REDUCED-DENSITY-MATRIX MECHANICS
WITH APPLICATION TO MANY-ELECTRON ATOMS AND MOLECULES

A SPECIAL VOLUME OF ADVANCES IN CHEMICAL PHYSICS

VOLUME 134

EDITORIAL BOARD

Bruce J. Berne, Department of Chemistry, Columbia University, New York, New York, U.S.A.
Kurt Binder, Institut für Physik, Johannes Gutenberg-Universität Mainz, Mainz, Germany
A. Welford Castlemena, Jr., Department of Chemistry, The Pennsylvania State University, University Park, Pennsylvania, U.S.A.
David Chandler, Department of Chemistry, University of California, Berkeley, California, U.S.A.
M. S. Child, Department of Theoretical Chemistry, University of Oxford, Oxford, U.K.
William T. Coffey, Department of Microelectronics and Electrical Engineering, Trinity College, University of Dublin, Dublin, Ireland
F. Fleming Crim, Department of Chemistry, University of Wisconsin, Madison, Wisconsin, U.S.A.
Ernest R. Davidson, Department of Chemistry, Indiana University, Bloomington, Indiana U.S.A.
Graham R. Fleming, Department of Chemistry, University of California, Berkeley, California, U.S.A.
Karl F. Freed, The James Franck Institute, The University of Chicago, Chicago, Illinois, U.S.A.
Pierre Gaspard, Center for Nonlinear Phenomena and Complex Systems, Brussels, Belgium
Eric J. Heller, Institute for Theoretical Atomic and Molecular Physics, Harvard-Smithsonian Center for Astrophysics, Cambridge, Massachusetts, U.S.A.
Robin M. Hochstrasser, Department of Chemistry, The University of Pennsylvania, Philadelphia, Pennsylvania, U.S.A.
R. Kosloff, The Fritz Haber Research Center for Molecular Dynamics and Department of Physical Chemistry, The Hebrew University of Jerusalem, Jerusalem, Israel
Rudolph A. Marcus, Department of Chemistry, California Institute of Technology, Pasadena, California, U.S.A.
G. Nicolis, Center for Nonlinear Phenomena and Complex Systems, Université Libre de Bruxelles, Brussels, Belgium
Thomas P. Russell, Department of Polymer Science, University of Massachusetts, Amherst, Massachusetts, U.S.A.
Donald G. Truhlar, Department of Chemistry, University of Minnesota, Minneapolis, Minnesota, U.S.A.
John D. Weeks, Institute for Physical Science and Technology and Department of Chemistry, University of Maryland, College Park, Maryland, U.S.A.
Peter G. Wolynes, Department of Chemistry, University of California, San Diego, California, U.S.A.

REDUCED-DENSITY-MATRIX MECHANICS

WITH APPLICATION TO MANY-ELECTRON ATOMS AND MOLECULES

ADVANCES IN CHEMICAL PHYSICS
VOLUME 134

Edited by

DAVID A. MAZZIOTTI

Series Editor

STUART A. RICE

Department of Chemistry
and
The James Franck Institute
The University of Chicago
Chicago, Illinois

WILEY-INTERSCIENCE
A JOHN WILEY & SONS, INC., PUBLICATION

Copyright © 2007 by John Wiley & Sons, Inc. All rights reserved

Published by John Wiley & Sons, Inc., Hoboken, New Jersey
Published simultaneously in Canada

No part of this publication may be reproduced, stored in a retrieval system, or transmitted in any form or by any means, electronic, mechanical, photocopying, recording, scanning, or otherwise, except as permitted under Section 107 or 108 of the 1976 United Stated Copyright Act, without either the prior written permission of the Publisher, or authorization through payment of the appropriate per-copy fee to the Copyright Clearance Center, Inc., 222 Rosewood Drive, Danvers, MA 01923, (978) 750-8400, fax (978) 750-4470, or on the web at www.copyright.com. Requests to the Publisher for permission should be addressed to the Permissions Department, John Wiley & Sons, Inc., 111 River Street, Hoboken, NJ 07030, (201) 748-6011, fax (201) 748-6008, or online at http://www.wiley.com/go/permission.

Limit of Liability/Disclaimer of Warranty: While the publisher and author have used their best efforts in preparing this book, they make no representations or warranties with respect to the accuracy or completeness of the contents of this book and specifically disclaim any implied warranties of merchantability or fitness for a particular purpose. No warranty may be created or extended by sales representatives or written sales materials. The advice and strategies contained herein may not be suitable for your situation. You should consult with a professional where appropriate. Neither the publisher nor author shall be liable for any loss of profit or any other commercial damages, including but not limited to special, incidental, consequential, or other damages.

For general information on our other products and services or for technical support, please contact our Customer Care Department within the United States at (800) 762-2974, outside the United States at (317) 572-3993 or fax (317) 572-4002.

Wiley also publishes its books in a variety of electronic formats. Some content that appears in print may not be available in electronic formats. For more information about Wiley products, visit our web site at www.wiley.com.

Wiley Bicentennial Logo: Richard J. Pacifico

Library of Congress Catalog Number: 58-9935

ISBN: 978-0-471-79056-3

Printed in the United States of America
10 9 8 7 6 5 4 3 2 1

CONTRIBUTORS TO VOLUME 134

D. R. ALCOBA Departamento de Física, Facultad de Ciencias Exactas y Naturales, Universidad de Buenos Aires, Ciudad Universitaria, 1428, Buenos Aires, Argentina

PAUL W. AYERS Department of Chemistry, McMaster University, Hamilton, Ontario L8S 4M1, Canada

BASTIAAN J. BRAAMS Department of Mathematics and Computer Science, Emory University, 400 Dowman Drive #W401, Atlanta, GA 30322 USA

GARNET KIN-LIC CHAN Department of Chemistry and Chemical Biology, Cornell University, Ithaca, New York 14853-1301 USA

A. JOHN COLEMAN Department of Mathematics and Statistics, Queen's University, Kingston, Ontario K7L 3N6, Canada

ERNEST R. DAVIDSON Department of Chemistry, University of Washington, Bagley Hall 303C/Box 351700, Seattle, WA 98195-1700 USA

ROBERT M. ERDAHL Department of Mathematics and Statistics, Queen's University, Kingston, Ontario K7L 3N6, Canada

MITUHIRO FUKUDA Department of Mathematical and Computing Sciences, Tokyo Institute of Technology, Japan

JOHN E. HARRIMAN Department of Chemistry, University of Wisconsin, Madison, WI 53706 USA

JOHN M. HERBERT Department of Chemistry, The Ohio State University, Columbus, OH 43210 USA

SABRE KAIS Department of Chemistry, Purdue University, West Lafayette, IN 47907 USA

WERNER KUTZELNIGG Lehrstuhl für Theoretische Chemie Ruhr-Universität Bochum, D-44780 Bochum, Germany

DAVID A. MAZZIOTTI Department of Chemistry and The James Franck Institute, The University of Chicago, Chicago, IL 60637 USA

DEBASHIS MUKHERJEE Department of Physical Chemistry, Indian Association for the Cultivation of Science, Calcutta 700 032, India and Jawaharlal Center for Advanced Scientific Research, Bangalore, India

MAHO NAKATA Department of Applied Chemistry, The University of Tokyo, Japan

JEROME K. PERCUS Courant Institute and Department of Physics, New York University, 251 Mercer Street, New York, NY 10012 USA

MARIO PIRIS Kimika Fakultatea, Euskal Herriko Unibertsitatea, and Donostia International Physics Center (DIPC) P.K. 1072, 20080 Donostia, Euskadi, Spain

B. C. RINDERSPACHER Department of Chemistry, The University of Georgia, Athens, GA 30602 USA

MITJA ROSINA Faculty of Mathematics and Physics, University of Ljubljana, Jadranska 19, P.O. Box 2964, 1001 Ljubljana, Slovenia and Jožef Stefan Institute, Ljubljana, Slovenia

C. VALDEMORO Instituto de Matemáticas y Física Fundamental, Consejo Superior de Investigaciones Científicas, Serrano 123, 28006 Madrid, Madrid, Spain

MAKOTO YAMASHITA Department of Information System Creation, Kanagawa University, Japan

TAKESHI YANAI Department of Chemistry and Chemical Biology, Cornell University, Ithaca, New York 14853-1301 USA

ZHENGJI ZHAO High Performance Computing Research Department, Lawrence Berkeley National Laboratory, 1 Cyclotron Road, Mail Stop 15F1650, Berkeley, CA 94720 USA

PREFACE

In the 1950s several scientists including Joseph Mayer, John Coleman, Per-Olov Lowdin, and Charles Coulson expounded on the possibilities for using the two-electron reduced density matrix to compute the energies and properties of atomic and molecular systems without the many-electron wavefunction. An N-electron density matrix may be assembled from an N-electron wavefunction by multiplying the wavefunction ψ by its adjoint ψ^* to obtain $\psi\psi^*$. Integrating the N-electron density matrix over all save two electrons yields the two-electron reduced density matrix (2-RDM). Because electrons interact with each other *in pairs* by Coulomb repulsion, the energies and other electronic properties of atoms and molecules can be computed directly from a knowledge of the 2-RDM. The fact that the 2-RDM is the repository of all the physically and chemically important information in the many-particle wavefunction suggests the tantalizing possibility that for a given molecular system the 2-RDM can be computed directly without constructing the many-electron wavefunction. For fifty years both scientists and mathematicians have pursued the goal of a 2-RDM approach to molecular electronic structure. Efforts, however, were stymied because the 2-RDM must be constrained by nontrivial conditions to ensure that the 2-RDM derives from an N-electron wavefunction. These restrictions on RDMs were given the appellation N-representability conditions by John Coleman. Ten years ago the calculation of the 2-RDM without the wavefunction seemed an impossibility. Dramatic progress, however, has been made since then, and today two complementary approaches to the direct calculation of the 2-RDM have emerged. The present book, with chapters from many of the scientists who contributed to these advances, provides a detailed yet pedagogical tour of modern 2-RDM theory and its present and potential applications to many-electron atoms and molecules.

I first became interested in 2-RDM theory when reading articles by Coleman and ter Haar in the chemistry library at Princeton University in the summer of 1995. For someone who had just graduated college, reading Coleman and ter Haar was rather "heady" material. However, despite some of the difficulties from N-representability it seemed apparent that a 2-RDM theory would offer a powerful bridge between the density functional methods, which were rapidly gaining popularity after many years of development, and the *ab initio* wavefunction methods. At Harvard under the splendid guidance of Dudley Herschbach I began to think about computing the 2-RDM without the wavefunction. In the summer of 1996 I began a long friendship with John Coleman through an innocuous e-mail about a reference in one of his articles. A few months later I found a typographical mistake in one of John's papers. With his typical humor John proclaimed this mistake to be the "Mazziotti typo" since I had brought it to his attention. (I suppose that I can accept greater responsibility for any typos in the present volume!) Recently, John admitted that he began to take my interest in the 2-RDM seriously after I had found a mistake in one of *his* papers. Because of the difficulties with minimizing the energy with respect to the 2-RDM, I began to think about what it would mean to contract the Schrödinger equation onto the space of two particles, and I found articles related to this idea by Hiroshi Nakatsuji and Carmela Valdemoro. Meanwhile, Valdemoro and Nakatsuji were making progress in solving the contracted Schrödinger equation (also known as the density equation).

The present book is divided into five related parts. Part I contains historical introductions by John Coleman and Mitja Rosina. Part II discusses the variational calculation of the 2-RDM including the development of a systematic hierarchy of N-representability conditions known as the positivity conditions and the design of effective semidefinite-programming algorithms for solving the 2-RDM optimization problem. In Part III the nonvariational calculation of the 2-RDM by the contracted Schrödinger equation (CSE) is presented including the reconstruction of the 3- and 4-RDMs from the 2-RDM by cumulant theory and the addition of N-representability conditions on the 2-RDM. Very recently developed methods that dramatically improve the accuracy from the CSE are presented in the last two chapters (12 and 13) of this part. Chapter 12, written by me, presents the anti-Hermitian CSE method in which the anti-Hermitian part of the CSE with cumulant reconstruction of the 3-RDM is solved for the ground-state energy and its 2-RDM. Chapter 13 by Garnet Chan and Takeshi Yanai incorporates RDM cumulant expansions into canonical diagonalization of the Hamiltonian. Chan's method, although it does not generate a 2-RDM, is included in the book because it can be interpreted as a solution of the anti-Hermitian CSE in the Heisenberg representation. The electronic energies from the anti-Hermitian CSE are competitive with the best wavefunction methods of similar computational expense. Both the second and third parts contain illustrative applications of the 2-RDM methods to a variety of atoms and molecules. Part IV of the book examines work on geminal, 1-RDM, and pair-density functional theories that explore the possibilities between using the 1-density as in density functional theory and using the 2-RDM. Each of these theories requires the development of a subset of the customary density functional. Finally, in Part V a parameterization of the 2-RDM with connections to quantum phase transitions is presented and the role of the 2-RDM in studying entanglement is examined.

I wish to extend my gratitude to each of the authors for contributing their ideas and enthusiasm to the present book. It was a great pleasure for me to work with everyone. I must also thank both Prof. Dudley R. Herschbach of Harvard University and Prof. Herschel A. Rabitz of Princeton University, whose support, encouragement, and example have been invaluable to me. I thank my father, Dr. Alexander R. Mazziotti, for encouraging and supporting me and sharing with me his love for science, especially chemistry. Finally, I would like to thank my colleagues and students at The University of Chicago. During the past four years I have had the pleasure of sharing my adventures in research with the following students: John Farnum, Daniel Jordan, Gergely Gidofalvi, Tamas Juhasz, Jeff Hammond, Eugene Kamarchik, Adam Rothman, Eugene De Prince, Marc Benayoun, Aiyan Lu, and Brittany Rohrman. The present volume in *Advances in Chemical Physics* surveys the recent advances in 2-RDM theory. The authors and I hope that the reader will view this book as a helpful guide in both understanding and exploring 2-RDM methods as the 2-RDM becomes increasingly important as a fundamental variable in the quantum computation of many-electron atoms and molecules.

DAVID A. MAZZIOTTI

Chicago, Illinois
October 2006

INTRODUCTION

DAVID A. MAZZIOTTI

*Department of Chemistry and The James Franck Institute,
The University of Chicago, Chicago, IL 60637 USA*

In a famous after-dinner address at a 1959 conference in Boulder, Colorado, Charles Coulson [1] discussed both the promises and the challenges of using the *two-electron reduced density matrix* (2-RDM) rather than the many-electron wavefunction as the primary variable in quantum computations of atomic and molecular systems. Integrating the N-electron density matrix,

$$^N D(1, 2, \ldots, N; 1', 2', \ldots, N') = \Psi(1, 2, \ldots, N)\Psi^*(1', 2', \ldots, N')$$

over coordinates 3 to N defines the 2-RDM:

$$^2 D(1, 2; 1', 2') = \int \Psi(1, 2, \ldots, N)\Psi^*(1', 2', \ldots, N) d3 \cdots dN$$

Because electrons are indistinguishable with only pairwise interactions, the energy of any atom or molecule may be expressed as a linear function of the 2-RDM rather than the many-electron wavefunction, that is,

$$E = \mathrm{Tr}(^2 K\, ^2 D)$$

where $^2 K$ is the two-electron reduced Hamiltonian matrix, which is the matrix representation of the operator

$$^2\hat{K} = \frac{1}{N-1}\left(-\frac{1}{2}\nabla_1^2 - \sum_j \frac{Z_j}{r_{1j}}\right) + \frac{1}{2}\frac{1}{r_{12}}$$

The expression of the energy for any N-electron system by a 2-RDM suggests the tantalizing possibility that the ground-state energy for any many-electron system can be computed through a *two-electron* calculation.

As Coleman describes eloquently in his introductory chapter, in the summer of 1951, as a young mathematician, with a derivation like the one above, he announced somewhat prematurely to a gathering of physicists at Chalk River that he had reduced the many-electron problem in quantum mechanics to a two-electron calculation. A simple variational calculation on lithium with the 2-RDM rather than the wavefunction produced a ground-state energy that was clearly too low. Coleman soon recognized that additional constraints must be added to the 2-RDM to guarantee that it derives from an N-electron density matrix (or wavefunction). Coleman called these constraints N-representability conditions—a term that became standard after Coleman's 1963 *Review of Modern Physics* article [2]. In 1955 both Joseph Mayer [3] and Per-Olov Löwdin [4] wrote papers for *Physical Review* that examined the expression of the energy as a function of the 2-RDM, and soon afterwards, several papers appeared that examined the need for additional conditions on the 2-RDM—a challenge that would become known as the N-representability problem. Coulson, Coleman, and others inspired several generations of chemists, physicists, and mathematicians to explore the 2-RDM conditions necessary for exploring many-particle quantum mechanics on the space of two particles, and Rosina, a nuclear physicist, describes in his chapter the unique interdisciplinary nature of the 2-RDM conferences at Queens University in the late 1960s and early 1970s. Despite significant efforts, however, the goal of computing the 2-RDM without the wavefunction remained elusive. Like John Coleman, Richard Feynman had been a first-year graduate student at Princeton in 1939 (in fact, a picture taken later by John of Feynman talking with Dirac is kindly reproduced with the permission of *Physics Today*, which published it on the cover of their August 1963 issue). In the late 1980s John wrote a letter to Feynman proposing that Feynman and he collaborate on the N-representability problem. The letter returned unopened, for Feynman had recently died. (At least, because the letter was unopened, we can surmise that Feynman was not troubled by the N-representability problem in the last days of his life!)

The present volume describes significant advances in 2-RDM mechanics (as distinct from conventional wave mechanics where the wavefunction is the primary variable of the calculation) that are realizing the direct calculation of the 2-RDM without the wavefunction. Two related and yet distinct approaches for computing the 2-RDM directly have emerged: (i) the variational calculation of the ground-state energy with the 2-RDM constrained by N-representability conditions, and (ii) the iterative solution of the contracted Schrödinger equation (CSE) for a nonvariational ground-state 2-RDM. The second part of this book explores the variational 2-RDM method while the third part of this book develops the nonvariational solution of the CSE including new research on solving only the anti-Hermitian part of the CSE. Historically, the present wave of advances in 2-RDM mechanics began with work by Carmela Valdemoro and her collaborators in the early 1990s on the solution of the CSE [5]. The CSE is a projection of the N-electron Schrödinger equation onto the space of two electrons. The CSE (also known as the density equation) was first formulated in separate papers by Cohen and Frishberg [6] and Nakatsuji [7]. However, because the CSE depends on not only the 2-RDM but also the 3- and 4-RDMs, the CSE cannot be solved for the 2-RDM without additional information. Valdemoro employed particle–hole duality to develop formulas for building the 3- and the 4-RDMs from a knowledge of the 2-RDM. Nakatsuji and Yasuda improved the formulas from their connection to Greens' functions [8], and I showed that the reconstruction of the 3- and 4-RDMs could be systematized through a cumulant theory for RDMs and improved by contraction relations for the cumulants [9–12]. Important techniques, known as correlated purification, have been developed for correcting the N-representability of the 2-RDM between iterations of the CSE [13, 14]. Very recently, a significant advance has been made in CSE theory with dramatic improvement in the accuracy of the energies and 2-RDMs. In Chapter 12, I develop a system of initial-value differential equations for solving the anti-Hermitian part of the CSE for the ground-state energy and its 2-RDM. The 3-RDM is reconstructed by its cumulant expansion including the second-order corrections by Yasuda and Nakatsuji [8] and Mazziotti [11, 12]. Molecular energies with the second-order corrections to the 3-RDM are as accurate as those from coupled cluster with single and double excitations. The anti-Hermitian CSE method also directly generates accurate 1- and 2-RDMs where the 2-RDMs very nearly satisfy known N-representability conditions. In Chapter 13, Chan and Yanai implement RDM cumulant expansions in the context of canonical diagonalization. Although the method does not produce a 2-RDM, it can be interpreted as a solution of the anti-Hermitian CSE in the Heisenberg representation. Because the method does not include second-order corrections to the 3-RDM, the energies computed from Hartree–Fock reference wavefunctions are not as accurate as those from coupled cluster singles–doubles. If multi–reference self-consistent-field wavefunctions are employed as references, however, the method generates very accurate energies at both equilibrium and nonequilibrium molecular geometries. An important advantage of the anti-Hermitian CSE methods is that their reference wavefunctions can readily be changed from Hartree–Fock to include multireference correlation effects. The concepts behind the CSE, the present-day CSE algorithms, the recently developed anti-Hermitian CSE algorithms, and future research directions are carefully developed in a series of chapters in Part III of this book by Valdemoro, Mazziotti, Alcoba, Herbert and Harriman, Kutzelnigg and Mukherjee, and Chan and Yanai.

The original goal of Coleman, Coulson, Mayer, and others was to develop a variational calculation of the ground-state energy as a functional of the 2-RDM. The difficult problem of identifying sufficiently stringent N-representability conditions on the 2-RDM had prevented this goal from being realized for fifty years.

With advances in theory and optimization in the late 1990s, however, the situation was prepared to change. First, with research on the CSE providing new insight into the structure of the necessary conditions, Erdahl and I developed a systematic hierarchy of the N-representability constraints known as p-positivity conditions [15]. The generalized uncertainty relations for all pairs of p-particle operators were shown to be enforced by the $2p$-positivity conditions. Second, in the 1990s, significant advances were made in a special form of optimization known as semidefinite programming for a variety of important problems in control theory, combinatorial optimization, and finance [16]. The minimization of the ground-state energy with respect to a 2-RDM constrained by p-positivity conditions constitutes a semidefinite program, where the process of solving a semidefinite program is known as semidefinite programming. In 2001 and 2002, the algorithms from control theory and combinatorial optimization were applied in separate works by Nakata et al. [17] and the author [18] to calculations of many-electron atoms and molecules (in minimal basis sets) with the 2-RDM constrained by 2-positivity conditions. The 2-RDM method with 2-positivity conditions yielded accurate shapes for potential energy surfaces without the multireference difficulties exhibited by many approximate wavefunction methods. (It should be mentioned that variational 2-RDM calculations with 2-positivity conditions had been performed in the 1970s with an early form of semidefinite programming by Rosina, Erdahl, Garrod, and their collaborators [19, 20], although these calculations were limited to four-electron atoms and molecules.) In 2004, Zhao et al. [21] applied a subset of the 3-positivity conditions, the T_1 and T_2 constraints, which had been proposed by Erdahl in 1978 [22], to closed- and open-shell molecules in minimal basis sets with coupled-cluster accuracy at equilibrium geometries. Later that year, I introduced a new, first-order algorithm for solving the 2-RDM semidefinite program [23, 24], which, by using a matrix factorization to enforce the positivity conditions, reduces the scaling for 2-positivity to r^6 in floating-point operations and r^4 in memory requirements (where r is the rank of the one-particle basis set). This first-order method enables the treatment of larger molecules and basis sets as well as the implementation of complete 3-positivity conditions [25–28]. These advances as well as other advances and applications are discussed in a series of chapters in Part II of this book by Mazziotti, Erdahl, Braams, Percus, and Zhao, and Fukuda, Nakata, and Yamashita.

In college I had a wonderful class in abstract algebra where the professor began each lecture with the last ten minutes of the previous lecture—a definite example of very useful *deja vu*. This book is organized around a similar principle. No effort has been made to keep each chapter strictly orthogonal to the preceding and proceeding chapters in the spirit that a certain degree of overlap is useful for the book to serve as both an instructional and a reference guide. Representing the frontiers of the 2-RDM mechanics, the book covers a great deal of new and exciting material, which, it is believed, will spur many future endeavors and explorations in 2-RDM theory. The book is organized into five parts with the second and third parts described above. The first part contains two historical perspectives, one by John Coleman and one by Mitja Rosina. The fourth part explores the connection between 2-RDM methods and one-particle and pair-density functional theories that aim to extend the well-known density functional theory. In Chapter 14, Piris describes the efforts since 1998 to develop a functional theory based on the one-particle RDM, and he shows how he has recently made progress by incorporating ideas from 2-RDM mechanics into the function. Rinderspacher reviews recent work on geminal functional theory, proposed by me in 2000 [29, 30], where the correlation energy is expressed as a function of a geminal. (A geminal denotes a two-electron function.) Rinderspacher extends geminal functional theory to consider the use of multiple geminal functions. In Chapter 16, Ayers and Davidson review the diagonal N-representability problem and its application to pair-density functional theory. The final, fifth part of the book, focusing on applications of 2-RDMs to entanglement and quantum phase transitions, contains two chapters: a chapter by Coleman on an exact parameterization of the 2-RDM, which was used by Coleman to discuss quantum phase transitions in a recent article [31], and a chapter by Kais that surveys the many applications of RDM theory to quantum entanglement, including a discussion of entanglement as a measure for correlation in electronic structure theory.

This volume in *Advances in Chemical Physics* provides a broad yet detailed survey of the recent advances and applications of reduced-density-matrix mechanics in chemistry and physics. With advances in theory and optimization, Coulson's challenge for the direct calculation of the 2-RDM has been answered. While significant progress has been made, as evident from the many contributions to this book, there remain many open questions and exciting opportunities for further development of 2-RDM methods and applications. It is the hope of the editor and the contributors that this book will serve as a guide for many further adventures and advancements in RDM mechanics.

References

1. C. A. Coulson, Present state of molecular structure calculations. *Rev. Mod. Phys.* **32**, 170 (1960).
2. A. J. Coleman, Structure of fermion density matrices. *Rev. Mod. Phys.* **35**, 668 (1963).
3. J. E. Mayer, Electron correlation. *Phys. Rev.* **100**, 1579 (1955).
4. P. O. Löwdin, Quantum theory of many-particle systems. 1. Physical interpretations by means of density matrices, natural spin-orbitals, and convergence problems in the method of configuration interaction. *Phys. Rev.* **97**, 1474 (1955).
5. F. Colmenero and C. Valdemoro, Approximating q-order reduced density-matrices in terms of the lower-order ones. 2. Applications. *Phys. Rev. A* **47**, 979 (1993).

6. L. Cohen and C. Frishberg, Hierarchy equations for reduced density matrices. *Phys. Rev. A* **13**, 927 (1976).
7. H. Nakatsuji, Equation for direct determination of density matrix. *Phys. Rev. A*, **14**, 41 (1976).
8. K. Yasuda and H. Nakatsuji, Direct determination of the quantum-mechanical density matrix using the density equation II. *Phys. Rev. A* **56**, 2648 (1997).
9. D. A. Mazziotti, Contracted Schrödinger equation: determining quantum energies and two-particle density matrices without wave functions. *Phys. Rev. A* **57**, 4219 (1998).
10. D. A. Mazziotti, Approximate solution for electron correlation through the use of Schwinger probes. *Chem. Phys. Lett.* **289**, 419 (1998).
11. D. A. Mazziotti, Pursuit of N-representability for the contracted Schrödinger equation through density-matrix reconstruction. *Phys. Rev. A* **60**, 3618 (1999).
12. D. A. Mazziotti, Complete reconstruction of reduced density matrices. *Chem. Phys. Lett.* **326**, 212 (2000).
13. D. A. Mazziotti, Correlated purification of reduced density matrices. *Phys. Rev. E* **65**, 026704 (2002).
14. D. R. Alcoba and C. Valdemoro, Spin structure and properties of the correlation matrices corresponding to pure spin states: controlling the S-representability of these matrices. *Int. J. Quantum Chem.* **102**, 629 (2005).
15. D. A. Mazziotti and R. M. Erdahl, Uncertainty relations and reduced density matrices: mapping many-body quantum mechanics onto four particles. *Phys. Rev. A* **63**, 042113 (2001).
16. L. Vandenberghe and S. Boyd, Semidefinite programming. *SIAM Rev.* **38**, 49 (1996).
17. M. Nakata, H. Nakatsuji, M. Ehara, M. Fukuda, K. Nakata, and K. Fujisawa, Variational calculations of fermion second-order reduced density matrices by semidefinite programming algorithm. *J. Chem. Phys.* **114**, 8282 (2001).
18. D. A. Mazziotti, Variational minimization of atomic and molecular ground-state energies via the two-particle reduced density matrix. *Phys. Rev. A* **65**, 062511 (2002).
19. C. Garrod, V. Mihailović, and M. Rosina, Variational approach to 2-body density matrix. *J. Math. Phys.* **10**, 1855 (1975).
20. R. M. Erdahl, Two algorithms for the lower bound method of reduced density matrix theory. *Reports Math. Phys.* **15**, 147 (1979).
21. Z. Zhao, B. J. Braams, H. Fukuda, M. L. Overton, and J. K. Percus, The reduced density matrix method for electronic structure calculations and the role of three-index representability conditions. *J. Chem. Phys.* **120**, 2095 (2004).
22. R. M. Erdahl, Representability. *Int. J. Quantum Chem.* **13**, 697 (1978).
23. D. A. Mazziotti, Realization of quantum chemistry without wavefunctions through first-order semidefinite programming. *Phys. Rev. Lett.* **93**, 213001 (2004).
24. D. A. Mazziotti, First-order semidefinite programming for the direct determination of two-electron reduced density matrices with application to many-electron atoms and molecules. *J. Chem. Phys.* **121**, 10957 (2004).
25. G. Gidofalvi and D. A. Mazziotti, Application of variational reduced-density-matrix theory to organic molecules. *J. Chem. Phys.* **122**, 094107 (2005).
26. G. Gidofalvi and D. A. Mazziotti, Application of variational reduced-density-matrix theory to the potential energy surfaces of the nitrogen and carbon dimers. *J. Chem. Phys.* **122**, 194104 (2005).
27. D. A. Mazziotti, Variational two-electron reduced-density-matrix theory for many-electron atoms and molecules: implementation of the spin- and symmetry-adapted T_2 condition through first-order semidefinite programming. *Phys. Rev. A* **72**, 032510 (2005).

28. J. R. Hammond and D. A. Mazziotti, Variational reduced-density-matrix calculations on radicals: an alternative approach to open-shell *ab initio* quantum chemistry. *Phys. Rev. A* **73**, 012509 (2006).
29. D. A. Mazziotti, Geminal functional theory: a synthesis of density and density matrix methods. *J. Chem. Phys.* **112**, 10125 (2000).
30. D. A. Mazziotti, Energy functional of the one-particle reduced density matrix: a geminal approach. *Chem. Phys. Lett.* **338**, 323 (2001).
31. A. J. Coleman, Kummer variety, geometry of N-representability, and phase transitions. *Phys. Rev. A* **66**, 022503 (2002).

CONTENTS

Part I	1
CHAPTER 1 N-REPRESENTABILITY	3
By A. John Coleman	
CHAPTER 2 HISTORICAL INTRODUCTION	11
By Mitja Rosina	
Part II	19
CHAPTER 3 VARIATIONAL TWO-ELECTRON REDUCED-DENSITY-MATRIX THEORY	21
By David A. Mazziotti	
CHAPTER 4 THE LOWER BOUND METHOD FOR DENSITY MATRICES AND SEMIDEFINITE PROGRAMMING	61
By Robert M. Erdahl	
CHAPTER 5 THE $T1$ AND $T2$ REPRESENTABILITY CONDITIONS	93
By Bastiaan J. Braams, Jerome K. Percus, and Zhengji Zhao	
CHAPTER 6 SEMIDEFINITE PROGRAMMING: FORMULATIONS AND PRIMAL–DUAL INTERIOR-POINT METHODS	103
By Mituhiro Fukuda, Maho Nakata, and Makoto Yamashita	
Part III	119
CHAPTER 7 THEORY AND METHODOLOGY OF THE CONTRACTED SCHRÖDINGER EQUATION	121
By C. Valdemoro	
CHAPTER 8 CONTRACTED SCHRÖDINGER EQUATION	165
By David A. Mazziotti	
CHAPTER 9 PURIFICATION OF CORRELATED REDUCED DENSITY MATRICES: REVIEW AND APPLICATIONS	205
By D. R. Alcoba	

CHAPTER 10 CUMULANTS, EXTENSIVITY, AND THE CONNECTED FORMULATION OF THE CONTRACTED SCHRÖDINGER EQUATION   261

By John M. Herbert and John E. Harriman

CHAPTER 11 GENERALIZED NORMAL ORDERING, IRREDUCIBLE BRILLOUIN CONDITIONS, AND CONTRACTED SCHRÖDINGER EQUATIONS   293

By Werner Kutzelnigg and Debashis Mukherjee

CHAPTER 12 ANTI-HERMITIAN FORMULATION OF THE CONTRACTED SCHRÖDINGER THEORY   331

By David A. Mazziotti

CHAPTER 13 CANONICAL TRANSFORMATION THEORY FOR DYNAMIC CORRELATIONS IN MULTIREFERENCE PROBLEMS   343

By Garnet Kin-Lic Chan and Takeshi Yanai

Part IV   385

CHAPTER 14 NATURAL ORBITAL FUNCTIONAL THEORY   387

By Mario Piris

CHAPTER 15 GEMINAL FUNCTIONAL THEORY   429

By B. C. Rinderspacher

CHAPTER 16 LINEAR INEQUALITIES FOR DIAGONAL ELEMENTS OF DENSITY MATRICES   443

By Paul W. Ayers and Ernest R. Davidson

Part V   485

CHAPTER 17 PARAMETERIZATION OF THE 2-RDM   487

By A. John Coleman

CHAPTER 18 ENTANGLEMENT, ELECTRON CORRELATION, AND DENSITY MATRICES   493

By Sabre Kais

AUTHOR INDEX   537
SUBJECT INDEX   551

PART I

CHAPTER 1

N-REPRESENTABILITY

A. JOHN COLEMAN

Department of Mathematics and Statistics, Queen's University, Kingston, Ontario K7L 3N6 Canada

CONTENTS

I. Introduction
II. Academic History
III. Summer 1951

I. INTRODUCTION

Few distinctions in quantum mechanics are as important as that between fermions and bosons. This distinction results from the fact that there are only two one-dimensional linear representations, on the space of wavefunctions, of the group, S_N, of permutations of $N \geq 2$ objects. For all groups there is the *identity* representation, which leaves the wavefunction fixed, and for indistinguishable particles there is one other that leaves the function fixed or changes its sign according to whether the permutation is even or odd. I do not have the authority to assert that God agrees with me as to the importance of this distinction, but I am sure that most happy humans will since, as noted by Eddington, if there were no fermions there would be no electrons, so no molecules, so no DNA, no humans!

What has this to do with *reduced density matrices*?

For a system of N identical fermions in a state ψ there is associated a reduced density matrix (RDM) of order p for each integer p, $1 \leq p \leq N$, which determines a Hermitian operator D^p, which we call a *reduced density operator* (RDO) acting on a space of antisymmetric functions of p particles. The case $p = 2$ is of particular interest for chemists and physicists who seldom consider

---

Reduced-Density-Matrix Mechanics: With Application to Many-Electron Atoms and Molecules,
A Special Volume of Advances in Chemical Physics, Volume 134, edited by David A. Mazziotti.
Series editor Stuart A. Rice. Copyright © 2007 John Wiley & Sons, Inc.

Hamiltonians involving more than two electron interactions. I shall use this to illustrate the general case of arbitrary p. The *second-order reduced density matrix* (2-RDM) of a pure state ψ, a function of four particles, is defined as follows:

$$D^2(12,1'2') = \int \psi(123\ldots N)\psi^*(1'2'3\ldots N)d(3\ldots N)$$

This we interpret as the kernel of an integral operator, D^2, which transforms an arbitrary function f on 2-space into a function $D^2 f$ on 2-space defined by

$$D^2 f(12) = \int D^2(12,1'2')f(1'2')d(1'2')$$

As we show later, the energy of the state of any system of N indistinguishable fermions or bosons can be expressed in terms of the Hamiltonian and $D^2(12,1'2')$ if its Hamiltonian involves at most two-particle interactions. Thus it should be possible to find the ground-state energy by variation of the 2-matrix, which depends on four particles. Contrast this with current methods involving direct use of the wavefunction that involves N particles. A principal obstruction for this procedure is the "N-representability" conditions, which ensure that the proposed RDM could be obtained from a system of N identical fermions or bosons.

II. ACADEMIC HISTORY

I will sketch briefly my personal academic history that prepared me to discuss these matters and then tell the story of how in the the summer of 1951 I hit upon and named "N-representability."

During my undergraduate years, 1935–1939, in Honors Mathematics and Physics at the University of Toronto, increasingly, I became interested in mathematical physics, picking up some elementary quantum mechanics and relativity. My first encounter with Einstein's *general relativity theory* (GRT) was in the substantial treatise of Levi-Civita on differential geometry, which ends with a 150-page introduction to GRT. This is a beautiful theory, which I presented in lectures from 1950 in Toronto because it had become the dominant orthodoxy everyone should know!

However, I never became a "True Believer" since by chance (but fortunately) I also read the *principle of relativity* (PR) by Whitehead, who pointed to a logical problem for Einstein, which, as far as I am aware, has never been dealt with adequately. Alfred North Whitehead (1860–1947) was a master of mathematical logic, which he showed as senior author of the famous three-volume work on the foundations of mathematics. As a mathematical Fellow and Tutor at Trinity

College, Cambridge, it was also his duty to keep abreast of developments in mathematical physics in the period 1885–1923. It was reported that when Bertrand Russell was asked "When did Whitehead become a Relativist?" he replied "At birth!"

In 1939, the first year of the Second World War, I had the delicious, difficult choice of graduate study at Harvard, Princeton, or St. John's College with Dirac! Because crossing the Atlantic was dangerous, and on the advice of the College Registrar, I regretfully turned down Harvard, chose Princeton and found myself sharing a first course in quantum mechanics with a student, nine days older than myself, from MIT of whom I and the rest of the world had never heard, named Richard Feynman!

The instructor in the first term was John A. Wheeler, who had begun his Princeton career the previous year; and in the second term, the famous authority on group theory, Eugene Wigner. All that I recall of this course was that one day in early January, Wigner arrived very excitedly saying that over the weekend he had learned from Lamb that there is a minute error in Dirac's formula for the spectrum of hydrogen. This was the "Lamb shift" and the harbinger of quantum field theory. At the time I had no idea why, but Wigner's excitement left no doubt for me, it was very important!

I assiduously attended all the lectures like a serious Torontonian even though there was little in the first term that I had not learned from Leopold Infeld at Toronto. I cannot remember seeing Feynman in class. He certainly knew more QM than I did. However, we did enjoy arguing vigorously on several occasions in the Discussion Rooms of Fine Hall Library. So much so that at least twice Miss Shields, who ruled the Library with an iron hand, ordered us to moderate our voices, which through the thick closed door were disturbing everyone in the Library! He told me about his engagement to a girl in New York City whose death from tuberculosis had been predicted and the opposition of his relatives and friends to his determination to keep his word to her. I sympathized but did not presume to advise. I became quite fond of him, admiring Feynman for simplicity and integrity of spirit. After I left Princeton we met only once, in July 1962 in Poland at a Conference about gravitation. It was there that I took the accompanying photo of him and Dirac, which was published on the cover of *Physics Today* in August 1963 (see Fig. 1).

I returned to the University of Toronto in the summer of 1940, having completed a Master's degree at Princeton, to enroll in a Ph.D. program under Leopold Infeld for which I wrote a thesis entitled: *A Study in Relativistic Quantum Mechanics Based on Sir A.S. Eddington's "Relativity Theory of Protons and Electrons."* This book summarized his thought about the constants of Nature to which he had been led by his shock that Dirac's equation demonstrated that a theory which was invariant under Lorentz transformation need not be expressed in terms of tensors.

Figure 1 Paul Dirac and Richard Feynman at the International Conference on Relativistic Theories of Gravitation, Warsaw, Poland, July 25–31, 1962. Photograph by A. John Coleman, courtesy AIP Emilio Segre Visual Archives, Physics Today Collection.

Eddington's final theory was dismissed by the physics establishment as philosophical and speculative nonsense. Though I found a serious error in Eddington's argument, the more errors I discovered the greater respect I developed for his physical insight. My admiration for Whitehead's gravitational theory and for Eddington's final work must cause orthodox physicists to dismiss me as espousing lost causes. However, as evidenced by this book, my pursuit of the second-order reduced density matrix appears in recent years to have gained some attention among chemists.

After obtaining a Ph.D. under Infeld at Toronto, I taught calculus and algebra at Queen's University for two years until the end of the War. Between 1945 and 1949, based in Geneva, I served as the University Secretary of the World's Student Christian Federation before joining the Mathematics Department of the University of Toronto until 1960, when I became Head of Mathematics and Statistics at Queen's University in Kingston, Ontario.

My years with the Federation provided a remarkable opportunity to broaden my understanding of international relations and to begin to understand sympathetically the diversity of religious and political experiences in Europe, Asia, Africa, and North America. But, what I had not expected, it also gave me a chance to meet distinguished scientists such as Hadamard, the French mathematician made famous for his discussion of the prime number theorem; Werner Heisenberg, with whom I enjoyed two evenings in Goettingen during the week in which I attended the funeral of Planck; W. Threlfall, the English topologist, who, protected by his Nazi student, Seifert, lived safely in Germany throughout the War during part of which he was housed inside a huge airplane factory in a splended cottage provided in case the Minister of the Luftwaffe came to inspect the factory, which he never did. He proudly told me that he was probably unique in Germany, lecturing on the "Jewish" relativity theories of Einstein during the courses he offered to engineers inside the factory!

III. SUMMER 1951

In the summer of 1951 it was my privilege to belong to the Research Institute of the Canadian Mathematical Congress, which later became the Canadian Mathematical Society. The Institute had been created by R. L. Jeffery to encourage young mathematicians to take time for research. I was working on Lie groups and algebras. But as a diversion I started to read about second quantization in Frenkel's advanced treatise on quantum mechanics. This was the only decent treatment of the topic in English available in 1951. I soon noticed that if the Hamiltonian, H, of a system of N electrons involves the electrons in at most two-particle interactions, the total energy of the ground state (GS) of the system can be expressed in terms of the *second-order reduced density matrix*. This mathematical object proved so important that it is also called *the second-order RDM*, or for the sake of brevity, simply *the 2-matrix*. Unsaid but assumed is the caveat "of the system."

We can justify the above conclusion as follows. If H involves at most two-particle interactions, it is expressible as

$$H = \sum_{klij} H_{klij} a_k^+ a_l^+ a_j a_i$$

where a_i and a_j denote *annihilators*, whereas a_k^+ and a_l^+ are *creators*. Therefore the energy of the state, ψ, is

$$E = \langle \psi | H | \psi \rangle$$
$$= \sum H_{klij} \langle \psi | a_k^+ a_l^+ a_j a_i | \psi \rangle$$

Thus the energy of the state is expressed in terms of coefficients of the Hamiltonian *and the quantities* $\langle \psi | a_k^+ a_l^+ a_j a_i | \psi \rangle$, which are coefficients of the 2-matrix which Dirac denoted by ρ_2.

I assume that the reader interprets the complex numbers $\langle\psi|a_k^+ a_l^+ a_j a_i|\psi\rangle$ as elements of a matrix representing a reduced density operator on two-particle space spanned by products of a fixed chosen complete set of orthonormal one-particle functions, ϕ_i, in terms of which ψ can also be expanded. Then a_i reduces the *occupancy* of ϕ_i to zero, while a_i^+ sets the *occupancy* at 1. For this reason some physicists consider a name such as "occupation number notation," used by many Russians, as preferable to "second quantization notation," which has an almost mystical connotation to my mind.

The RDO, ρ_2, is defined by

$$\rho_2 = N(N-1)D^2(12, 1'2')$$
$$= \int \psi(123\ldots N)\psi^*(1'2'3\ldots N)d(3\ldots N)$$

Since I have assumed that ψ is normalized to 1, the trace of D^2 is also 1 and the trace of ρ_2 is $N(N-1)$. We now define the *reduced Hamiltonian operator*

$$K = H(1) + H(2) + (N-1)H(12)$$

where $H(i)$ denotes the interaction between particle i and the fixed environment, while $H(ij)$ denotes the interaction between the ith and jth particles. Note that $\langle\psi|H(ij)|\psi\rangle$ is independent of which pair of distinct integers (ij) denotes. Similarly, $\langle\psi|H(i)|\psi\rangle$ is independent of i. It is then merely a question of counting to show that the energy, E, of the system is given by

$$E = \langle\psi|H|\psi\rangle$$
$$= \tfrac{1}{2}N(KD^2)$$

Taking a hint from the treatment of helium by Hylleraas, I realized that one *merely had to choose* $D^2(12, 1'2')$ to minimize the above expression for fixed N and with K appropriate for any quantum system of N identical fermions to obtain the ground-state energy level.

To impress physicists one needed to do this for a system more complicated than helium. So I tried to find the ground state of lithium assuming that my guess for $D^2(12, 1'2')$ was restricted only by the conditions that it be antisymmetric in 12 and $1'2'$ and change these pairs under complex conjugation. I did *too well*, obtaining a level about 10% BELOW the observed ground-state energy!

Impossible!

It did not take long, perhaps a day, to realize that I had not imposed some limitation on the allowed 2-RDM additional to those mentioned above.

I have no record of how long it took for me to realize that the needed conditions were that the 2-matrix be derived from a function that is antisymmetric in N particles. This led me to invent the term "N-representable" to point to a key obstruction to solving N-electron problems by variation if N is larger than 2. I believed that I had made a huge step forward and later in that summer brashly claimed to a group of physicists at Chalk River that I had reduced the problem for arbitrary N to a 2.5 particle problem. This claim is so intriguing that it attracted several scientists, especially chemists, to attempt to use my approach. I assured my audience at Chalk River that the obstacle of N-representability would quickly be overcome by an able mathematician—presumably, like myself! This proved the arrogant idea of a brash young scientist since the search for a neat easy solution has not ended after 55 years.

The search was first pursued in a series of conferences organized by Bob Erdahl, Hans Kummer, the late Vedene Smith, Jr., and myself. However, many others have been involved, notably Prof. Valdemoro and her colleagues in Spain, Prof. Nakatsuji and his associates in Japan, and since completing his Ph.D. at Harvard, Prof. Mazziotti in Chicago.

This book shows that great progress has been made in using the 2-matrix effectively, especially in chemistry. I believe that the role of RDM for condensed matter physics is just as important as in chemistry. Some of these connections will be explored in later chapters.

CHAPTER 2

HISTORICAL INTRODUCTION

MITJA ROSINA

Faculty of Mathematics and Physics, University of Ljubljana, Jadranska 19, P.O. Box 2964, 1001 Ljubljana, Slovenia and Jožef Stefan Institute, Ljubljana, Slovenia

CONTENTS

I. A Short Chronicle
II. The Seven International Conferences/Workshops on Reduced Density Matrices
III. Recollections of a Nuclear Physicist
References

I. A SHORT CHRONICLE

In the Pre-RDM Era, hominids made impressive progress in understanding the principles of quantum mechanics and they were solving a limited set of relevant physical problems at an unprecedented rate. The main tool became the wavefunction. However, the problems were more or less of a single-particle type (independent particles in a potential, possibly a mean potential approximating the influence of other particles). When many-body problems appeared, the single-particle picture was no longer accurate enough and the calculation of a many-body wavefunction was difficult, indeed.

Then, in the Old Ages (1940 or 1951–1967) some ingenious people became aware that, in the case of two-body interactions, it is the two-particle reduced density matrix (2-RDM) that carries in a compact way all the relevant information about the system (energy, correlations, etc.). Early insight by Husimi (1940) and challenges by Charles Coulson were completed by a clear realization and formulation of the N-representability problem by John Coleman in 1951 (for the history, see his book [1] and Chapters 1 and 17 of the present book). Then a series of theorems on N-representability followed, by John Coleman and many

---

Reduced-Density-Matrix Mechanics: With Application to Many-Electron Atoms and Molecules,
A Special Volume of Advances in Chemical Physics, Volume 134, edited by David A. Mazziotti.
Series editor Stuart A. Rice. Copyright © 2007 John Wiley & Sons, Inc.

others (e.g., Fukashi Sasaki's upper bound on eigenvalues of 2-RDM [2]). Claude Garrod and Jerome Percus [3] formally wrote the necessary and sufficient N-representability conditions. Hans Kummer [4] provided a generalization to infinite spaces and a nice review. Independently, there were some clever practical attempts to reduce the three-body and four-body problems to a reduced two-body problem without realizing that they were actually touching the variational 2-RDM method: Fritz Bopp [5] was very successful for three-electron atoms and Richard Hall and H. Post [6] for three-nucleon nuclei (if assuming a fully attractive nucleon–nucleon potential).

The Middle Ages (1967–1985) can be characterized by the six RDM conferences and workshops, which are listed and discussed in the next section. These meetings turned out to be a great catalyst among participants from different branches of science: mathematicians, physicists, and chemists (at that time, computer scientists were still missing!). They became aware that they were not alone in RDM research and many collaborations started. In the Middle Ages, the list of useful N-representability conditions was still insufficient and computer power too weak. Regarding numerical results, RDMs could hardly compete with wavefunctions. There were, however, many new conceptual insights in atomic and nuclear many-body systems, such as natural orbitals and geminals, the role of symmetries, characterization of correlations, and pairing. In parallel, the highly successful *density functional method* was evolving.

In the 1990s, the New Ages started, with the breakthrough of Hiroshi Nakatsuji, Carmela Valdemoro, and David Mazziotti. They introduced improved N-representability by means of a hierarchy of equations connecting p-RDMs with $(p + 2)$-RDMs (e.g., the *contracted Schrödinger equation*). Also, increased computer power and improved algorithms in *semidefinite programming* allowed very promising practical atomic and molecular calculations. References are given in later chapters.

II. THE SEVEN INTERNATIONAL CONFERENCES/ WORKSHOPS ON REDUCED DENSITY MATRICES

1. **Density Matrix Conference**, Kingston, August 28–September 1, 1967. Sponsored by: U.S. Air Force, Office of Scientific Research; U.S. Office of Naval Research; National Research Council of Canada; Queen's University. Co-organizers: A. J. Coleman and R. M. Erdahl. Proceedings: A. J. Coleman and R. M. Erdahl, editors, *Reduced Density Matrices with Applications to Physical and Chemical Systems*, Queen's Papers in Pure and Applied Mathematics No. 11 (1967), 434 pp.

This was a great "coming together" of mathematicians, physicists, and quantum chemists, and an exciting review of the progress already achieved with the

RDM method. The participants learned about other researchers, the interdisciplinary nature of the problem, the existence of *N*-representability theorems, and the availability of computational methods. Many collaborations started.

2. **Density Matrix Seminar**, Kingston, June 17–July 12, 1968. Sponsored by: U.S. Air Force, Office of Scientific Research; U.S. Office of Naval Research; Queen's University. Co-organizers: A. J. Coleman and R. M. Erdahl. Proceedings: A. J. Coleman and R. M. Erdahl, editors, *Report of the Density Matrix Seminar*, Queen's Press (1968), 78 pp.

A beautiful month on Lake Ontario in a very friendly atmosphere. John Coleman, Bob Erdahl, Claude Garrod, Richard Hall, Hans Kummer, J. Lindenberg, R. McWeeny, Yngve Ohrn, David Peat, Mitja Rosina, Mary-Beth Ruskai, Darwin Smith, George Warsket, Antonio Ciampi, Ernest Davidson, and others discussed *N*-representability, the interpretation of RDMs, and other unsolved problems.

3. **Density Matrix Seminar II**, Kingston, August 4–29, 1969. Sponsored by: U.S. Air Force, Office of Scientific Research; Queen's University. Co-organizers: A. J. Coleman and R. M. Erdahl. Proceedings: A. J. Coleman and R. M. Erdahl, editors, *Report of the Density Matrix Seminar*, Queen's Press (1969), 151 pp.

Some previous and some new participants continued to explore the characterization of RDMs, related conceptual problems, as well as practical problems to incorporate correlations.

4. **Reduced Density Operators Conference**, Kingston, June 20–22, 1974. Sponsored by: National Research Council of Canada; Queen's University. Organizer: R. M. Erdahl. Proceedings: R. M. Erdahl, editor, *Reduced Density Operators with Applications to Physical and Chemical Systems—II*, Queen's Papers in Pure and Applied Mathematics No. 40 (1974), 234 pp.

The Conference was followed by an extended workshop. The *Density Matrix Club* had increased. The structure and symmetries of RDM were further studied, and direct variational calculations were encouraged. Some new names were Hubert Grudzinski, Everett Larson, and Vedene Smith. The lively workshop encouraged the initiation of an informal newsletter to be distributed to old and new participants. Three Editions of *RDO News* followed, edited by Bob Erdahl (1975, 1976, 1977).

5. **Reduced Density Matrices Conference**, Université de Moncton, New Brunswick, June 1977. Sponsored by: National Research Council of Canada; Queen's University. Co-organizers: A. J. Coleman, R. M. Erdahl, and V. H. Smith, Jr. Proceedings: A. J. Coleman, R. M. Erdahl, and V. H. Smith, Jr., editors, *Proceedings of the Reduced Density Matrix Conference at Moncton, New Brunswick, International Journal of Quantum Chemistry* **13** (1978), 204 pp.

Again a rich harvest of new ideas.
6. **Density Matrices and Density Functionals, the A. John Coleman Symposium**, Kingston, August 25–29, 1985. Sponsored by: National Sciences and Engineering Research Council of Canada; Queen's University. Co-organizers: R. M. Erdahl and V. H. Smith, Jr. Proceedings: R. M. Erdahl and V. H. Smith, Jr., editors, *Density Matrices and Density Functionals*, Reidel (1987), 600 pp.

During the lively symposium the history and concepts of RDM were reviewed. Even small steps toward N-representability were welcomed. Many details in calculated densities and correlations emerged. The comparison (or competition) between RDM and density functionals was interesting. The symposium was followed by a seminar including a few enthusiasts.

7. **Reduced Density Matrix Workshop**, Kingston, August 29–31, 1999. Sponsored by: Queen's University. Organizer: A. J. Coleman. Monograph (Instead of Proceedings): Jerzy Cioslowski, editor, *Many-Electron Densities and Reduced Density Matrices*, Kluwer Academic/Plenum (2000), 301 pp.

New optimism was brought into the field of RDMs by Hiroshi Nakatsuji, Carmela Valdemoro, and David Mazziotti with their cumulant expansion, the hierarchy of equations connecting the 2-RDM with 4-RDMs, and the contracted Schrödinger equation. John Coleman continues to be the "motor" for further progress.

III. RECOLLECTIONS OF A NUCLEAR PHYSICIST

In the 1950s, many basic nuclear properties and phenomena were qualitatively understood in terms of single-particle and/or collective degrees of freedom. A hot topic was the study of collective excitations of nuclei such as giant dipole resonance or shape vibrations, and the state-of-the-art method was the *nuclear shell model* plus *random phase approximation* (RPA). With improved experimental precision and theoretical ambitions in the 1960s, the *nuclear many-body problem* was born. The importance of the ground-state correlations for the transition amplitudes to excited states was recognized.

In Ljubljana, we participated in the measurements of the giant dipole resonance in light nuclei (1958–1960) and we discovered its rich structure. To go beyond RPA, we introduced a configuration interaction of the two-particle–two hole type, which indeed split the resonance in rough agreement with experiment (1962). However, for improvement, we needed some ground-state correlations and we expressed [7] the ground-state energy (actually the G-matrix) in terms of bilinear products of transition amplitudes to chosen excited states n:

$$G_{c,d}^{a,b} = \sum_n A_{a,b}^{(n)*} A_{c,d}^{(n)} = \sum_n \langle g|(\hat{a}_a^\dagger \hat{a}_b)^\dagger |n\rangle \langle n|\hat{a}_c^\dagger \hat{a}_d|g\rangle$$

When applying the variational principle to the ground-state energy, we realized that we have to satisfy some subsidiary relations for our variational parameters A. And there we were, involved in the N-representability problem! We were very excited, hoping to find an alternative to wavefunction calculations. Soon we became aware that some other people such as Fritz Bopp, Fukashi Sasaki, and, above all, John Coleman were already ahead of us, attacking this fortress. My optimism suggested: let the mathematicians rigorously solve the mathematical problem, and we physicists (and quantum chemists) will fruitfully apply it. Fortunately, I was impatient to wait and I soon realized that we have to collaborate. I was extremely lucky that John Coleman invited me to the very first RDM Conference (1967), where I presented our variational calculation with transition amplitudes (precursor of the RDM approach) and the one-to-one mapping from the 2-RDM to the N-particle wavefunction in the case of the ground state of a Hamiltonian with at most two-body interactions [8]. I participated in six of the seven RDM conferences or seminars [9–10]. The search for N-representability conditions on the 2-RDM and viable algorithms to implement the conditions became team work (which I enjoyed very much). The progress was exciting but slow, with its ups and downs. It was the infectious optimism of John Coleman (and of my senior collaborator in Ljubljana, Miodrag Mihailović) that kept me up [17, 18].

It was a fruitful period when I collaborated with Claude Garrod [19, 20]. He had also been excited by RDMs, and he had discovered (with Jerome Percus) the importance of the G-matrix nonnegativity condition [3]; whichever simple many-body problem they tried, they obtained exact results—they had thought they had resolved the N-representability. However, with more complicated interactions, the results were poorer, and Claude Garrod started the search for new conditions. He realized, however, that the method should be computationally viable and suggested a type of variational calculation in which D- and G-matrix nonnegativity would be imposed iteratively by a converging infinite series of linear inequalities. This algorithm later became known as the *cutting plane method*, an extension of *dual simplex*. It is amusing that the method has been rediscovered recently under the name of *semidefinite programming*. Together, we developed the code and implemented the method to some atoms (Be) and light nuclei (^{15,16,17}O, ^{20}Ne, ^{24}S, ^{28}Si). While it was excellent for the atom, it gave only 10% precision for nuclei (compared to configuraton interaction calculations) [17, 21–24]. Also, the imprecision and convergence time increased with the number of particles. This means that the method *is clever, and it is aware* of the type of interaction (Coulombic versus nuclear) as well as the number of particles (even if it enters only as a parameter—the trace of the D-matrix).

An interesting application of the 2-RDM was the calculation of excited states in the space of one-particle one-hole excitations [18, 25–30] with reasonably good results, as well as the study of wavepacket dynamics [31].

The partial sucess/failure slowed down the applications, especially as the computers at that time were too slow to manage larger model spaces and additional, more complicated, N-representability conditions. Some hope was offered by applying symmetries—orbital rotation, spin, isobaric spin—and it was stimulating to explore them with Bob Erdahl and my younger collaborator Bojan Golli [32]. However, new ideas were needed.

The final breakthrough came with the advent of powerful computers, which enabled the algorithms of Hiroshi Nakatsuji, Carmela Valdemoro, and David Mazziotti to come to life. I feel very happy about the revival of the RDM approach to many-body problems.

I apologize to all those RDM contributors whose worthy works I did not mention due to the limited space.

References

1. A. J. Coleman and V. I. Yukalov, *Reduced Density Matrices, Coulson's Challenge*, Springer-Verlag, New York, (2000).
2. F. Sasaki, *Phys. Rev.* **138B**, 1338–1342 (1965).
3. C. Garrod and J. K. Percus, *J. Math. Phys.* **5**, 1756 (1964).
4. H. Kummer, N-Representability problem for reduced density matrices. *J. Math. Phys.* **8**, 2063–2081 (1967).
5. F. Bopp, *Z. Phys.* **56**, 348 (1959).
6. R. L. Hall and H. R. Post, *Proc. Phys. Soc.* **A69**, 936 (1956); **A79**: 819 (1962); **A90**: 381 (1967); **A91**: 16 (1967).
7. M. V. Mihailović and M. Rosina, Excitations as ground state variational parameters. *Nucl. Phys.* **A130**, 386–400 (1969).
8. M. Rosina, Transition amplitudes as ground state variational parameters, in *Reduced Density Matrices with Applications to Physical and Chemical Systems* (A. J. Coleman and R. M. Erdahl, eds.), Queen's Papers in Pure and Applied Mathematics No. 11, Queen's University, Kingston, Ontario, 1967, p. 369.
9. M. Rosina, Large eigenvalues of the G-matrix and collective states, in *Report of the Density Matrix Seminar* (A. J. Coleman and R. M. Erdahl, eds.), Queen's Press, Kingston, Ontario, 1968, p. 51.
10. M. Rosina, A lower bound on the ground state energy of some few-nucleon systems, in *Report of the Density Matrix Seminar* (A. J. Coleman and R. M. Erdahl, eds.), Queen's Press, Kingston, Ontario, 1969, p. 82.
11. M. Rosina, The characterization of collective states by means of density matrices, in *Report of the Density Matrix Seminar* (A. J. Coleman and R. M. Erdahl, eds.), Queen's Press, Kingston, Ontario, 1969, p. 89.
12. C. Garrod and M. Rosina, The nonnegativity of the G-matrix and saturation, in *Report of the Density Matrix Seminar* (A. J. Coleman and R. M. Erdahl, eds.), Queen's Press, Kingston, Ontario, 1969, p. 76.
13. R. M. Erdahl and M. Rosina, The B-condition is implied by the G-condition, in *Reduced Density Operators with Applications to Physical and Chemical Systems—II* (R. M. Erdahl, ed.), Queen's Papers in Pure and Applied Mathematics No. 40, Queen's University, Kingston, Ontario, 1974, p. 36.

14. M. Rosina, (a) Direct variational calculation of the two-body density matrix; (b) On the unique representation of the two-body density matrices corresponding to the AGP wave function; (c) The characterization of the exposed points of a convex set bounded by matrix nonnegativity conditions; (d) Hermitian operator method for calculations within the particle-hole space; in *Reduced Density Operators with Applications to Physical and Chemical Systems—II* (R. M. Erdahl, ed.), Queen's Papers in Pure and Applied Mathematics No. 40, Queen's University, Kingston, Ontario, 1974, (a) p. 40, (b) p. 50, (c) p. 57, (d) p. 126.

15. M. Rosina, B. Golli, and R. M. Erdahl, A lower bound to the ground state energy of a boson system with fermion source, in *Density Matrices and Density Functionals* (R. M. Erdahl and V. H. Smith, Jr., eds.), Reidel, Dordrecht, 1987, p. 231.

16. M. Rosina, Some theorems on uniqueness and reconstruction of higher-order density matrices, in *Many-Electron Densities and Reduced Density Matrices* (J. Cioslowski, ed.), Kluwer Academic/Plenum Publishers, New York, 2000, pp. 19–32.

17. M. V. Mihailović and M. Rosina, The variational approach to the density matrix for light nuclei. *Nucl. Phys.* **A237**, 221–228 (1975).

18. M. V. Mihailović and M. Rosina, The particle–hole states in some light nuclei calculated with the two-body density matrix of the ground state. *Nucl. Phys.* **A237**, 229–234 (1975).

19. M. Rosina, J. K. Percus, L. J. Kijewski, and C. Garrod, Reduced density matrices of energy eigenstates. *J. Math. Phys.* **10**, 1761–1763 (1969).

20. M. Rosina and C. Garrod, The particle–hole matrix: its connection with the symmetries and collective features of the ground state. *J. Math. Phys.* **10**, 1855 (1969).

21. C. Garrod, M. V. Mihailović, and M. Rosina, The variational approach to the two-body density matrix. *J. Math. Phys.* **16**, 868–874 (1975).

22. M. Rosina and C. Garrod, The variational calculation of reduced density matrices. *J. Computational Phys.* **18**, 300–310 (1975).

23. M. Rosina, Direct variational calculation of the two-body density matrix, in *The Nuclear Many-Body Problem (Proceedings of the Symposium on Present Status and Novel Developments in the Nuclear Many-Body Problem*, Rome 1972, (F. Calogero and C. Ciofi degli Atti, eds.), Editrice Compositori, Bologna, 1973.

24. M. Rosina, Lower bounds to ground state energy obtained by variational calculation of density matrices, in *Sur des Récents Developpements en Physique Nucléaire Théorique* (E. El Baz, ed.), Session d'Etudes de La Toussuire, 1973.

25. M. Rosina and M. V. Mihailović, The determination of the particle–hole excited states by using the variational approach to the ground state two-body density matrix, in *International Conference on Properties of Nuclear States*, Montreal 1969, Les Presses de l'Université de Montreal, 1969.

26. M. Bouten, P. Van Leuven, M. V. Mihailović, and M. Rosina, A new particle–hole approach to collective states. *Nucl. Phys.* **A202**, 127–144 (1973).

27. M. Bouten, P. Van Leuven, M. V. Mihailović, and M. Rosina, Two exactly soluble models as a test of the Hermitian operator method. *Nucl. Phys.* **A221**, 173–182 (1974).

28. M. Rosina, The calculation of excited states in the particle–hole space using the two-body density matrix of the ground state, in *Proceedings of the International Conference on Nuclear Structure and Spectroscopy*, Vol. 1 (H. P. Blok and A. E. L. Dieperink, eds.), North-Holland, Amsterdam, 1974.

29. M. V. Mihailović and M. Rosina, Particle–hole states in light nuclei, in *Proceedings of the International Conference on Nuclear Self-Consistent Fields*, Trieste 1975 (G. Ripka and M. Porneuf, eds.), North-Holland, Amsterdam, 1975, p. 37.

30. M. Rosina, Application of the two-body density matrix of the ground state for calculations of some excited states. *Int. J. Quantum Chem.* **13**, 737–742 (1978).
31. M. Rosina and P. Van Leuven, Density-matrix approach to wave-packet dynamics. *Phys. Rev. A* **45**, 64–69 (1992).
32. R. M. Erdahl, C. Garrod, B. Golli, and M. Rosina, The application of group theory to generate new representability conditions for rotationally invariant density matrices. *J. Math. Phys.* **20**, 1366–1374 (1979).

PART II

CHAPTER 3

VARIATIONAL TWO-ELECTRON REDUCED-DENSITY-MATRIX THEORY

DAVID A. MAZZIOTTI

Department of Chemistry and The James Franck Institute, The University of Chicago, Chicago, IL 60637 USA

CONTENTS

I. Introduction
II. Theory
    A. Energy as a 2-RDM functional
    B. 2-Positivity conditions
    C. 3-Positivity conditions
    D. Partial 3-positivity conditions
        1. Lifting conditions
        2. T_1/T_2 conditions
    E. Convex set of two-particle reduced Hamiltonian matrices
        1. Convex Set of N-representable 2-RDMs
        2. Positivity and the 1-RDM
        3. Positivity and the 2-RDM
        4. Strength of positivity conditions
    F. Spin and spatial symmetry adaptation
        1. Spin adaptation and S-representability
    G. Open-shell molecules
III. Semidefinite programming
IV. Applications
V. A Look Ahead
Acknowledgments
References

I. INTRODUCTION

In 1927 Landau [1] and von Neumann [2] introduced the *density matrix* into quantum mechanics. The density matrix for the N-electron ground-state

---

Reduced-Density-Matrix Mechanics: With Application to Many-Electron Atoms and Molecules,
A Special Volume of Advances in Chemical Physics, Volume 134, edited by David A. Mazziotti.
Series editor Stuart A. Rice. Copyright © 2007 John Wiley & Sons, Inc.

wavefunction $\Psi(1, 2, \ldots, N)$, where the numbers represent the spatial and spin coordinates for each electron, is given by

$$^ND(1, 2, \ldots, N; 1', 2', \ldots, N') = \Psi(1, 2, \ldots, N)\Psi^*(1', 2', \ldots, N') \quad (1)$$

Integrating the N-electron density matrix over coordinates 3 to N generates the *two-electron density matrix* (2-RDM):

$$^2D(1, 2; 1', 2') = \int \Psi(1, 2, \ldots, N)\Psi^*(1', 2', \ldots, N) d3 \cdots dN \quad (2)$$

Because electrons are indistinguishable with only pairwise interactions, the energy of any atom or molecule may be expressed as a linear functional of the 2-RDM [3, 4]. Formulating the energy as a linear functional of the 2-RDM, however, suggests the tantalizing possibility of employing the 2-RDM rather than the many-electron wavefunction to compute the ground-state energy of atoms and molecules. In 1955 Mayer [4] performed an encouraging pencil-and-paper calculation, but Tredgold [5] soon discovered that the energy for a simple system from a trial 2-RDM could be optimized substantially *below the exact ground-state energy*. Why did the Rayleigh–Ritz variational principle not hold for the 2-RDM expression of the energy? Tredgold [5], Coleman [6], Coulson [7], and others realized that for an N-electron problem the trial 2-RDM was assuming a form that did not correspond to an N-electron wavefunction: that is, the trial 2-RDM at the minimum energy could not be obtained from the integration of an N-electron density matrix. The 2-RDM must be constrained by additional rules (or conditions) to derive from an N-electron wavefunction. Coleman described these necessary and sufficient rules as *N-representability conditions* [6].

The unsuccessful back-of-the-envelope 2-RDM calculations of Mayer and Tredgold already employed four basic requirements for a density matrix of indistinguishable fermions [6]: the matrix should be (i) normalized to conserve particle number, (ii) Hermitian, (iii) antisymmetric under particle exchange, and (iv) positive semidefinite to keep probabilities nonnegative. A matrix is *positive semidefinite* if and only if all of its eigenvalues are nonnegative. These conditions are sufficient to guarantee that 2-RDM is a density matrix but not sufficient for the matrix to be representable by an N-electron density matrix, or N-representable. What additional conditions must be imposed on a 2-RDM to restrict it to be N-representable? While a considerable research effort was initially made to understand these conditions, interest in the 2-RDM approach to many-electron atoms and molecules began to wane as the N-representability problem appeared intractable.

Interest in the 2-RDM and its N-representability returned in the 1990s with the direct calculation of the ground-state 2-RDM without the many-electron wavefunction from a self-consistent solution of the contracted Schrödinger equation [8–15]. Recent progress has revealed the importance of a class of N-representability constraints, called positivity conditions [16, 17]. Erdahl and Jin [16] and Mazziotti and Erdahl [17] generalized these conditions, originally discussed by Coleman [6] and Garrod and Percus [18], to a hierarchy of N-representability conditions, and Mazziotti and Erdahl [17] showed that each level of the hierarchy corresponds to enforcing the generalized uncertainty relations for a class of operators. With the positivity conditions, an accurate *lower bound* on the ground-state energy of many-electron atoms and molecules can be computed through a variational calculation in which the energy is directly parameterized as a linear functional of the 2-RDM [16, 17, 19–37]. The method produces realistic energies and RDMs even when the wavefunction becomes difficult to parameterize, as in transition-state structures or other stretched geometries of a potential energy surface [21, 22, 28, 29, 31, 35, 37]. Variational solution of the 2-RDM with positivity constraints requires a special constrained optimization known as *semidefinite programming*, which also has applications in control theory, combinatorial optimization, and even finance.

II. THEORY

After the energy is expressed as a functional of the 2-RDM, a systematic hierarchy of N-representability constraints, known as p-positivity conditions, is derived [17]. We develop the details of the 2-positivity, 3-positivity, and partial 3-positivity conditions [21, 27, 34, 33]. In Section II.E the formal solution of N-representability for the 2-RDM is presented through a convex set of two-particle reduced Hamiltonian matrices [7, 21]. It is shown that the positivity conditions correspond to certain classes of reduced Hamiltonian matrices, and consequently, they are exact for certain classes of Hamiltonian operators at any interaction strength. In Section II.F the size of the 2-RDM is reduced through the use of spin and spatial symmetries [32, 34], and in Section II.G the variational 2-RDM method is extended to open-shell molecules [35].

A. Energy as a 2-RDM Functional

Because electrons interact pairwise, the many-electron Hamiltonian for any atom or molecule can be written

$$\hat{H} = \sum_{i,j,k,l} {}^2K^{i,j}_{k,l} a^\dagger_i a^\dagger_j a_l a_k \qquad (3)$$

where the $a^\dagger$ and the a are the second-quantized creation and annihilation operators, the indices refer to members of a spin-orbital basis set, and the two-electron reduced Hamiltonian matrix ^2K is the matrix representation of the operator

$$^2\hat{K} = \frac{1}{N-1}\left(-\frac{1}{2}\nabla_1^2 - \sum_j \frac{Z_j}{r_{1j}}\right) + \frac{1}{2}\frac{1}{r_{12}} \qquad (4)$$

The expectation value of the Hamiltonian operator yields the many-electron energy

$$E = \sum {}^2K_{k,l}^{i,j}\,{}^2D_{k,l}^{i,j} \qquad (5)$$

$$E = \text{Tr}(^2K\,{}^2D) \qquad (6)$$

as a functional of the reduced Hamiltonian matrix and the two-electron reduced density matrix (2-RDM), where

$$^2D_{k,l}^{i,j} = \langle \Psi | a_i^\dagger a_j^\dagger a_l a_k | \Psi \rangle \qquad (7)$$

Both the energy as well as the one- and two-electron properties of an atom or molecule can be computed from a knowledge of the 2-RDM. To perform a variational optimization of the ground-state energy, we must constrain the 2-RDM to derive from integrating an N-electron density matrix. These necessary yet sufficient constraints are known as N-representability conditions.

B. 2-Positivity Conditions

General p-particle N-representability conditions on the 2-RDM are derivable from metric (or overlap) matrices. From the ground-state wavefunction $|\Psi\rangle$ and a set of p-particle operators $\{\hat{C}_{i_1,i_2,\ldots,i_p}\}$, a set of basis functions can be defined,

$$\langle \Phi_{i_1,i_2,\ldots,i_p}| = \langle\Psi|\hat{C}_{i_1,i_2,\ldots,i_p} \qquad (8)$$

for which the metric (or overlap) matrix M with elements

$$M_{j_1,j_2,\ldots,j_p}^{i_1,i_2,\ldots,i_p} = \langle \Phi_{i_1,i_2,\ldots,i_p}|\Phi_{j_1,j_2,\ldots,j_p}\rangle \qquad (9)$$

$$= \langle\Psi|\hat{C}_{i_1,i_2,\ldots,i_p}\hat{C}_{j_1,j_2,\ldots,j_p}^\dagger|\Psi\rangle \qquad (10)$$

must be positive semidefinite. We indicate that a matrix has this property by the notation $M \geq 0$. For a p-RDM that is parameterized by a wavefunction, these vector-space restrictions are *always* satisfied. More generally, however, these conditions, known as *p-positivity conditions*, offer a systematic approach for imposing N-representability conditions on an RDM *without* using the wavefunction.

When $p = 2$, we may choose the $\hat{C}_{i,j}$ in three distinct ways: (i) to create one particle in the jth orbital and one particle in the ith orbital, that is, $\hat{C}_{i,j} = a_i^\dagger a_j^\dagger$; (ii) to annihilate one particle in the jth orbital and one particle in the ith orbital (or create holes in each of these orbitals), $\hat{C}_{i,j} = a_i a_j$; and (iii) to annihilate one particle in the jth orbital and create one particle in the ith orbital, that is, $\hat{C}_{i,j} = a_i^\dagger a_j$. These three choices for the $\hat{C}_{i,j}$ produce the following *three* different metric matrices for the 2-RDM:

$$^2D_{k,l}^{i,j} = \langle \Psi | a_i^\dagger a_j^\dagger a_l a_k | \Psi \rangle \tag{11}$$

$$^2Q_{k,l}^{i,j} = \langle \Psi | a_i a_j a_l^\dagger a_k^\dagger | \Psi \rangle \tag{12}$$

$$^2G_{k,l}^{i,j} = \langle \Psi | a_i^\dagger a_j a_l^\dagger a_k | \Psi \rangle \tag{13}$$

which must be positive semidefinite if the 2-RDM is N-representable [6, 17, 18]. All three matrices contain equivalent information in the sense that rearranging the creation and annihilation operators produces linear mappings between the elements of the three matrices; particularly, the two-hole RDM ^2Q and the particle–hole RDM ^2G may be written in terms of the two-particle RDM ^2D as follows

$$^2Q_{k,l}^{i,j} = 2\,^2I_{k,l}^{i,j} - 4\,^1D_k^i \wedge {}^1I_l^j + {}^2D_{k,l}^{i,j} \tag{14}$$

and

$$^2G_{k,l}^{i,j} = {}^1I_l^j\,{}^1D_k^i - {}^2D_{k,j}^{i,l} \tag{15}$$

While all three matrices are interconvertible, the nonnegativity of the eigenvalues of one matrix does not imply the nonnegativity of the eigenvalues of the other matrices, and hence the restrictions $^2Q \geq 0$ and $^2G \geq 0$ provide two important N-representability conditions in addition to $^2D \geq 0$. These conditions physically restrict the probability distributions for two particles, two holes, and one particle and one hole to be nonnegative with respect to all unitary transformations of the two-particle basis set. Collectively, the three restrictions are known as the *2-positivity* conditions [17].

Because $^2D \geq 0$ and $^2Q \geq 0$ imply $^1D \geq 0$ and $^1Q \geq 0$ by contraction

$$^1D_k^i = \frac{1}{N-1} \sum_j {}^2D_{k,j}^{i,j} \tag{16}$$

$$^1Q_k^i = \frac{1}{r-N-1} \sum_j {}^2Q_{k,j}^{i,j} \tag{17}$$

the 2-positivity conditions imply the 1-positivity conditions. The r in the contraction of the two-hole RDM denotes the rank of the one-particle basis set. In general, the p-positivity conditions imply the q-positivity conditions for $q \leq p$. The 1-positivity conditions from the metric matrices for the one-particle and one-hole RDMs, ^1D and ^1Q, restrict the occupation numbers n_i (or eigenvalues) of the 1-RDM to lie in the interval $n_i \in [0, 1]$. Coleman showed this condition on the eigenvalues to be both *necessary and sufficient* for the N-representability of the 1-RDM [6].

C. 3-Positivity Conditions

The conditions that a 3-RDM be 3-positive follow from writing the operators in Eq. (8) as products of *three* second-quantized operators [16, 17]. The resulting basis functions lie in four vector spaces according to the number of creation operators in the product; the four sets of operators defining the basis functions in Eq. (8) are

$$\hat{C}^D_{i,j,k} = \hat{a}^\dagger_i \hat{a}^\dagger_j \hat{a}^\dagger_k \tag{18}$$

$$\hat{C}^E_{i,j,k} = \hat{a}^\dagger_i \hat{a}^\dagger_j \hat{a}_k \tag{19}$$

$$\hat{C}^F_{i,j,k} = \hat{a}_i \hat{a}_j \hat{a}^\dagger_k \tag{20}$$

$$\hat{C}^Q_{i,j,k} = \hat{a}_i \hat{a}_j \hat{a}_k \tag{21}$$

Basis functions between these vector spaces are orthogonal because they are contained in Hilbert spaces with different numbers of particles. Hence the four metric matrices that must be constrained to be positive semidefinite for 3-positivity [17] are given by

$$^3D^{i,j,k}_{p,q,r} = \langle \Psi | \hat{a}^\dagger_i \hat{a}^\dagger_j \hat{a}^\dagger_k \hat{a}_r \hat{a}_q \hat{a}_p | \Psi \rangle \tag{22}$$

$$^3E^{i,j,k}_{p,q,r} = \langle \Psi | \hat{a}^\dagger_i \hat{a}^\dagger_j \hat{a}_k \hat{a}^\dagger_r \hat{a}_q \hat{a}_p | \Psi \rangle \tag{23}$$

$$^3F^{i,j,k}_{p,q,r} = \langle \Psi | \hat{a}_i \hat{a}_j \hat{a}^\dagger_k \hat{a}_r \hat{a}^\dagger_q \hat{a}^\dagger_p | \Psi \rangle \tag{24}$$

$$^3Q^{i,j,k}_{p,q,r} = \langle \Psi | \hat{a}_i \hat{a}_j \hat{a}_k \hat{a}^\dagger_r \hat{a}^\dagger_q \hat{a}^\dagger_p | \Psi \rangle \tag{25}$$

As in Eqs.(14) and (15) for the 2-positive metric matrices, the 3-positive metric matrices are connected by linear mappings, which can be derived by rearranging the second-quantized operators. A 2-RDM is defined to be 3-positive if it arises from the *contraction* of a 3-positive 3-RDM:

$$^2D^{i,j}_{p,q} = \frac{1}{N-2} \sum_k {}^3D^{i,j,k}_{p,q,k} \tag{26}$$

Physically, the 3-positivity conditions restrict the probability distributions for "three particles," "two particles and one hole," "one particle and two holes," and "three holes" to be nonnegative with respect to all unitary transformations of the one-particle basis set. These conditions have been examined in variational 2-RDM calculations on spin systems in the work of Erdahl and Jin [16], Mazziotti and Erdahl [17], and Hammond and Mazziotti [33], where they give highly accurate energies and 2-RDMs.

D. Partial 3-Positivity Conditions

Two different *partial 3-positivity* conditions have been proposed: (i) the lifting conditions of Mazziotti [21, 33], and (ii) the T_1/T_2 conditions of Erdahl [27, 34, 38]. The T_1/T_2 conditions have been implemented for molecules by Zhao et al. [27] and Mazziotti [34].

1. Lifting Conditions

The lifted 3-RDMs [21] are defined by taking the expectation values of particle (or hole) projection operators $\hat{n}_k = \hat{a}_k^\dagger \hat{a}_k$ (or $1 - \hat{n}_k = \hat{a}_k \hat{a}_k^\dagger$) over the space spanned by the basis functions in the three metric matrices for 2-positivity:

$$^2D_{kl}^{ij} = \langle \Phi_{ij}^D | \Phi_{kl}^D \rangle = \langle \Psi | \hat{a}_i^\dagger \hat{a}_j^\dagger \hat{a}_l \hat{a}_k | \Psi \rangle \tag{27}$$

$$^2Q_{kl}^{ij} = \langle \Phi_{ij}^Q | \Phi_{kl}^Q \rangle = \langle \Psi | \hat{a}_i \hat{a}_j \hat{a}_l^\dagger \hat{a}_k^\dagger | \Psi \rangle \tag{28}$$

$$^2G_{lj}^{ik} = \langle \Phi_{ik}^G | \Phi_{lj}^G \rangle = \langle \Psi | \hat{a}_i^\dagger \hat{a}_k \hat{a}_j^\dagger \hat{a}_l | \Psi \rangle \tag{29}$$

where $|\Phi_{kl}^D\rangle$, $|\Phi_{kl}^Q\rangle$, and $|\Phi_{kl}^G\rangle$, are $(N-2)$-, $(N+2)$-, and N-particle basis functions, respectively. An example of this type of expectation value is

$$\langle \Phi_{ij}^D | (1 - \hat{n}_k) | \Phi_{lm}^D \rangle = \langle \Psi | \hat{a}_i^\dagger \hat{a}_j^\dagger \hat{a}_k \hat{a}_k^\dagger \hat{a}_m \hat{a}_l | \Psi \rangle \tag{30}$$

which is the ^3E matrix in Eq. (22) with an upper index set equal to a lower set. Summing over the particle projection operators for all orbital basis functions gives the number operator $\hat{N}_k = \sum_k \hat{n}_k$. Hence, because Eq. (30) contracts to the G-condition, it includes the N-representability restrictions from the G-condition as well as additional constraints [21]. The lifted conditions are part of the four 3-positivity conditions since every principal submatrix of a positive semidefinite matrix must also be positive semidefinite. By inserting either the particle or hole projection (or lifting) operator between the basis functions $|\Phi^D\rangle$, we generate two lifted metric matrices ^3D and ^3E. Similarly, from the basis functions for ^2Q and ^2G, we generate four more conditions for a total of *six* partial 3-positive conditions:

$$\langle \Phi^D_{ij}|\hat{a}^\dagger_k\hat{a}_k|\Phi^D_{lm}\rangle = \langle \Psi|\hat{a}^\dagger_i\hat{a}^\dagger_j\hat{a}^\dagger_k\hat{a}_k\hat{a}_m\hat{a}_l|\Psi\rangle = {}^3D^{ijk}_{lmk} \tag{31}$$

$$\langle \Phi^D_{ij}|\hat{a}_k\hat{a}^\dagger_k|\Phi^D_{lm}\rangle = \langle \Psi|\hat{a}^\dagger_i\hat{a}^\dagger_j\hat{a}_k\hat{a}^\dagger_k\hat{a}_m\hat{a}_l|\Psi\rangle = {}^3E^{ijk}_{lmk} \tag{32}$$

$$\langle \Phi^G_{ij}|\hat{a}^\dagger_k\hat{a}_k|\Phi^G_{lm}\rangle = \langle \Psi|\hat{a}^\dagger_i\hat{a}_j\hat{a}^\dagger_k\hat{a}_k\hat{a}^\dagger_m\hat{a}_l|\Psi\rangle = {}^3\widetilde{E}^{ijk}_{lmk} \tag{33}$$

$$\langle \Phi^G_{ij}|\hat{a}_k\hat{a}^\dagger_k|\Phi^G_{lm}\rangle = \langle \Psi|\hat{a}^\dagger_i\hat{a}_j\hat{a}_k\hat{a}^\dagger_k\hat{a}^\dagger_m\hat{a}_l|\Psi\rangle = {}^3\widetilde{F}^{ijk}_{lmk} \tag{34}$$

$$\langle \Phi^Q_{ij}|\hat{a}^\dagger_k\hat{a}_k|\Phi^Q_{lm}\rangle = \langle \Psi|\hat{a}_i\hat{a}_j\hat{a}^\dagger_k\hat{a}_k\hat{a}^\dagger_m\hat{a}^\dagger_l|\Psi\rangle = {}^3F^{ijk}_{lmk} \tag{35}$$

$$\langle \Phi^Q_{ij}|\hat{a}_k\hat{a}^\dagger_k|\Phi^Q_{lm}\rangle = \langle \Psi|\hat{a}_i\hat{a}_j\hat{a}_k\hat{a}^\dagger_k\hat{a}^\dagger_m\hat{a}^\dagger_l|\Psi\rangle = {}^3Q^{ijk}_{lmk} \tag{36}$$

Three distinct sets of linear mappings for the partial 3-positivity matrices in Eqs. (31)–(36) are important: (i) the contraction mappings, which relate the lifted metric matrices to the 2-positive matrices in Eqs. (27)–(29); (ii) the linear interconversion mappings from rearranging creation and annihilation operators to interrelate the lifted metric matrices; and (iii) antisymmetry (or symmetry) conditions, which enforce the permutation of the creation operators for fermions (or bosons). Note that the correct permutation of the annihilation operators is automatically enforced from the permutation of the creation operators in (iii) by the Hermiticity of the matrices.

2. T_1/T_2 Conditions

Because the addition of any two positive semidefinite matrices produces a positive semidefinite matrix, the four 3-positivity conditions [17] imply the following two less stringent constraints:

$$T_1 = {}^3D + {}^3Q \geq 0 \tag{37}$$

$$T_2 = {}^3E + {}^3F \geq 0 \tag{38}$$

known as the T_1 and T_2 conditions [27, 34, 38]. These conditions can be written as explicit linear functionals of the 2-RDM because the terms with six creation and/or annihilation operators in ^3D and ^3Q (as well as ^3E and ^3F) cancel upon addition due to opposite signs. If the metric matrices are expressed as cumulant expansions [14, 15, 39, 40], it can be shown that it is precisely the connected (or cumulant) parts of the metric matrices that cancel upon addition [33]. Hence the T_1 and T_2 matrices are unconnected. The T_1 and T_2 matrices correspond to metric matrices, where the operators $\hat{C}_{i,j,k}$ in Eq. (9) are defined as

$$\hat{C}^{T_1}_{i,j,k} = a^\dagger_i a^\dagger_j a^\dagger_k + a_i a_j a_k \tag{39}$$

$$\hat{C}^{T_2}_{i,j,k} = a^\dagger_i a^\dagger_j a_k + a_i a_j a^\dagger_k \tag{40}$$

respectively.

In contrast to the ^3E and ^3F metric matrices in 3-positivity, the strength of the T_2 matrix as a 2-RDM N-representability condition is not completely invariant upon altering the order of the second-quantized operators in $\hat{C}^{T_2}_{i,j,k}$. For example, a slightly different metric matrix $\tilde{T}_2$ can be defined by exchanging the operators a_i and $a_k^\dagger$ in Eq. (40) to obtain

$$\hat{C}^{\tilde{T}_2}_{i,j,k} = a_i^\dagger a_j^\dagger a_k + a_k^\dagger a_j a_i \tag{41}$$

This dependence on ordering occurs because, unlike the set of operators $\hat{C}^E_{i,j,k}$ and $\hat{C}^F_{i,j,k}$ in the 3-positivity conditions, the operators $\hat{C}^{T_2}_{i,j,k}$ do not include the set of single-particle excitation and deexcitation operators, that is,

$$\left\{ a_j^\dagger, a_j \right\} / \subset \left\{ \hat{C}^{T_2}_{i,j,k} \right\} \tag{42}$$

To demonstrate the reason for this difference between ^3E and T_2, we note that

$$\sum_i \hat{C}^E_{i,j,i} = -a_j^\dagger \hat{N} \tag{43}$$

where $\hat{N}$ is the number operator, while

$$\sum_i \hat{C}^{T_2}_{i,j,i} = -\left(a_j^\dagger + a_j \right) \hat{N} \tag{44}$$

Rearranging the operators $\hat{C}^{T_2}_{i,j,i}$ to $\hat{C}^{\tilde{T}_2}_{i,j,i}$ produces a term with a single annihilation operator. Because this term cannot be expressed in terms of the set of operators $\{\hat{C}^{T_2}_{i,j,k}\}$, the space spanned by the basis functions in the metric matrix T_2 differs slightly from the space spanned by the basis functions in the metric matrix $\tilde{T}_2$.

A *generalized* metric matrix $\bar{T}_2$, however, can be obtained by supplementing the $\hat{C}^{T_2}_{i,j,k}$ operators from Eq. (40) with the set of single-particle excitation and deexcitation operators:

$$\hat{C}^{\bar{T}_2} \in \left\{ a_i^\dagger, a_i, a_i^\dagger a_j^\dagger a_k + a_i a_j a_k^\dagger \right\} \tag{45}$$

The generalized $\bar{T}_2$ matrix is contained in Erdahl's original theoretical treatment of these conditions [38], although the recent applications to atoms and molecules [27, 34] employ either T_2 or $\tilde{T}_2$. The condition $\bar{T}_2 \geq 0$ implies both $T_2 \geq 0$ and $\tilde{T}_2 \geq 0$ as well as any other conditions from different orderings of the second-quantized operators.

E. Convex Set of Two-Particle Reduced Hamiltonian Matrices

The formal solution of N-representability for the 2-RDM is developed in terms of a convex set of two-particle reduced Hamiltonian matrices. To complement the derivation of the positivity conditions from the metric matrices, we derive them from classes of these two-particle reduced Hamiltonian matrices. This interpretation allows us to demonstrate that the 2-positivity conditions are exact for certain classes of Hamiltonian operators for any interaction strength. In this section all of the RDMs are normalized to unity. Much of this discussion appeared originally in Refs. [21, 29].

1. Convex Set of N-Representable 2-RDMs

The energy for a system of N fermions with p-particle interactions may be written as a linear functional of the p-RDM:

$$E = \text{Tr}[H^N D] = \text{Tr}[^p K\, ^p D] \tag{46}$$

where $^p K$ is the p-particle reduced Hamiltonian matrix. A contraction operator L_N^p may be defined to integrate (or sum) the N-particle density matrix to the p-RDM, where in this section we assume that all density matrices are normalized to unity. Employing the contraction operator in Eq. (46) and taking its adjoint

$$E = \text{Tr}[^p K\, L_N^p(^N D)] = \text{Tr}[\Gamma_p^N(^p K)\,^N D] \tag{47}$$

defines a *lifting operator* Γ_p^N, which by comparison with Eq. (46) maps the p-particle reduced Hamiltonian matrix to the N-particle Hamiltonian H [41, 42]. The lifting operator may be evaluated with a Grassmann *wedge product* of the p-particle reduced Hamiltonian matrix with the $(N-p)$-particle identity matrix

$$H = \Gamma_p^N(^p K) = {}^p K \wedge {}^{(N-p)} I \tag{48}$$

where the Grassmann wedge $\wedge$ denotes the antisymmetric tensor product [13, 43]. The wedge product is computed by summing all distinct antisymmetric permutations of the upper and lower indices and dividing by the total number of permutations.

Direct minimization of the energy as a functional of the p-RDM may be achieved if the p-particle density matrix is restricted to the set of N-representable p-matrices, that is, p-matrices that derive from the contraction of at least one N-particle density matrix. The collection of ensemble N-representable p-RDMs forms a *convex set*, which we denote as P_p^N. To define P_p^N, we first consider the convex set B_p^N of p-particle reduced Hamiltonians, which are positive semidefinite (nonnegative eigenvalues) *after* being lifting to the N-particle space:

$$B_p^N = \left\{ {}^pB | {}^pB \wedge {}^{(N-p)}I \geq 0 \right\} \tag{49}$$

When $N = p$, the set B_p^p simply contains the p-particle reduced Hamiltonians, which are positive semidefinite, but when $N = p + 1$, because the lifting process raises the lowest eigenvalue of the reduced Hamiltonian, the set B_p^{p+1} also contains p-particle reduced Hamiltonians that are *lifted* to positive semidefinite matrices. Consequently, the number of N-representability constraints must increase with N, that is, $B_p^N \subset B_p^{N+1}$. To constrain the p-RDMs, we do not actually need to consider all ${}^pB$ in B_p^N, but only the members of the convex set B_p^N, which are *extreme* A member of a convex set is *extreme* if and only if it cannot be expressed as a positively weighted ensemble of other members of the set (i.e., the extreme points of a square are the four corners while every point on the boundary of a circle is extreme). These extreme constraints form a *necessary* and *sufficient* set of N-representability conditions for the p-RDM [18, 41, 42], which we can formally express as

$$P_p^N = \left\{ {}^pD | \mathrm{Tr}[{}^pB {}^pD] \geq 0, \forall {}^pB \in B_p^N \right\} \tag{50}$$

The set of N-representable p-RDMs becomes smaller as N increases, that is, $P_p^{N+1} \subset P_p^N$.

A significant class of p-particle reduced Hamiltonians in the set B_p^N includes those that are positive semidefinite ${}^pB \geq 0$, and the extreme Hamiltonian matrices that satisfy this positivity constraint may be parameterized as follows:

$${}^pB_j^i = c_i c_j^* \tag{51}$$

Substitution of this class of reduced Hamiltonians into Eq. (50) gives

$$\sum_{i,j} c_i c_j^* \, {}^pD_j^i \geq 0 \tag{52}$$

which is one definition for restricting the p-RDM to be *positive semidefinite* ${}^pD \geq 0$. However, for $p < N$ this does not exhaust the extreme elements of the set B_p^N. The task of determining the complete set of ${}^pB$ *without* checking the conditions in Eq. (49) may appear daunting or even impossible. However, in the next sections this set is described exactly for the 1-RDM at the one-particle level and approximately for the 2-RDM at the two-particle level. Special emphasis is placed on the interpretation of positivity conditions as testing significant classes of extreme reduced Hamiltonians.

2. Positivity and the 1-RDM

A quantum system of N particles may also be interpreted as a system of $(r-N)$ holes, where r is the rank of the one-particle basis set. The complementary nature of these two perspectives is known as the *particle–hole duality* [13, 44, 45]. Even though we treated only the N-representability for the particles in the formal solution, any p-hole RDM must also be derivable from an $(r-N)$-hole density matrix. While the development of the formal solution in the literature only considers the particle reduced Hamiltonian, both the particle and the hole representations for the reduced Hamiltonian are critical in the practical solution of N-representability problem for the 1-RDM [6, 7]. The hole definitions for the sets B_p^{r-N} and P_p^{r-N} are analogous to the definitions for particles except that the number $(r-N)$ of holes is substituted for the number of particles. In defining the hole RDMs, we assume that the rank r of the one-particle basis set is finite, which is reasonable for practical calculations, but the case of infinite r may be considered through the limiting process as $r \to \infty$.

The ground-state energy for the N-particle Hamiltonian defined with ^1K in Eq. (48) may be expressed from either the particle or the hole perspective:

$$E = \text{Tr}[^1K\,^1D] \tag{53}$$
$$= \text{Tr}[^1\bar{K}\,^1\bar{D}]$$

where the 1-hole RDM and reduced Hamiltonian may be written in terms of the 1-RDM and the 1-particle reduced Hamiltonian through the rearrangement of the creation and the annihilation operators:

$$(r-N)\,^1\bar{D} + N\,^1D = {}^1I \tag{54}$$

and

$$^1\bar{K}_j^i = \frac{1}{N}(\text{Tr}[^1K]\,^1I_j^i - (r-N)\,^1K_j^i) \tag{55}$$

Any arbitrary one-particle reduced Hamiltonian shifted by its N-particle ground-state energy must be expressible by the extreme Hamiltonian elements in the convex set B_1^N. As we showed in Eq. (52), keeping the 1-RDM positive semidefinite is equivalent to applying the N-representability constraints in Eq. (50) for the class of extreme positive semidefinite ^1B, which may be parameterized by

$$^1B_j^i = c_i c_j^* \tag{56}$$

Each *extreme* ^1B matrix is a projector onto an orbital defined by the set of expansion coefficients $\{c_i\}$. This class of Hamiltonians, however, is not complete, as may be seen by shifting an arbitrary Hamiltonian ^1K by its N-particle ground-state energy E_N and then expanding the resulting matrix ^1C in terms of its eigenvalues $\{\epsilon_i\}$ and eigenvectors $\{c_i\}$:

$$^1C = {}^1K - E_N\,{}^1I = \sum_i \epsilon_i c_i c_i^* \tag{57}$$

Because E_N is the N-particle energy and not the lowest eigenvalue of ^1K, some of the eigenvalues of ^1C will be *negative*, and this portion of the reduced Hamiltonian cannot be represented by the positive semidefinite Hamiltonians in Eq. (56).

A similar argument, however, may also be made from the perspective of the holes. Restricting the one-hole RDM to be positive semidefinite corresponds to applying the N-representability constraints in Eq. (50) to the class of extreme positive semidefinite $^1\bar{B}$,

$$^1\bar{B}_j^i = c_i c_j^* \tag{58}$$

or, after being mapped to the particle reduced Hamiltonian,

$$^1B_j^i = \frac{1}{r-N}({}^1I_j^i - N\,{}^1c_i c_j^*) \tag{59}$$

While the extreme Hamiltonians in either Eq. (56) or Eq. (59) alone are not sufficient, together they provide all of the extreme Hamiltonians in the set B_1^N. Independent of the correlation present in the 1-RDM, an ensemble of the extreme elements in Eqs. (56) and (59) may generate any energy-shifted Hamiltonian ^1C—both the positive and the negative parts of its spectrum. Proof of this important idea was first given by Coleman [6, 7]. From the formal definition of the N-representability constraints in Eq. (50), therefore, the positivity of the one-particle and the one-hole RDMs is *necessary* and *sufficient* for the ensemble N-representability of the 1-RDM [6]. This result highlights the importance of examining different representations of the reduced Hamiltonian. Without wedging to the N-particle space the particle- and the hole-reduced Hamiltonians provide a complete solution of ensemble representability for the 1-RDM on the one-particle space.

3. Positivity and the 2-RDM

The ground-state energies of atoms and molecules where the N-particle Hamiltonian is defined by Eq. (48) may be expressed through *three* different representations of the 2-RDM and the two-particle reduced Hamiltonian:

$$\begin{aligned} E &= \text{Tr}[^2K\,{}^2D] \\ &= \text{Tr}[^2\bar{K}\,{}^2Q] \\ &= \text{Tr}[^2\tilde{K}\,{}^2G] \end{aligned} \tag{60}$$

where the elements of the two-hole RDM are given by

$$^2Q^{i,j}_{k,l} = \frac{1}{n_q}\langle\Psi|a_i a_j a^\dagger_l a^\dagger_k|\Psi\rangle \geq 0 \tag{61}$$

or in terms of the 1- and the 2-RDMs

$$\frac{n_q}{2}{}^2Q^{i,j}_{k,l} = {}^2I^{i,j}_{k,l} - N\,{}^1D^i_k \wedge {}^1I^j_l + \frac{n_d}{2}{}^2D^{i,j}_{k,l} \tag{62}$$

and the elements of the two-particle G-matrix are given by

$$^2G^{i,k}_{l,j} = \frac{1}{n_g}\langle\Psi|a^\dagger_i a_k a^\dagger_j a_l|\Psi\rangle \geq 0 \tag{63}$$

or in terms of the 1- and the 2-RDMs

$$n_g\,{}^2G^{i,k}_{l,j} = N\delta^j_k\,{}^1D^i_l + n_d\,{}^2D^{i,j}_{k,l} \tag{64}$$

The factors n_d, n_q, and n_g, which normalize the D-, the Q- and the G-matrices to unity, are given by $N(N-1)$, $(r-N)(r-N-1)$, and $N(r-N+1)$, respectively. Because the D-, the Q-, and the G-matrices are expressible as metric (or overlap) matrices M,

$$M^i_j = \langle\Phi_i|\Phi_j\rangle = \langle\Psi|\hat{C}_i\hat{C}^\dagger_j|\Psi\rangle \geq 0 \tag{65}$$

each of them must be positive semidefinite for the 2-RDM to be N-representable. These positivity conditions were originally proposed by Garrod and Percus [18].

Using the second-quantized definitions of the D-, the Q-, and the G- matrices, the Q- and the G-reduced Hamiltonians $^2\bar{K}$ and $^2\tilde{K}$ in Eq. (60) may be expressed in terms of the usual D-representation of the reduced Hamiltonian:

$$^2\bar{K}^{i,j}_{k,l} = \frac{2}{n_d}\left(\text{Tr}[^2K]\,{}^2I^{i,j}_{k,l} - (r-N)L^1_2[^2K]^i_k \wedge {}^1I^j_l + \frac{n_q}{2}{}^2K^{i,j}_{k,l}\right) \tag{66}$$

and

$$^2\tilde{K}^{i,k}_{l,j} = \frac{n_g}{n_d}\left(^2K^{i,j}_{k,l} + \delta^j_k/(r-N+1)\sum_m {}^2K^{i,m}_{l,m}\right) \tag{67}$$

To the Q- and the G-reduced Hamiltonians, we may wish to affix the name *generalized* reduced Hamiltonians. Just as the D-form of the reduced Hamiltonian may be lifted to an N-particle Hamiltonian, the Q-form of the reduced

Hamiltonian may be lifted by Eq. (48) to an $(r-N)$-hole Hamiltonian, which shares the same ground-state as the N-particle Hamiltonian. A similar lifting may be extended to the G-form of the reduced Hamiltonian, but the procedure is slightly more subtle since the G-matrix combines the particle and the hole perspectives.

The *three* complementary representations of the reduced Hamiltonian offer a framework for understanding the D-, the Q-, and the G-positivity conditions for the 2-RDM. Each positivity condition, like the conditions in the one-particle case, correspond to including a different class of two-particle reduced Hamiltonians in the N-representability constraints of Eq. (50). The positivity of ^2D arises from employing all positive semidefinite ^2B in Eq. (50) while the Q- and the G-conditions arise from positive semidefinite $^2\bar{B}$ and $^2\tilde{B}$, respectively. To understand these positivity conditions in the particle (or D-matrix) representation, we define the D-form of the reduced Hamiltonian in terms of the Q- and the G-representations:

$$^2B^{i,j}_{k,l} = \frac{2}{n_q}\left(\text{Tr}[^2\bar{B}]\,^2I^{i,j}_{k,l} - NL^1_2[^2\bar{B}]^i_k \wedge {}^1I^j_l + \frac{n_d}{2}\,^2\bar{B}^{i,j}_{k,l}\right) \tag{68}$$

and

$$^2B^{i,j}_{k,l} = \frac{n_d}{n_g}\left(^2\tilde{B}^{i,k}_{l,j} - \delta^j_k/(N-1)\sum_m \tilde{B}^{i,m}_{l,m}\right) \tag{69}$$

The Q- and the G-conditions are thus equivalent to the constraints in Eq. (50) with the two-particle reduced Hamiltonians in Eqs. (68) and (69), where $^2\bar{B} \geq 0$ and $^2\tilde{B} \geq 0$. Unlike the one-particle case, these reduced Hamiltonians do not exhaust all of the extreme constraints in Eq. (50), and yet the explicit forms of the Hamiltonians give us insight into the variety of correlated Hamiltonians that can be treated accurately.

4. Strength of Positivity Conditions

Many methods in chemistry for the correlation energy are based on a form of perturbation theory, but the positivity conditions are quite different. Traditional perturbation theory performs accurately for all kinds of two-particle reduced Hamiltonians, which are *close* enough to a mean-field (Hartree–Fock) reference. There are a myriad of chemical systems, however, where the correlated wavefunction (or 2-RDM) is not sufficiently close to a statistical mean field. Different from perturbation theory, the positivity conditions function by increasing the number of extreme two-particle Hamiltonians in B^N_2, which are employed as constraints upon the 2-RDM in Eq. (50) and, hence, they exactly treat a certain convex set of reduced Hamiltonians to all orders of perturbation theory. For the

D-, the Q-, and the G-conditions we have: *If the two-particle reduced Hamiltonian shifted by its N-particle ground-state energy can be written as an ensemble of the reduced Hamiltonians in the set $\{^2B \geq 0\}$ as well as the Q- and the G-reduced Hamiltonians parameterized in Eqs. (68) and (69), then the energy for an N-particle system may be computed exactly.*

To gain an understanding of this mechanism, consider the Hamiltonian operator $(\hat{H} - E_g\hat{I})$ with only two-body interactions, where E_g is the lowest energy for an N-particle system with Hamiltonian $\hat{H}$ and the identity operator $\hat{I}$. Because E_g is the lowest (or ground-state) energy, the Hamiltonian operator is positive semidefinite on the N-electron space; that is, the expectation values of $\hat{H}$ with respect to all N-particle functions are nonnegative. Assume that the Hamiltonian may be expanded as a sum of operators $\hat{O}_i\hat{O}_i^\dagger$

$$\hat{H} - E_g\hat{I} = \sum_i w_i \hat{O}_i \hat{O}_i^\dagger \tag{70}$$

where each $\hat{O}_i$ operator is a sum of products of two creation and/or annihilation operators and the weights w_i are defined to be nonnegative. If the 2-RDM is constrained to be *2-positive*, then upon evaluation with the 2-RDM every term in the sum in Eq. (70) is nonnegative. For example, if the $\hat{O}_i$ operator is assembled from products like $a^\dagger a$, then the ith term may be written in terms of the G-matrix:

$$\langle \Psi | \hat{O}_i \hat{O}_i^\dagger | \Psi \rangle = \left\langle \Psi \left| \sum_{i,j} o_{i,j} a_i^\dagger a_j \sum_{k,l} o_{k,l}^* a_l^\dagger a_k \right| \Psi \right\rangle \tag{71}$$

$$= \sum_{i,j,k,l} o_{i,j} o_{k,l}^* \langle \Psi | a_i^\dagger a_j a_l^\dagger a_k | \Psi \rangle \tag{72}$$

$$= \sum_{i,j,k,l} o_{i,j} o_{k,l}^* {}^2G_{k,l}^{i,j} \tag{73}$$

If the G-matrix is positive semidefinite, then the above expectation value of the G-matrix with respect to the vector of expansion coefficients $o_{i,j}$ must be nonnegative. Similar analysis applies to $\hat{O}_i$ operators expressible with the D- or Q-matrix or any combination of D, Q, and G. Therefore variationally minimizing the ground-state energy of a $(\hat{H} - E_g\hat{I})$ operator, consistent with Eq. (70), as a function of the *2-positive* 2-RDM cannot produce an energy less than zero. For this class of Hamiltonians, we conclude, the 2-positivity conditions on the 2-RDM are sufficient to compute the *exact* ground-state N-particle energy on the two-particle space.

The G-reduced Hamiltonians are necessary and sufficient for at least three important classes of Hamiltonians: (i) all one-particle Hamiltonians, (ii) bosons or fermions with harmonic interactions [24], and (iii) all Hamiltonians with antisymmetrized geminal power (AGP) ground states. For ^2G the operators $\hat{C}$ in Eq. (65) are chosen to be $a_i^\dagger a_j$. Selecting the operators $\hat{C}$ to be $a_i a_j^\dagger$ produces

$$^2\tilde{G} = \frac{1}{\tilde{n}_g} \langle \Psi | a_i a_j^\dagger a_l a_k^\dagger | \Psi \rangle \geq 0 \tag{74}$$

where $\tilde{n}_g = (N+1)(r-N)$. Because these two sets of basis functions may be interconverted by the anticommutation relation, however, they merely represent a different organization of the same functions. Hence positivity of one metric matrix implies positivity of the other. Contraction of these two G-matrices (^2G and $^2\tilde{G}$) yields the positivity of the one-particle and the one-hole RDMs, respectively, which proves that the G-condition alone forces the 1-RDM to be N-representable. This proof is *stronger* than Rosina's proof that the generalized G-condition produces an N-representable 1-RDM [46]. While neither the D- nor the Q-condition alone restricts the 1-RDM to be N-representable, it is well-known that the combination of these conditions enforces N-representability of the 1-RDM [6].

Gidofalvi and Mazziotti [24] have examined a harmonically interacting system of bosons with application to Bose condensation. At any interaction strength it was shown numerically that energy minimization with respect to a 2-RDM constrained by 2-positivity conditions yielded the exact ground-state energy, and theoretically, the Hamiltonian was proved to belong to the family characterized by Eq. (70). In contrast, other many-body methods including perturbation theory and the connected-moments expansion [47] failed to give more than half of the correlation energy at large interactions.

An AGP wavefunction [48, 49] is generated from wedge products involving a single geminal $g(1,2)$:

$$\Psi_{AGP} = g(1,2) \wedge g(3,4) \wedge \cdots \wedge g(N-1, N) \tag{75}$$

Coleman [50] has shown that the Hamiltonians $b(g) \in B_2^N$, which have an AGP ground-state wavefunction, are given by

$$b(g) = {}^2I - (N-2)\,{}^1D(g) \wedge {}^1I - (N-1)gg^* \tag{76}$$

where

$$^1D(g) = L_2^1[gg^*] \tag{77}$$

Erdahl and Rosina [51] have demonstrated that the set of $b(g)$ Hamiltonians is contained in the convex set defined by the extreme reduced Hamiltonians from the G-condition, and hence we conclude that the G-condition is sufficient to obtain the correct ground-state energy for any system with an AGP ground-state wavefunction.

Because an AGP wavefunction may be *highly correlated*, the G-condition's treatment of AGP Hamiltonians illustrates how the positivity conditions have the ability to transcend the limitations of perturbation theory. For a mean-field state the occupation numbers of the 2-RDM are equal to $1/[N(N-1)]$ or 0. In early work on N-representability, Bopp assumed that the occupation numbers of the 2-RDM were bounded from above by $1/[N(N-1)]$. Yang, Sasaki, and Coleman, however, showed that the maximum occupation number is actually $1/(N-1)$, which is achieved by a certain class of AGP wavefunctions [6, 52, 53]. The appearance of a large occupation number in the 2-RDM may be associated with long-range order and pair formation in superconductivity. Although these highly correlated phenomena easily cause single-reference perturbation theory to fail, they may be treated exactly within the framework of G-positivity.

Our discussion may readily be extended from 2-positivity to p-positivity. The class of Hamiltonians in Eq. (70) may be expanded by permitting the $\hat{O}_i$ operators to be sums of products of p creation and/or annihilation operators for $p > 2$. If the p-RDM satisfies the p-positivity conditions, then expectation values of this expanded class of Hamiltonians with respect to the p-RDM will be nonnegative, and a variational RDM method for this class will yield exact energies. Geometrically, the convex set of 2-RDMs from p-positivity conditions for $p > 2$ is contained within the convex set of 2-RDMs from 2-positivity conditions. In general, the p-positivity conditions imply the q-positivity conditions, where $q < p$. As a function of p, experience implies that, for Hamiltonians with two-body interactions, the positivity conditions converge rapidly to a computationally sufficient set of representability conditions [17].

F. Spin and Spatial Symmetry Adaptation

While previous variational 2-RDM calculations for electronic systems have employed the above formulation [20–31], the size of the largest block diagonal matrices in the 2-RDMs may be further reduced by using *spin-adapted* operators $\hat{C}_i$ in Eq. (9). Spin-adapted operators are defined to satisfy the following mathematical relations [54, 55]:

$$[\hat{S}_z, \hat{C}^{s,m}] = m\hat{C}^{s,m} \tag{78}$$

and

$$[\hat{S}_\pm, \hat{C}^{s,m}] = \sqrt{s(s+1) - m(m \pm 1)}\hat{C}^{s,m\pm 1} \tag{79}$$

where

$$\hat{S}_+ = \sum_i \hat{a}^\dagger_{p,\alpha} \hat{a}_{p,\beta} \tag{80}$$

$$\hat{S}_- = \sum_i \hat{a}^\dagger_{p,\beta} \hat{a}_{p,\alpha} \tag{81}$$

$$\hat{S}_z = \frac{1}{2} \sum_i (\hat{a}^\dagger_{i,\alpha} \hat{a}_{i,\alpha} - \hat{a}^\dagger_{i,\beta} \hat{a}_{i,\beta}) \tag{82}$$

$\hat{C}^{s,s+1}_{i,j} = 0$ and $\hat{C}^{s,-s-1}_{i,j} = 0$, and the upper right superscripts s and m in $C^{s,m}_{i,j}$ denote the square of the total spin and the z-component of the total spin for the operators. The operators $\hat{C}_i$ employed in the previous section as well as earlier work satisfy only Eq. (78). The spin-adapted products of two creation operators are:

$$\hat{C}^{0,0}_{i,j} = \frac{1}{\sqrt{2}} (\hat{a}^\dagger_{i,\alpha} \hat{a}^\dagger_{j,\beta} + \hat{a}^\dagger_{j,\alpha} \hat{a}^\dagger_{i,\beta}) \tag{83}$$

$$\hat{C}^{1,0}_{i,j} = \frac{1}{\sqrt{2}} (\hat{a}^\dagger_{i,\alpha} \hat{a}^\dagger_{j,\beta} - \hat{a}^\dagger_{j,\alpha} \hat{a}^\dagger_{i,\beta}) \tag{84}$$

$$\hat{C}^{1,1}_{i,j} = \hat{a}^\dagger_{i,\alpha} \hat{a}^\dagger_{j,\alpha} \tag{85}$$

$$\hat{C}^{1,-1}_{i,j} = \hat{a}^\dagger_{i,\beta} \hat{a}^\dagger_{j,\beta} \tag{86}$$

Inserting these four operators into Eq. (8), we can generate four sets of $(N+2)$-electron basis functions for the two-hole RDM (^2Q matrix), $|{^Q\Phi^{0,0}_{i,j}}\rangle$, $|{^Q\Phi^{1,0}_{i,j}}\rangle$, $|{^Q\Phi^{1,1}_{i,j}}\rangle$, and $|{^Q\Phi^{1,-1}_{i,j}}\rangle$, respectively. For a ground-state wavefunction with a definite total S and z-component M spin quantum numbers, the $(N+2)$-electron basis functions with different m are orthogonal. Furthermore, whenever the ground-state wavefunction has $M = 0$ for any definite S, it is readily shown that the basis functions generated from operators with different s but the same m are also orthogonal. Therefore, when $M = 0$ in the ground-state wavefunction, the spin-adapted two-hole RDM has four blocks, with the singlet and triplet blocks scaling as $r_s(r_s + 1)/2$ and $r_s(r_s - 1)/2$, respectively [56, 57]. The singlet block is symmetric in its spatial indices while each triplet block is antisymmetric in its indices. If the ground-state wavefunction is also a singlet ($S = 0$), the three triplet blocks are equivalent, and hence only two distinct blocks must be constrained to be positive semidefinite.

By particle–hole duality, the same block structure appears in the spin-adapted two-electron RDM. The four blocks of the 2-RDM have the following traces [57]:

$$\mathrm{Tr}(^2D^{0,0}) = \frac{N(N+2)}{4} - S(S+1) \tag{87}$$

$$\mathrm{Tr}(^2D^{1,0}) = \frac{N(N-2)}{4} - 2M^2 + S(S+1) \tag{88}$$

$$\text{Tr}(^2D^{1,1}) = \left(\frac{N}{2}+M\right)\left(\frac{N}{2}+M-1\right) \tag{89}$$

$$\text{Tr}(^2D^{1,-1}) = \left(\frac{N}{2}-M\right)\left(\frac{N}{2}-M-1\right) \tag{90}$$

When the ground-state wavefunction is a singlet, the three triplet blocks have the same traces. In the variational 2-RDM calculations, these trace restrictions are enforced as constraints. Because the ^2Q and ^2G matrices are related to ^2D by linear mappings, these trace conditions also produce the correct traces of the spin-adapted ^2Q and ^2G matrices. As will be shown later, these trace conditions also enforce the correct expectation value of the $\hat{S}^2$ operator.

To generate the spin-adapted ^2G matrix, we spin-adapt the products of one creation operator and one annihilation operator:

$$\hat{C}_{i,j}^{0,0} = \frac{1}{\sqrt{2}}(\hat{a}_{i,\alpha}^\dagger \hat{a}_{j,\alpha} + \hat{a}_{i,\beta}^\dagger \hat{a}_{j,\beta}) \tag{91}$$

$$\hat{C}_{i,j}^{1,-1} = \hat{a}_{i,\beta}^\dagger \hat{a}_{j,\alpha} \tag{92}$$

$$\hat{C}_{i,j}^{1,0} = \frac{1}{\sqrt{2}}(\hat{a}_{i,\alpha}^\dagger \hat{a}_{j,\alpha} - \hat{a}_{i,\beta}^\dagger \hat{a}_{j,\beta}) \tag{93}$$

$$\hat{C}_{i,j}^{1,1} = \hat{a}_{i,\alpha}^\dagger \hat{a}_{j,\beta} \tag{94}$$

These operators satisfy the formal definition for spin-adapted operators in Eqs. (79) and (78). Inserting these four operators into Eq. (8), we can generate four sets of N-electron basis functions for the ^2G matrix, $|^G\Phi_{i,j}^{0,0}\rangle$, $|^G\Phi_{i,j}^{1,0}\rangle$, $|^G\Phi_{i,j}^{1,1}\rangle$, and $|^G\Phi_{i,j}^{1,-1}\rangle$, respectively. As in the case of the ^2Q matrix, when $M=0$ in the ground-state wavefunction, the spin-adapted two-electron RDM has four diagonal blocks, scaling as r_s^2. The blocks of the ^2G matrix are neither symmetric nor antisymmetric in the permutation of the spatial indices. If the ground-state wavefunction is also a singlet ($S=0$), the three triplet blocks are equivalent, and hence only two distinct blocks must be constrained to be positive semidefinite.

Similar to spin adaptation each 2-RDM spin block may further be divided upon considering the spatial symmetry of the basis functions. Here we assume that the 2-RDM has already been spin-adapted and consider only the spatial symmetry of the basis function for the 2-RDM. Denoting the irreducible representation of orbital i as Γ_i, the 2-RDM matrix elements are given by

$$^2D_{k\Gamma_k,l\Gamma_l}^{i\Gamma_i,j\Gamma_j} = \langle\Psi|\hat{a}_{j,\Gamma_j}^\dagger \hat{a}_{i,\Gamma_i}^\dagger \hat{a}_{k,\Gamma_k} \hat{a}_{l,\Gamma_l}|\Psi\rangle \tag{95}$$

These matrix elements are nonzero by spatial symmetry only if the direct products $\Gamma_i \otimes \Gamma_j$ and $\Gamma_k \otimes \Gamma_l$ share a common irreducible representation [58].

Hence the 2-RDM is further divided into blocks according to the spatial symmetry of the orbitals.

To illustrate the advantage of spin- and spatial-symmetry adaptation, consider the BH molecule in a minimal basis set. If only $\hat{S}_z$ is considered, the largest block of the two-electron RDM (i.e., $^2D_{\alpha,\beta}^{\alpha,\beta}$) is of dimension 36. Spin adaptation divides ^2D into two blocks, $^2D_{\alpha,\beta}^{\alpha,\beta} = \frac{1}{2}(^2D^{1,0} + {}^2D^{0,0})$, with sizes 15 and 21, respectively. Furthermore, because there are three molecular orbitals (MOs) with A_1, one MO with B_2, and one MO with B_1 spatial symmetry, each spin block is divided into four spatial blocks. In particular, the singlet spin block has the structure

$$^2D^{1,0} = \begin{pmatrix} D_{A_1,A_1}^{A_1,A_1} & D_{B_2,B_2}^{A_1,A_1} & D_{B_1,B_1}^{A_1,A_1} & 0 & 0 & 0 \\ D_{A_1,A_1}^{B_2,B_2} & D_{B_2,B_2}^{B_2,B_2} & D_{B_1,B_1}^{B_2,B_2} & 0 & 0 & 0 \\ D_{A_1,A_1}^{B_1,B_1} & D_{B_2,B_2}^{B_1,B_1} & D_{B_1,B_1}^{B_1,B_1} & 0 & 0 & 0 \\ 0 & 0 & 0 & D_{A_1,B_2}^{A_1,B_2} & 0 & 0 \\ 0 & 0 & 0 & 0 & D_{A_1,B_1}^{A_1,B_1} & 0 \\ 0 & 0 & 0 & 0 & 0 & D_{B_1,B_2}^{B_1,B_2} \end{pmatrix}$$

where the blocks have dimensions 12, 4, 4, and 1, respectively. The triplet block has a similar structure, except that the largest block is of a smaller dimension (i.e., dimension 6) because of the restriction ($i < j$) on the upper (and lower) indices. By particle–hole duality, the two-hole or ^2Q matrix has precisely the same block structure. Similar arguments may be employed to show that the electron–hole or ^2G matrix is partitioned into four blocks with sizes 18, 8, 8, and 2.

1. Spin Adaptation and S-Representability

An N-representable RDM is also defined to be *S-representable* if it derives from an N-particle wavefunction or an ensemble of N-particle wavefunctions with a definite spin quantum number S [57]. By definition, an S-representable two-electron RDM yields the correct expectation value

$$\langle \Psi^{S,M} | \hat{S}^2 | \Psi^{S,M} \rangle = S(S+1) \tag{96}$$

of the total spin angular momentum operator

$$\hat{S}^2 = \hat{S}_z + \hat{S}_z^2 + \hat{S}_- \hat{S}_+ \tag{97}$$

For an N-electron wavefunction with quantum numbers S and M, Eq. (96) reduces to

$$\sum_{i,j} {}^2D_{j\alpha,i\beta}^{i\alpha,j\beta} = N/2 + M^2 - S(S+1) \tag{98}$$

where the $\alpha\beta$ block of the unadapted 2-RDM block is normalized according to its definition in second quantization. Because the summation in Eq. (98) is not a simple trace, if the basis functions for the 2-RDM are only eigenfunctions of $\hat{S}_z$, the S-representability condition in Eq. (98) must be enforced in addition to the 2-positivity conditions of Eq. (11) as in previous 2-RDM calculations.

Using the definition of the spin-adapted 2-RDMs, however, we have the following relation between the $\alpha\beta$ block of the 2-RDM and the spin-adapted 2-RDMs:

$$^2D^{i\alpha,j\beta}_{k\alpha,l\beta} + {}^2D^{j\alpha,i\beta}_{l\alpha,k\beta} = {}^2D^{0,0}_{i,j;k,l} + {}^2D^{1,0}_{i,j;k,l} \qquad (99)$$

Dividing this equation by 2, setting $k = j$ and $l = i$, and then summing over i and j yields

$$\sum_{i,j} {}^2D^{i\alpha,j\beta}_{j\alpha,i\beta} = \frac{1}{2}\sum_{i,j}\left({}^2D^{0,0}_{i,j;j,i} + {}^2D^{1,0}_{i,j;j,i}\right) \qquad (100)$$

Because the singlet and triplet blocks are symmetric and antisymmetric in their indices, respectively, we have

$$\sum_{i,j} {}^2D^{i\alpha,j\beta}_{j\alpha,i\beta} = \frac{1}{2}\sum_{i,j}\left({}^2D^{0,0}_{i,j;i,j} - {}^2D^{1,0}_{i,j;i,j}\right) \qquad (101)$$

$$= \frac{1}{2}[\mathrm{Tr}({}^2\mathrm{D}^{0,0}) - \mathrm{Tr}({}^2\mathrm{D}^{1,0})] \qquad (102)$$

$$= N/2 + M^2 - S(S+1) \qquad (103)$$

where the traces for the singlet and triplet blocks are evaluated from Eqs. (87) and (88). Hence, as also discussed in reference [57], spin adaptation of the 2-RDM automatically enforces the S-representability condition in Eq. (96) for a general wavefunction with definite quantum numbers S and M. This result is especially important for the variational 2-RDM method because it proves that the constraint on the expectation value of $\hat{S}^2$ may be eliminated from the optimization if the 2-RDM is spin-adapted.

G. Open-Shell Molecules

Open-shell molecules, or radicals, may readily be treated within the variational 2-RDM theory. Here we compute the radical's energy and 2-RDM as the limit of dissociating one or more hydrogen atoms from a molecule in a singlet state. Calculation of the dissociated molecule's energy yields the energy of the radical plus the energy of one or more hydrogen atoms. The energy of the radical is then readily determined by subtracting the energy of the one-or-more hydrogen atoms from the energy of the total dissociated system. In a complete basis set the energy of each hydrogen atom would be -0.5 au, but in a finite Gaussian-orbital basis set the energy is slightly higher. With this approach the energy and properties of a radical may be computed through a singlet calculation.

The spin of the radical is characterized by two spin quantum numbers, the total spin S and the component of the total spin along the z-axis M. The simplest type of radical has one unpaired electron, and hence $S = \frac{1}{2}$ and $M = \pm\frac{1}{2}$, where the sign of M indicates the orientation of the electron spin in the z-direction. The dissociated singlet molecule, described by the $(N+1)$-electron wavefunction, consists of the radical and a hydrogen atom in orbital ϕ at "infinity,"

$$\Psi_{N+1}^{0,0} = c_\alpha \phi_\alpha \wedge \Psi_N^{1/2,-1/2} + c_\beta \phi_\beta \wedge \Psi_N^{1/2,1/2} \tag{104}$$

where c_α and c_β are expansion coefficients such as that $c_\alpha^2 + c_\beta^2 = 1$, the right superscripts on Ψ_{N+1} and Ψ_N denote S and M, and the symbol $\wedge$ denotes the antisymmetric tensor product known as the Grassmann (or wedge) product [13]. How are the spin quantum numbers S and M of the radical determined in the variational 2-RDM calculation of the singlet dissociated molecule?

The total spin of the radical is constrained *implicitly* by the search for the minimum energy. As long as the hydrogen atom has one electron, the radical will be in a doublet state $S = \frac{1}{2}$. The doublet state of the radical will not be violated as long as neither of the following two events occurs: (i) the hydrogen atom donates its electron to the radical, or (ii) the hydrogen atom takes the unpaired electron of the radical. These two events, however, are energetically unfavorable because the ionization energy of the hydrogen atom is much higher than the electron affinities of radicals and the electron affinity of the hydrogen atom is lower than the ionization energies of radicals. Hence, by the variational principle, minimizing the energy of the dissociated singlet molecule yields the radical in its doublet state $S = \frac{1}{2}$.

The 2-RDM *for the radical* may be computed from the $(N+1)$-electron density matrix for the dissociated molecule by integrating over the spatial orbital and spin associated with the hydrogen atom and then integrating over $N-2$ electrons. Because the radical in the dissociated molecule can exist in a doublet state with its unpaired electron either up or down, that is, $M = \pm\frac{1}{2}$, the 2-RDM for the radical is an arbitrary convex combination

$$^2D_{\text{radical}}^{S=1/2} = w_\alpha \, ^2D_{\text{radical}}^{1/2,-1/2} + w_\beta \, ^2D_{\text{radical}}^{1/2,+1/2} \tag{105}$$

where the right superscripts on the ^2D denote S and M, respectively, $w_\alpha = c_\alpha^2$, $w_\beta = c_\beta^2$, and $w_\alpha + w_\beta = 1$. While we have examined the doublet case, a similar analysis is valid for radicals with higher spin states. If the total spin S of the radical is greater than $\frac{1}{2}$, the singlet calculation with more-than-one dissociated hydrogen atom yields a 2-RDM for the radical that is a convex combination

$$^2D^S_{\text{radical}} = \sum_{M=-S}^{S} w_M \, ^2D^{S,M}_{\text{radical}} \tag{106}$$

of the allowed S_z eigenvalues $M = -S \ldots S$, where $w_M \geq 0$. For example, if the ground-state of the radical is a triplet state, the variational principle will produce a triplet radical with the pair of "removed" hydrogen atoms also in a triplet state to preserve the overall singlet symmetry of the dissociated molecule.

Treating a doublet radical by a singlet calculation requires the placement of a hydrogen atom at "infinity." Two approaches are: (i) computing the electron integrals of the parent molecule with its hydrogen atom stretched to a large distance (10^9 Å), and (ii) computing the electron integrals of the radical first and then adding integrals for the hydrogen atom that, due to the "infinite" separation, do not couple with the radical. The second approach has several advantages: (i) full spatial symmetry of the radical may be exploited in the integrals, (ii) any roundoff coupling in the integrals between the hydrogen atom and the radical is eliminated, and (iii) comparison with wavefunction methods is facilitated. Because the radical and the hydrogen atom do not couple in the electron integrals, each 2-RDM spin block subdivides into a block for the radical and an extremely small block for the spin entanglement of the hydrogen atom and the radical. Consequently, the calculation of the radical by a singlet calculation, dominated by the size of the largest block, is computationally less expensive than the calculation of the parent molecule.

Although the 2-RDM of an open-shell molecule (radical) may be determined variationally without considering a singlet parent molecule [20, 27], the present approach has several advantages: (i) greater computational efficiency, (ii) more consistent accuracy for the radical and its molecule, and (iii) easy implementation within 2-RDM code for singlet molecules. The 2-RDM for $M \neq 0$ has three blocks [21] with the largest scaling as r_s^2, but the 2-RDM for a singlet state has only two distinct spin-adapted blocks [32, 56, 57, 59–62] scaling as $r_s(r_s - 1)/2$ and $r_s(r_s + 1)/2$. Similar savings exist for the other metric matrices ^2Q and ^2G as well as the T_2 matrix [32, 34]. The singlet calculation of a radical by dissociation, therefore, uses one-quarter the memory and one-eighth the number of floating-point operations as a direct calculation of the radical with $M = S$. In addition, because the radical 2-RDM from the dissociation limit in Eq. (106) is more consistent with the spin of the radical moiety of the parent molecule than the 2-RDM with a fixed $M = S$, calculation of the radical by hydrogen dissociation may improve the relative accuracy of the molecule/radical energies.

Also the variational 2-RDM method for computing the 2-RDM of a singlet state may be applied directly to open-shell molecules without significant modification.

III. SEMIDEFINITE PROGRAMMING

Variational calculation of the energy with respect to the 2-RDM constrained by 2-positivity conditions requires minimizing the energy in Eq. (46) while restricting the ^2D, ^2Q, and ^2G to be not only positive semidefinite but also interrelated by the linear mappings in Eqs. (14)–(16). This is a special optimization problem known as a *semidefinite program*. The solution of a semidefinite program is known as *semidefinite programming* [63–65].

A semidefinite program may be written in two complementary formulations, which are known as the *primal* and *dual programs*. For convenience we define the map M that transforms any vector $|x\rangle$ of length n^2 into an $n \times n$ matrix $M(x)$ by creating each column of the matrix sequentially with the elements of the vector. The *primal* formulation of the semidefinite program may be expressed in general notation as

$$\begin{aligned} & \text{minimize } \langle c|x\rangle \\ & \text{such that } A|x\rangle = |b\rangle \\ & M(x) \geq 0 \end{aligned} \qquad (107)$$

where the vector $|c\rangle$ defines the system, the vector $|x\rangle$ denotes the primal solution, the $m \times n$ matrix A and the m-dimensional vector $|b\rangle$ enforce m linear constraints upon the solution $|x\rangle$, and the matrix $M(x)$ is restricted to be positive semidefinite. Similarly, the *dual* formulation of the semidefinite program may be expressed generally as

$$\begin{aligned} & \text{maximize } \langle b|y\rangle \\ & \text{such that } |z\rangle = |c\rangle - A^T|y\rangle \\ & M(z) \geq 0 \end{aligned} \qquad (108)$$

where the vector $|y\rangle$ of length m is the dual solution, A^T is the $n \times m$ transpose of the matrix A, and the $n \times n$ matrix $M(z)$ is constrained to be positive semidefinite.

In the variational 2-RDM method with 2-positivity the solution $|x\rangle$ of the primal program is a vector of the three metric matrices from the 2-RDM, the ^2D, the ^2Q, and the ^2G matrices

$$M(x) = \begin{pmatrix} ^2D & 0 & 0 \\ 0 & ^2Q & 0 \\ 0 & 0 & ^2G \end{pmatrix} \qquad (109)$$

the vector $|c\rangle$ holds specific information about the quantum system in the form of the two-particle reduced Hamiltonian

$$M(c) = \begin{pmatrix} {}^2K & 0 & 0 \\ 0 & 0 & 0 \\ 0 & 0 & 0 \end{pmatrix} \qquad (110)$$

and the matrix A and the vector $|b\rangle$ contain the linear mappings among the ${}^2D$, ${}^2Q$, and ${}^2G$ matrices in Eqs. (14) and (15), the contraction (Eq. (16)), and trace conditions $(\text{Tr}({}^1D) = N)$, as well as any spin constraints. The constraint $M(x) \geq 0$ in Eq. (107) restricts the ${}^2D$, ${}^2Q$, and ${}^2G$ matrices to be positive semidefinite.

Feasible $|x\rangle$ and $|y\rangle$ give upper and lower bounds on the optimal value of the objective function, which in the 2-RDM problem is the ground-state energy in a finite basis set. The primal and dual solutions, $|x\rangle$ and $|y\rangle$, are *feasible* if they satisfy the primal and dual constraints in Eqs. (107) and (108), respectively. The difference between the feasible primal and the dual objective values, called the *duality gap* μ, which equals the inner product of the vectors $|x\rangle$ and $|z\rangle$,

$$\mu = \langle x|z\rangle \geq 0 \qquad (111)$$

vanishes if and only if the solution is a global extremum. This important result was first proved by Erdahl in 1979 in the context of 2-RDM theory. For the variational 2-RDM method the duality gap furnishes us with a mathematical guarantee that we have determined the optimal energy within the convex set defined by the positivity conditions. With necessary N-representability conditions the optimal energy is a *lower bound* to the energy from full configuration interaction in the selected basis set.

In the mid-1990s a powerful family of algorithms, known as *primal–dual interior-point algorithms*, was developed for solving semidefinite programs [63]. The phrase *interior point* means that the method keeps the trial primal and dual solutions on the *interior* of the feasible set throughout the solution process. In these algorithms a good initial guess for the 2-RDM is a scalar multiple of the two-particle identity matrix. Advantages of the interior-point methods are: (i) rapid quadratic convergence from the identity matrix to the optimal 2-RDM for a set of positivity conditions, and (ii) a rigorous criterion in the duality gap for convergence to the global minimum. These benefits, however, are accompanied by large memory requirements and a significant number of floating-point operations per iteration, specifically $O(nm^3 + n^2m^2)$, where n is the number of variables and m is the number of constraints. With m and n proportional to the number of elements in the 2-RDM ($\approx r^4$), the method scales approximately as r^{16}, where r is the rank of the one-particle basis set [21, 23]. The variational 2-RDM method has been explored for minimal basis sets with the primal–dual interior-point algorithm, but the computational scaling significantly limits both the number of active electrons and the size of the basis set [20–23, 26, 27].

The author has recently developed a large-scale semidefinite programming algorithm for solving the semidefinite program in the variational 2-RDM method [28, 29]. The optimization challenge in the 2-RDM method is to constrain the metric matrices to be positive semidefinite while the ground-state energy is minimized. The algorithm constrains the solution matrix M to be positive semidefinite by a matrix factorization

$$M = RR^* \tag{112}$$

where for the 2-positivity conditions M is given in Eq. (109). Such a matrix factorization was previously considered in the context of 2-RDM theory by Mihailović and Rosina [66] Harriman [67], and the author [13], and it was recently employed for solving large-scale semidefinite programs in combinatorial optimization [68]. The linear constraints, including the trace, the contraction, and the interrelations between the metric matrices, become quadratic in the new independent variables R. Therefore the factorization in Eq. (112) converts the semidefinite program into a *nonlinear program,* where the energy must be minimized with respect to R while nonlinear constraint equalities are enforced.

We solve the nonlinear formulation of the semidefinite program by the augmented Lagrange multiplier method for constrained nonlinear optimization [28, 29]. Consider the augmented Lagrangian function

$$L(R) = E(R) - \sum_i \lambda_i c_i(R) + \frac{1}{\mu} \sum_i c_i(R)^2 \tag{113}$$

where R is the matrix factor for the solution matrix M, $E(R)$ is the ground-state energy as a function of R, $\{c_i(R)\}$ is the set of equality constraints, $\{\lambda_i\}$ is the set of Lagrange multipliers, and μ is the penalty parameter. For an appropriate set of multipliers $\{\lambda_i\}$ the minimum of the Lagrangian function with respect to R corresponds to the minimum of the energy $E(R)$ subject to the nonlinear constraints $c_i(R)$. The positive third term in the augmented Lagrangian function, known as the quadratic penalty function, tends to zero as the constraints are satisfied.

The augmented Lagrange multiplier algorithm finds the energy minimum of the constrained problem with an iterative, three-step procedure:

Step 1. For a given set of Lagrange multipliers $\{\lambda_i^{(n)}\}$ and penalty parameter $\mu^{(n)}$, minimize the Lagrangian function $L(R)$ to obtain an improved estimate R_{n+1} of the factorized 2-RDM at the energy minimum.

Step 2. If the maximum absolute error in the constraints $\max\{c_i(R_{n+1})\}$ is below a chosen threshhold (i.e., $0.25\max\{c_i(R_n)\}$), then the Lagrange multipliers are updated by a first-order correction

$$\lambda_i^{(n+1)} = \lambda_i^{(n)} - c_i(R_{n+1})/\mu^{(n)}$$

while the penalty parameter remains the same

$$\mu^{(n+1)} = \mu^{(n)}$$

Step 3. If Step 2 is not executed, then the penalty parameter is decreased to better enforce the constraints

$$\mu^{(n+1)} = 0.1\mu^{(n)}$$

while the Lagrange multipliers remain the same

$$\lambda^{(n+1)} = \lambda^{(n)}$$

Steps 1–3 are repeated until the maximum absolute error in the constraints falls below a target threshhold. Before the first iteration the Lagrange multipliers may be initialized to zero and the penalty parameter set to 0.1. The constraints are not fully enforced until convergence, and the energy in the primal program approaches the optimal value from below.

The cost of the algorithm is dominated by r^6 floating-point operations [28], mainly from the matrix multiplication of the block-diagonal R matrix with itself, where r is the rank of the one-particle basis set. Storage of the factorized 2-RDM, several copies of its gradient, and the Lagrange multipliers scales as r^4. In comparison with the primal–dual interior-point approach, which scales as r^{16} and r^8 in floating-point operations and memory storage, the first-order nonlinear algorithm for the variational 2-RDM method [28, 29] provides a significant improvement in computational efficiency.

IV. APPLICATIONS

The variational 2-RDM method has been applied to a variety of atoms and molecules at both equilibrium and stretched geometries. We will summarize calculations on a variety of molecules: (i) the nitrogen molecule [31], (ii) carbon monoxide with and without an electric field [37], (iii) a set of inorganic molecules [34], (iv) the hydroxide radical [35], and (v) a hydrogen chain [28].

A challenging correlation problem is the accurate description of the stretching and dissociation of the triple bond in nitrogen. Six-to-eight-particle excitations from the Hartree–Fock determinant are required to treat the nitrogen dissociation correctly. Using a correlation-consistent polarized double-zeta basis set, we compare in Fig. 1a and 1b the shape of the potential curve for nitrogen from the variational 2-RDM method with the curves from several wavefunction methods including full configuration interaction (FCI) [31]. The 2-RDM energies are

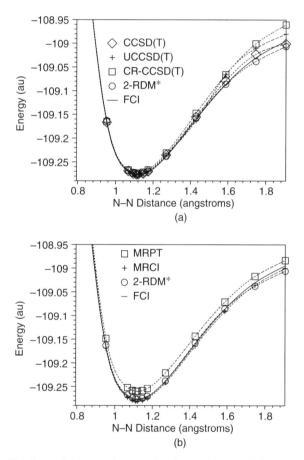

Figure 1. The shape of the potential curve for nitrogen in a correlation-consistent polarized double-zeta basis set is presented for the variational 2-RDM method as well as (a) single-reference coupled cluster, (b) multireference second-order perturbation theory (MRPT) and single–double configuration interaction (MRCI), and full configuration interaction (FCI) wavefunction methods. The symbol 2-RDM* indicates that the potential curve was shifted by the difference between the 2-RDM and CCSD(T) energies at equilibrium.

TABLE I
Equilibrium Bond Distance and the Harmonic Frequency for N_2 from the 2-RDM Method with 2-Positivity (DQG) Conditions Compared with Their Values from Coupled-Cluster Singles–Doubles with Perturbative Triples (CCD(T)), Multireference Second-Order Perturbation Theory (MRPT), Multireference Configuration Interaction with Single–Double Excitations (MRCI), and Full Configuration Interaction (FCI)[a].

Method	R_{eq}(Å)	$\omega(\text{cm}^{-1})$
CCSD(T)	1.1185	2344
MRPT	1.1176	2309
MRCI	1.1184	2311
2-RDM	1.1167	2311
FCI	1.1172	2321

[a]All methods employ a correlation-consistent polarized double-zeta basis set.

consistent lower bounds to the FCI energies throughout the stretch. In the figures we present the 2-RDM curve shifted by the difference between the 2-RDM and CCSD(T) energies at equilibrium. (The symbol 2-RDM* indicates that the potential curve was shifted by the difference between the 2-RDM and CCSD(T) energies at equilibrium.) The 2-RDM* method yields a potential energy curve that is more accurate than the single-reference methods in Fig. 1a and equally accurate as the multireference methods in Fig. 1b. The equilibrium bond distance and the harmonic frequency from the 2-RDM method are 1.1167 Å and 2311 cm^{-1}, which is in good agreement with the FCI numbers, 1.1172 Å and 2321 cm^{-1}. Multireference configuration interaction with single–double excitations yields 1.1184 Å and 2311 cm^{-1} (see Table I).

The variational 2-RDM method with 2-positivity conditions, implemented by a first-order nonlinear algorithm for semidefinite programming [28, 29], is applied to compute the ground-state potential energy surface of the carbon monoxide molecule in the absence and in the presence of electric fields. Even without an electric field, the calculation of the potential energy surface of the carbon monoxide molecule is a challenging task because proper treatment of the triple bond requires six-to-eight-particle excitations from a single Slater determinant or Hartree–Fock reference. We find that solving for the electronic structure of carbon monoxide in the presence of an electric field can either diminish or enhance the effects of the correlation along the bond dissociation curve. Modeling molecules within electric fields, therefore, provides a stringent test for electronic structure methods since we can increase the effects of correlation beyond their role in the absence of the field. In the absence of an electric field, Fig. 2a compares the 2-RDM*, coupled-cluster, MRPT2, and FCI potential energy surfaces of CO, where all valence electrons are correlated. 2-RDM* and MRPT2 accurately describe the features of the FCI potential energy surface. Figure 2b shows the potential energy curve for carbon monoxide in an electric field of strength 0.10 a.u. applied in the direction of the permanent dipole moment. The direction of the electric field affects the accuracy of the 2-RDM* energies much less than it affects the accuracy of the coupled-cluster methods.

The N-representability conditions on the 2-RDM can be systematically strengthened by adding some of the 3-positivity constraints to the 2-positivity conditions. For three molecules in valence double-zeta basis sets Table II shows

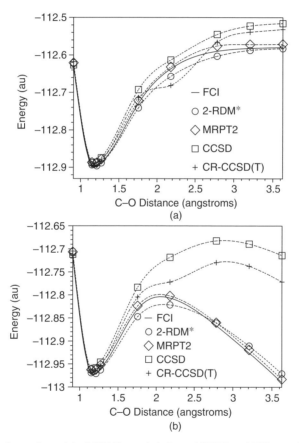

Figure 2. Comparison of the 2-RDM*, coupled-cluster, MRPT2, and FCI potential energy surfaces of CO in a valence double-zeta basis set, where all valence electrons are correlated (a) without an electric field and (b) with an electric field of strength 0.10 au applied in the direction of the permanent dipole moment. The 2-RDM* and MRPT2 methods accurately describe the features of the FCI potential energy surface.

TABLE II

For Three Molecules in Valence Double-Zeta Basis Sets, a Comparison of Energies in Hartrees (H) from the 2-RDM Method with the T_2 Condition (DQGT2) with the Energies from Second-Order Many-Body Perturbation Theory (MP2), Coupled-Cluster Method with Single–Double Excitations and a Perturbative Triples Correction (CCSD(T)), and Full Configuration Interaction (FCI)

	Total	Error in (mH)		
Molecule	FCI Energy (H)	MP2	CCSD(T)	2POS+T_2
CH_2	−38.9465	+23.3	+0.6	−0.1
BeH_2	−15.8002	+11.8	+0.2	−0.2
H_2O	−76.1411	+8.0	+0.5	−1.8

that the 2-RDM method with the T_2 condition (DQGT2) yields energies at equilibrium geometries that are similar in accuracy as the coupled-cluster method with single–double excitations and a perturbative triples correction. The error is reported in millihartrees (mH). Table III displays the ground-state energy of the nitrogen molecule as a function of bond length for the 2-RDM method with 2-positivity (2POS), 2-positivity plus T_1 and the generalized T_2 (denoted $\bar{T}_2$), and 3-positivity (3POS) as well as both configuration interaction and coupled-cluster wavefunction methods. The 2-RDM method with 3-positivity (3POS) has a maximum error of −1.4 mH at $R = 1.7$ Å. Around equilibrium the 3-positivity (3POS) improves the energies from 2-positivity plus $T_1\bar{T}_2$ and 2-positivity (2POS) by one and two orders of magnitude, respectively; it is an order of magnitude more accurate than CCSDT near equilibrium. Both

TABLE III

Ground-State Energy of the Nitrogen Molecule as a Function of Bond Length Examined with 2-RDM and Wavefunction Methods[a]

		Error in the Ground-State Energy (mH)						
		Wavefunction Methods				2-RDM Methods		
R	Total FCI Energy (H)	HF	CISD	CISDT	CCSD	CCSDT	2POS	2POS + $T_1\bar{T}_2$	3POS
1.0	−108.59599	131.0	8.2	6.6	2.7	1.1	−9.4	−1.2	−0.0
1.2	−108.72686	191.2	19.5	17.3	6.0	3.8	−14.9	−1.8	−0.1
1.5	−108.63545	311.3	55.6	53.2	13.8	11.7	−22.6	−4.4	−0.4
2.0	−108.81776	585.6	170.4	169.4	−103.5	−104.5	−15.9	−1.3	−0.5

[a] The 2-RDM method with 3-positivity (3POS) has a maximum error of −1.4 mH at $R = 1.7$ Å. Around equilibrium the 3-positivity (3POS) improves the energies from 2-positivity plus $T_1\bar{T}_2$ and 2-positivity (2POS) by one and two orders of magnitude, respectively; it is an order of magnitude more accurate than CCSDT near equilibrium. Both configuration interaction and coupled-cluster methods have difficulty at stretched geometries, where multiple Slater determinants contribute to the wavefunction.

configuration interaction and coupled-cluster methods have difficulty at stretched geometries, where multiple Slater determinants contribute to the wavefunction.

Computation of open-shell energies and properties is important in many areas of chemistry from combustion and atmospheric chemistry to medicine, and yet such molecules are often challenging due to the appearance of multireference spin effects. We have recently extended the variational 2-RDM method from closed-shell to open-shell molecules [33]. The *shape* of the potential energy curve of the OH radical is shown in Fig. 3 from the 2-RDM methods with 2-positivity (DQG) and 2-positivity plus T_2 (DQGT2) conditions as well as the wavefunction methods unrestricted second-order many-body perturbation theory (MBPT2), unrestricted coupled-cluster singles–doubles (UCCSD), and full configuration interaction (FCI). The potential energy curves of the approximate methods have been shifted by a constant to make them agree with the FCI curve at equilibrium. In the bonding region, the 2-RDM/DQG, UCCSD, and FCI curves are nearly indistinguishable, whereas in the stretched-bond region the 2-RDM/DQG and UCCSD curves move away slightly from the FCI solution in opposite directions. Recently, we have implemented a spin- and symmetry-

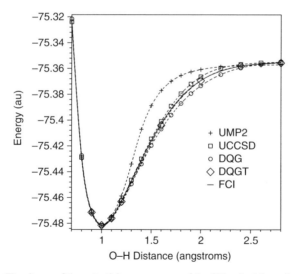

Figure 3. The *shapes* of the potential energy curves of the OH radical from the 2-RDM methods with DQG and DQGT2 conditions as well as the approximate wavefunction methods UMP2 and UCCSD are compared with the shape of the FCI curve. The potential energy curves of the approximate methods are shifted by a constant to make them agree with the FCI curve at equilibrium or 1.00 Å. The 2-RDM method with the DQGT2 conditions yields a potential curve that within the graph is indistinguishable in its contour from the FCI curve.

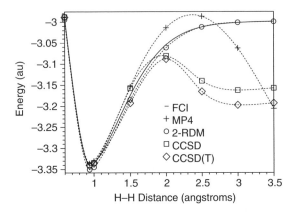

Figure 4. Ground-state potential energy curves of H_6 from 2-RDM and wavefunction methods are shown. MP2 and MP4 denote second- and fourth-order perturbation theories, while CCSD and CCSD(T) represent coupled cluster methods.

adapted form of the T_2 constraint within the large-scale semidefinite-programming algorithm for the 2-RDM method [34]. For OH the 2-RDM method with the DQGT2 conditions yields potential curves whose shapes in the figures are indistinguishable from the shapes of the FCI curves. Metallic hydrogen is an infinite chain of equally spaced hydrogen atoms. It can serve as a simple model for polymers and crystals. We consider the equally spaced, finite chain H_6, where the hydrogen atoms are described by the valence triple-zeta basis set. A potential energy curve may be formed by equally stretching the five bonds in H_6. Ground-state energies from the variational 2-RDM method and a variety of wavefunction techniques are shown in Fig. 4 as functions of the distance R between adjacent hydrogen atoms [28]. The 2-RDM method yields consistent energies with a maximum error of -10.8 mH at 1.5 Å. While the coupled-cluster methods are - accurate near the equilibrium geometry with errors at 1 Å of 1.3 mH (CCSD) and 0.2 mH (CCSD(T)), their performance rapidly degrades as the bonds are stretched. At 3.5 Å each of the coupled-cluster methods has an energy error of at least -160 mH while the 2-RDM method has an error of only -0.4 mH.

V. A LOOK AHEAD

Since the time that Coulson [7] discussed the promise and challenges of computing the energies and properties of atoms and molecules without the many-electron wavefunction, quantum chemistry has experienced many important advances toward the accurate treatment of electron correlation including the development of density functional theory, coupled-cluster theory, Monte Carlo methods, and multireference perturbation theory. The recent progress in the 2-RDM methods contributes both a new perspective and tool for describing energies and properties of atoms and molecules in which correlation effects are important. In this chapter we have developed the variational calculation of the ground-state energy as a function of the 2-RDM constrained by N-representability conditions [16, 17, 19–37].

A critical part of realizing a "quantum chemistry without wavefunctions" through the variational 2-RDM method is the development of robust algorithms for large-scale semidefinite programming. As discussed in the chapter a large-scale algorithm, developed by the author, reduces the computational scaling of the 2-RDM method by orders of magnitude in both floating-point operations and memory. A key feature of the algorithm is the expression of the semidefinite program as nonlinear constrained optimization, which is then solved by the method of augmented Lagrange multipliers. Zhao et al. [27] have examined improving the performance of the primal–dual interior-point methods by redefining the statement of the semidefinite program. Further advances in large-scale semidefinite programming will have an important impact on the variational 2-RDM method, and similarly, the problems in electronic structure offer fertile ground for testing and benchmarking new large-scale algorithms. Advances will also have a broad impact on many other scientific problems in areas like control theory, combinatorial optimization, quantum information, and finance.

The variational calculation of the 2-RDM with necessary N-representability conditions yields a *lower bound* on the ground-state energy in a given finite basis set. The strict lower bound occurs because the 2-RDM is optimized over a set that contains all correlated N-electron wavefunctions. Within wave mechanics the challenge is to introduce sufficient variational flexibility into the wavefunction, but in reduced-density-matrix mechanics the challenge is to limit the 2-RDM, which has the flexibility to model all correlated wavefunctions, to represent only realistic N-electron wavefunctions. A practical consequence is that the 2-RDM method has the potential to produce ground-state energies with useful accuracy even if the wavefunction is challenging to parameterize as in transition-state structures or other stretched geometries of a potential energy surface [21, 22, 28, 29, 31]. The calculation of the 2-RDM has important applications in chemistry to studying reactivity [30, 31] as well as in other areas of correlation such as spin systems like the Hubbard model, Bose condensation [24], and molecular conductivity. The 2-RDM methods may be especially well suited for the use of explicitly correlated basis sets for enhancing basis set convergence. While still in its early stages, the 2-RDM method for computing energies and properties without the many-electron wavefunction represents a new approach to investigating the electronic structure of atomic and molecular systems.

Acknowledgments

The author expresses his appreciation to Dudley Herschbach, Herschel Rabitz, John Coleman, and Alexander Mazziotti for their support and encouragement. The author thanks the NSF, the Henry-Camille Dreyfus Foundation, the Alfred P. Sloan Foundation, and the David-Lucile Packard Foundation for their support.

References

1. L. D. Landau, Das dämpfungsproblem in der wellenmechanik. *Z. Phys.* **45**: 430 (1927).
2. J. von Neumann, *Mathematical Foundations of Quantum Mechanics*, Princeton University Press, Princeton, 1995.
3. P. O. Löwdin, Quantum theory of many-particle systems. 1. Physical interpretations by means of density matrices, natural spin-orbitals, and convergence problems in the method of configuration interaction. *Phys. Rev.* **97**, 1474 (1955).
4. J. E. Mayer, Electron correlation. *Phys. Rev.* **100**, 1579 (1955).
5. R. H. Tredgold, Density matrix and the many-body problem. *Phys. Rev.* **105**, 1421 (1957).
6. A. J. Coleman, Structure of fermion density matrices. *Rev. Mod. Phys.* **35**, 668 (1963).
7. A. J. Coleman and V. I. Yukalov, *Reduced Density Matrices: Coulson's Challenge*, Springer-Verlag, New York, 2000.
8. F. Colmenero and C. Valdemoro, Approximating q-order reduced density-matrices in terms of the lower-order ones. 2. Applications. *Phys. Rev. A* **47**, 979 (1993).
9. F. Colmenero and C. Valdemoro, Self-consistent approximate solution of the 2nd-order contracted Schrödinger equation. *Int. J. Quantum Chem.* **51**, 369 (1994).
10. H. Nakatsuji and K. Yasuda, Direct determination of the quantum-mechanical density matrix using the density equation. *Phys. Rev. Lett.* **76**, 1039 (1996).
11. K. Yasuda and H. Nakatsuji, Direct determination of the quantum-mechanical density matrix using the density equation II. *Phys. Rev. A* **56**, 2648 (1997).
12. C. Valdemoro, L. M. Tel, and E. Perez-Romero, The contracted Schrödinger equation: some results. *Adv. Quantum Chem.* **28**, 33 (1997).
13. D. A. Mazziotti, Contracted Schrödinger equation: determining quantum energies and two-particle density matrices without wave functions. *Phys. Rev. A* **57**, 4219 (1998).
14. D. A. Mazziotti, Approximate solution for electron correlation through the use of Schwinger probes. *Chem. Phys. Lett.* **289**, 419 (1998).
15. D. A. Mazziotti, 3,5-Contracted Schrödinger equation: determining quantum energies and reduced density matrices without wave functions. *Int. J. Quantum Chem.* **70**, 557 (1998).
16. R. M. Erdahl and B. Jin, The lower bound method for reduced density matrices. *J. Mol. Struc. (Theochem.)* **527**, 207 (2000).
17. D. A. Mazziotti and R. M. Erdahl, Uncertainty relations and reduced density matrices: mapping many-body quantum mechanics onto four particles. *Phys. Rev. A* **63**, 042113 (2001).
18. C. Garrod and J. Percus, Reduction of N-particle variational problem. *J. Math. Phys.* **5**, 1756 (1964).
19. C. Garrod, V. Mihailović, and M. Rosina, Variational approach to 2-body density matrix. *J. Math. Phys.* **10**, 1855 (1975).
20. M. Nakata, H. Nakatsuji, M. Ehara, M. Fukuda, K. Nakata, and K. Fujisawa, Variational calculations of fermion second-order reduced density matrices by semidefinite programming algorithm. *J. Chem. Phys.* **114**, 8282 (2001).

21. D. A. Mazziotti, Variational minimization of atomic and molecular ground-state energies via the two-particle reduced density matrix. *Phys. Rev. A* **65**, 062511 (2002).
22. M. Nakata, M. Ehara, and H. Nakatsuji, Density matrix variational theory: application to the potential energy surfaces and strongly correlated systems. *J. Chem. Phys.* **116**, 5432 (2002).
23. D. A. Mazziotti, Solution of the 1,3-contracted Schrödinger equation through positivity conditions on the 2-particle reduced density matrix. *Phys. Rev. A* **66**, 062503 (2002).
24. G. Gidofalvi and D. A. Mazziotti, Boson correlation energies via variational minimization with the two-particle reduced density matrix: exact N-representability conditions for harmonic interactions. *Phys. Rev. A* **69**, 042511 (2004).
25. G. Gidofalvi and D. A. Mazziotti, Variational reduced-density-matrix theory: strength of Hamiltonian-dependent positivity conditions. *Chem. Phys. Lett.* **398**, 434 (2004).
26. T. Juhász and D. A. Mazziotti, Perturbation theory corrections to the two-particle reduced density matrix variational method. *J. Chem. Phys.* **121**, 1201 (2004).
27. Z. Zhao, B. J. Braams, H. Fukuda, M. L. Overton, and J. K. Percus, The reduced density matrix method for electronic structure calculations and the role of three-index representability conditions. *J. Chem. Phys.* **120**, 2095 (2004).
28. D. A. Mazziotti, Realization of quantum chemistry without wavefunctions through first-order semidefinite programming. *Phys. Rev. Lett.* **93**, 213001 (2004).
29. D. A. Mazziotti, First-order semidefinite programming for the direct determination of two-electron reduced density matrices with application to many-electron atoms and molecules. *J. Chem. Phys.* **121**, 10957 (2004).
30. G. Gidofalvi and D. A. Mazziotti, Application of variational reduced-density-matrix theory to organic molecules. *J. Chem. Phys.* **122**, 094107 (2005).
31. G. Gidofalvi and D. A. Mazziotti, Application of variational reduced-density-matrix theory to the potential energy surfaces of the nitrogen and carbon dimers. *J. Chem. Phys.* **122**, 194104 (2005).
32. G. Gidofalvi and D. A. Mazziotti, Spin and symmetry adaptation of the variational two-electron reduced-density-matrix method. *Phys. Rev. A* **72**, 052505 (2005).
33. J. R. Hammond and D. A. Mazziotti, Variational two-electron reduced-density-matrix theory: partial 3-positivity conditions for N-representability. *Phys. Rev. A* **71**, 062503 (2005).
34. D. A. Mazziotti, Variational two-electron reduced-density-matrix theory for many-electron atoms and molecules: implementation of the spin- and symmetry-adapted T_2 condition through first-order semidefinite programming. *Phys. Rev. A* **72**, 032510 (2005).
35. J. R. Hammond and D. A. Mazziotti, Variational reduced-density-matrix calculations on radicals: an alternative approach to open-shell *ab initio* quantum chemistry. *Phys. Rev. A* **73**, 012509 (2006).
36. D. A. Mazziotti, Quantum chemistry without wave functions: two-electron reduced density matrices. *Acc. Chem. Res.* **39**, 207 (2006).
37. G. Gidofalvi and D. A. Mazziotti, Variational reduced-density-matrix theory applied to the potential energy surfaces of carbon monoxide in the presence of electric fields. *J. Phys. Chem. A* **110**, 5481 (2006).
38. R. M. Erdahl, Representability. *Int. J. Quantum Chem.* **13**, 697 (1978).
39. W. Kutzelnigg and D. Mukherjee, Cumulant expansion of the reduced density matrices. *J. Chem. Phys.* **110**, 2800 (1999).
40. D. A. Mazziotti, Complete reconstruction of reduced density matrices. *Chem. Phys. Lett.* **326**, 212 (2000).

41. H. Kummer, N-representability problem for reduced density matrices. *J. Math. Phys.* **8**, 2063 (1967).
42. A. J. Coleman, Convex structure of electrons. *Int. J. Quantum Chem.* **11**, 907 (1977).
43. W. Slebodziński, *Exterior Forms and their Applications*, Polish Scientific Publishers, Warsaw, 1970.
44. M. B. Ruskai, N-representability problem: particle–hole equivalence. *J. Math. Phys.* **11**, 3218 (1970).
45. R. M. Erdahl, Convex structure of the set of N-representable reduced 2-matrices. *J. Math. Phys.* **13**, 1608 (1972).
46. M. V. Mihailovic and M. Rosina, Excitations as ground-state variational parameters. *Nucl. Phys. A* **130**, 386 (1969).
47. J. Cioslowski, Connected-moments expansion — a new tool for quantum many-body theory. *Phys. Rev. Lett.* **58**, 83 (1987).
48. A. J. Coleman, Structure of fermionic density matrices.2. antisymmetrized geminal powers. *J. Math. Phys.* **6**, 1425 (1965).
49. D. A. Mazziotti, Geminal functional theory: a synthesis of density and density matrix methods. *J. Chem. Phys.* **112**, 10125 (2000). D. A. Mazziotti, Energy functional of the one-particle reduced density matrix: a geminal approach, *Chem. Phys. Lett.* **338**, 323 (2001).
50. A. J. Coleman, Necessary conditions for N-representability of reduced density matrices. *J. Math. Phys.* **13**, 214 (1972).
51. R. M. Erdahl and M. Rosina, in *Reduced Density Operators with Applications to Physical and Chemical Systems II* (R.M. Erdahl, ed.), Queen's Papers in Pure and Applied Mathematics No. 40, Queen's University, Kingston, Ontario, 1974, p. 131.
52. C. N. Yang, Concept of off-diagonal long-range order and quantum phases of liquid helium and superconductors. *Rev. Mod. Phys.* **34**, 694 (1962).
53. F. Sasaki, Eigenvalues of fermion density matrices. *Phys. Rev.* **138**, B1338 (1965).
54. D. M. Brink and G. R. Satchler, *Angular Momentum*, Clarendon Press, New York, 1968.
55. T. Helgaker, P. Jorgensen, and J. Olsen, *Molecular Electronic-Structure Theory*, John Wiley & Sons, Hoboken, NJ, 2000.
56. W. A. Bingel and W. Kutzelnigg, Symmetry of the first-order density matrix and its natural orbitals for linear molecules. *Adv. Quantum Chem.* **5**, 201 (1970).
57. E. Perez-Romero, L. M. Tel, and C. Valdemoro, Traces of spin-adapted reduced density matrices. *Int. J. Quantum Chem.* **61**, 55 (1997).
58. F. A. Cotton, *Chemical Applications of Group Theory*, 2nd ed. John Wiley & Sons, Hoboken, NJ, 1971.
59. R. McWeeny and Y. Mizuno, Density matrix in many-electron quantum mechanics. 2. Separation of space and spin variables — spin coupling problems. *Proc. R. Soc. London* **259**, 554 (1961).
60. W. Kutzelnigg, Uber die symmetrie-eigenschaften der reduzierten dichtermatrizen und der naturlichen spin-orbitale und spin geminale. *Z. Naturforsch. A* **18A**, 1058 (1963).
61. D. J. Klein, N. H. March, and A. K. Theophilou, Form of spinless first- and second-order density matrices in atoms and molecules, derived from eigenfunctions of S^2 and S_z. *J. Math. Chem.* **21**, 261 (1997).
62. L. M. Tel, E. Perez-Romero, F. J. Casquero, and C. Valdemoro, Compact forms of reduced density matrices. *Phys. Rev. A* **67**, 052504 (2003).
63. L. Vandenberghe and S. Boyd, Semidefinite programming. *SIAM Rev.* **38**, 49 (1996).

64. S. Wright, *Primal–Dual Interior-Point Methods*, SIAM, Philadelphia, 1997.
65. Y. Nesterov and A. S. Nemirovskii, *Interior Point Polynomial Method in Convex Programming: Theory and Applications*, SIAM, Philadelphia, 1993.
66. M. V. Mihailović and M. Rosina, Excitations as ground-state variational parameters. *Nucl. Phys. A* **130**, 386 (1971).
67. J. E. Harriman, Geometry of density matrices. II. Reduced density matrices and N-representability. *Phys. Rev. A* **17**, 1257 (1978).
68. S. Burer and R. D. C. Monteiro, A nonlinear programming algorithm for solving semidefinite programs via low-rank factorization. *Math. Programming Ser. B* **95**, 329 (2003).

CHAPTER 4

THE LOWER BOUND METHOD FOR DENSITY MATRICES AND SEMIDEFINITE PROGRAMMING

ROBERT M. ERDAHL

Department of Mathematics and Statistics, Queen's University, Kingston, Ontario K7L 3N6, Canada

CONTENTS

I. Introduction
   A. Brief History of the Lower Bound Method
   B. Strong Two-Body Forces
II. Kth-Order Approximations for States
   A. Approximating States by k-Densities
   B. Matrix Representations
   C. Expectation Values
   D. The Pauli Subspace
   E. Additional Properties of Matrix Representations
   F. Energy Lower Bounds
III. Semidefinite Programming
   A. Lagrange Duality
   B. The Gap Formula
IV. The Fundamental Theorem
   A. Fundamental Theorem
   B. Existence Theorems
V. Algorithms
   A. Standard Formulation
   B. Alternate Formulations
   C. Algorithms
      1. Primal–Dual Interior Point Methods
      2. Boundary Methods
   D. Interpreting the Solution
VI. Modeling a One-Dimensional Superconductor
   A. Details of the Model
   B. Results

Reduced-Density-Matrix Mechanics: With Application to Many-Electron Atoms and Molecules,
A Special Volume of Advances in Chemical Physics, Volume 134, edited by David A. Mazziotti.
Series editor Stuart A. Rice. Copyright © 2007 John Wiley & Sons, Inc.

C. Convergence with Order k
D. The Lipkin Model
VII. Concluding Remarks
References

There are two recent developments that have turned the *lower bound method* of density matrix theory into a powerful computational tool for electronic structure theory, and thereby solved John Coleman's N-representability problem [1]. These two developments are discussed here. First, it is now understood how to achieve accurate results, even when two-body forces dominate; the picture that has emerged through computational experiments by several authors is described. Second, the central energy minimization problem of the lower bound method belongs to a class called semidefinite programs, and effective algorithms are now available to solve such problems. How the lower bound method and semidefinite programming have come together is described, and at the same time a self-contained treatment of the mathematical results at the core of semidefinite programming is given. The treatment includes a new proof of the fundamental theorem of semidefinite programming.

I. INTRODUCTION

By replacing the wavefunction with a density matrix, the electronic structure problem is reduced in size to that for a two- or three-electron system. Rather than solve the Schrödinger equation to determine the wavefunction, the *lower bound method* is invoked to determine the density matrix; this requires adjusting parameters so that the energy content of the density matrix is minimized. More precisely, the lower bound method requires finding a solution to the *energy problem*,

$$\min_{\mathbf{P} \in \mathcal{S}^\perp \cap \mathbb{P}_0^k} \langle \mathbf{P}, \mathbf{H} \rangle_k$$

where $\mathbf{P}$ is a matrix representation of the quantum state called a k-matrix, $\mathbf{H}$ is a matrix representation of the Hamiltonian, and $\langle \mathbf{P}, \mathbf{H} \rangle_k$ is the energy—the trace scalar product of $\mathbf{P}$ with $\mathbf{H}$. The k-matrix that minimizes the energy is an estimate of the von Neumann density for the ground state, and the energy content of this k-matrix is a lower bound to the ground-state energy. The variation is over the set of k-matrices, $\mathcal{S}^\perp \cap \mathbb{P}_0^k$, which is a section of the cone of positive semidefinite matrices; $\mathbb{P}_0^k$ is the convex set of positive semidefinite matrices with unit trace, and $\mathcal{S}$ is the *Pauli subspace*—a subspace of symmetric matrices that encodes the conditions imposed by the Pauli Principle. The compact convex set $\mathcal{S}^\perp \cap \mathbb{P}_0^k$ is a kth-order approximation of the set of von Neumann density matrices, and a variant of the set of kth-order reduced density matrices. Included in the definition of

k-matrix is the condition that $\mathbf{P}$ be k-positive, a condition that rapidly becomes stringent as k increases. Details on how the sets $\mathcal{S}^\perp \cap \mathbb{P}_0^k, k = 2, 3, 4, \ldots$, are constructed and the definition of k-positive are given in Section II.

One purpose here is to report on some recent computational experiments that show that the rate of convergence to exact solutions with the order parameter k is extraordinarily rapid—much faster than anticipated. This was first uncovered in the doctoral dissertation of B. Jin, who studied a solid model where the Hamiltonian included two-body interactions alone. In his thesis, Jin [2] reported a striking improvement in accuracy in going from second- to third-order estimates: second-order estimates contained no useful information, even predicting wrong trends in certain instances, but with third-order estimates four-figure accuracy was achieved. It came as a surprise that second-order approximations were so poor, and it was equally surprising that third-order approximations were so good. The significance was immediately obvious, that convergence with k was extremely rapid, and accurate solutions could be achieved while complexity was kept within bounds. It is the condition that $\mathbf{P}$ be k-positive that accounts for this rapid convergence and makes possible the characterization of increasingly complex correlations. Several papers followed that filled out the picture on how accurate solutions could be achieved [3–6]. These computational experiments showed that when two-body interactions play a relatively minor role, as they do in atoms and molecules, approximations to second order and 2-positivity do an adequate job for most purposes. It was also discovered that the accuracy of third-order estimates in these cases exceeds that for all other approximate methods currently used. However, when two-body forces become more important, as they do in nuclear and solid state problems, third-order approximations and 3-positivity are required.

The minimum energy problem is a *semidefinite program*, a new class of optimization problems that emerged in the late 1980s and has been intensively studied ever since. Semidefinite programming has applications that extend far beyond the electronic structure problem we are considering, covering vast new areas in applied mathematics; linear programming is a special case. That the variation for the energy problem is over a section of the cone of positive semidefinite matrices is the characteristic that identifies this problem as a semidefinite program—and is the origin of the name. The recent development of effective algorithms to solve semidefinite programming problems gives a second reason why the lower bound method is emerging as an important computational tool. For strongly interacting systems of electrons, approximations must be carried through to at least third order, and these are typically large problems. It is the recent and ongoing development of algorithms for semidefinite programming that has made large-scale electronic structure problems feasible.

The energy problem is accompanied by the dual *spectral optimization problem*,

$$\max_{\mathbf{S} \in \mathcal{S}} \quad \lambda_0(\mathbf{H} + \mathbf{S})$$

where $\lambda_0(\mathbf{H}+\mathbf{S})$ is the bottom eigenvalue of the matrix $\mathbf{H}+\mathbf{S}$; the variation is over the Pauli subspace $\mathcal{S}$, which is defined in Section II. Solutions of the energy and spectral optimization problem occur simultaneously, which proves to be an enormous asset when interpreting solutions. We will see that the dual spectral optimization problem serves as a sharp tool in analyzing the correlations in the ground state as it responds to the interactive forces experienced by the electrons. The energy and spectral optimization problems are tied together by the fundamental Euler equation for these problems,

$$\mathbf{PQ} = \mathbf{0}$$

where $\mathbf{Q} = \mathbf{H} + \mathbf{S} - \lambda_0(\mathbf{H}+\mathbf{S})\mathbf{I}$; the Euler equation is a stability equation that the optimal matrices $\mathbf{P}, \mathbf{S}$ must satisfy. Any two matrices $\mathbf{P}, \mathbf{Q}$ that satisfy this equation provide optimal solutions to the energy and spectral optimization problems, as long as they have the correct form: $\mathbf{P}$ must be contained in the convex set $\mathcal{S}^\perp \cap \mathbb{P}_0^k$, and $\mathbf{Q}$ must have the form $\mathbf{Q} = \mathbf{H} + \mathbf{S} - \lambda_0(\mathbf{H}+\mathbf{S})\mathbf{I}$. A self-contained treatment is given of the theoretical results at the core of semidefinite programming, which includes a new proof of the result that solution of the spectral optimization problem is equivalent to solution of the Euler equation $\mathbf{PQ} = \mathbf{0}$. The treatment uses only elementary matrix theory and convexity theory and has the virtue of being brief. Moreover, the results are formulated in terms of the energy problem for density matrices.

A. Brief History of the Lower Bound Method

The first accurate estimate of a 2-matrix using the lower bound method was made by M. A. Fusco in his doctoral dissertation, directed by Claude Garrod at the University of California, Davis; in his 1974 thesis [7], Fusco reported on his work on the beryllium atom. There were two papers that quickly followed, both reporting accurate lower bound calculations on the beryllium atom. In their 1975 paper, Garrod, Mihailovic, and Rosina [8] reported a lower bound to the energy that was only slightly below the exact ground-state energy calculated using a complete configuration interaction treatment; the lower bound they calculated was -14.60999 atomic units, and the configuration interaction energy was -14.609987; which represented seven-figure accuracy. In this calculation the convex set of density matrices was approximated by a superscribing polytope, which was updated at each step. The energy minimization problem was converted to a linear programming problem, and successively tighter lower bounds were computed as the polytope was adjusted to match the data near the optimal solution. The thesis work of Fusco was reported in the 1976 paper of Garrod and Fusco [9]; in this work a penalty function method was applied to minimize the energy. The conditions imposed by these authors were that the associated P-, Q-, and G-matrices be positive semidefinite; these conditions did not ensure that the 2-matrix was N-representable, but were sufficiently effective that seven-figure accuracy was achieved.

The second test of the *lower bound method*, and the first test for a molecule, appeared in 1979 when Erdahl calculated a lower bound to the energy for a pair of weakly bound helium atoms [10]. By requiring that the density matrix be 2-positive, a lower bound was calculated that was accurate to five figures when compared to a complete configuration interaction treatment (from now on we refer to the condition that the P-, Q-, and G-matrices be positive semidefinite as the condition that **P** be 2-*positive*; the notion of 2-positive, and more generally k-positive, is defined in Section II). The Be and He^2 calculations showed that accurate lower bounds could be achieved for atoms and molecules and, in particular, showed the importance of the condition that **P** be 2-positive. The calculation on the weakly bound helium molecule was a side issue in Ref. [10], only introduced to test a new method for solving the energy problem. The main result was the derivation of the Euler equation $\mathbf{PQ} = \mathbf{0}$, which was then solved to obtain the optimal solution of the energy problem. Thus this second test of the lower bound method anticipated the development of semidefinite programming by over ten years. The equation $\mathbf{PQ} = \mathbf{0}$ and its relation to the spectral optimization problem are rediscovered by M. L. Overton and reported in his 1988 paper [11]. It is this paper that served as a precursor for the rapid development of semidefinite programming that started in the early 1990s.

The promise of the early work on Be and He^2 has recently been confirmed in the work of Nakatsuji and Mazziotti, which started to appear in 2001. This work showed that the lower bound method combined with second-order approximations yields accurate information for atoms and molecules. Nakatsuji and his co-workers [12] did a series of computational experiments where accuracies of between four and five figures were typically achieved. More precisely, they reported the correlation energy as a percentage of the exact correlation energy for a variety of atoms and molecules. They found these percentages ranged between 100% and 110% for atoms and diatomic molecules, and between 110% and 120% for triatomic molecules; since these percentages are for lower bounds they never go below 100%.

An even more exacting test of 2-positivity appeared in the work of Mazziotti [13], who computed binding energy curves for LiH and H_2O. These computational experiments showed that accurate binding energy curves could be calculated when second-order approximations are invoked; the curves he calculated compared favorably with curves generated by a full configuration interaction treatment. These results were compared with curves generated using second- and fourth-order perturbation theory, and these deteriorated rapidly with increasing bond length. These experiments showed that lower bounds can successfully track energy shifts over a range of molecular geometries, subtle shifts that cannot be tracked by perturbation theory.

Another significant theme was reported on in the Mazziotti paper. Experiments were made by selectively adding higher-order conditions to tighten the lower bounds computed using 2-positivity. These were a portion of the 3- *positivity conditions* selected on the basis of physical intuition about correlations and experience gained with second-order conditions. This strategy adds flexibility and strength to the lower bound method in much the way the selection of configurations adds to the configuration interaction method. These computational experiments were first steps in developing a systematic approach to handling the overwhelming array of higher-order conditions, the strategies used being those proposed by Erdahl and Jin [3, 4] under the heading *dual configuration interaction*.

The inclusion of some 3-positivity conditions along with the 2-positivity was also used by Zhao and co-workers in their recent computational experiments with the lower bound method [6, 14]; in addition to the P-, Q-, and G-conditions they added the third-order conditions formulated in Section 8 of Ref. [15], which they referred to as the $T1$ and $T2$ conditions. They compared the strength of the various conditions they imposed by estimating the ground-state density matrix for 38 small molecules. They noticed a "spectacular increase in accuracy" when these third-order conditions were imposed in their calculations: "We find that including the $T1$ and $T2$ conditions results in a spectacular increase in the accuracy of the results, and gives in the cases studied an accuracy better than that of other more familiar approximate methods: singly and doubly substituted configuration interaction (SDCI), Brueckner doubles (with triples) (BD(T)) and coupled cluster singles and doubles with perturbational treatment of triples (CCSD(T))."

B. Strong Two-Body Forces

The optimism generated by the Be and He^2 calculations was tempered by another line of investigation in the 1970s. Mihailovic and Rosina applied the methods they developed for the beryllium atom to light nuclei [16] and found lower bounds falling below exact values by as much as 15%. For example, the series of nuclei ^{15}O, ^{16}O, ^{17}O, and ^{18}O have 3, 4, 5, and 6 valence nucleons above a ^{12}C core. The greatest errors they observed were for the last in this series, where they computed a lower bound of -60.06 MeV, which was 8.55 MeV below the configuration interaction energy of -53.91 MeV, representing an error of 15.86%. For this series the percentage errors are given, respectively, by 5.82%, 3.76%, 14.93% and 15.86%, showing that errors increase as half-filled structures are approached. The results for ^{15}O and ^{16}O were reasonable, but for ^{17}O and ^{18}O they commented that "new relevant conditions" should be found to supplement 2-positivity. In response to the dominating two-body terms, the nucleons in a shell form highly correlated arrangements that are inadequately characterized by second-order approximations. That second-order approximations can effectively characterize electron correlations but not nucleon correlations can be explained as follows. The electrons in atoms and molecules configure themselves largely in response to the nuclear attraction terms in the Hamiltonian, but protons and neutrons in the nucleus configure themselves in response to the nucleon–nucleon attraction terms. That is, for atoms and molecules correlations are largely driven by one-body operators, whereas for nuclei correlations are driven by two-body operators. The simpler correlations in atoms and molecules are effectively characterized by second-order approximations, but the more complex nuclear correlations are not. In Section VI a review is given of recent work on strongly interacting fermion systems that resolve the difficulties that Mihailovic and Rosina faced. Jin and Erdahl [2–4] studied a lattice model, where electrons interact through two-body forces alone, and Mazziotti and Erdahl [5] studied the Lipkin model for electrons, where the strength of the two-body forces can be turned on by adjusting a parameter V. There is convincing evidence [2–4] supporting the comment of Mihailovic and Rosina that "new relevant conditions" are required beyond 2-positivity. The articles make it abundantly clear that 2-positivity cannot effectively characterize the correlations induced by strong two-body forces. These three papers systematically explore higher-order conditions and show that by requiring the reduced density matrix to be 3-positive the problems faced by Mihailovic and Rosina are completely resolved. Both 3-positivity and 4-positivity were explored, and when the reduced density matrix is required to be 4-positive the lower bound estimates are accurate to ten figures. Very recently, Mazziotti [17] has applied the *complete* 3-positivity conditions to atoms and molecules with spectacular accuracy in energy and properties that substantially improves upon even the highly accurate $T1$ and $T2$ conditions. At the equilibrium geometry of N_2 the lower bound method with 2-positivity, 2-positivity plus $T1$ and $T2$, and 3-positivity deviates from full configuration interaction by 0.0149, 0.0018, and 0.0001 atomic units, respectively. The 3-positivity conditions, therefore, improve upon the accuracy of 2-positivity plus $T1$ and $T2$ by an order of magnitude. As a side issue, the work of Jin and Erdahl involved developing methods for solving the Euler equation $\mathbf{PQ} = \mathbf{0}$ of semidefinite programming.

II. Kth-ORDER APPROXIMATIONS FOR STATES

A Hermitian operator p is a von Neumann density if it is nonnegative and has unit trace. In more concrete terms, if $\mathfrak{F}$ is the finite-dimensional Fock space for a quantum model where electrons are distributed over a finite number of states, then p is a von Neumann density if (i) $\langle v, pv \rangle_{\mathfrak{F}} \geq 0$ *for all operators v on $\mathfrak{F}$*; and (ii) $\langle p, 1 \rangle_{\mathfrak{F}} = 1$. By the formula $\langle v, pv \rangle_{\mathfrak{F}}$ we mean the trace scalar product of the operators v and pv, that is, $\langle v, pv \rangle_{\mathfrak{F}} = \text{trace}_{\mathfrak{F}}(v^* pv)$; since $\langle p, 1 \rangle_{\mathfrak{F}} = \text{trace}_{\mathfrak{F}} p = 1$ we have used this scalar product to express the trace condition. More generally, if x, y are operators on $\mathfrak{F}$, then $\langle x, y \rangle_{\mathfrak{F}} = \text{trace}(x^* y)$. We denote by $\mathfrak{P}$ the cone of nonnegative operators on $\mathfrak{F}$, and by $\mathfrak{P}_0$ the convex set of all von Neumann densities. The elements of $\mathfrak{P}_0$ are the quantum states for the model.

We discuss a lattice model where spin-up, spin-down electrons move on a one-dimensional lattice Λ of size $|\Lambda| = r$, so that $\dim \mathfrak{F} = 2^{2r}$. An annihilator for a spin-up electron on lattice site $\mu \in \Lambda$ is denoted by a_μ, and that for a spin-down electron by b_μ. An arbitrary operator on $\mathfrak{F}$ can be written as a polynomial in the $2r$ annihilation and $2r$ creation operators.

A. Approximating States by k-Densities

By relaxing the condition that a von Neumann density be positive semidefinite, a graded family of approximations can be constructed. Since an operator can be represented as a polynomial in the annihilation and creation operators, it can be assigned a degree; for example, if $v = \sum_\mu v_\mu a_\mu^\dagger + \sum_{\mu,\nu,\rho} v_{\mu\nu\rho} a_\mu^\dagger b_\nu^\dagger b_\rho$, then $\deg(v) = 3$. We say that a Hermitian operator p is k-*positive* if it satisfies the condition that $\langle v, pv \rangle_{\mathfrak{F}} \geq 0$ *for all operators v where* $\deg(v) \leq k$.

Definition 1 A Hermitian operator p is a k-density if (i) $\langle v, pv \rangle_{\mathfrak{F}} \geq 0$ for all operators v, where $\deg v \leq k$; and (ii) $\langle p, 1 \rangle_{\mathfrak{F}} = 1$.

The k-densities approximate von Neumann densities to kth order; they are k-positive with unit trace. We denote the cone of all k-positive operators by $\mathfrak{P}^k$, and the convex set of all k-densities by $\mathfrak{P}_0^k$. The k-densities satisfy the relations $\mathfrak{P}_0 \subset \cdots \subset \mathfrak{P}_0^3 \subset \mathfrak{P}_0^2 \subset \mathfrak{P}_0^1$ and $\bigcap_k \mathfrak{P}_0^k = \mathfrak{P}_0$.

B. Matrix Representations

Let $\mathcal{V}^k$ be the linear space of operators v satisfying the condition $\deg(v) \leq k$. We take the monomial basis $\mathcal{M} = \{m_1, m_2, m_3, \ldots\}$ for $\mathcal{V}^k$, namely, all possible monomials in the factors $a_\mu^* + a_\mu, a_\mu^* - a_\mu, b_\mu^* + b_\mu, b_\mu^* - b_\mu, \mu \in \Lambda$, so that $\deg(m_i) \leq k$; we append the scale factor $1/\sqrt{\dim \mathcal{V}^k}$ to each of the monomials. Since $(a_\mu^* + a_\mu)^* (a_\mu^* + a_\mu) = 1$, and similarly for the other factors, we choose monomials with no repeated factors. It is easy to deduce that this monomial basis has the following properties: (1) $\sum_{i=1}^{\dim \mathcal{V}^k} m_i m_i^* = 1$, which follows from the equality $m_i m_i^* = 1/(\dim \mathcal{V}^k)$; (2) $\langle m_i, m_j \rangle_{\mathfrak{F}} = (\dim \mathfrak{F}/\dim \mathcal{V}^k) \delta_{ij}$.

Definition 2 Suppose that p is a k-density. Then the k-matrix for p, relative to the monomial basis $\mathcal{M} = \{m_1, m_2, m_3, \ldots\}$, is the matrix **P** with entries $P_{ij} = \langle m_i, pm_j \rangle_{\mathfrak{F}}$.

The k-matrix is a $\dim \mathcal{V}^k \times \dim \mathcal{V}^k$ matrix representation for p and is a variant of the kth-order reduced density matrix.

Proposition 3 *Let* **P** *be the k-matrix for the k-density p. Then* **P** *is Hermitian, positive semidefinite, and has unit trace.*

Proof. $P_{ji} = \langle m_j, pm_i \rangle_{\widetilde{\mathfrak{F}}} = \text{trace}(m_j^* P m_i) = (\text{trace}(m_i^* P m_j))^* = \langle m_i, pm_j \rangle_{\widetilde{\mathfrak{F}}}^* = P_{ij}^*$. Therefore **P** is Hermitian. Let $v = \sum_{i=1}^{\dim \mathcal{V}^k} v_i m_i$ be an arbitrary element of $\mathcal{V}^k$. Then the inequality $0 \leq \langle v, pv \rangle_{\widetilde{\mathfrak{F}}} = \sum_{i,j=1}^{\dim \mathcal{V}^k} v_i^* \langle m_i, pm_j \rangle_{\widetilde{\mathfrak{F}}} v_j = \mathbf{v}^* \mathbf{P} \mathbf{v}$ shows that **P** is positive semidefinite; **v** is the column vector of coefficients in the expansion. The equalities $1 = \langle p, 1 \rangle_{\widetilde{\mathfrak{F}}} = \langle p, \sum_{i=1}^{\dim \mathcal{V}^k} m_i m_i^* \rangle_{\widetilde{\mathfrak{F}}} = \sum_{i=1}^{\dim \mathcal{V}^k} \langle m_i, pm_i \rangle_{\widetilde{\mathfrak{F}}} = \sum_{i=1}^{\dim \mathcal{V}^k} P_{ii}$ establish that **P** has unit trace; we have used the identity $\sum_{i=1}^{\dim \mathcal{V}^k} m_i m_i^* = 1$, which was established above. ∎

C. Expectation Values

If x is an operator on $\widetilde{\mathfrak{F}}$, and if x can be written as $x = \sum_{i,j=1}^{\dim \mathcal{V}^k} X_{ij} m_i m_j^*$, then we take the matrix **X**, with entries X_{ij}, to be the matrix representation of **x**. With this definition, expectation values can be written

$$\langle p, x \rangle_{\widetilde{\mathfrak{F}}} = \langle \mathbf{P}, \mathbf{X} \rangle_k$$

where $\langle \mathbf{P}, \mathbf{X} \rangle_k = \text{trace } \mathbf{P}^* \mathbf{X}$ is the trace scalar product of the two matrices **P**, **X**; the subscript k is added to indicate that the trace is over the linear space of coefficients of the elements of $\mathcal{V}^k$. This equality is achieved by the following sequence of steps: $\langle p, x \rangle_{\widetilde{\mathfrak{F}}} = \langle p, \sum_{i,j=1}^{\dim \mathcal{V}^k} X_{ij} m_i m_j^* \rangle_{\widetilde{\mathfrak{F}}} = \sum_{i,j=1}^{\dim \mathcal{V}^k} \langle m_j, pm_i \rangle_{\widetilde{\mathfrak{F}}} X_{ij} = \sum_{i,j=1}^{\dim \mathcal{V}^k} P_{ji} X_{ij} = \sum_{i,j=1}^{\dim \mathcal{V}^k} P_{ij}^* X_{ij} = \langle \mathbf{P}, \mathbf{X} \rangle_k$.

D. The Pauli Subspace

It follows from the fermion commutation relations that the entries of a k-matrix are related by a system of linear equalities. For example, consider the pair transport operator $T_{\mu\nu}^2 = 2(b_\mu^* a_\mu^* a_\nu b_\nu + b_\nu^* a_\nu^* a_\mu b_\mu)$, which moves a spin-up, spin-down pair of electrons between sites μ, ν of Λ. If we define $v_{\pm 0} = a_\mu b_\mu \pm a_\nu b_\nu$, $v_{\pm 1} = a_\mu^* b_\nu \pm b_\mu a_\nu^* r$, $v_{\pm 2} = a_\mu b_\nu^* \pm b_\mu^* a_\nu$, $v_{\mp 3} = a_\mu^* a_\nu \mp b_\mu b_\nu^*$, and $v_{\mp 4} = a_\mu a_\nu^* \mp b_\mu^* b_\nu$, then, by the commutation relations, it follows that

$$T_{\mu\nu}^2 = v_{+i} v_{+i}^* - v_{-i} v_{-i}^*, \quad i = 0, 1, 2$$
$$= v_{-i} v_{-i}^* - v_{+i} v_{+i}^*, \quad i = 3, 4$$

This system of equalities is equivalent to the following linear conditions on the k-matrix:

$$\mathbf{v}_{+0}^* \mathbf{P} \mathbf{v}_{+0} - \mathbf{v}_{-0}^* \mathbf{P} \mathbf{v}_{-0} = \mathbf{v}_{+i}^* \mathbf{P} \mathbf{v}_{+i} - \mathbf{v}_{-i}^* \mathbf{P} \mathbf{v}_{-i}, \quad i = 1, 2$$
$$= \mathbf{v}_{-i}^* \mathbf{P} \mathbf{v}_{-i} - \mathbf{v}_{+i}^* \mathbf{P} \mathbf{v}_{+i}, \quad i = 3, 4$$

The column vector $\mathbf{v}_{\pm i}$ is the coordinate vector for $v_{\pm i}$ referred to the fixed basis for $\mathcal{V}^k$, and the equivalence is established using the identities $\langle v_{\pm i}, pv_{\pm i}\rangle_{\tilde{\mathfrak{F}}} = \mathbf{v}_{\pm i}^* \mathbf{P} \mathbf{v}_{\pm i}$. The coordinate vectors $\mathbf{v}_{\pm i}$ are linearly independent by the independence of the operators $v_{\pm i}$. It follows that these equalities represent four independent conditions on the k-matrix $\mathbf{P}$.

The linear conditions on $\mathbf{P}$ can be rewritten

$$\mathbf{P} \perp \mathbf{v}_{+0}\mathbf{v}_{+0}^* - \mathbf{v}_{-0}\mathbf{v}_{-0}^* - \mathbf{v}_{+i}\mathbf{v}_{+i}^* + \mathbf{v}_{-i}\mathbf{v}_{-i}^*, \quad i = 1, 2$$
$$\perp \mathbf{v}_{+0}\mathbf{v}_{+0}^* - \mathbf{v}_{-0}\mathbf{v}_{-0}^* - \mathbf{v}_{-i}\mathbf{v}_{-i}^* + \mathbf{v}_{+i}\mathbf{v}_{+i}^*, \quad i = 3, 4$$

so that $\mathbf{P}$ is orthogonal to a four-dimensional subspace of Hermitian matrices with respect to the trace scalar product. There are many other such linear conditions on $\mathbf{P}$, and taken together they are equivalent to requiring that $\mathbf{P}$ lie in the orthogonal complement of a real linear space of Hermitian matrices, which we denote by $\mathcal{S}$. We call $\mathcal{S}$ the *Pauli subspace* since it encodes the content of the Pauli principle.

Since an arbitrary k-matrix is orthogonal to $\mathcal{S}$, the matrix representation for a Hermitian operator is far from unique. Suppose that $\mathbf{H}$ is the matrix representation of some Hamiltonian, and that $\mathbf{S}$ is an arbitrary matrix in $\mathcal{S}$. Then $\mathbf{H} + \mathbf{S}$ is an equally valid representation since the identity $\langle p, h\rangle_{\tilde{\mathfrak{F}}} = \langle \mathbf{P}, \mathbf{H}\rangle_k = \langle \mathbf{P}, \mathbf{H} + \mathbf{S}\rangle$ clearly holds for all k-matrices $\mathbf{P}$. The explanation for such a large number of representations is straightforward: *the operator s corresponding to a matrix $\mathbf{S} \in \mathcal{S}$ is equal to the zero operator*. The operator s is constructed by taking the matrix elements to be coefficients in an expansion but can then be reduced to the zero operator using the commutation relations.

We summarize this discussion with the following theorem characterizing the convex set of k-matrices. In the statement of this theorem we introduce $\mathbb{P}_0^k$, the symbol we use to denote the convex set of positive semidefinite matrices with unit trace on the linear space of coefficients of the elements of $\mathcal{V}^k$; similarly, we use $\mathbb{P}^k$ to denote the cone of positive semidefinite matrices.

Theorem 4 *The set of k-matrices is given by $\mathbb{P}_0^k \cap \mathcal{S}^\perp$.*

The convex set of k-matrices $\mathbb{P}_0^k \cap \mathcal{S}^\perp$ is compact, but the corresponding convex set of k-densities $\mathfrak{P}_0^k$ is not; each k-matrix corresponds to an affine space of k-densities.

E. Additional Properties of Matrix Representations

At this point we list several useful properties of the matrix representation we have just constructed; these properties will be used in subsequent sections.

R1. *The k-matrix for the von Neumann density $(1/dim\,\tilde{\mathfrak{F}})1$ is $(1/dim\,\mathcal{V}^k)\,\mathbf{I}$*: Using the properties of the monomial basis for $\mathcal{V}^k$, the entries for the matrix representation are given by

$$\left\langle m_i, \frac{1}{dim\,\tilde{\mathfrak{F}}} 1 m_j \right\rangle_{\tilde{\mathfrak{F}}} = \frac{1}{dim\,\tilde{\mathfrak{F}}} \langle m_i, m_j \rangle_{\tilde{\mathfrak{F}}} = \frac{1}{dim\,\tilde{\mathfrak{F}}} \frac{dim\,\tilde{\mathfrak{F}}}{dim\,\mathcal{V}^k} \delta_{ij} = \frac{1}{dim\,\mathcal{V}^k} \delta_{ij}$$

R2. *The operator 1 is represented by $\mathbf{I}$*: This follows from the formula $1 = \sum_{i=1}^{dim\,\mathcal{V}^k} m_i m_i^*$.

R3. *Each Hermitian operator x admitting an expansion of the form*

$$x = \sum_{i,j=1}^{dim\,\mathcal{V}^k} X_{ij} m_i m_j^*$$

has a unique representation that is orthogonal to $\mathcal{S}$: If $\mathbf{X}$ is an arbitrary representation, and $\pi_{\mathcal{S}}$ is an orthogonal projection onto the Pauli space, then $\mathbf{X} - \pi_{\mathcal{S}}(\mathbf{X}) = \pi_{\mathcal{S}\perp}(\mathbf{X})$ is an equally valid representation that is orthogonal to $\mathcal{S}$ and uniquely determined by x.

R4. *Assume that $\mathbf{X}$ is an arbitrary representation of the traceless operator x; then* trace$(\mathbf{X}) = 0$: This follows from the equalities $0 =$ trace$(x) = \langle 1, x \rangle_{\tilde{\mathfrak{F}}} = dim\,\tilde{\mathfrak{F}} \langle (1/dim\,\tilde{\mathfrak{F}})1, x \rangle_{\tilde{\mathfrak{F}}} = dim\,\tilde{\mathfrak{F}} \langle (1/dim\,\mathcal{V}^k)\mathbf{I}, \mathbf{X} \rangle_k = (dim\,/\tilde{\mathfrak{F}} dim\,\mathcal{V}^k) \langle \mathbf{I}, \mathbf{X} \rangle_k = (dim\,/\tilde{\mathfrak{F}} dim\,\mathcal{V}^k)$ trace$(\mathbf{X})$.

R5. *If $\mathbf{S} \in \mathcal{S}$, then* trace$(\mathbf{S}) = 0$: This follows directly from R4 since $\mathbf{S}$ is a matrix representation of the zero operator, which is traceless.

F. Energy Lower Bounds

Since the convex set of k-densities contains the convex set of von Neumann densities, the following inequality holds:

$$\min_{p \in \mathfrak{P}_0^k} \langle p, h \rangle_{\tilde{\mathfrak{F}}} \leq \min_{p \in \mathfrak{P}_0} \langle p, h \rangle_{\tilde{\mathfrak{F}}}$$

where h is the Hamiltonian. An energy lower bound and a kth order approximation for the state are found by varying over the set of k-densities. Since the k-densities satisfy the relations $\mathfrak{P}_0 \subset \cdots \subset \mathfrak{P}_0^3 \subset \mathfrak{P}_0^2 \subset \mathfrak{P}_0^1$ and $\bigcap_k \mathfrak{P}_0^k = \mathfrak{P}_0$, these lower bounds converge to the exact ground-state energy from below.

$$\min_{p \in \mathfrak{P}_0^2} \langle p, h \rangle_{\tilde{\mathfrak{F}}} \leq \min_{p \in \mathfrak{P}_0^3} \langle p, h \rangle_{\tilde{\mathfrak{F}}} \leq \min_{p \in \mathfrak{P}_0^4} \langle p, h \rangle_{\tilde{\mathfrak{F}}} \leq \cdots \leq \min_{p \in \mathfrak{P}_0} \langle p, h \rangle_{\tilde{\mathfrak{F}}}$$

If **P** is a k-matrix and **H** is a matrix representation of the Hamiltonian, as described above, then the equality

$$\langle \mathbf{P}, \mathbf{H} \rangle_k = \langle p, h \rangle_{\tilde{\mathfrak{F}}}$$

holds. It follows that the lower bounds can be computed using k-matrices, and the sequence of lower bounds can be written

$$\min_{\mathbf{P} \in \mathbb{P}_0^2 \cap \mathcal{S}^\perp} \langle \mathbf{P}, \mathbf{H} \rangle_{\tilde{\mathfrak{F}}} \leq \min_{\mathbf{P} \in \mathbb{P}_0^3 \cap \mathcal{S}^\perp} \langle \mathbf{P}, \mathbf{H} \rangle_{\tilde{\mathfrak{F}}} \leq \min_{\mathbf{P} \in \mathbb{P}_0^4 \cap \mathcal{S}^\perp} \langle \mathbf{P}, \mathbf{H} \rangle_{\tilde{\mathfrak{F}}} \leq \cdots$$

and the corresponding sequence of k-matrices converge to k-matrices for the von Neumann density for the ground-state. A question of great practical importance is the speed of convergence of these lower bounds to the ground-state energy, and this will be thoroughly discussed later.

III. SEMIDEFINITE PROGRAMMING

The central problem in electronic structure theory is to determine the ground state of a system of electrons, which is typically done variationally by minimizing the energy. The lower bound method can be invoked to achieve a kth-order approximation by replacing the variation $\min_{p \in \mathfrak{P}_0} \langle p, h \rangle_{\tilde{\mathfrak{F}}}$ by the *semidefinite program*

$$\min_{P \in \mathbb{P}_0^k \cap \mathcal{S}^\perp} \langle \mathbf{P}, \mathbf{H} \rangle_k$$

where the minimum is taken over the the set of k-matrices that represent quantum states; **H** is a matrix representation of the Hamiltonian h. This is called a semidefinite program because the variation is over a section of the cone of positive semidefinite matrices, namely, $\mathbb{P}_0^k \cap \mathcal{S}^\perp$.

A. Lagrange Duality

Rather than minimize the energy function $E_0(\mathbf{P}) = \langle \mathbf{P}, \mathbf{H} \rangle_k$ by varying over the set of k-matrices, there is a dual formulation where the bottom eigenvalue $\lambda_0(\mathbf{H} + \mathbf{S})$ of the matrix $\mathbf{H} + \mathbf{S}$ is maximized over the set of Pauli matrices $\mathbf{S} \in \mathcal{S}$. The dual formulation can be derived using Lagrange's method, which requires converting the constrained *energy problem* to an unconstrained one. If

$\{\mathbf{S}_1, \mathbf{S}_2, \mathbf{S}_3, \ldots\}$ is a basis for the Pauli space $\mathcal{S}$, then

$$\min_{\mathbf{P} \in \mathbb{P}_0^k \cap \mathcal{S}^\perp} E_0(\mathbf{P}) = \min_{\mathbf{P} \in \mathbb{P}_0^k \cap \mathcal{S}^\perp} \langle \mathbf{P}, \mathbf{H} \rangle_k$$

$$\geq \max_\beta \min_{\mathbf{P} \in \mathbb{P}^k} \left\{ \langle \mathbf{P}, \mathbf{H} \rangle_k + \beta_0(\langle \mathbf{P}, \mathbf{I} \rangle_k - 1) + \sum_i \beta_i \langle \mathbf{P}, \mathbf{S}_i \rangle_k \right\}$$

$$= \max_\beta \left\{ -\beta_0 + \min_{\mathbf{P} \in \mathbb{P}^k} \left\langle \mathbf{P}, \mathbf{H} + \beta_0 \mathbf{I} + \sum_i \beta_i \mathbf{S}_i \right\rangle_k \right\}$$

$$= \max_\beta \left\{ \lambda_0 + \min_{\mathbf{P} \in \mathbb{P}^k} \left\langle \mathbf{P}, \mathbf{H} + \sum_i \beta_i \mathbf{S}_i - \lambda_0 \mathbf{I} \right\rangle_k \right\}$$

$$= \sup_{\mathbf{S} \in \mathcal{S}} \lambda_0(\mathbf{H} + \mathbf{S})$$

where $\lambda_0(\mathbf{H} + \mathbf{S})$ is the bottom eigenvalue of the matrix $\mathbf{H} + \mathbf{S}$. The last two steps require first noting that the value of

$$\min_{\mathbf{P} \in \mathbb{P}^k} \left\langle \mathbf{P}, \mathbf{H} + \beta_0 \mathbf{I} + \sum_i \beta_i \mathbf{S}_i \right\rangle_k$$

is negative infinity unless $\mathbf{H} + \beta_0 \mathbf{I} + \sum_i \beta_i \mathbf{S}_i$ is positive semidefinite, and then observing that the maximum over the vector of parameters β is achieved when $-\beta_0$ is the bottom eigenvalue λ_0 of the matrix $\mathbf{H} + \sum_i \beta_i \mathbf{S}_i$. By passing to the dual the energy minimization problem is converted to the problem of *optimizing the spectrum*, namely, the problem of maximizing the bottom eigenvalue. If $\mathbf{P} \in \mathbb{P}_0^k \cap \mathcal{S}$ is an arbitrary k-matrix, and $\mathbf{S} \in \mathcal{S}$ an arbitrary element of the Pauli space, Lagrange's argument shows that

$$\Delta = E_0(\mathbf{P}) - \lambda_0(\mathbf{H} + \mathbf{S}) \geq 0$$

The *gap* between the energy and bottom eigenvalue is nonnegative. If the k-matrix $\mathbf{P}$ moves to further decrease the energy, and if the Pauli matrix $\mathbf{S}$ moves to further increase the bottom eigenvalue, the gap narrows and possibly shrinks to zero. It is important to note that there are semidefinite programs where this gap cannot shrink to zero; we discuss such an example later. However, In our special case where we vary k-matrices and Pauli matrices, as we have defined them, the gap shrinks to zero. This is an important result for both theoretical and practical reasons; a proof is supplied below.

B. The Gap Formula

The semidefinite program we are considering is formulated in terms of the symmetric matrix $\mathbf{H}$ and the linear subspace of symmetric matrices $\mathcal{S}$. For the moment we consider a general situation where the matrix $\mathbf{H}$ and the subspace $\mathcal{S}$ are specified, but not as defined in Section II where kth-order approximations to the von Neumann density were considered. We only impose the condition that $\mathbf{I}, \mathbf{H}, \mathcal{S}$ be linearly independent, so that there are no nontrivial linear relations of the form $\alpha \mathbf{I} + \beta \mathbf{H} + \mathbf{S} = 0$, where $\mathbf{S} \in \mathcal{S}$. In such a general setting, it is possible that the energy problem does not have an optimal solution, or that the spectral optimization problem does not have an optimal solution.

Theorem 5 *If* $\mathbf{P} \in \mathbb{P}_0^k \cap \mathcal{S}$, *and* $\mathbf{S} \in \mathcal{S}$ *are chosen arbitrarily, then*

$$\Delta = E_0(\mathbf{P}) - \lambda_0(\mathbf{H} + \mathbf{S}) \geq 0$$

This gap inequality follows from Lagrange's derivation of the dual spectral optimization problem, but there is a more direct proof that we now present.

Proof. Define

$$\mathbf{Q} = \mathbf{H} + \mathbf{S} - \lambda_0(\mathbf{H} + \mathbf{S})\mathbf{I}$$

where $\lambda_0(\mathbf{H} + \mathbf{S})$ is the bottom eigenvalue of the matrix $\mathbf{H} + \mathbf{S}$. Then the equality $\langle \mathbf{P}, \mathbf{Q} \rangle_k = \langle \mathbf{P}, \mathbf{H} \rangle_k - \lambda_0(\mathbf{H} + \mathbf{S}) = \Delta$ follows from the conditions $\langle \mathbf{P}, \mathcal{S} \rangle_k = 0, \langle \mathbf{P}, \mathbf{I} \rangle_k = 1$, which are assumed to hold. Since both $\mathbf{P}$ and $\mathbf{Q}$ are positive semidefinite, this scalar product $\langle \mathbf{P}, \mathbf{Q} \rangle_k$ is nonnegative. ∎

The search for optimal solutions to both the energy problem and the spectral optimization problem typically starts with matrices $\mathbf{P}$ and $\mathbf{S}$ that have a positive gap. Iterations are designed to move $\mathbf{P}$ so that the energy is decreased, and to move $\mathbf{S}$ so that the bottom eigenvalue is increased, and such motions cause the gap to narrow. It is important that there are semidefinite programs where this gap cannot shrink to zero, and we discuss such an example later. However, in our special case where we vary k-matrices and Pauli matrices, as we have defined them, the gap shrinks to zero. This is an important result for both theoretical and practical reasons; a proof is supplied below.

Corollary 6 *Assume that for* $\mathbf{P} \in \mathbb{P}_0^k \cap \mathcal{S}$, *and* $\mathbf{S} \in \mathcal{S}$, *the gap* $\Delta = E_0(\mathbf{P}) - \lambda_0(\mathbf{H} + \mathbf{S}) = 0$. *Then* $\mathbf{P}$ *solves the energy problem, and* $\mathbf{S}$ *solves the spectral optimization problem.*

Proof. This follows immediately from the gap inequality $\Delta = E_0(\mathbf{P}) - \lambda_0(\mathbf{H} + \mathbf{S}) \geq 0$. ∎

Corollary 7 *Let* $\mathbf{P} \in \mathbb{P}_0^k \cap \mathcal{S}$, *and let* $\mathbf{S} \in \mathcal{S}$. *Then* $\Delta = E_0(\mathbf{P}) - \lambda_0(\mathbf{H} + \mathbf{S}) = 0$ *if and only if* $\mathbf{PQ} = \mathbf{0}$, *where* $\mathbf{Q} = \mathbf{H} + \mathbf{S} - \lambda_0(\mathbf{H} + \mathbf{S})\mathbf{I}$.

Proof. For this result we use the identity $\Delta = \langle \mathbf{P}, \mathbf{Q} \rangle_k$. Since $\mathbf{P}$ and $\mathbf{Q}$ are positive semidefinite, $\langle \mathbf{P}, \mathbf{Q} \rangle_k = 0$ if and only if $\mathbf{PQ} = \mathbf{0}$. ■

The equation $\mathbf{PQ} = \mathbf{0}$ requires that the range spaces for $\mathbf{P}$ and $\mathbf{Q}$ lie in complementary orthogonal subspaces.

The energy and spectral optimization problems are convex programs so when there are multiple solutions the solution sets form a convex set. The following corollary characterizes how these convex sets of solutions relate to solutions of the Euler equation. In the formulation of this corollary we use the notion of *optimal gap* Δ_0—the gap achieved by optimal $\mathbf{P}$ and $\mathbf{S}$. The optimal gap is a characteristic of the energy problem, depending only on $\mathbf{H}$ and $\mathcal{S}$.

Corollary 8 *Let $\mathcal{P}_0$ be the convex set of solutions of the energy problem, and let $\mathcal{S}_0$ be the convex set of solutions of the spectral optimization problem. If the optimal gap $\Delta_0 = 0$, then $\mathbf{P} \in \mathcal{P}_0$, $\mathbf{S} \in \mathcal{S}_0$ if and only if $\mathbf{PQ} = \mathbf{0}$, where $\mathbf{Q} = \mathbf{H} + \mathbf{S} - \lambda_0(\mathbf{H} + \mathbf{S})\mathbf{I}$.*

Proof. This follows from Corollary 7. ■

IV. THE FUNDAMENTAL THEOREM

In this section we continue the discussion of the energy and spectral optimization problems,

$$\min_{\mathbf{P} \in \mathbb{P}_0^k \cap \mathcal{S}^\perp} E_0(\mathbf{P}), \qquad \max_{\mathbf{S} \in \mathcal{S}} \lambda_0(\mathbf{H} + \mathbf{S})$$

giving conditions that characterize the class of problems where optimal solutions can be found by solving the *Euler equation*

$$\mathbf{PQ} = \mathbf{0} \quad \text{where} \quad \mathbf{Q} = \mathbf{H} + \mathbf{S} - \lambda_0(\mathbf{H} + \mathbf{S})\mathbf{I}$$

This class is important since the most effective algorithms for solving semidefinite programs take as their starting point the Euler equation. As in the previous section the setting is general, the only condition initially imposed being that $\mathbf{I}, \mathbf{H}, \mathcal{S}$ are linearly independent. Throughout our discussion the positive semidefinite matrix $\mathbf{P}$ is required to satisfy the conditions $\langle \mathbf{P}, \mathcal{S} \rangle_k = 0$, $\langle \mathbf{P}, \mathbf{I} \rangle_k = 1$, and the positive semidefinite matrix $\mathbf{Q}$ is required to have the form $\mathbf{H} + \mathbf{S} - \lambda_0(\mathbf{H} + \mathbf{S})\mathbf{I}$, where $\mathbf{S} \in \mathcal{S}$ and $\lambda_0(\mathbf{H} + \mathbf{S})$ is the bottom eigenvalue of $\mathbf{H} + \mathbf{S}$.

A. Fundamental Theorem

To prove our main theorem, establishing that the existence of a solution of the spectral optimization problem is equivalent to the existence of a solution of the

Euler equation, we need the following lemma. In the statement of the lemma K^* denotes the polar cone of the cone $K \subset \mathbb{R}^p$, which is given by

$$K^* = \{\mathbf{v} \in \mathbb{R}^p | \mathbf{k}^T \mathbf{v} \geq 0, \forall \mathbf{k} \in K\}$$

$\mathbf{k}^T \mathbf{v}$ is the Euclidean scalar product of $\mathbf{k}$ with $\mathbf{v}$.

Lemma 9 *Let $L \subset \mathbb{R}^n$ be a subspace and let $K \subset \mathbb{R}^n$ be a closed pointed cone with nonempty interior and vertex at the origin. Assume that $\mathbf{b} \in \partial K$. Then exactly one of two alternatives must hold: (A1) there is a nonzero element $\mathbf{u} \in (\mathbf{b} + L) \cap \mathrm{int}\, K$; (A2) there is a nonzero element $\mathbf{v} \in \{\mathbf{b}, L\}^\perp \cap K^*$.*

Proof. We prove this result by showing that alternative A2 is equivalent to the statement $(\mathbf{b} + L) \cap \mathrm{int}\, K$ is empty. Suppose that $(\mathbf{b} + L) \cap \mathrm{int}\, K$ is empty. By the celebrated separation theorem of convex analysis there is a hyperplane H separating $\mathbf{b} + L$ and $\mathrm{int}\, K$. Since $\mathbf{b} \in \partial K, \mathbf{b} \in H$. It follows that H is a supporting hyperplane of K and that $\mathbf{b} + L \subset H$. Since K is a cone, more is true: H passes through the vertex of K, the origin, and is a subspace. For this reason $\mathbf{b}, L \subset H$. Let $\mathbf{v}$ be nonzero, normal to H, and pointing into the closed half-space containing K. Since $\mathbf{v}^T \mathbf{k} \geq 0$ for all $\mathbf{k} \in K$, it follows that $\mathbf{v} \in K^*$; since $\mathbf{v} \perp \{\mathbf{b}, L\}$ alternative A2 holds.

Now assume that statement A2 holds. Since $\mathbf{v} \perp \mathbf{b}, L$, and since $\mathbf{v}^T \mathbf{k} > 0$ for all $\mathbf{k} \in \mathrm{int}\, K$, it follows that $(\mathbf{b} + L) \cap \mathrm{int}\, K$ is empty. ∎

The preceeding lemma does most of the work in establishing the following fundamental theorem of semidefinite programming.

Theorem 10 *Let $\mathbf{S} \in \mathcal{S}$, and let $\mathbf{Q} = \mathbf{H} + \mathbf{S} - \lambda_0(\mathbf{H} + \mathbf{S})\mathbf{I}$, where $\lambda_0(\mathbf{H} + \mathbf{S})$ is the bottom eigenvalue of $\mathbf{H} + \mathbf{S}$. Then $\mathbf{S}$ maximizes λ_0 if and only if there is a positive semidefinite matrix $\mathbf{P}$ satisfying the following conditions: (E1) $\mathbf{PQ} = \mathbf{0}$; (E2) $\langle \mathbf{P}, \mathcal{S} \rangle_k = 0$; (E3) $\langle \mathbf{P}, \mathbf{I} \rangle_k = 1$.*

Proof. It is clear that $\mathbf{S} \in \mathcal{S}$ maximizes λ_0 if and only if $(\mathbf{Q} + \mathcal{S}) \cap \mathrm{int}\, \mathbb{P}^k$ is empty, where $\mathbf{Q} = \mathbf{H} + \mathbf{S} - \lambda_0(\mathbf{H} + \mathbf{S})\mathbf{I}$. By making the identifications $\mathbf{b} = \mathbf{Q}, L = \mathcal{S}, K = K^* = \mathbb{P}^k$ (since $\mathbb{P}^k$ is self-polar), and applying the above lemma a more useful characterization is achieved: $\mathbf{S} \in \mathcal{S}$ maximizes λ_0 if and only if there is a nonzero element $\mathbf{P} \in \{\mathbf{Q}, \mathcal{S}\}^\perp \cap \mathbb{P}^k = (\mathbf{Q}^\perp \cap \mathbb{P}^k) \cap (\mathcal{S}^\perp \cap \mathbb{P}^k)$. Since $\mathbf{Q} \in \mathbb{P}^k$ the condition $\mathbf{P} \in \mathbf{Q}^\perp \cap \mathbb{P}^k$ is equivalent to the condition $\mathbf{PQ} = \mathbf{0}$, which is condition E1. The condition $\mathbf{P} \in \mathcal{S}^\perp \cap \mathbb{P}^k$ is equivalent to condition E2. Since $\mathbf{P}$ is nonzero it can be scaled so that condition E3 is satisfied. ∎

The virtue of this theorem is that it reduces the dual problem to the question of solving the Euler equation $\mathbf{PQ} = \mathbf{0}$, a second-order algebraic equation for the entries of **P** and **Q**. Earlier proofs [10, 11] of this result made use of elements of nonsmooth analysis, but the proof given here draws only on standard ideas from matrix theory and convexity theory.

Corollary 11 *If the spectral optimization problem has an optimal solution* **S**, *then there is an optimal solution* **P** *of the energy problem, and* **PQ** = **0**, *where* **Q** = **H** + **S** − λ_0(**H** + **S**)**I**.

Proof. This result follows directly from Theorem 10 and Corollaries 6 and 7. ■

B. Existence Theorems

If the subspace $\mathcal{S}$ contains a positive definite element $\mathbf{S}^+$, neither the energy problem nor the spectral optimization problem has an optimal solution: since there is no positive semidefinite matrix **P** satisfying both the conditions $\langle \mathbf{P}, \mathbf{S}^+ \rangle_k = 0$, $\langle \mathbf{P}, \mathbf{I} \rangle_k = 1$, the convex set $\mathbb{P}_0^k \cap \mathcal{S}$ is empty, and the energy problem has no solution. Since the values of $\lambda_0(\mathbf{H} + \alpha \mathbf{S}^+)$ increase indefinitely as $\alpha \in \mathbb{R}$ goes to positive infinity, the spectral optimization problem has no solution. When $\mathcal{S}$ contains positive semidefinite elements, but no positive definite elements, the spectral optimization problem can have a solution, but need not, as is illustrated by the following example. Consider the data

$$\mathbf{H} = \begin{bmatrix} 1 & 1 \\ 1 & 0 \end{bmatrix}; \quad \mathcal{S} = \mathrm{span}\left\{ \begin{bmatrix} 0 & 0 \\ 0 & 1 \end{bmatrix} \right\}$$

and the associated energy problem. The conditions $\langle \mathbf{P}, \mathcal{S} \rangle_k = 0$, $\langle \mathbf{P}, \mathbf{I} \rangle_k = 1$ uniquely determine **P** to be

$$\mathbf{P} = \begin{bmatrix} 1 & 0 \\ 0 & 0 \end{bmatrix}$$

The minimum energy is then given by $E_0(\mathbf{P}) = \langle \mathbf{P}, \mathbf{H} \rangle_k = 1$.

For the spectral optimization problem we must calculate the bottom eigenvalue of

$$\mathbf{H} + \alpha \mathbf{S} = \begin{bmatrix} 1 & 1 \\ 1 & \alpha \end{bmatrix}$$

It is easy to see that inequality $\lambda_0(\mathbf{H} + \alpha \mathbf{S}) < 1$ holds for all $\alpha \in \mathbb{R}$. For $\alpha > 1$

$$\lambda_0(\mathbf{H} + \alpha \mathbf{S}) = \frac{1}{2}\left[1 + \sqrt{1 + \left(\frac{2}{\alpha - 1}\right)^2} \right] + \frac{\alpha}{2}\left[1 - \sqrt{1 + \left(\frac{2}{\alpha - 1}\right)^2} \right]$$

$$\sim -\frac{1}{\alpha - 1}$$

and $\lambda_0(\mathbf{H} + \alpha\mathbf{S})$ approaches 1 asymtotically with increasing α. Therefore the spectral optimization problem has no solution. The gap is given by

$$\Delta = E_0(\mathbf{P}) - \lambda_0(\mathbf{H} + \alpha\mathbf{S})$$
$$\sim 1 - \left(1 - \frac{1}{\alpha - 1}\right)$$
$$\sim \frac{1}{\alpha - 1}$$

which is always positive, but decreases asymtotically to zero as α increases.

Theorem 12 *The energy minimization problem has an optimal solution if and only if $\mathcal{S} \cap \mathrm{int}\mathbb{P}^k = \emptyset$.*

Proof. In our introductory comments to this subsection we have argued that the energy problem has no solution when $\mathcal{S} \cap \mathrm{int}\,\mathbb{P}^k \neq \emptyset$. It remains to argue that the energy problem has a solution when $\mathcal{S} \cap \mathrm{int}\,\mathbb{P}^k = \emptyset$. After making the identifications $\mathbf{b} = \mathbf{0}, L = \mathcal{S}, K = K^* = \mathbb{P}^k$, we apply Lemma 9 to show that there is a nonzero element $\mathbf{P}$ in $\mathcal{S} \cap \mathbb{P}^k$. We can then scale $\mathbf{P}$ so that it has unit trace and conclude that the convex set determined by the two conditions $\langle \mathbf{P}, \mathcal{S} \rangle_k = 0$, $\langle \mathbf{P}, \mathbf{I} \rangle_k = 1$ is not empty. Since this set is also compact, the energy problem necessarily has a solution. ∎

Theorem 13 *The spectral optimization problem has an optimal solution if $\mathcal{S} \cap \mathbb{P}^k = \mathbf{0}$.*

Proof. We first consider the condition $\mathcal{S} \cap \mathbb{P}^k = \mathbf{0}$. After making the identifications $\mathbf{b} = \mathbf{0}, L = \mathcal{S}, K = K^* = \mathbb{P}^k$, we apply Lemma 9 to show that this condition is equivalent to the existence of a nonzero element in $\mathcal{S} \cap \mathrm{int}\,\mathbb{P}^k$, which we scale to yield the matrix $\mathbf{I}^*$ satisfying the condition $\langle \mathbf{I}^*, \mathbf{I} \rangle_k = 1$. We then replace $\mathbf{H}$ by a translate $\mathbf{H}_0 = \mathbf{H} - \kappa\mathbf{I}$ so that the condition $\langle \mathbf{I}^*, \mathbf{H}_0 \rangle_k = 0$ is satisfied. Since we have assumed that $\mathbf{I}, \mathbf{H}, \mathcal{S}$ are linearly independent, the projection of $\mathbf{H}_0$ onto the orthogonal complement of $\{\mathbf{I}, \mathcal{S}\}$ is nonzero, so it can then be scaled to give the matrix $\mathbf{H}_0^*$ satisfying the conditions $\langle \mathbf{H}_0^*, \mathbf{I} \rangle_k = 0, \langle \mathbf{H}_0^*, \mathbf{H}_0 \rangle_k = 1$. We next introduce the subspace $\mathcal{K} = \{\mathbf{I}, \mathbf{H}_0, \mathcal{S}\}^{\perp}$ to achieve the following symmetrical situation: just as the matrices $\mathbf{I}, \mathbf{H}_0, \mathcal{S}$ span $\mathcal{K}^{\perp}$, the matrices $\mathbf{I}^*, \mathbf{H}_0^*, \mathcal{K}$ span $\mathcal{S}$. Moreover, $\mathbf{I}^*, \mathbf{I}$ are positive definite and the conditions $\langle \mathbf{I}^*, \mathbf{I} \rangle_k = 1, \langle \mathbf{I}^*, \mathbf{H}_0 \rangle_k = 0, \langle \mathbf{H}_0^*, \mathbf{I} \rangle_k = 0, \langle \mathbf{H}_0^*, \mathbf{H}_0 \rangle_k = 1$ hold.

By Theorem 12 the condition $\mathcal{S} \cap \mathbb{P}^k = \mathbf{0}$ ensures that the energy problem has an optimal solution $\mathbf{P}$, which we write as $\mathbf{P} = \mathbf{I}^* - \alpha\mathbf{H}_0^* + \mathbf{K}$, where $\mathbf{K} \in \mathcal{K}$. Since $\mathbf{P}$ minimizes the energy it follows that $\alpha > 0$ in this expansion. Since $\mathbf{P}$

is optimal we also conclude that the condition $(\mathbf{P} + \mathcal{K}) \cap \text{int } \mathbb{P}^k = \emptyset$ must hold. Otherwise, there would be an element $\Delta \mathbf{K} \in \mathcal{K}$, and a positive real number ε, so that $\mathbf{P}' = \mathbf{P} - \varepsilon \mathbf{H}_0^* + \Delta \mathbf{K} \in \mathcal{S} \cap \mathbb{P}_0^k$. It would then follow that $\langle \mathbf{P}' - \mathbf{P}, \mathbf{H}_0 \rangle_k = \langle -\varepsilon \mathbf{H}_0^* + \Delta \mathbf{K}, \mathbf{H}_0 \rangle_k = -\varepsilon \langle \mathbf{H}_0^*, \mathbf{H}_0 \rangle_k = -\varepsilon < 0$, which cannot hold since $\mathbf{P}$ is optimal.

After making the identifications $\mathbf{b} = \mathbf{P}, L = \mathcal{K}, K = K^* = \mathbb{P}^k$, we apply Lemma 9 to convert the condition $(\mathbf{P} + \mathcal{K}) \cap \text{int } \mathbb{P}^k = \emptyset$ to a more useful form: there is a nonzero element $\mathbf{Q}$ in $\{\mathbf{P}, \mathcal{K}\}^\perp \cap \mathbb{P}^k = \mathbf{P}^\perp \cap \{\mathbf{I}, \mathbf{H}, \mathcal{S}\} \cap \mathbb{P}^k$. Since $\langle \mathbf{I}_0^*, \mathbf{Q} \rangle_k > 0$, we initially scale $\mathbf{Q}$ so that it can be written $\mathbf{Q} = \mathbf{I} + \gamma \mathbf{H}_0 + \mathbf{S}'$, where $\mathbf{S}' \in \mathcal{S}$. The condition $\langle \mathbf{P}, \mathbf{Q} \rangle_k = \langle \mathbf{I}^* - \alpha \mathbf{H}_0^* + \mathbf{K}, \mathbf{I} + \gamma \mathbf{H}_0 + \mathbf{S}' \rangle_k = 1 - \alpha\gamma = 0$ requires that $\gamma = 1/\alpha > 0$. Rescaling by α converts $\mathbf{Q}$ to the form $\mathbf{H}_0 + \alpha \mathbf{S}' + \alpha \mathbf{I} = \mathbf{H} + \alpha \mathbf{S}' + (\alpha - \kappa)\mathbf{I}$. The final form is $\mathbf{Q} = \mathbf{H} + \mathbf{S} - \lambda_0(\mathbf{H} + \mathbf{S})\mathbf{I}$, where $\mathbf{S} = \alpha \mathbf{S}' \in \mathcal{S}$, and $\lambda_0(\mathbf{H} + \mathbf{S}) = -(\alpha - \kappa)$ is the bottom eigenvalue of $\mathbf{H} + \mathbf{S}$; this is the form required by the spectral optimization problem.

Since $\Delta = \langle \mathbf{P}, \mathbf{Q} \rangle_k = E_0(\mathbf{P}) - \lambda_0(\mathbf{H} + \mathbf{S}) = 0$, an application of Corollary 6 shows that $\mathbf{Q}$ is a solution of the spectral optimization problem. ∎

Corollary 14 *Assume that $\mathcal{S}$ is the Pauli subspace as defined in Section II. Then the energy minimization and spectral optimization problems have optimal solutions $\mathbf{P}, \mathbf{S}$, and $\mathbf{PQ} = \mathbf{O}$, where $\mathbf{Q} = \mathbf{H} + \mathbf{S} - \lambda_0(\mathbf{H} + \mathbf{S})\mathbf{I}$.*

Proof. By property R5 listed at the end of Section II, the elements of the Pauli subspace $\mathcal{S}$ are traceless, from which we infer by Theorems 12 and 13 that the energy problem and the spectral optimization problem have optimal solutions. By Theorem 10 these solutions are characterized by the Euler equation $\mathbf{PQ} = \mathbf{0}$. ∎

V. ALGORITHMS

In order to further develop the lower bound method, more effective algorithms need to be devised to solve the Euler equation $\mathbf{PQ} = \mathbf{0}$, and attention is now focused in this direction. For electronic structure calculations the matrices $\mathbf{P}, \mathbf{Q}$ are large, and in many cases out of range for the current generation of algorithms—the next generation will make use of particular features inherent in electronic structure theory so that larger problems can be accommodated. In this section we give a brief treatment of algorithms for semidefinite programming.

A. Standard Formulation

We now return to the original formulation of the energy problem,

$$\min_{\mathbf{P} \in \mathcal{S}^\perp \cap \mathbb{P}_0^k} E_0(\mathbf{P})$$

where $E_0(\mathbf{P}) = \langle \mathbf{P}, \mathbf{H} \rangle_k$, and the k-matrix $\mathbf{P}$ and Hamiltonian matrix $\mathbf{H}$ are referred to the original monomial basis $\mathcal{M}$. This representation has many nice properties, including the five properties listed toward the end of Section II. We first note that by property R5 the elements $\mathbf{S}$ of the Pauli space $\mathcal{S}$ are traceless. It is convenient to assume that the original Hamiltonian h on Fock space is traceless, in which case the matrix representation $\mathbf{H}$ is traceless by property R4. It is also convenient to choose a representation $\mathbf{H}$ that is orthogonal to the Pauli space, which is possible by property R3. By introducing the subspace $\mathcal{K} = \{\mathbf{I}, \mathbf{H}, \mathcal{S}\}^\perp$, we have the following orthogonal decomposition of the set of symmetric matrices:

$$\mathbf{I} \perp \mathbf{H} \perp \mathcal{S} \perp \mathcal{K}$$

The matrices $\mathbf{P}$ and $\mathbf{Q}$ then belong to the following subspaces of symmetric matrices:

$$\mathbf{P} \in \mathcal{S}^\perp = \mathrm{span}\{\mathbf{I}, \mathbf{H}, \mathcal{K}\}$$
$$\mathbf{Q} \in \mathcal{K}^\perp = \mathrm{span}\{\mathbf{I}, \mathbf{H}, \mathcal{S}\}$$

By introducing bases $\{\mathbf{S}_1, \mathbf{S}_2, \mathbf{S}_3, \ldots\}, \{\mathbf{K}_1, \mathbf{K}_2, \mathbf{K}_3, \ldots\}$ for $\mathcal{S}, \mathcal{K}$ the matrices $\mathbf{P}$ and $\mathbf{Q}$ can be written

$$\mathbf{P} = \frac{1}{|\mathbf{I}|^2}\mathbf{I} + \frac{E_0}{|\mathbf{H}|^2}\mathbf{H} + \sum_i \alpha_i \mathbf{K}_i \quad \text{and} \quad \mathbf{Q} = \mathbf{H} + \sum_i \beta_i \mathbf{S}_i - \lambda_0 \mathbf{I}$$

The coefficients of $\mathbf{I}$ and $\mathbf{H}$ are treated separately in $\mathbf{P}$ so that the trace condition, $\langle \mathbf{P}, \mathbf{I} \rangle_k = 1$, holds and so that the energy is given by $\langle \mathbf{P}, \mathbf{H} \rangle_k = E_0$; the coefficient of $\mathbf{I}$ in $\mathbf{Q}$ becomes the bottom eigenvalue λ_0 when parameters are adjusted so that $\mathbf{Q}$ is optimal. The optimal values for E_0 and λ_0 are equal.

Finding a solution for the energy problem and spectral optimization problem is reduced to solving a system of quadratic equations obtained by substituting $\mathbf{P}$ and $\mathbf{Q}$ into the Euler equation $\mathbf{PQ} = \mathbf{0}$. It is easy to devise algorithms to determine the parameters $E_0, \alpha_1, \alpha_2, \ldots, \lambda_0, \beta_1, \beta_2, \ldots$, but there is an important caveat—the only meaningful solutions are ones where the parameters determine positive semidefinite $\mathbf{P}$ and $\mathbf{Q}$. The goal of current research is to develop algorithms that both converge superlinearly and converge to positive semidefinite solutions $\mathbf{P}$ and $\mathbf{Q}$; an enormous effort has been expended in this direction. Moreover, since the applications of semidefinite programming are wide ranging, the efforts expended in algorithm development are broad based. The observation that linear programming is a special case of semidefinite programming, and that the difficulties faced in developing interior point methods are common to both theories, has served to intensify efforts in algorithm development.

B. Alternate Formulations

With the basic problem in standard form it is easy to switch to an alternate formulation by scaling the matrices $\mathbf{P}$ and $\mathbf{Q}$. By scaling $\mathbf{P}$ by $|\mathbf{H}|^2/E_0$ and scaling $\mathbf{Q}$ by $1/\lambda_0|\mathbf{I}|^2$, we obtain

$$\mathbf{P}' = \mathbf{H} + \frac{|\mathbf{H}|^2}{E_0}\sum_i \alpha_i \mathbf{K}_i - \frac{|\mathbf{H}|^2}{E_0|\mathbf{I}|^2}\mathbf{I}, \quad \mathbf{Q}' = \frac{1}{|\mathbf{I}|^2}\mathbf{I} + \frac{1}{\lambda_0|\mathbf{I}|^2}\mathbf{H} + \sum_i \beta_i \mathbf{S}_i$$

so that $\mathbf{P}'$ has the required form for a spectral optimization problem, and $\mathbf{Q}'$ the form for an energy minimization problem. If $\mathbf{P}', \mathbf{Q}'$ are optimal, so that $\mathbf{P}'\mathbf{Q}' = \mathbf{0}$, the corresponding optimal values of the bottom eigenvalue and energy are given by $|\mathbf{H}|^2/E_0|\mathbf{I}|^2$ and $|\mathbf{H}|^2/\lambda_0|\mathbf{I}|^2$.

That we can so easily pass between the two forms of optimization problem without worrying about the existence of optimal solutions follows from the fact that the matrices in the two subspaces $\mathcal{S}$ and $\mathcal{K}$ are traceless: Theorems 12 and 13 can then be applied to ensure existence of solutions. This can also be seen directly by noting that the optimal values for E_0 and λ_0 are negative, and therefore nonzero, so that the scale factors $|\mathbf{H}|^2/E_0, 1/\lambda_0|\mathbf{I}|^2$, used in passing to the primed matrices $\mathbf{P}', \mathbf{Q}'$, are well defined.

C. Algorithms

Here we briefly sketch two directions in research on algorithms for semidefinite programming. A more complete discussion can be found in M. Todd's *Semidefinite Optimization* [18], or in the *Handbook of Semidefinite Programming* edited by Wolkowicz et al. [19].

The bottom eigenvalues for optimal $\mathbf{P}$ and $\mathbf{Q}$ are typically multiple, with many additional eigenvalues concentrated in a narrow band at the bottom end of the spectrum, and this structure is particularly difficult for algorithms to resolve. Although solutions of the equation $\mathbf{PQ} = \mathbf{0}$ are sought where all eigenvalues are nonnegative, there are many nearby solutions with small negative eigenvalues. As semidefinite programming emerged in the early 1990s increased efforts were made, that are ongoing, to devise algorithms that avoid these spurious solutions. That the bottom end of the spectrum for $\mathbf{P}$ and $\mathbf{Q}$ is compressed is easily understood by considering the spectral optimization problem. If $\mathbf{S} = \beta_1 \mathbf{S}_1 + \beta_2 \mathbf{S}_2 + \cdots + \beta_s \mathbf{S}_s$, then the function $\lambda_0(\mathbf{H} + \mathbf{S})$ can be considered a function $\lambda_0(\beta)$ of vector variable $\beta = [\beta_1, \beta_2, \ldots, \beta_s] \in \mathbb{R}^s$. As β moves toward the position of the maximum, the bottom eigenvalue is pushed up, compressing the bottom part of the spectrum. Thus at the maximum there are typically multiple eigenvalues with the same maximal value, and many additional eigenvalues nearby.

1. Primal–Dual Interior Point Methods

Starting with Karmarker's 1984 paper where he proposed an interior point method for linear programming, a new line of research on algorithms was started. Rather than deal with the detailed structure of the boundary of a polytope, Karmarker devised a search method that proceeds through the interior, making contact with the boundary at only the last moment, when the optimal solution is reached. Search directions need not be informed of details of the boundary structure, but must be devised so that the boundary is avoided. The ideas of Karmarker were steadily improved upon until the primal–dual path-following methods emerged as the dominant class of algorithms. These algorithms can be applied equally to semidefinite programming problems and linear programming problems. For semidefinite programming the primal–dual algorithms replace the equation $\mathbf{PQ} = \mathbf{0}$ by the equation $\mathbf{PQ} = \tau\mathbf{I}$, where τ is a positive number. When the right-hand side is $\mathbf{0}$ the equation determines $\mathbf{P}, \mathbf{Q}$ on the boundary of the cone of positive semidefinite matrices, but when the right-hand side is set equal to $\tau\mathbf{I}$ the equation determines positive definite $\mathbf{P}, \mathbf{Q}$. As τ approaches zero the solutions $\mathbf{P}_\tau, \mathbf{Q}_\tau$ track along the interior of the cone of positive semidefinite matrices, arriving at the boundary at the last moment, when τ converges to zero. When this happens we arrive at a solution of the Euler equation $\mathbf{PQ} = \mathbf{0}$.

2. Boundary Methods

The recent papers of Mazziotti show that boundary methods can be effective in solving the semidefinite program that accompanies electronic structure theory. His algorithm seems to outperform standard primal–dual path-following algorithms in terms of both the complexity of the basic step and number of steps to convergence [20–22]. Central to boundary methods, and to Mazziotti's algorithm, is a parameterization of $\mathbf{P}$, or $\mathbf{Q}$, or possibly both, that ensures the positive semidefinite property. For example, to ensure that $\mathbf{P}$ remains positive semidefinite, it is simply written $\mathbf{P} = \mathbf{RR}^*$, and the entries of $\mathbf{R}$ become the variational parameters. By imposing the positive semidefinite condition in this way, the algorithm can track along the boundary by directly invoking the equation $\mathbf{PQ} = \mathbf{0}$.

Information on the ranks of the optimal matrices $\mathbf{P}$ and $\mathbf{Q}$ can be used to gain efficiency since then the factor $\mathbf{R}$ need not have full rank, and the number of parameters is reduced accordingly. In the problem of quantum phases discussed in the following sections, the ranks of both $\mathbf{P}$ and $\mathbf{Q}$ can be predicted, which allows such efficiencies to be deployed in the solution process.

D. Interpreting the Solution

By the *spectral theorem* the optimal matrix $\mathbf{Q}$ can be written

$$\mathbf{Q} = \mathbf{H} + \mathbf{S} - \lambda_0(\mathbf{H} + \mathbf{S})\mathbf{I} = \sum \mathbf{g}_i \mathbf{g}_i^*$$

which by passing to the operator representation becomes

$$h - \lambda_0 1 = \sum g_i g_i^*$$

$\mathbf{g}_i$ is the coordinate vector for the operator g_i referred to the fixed basis for Q^k. The vectors $\mathbf{g}_i$ belong to the kernel of the k-matrix $\mathbf{P}$, and the operators g_i are annihilators for the corresponding approximate von Neumann density: $g_i p = 0$. Both $\mathbf{g}_i$ and g_i give a dual description of a type of correlation present in the ground state; these are the "killers" of the ground state. The $\mathbf{P}$ and $\mathbf{Q}$ matrices are typically block diagonal, with the blocks varying in size and type—the blocks are labeled by quantum numbers, which serve to classify the blocks. For symmetrical problems the number of labels increases, and the blocks become smaller and more numerous. The coordinate vectors $\mathbf{g}_i$ are associated with a particular block and therefore can be labeled by such quantum numbers. These serve to classify the possible types of correlations present in the ground state. It is important that the number of types of blocks that appear when k-matrices are used to represent the ground state is far greater than when a wavefunction is used, so k-matrices add precision to the discussion of correlations. The example we conclude with is the calculation of the ground-state k-matrix for the superconducting phase. This is a very symmetrical problem, so the number of types of correlations is large. As model parameters are varied, and the superconducting phase is traversed, the number and types of "killers" g_i remain constant, so the types of correlation that characterize the ground state remain constant. The coordinate vectors themselves vary smoothly, but the ranks of the various blocks remain constant. This property, that the ranks remain constant, vividly illustrates the stability that accompanies quantum phases and even serves to characterize quantum phases.

The electronic structure of atoms shows a similar stability since the shell structure remains constant over a wide range of experimental environments. However, with molecules this picture must be modified. The electronic structure of a diatomic molecule varies with bond length, the limit being that of a pair of separated atoms. Accordingly, the ranks of the blocks in the k-matrix description vary with bond length.

VI. MODELING A ONE-DIMENSIONAL SUPERCONDUCTOR

We first review the two papers [3, 4] where, using semidefinite programming, the 3-matrix of a one-dimensional superconductor was calculated for a two-parameter family of Hamiltonians. This work was the first to reveal the importance of the order of accuracy k, and to give a clear picture of how order of accuracy relates to the balance between one- and two-body forces in the Hamiltonian. It was established that convergence to exact values is extremely rapid with k, with third-order estimates being sufficiently accurate for most purposes. More specifically, these papers showed poor results when calculations are carried through to second order but showed three-figure accuracy when calculations are carried through to third order. The improvement with order of accuracy k was dramatic, suggesting that the accuracy of fourth-order estimates would be in the range of ten figures; this was later confirmed [5]. It is this characteristic, rapid convergence with k that converts the lower bound method into an effective computational tool, even when the correlations are induced by strong two-body forces. These papers also showed how order of accuracy k relates to the balance between one- and two-body forces in the Hamiltonian. The Hamiltonian for the superconducting model contains two-body forces alone—the opposite extreme from atoms and molecules where one-body forces dominate. Two-body forces alone provided a severe test—the second-order estimates contained no useful information. It came as somewhat of a surprise that the second-order estimates were so poor, and it was equally surprising that the third-order estimates were so good.

A. Details of the Model

In the superconducting model studied by Erdahl and Jin [2–4] spin-up, spin-down pairs of electrons wander on a one-dimensional periodic lattice Λ. The *local Hamiltonian*,

$$h_{\mu\nu} = \alpha_E^2 E_{\mu\nu} + \alpha_T^2 T_{\mu\nu}$$

where $\mu, \nu \in \Lambda$ are nearest neighbors, determines how the pairs interact along a nearest neighbor bond and the Hamiltonian is formed by summing these bond contributions:

$$h = \sum_{|\mu-\nu|=1} h_{\mu\nu}$$

There are an even number of lattice points equispaced on a ring, which are represented by the integers $\{1, 2, \ldots, |\Lambda|\}$. The restriction $|\mu - \nu| = 1$ limits the summation to nearest neighbor contributions, and since Λ is periodic the identity $|\Lambda| + 1 = 1$ holds. Along a bond the pairs interact through an electrostatic force $\alpha_E^2 E_{\mu\nu}$, the magnitude and sign determined by the coefficient α_E. The pairs also respond to a transport term $\alpha_T^2 T_{\mu\nu}$, which moves pairs between adjacent lattice sites. The details of these interactions are given by the expressions

$$^2E_{\mu\nu} = \tfrac{1}{4}(e_{a_\mu}e_{a_\nu} + e_{b_\mu}e_{b_\nu} + e_{a_\mu}e_{b_\nu} + e_{b_\mu}e_{a_\nu})$$
$$^2T_{\mu\nu} = 2(b_\mu^\dagger a_\mu^\dagger a_\nu b_\nu + b_\nu^\dagger a_\nu^\dagger a_\mu b_\mu)$$

The operator a_μ annihilates a spin-up electron at $\mu \in \Lambda$, and the operator b_μ annihilates a spin-down electron; $e_{a_\mu} = a_\mu^\dagger a_\mu - a_\mu a_\mu^\dagger$. The electrostatic operator has expected value $+1$ for states when sites μ, ν are either both occupied or unoccupied, and expected value -1 in the other case where one site is occupied, the other not.

The superscript 2 is added to emphasize that these interactions are *two-body*. That is, *these operators are orthogonal to all scalar and one-body operators with respect to the trace scalar product.*

Definition 15 A k-body operator is a Hermitian operator that can be represented as a polynomial of degree $2k$ in the annihilation and creation operators, and is of even degree in these operators. In addition, a k-body operator must be orthogonal to all $(k-1)$-body operators, all $(k-2)$-body operators, ..., and all scalar operators, with respect to the trace scalar product.

B. Results

As p varies over the set of von Neumann densities $\mathfrak{P}_0$, the vector of matrix elements $\beta = [\beta_T, \beta_E] = [\langle p^2, T_{\mu\nu}\rangle, \langle p^2, E_{\mu\nu}\rangle]$ fills in the *representable region* $\mathcal{R}$ pictured in Fig. 1. Each von Neumann density $p \in \mathfrak{P}_0$ is represented by

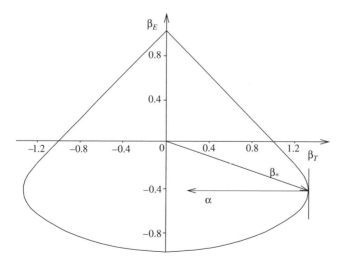

Figure 1. The representable region $\mathcal{R}$.

a point in $\mathcal{R}$, and each point in $\mathcal{R}$ represents at least one von Neumann density.

Each point $\beta \in \mathcal{R}$ corresponds to a quantum state with total energy given by $\epsilon = \langle p, h \rangle = |\Lambda| \langle p, h_{\mu\nu} \rangle = |\Lambda|(\alpha_T \langle p^2, T_{\mu\nu} \rangle + \alpha_E \langle p^2, E_{\mu\nu} \rangle) = |\Lambda| \alpha \cdot \beta$, where $\alpha = [\alpha_E, \alpha_T]$; the number of lattice sites $|\Lambda|$ is equal to the number of bonds in the ring, and therefore equal to the number of like contributions of bond energy. The minimum energy state for a Hamiltonian with coefficients $\alpha = [0, \alpha_T]$, $\alpha_T < 0$, is represented by the point $\beta_* \in \mathcal{R}$ that is pictured. This follows since β_* is a point of tangency for a line that is tangent to $\mathcal{R}$, and perpendicular to α; the energy is constant along the tangent line, and the vector α points in the direction of increasing energy.

The open circles slightly outside $\mathcal{R}$ in Fig. 2 are third-order estimates of boundary points obtained by the lower bound method and are accurate to three figures. These results are for three spin-up, spin-down pairs of electrons wandering on a ring with six lattice sites.

As the number of lattice sites increases, the electrons experience additional correlations, so the representable region shrinks. That is, if $\mathcal{R}_i$ is the representable region for a lattice with $i = |\Lambda|$ sites, then $\mathcal{R}_4 \supset \mathcal{R}_6 \supset \mathcal{R}_8 \supset \mathcal{R}_{10} \supset \cdots$. This phenomenon is accurately tracked by the third-order estimates, and Fig. 3 shows that convergence to the limiting case where $|\Lambda| \to \infty$ is rapid.

C. Convergence with Order k

Second-order estimates of the representable region $\mathcal{R}$ were also made for the cases $|\Lambda| = 4, 6$. These are the *pentagonal* regions $\mathcal{R}_4^2 \subset \mathcal{R}_6^2$ that appear in Fig. 4. Not only do the pentagonal regions poorly represent the corresponding representable regions, they *increase in size* in going from four to six lattice sites

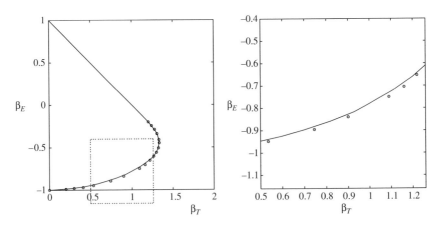

Figure 2. Accuracy of the lower bound method.

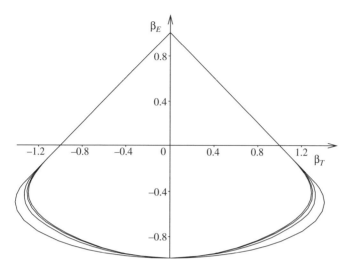

Figure 3. The representable regions $\mathcal{R}_4 \supset \mathcal{R}_6 \supset \mathcal{R}_8 \supset \mathcal{R}_{10}$.

rather than shrink! The limiting case, when $|\Lambda| \to \infty$, is the triangle with vertices at $[0, 1], [2, -1], [-2, -1]$. Thus the second-order estimates completely misrepresent the physical situation and are of little value. The two representable regions $\mathcal{R}_4 \supset \mathcal{R}_6$ have been added for comparison.

These results confirm the observation of Mihailovic and Rosina that second-order methods cannot characterize the correlations induced by two-body forces,

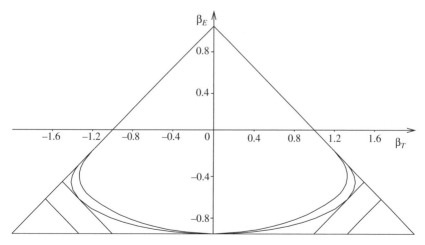

Figure 4. The estimates $\mathcal{R}_4^2 \subset \mathcal{R}_6^2$.

and show that third order can. The vast difference between the second- and third-order estimates shows that convergence to the exact representable region is rapid in k—it is clear that fourth-order estimates would give a very precise description of these correlations.

D. The Lipkin Model

The extraordinary speed of convergence with k was confirmed in the recent work of Mazziotti and Erdahl [5], where the Lipkin model was studied using the lower bound method. Second-, third-, and fourth-order approximations were studied for a range of strengths of the two-body forces. The jump in accuracy in going from second- to third-order approximations was again impressive, and the fourth-order estimates effectively removed all remaining errors. The accuracy goes from about four to ten figures when third-order estimates are replaced by fourth. A significant observation was that fourth-order estimates were an order of magnitude more accurate than estimates calculated using other many-body methods, showing that convergence in k is rapid enough to be of enormous practical significance. These results show that 4-positivity gives a solution to the correlation problem for strong two-body forces. The Lipkin model was introduced to study the strong two-body forces that dominate the dynamics of protons and neutrons in the nucleus. This is a two-level system where the energy levels -1 and $+1$ are each N-fold degenerate, and the strength of a two-body term can be adjusted by varying the parameter V. When V is zero all N fermions occupy the ground state, but when V is turned on pairs are promoted to the excited state; the action of the two-body term is to move pairs of fermions between the ground and excited states. For large interaction strengths both levels are nearly equally occupied; there are two half-filled shells, which is the configuration Mihailovic and Rosina found most difficult. In modeling nuclear systems V is chosen to be negative so the two-body forces are attractive, but in the work of Mazziotti and Erdahl V was positive, the two-body forces were repulsive, and an electronic system was modeled.

Convergence in order k was tracked for three different particle numbers, $N = 10, 30, 75$, and for two values of the interaction strength, $V = 0.8, 1.6$. Comparisons were made by recording the percentage of the correlation accounted for using second-, third-, and fourth-order approximations, with the results recorded in Table I.

The third column contains the exact correlation energy computed using a full configuration interaction treatment. As anticipated, convergence in k is very fast: the accuracy of the fourth-order approximations of the energy exceeded ten figures in some instances. A significant comparison is with other wavefunction and many-body perturbation methods currently employed in quantum theory. These comparisons are made in Table II, where correlation energies are tabulated for a variety of methods.

TABLE I
Convergence of Correlation Energy with k

N	V	E_{corr}	$\mathfrak{P}_0^2$	$\mathfrak{P}_0^3$	$\mathfrak{P}_0^4$
10	0.8	−0.0384	104.6640	100.0376	100.000103
30	0.8	−0.0130	102.4673	100.0103	100.0000670
75	0.8	−0.00526	101.1621	100.00167	100.0000113
10	1.6	−0.186	122.3515	101.85	100.0958
30	1.6	−0.128	119.6154	102.86	101.90
75	1.6	−0.117	108.9054	100.644	100.428

Source: Mazziotti and Erdahl [5].

Comparisons are made with: (i) a configuration interaction wavefunction calculation with single, double, triple, and quadruple excitations (*SDTQCI*); (ii) fourth-order many-body perturbation theory (*MP*4); (iii) solution of the single–double coupled-cluster equations (*CCSD*); and (iv) the contracted Schrödinger equation (*CSE*). Of the comparison methods the single–double coupled-cluster treatment gives the best results, but failed to converge in the strong interaction case where $V = 1.6$. The energies computed using the lower bound method are four orders of magnitude more accurate than those computed using single–double coupled clusters. These comparisons show that the rate of convergence of the lower bound method in k is fast enough that when $k = 4$ accuracies are achieved they are superior to other many-body techniques by several orders of magnitude. The correlations induced by two-body interactions are given a precise characterization by 4-positivity.

VII. CONCLUDING REMARKS

It is now firmly established that the lower bound method can be relied on for accurate electronic structure calculations for atoms and molecules. The recent

TABLE II
Comparison with Other Many-Body Methods

N	V	E_{corr}	*SDTQCI*	*MP*4	*CCSD*	*CSE*	$\mathfrak{P}_0^4$
10	0.8	−0.0384	98.85	99.703	101.07	99.9856	100.000103
30	0.8	−0.0130	95.94	96.43	101.08	99.9860	100.0000670
75	0.8	−0.00526	94.25	94.5	100.642	99.9092	100.0000113
10	1.6	−0.186	88.9	100.224	*	100.0336	100.0958
30	1.6	−0.128	47.5	53.9	*	100.486	101.90
75	1.6	−0.117	20.9	23.6	*	99.9657	100.428

Source: Mazziotti and Erdahl [5].

computational experiments [5, 6, 12, 14] have shown that second-order approximations are adequate for most purposes, even giving accurate binding energy curves in many instances. This last observation indicates that the correlations are treated uniformly over a range of molecular geometries, a characteristic that is particularly valuable and not shared by other approximation methods. Moreover, these experiments establish that if approximations are carried through to third order, the lower bound method is more accurate than other approximation methods currently used. This picture was filled out by studies of model systems where two-body forces dominate [2–5], providing a more severe test for the lower bound method. This second group of experiments show that with strong two-body forces second-order approximations fall far short of acceptable accuracies, but that with third-order approximations four-figure accuracy can be achieved. That is, these experiments show that convergence in the order k is so rapid that target accuracies can be achieved in all cases while complexity is kept within bounds. It is this rapid convergence, requiring that for kth-order approximations be k-positive, that stands behind the claim that the N-representability problem now has a satisfactory solution. The second theme we pursued is how semidefinite programming and the lower bound method have come together so that the calculations required for the energy minimum problem are "within grasp." Within grasp means that calculations carried to third order are now possible for small molecules. Attention is now moving toward semidefinite algorithms that exploit special features of the electronic structure problem so that larger systems can be studied [20–22], and this is where a rapid advance is possible.

We have focused on the lower bound method, but density matrix research has moved forward on a much broader front than that. In particular, work on the contracted Schrödinger equation played an important role in developments. A more complete picture can be found in Coleman and Yukalov's book [23]. It has taken 55 years and work by many scientists to fulfill Coleman's 1951 claim at Chalk River that "except for a few details which would be easily overcome in a couple of weeks—the N-body problem has been reduced to a 2.5-body problem!"

REFERENCES

1. A. J. Coleman, The structure of fermion density matrices. *Rev. Mod. Phys.* **35**, 668–687 (1963).
2. Beiyan Jin, *Quantum Phases for Two-Body Spin-Invariant Nearest Neighbor Interactions*, doctoral dissertation, Queens University, Kingston, Ontario, 1998.
3. R. M. Erdahl and B. Jin, *The lower bound method for reduced density matrices*. *J. Mol. Struct. Theochem.* **527**, 207–220 (2000).
4. R. M. Erdahl and B. Jin, On calculating approximate and exact density matrices, in Chapter 4 of *Many-Electron Densities and Density Matrices* (J. Cioslowski, ed.), Kluwer, Boston, 2000.
5. D. A. Mazziotti and R. M. Erdahl, Uncertainty relations and reduced density matrices: mapping many-body quantum mechanics onto four particles. *Phys. Rev. A* **63**, 042113 (2001).

6. Z. Zhao, B. J. Braams, M. Fukuda, M. L. Overton, and J. K. Percus, The reduced density matrix method for electronic structure calculations and the role of three-index representability. *J. Chem. Phys.* **120**, 2095 (2004).
7. M. A. Fusco, *Density Matrices and the Atomic Structure Problem*, doctoral dissertation, University of California, Davis, 1974.
8. C. Garrod, M.V. Mihailovic, and M. Rosina, The variational approach to the two-body density matrix. *J. Math. Phys.* **16**, 868–874 (1975).
9. C. Garrod and M. A. Fusco, A density matrix variational calculation for atomic Be. *Int. J. Quantum Chem.* **10**, 495 (1976).
10. R. M. Erdahl, Two algorithms for the lower bound method of reduced density matrix theory. *Reports Math. Phys.* **15**, 147–162 (1979).
11. M. L. Overton, On minimizing the maximum eigenvalue of a symmetric matrix. *SIAM J. Matrix Anal. Appl.* **9**, 256–268 (1988).
12. M. Nakata, H. Nakatsuji, M. Ehara, M. Fukuda, K. Nakata, and K. Fujisawa, Variational calculations of fermion second-order reduced density matrices by semi-definite programming algorithm. *J. Chem. Phys.* **114**, 8282–8202 (2001).
13. D. A. Mazziotti, Variational minimization of atomic and molecular ground-state energies via the two-particle reduced density matrix. *Phys. Rev. A* **65**, 062511 (2002).
14. M. Fukuda, B. J. Braams, M. Nakata, M. L. Overton, J. K. Percus, M. Yamashita, and Z. Zhao, Large-scale semidefinite programs in electronic structure calculation, *Math. Programming Ser. B* (in press).
15. R. M. Erdahl, Representability. *Int. J. Quantum Chem.* **13** 697–718 (1978).
16. M.V. Mihailovic and M. Rosina, The variational approach to the density matrix for light nuclei. *Nucl. Phys. A* **237**, 221–228 (1975).
17. D. A. Mazziotti, Variational reduced-density-matrix method using three-particle N-representability conditions with application to many-electron molecules. *Phys. Rev. A* **74**, 032501 (2006).
18. M. J. Todd, Semidefinite optimization. *Acta Numer.* **10**, 515 (2001).
19. H. Wolkowicz, R. Saigal, and L. Vandenberghe (eds.), *Handbook of Semidefinite Programming: Theory, Algorithms, and Applications*, Kluwer Academic Publishers, Nowell, MA, 2000.
20. D. A. Mazziotti, Realization of quantum chemistry without wave functions through first-order semidefinite programming. *Phys. Rev. Lett.* **93**, 213001 (2004).
21. D. A. Mazziotti, First-order semidefinite programming for the direct determination of two-electron reduced density matrices with application to many-electron atoms and molecules. *J. Chem. Phys.* **121**, 10957 (2004).
22. D. A. Mazziotti, Variational two-electron reduced density matrix theory for many-electron atoms and molecules: implementation of the spin- and symmetry-adapted T_2 condition through first-order semidefinite programming. *Phys. Rev. A* **72**, 032510 (2005).
23. A. J. Coleman and V. I. Yukalov, *Reduced Density Matrices—Coulson's Challenge*, Springer-Verlag, New York, 2000.

CHAPTER 5

THE $T1$ AND $T2$ REPRESENTABILITY CONDITIONS[‡]

BASTIAAN J. BRAAMS[*]

Department of Mathematics and Computer Science, Emory University, 400 Dowman Drive #W401, Atlanta, GA 30322 USA

JEROME K. PERCUS[†]

Courant Institute and Department of Physics, New York University, 251 Mercer Street, New York, NY 10012 USA

ZHENGJI ZHAO[‡]

High Performance Computing Research Department, Lawrence Berkeley National Laboratory, 1 Cyclotron Road, Mail Stop 15F1650, Berkeley, CA 94720 USA

CONTENTS

I. Introduction and background
II. Erdahl's $T1$ and $T2$ conditions
III. Numerical results
IV. Conclusion
Acknowledgment
References

---

[*]Email: braams@mathcs.emory.edu.
[†]Email: percus@cims.nyu.edu.
[‡]Email: zzhao@lbl.gov.
[‡] Submitted on April 2, 2006.

Reduced-Density-Matrix Mechanics: With Application to Many-Electron Atoms and Molecules,
A Special Volume of Advances in Chemical Physics, Volume 134, edited by David A. Mazziotti.
Series editor Stuart A. Rice. Copyright © 2007 John Wiley & Sons, Inc.

I. INTRODUCTION AND BACKGROUND

The starting point for all investigations of the classical RDM method (determination of the 1-RDM and 2-RDM by constrained optimization of the energy as a linear functional of these RDMs) are the representability conditions obtained from positivity of A^+A, where A is any one-body or two-body operator [1, 2]. Positivity of A^+A when A is a p-body operator likewise provides constraints on the p-RDM, and they have been called p-positivity conditions [3]. The $T1$ and $T2$ conditions that are the subject of this chapter are obtained as special positive linear combinations of 3-positivity conditions in which terms involving the 3-RDM cancel out. These conditions were introduced by Erdahl in the concluding section of his 1978 survey paper [4]. He presents them in a general Fock-space setting in the concise form $0 \preceq y^+y$, where the operator y is a polynomial containing terms of degrees 1 and 3 in the annihilation and creation operators and having the property that y^+y involves terms only up to degree 4. Erdahl's presentation is clear, but no further reference to these conditions is found in the 25 years following Ref. [4]. Part of the reason may be that a computationally efficient implementation of the classical 2-positivity conditions presented enough of a challenge already, and another part of the reason may be that Erdahl did not spell out the precise semidefinite conditions obtained after specialization to fixed particle number.

The conditions were rederived and explored by us in a 2004 paper [5] and they have been further investigated in Refs. [6–8]. In the present chapter we present these conditions, including a slight modification and strengthening of the $T2$ condition as compared to Ref. [5]. We also make precise the relation between the present $T1$ and $T2$ conditions and the Fock-space positivity condition given by Erdahl, and we show the effect of the strengthening of the $T2$ condition relative to our earlier paper [5] by recalculating the most difficult cases of that work. A brief review of the derivation of the classical positivity conditions of Refs. [1, 2] and an overview of the well-known diagonal conditions [9, 10] will set the stage.

In analytical investigations it is often desirable to leave the particle number free and consider operators that fix only the parity, but in applications to electronic structure theory one deals with fixed particle number and one may restrict A to have a definite action on the particle number N, so that A^+A is particle conserving. There are then two cases for the one-body operator A: consideration of $A = \sum_i f_i a_i$ with undetermined coefficients f_i gives rise to the condition $0 \preceq \gamma$, while consideration of $A = \sum_i f_i a_i^+$ gives rise to the condition $0 \preceq I - \gamma$; here the a_i form a (finite-dimensional) basis of annihilation operators, γ is the 1-RDM, and I is the identity matrix of the appropriate size. If the density matrices are known to be real symmetric then the f_i may be assumed real, otherwise they should be assumed complex. For fixed particle number there are three kinds of two-body operators to consider: letting $A = \sum_{i,j} f_{i,j} a_i a_j$ one obtains the condition $0 \preceq \Gamma$, consideration of $A = \sum_{i,j} f_{i,j} a_i^+ a_j^+$ leads to the condition $0 \preceq Q$, and consideration of $A = \sum_{i,j} f_{i,j} a_i^+ a_j$ leads to the condition $0 \preceq G$. Here, Γ is the 2-RDM, and matrices Q and G are expressed in terms of γ and Γ by

$$Q^{i,j}_{k,l} = \Gamma^{i,j}_{k,l} - \delta^i_k \gamma^j_l - \delta^j_l \gamma^i_k + \delta^i_l \gamma^j_k + \delta^j_k \gamma^i_l + \delta^i_k \delta^j_l - \delta^i_l \delta^j_k \qquad (1)$$

$$G^{i,j}_{k,l} = \Gamma^{j,k}_{i,l} + \delta^i_k \gamma^j_l \qquad (2)$$

(where δ denotes the Dirac delta function). The Q matrix is antisymmetric in each pair of indices and is understood as a matrix of size $\binom{r}{2} \times \binom{r}{2}$, where r is the dimension of the one-particle basis. The G matrix is of dimension $r^2 \times r^2$.

Conditions $0 \preceq \gamma$, $0 \preceq I - \gamma$, $0 \preceq \Gamma$, and $0 \preceq Q$ are found in Ref. [1] and the condition $0 \preceq G$ is from Ref. [2], where the unity of this family of conditions is emphasized. (For the history of the subject the less well-known paper by Mayer [11] deserves notice, and for reviews we mention Refs. [4, 12].) Garrod and Percus [2] developed condition $0 \preceq G$ in a more general form by considering the positivity of $A^+ A$, where $A = c + \sum_{i,j} f_{i,j} a_i^+ a_j$; however, for the case of fixed particle number N the free constant c can be absorbed into the sum using the operator identity $\sum_{i,j} \delta^i_j a_i^+ a_j = \hat{N}$, and this has always been done in computational work. The implementation of the RDM method subject to these positivity conditions belongs to semidefinite programming and this is shown in a beautiful way by Rosina and Garrod [13], which presents—even before the name semidefinite programming was in use—a cutting-plane method applied to a linear programming relaxation and also a barrier function approach that looks ahead to present-day interior-point methods.

Beyond the semidefinite conditions on γ, $I - \gamma$, Γ, Q, and G, one other class of representability conditions for the 2-RDM has long been studied: these are the "diagonal" conditions, which include the three-index conditions due to Weinhold and Wilson [9] and which were systematically investigated by McRae and Davidson [10]. The simplest, one-index, diagonal conditions are $0 \leq \gamma^i_i$ and $0 \leq 1 - \gamma^i_i$ (for all i), and these obviously specialize the conditions $0 \preceq \gamma$ and $0 \preceq I - \gamma$. The two-index diagonal conditions are $0 \leq \Gamma^{i,j}_{i,j}$, $0 \leq 1 - \gamma^i_i - \gamma^j_j + \Gamma^{i,j}_{i,j}$, and $0 \leq \gamma^i_i - \Gamma^{i,j}_{i,j}$ (for all i, j; $i \neq j$), and these specialize $0 \preceq \Gamma$, $0 \preceq Q$, and $0 \preceq G$. Next in the hierarchy are the three-index (Weinhold–Wilson) diagonal conditions: $0 \leq 1 - \gamma^i_i - \gamma^j_j - \gamma^k_k + \Gamma^{i,j}_{i,j} + \Gamma^{i,k}_{i,k} + \Gamma^{j,k}_{j,k}$ and $0 \leq \gamma^i_i + \Gamma^{j,k}_{j,k} - \Gamma^{i,j}_{i,j} - \Gamma^{i,k}_{i,k}$ (with i, j, k all distinct). The hierarchy continues, but as observed by McRae and Davidson the conditions become progressively more unwieldy. The 3-index and higher diagonal conditions have not played much of a role in RDM computations, although it would not be hard to incorporate any number of them into an SDP formulation via a cutting-plane approach. They depend on the choice of basis, and an invariant choice would be the basis that makes the 1-RDM diagonal. (It is a bit more challenging if one insists that the conditions must hold after *any* transformation of the one-electron basis.) In Ref. [14] violation of the 3-index conditions was tested, but the conditions were not then used to improve the solution, and in Ref. [15] the same authors incorporated the diagonal conditions but found only a weak improvement for the systems studied.

The complexity of the diagonal representability problem is now understood in a manner that was not available to McRae and Davidson: the diagonal conditions of RDM theory are recognized in combinatorial optimization as conditions characterizing the *Boolean quadric polytope* (BQP) and, after a simple transformation, the *cut polytope* [16]. Optimization over BQP is *NP*-hard and a concise characterization of the polytope is not available unless $P = NP$ [16, p. 397; 17]. These fundamental complexity results tell us that we should not look for a concise complete solution to the fermion representability problem. On the other hand, the original full configuration interaction eigenvalue problem looks exponentially hard in any case, and therefore the analytical challenge for the RDM method is to develop representability conditions that provide high accuracy even if they will not form a complete family.

II. ERDAHL'S *T*1 AND *T*2 CONDITIONS

Following Ref. [5] the $T1$ condition is obtained by considering an operator $A = \sum_{i,j,k} g_{i,j,k} a_i a_j a_k$, where the $g_{i,j,k}$ are arbitrary real or complex coefficients totally antisymmetric in the three indices. (We view g as a vector of dimension $\binom{r}{3}$, where r is the size of the one-electron basis.) The contractions $\langle \Psi | A^+ A | \Psi \rangle$ and $\langle \Psi | A A^+ | \Psi \rangle$ both involve the 3-RDM, but with opposite sign, and so the nonnegativity of $\langle \Psi | A^+ A + A A^+ | \Psi \rangle$ for all three-index functions g provides a representability condition involving only the 1-RDM and 2-RDM. In explicit form the condition is of semidefinite form, $0 \preceq T1$, where the Hermitian matrix $T1$ is defined in terms of γ and Γ by

$$T1_{i',j',k'}^{i,j,k} = \mathcal{A}[i,j,k]\mathcal{A}[i',j',k']\left(\tfrac{1}{6}\delta_{i'}^{i}\delta_{j'}^{j}\delta_{k'}^{k} - \tfrac{1}{2}\delta_{i'}^{i}\delta_{j'}^{j}\gamma_{k'}^{k} + \tfrac{1}{4}\delta_{i'}^{i}\Gamma_{j',k'}^{j,k}\right) \quad (3)$$

We are using $\mathcal{A}[i,j,k]f_{i,j,k}$ to denote anti symmetrization with respect to (i,j,k): $f_{i,j,k}$ summed over all permutations of the indices with each term multiplied by the sign of the permutation. The dimension of the $T1$ matrix is $\binom{r}{3} \times \binom{r}{3}$.

The $T2$ condition, slightly strengthened from Ref. [5], but in the real case already contained in Ref. [4], is obtained by considering operators $A = \sum_{i,j,k} g_{i,j,k} a_i^+ a_j a_k$ and $B = \sum_l h_l a_l$ for arbitrary real or complex coefficients $g_{i,j,k}$ and h_l, with $g_{i,j,k}$ antisymmetric in (j,k). (So this g may be compressed to a vector of dimension $r\binom{r}{2}$.) Operators A and B both lower the particle number by 1. Similar to the case above, the contractions $\langle\Psi|A^+A|\Psi\rangle$ and $\langle\Psi|AA^+|\Psi\rangle$ involve the 3-RDM with opposite sign, and so the combination $\langle\Psi|(A+B)^+(A+B)+AA^+|\Psi\rangle$ involves again only the 1-RDM and 2-RDM and is, of course, nonnegative and also particle conserving. This provides a semi-definite representability condition $0 \preceq T2'$, where the matrix $T2'$ has the block form

$$T2' = \begin{pmatrix} T2 & X \\ X^+ & \gamma \end{pmatrix} \qquad (4)$$

The $T2$ diagonal subblock has dimension $r\binom{r}{2} \times r\binom{r}{2}$ and acts on the coefficients $g_{i,j,k}$, the 1-RDM is the other diagonal subblock, of dimension $r \times r$, and acts on the coefficients h_l, and the off-diagonal blocks X and X^+ mix the two sets of coefficients; X^+ is, of course, the Hermitian adjoint of X. Specifically,

$$T2^{i,j,k}_{i',j',k'} = \mathcal{A}[j,k]\mathcal{A}[j',k']\left(\tfrac{1}{2}\delta^i_{j'}\delta^k_{k'}\gamma^{i'}_i + \tfrac{1}{4}\delta^i_{i'}\Gamma^{j,k}_{j',k'} - \delta^j_{j'}\Gamma^{i',k}_{i,k'}\right) \qquad (5)$$

$$X^l_{i',j',k'} = \Gamma^{i',l}_{j',k'} \qquad (6)$$

In our earlier work [5] we did not include the contribution from the one-body operator B, and so we obtained only the main $T2$ subblock. It might be thought, briefly, that the condition could be further strengthened by considering an independent pair of one-body annihilator operators B and C and then developing the positivity of $(A+B)^+(A+B)+(A+C)(A+C)^+$. However, this yields nothing stronger in the case of fixed particle number, because if $N \geq 2$ then the operator C can be absorbed into A, much like the constant term c could be absorbed into the two-body operator used in defining the G condition, while if $N = 1$ then effectively $A = 0$ and again the extreme conditions require only the operator B.

Let us make clear now the correspondence between our treatment here and Erdahl's 1978 treatment [4, Sec. 8]. Erdahl works in general Fock space and his operators conserve only the parity of the number of nuclei. He exhibits two families of operators that are polynomials in the annihilation and creation operators containing a three-body and a one-body term. Generic instances of these operators are denoted y and w. The coefficients are real, and Erdahl stresses that this is essential for his treatment. The one-body term is otherwise unrestricted, but the three-body term must satisfy conditions to guarantee that y^+y or w^+w does not contain a six-body term. For the first family the conditions amount to the three-body term being even under taking the adjoint, and for the second family the three-body term must be odd under taking the adjoint. Writing out the polynomials one obtains the general form

$$y = \sum_{i,j,k} f1_{i,j,k} a_i a_j a_k + \sum_{i,j,k} (f1_{i,j,k} a_i a_j a_k)^+$$
$$+ \sum_{i,j,k} f2_{i,j,k} a_i^+ a_j a_k + \sum_{i,j,k} (f2_{i,j,k} a_i^+ a_j a_k)^+$$
$$+ \sum_i g_i a_i + \sum_i h_i a_i^+ \quad (7)$$

$$w = \sum_{i,j,k} f1_{i,j,k} a_i a_j a_k - \sum_{i,j,k} (f1_{i,j,k} a_i a_j a_k)^+$$
$$+ \sum_{i,j,k} f2_{i,j,k} a_i^+ a_j a_k - \sum_{i,j,k} (f2_{i,j,k} a_i^+ a_j a_k)^+$$
$$+ \sum_i g_i a_i + \sum_i h_i a_i^+ \quad (8)$$

One may now proceed to write out y^+y and w^+w and, to make the connection to the present work, retain only the terms that are particle conserving. The result are representability conditions, and they include terms quadratic in $f1$ and terms quadratic in $f2$, but no mixed terms. It will be clear then that the extreme conditions—and they are all that matter—involve either $f1$ or $f2$, but not both; moreover, one will observe that $0 \preceq y^+y$ and $0 \preceq w^+w$ lead to the same conditions, which are the real cases of the $T1$ and the strengthened $T2$ conditions.

In summary, Erdahl's treatment is more general and allows a more concise formulation because he works in Fock space, conserving only the parity of the number of particles; however, he finds it necessary to restrict the coefficients to be real. We work at fixed particle number and have no reason for the restriction to real coefficients. If the Hamiltonian should be general Hermitian, in which case the RDM must likewise be assumed to be general Hermitian, then our approach leads to Hermitian semidefinite conditions.

III. NUMERICAL RESULTS

In our first exploration of the $T1$ and $T2$ conditions [5] we obtained results of the RDM method for the ground-state energy and dipole moment for a collection of small molecules and molecular ions, both closed-shell and open-shell systems. (We don't mean "closed shell" in a strict sense, and we only constrained the spin and spin multiplicity eigenvalues, not the elements of the RDM.) The choice of molecules and configurations largely followed Ref. [18]—a paper that, we think, reinvigorated the classical RDM approach. We showed that the addition of the $T1$ and $T2$ conditions ($T2$ without the off-diagonal block X)

provided a significant improvement in accuracy over that obtained using only the classical 2-positivity conditions denoted P, Q, and G. The worst cases for the energy in our sample were the O_2^+ molecule, for which the RDM method subject to the P, Q, G, $T1$, and $T2$ conditions gave an error of 2.8 mH relative to the full configuration interaction (FCI) benchmark, and the CF molecule, giving an error of 0.9 mH relative to FCI. For the dipole moment the worst case was the CF molecule, for which the error was 0.0045 a.u. relative to FCI. All our results, including these worst cases, were competitive in accuracy with those obtained from the best standard methods: coupled-cluster singles and doubles with perturbative correction of triples [CCSD(T)] for the energy and singly and doubly substituted configuration interaction (SDCI) for the dipole. Further demonstrations of the strength of these conditions are provided in Refs. [6, 7]. Reference [8] provides a detailed description of our present implementation of the SDP approach.

The strengthened $T2$ condition presented here involves only a very slightly larger matrix: matrix $T2'$ of Eq. (4) is of size $(r\binom{r}{2} + r) \times (r\binom{r}{2} + r)$, whereas the $T2$ subblock is of size $r\binom{r}{2} \times r\binom{r}{2}$, and so really this $T2'$ matrix should replace $T2$ in any use of these conditions. (We thought of changing the name and calling the strengthened conditions the $T2'$ conditions, but for the long haul that seems a poor choice.) We recalculated five cases from Ref. [5] for which the worst accuracy was obtained and found that the strengthening of $T2$ to the $T2'$ matrix improved the accuracy further, as displayed in Tables I and II; in particular, the O_2^+ energy error decreased from 2.8 mH to 2.1 mH, the CF energy error went down from 0.9 mH to 0.5 mH, and the CF dipole error decreased from 0.0045 au to 0.0037 a.u. In the table, subscript PQG refers to the three classical 2-positivity conditions, $PQGT$ refers to those conditions and also the $T1$ and $T2$ conditions as used in Ref. [5], and $PQGT'$ refers to the classical 2-positivity conditions and the $T1$ and strengthened $T2$ condition as presented here.

TABLE I
Ground-State Energies (in Hartree Units) Relative to the Full CI Result, Calculated by the RDM Method Subject to Three Sets of Representability Conditions (Column 6: P, Q, G; Column 7: PQG and $T1$, $T2$ as in Ref. [5]; Column 7: PQG and $T1$, $T2$ as in Eq. (4) and Calculated by CCSD(T) Using *Gaussian 98* [19] (Column 9), and the Full CI Reference Value (Last Column)[a]

System	State	$N(N_\alpha)$	r	$2S+1$	ΔE_{PQG}	ΔE_{PQGT}	$\Delta E_{PQGT'}$	$\Delta E_{CCSD(T)}$	E_{FCI}
NH_3	$^1A_1$	10(5)	16	1	−0.0109	−0.0003	−0.0002	+0.0018	−56.0142
H_3O^+	$^1A_1$	10(5)	16	1	−0.0073	−0.0002	−0.0002	+0.0002	−76.1046
CF	$^2\Pi$	15(8)	20	2	−0.0076	−0.0009	−0.0005	+0.0010	−136.6775
O_2^+	$^2\Pi g$	15(8)	20	2	−0.0167	−0.0028	−0.0021	+0.0033	−148.7933

[a]Here r is the basis size, N is the electron number, N_α the number of spin-up electrons, and $2S+1$ is the multiplicity. The geometries used are the experimental ones from Ref. [20]. The basis set is STO-6G for all systems.

TABLE II

Dipole Moments (in a.u.) Calculated by the RDM Method Subject to Three Sets of Representability Conditions (Columns 6–8) and Calculated by SDCI Using *Gaussian 98* and Full CI (Last Two Columns)[a]

System	State	$N(N_\alpha)$	r	$2S+1$	D_{PQG}	D_{PQGT}	$D_{PQGT'}$	D_{SDCI}	D_{FCI}
NH_3	$^1A_1$	10(5)	16	1	0.0748	0.0799	0.0799	0.0803	0.0800
H_3O^+	$^1A_1$	10(5)	16	1	0.7106	0.7201	0.7202	0.7216	0.7203
CF	$^2\Pi$	15(8)	20	2	0.4505	0.4255	0.4247	0.3929	0.4210

[a]Other conventions as in Table I.

IV. CONCLUSION

In this chapter we reviewed the semidefinite extension of the three-index diagonal conditions which is due originally to Erdahl [4], but was long ignored. The RDM method subject to these conditions appears to provide an accuracy very much better than that obtained using only the traditional 2-positivity conditions and competitive with that of the best standard methods of ab initio theory. A strengthening of the $T2$ condition was described here relative to our earlier derivation [5], and the $T1$ and $T2'$ pair is, in the real and particle conserving case, equivalent to Erdahl's condition. The semidefinite matrix associated with the revised $T2$ condition is only very slightly larger than the original $T2$, and the revised condition provides a further noticeable improvement in accuracy.

Acknowledgments

This work was supported by the National Science Foundation under Grant No. ITR-CHE-0219331 (BJB), by the Department of Energy under Grant No. DE-FG02-02ER15292 (JKP), and by the Director, Office of Science, of the U.S. Department of Energy under Contract No. DE-AC02-05CH11231 (ZZ).

References

1. A. J. Coleman, Structure of fermion density matrices. *Rev. Mod. Phys.* **35**, 668 (1963).
2. C. Garrod and J. K. Percus, Reduction of the N-particle variational problem. *J. Math. Phys.* **5**, 1756 (1964).
3. D. A. Mazziotti and R. M. Erdahl, Uncertainty relations and reduced density matrices: mapping many-body quantum mechanics onto four particles. *Phys. Rev. A* **63**, 042113 (2001).
4. R. M. Erdahl, Representability. *Int. J. Quantum Chem.* **13**, 697 (1978).
5. Z. Zhao, B. J. Braams, M. Fukuda, M. L. Overton, and J. K. Percus, The reduced density matrix method for electronic structure calculations and the role of three-index representability conditions. *J. Chem. Phys.* **120**, 2095–2104 (2004).
6. J. R. Hammond and D. A. Mazziotti, Variational two-electron reduced-density-matrix theory: partial 3-positivity conditions for N-representability. *Phys. Rev. A* **71**, 062503 (2005).

7. D. A. Mazziotti, Variational two-electron reduced density matrix theory for many-electron atoms and molecules: implementation of the spin- and symmetry-adapted T_2 condition through first-order semidefinite programming. *Phys. Rev. A* **72**, 032510 (2005).
8. M. Fukuda, B. J. Braams, M. Nakata, M. L. Overton, J. K. Percus, M. Yamashita, and Z. Zhao, Large-scale semidefinite programs in electronic structure calculation. *Math. Programming Ser. B*, in press.
9. F. Weinhold and E. Bright Wilson, Jr., Reduced density matrices of atoms and molecules. II. On the N-representability problem. *J. Chem. Phys.* **47**, 2298 (1967).
10. W. B. McRae and E. R. Davidson, Linear inequalities for density matrices II. *J. Math. Phys.* **13**, 1527 (1972).
11. J. E. Mayer, Electron correlation. *Phys. Rev.* **100**, 1579 (1955).
12. A. J. Coleman, Reduced density matrices: 1929–1989, in *Density Matrices and Density Functionals, Proceedings of the A. John Coleman Symposium* (R. M. Erdahl and V. H. Smith, Jr., eds.), D. Reidel Publishing Company, Dordrecht, 1987.
13. M. Rosina and C. Garrod, The variational calculation of reduced density matrices. *J. Comput. Phys.* **18**, 300 (1975).
14. M. Nakata, M. Ehara, and H. Nakatsuji, Density matrix variational theory: application to the potential energy surfaces and strongly correlated systems. *J. Chem. Phys.* **116**, 5432 (2002).
15. M. Nakata, M. Ehara, and H. Nakatsuji, Density matrix variational theory: strength of Weinhold–Wilson inequalities, in *Fundamental World of Quantum Chemistry*, Vol. I, (Erkki J. Brändas and Eugene S. Kryachko, eds.), Kluwer Academic, Norwell, MA, 2003.
16. M. M. Deza and M. Laurent, *Geometry of Cuts and Metrics*, Springer-Verlag, Berlin, 1997.
17. R. M. Karp and C. H. Papadimitriou, On linear characterizations of combinatorial optimization problems. *SIAM J. Comput.* **11**, 620 (1982).
18. M. Nakata, H. Nakatsuji, M. Ehara, M. Fukuda, K. Nakata, and K. Fujisawa, Variational calculations of fermion second-order reduced density matrices by semidefinite programming algorithm. *J. Chem. Phys.* **114**, 8282 (2001).
19. M. J. Frisch, G. W. Trucks, H. B. Schlegel, G. E. Scuseria, M. A. Robb, J. R. Cheeseman, V. G. Zakrzewski, J. A. Montgomery, Jr., R. E. Stratmann, J. C. Burant, S. Dapprich, J. M. Millam, A. D. Daniels, K. N. Kudin, M. C. Strain, O. Farkas, J. Tomasi, V. Barone, M. Cossi, R. Cammi, B. Mennucci, C. Pomelli, C. Adamo, S. Clifford, J. Ochterski, G. A. Petersson, P. Y. Ayala, Q. Cui, K. Morokuma, D. K. Malick, A. D. Rabuck, K. Raghavachari, J. B. Foresman, J. Cioslowski, J. V. Ortiz, A. G. Baboul, B. B. Stefanov, G. Liu, A. Liashenko, P. Piskorz, I. Komaromi, R. Gomperts, R. L. Martin, D. J. Fox, T. Keith, M. A. Al-laham, C. Y. Peng, A. Nanayakkara, M. Challacombe, P. M. W. Gill, B. Johnson, W. Chen, M. W. Wong, J. L. Andres, C. Gonzalez, M. Head-Gordon, E. S. Replogle, and J. A. Pople, *Gaussian 98, Revision A.9*, Gaussian, Inc., Pittsburgh, PA, 1998.
20. http://www.emsl.pnl.gov/proj/crdb/.

CHAPTER 6

SEMIDEFINITE PROGRAMMING: FORMULATIONS AND PRIMAL–DUAL INTERIOR-POINT METHODS

MITUHIRO FUKUDA

Department of Mathematical and Computing Sciences, Tokyo Institute of Technology, Japan

MAHO NAKATA

Department of Applied Chemistry, The University of Tokyo, Japan

MAKOTO YAMASHITA

Department of Information System Creation, Kanagawa University, Japan

CONTENTS

I. Introduction
II. Formulation as an SDP Problem
III. The Primal–Dual Interior-Point Method
IV. Other Methods for SDP Problems
V. Solving SDP Problems in Practice
Acknowledgments
References

I. INTRODUCTION

In 2001, Nakata and co-workers presented the results of realistic fermionic systems, like atoms and molecules, larger than previously reported for the variational calculation of the second-order reduced density matrix (2-RDM) [1].

Reduced-Density-Matrix Mechanics: With Application to Many-Electron Atoms and Molecules,
A Special Volume of Advances in Chemical Physics, Volume 134, edited by David A. Mazziotti.
Series editor Stuart A. Rice. Copyright © 2007 John Wiley & Sons, Inc.

They employed a general-purpose semidefinite programming (SDP) software [2] for these calculations.

Considering such recent relevance of SDP in quantum chemistry, this chapter discusses some practical aspects of this variational calculation of the 2-RDM formulated as an SDP problem. We first present the definition of an SDP problem, and then the primal and dual SDP formulations of the variational calculation of the 2-RDM as SDP problems (Section II), an efficient algorithm to solve the SDP problems: the primal–dual interior-point method (Section III), a brief section about alternative and also efficient augmented Lagrangian methods (Section IV), and some computational aspects when solving the SDP problems (Section V).

The SDP problem is a convex optimization problem that stimulated intensive research since the 1990s for two main reasons: generalization of the exciting new method called the interior-point method for linear programming, and its wide-range applications in many diverse areas [3–7]. The SDP problem consists in computing a maximization or a minimization solution of a real-valued linear function defined on the space of positive semidefinite Hermitian matrices (i.e., Hermitian matrices with nonnegative eigenvalues) restricted by linear equality and/or inequality constraints on the same space. This mathematical description of an SDP problem is quite general. However, due to historical reasons, there is a preferred formulation in which it has been "popularized" in mathematical programming.

Let us denote by $\mathbb{S}$ the space of block-diagonal real symmetric matrices (i.e., multiple symmetric matrices arranged diagonally in a unique large matrix) with prescribed dimensions, and by $\mathbb{R}^m$ the m-dimensional real space. Given the constants $C, A_1, A_2, \ldots, A_m \in \mathbb{S}$, and $b \in \mathbb{R}^m$, an SDP problem is usually defined either as the *primal SDP problem*,

$$\begin{cases} \max & \langle C, X \rangle \\ \text{subject to} & \langle A_p, X \rangle = [b]_p, \quad p = 1, 2, \ldots, m \\ & X \succeq O \end{cases} \quad (1)$$

or equivalently (under mild conditions, e.g., the Slater condition) as the *dual SDP problem*,

$$\begin{cases} \min & b^t y \\ \text{subject to} & \sum_{p=1}^{m} A_p [y]_p - C \succeq O \\ & y \in \mathbb{R}^m \end{cases} \quad (2)$$

Here $\langle C, X \rangle$ denotes the inner product $\sum_{ij} C_{ij} X_{ij}$, b^t the transpose of the vector b, $[y]_p$ the pth coordinate of the vector y, and $X \succeq O$ means that the matrix X is a positive semidefinite symmetric matrix. The variables for the primal SDP problem in Eq. (1) and the dual SDP problem in Eq. (2) are $X \in \mathbb{S}$ and $y \in \mathbb{R}^m$, respectively. Therefore the size of an SDP problem depends on the size of each block-diagonal matrix of X and m. We should also mention that the problem as represented by Eq. (1) is the preferred "format" for the primal SDP formulation of the variational calculation, which we present in the next section.

We can also define a more general SDP problem as follows. Let us consider now the constants $(C, c), (A_1, a_1), (A_2, a_2), \ldots, (A_m, a_m) \in \mathbb{S} \times \mathbb{R}^s$, and $b \in \mathbb{R}^m$. Then we can define the *primal SDP problem with free variables*,

$$\begin{cases} \max & \langle C, X \rangle + c^t x \\ \text{subject to} & \langle A_p, X \rangle + a_p^t x = [b]_p, \quad p = 1, 2, \ldots, m \\ & X \succeq O, \ x \in \mathbb{R}^s \end{cases} \quad (3)$$

and equivalently (under mild conditions, e.g., the Slater condition), the *dual SDP problem with equality constraints*,

$$\begin{cases} \min & b^t y \\ \text{subject to} & \sum_{p=1}^{m} A_p [y]_p - C \succeq O \\ & \sum_{p=1}^{m} a_p [y]_p = c, \quad y \in \mathbb{R}^m \end{cases} \quad (4)$$

In this case, the variables for the primal SDP problem with free variables (Eq. (3)) and the dual SDP problem with equality constraints (Eq. (4)) are $(X, x) \in \mathbb{S} \times \mathbb{R}^s$ and $y \in \mathbb{R}^m$, respectively. Therefore the size of an SDP problem depends now on the size of each block-diagonal matrix of X, m, and s. We should also mention that the problem as represented by Eq. (4) is the preferred "format" for the dual SDP formulation of the variational calculation, which we present in the next section, too.

The primal–dual pair of SDP problems Eqs. (1)–(2) or Eqs. (3)–(4) is a natural extension of linear programming problems [3–6]. Therefore, owing to the primal–dual "nature," if $(\bar{X}, \bar{x})$ satisfies the constraints in Eq. (3) and $\bar{y}$ satisfies the constraints in Eq. (4), we can simply verify that $b^t \bar{y} - (\langle C, \bar{X} \rangle + c^t \bar{x}) = \langle \bar{X}, \bar{S} \rangle \geq 0$, where $\bar{S} = \sum_{p=1}^{m} A_p [\bar{y}]_p - C$. Furthermore, $\langle \bar{X}, \bar{S} \rangle = 0$ (or equivalently $\bar{X}\bar{S} = O$) holds if and only if the maximum and minimum values of the respective problems, Eqs. (3) and (4), are attained (and they are the same). The same results are valid for the particular case of Eqs. (1)–(2).

II. FORMULATION AS AN SDP PROBLEM

In this section, we focus on how to formulate the variational calculation of the 2-RDM as an SDP problem. In fact, it can always be formulated as a primal SDP problem (Eq. (1)) [1, 8–13] or as a dual SDP problem with equality constraints

(Eq. (4)) [14–16]. A key point here to understand the difference between these two formulations is that the dual SDP formulation (Eq. (4)) is *not* the dual of the primal SDP formulation (Eq. (1)). Both formulations produce two distinct pairs of primal and dual SDP problems, which mathematically describe the same fermionic system. Since their mathematical formulations differ, this implies differences in the computational effort to solve them.

As an instructive example, we consider the primal SDP formulation in detail. First, we show that the variational minimization of a two-particle system can be trivially formulated as a primal SDP problem. Next, we show how we constrain the eigenvalues of the 1-RDM between zero and one, and finally, how we set the SDP constraints to satisfy P and Q conditions simultaneously.

First consider a two-particle system. In this simplest case, the 2-RDM Γ is N-representable if it is positive semidefinite and the number of particles is fixed to two. It can easily be cast as the following SDP problem:

$$\begin{cases} \min & \langle H, \Gamma \rangle \\ \text{subject to} & \langle \hat{N}, \Gamma \rangle = 2 \\ & \Gamma \succeq O \end{cases}$$

where H is the Hamiltonian of the system, and $\hat{N}$ is the number operator.

For the next step, we show how we consider the N-representability conditions for the 1-RDM γ for a system with N particles; that is all of its eigenvalues should be between zero and one [17]. In other words, this condition is equivalent to saying that γ and $I - \gamma$ are positive semidefinite, where I is the identity matrix. Assuming that H_1 is the one-body Hamiltonian, we have

$$\begin{cases} \min & \langle H_1, \gamma \rangle \\ \text{subject to} & \langle \hat{N}, \gamma \rangle = N \\ & \gamma \succeq O \text{ and } I - \gamma \succeq O \end{cases} \quad (5)$$

The difficulty here is how to simultaneously constrain γ and $I - \gamma$ to be positive semidefinite. To formulate it as a primal SDP problem (Eq. (1)), we should express these two conditions as a positive semidefinite constraint over a single matrix: let $\widetilde{\gamma}$ be a block-diagonal matrix in which two symmetric matrices $\widetilde{\gamma}_1$ and $\widetilde{\gamma}_2$ are arranged diagonally, and let us express the interrelation between these two matrices via linear constraints defined by the matrices A_p and the constants $[b]_p$ as in Eq. (1). That is,

$$\widetilde{\gamma} = \begin{pmatrix} \widetilde{\gamma}_1 & O \\ O & \widetilde{\gamma}_2 \end{pmatrix}$$

where $\widetilde{\gamma}_1$ and $\widetilde{\gamma}_2$ should satisfy $\widetilde{\gamma}_1 = \gamma$ and $\widetilde{\gamma}_2 = I - \gamma$. These conditions can be equivalently rewritten as $[\widetilde{\gamma}_1]_{ij} + [\widetilde{\gamma}_2]_{ij} = \delta^i_j$ for $i, j = 1, 2, \ldots, r$, where δ^i_j denotes the Kronecker delta and r is the dimension of γ. Now for $i,j = 1, 2, \ldots, r$ and $i \leq j$, let us define the $r \times r$ symmetric matrix E_{ij} whose (i,i) element is one, (i,j) and (j,i) elements are $\frac{1}{2}$ for $i < j$, and zero otherwise. By taking the inner product of E_{ij} and an arbitrary symmetric matrix X, we can pull out the (i,j) element (or the (j,i) element) of X, that is, $\langle E_{ij}, X \rangle = [X]_{ij}$. Additionally, we define the matrices

$$\widetilde{H}_1 = \begin{pmatrix} H_1 & O \\ O & O \end{pmatrix}, \quad \widetilde{N} = \begin{pmatrix} \hat{N} & O \\ O & O \end{pmatrix}, \quad \text{and} \quad A_{ij} = \begin{pmatrix} E_{ij} & O \\ O & E_{ij} \end{pmatrix}$$

Notice that by taking the trace of $\widetilde{\gamma}$ with A_{ij} it becomes $\langle A_{ij}, \widetilde{\gamma} \rangle = \langle E_{ij}, \widetilde{\gamma}_1 \rangle + \langle E_{ij}, \widetilde{\gamma}_2 \rangle = [\widetilde{\gamma}_1]_{ij}/2 + [\widetilde{\gamma}_1]_{ji}/2 + [\widetilde{\gamma}_2]_{ij}/2 + [\widetilde{\gamma}_2]_{ji}/2 = [\widetilde{\gamma}_1]_{ij} + [\widetilde{\gamma}_2]_{ij} = \delta_j^i$. Finally, Eq. (5) can be rewritten as the following primal SDP problem (Eq. (1)):

$$\begin{cases} \max & \langle -\widetilde{H}_1, \widetilde{\gamma} \rangle \\ \text{subject to} & \langle \widetilde{N}, \widetilde{\gamma} \rangle = N \\ & \langle A_{ij}, \widetilde{\gamma} \rangle = \delta_j^i, \quad i,j = 1, 2, \ldots, r, \quad i \leq j \\ & \widetilde{\gamma} \succeq O \end{cases}$$

where $\widetilde{N}, A_{ij}$ and N, δ_j^i take the role for the A_p and $[b]_p$ in Eq. (1), respectively.

Summing up, the main points when formulating this variational calculation as a primal SDP problem are:

- Prepare a block-diagonal matrix in which γ and $I - \gamma$ are placed diagonally.
- Define the constraint matrices A_p and the constants $[b]_p$ for *each element* of γ and $I - \gamma$ to satisfy the linear relations between these two matrices.

Finally, we show how to consider the Q condition [17]:

$$\begin{cases} \min & \langle H, \Gamma \rangle \\ \text{subject to} & \langle \hat{N}, \Gamma \rangle = N \\ & \Gamma \succeq O \text{ and } Q \succeq O \end{cases} \quad (6)$$

and convert it into the primal SDP problem (Eq. (1)). The relation between γ, Γ, and Q is as follows:

$$Q_{j_1 j_2}^{i_1 i_2} + (\delta_{j_1}^{i_1} \gamma_{j_2}^{i_2} + \delta_{j_2}^{i_2} \gamma_{j_1}^{i_1}) - \delta_{j_2}^{i_1} \gamma_{j_1}^{i_2} + \delta_{j_1}^{i_2} \gamma_{j_2}^{i_1}) - \Gamma_{j_1 j_2}^{i_1 i_2} = (\delta_{j_1}^{i_1} \delta_{j_2}^{i_2} - \delta_{j_2}^{i_1} \delta_{j_1}^{i_2}),$$
$$i_1, i_2, j_1, j_2 = 1, 2, \ldots, r \quad (7)$$

Let $\widetilde{\Gamma}$ be a block-diagonal matrix where Γ and Q are diagonally arranged:

$$\widetilde{\Gamma} = \begin{pmatrix} \Gamma & O \\ O & Q \end{pmatrix}$$

Γ and Q have four indices and they should be mapped to matrices with two indices, respectively. This mapping is clear from the context. For $i_1, i_2, j_1, j_2 = 1, 2, \ldots, r$, $i_i \leq j_1$ and $i_2 \leq j_2$, let us define the $r^2 \times r^2$ symmetric matrix $E_{i_1 i_2, j_1 j_2}$ whose $(i_1 + (i_2 - 1)r, i_1 + (i_2 - 1)r)$ element is one, $(i_1 + (i_2 - 1)r, j_1 + (j_2 - 1)r)$ and $(j_1 + (j_2 - 1)r, i_1 + (i_2 - 1)r)$ elements are $\frac{1}{2}$ for $i_1 < j_1$ or $i_2 < j_2$, and zero otherwise. Additionally, define the matrices with the same dimension $\widetilde{E}_{i_2 i_2, j_1 j_2} = \sum_{k=1}^{r} (\delta_{j_1}^{i_1} E_{i_2 k, j_2 k} + \delta_{j_2}^{i_2} E_{i_1 k, j_1 k} - \delta_{j_2}^{i_1} E_{i_2 k, j_1 k} - \delta_{j_1}^{i_2} E_{i_1 k, j_2 k}) /(N - 1)$. Letting

$$\widetilde{H} = \begin{pmatrix} H & 0 \\ 0 & 0 \end{pmatrix}, \quad \widetilde{N} = \begin{pmatrix} \hat{N} & 0 \\ 0 & 0 \end{pmatrix}$$

$$A_{i_1 i_2, j_1 j_2} = \begin{pmatrix} \widetilde{E}_{i_1 i_2, j_1 j_2} - E_{i_1 i_2, j_1 j_2} & 0 \\ 0 & E_{i_1 i_2, j_1 j_2} \end{pmatrix}$$

and $b_{i_1 i_2, j_1 j_2} = \delta_{j_1}^{i_1} \delta_{j_2}^{i_2} - \delta_{j_2}^{i_1} \delta_{j_1}^{i_2}$, we obtain the linear constraints

$$\langle A_{i_1 i_2, j_1 j_2}, \widetilde{\Gamma} \rangle = b_{i_1 i_2, j_1 j_2}, \quad i_1, i_2, j_1, j_2 = 1, 2, \ldots, r, \quad i_1 \leq j_1, \ i_2 \leq j_2$$

which express the Q condition (Eq. (7)) since $\gamma_j^i = \sum_{k=1}^{r} \Gamma_{jk}^{ik}/(N-1)$.

Finally, the problem represented by Eq. (6) can be reduced into the primal SDP problem (Eq. (1)):

$$\begin{cases} \max & \langle -\widetilde{H}, \widetilde{\Gamma} \rangle \\ \text{subject to} & \langle \widetilde{N}, \widetilde{\Gamma} \rangle = N \\ & \langle A_{i_1 i_2, j_1 j_2}, \widetilde{\Gamma} \rangle = b_{i_1 i_2, j_1 j_2}, \quad i_1, i_2, j_1, j_2 = 1, 2, \ldots, r, \\ & \qquad \qquad \qquad \qquad \qquad \qquad \ i_1 \leq j_1, i_2 \leq j_2 \\ & \widetilde{\Gamma} \succeq O \end{cases}$$

Furthermore, we can make use of the antisymmetric properties and generic spin symmetries to reduce the sizes of above problems [1, 9, 14].

The inclusion of other known N-representability conditions like G, $T1$, and $T2$ [14] in the variational calculation can be embedded into the primal SDP problem in a similar way.

Table I (which can be deduced from Ref. [15]) shows the dimensions of the block-diagonal matrices of X and the number of linear equalities m in Eq. (1) relative to the number r of spin orbitals of a generic reference basis when employing the primal SDP formulation. It also considers conditions on α electron number, total spin, and spin symmetries of the N-representability. In the table

$$\binom{a}{b} = \frac{a!}{b!(a-b)!} \quad \text{for integers } a \geq b > 0$$

For example, the eigenvalue restrictions on the 1-RDM, $\gamma \succeq O$ and $I - \gamma \succeq O$, correspond to four block-diagonal matrices of dimensions $r/2 \times r/2$ in X (see

TABLE I
Dimensions of the Primal SDP Formulation for the Variational Calculation of the 2-RDM

N-Representability Conditions	Dimensions of Each Block-Diagonal Matrix of X
Restrictions on 1-RDM	$r/2 \times r/2$ (4 blocks)
P condition	$(r/2)^2 \times (r/2)^2$ (1 block), $\binom{r/2}{2} \times \binom{r/2}{2}$ (2 blocks)
Q condition	$(r/2)^2 \times (r/2)^2$ (1 block), $\binom{r/2}{2} \times \binom{r/2}{2}$ (2 blocks)
G condition	$2(r/2)^2 \times 2(r/2)^2$ (1 block), $(r/2)^2 \times (r/2)^2$ (2 blocks)
T1 condition	$\frac{r}{2}\binom{r/2}{2} \times \frac{r}{2}\binom{r/2}{2}$ (2 blocks), $\binom{r/2}{3} \times \binom{r/2}{3}$ (2 blocks)
T2 condition	$\frac{r}{6}\binom{3r/2}{2} \times \frac{r}{6}\binom{3r/2}{2}$ (2 blocks), $\frac{r}{2}\binom{r/2}{2} \times \frac{r}{2}\binom{r/2}{2}$ (2 blocks)

Conditions Considered in the Formulation	Size m in Eq. (1)
P, Q, G conditions	$5 + 6\binom{r/2+1}{2} + 3\binom{r^2/4+1}{2} + 2\binom{r(r/2-1)/4+1}{2} + \binom{r^2/2+1}{2}$
P, Q, G, T1 conditions	Above line $+2\binom{r^2(r/2-1)/8+1}{2} + 2\binom{r(r/2-1)(r/2-2)/12+1}{2}$
P, Q, G, T1, T2 conditions	Above line $+2\binom{r^2(3r/2-1)/8+1}{2} + 2\binom{r^2(r/2-1)/8+1}{2}$

first line in Table I). The largest block-diagonal matrices in X correspond to the T2 condition and its number of rows/columns scales as $3r^3/16$. The number of equality constraints m in Eq. (1) depends on the N-representability conditions considered in the variational formulation. For instance, if we employ the P, Q, and G conditions, m scales as $15r^4/64$, while if we further add the T1 and T2 conditions, it will scale as $25r^6/576$. Furthermore, these sizes can be reduced if spatial and spin symmetries peculiar to each atom or molecule are incorporated [12, 13, 18], but even these symmetries will not change the order of magnitude of the SDP problems.

The dual SDP formulation [14–16] follows a similar spirit but seems less obvious at a first sight. In general lines, it can be briefly described as follows. Let us represent all nonrepeating elements of the 2-RDM Γ (after considering the antisymmetric condition on it) by a vector $y \in \mathbb{R}^m$. Then, defining an appropriate vector $b \in \mathbb{R}^m$ from the Hamiltonian for the corresponding system, the ground-state energy of the fermionic system can be computed by $\min b^t y$ restricted to Eq. (4). In this case, the P, Q, G, T1, and T2 conditions will correspond to the block-diagonal matrices of $\sum_{p=1}^m A_p[y]_p - C$ when $C, A_1, A_2, \ldots, A_m$ are appropriately defined [14–16]. The other equalities as the restriction on the α electron number, total spin, and so on will be defined by

TABLE II
Dimensions of the Dual SDP Formulation for the Variational Calculation of the 2-RDM

N-Representability Conditions	Dimensions of Each Block-Diagonal Matrix of $\sum_{p=1}^{m} A_p[y]_p - C$	
Restrictions on 1-RDM and P, Q, G, T1, T2 conditions	Same as Table I	
Conditions Considered in the Formulation	Size m in Eq. (4)	Size s in Eq. (4)
m and s do not depend on the N-representability conditions considered in the formulation	$\binom{r^2/4+1}{2} + 2\binom{r(r/2-1)/4+1}{2}$	$5 + 2\binom{r/2+1}{2}$

$c, a_1, a_2, \ldots, a_m \in \mathbb{R}^s$ in Eq. (4) [15, 16]. We omit its details here. More details of the formulation can be found in Refs. [14–16].

Table II shows the dimensions of the block-diagonal matrices of $\sum_{p=1}^{m} A_p[y]_p - C$, the dimension m of the variable vector y, and the number of equality constraints s in Eq. (4) relative to the number r of spin orbitals of a generic reference basis [15].

If we employ the dual SDP formulation and include the P, Q, G, $T1$, and $T2$ conditions, the number of rows/columns of the largest block-diagonal matrices scale as $3r^3/16$ again, while m scales as $3r^4/64$ and s as $r^2/4$.

The advantages of the dual SDP formulation are clear when comparing Tables I and II. First, notice that the sizes of the block-diagonal matrices are unchanged in both formulations. There is also an additional constraint $\sum_{p=1}^{m} a_p[y]_p = c$ in the dual SDP formulation, which is absent in the primal SDP formulation. Then, while the size m of equality constraint in the primal SDP formulation (see Eq. (1)) corresponds to the dimensions of the $Q, G, T1$, and $T2$ matrices included in the formulation and scales as $25r^6/576$, the dimension m of the variable vector $y \in \mathbb{R}^m$ in the dual SDP formulation (see Eq. (4)) corresponds to the dimension of the 2-RDM and scales merely as $3r^4/64$. The difference becomes more remarkable when more N-representability conditions are considered in these primal or dual SDP formulations. Computational implications when solving the SDP problems employing the primal and dual SDP formulations are discussed in Section V.

III. THE PRIMAL–DUAL INTERIOR-POINT METHOD

Interior-point methods for SDPs were independently proposed by Nesterov and Nemirovskii [19] and Alizadeh [20] in the early 1990s. These methods were primal or dual only interior-point methods. Several variants of interior-point methods have been proposed so far, but after a decade of theoretical maturation followed by successful implementations, the widely accepted and most efficient variant is the *infeasible primal–dual path-following Mehrotra-type predictor–corrector interior-point method*. Henceforth, we restrict ourselves to present its basic idea. See Refs. [4–7] and references therein for more details and a partial list of other variants.

In order to simplify the discussion, we consider, for a while, the primal–dual pair of SDPs Eqs. (1)–(2) instead of Eqs. (3)–(4). We also assume that the space $\mathbb{S}$ is formed by single block-diagonal symmetric matrices with size $n \times n$.

Let us assume henceforth that the *Slater condition* is valid for Eqs. (1)–(2); that is there exist $\hat{X} \in \mathbb{S}$ and $\hat{y} \in \mathbb{R}^m$ such that $\langle A_p, \hat{X} \rangle = [b]_p$ $(p = 1, 2, \ldots, m)$, $\hat{X} \succ 0$, and $\sum_{p=1}^m A_p[\hat{y}]_p - C \succ O$ (i.e., all of their eigenvalues are positive). Under this assumption, $(X, y) \in \mathbb{S} \times \mathbb{R}^m$ must be an optimal solution; that is, the solution that maximizes and minimizes the functions, respectively, for Eqs. (1)–(2) if and only if it satisfies the *Karush–Kuhn–Tucker condition*:

$$\begin{cases} \langle A_p, X \rangle = [b]_p, & p = 1, 2, \ldots, m \\ S \equiv \sum_{p=1}^m A_p[y]_p - C \\ XS = O \\ X \succeq 0, \quad S \succeq 0 \end{cases} \quad (8)$$

Note that only the third equation is nonlinear. Let us now perturb this nonlinear equation by introducing a positive parameter $\mu \in \mathbb{R}$:

$$\begin{cases} \langle A_p, X \rangle = [b]_p, & p = 1, 2, \ldots, m \\ S = \sum_{p=1}^m A_p[y]_p - C \\ XS = \mu I \\ X \succ 0, \quad S \succ 0 \end{cases} \quad (9)$$

where $I \in \mathbb{S}$ is the identity matrix. The basic idea of primal–dual interior-point methods is to apply the damped (or modified) Newton's method to the perturbed system in Eq. (9) with a fixed μ from an initial guess $(\bar{X}, \bar{y}, \bar{S})$ such that $\bar{X} \succ 0$ and $\bar{S} \succ 0$ (which does not necessary satisfy the two linear equations in Eq. (9)). Along the major iterations, the parameter μ is decreased until zero in order to obtain Eq. (8), and the variables X and S are maintained positive definite to have the Newton's system solvable (that is why it is called interior-point method).

The first-order approximations for the first two equations of Eq. (9) for a fixed (X, y, S) are

$$\begin{cases} \langle A_p, \Delta X \rangle = [r]_p \equiv [b]_p - \langle A_p, X \rangle, & p = 1, 2, \ldots, m \\ \sum_{p=1}^m A_p[\Delta y]_p - \Delta S = R \equiv C - \sum_{p=1}^m A_p[y]_p + S \end{cases} \quad (10)$$

where $(\Delta X, \Delta y, \Delta S)$, is the first-order contribution for the linearization, called *search direction*. Furthermore, a naive linearization of the third equation in Eq. (9) gives

$$X\Delta S + \Delta XS = \mu I - XS \tag{11}$$

and together with Eq. (10) are clearly an undefined system since there are more equations than variables (notice that ΔX must be a symmetric matrix belonging to $\mathbb{S}$ and not to $\mathbb{R}^{n \times n}$). This arbitrariness to choose a specific solution for the search direction gives rise to the existence of different search directions in SDP theory. One of the several alternatives to remedy it is to introduce a scaling nonsingular matrix $E \in \mathbb{R}^{n \times n}$ and replace Eq. (11) by

$$\frac{1}{2}(EX\Delta SE^{-1} + E\Delta XSE^{-1} + E^{-t}\Delta SXE^t + E^{-t}S\Delta XE^t)$$
$$= \mu I - \frac{1}{2}(EXSE^{-1} + E^{-t}SXE^t) \tag{12}$$

Different choices for E result in different search directions. However, the most successful implementations are the NT (search) direction with $E = W^{-1/2}$, where $W = X^{1/2}(X^{1/2}SX^{1/2})^{-1/2}X^{1/2}$, and the HRVW/KSH/M (search) direction with $E = S^{1/2}$. Here $S^{1/2}$ is the unique positive definite decomposition of the positive definite matrix S such that $S = S^{1/2}S^{1/2}$.

Once again, we restrict ourselves to the HRVW/KSH/M direction for simplicity, and then Eq. (12) can be rewritten

$$\begin{cases} \widehat{\Delta X}S + X\Delta S = K \equiv \mu I - XS \\ \Delta X = (\widehat{\Delta X} + \widehat{\Delta X^t})/2 \end{cases} \tag{13}$$

where $\widehat{\Delta X} \in \mathbb{R}^{n \times n}$ is an auxiliary matrix not necessarily symmetric. Under the linear independence assumption of the data matrices $\{A_1, A_2, \ldots, A_m\}$, and for arbitrary $X \succ O, S \succ O$, and K, the system of linear equations (10) and (13) has a unique solution $(\Delta X, \Delta y, \Delta S)$. This system can be further reduced to the recursive system of linear equations

$$\begin{cases} B\Delta y = g \\ \Delta S = \sum_{p=1}^{m} A_p \Delta y - R \\ \widehat{\Delta X} = (K - X\Delta S)S^{-1}, \ \Delta X = (\widehat{\Delta X} + \widehat{\Delta X^t})/2 \end{cases} \tag{14}$$

where

$$\begin{aligned} B_{pq} &\equiv \langle XA_p S^{-1}, A_q \rangle, \ p, q = 1, 2, \ldots, m \\ [g]_p &\equiv \langle (K + XR)S^{-1}, A_p \rangle - [r]_p, \ p = 1, 2, \ldots, m \end{aligned} \tag{15}$$

Finally, the general algorithm framework of the infeasible primal–dual path-following Mehrotra-type predictor–corrector interior-point method is the following.

Primal–Dual Interior-Point Method

Input: $(\bar{X}, \bar{y}, \bar{S})$ with $\bar{X} \succ O$ and $\bar{S} \succ O$.
Set $\varepsilon > 0$, $k = 0$, and $(X^k, y^k, S^k) = (\bar{X}, \bar{y}, \bar{S})$.

Convergence Test. We iterate the following sequential computations until (X^k, y^k, S^k) satisfies the constraints in Eqs. (1)–(2) with $S^k = \sum_{p=1}^{m} A_p[y^k]_p - C$, and $b^t y^k - \langle C, X^k \rangle < \varepsilon$.

Predictor Step. Set $0 \leq \beta_p \leq 1$, and solve Eqs. (14) and (15) for $K = \beta_p (\langle X^k, S^k \rangle / n) I - XS$ and obtain $(\widetilde{\Delta X}, \widetilde{\Delta y}, \widetilde{\Delta S})$.

Corrector Step. Set $0 \leq \beta_c \leq 1$ such that $\beta_c > \beta_p$, and solve Eqs. (14) and (15) for $K = \beta_c (\langle X^k, S^k \rangle / n) I - X^k S^k - \widetilde{\Delta X}\widetilde{\Delta S}$ to obtain $(\Delta X, \Delta y, \Delta S)$.

Step Length. Take $\alpha_x \in \{\alpha \in [0,1] : X^k + \alpha \Delta X \succeq O\}$, $\alpha_s \in \{\alpha \in [0,1] : S^k + \alpha \Delta S \succeq O\}$ and set $X^{k+1} = X^k + \alpha_x \Delta X$, $y^{k+1} = \alpha_s \Delta y$, $S^{k+1} = S^k + \alpha_s \Delta S$, and $k = k + 1$.

Note that the parameter μ has been replaced by $\langle X^k, S^k \rangle / n$ above because μ is equal to $\langle X^k, S^k \rangle / n$ whenever (X^k, y^k, S^k) satisfies the constraints in Eqs. (1)–(2) with $S^k = \sum_{p=1}^{m} A_p[y^k]_p - C$, and $X^k S^k = \mu I$ is valid.

We can prove that the above algorithm converges in polynomial time (i.e., the number of floating-point operations is proportional to a polynomial in the problem sizes m and n) by choosing appropriately β_p, β_c, α_x, and α_s.[1] See Refs. [4–7]. The computational cost of each major iteration is at most proportional to $mn^3 + m^2 n^2 + m^3 + n^3$ floating-point operations, and the maximum number of iterations is proportional to $\sqrt{n} \ln \varepsilon^{-1}$.

The above analysis is just of theoretical interest and real implementation codes including the ones listed in Section V perform much faster in practice. These codes frequently ignore conservative choices for the above parameters, taking ambitious values, and compute the solution in a much shorter time. Also, the use of efficient numerical linear algebra libraries and the exploration of several sparsity properties of $C, A_1, A_2, \ldots, A_m$ [21] have tremendously reduced the computational time of these codes every few months.

Primal–Dual interior-point methods always compute the desired solution within a guaranteed time complexity framework. Moreover, we can always

---

[1] Strictly speaking, we additionally need to "control" the distance between the initial guess $(\bar{X}, \bar{y}, \bar{S})$ and the region formed by the variables (X, y, S), which satisfies the constraints in eqs. (1)–(2) along the major iterations.

check the reliability of the computed solution (X, y, S) by substituting in Eq. (8). Another great advantage of these methods is the possibility to parallelize their computation. It is known that two major routines are the most time consuming ones in these algorithms: forming the matrix B in Eq. (15) and solving the system of linear equations $B\Delta y = g$ in Eq. (14). The parallel versions of the Primal–Dual interior-point methods explore the parallelization of these two major routines and have given notable results [14, 15, 22, 23].

Finally, all of the above algorithm can be extended trivially to matrices with several diagonal blocks (matrices) without the restriction to a single diagonal block (matrix) as we did at the beginning of this section. Problems with extra data $c, a_1, a_2, \ldots, a_m \in \mathbb{R}^s$, and the variable $x \in \mathbb{R}^s$ as in Eqs. (3)–(4) can be reduced to the form of Eqs. (1)–(2) in the following way. The variable $x \in \mathbb{R}^s$ in Eq. (3) can be replaced by a difference of two nonnegative variables $x^+, x^- \in \mathbb{R}^s$, with $x = x^+ - x^-$ and $x^+, x^- \geq 0$. This is also equivalent to replacing the equality $\sum_{p=1}^{m} a_p [y]_p - c = 0$ in Eq. (4) by two inequalities $\sum_{p=1}^{m} a_p [y]_p - c \geq 0$ and $\sum_{p=1}^{m} a_p [y]_p - c \leq 0$. Then Eqs. (3)–(4) can be reduced to Eqs. (1)–(2) as

$$\begin{cases} \max & \left\langle \begin{pmatrix} C & 0 & 0 \\ 0 & c & 0 \\ 0 & 0 & -c \end{pmatrix}, \begin{pmatrix} X & 0 & 0 \\ 0 & x^+ & 0 \\ 0 & 0 & x^- \end{pmatrix} \right\rangle \\ \text{subject to} & \left\langle \begin{pmatrix} A_p & 0 & 0 \\ 0 & a_p & 0 \\ 0 & 0 & -a_p \end{pmatrix}, \begin{pmatrix} X & 0 & 0 \\ 0 & x^+ & 0 \\ 0 & 0 & x^- \end{pmatrix} \right\rangle = [b]_p, \\ & \qquad\qquad\qquad\qquad\qquad\qquad p = 1, 2, \ldots, m \\ & \begin{pmatrix} X & 0 & 0 \\ 0 & x^+ & 0 \\ 0 & 0 & x^- \end{pmatrix} \succeq O \end{cases}$$

and

$$\begin{cases} \min & b^t y \\ \text{subject to} & \sum_{p=1}^{m} \begin{pmatrix} A_p & 0 & 0 \\ 0 & a_p & 0 \\ 0 & 0 & -a_p \end{pmatrix} [y]_p - \begin{pmatrix} C & 0 & 0 \\ 0 & c & 0 \\ 0 & 0 & -c \end{pmatrix} \succeq O \quad (16) \\ & y \in \mathbb{R}^m \end{cases}$$

A careful reader will observe that this algebraic transformation will produce a dual SDP problem that does not have $y \in \mathbb{R}^m$ such that the matrix in Eq. (16) has all of its eigenvalues positive and, therefore, will not satisfy the Slater conditions. However, numerical experiments have shown that practical algorithms still can solve these problems efficiently [16].

IV. OTHER METHODS FOR SDP PROBLEMS

The success of Primal–Dual interior-point methods is due to its feature of computing reliable and highly precise solutions in a guaranteed time framework, although its computational cost can become prohibitively expensive for large-scale SDP problems.

Another recent approach that has demonstrated effectiveness is based on the augmented Lagrangian method for SDPs and was successfully implemented by Kočvara and Stingl [24]. The basic difference from the standard augmented Lagrangian methods [25] is in the definition of the augmented Lagrangian function, which is defined employing a penalty/barrier function with special properties. We omit its details here.

A more recent approach was described by Mazziotti [10, 11], which reformulates the SDP problem (Eq. (1)) as a nonlinear and nonconvex optimization problem and applies a combination of the augmented Lagrangian method with the quasi-Newton method [25]. He calls this method the first-order method and it is implemented in RRSDP. In this reformulation, the variable $X \in \mathbb{S}$ in Eq. (1) is replaced by a full-rank factorization RR^t, where R is a nonsymmetric matrix and has the same number of rows/columns as X. In this sense, it can be viewed as a special case of the Burer–Monteiro low-rank factorization method [26–28] since this latter employs a low-rank factorization $VV^t = X \in \mathbb{S}$, where V can have fewer columns than R. Since both reformulations produce nonconvex optimization problems, there was no guarantee that these algorithms could find an optimal solution of an SDP problem. However, Burer and Monteiro further showed that these algorithms indeed converge and find the exact solution certifying the validity of these methods [27]; although it is not proved so far that these algorithms have theoretical bounds on the number of iterations required to converge as interior-point methods do.

V. SOLVING SDP PROBLEMS IN PRACTICE

Currently, there are several open source free software packages that can solve SDP problems in the form of Eqs. (1)–(2) and/or Eqs. (3)–(4) by Primal–Dual interior-point methods [3–7]: SDPA [2] is written in C++, CSDP [29] is written in C, and SeDuMi [30] and SDPT3 [31] have interfaces in MATLAB. Furthermore, SDPA and CSDP have their respective parallel versions: SDPARA [22, 23] which can solve larger problems in a more reasonable time. It is also possible to solve SDP problems without installing these software packages in your own computer. *NEOS Server* [32] and *CaNEOS Server for Optimization* [33] provide free services in solving these SDP problems submitted through a web browser, and *SDPA Online for Your Future* [34] allows one even to solve larger problems by the parallel SDP solver SDPARA [22] on a PC cluster.

Software based on the augmented Lagrangian method (Section IV) is also available: PENSDP by Kočvara and Stingl [24] (unique commercial code) and SDPLR by Burer, Monteiro, and Choi [26–28].

For the SDP problems arising from the variational calculation, in which we are interested, the theoretical number of floating-point operations required by parallel Primal–Dual interior-point method-based software scales as $m^2u^2/d + m^3/d + n^3 + mn^2$ per iteration [15], while the number of major iterations is at most proportional to $\sqrt{n}\ln\varepsilon^{-1}$ (but requires many fewer iterations in practice). Here u denotes the maximum number of nonzero elements of each A_p ($p = 1, 2, \ldots, m$), n the number of rows/columns of the largest block-diagonal matrix of the above matrices, d the number of processors used by the parallel code, and ε the difference between the approximate value of the primal and dual functions in Eqs. (1) and (2) (or Eqs. (3) and (4)). Notice that this number of floating-point operations per iteration is less than $mn^3 + m^2n^2 + m^3 + n^3$ (see Section III) because we can explore the sparsity of the data [21]. Let r be the number of spin orbitals of a generic reference basis. Since u is constant and it scales as r^2 for the primal and dual SDP formulations, respectively [15], we obtain Table III based on the information in Tables I and II. Table III also shows the memory usage of parallel Primal–Dual interior-point methods.

Let us analyze now the first-order method: RRSDP [10, 11]. This method usually requires a number of floating-point operations that scale as $n^3 + mu$ per iteration. However, as we mentioned before, there is no theoretical bound on the number of major iterations required for its convergence. Once again, let r be the number of spin orbitals. Considering the information from Tables I and II, and remembering that u is constant and scales as r^2 for the primal and dual SDP formulations, respectively, we can obtain the following number of floating-point operations and the memory usage for the first-order method and presented in Table III.

TABLE III

Theoretical Number of Floating-Point Operations per Iteration (FLOPI), Maximum Number of Major Iterations, and Memory Usage for the Parallel Primal–Dual Interior-Point Method (pPDIPM) and for the First-Order Method (RRSDP) Applied to Primal and Dual SDP Formulations[a].

N-Representability Conditions		P, Q, G			P, Q, G, T1 or P, Q, G, T1, T2		
Formulation	algorithm	FLOPI	# Iterations	Memory	FLOPI	# Iterations	Memory
Primal SDP	pPDIPM	r^{12}/d	$r\ln\varepsilon^{-1}$	r^8	r^{18}/d	$r^{3/2}\ln\varepsilon^{-1}$	r^{12}
formulation	RRSDP	r^6	—	r^4	r^9	—	r^6
Dual SDP	pPDIPM	r^{12}/d	$r\ln\varepsilon^{-1}$	r^8	r^{12}/d	$r^{3/2}\ln\varepsilon^{-1}$	r^8
formulation	RRSDP	r^6	—	r^4	r^9	—	r^6

[a]r denotes the number of spin orbitals, and d the number of processors used for the parallel computation.

From the table, we can see that the first-order method usually requires fewer floating-point operations and memory storage if compared with the Primal–Dual interior-point method. The unique drawback of the former method is that we cannot guarantee a convergence of the method in a certain time frame.

We can also conclude that if we employ the Primal–Dual interior-point method, the dual SDP formulation provides a more "reduced" mathematical description of the variational calculation of the 2-RDM than employing the primal SDP formulation. The former formulation also allows us to reach a faster computational solution. On the other hand, the number of floating-point operations and the memory storage of RRSDP do not depend on the primal or dual SDP formulations.

Even with the existence of these efficient methods to solve SDP problems, we recognize that we still need to pursue the development of new methods to solve these problems and provide low-cost computations for variational calculations involving RDMs.

ACKNOWLEDGMENTS

The authors are very thankful to the editor of the volume, D. A. Mazziotti, for the encouragement to write this chapter, and also to B. J. Braams for providing valuable comments. The first two authors were supported by fellowships from the Japan Society for the Promotion of Science (JSPS) and Grant-in-Aids for Scientific Research from the Japanese Ministry of Education, Culture, Sports, Science, and Technology.

References

1. M. Nakata, H. Nakatsuji, M. Ehara, M. Fukuda, K. Nakata, and K. Fujisawa, *J. Chem. Phys.* **114**, 8282 (2001).
2. M. Yamashita, K. Fujisawa, and M. Kojima, *Optimization Methods & Software* **18**, 491 (2003). Available at http://grid.r.dendai.ac.jp/sdpa.
3. L. Vandenberghe and S. Boyd, *SIAM Rev.* **38**, 49 (1996).
4. H. Wolkowicz, R. Saigal, and L. Vandenberghe (eds.), *Handbook of Semidefinite Programming: Theory, Algorithms, and Applications*, Kluwer Academic, Norwell, MA, 2000.
5. A. Ben-Tal and A. Nemirovski, *Lectures on Modern Convex Optimization: Analysis, Algorithms, and Engineering Applications*, SIAM, Philadelphia, 2001.
6. M. J. Todd, *Acta Numerica* **10**, 515 (2001).
7. R. D. C. Monteiro, *Math. Programming Ser. B* **97**, 209 (2003).
8. M. Nakata, M. Ehara, and H. Nakatsuji, *J. Chem. Phys.* **116**, 5432 (2002).
9. D. A. Mazziotti, *Phys. Rev. A* **65**, 062511 (2002).
10. D. A. Mazziotti, *Phys. Rev. Lett.* **93**, 213001 (2004).
11. D. A. Mazziotti, *J. Chem. Phys.* **121**, 10957 (2004).
12. D. A. Mazziotti, *Phys. Rev. A* **72**, 032510 (2005).
13. G. Gidofalvi and D. A. Mazziotti, *Phys. Rev. A* **72**, 052505 (2005).
14. Z. Zhao, B. J. Braams, M. Fukuda, M. L. Overton, and J. K. Percus, *J. Chem. Phys.* **120**, 2095 (2004).

15. M. Fukuda, B. J. Braams, M. Nakata, M. L. Overton, J. K. Percus, M. Yamashita, and Z. Zhao, *Math. Programming Ser.* B (in press).
16. M. Fukuda, B. J. Braams, M. Nakata, M. L. Overton, J. K. Percus, M. Yamashita, and Z. Zhao, *Sūri Kaiseki Kenkyūjyo Koukyūroku* **1461**, 15 (2005). Also, Research Report B-420, Dept. Mathematical and Computing Sciences, Tokyo Institute of Technology, 2005.
17. A. J. Coleman, *Rev. Mod. Phys.* **35**, 668 (1963).
18. F. Weinhold and E. B. Wilson, Jr., *J. Chem. Phys.* **47**, 2298 (1967).
19. Yu. Nesterov and A. Nemirovskii, *Interior-Point Polynomial Algorithms in Convex Programming*, SIAM, Philadelphia, 1993.
20. F. Alizadeh, *SIAM J. Optimization* **5**, 13 (1995).
21. K. Fujisawa, M. Kojima, and K. Nakata, *Math. Programming Ser.* B **79**, 235 (1997).
22. M. Yamashita, K. Fujisawa, and M. Kojima, *Parallel Computing* **29**, 1053 (2003). Available at http://grid.r.dendai.ac.jp/sdpa.
23. B. Borchers and J. Young, *Implementation of a Primal–Dual Method for SDP on a Parallel Architecture*, Research Report, 2005. Available at http://infohost.nmt.\ edu/~borchers/csdp.html.
24. M. Kočvara and M. Stingl, *Optimization Methods & Software* **18**, 317 (2003). Available at http://www.penopt.com.
25. J. Nocedal and S. J. Wright, *Numerical Optimization*, Springer-Verlag, New York, 1999.
26. S. Burer and R. D. C. Monteiro, *Math. Programming Ser.* B **95**, 329 (2003).
27. S. Burer and R. D. C. Monteiro, *Math. Programming Ser.* A **103**, 427 (2005).
28. S. Burer and C. Choi, *Optimization Methods & Software* **21**, 493 (2006). Available at http://dollar.biz.uiowa.edu/~sburer/software/SDPLR.
29. B. Borchers, *Optimization Methods & Software* **11–12**, 597 (1999). Available at http://infohost.nmt.edu/~borchers/csdp.html.
30. J. F. Sturm, *Optimization Methods & Software* **11–12**, 625 (1999). Available at http://sedumi.mcmaster.ca.
31. K.-C. Toh, M. J. Todd, and R. H. Tütüncü, *Optimization Methods & Software* **11–12**, 545 (1999). Available at http://www.math.nus.edu.sg/~mattohkc/sdpt3.html.
32. J. Moré, T. Munson, J. Sarich, et al., *NEOS Server*. Available at http://www-neos.mcs.anl.gov/neos/solvers.
33. T. Terlaky, O. Romanko, and L. Zhu, *CaNEOS Server for Optimization*. Available at http://caneos.mcmaster.ca/server-solvers.html\#SDSOCP.
34. K. Fujisawa, M. Fukuda, Y. Futakata, K. Kobayashi, M. Kojima, K. Nakata, M. Nakata, and M. Yamashita, *SDPA Online for Your Future*. Available at http://grid.r.dendai.ac.jp/sdpa.

PART III

CHAPTER 7

THEORY AND METHODOLOGY OF THE CONTRACTED SCHRÖDINGER EQUATION

C. VALDEMORO

Instituto de Matemáticas y Física Fundamental, Consejo Superior de Investigaciones Científicas, Serrano 123, 28006 Madrid, Madrid, Spain

CONTENTS

I. Introduction
II. Theoretical Background
   A. Notation and Definitions
      1. The Hamiltonian
      2. The Reduced Density Matrices
   B. Properties of the 2-RDM and the N-Representability Problem
   C. The Matrix Contracting Mapping
III. The Contracted Schrödinger Equation
   A. Matrix Representation of the Schrödinger Equation and its Contraction
   B. The Role of the Spin
      1. The Spin Contracted Equation
      2. The Singlet Case
   C. Iterative Solution of the Contracted Schrödinger Equation
IV. The Reduced Density Matrices Construction Algorithms
   A. The 2-RDM Construction Algorithm
   B. Higher-Order RDMs Construction Algorithms
   C. Other Approaches
   D. A Unifying Algorithm
      1. A Criterion for Selecting the Parameter Value for the 4-RDM
   E. Estimating the Error Matrix $^3\Delta$
      1. Analytical Contraction of the Unifying Algorithm
   F. The N-Representability Problem: Introducing Bounds Correction
   G. Some Comparative Results
V. Factors Affecting the Convergence of the Iterative Solution of the CSE
   A. Influence of the Algorithms on the 2-CSE Convergence
   B. Influence of the N- and S-Representability of the 2-RDM on the Convergence of the 2-CSE Iterative Process

---

Reduced-Density-Matrix Mechanics: With Application to Many-Electron Atoms and Molecules, A Special Volume of Advances in Chemical Physics, Volume 134, edited by David A. Mazziotti. Series editor Stuart A. Rice. Copyright © 2007 John Wiley & Sons, Inc.

VI. An Exact Formal Solution to the Contracted Schrödinger Equation's Indeterminacy
  A. Decomposition of High-Order RDMs and a Basic Cancellation Relation
  B. Derivation of the Fourth-Order MCSE
  C. Some Significant Results
VII. Some Final Remarks
Acknowledgments
References

I. INTRODUCTION

The many-body Hamiltonian operator is the sum of one- and two-electron operators, which is the reason why the energy of an N-electron system can be expressed as a functional of a mathematical object which only depends on the variables of two electrons, the second-order reduced density matrix (2-RDM). The quest for a method of studying the structure of electronic systems by determining the 2-RDM instead of using the N-electron wavefunction dates from the 1950s [1–5]. Since then, a rich bibliography on the 2-RDM theory has been developed, in particular, several books and proceedings [6–12] describe the progress achieved. The incomplete knowledge of the 2-RDM mathematical properties has greatly hindered this progress but, in spite of it, recent new developments have reawakened this line of research.

One question that several people, independently and at different times, have asked themselves may probably be stated as: Can the Schrödinger equation be mapped into the two-electron space? And what would be the properties of the resulting equation? The answer to this double question was obtained by following two essentially different approaches. Thus Cho [13], Cohen and Frishberg [14,15], and Nakatsuji [16], by integrating the Schrödinger equation, obtained in first quantization the density equation; and Valdemoro [17], by applying a contracting mapping to the matrix representation of the Schrödinger equation, obtained the contracted Schrödinger equation (CSE). Although these two equations are apparently very different, they are in fact equivalent. An important feature of these equations is that they constitute a hierarchy of equations. Thus the contraction of the Schrödinger equation to the p-electron space generates a p-order CSE that depends on the $(p+1)$-CSE and on the $(p+2)$-CSE. This hierarchy dependence causes the p-CSE to depend not only on the p-RDM but also on higher-order RDMs, which renders the equation indeterminate [18]. In 1992 Valdemoro [19] showed that a reasonable approximation of the 2-RDM could be obtained in terms of the 1-RDM by a method that could be extended to higher-order RDMs [20, 21]. In this way, Colmenero and Valdemoro solved approximatly the CSE [21]. The aim of this chapter is to give an overview of the CSE theory and of the construction algorithms for the high-order RDMs, which are the basic part of the methodology that has been developed, in order to obtain a good approximate solution of the 2-CSE.

II. THEORETICAL BACKGROUND

In order to have a self-contained chapter, the well-known definitions and concepts that are used in the rest of the chapter are recalled in this section. Other matters, not so widely known but also basic for later developments, are also described here.

A. Notation and Definitions

In what follows, the number N of particles is assumed to be a constant. The one-electron basis is assumed to be finite and formed by $2\mathcal{K}$ orthonormal spin orbitals denoted by the italic letters $i, j, k, l \ldots$ or, when the spin is considered explicitly by i_σ or $i_\alpha \ldots$.

1. The Hamiltonian

The electronic many-body Hamiltonian is

$$\hat{H} = \sum_{i,j} h_{i;j} a_i^\dagger a_j + \frac{1}{2} \sum_{i,j,k,l} \langle ij|kl \rangle a_i^\dagger a_j^\dagger a_l a_k \quad (1)$$

where h is a matrix formed by the one-electron integrals and $\langle ij|kl \rangle$ is the two-electron integral matrix in the Condon and Shortley notation. It is useful to transform the form of the Hamiltonian to

$$\hat{H} = \frac{1}{2} \sum_{i,j,k,l} K_{i,j;k,l} a_i^\dagger a_j^\dagger a_l a_k \quad (2)$$

where

$$K_{i,j;k,l} = \left[\frac{1}{N-1} (h_{i;k} \delta_{j,l} + h_{j;l} \delta_{i,k}) + \langle ij|kl \rangle \right] \quad (3)$$

The matrix K is the reduced Hamiltonian [22, 25] and has the same symmetry properties as the two-electron matrix; that is,

$$K_{i,j;k,l} = K_{j,i;l,k} = K^*_{k,l;i,j} = K^*_{l,k;j,i} \quad (4)$$

2. The Reduced Density Matrices

The p-order reduced density matrix (p-RDM) is defined as

$$^p D^{\Psi\Psi'}_{i_1,i_2,\ldots,i_p;j_1,j_2,\ldots,j_p} = \frac{1}{p!} \langle \Psi | a^\dagger_{i_1} a^\dagger_{i_2} \cdots a^\dagger_{i_p} a_{j_p} \cdots a_{j_2} a_{j_1} | \Psi' \rangle \qquad (5)$$

When $\Psi \neq \Psi'$, this expression defines an element of the p-order transition density matrix (p-TRDM) [2]. In what follows when $\Psi = \Psi'$, one instead of two upper indices denoting Ψ will be used.

The complementary matrix to the p-RDM is the p-order holes reduced density matrix

$$^p \bar{D}^{\Psi}_{i_1,i_2,\ldots,i_p;j_1,j_2,\ldots,j_p} = \frac{1}{p!} \langle \Psi | a_{j_p} \cdots a_{j_2} a_{j_1} a^\dagger_{i_1} a^\dagger_{i_2} \cdots a^\dagger_{i_p} | \Psi \rangle \qquad (6)$$

The concept of *hole* here implies that Ψ itself is the state of reference, not the Fermi sea or other state models.

For some purposes it is convenient to use the following *global* operators [26]:

$$^p B^\dagger_\Lambda = a^\dagger_{i_1} a^\dagger_{i_2} \cdots a^\dagger_{i_p} \qquad (7)$$

and

$$^p B_\Omega = a_{j_p} \cdots a_{j_2} a_{j_1} \qquad (8)$$

where the indices must have a unique ordering: that is, $i_1 < i_2 < \cdots < i_p$ and $j_1 < j_2 < \cdots < j_p$. That is,

$$^p B^\dagger_\Lambda |0\rangle \equiv |\Lambda\rangle \equiv |i_1 i_2 \cdots i_p\rangle \qquad (9)$$

and the p-RDM can be written

$$^p D^\Psi_{\Lambda;\Omega} = \langle \Psi |^p B^\dagger_\Lambda {}^p B_\Omega | \Psi \rangle \qquad (10)$$

B. Properties of the 2-RDM and the N-Representability Problem

The N-representability problem was defined in a remarkable paper by Coleman in 1963 [27]. This problem asks about the necessary and sufficient conditions that a matrix represented in a p-electron space must satisfy in order to be N-representable; that is, the conditions that must be imposed to ensure that there exists an N-electron wavefunction from which this matrix may be obtained by integration over $N-p$ electron variables. All the relations and properties that will now be described are the basis of a set of important necessary

N-representability conditions [28]. Let us start this description by focusing on the RDM's properties, which may be deduced from their definition as expectation values of density fermion operators. Thus the RDMs are Hermitian, are positive semidefinite, and contract to finite values that depend on the number of electrons, N, and in the case of the HRDMs on the size of the one-electron basis of representation, $2\mathcal{K}$. Thus

$$\mathrm{Tr}\,({}^p\underline{D}^\Psi) = \binom{N}{p} \tag{11}$$

$$\mathrm{Tr}\,({}^p\underline{\bar{D}}^\Psi) = \binom{2\mathcal{K}-N}{p} \tag{12}$$

Also, the fermion anticommutation rules interrelate the RDMs with the HRDMs; they render these matrices antisymmetric with respect to odd permutations of the row or column indices; and, finally, they interrelate them with two other families of matrices: the G-matrices and the correlation matrices.

Let us recall here the less obvious of these properties

- The anticommutator of a creator with an annihilator is

$$[a_j, a_i^\dagger]_+ = \delta_{i,j} \tag{13}$$

and its expectation value gives

$$ {}^1\bar{D}^\Psi_{i;k} + {}^1 D^\Psi_{i;k} = \delta_{i,k} \tag{14}$$

which relates the value of an element of the 1-RDM to the same element of the 1-HRDM. Since both the 1-RDM and the 1-HRDM are positive semidefinite, relation (14) imposes that the eigenvalues of these matrices are bounded by the numbers 0 and 1. To be positive semidefinite, together with these bounds, constitutes a necessary and sufficient condition for the 1-RDM to be ensemble N-representable [27].

- The second-order commutator is

$$[a_l a_k, a_i^\dagger a_j^\dagger]_- = \delta_{l,j}\delta_{k,i} - \delta_{l,i}\delta_{k,j} - \delta_{l,j}a_i^\dagger a_k \\ - \delta_{k,i}a_j^\dagger a_l + \delta_{k,j}a_i^\dagger a_l + \delta_{l,i}a_j^\dagger a_k \tag{15}$$

and its expectation value gives the second-order fermion relation [26]

$$ {}^2\bar{D}^\Psi_{ij;kl} - {}^2 D^\Psi_{ij;kl} = \delta_{l,j}\delta_{k,i} -, \delta_{l,i}\delta_{k,j} - \delta_{l,j}\,{}^1 D^\Psi_{i;k} \\ - \delta_{k,i}\,{}^1 D^\Psi_{j;l} + \delta_{k,j}\,{}^1 D^\Psi_{i;l} + \delta_{l,i}\,{}^1 D^\Psi_{j;k} \tag{16}$$

which interrelates the 2-HRDM with the 2-RDM. The condition that the 2-HRDM, as defined by this relation in terms of the 1- and 2-RDMs, be positive semidefinite constitutes what is called the necessary N-representability Q-condition [28–33].

- Let us reorder the fermion operators of a 2-RDM element

$$2!\,{}^2D^\Psi_{i,j;k,l} \equiv \langle\Psi|a_i^\dagger a_j^\dagger a_l a_k|\Psi\rangle$$
$$= -\delta_{k,j}\langle\Psi|a_i^\dagger a_l|\Psi\rangle + \langle\Psi|a_i^\dagger a_k a_j^\dagger a_l|\Psi\rangle \quad (17)$$

and let us now insert the unit operator in the middle of the second term

$$2!\,{}^2D^\Psi_{i,j;k,l} = -\delta_{k,j}\langle\Psi|a_i^\dagger a_l|\Psi\rangle$$
$$+ \langle\Psi|a_i^\dagger a_k|\Psi\rangle\langle\Psi|a_j^\dagger a_l|\Psi\rangle \quad (18)$$
$$+ \sum_{\Psi'\neq\Psi}\langle\Psi|a_i^\dagger a_k|\Psi'\rangle\langle\Psi'|a_j^\dagger;a_l|\Psi\rangle$$

This may be rewritten as

$$2!\,{}^2D^\Psi_{i,j;k,l} = {}^1D^\Psi_{i;k}\,{}^1D^\Psi_{j;l} - \delta_{k,j}\,{}^1D^\Psi_{i;l} + {}^2\mathcal{C}^\Psi_{i,j;k,l} \quad (19)$$

where $\mathcal{C}$ was defined by Valdemoro et al. [34, 35] as correlation matrix (CM) because it cannot be factorized in terms of the 1-RDM. It can be interpreted as describing the virtual excitations undergone by the electrons in order to avoid each other. Note that the row and column labels of the CM coincide with those of the 2-RDM from which it is derived. The properties of this matrix have recently been studied in detail [36–38].

The same elements appearing in the CM, but in a different ordering, form the G-matrix; thus

$${}^2\mathcal{C}^\Psi_{i,j;k,l} \equiv {}^2\mathcal{G}^\Psi_{i,k;l,j} \quad (20)$$

An equivalent equation to Eq. (19) in terms of the G-matrix was first published in 1969 by Garrod and Rosina [39] and also later reported by Valdemoro et al. [35].

The G-matrix that has been obtained here by decomposing the 2-RDM was defined by Garrod and Percus [29] as

$${}^2\mathcal{G}^\Psi_{i,j;k,l} = \langle\Psi|a_i^\dagger a_j a_l^\dagger a_k|\Psi\rangle - \langle\Psi|a_i^\dagger a_j|\Psi\rangle\langle\Psi|a_l^\dagger a_k|\Psi\rangle \quad (21)$$

The G-matrix is also Hermitian and positive semidefinite. The condition

$${}^2\underline{\mathcal{G}}^\Psi \geq 0 \quad (22)$$

is called the G-condition.

THEORY AND METHODOLOGY   127

The 2-RDM, the 2-HRDM, and the G-matrix are the only three second-order matrices which (by themselves) are Hermitian and positive semidefinite; thus they are at the center of the research in this field. Recently, a formally exact solution of the N-representability problem was published [12] but this solution is unfeasable in practice [40].

C. The Matrix Contracting Mapping

The contraction of a q-RDM to get a p-RDM with $p < q$ was formally defined by Kummer [41] as

$$^p\underline{D}^\Psi \equiv L_q^p \, ^q\underline{D}^\Psi \tag{23}$$

where L_q^p represents the contraction operation and $N \geq q > p$.

It is simple to contract an RDM by applying the expression of the $\hat{N}$ operator. Thus

$$\langle \Psi | a_i^\dagger a_j | \Psi \rangle \equiv \langle \Psi | a_i^\dagger \left(\sum_l \frac{a_l^\dagger a_l}{N-1} \right) a_j | \Psi \rangle \tag{24}$$

but when the matrix that must be contracted represents an operator that, while being related to the density operator, is a different one (i.e., the Hamiltonian), the question is more complicated. This general case was solved in 1983 by Valdemoro [17, 24, 26, 42, 43], who obtained the general matrix contracting mapping (MCM)

$$^p\mathcal{M}_{\lambda;\omega} \equiv \frac{\binom{N}{p}}{\binom{q}{p}\binom{N}{q}} \sum_{\Pi,\Gamma} {}^p D_{\lambda;\omega}^{\Pi\,\Gamma} \, ^q\mathcal{M}_{\Pi;\Gamma} \tag{25}$$

where $^q\mathcal{M}$ is the q-order matrix that must be contracted into the p-electron space and where the Π and Γ letters represent q-electron configurations.

III. THE CONTRACTED SCHRÖDINGER EQUATION

As mentioned in Section I, Cho [13], Cohen and Frishberg [14, 15], and Nakatsuji [16] integrated the Schrödinger equation and obtained an equation that they called the *density equation*. This equation was at the time also studied by Schlosser [44] for the 1-TRDM. In 1986 Valdemoro [17] applied a contracting mapping to the matrix representation of the Schrödinger equation and obtained the *contracted Schrödinger equation* (CSE). In 1986, at the Coleman Symposium where the CSE was first reported, Löwdin asked whether there was a connection between the CSE and the Nakatsuji's density equation. It came out that both equations, although different, are completely equivalent. In 1998, Mazziotti [45] showed that one could obtain the same equation directly by considering the density operator as a probe. The direct connection of the CSE with the Schrödinger equation is, however, best understood by contracting the matrix representation of this fundamental equation and this is the derivation that will now be given [46].

A. Matrix Representation of the Schrödinger Equation and Its Contraction

Let us consider the Schrödinger equation

$$\hat{H}|\Psi\rangle = E_\Psi |\Psi\rangle \tag{26}$$

or equivalently

$$\hat{H}|\Psi\rangle\langle\Psi| = E_\Psi |\Psi\rangle\langle\Psi| \tag{27}$$

and let us represent this operatorial equation in a basis of N-electron functions, that is, Slater determinants

$$\langle\Lambda|\hat{H}|\Psi\rangle \langle\Psi|\Omega\rangle = E_\Psi \langle\Lambda|\Psi\rangle \langle\Psi|\Omega\rangle \tag{28}$$

That is,

$$\langle\Lambda|\hat{H}|\Psi\rangle \langle\Psi|\Omega\rangle = E_\Psi \,^N D^\Psi_{\Lambda\Omega} \tag{29}$$

This equation is the matrix representation of the Schrödinger equation in the N-electron space. In order to contract it into the two-electron space, we will apply the MCM to both sides of the equation and get

$$\sum_{\Lambda\Omega} {}^2 D^{\Lambda\Omega}_{i_1,i_2;j_1,j_2} (\underline{H}\,^N\underline{D}^\Psi)_{\Lambda\Omega} = E_\Psi \,^2 D^\Psi_{i_1,i_2;j_1,j_2} \tag{30}$$

When this equation is developed, one obtains

$$\langle\Psi|\hat{H}\, a^\dagger_{i_1} a^\dagger_{i_2} a_{j_2} a_{j_1} |\Psi\rangle = E_\Psi \,^2 D^\Psi_{i_1,i_2;j_1,j_2} \tag{31}$$

This equation is the CSE in compact form, which was the starting point of Mazziotti's derivation [45]. By replacing $\hat{H}$ by relation (2), and transforming the string of operators into its normal form, one obtains one of the usual forms of the CSE:

$$\boxed{\begin{aligned}(\underline{K}^2\underline{D}^\Psi)_{r,s;t,u} + 3\sum_{i,k,l}(K_{i,s;k,l}\,^3D^\Psi_{t,u,i;r,l,k} + K_{i,r;k,l}\,^3D^\Psi_{t,u,i;l,s,k}) \\ + 6\sum_{i,j,k,l} K_{i,j;k,l}\,^4D^\Psi_{t,u,i,j;r,s,k,l} = E_\Psi \,^2D^\Psi_{r,s;t,u}\end{aligned}} \tag{32}$$

A simple inspection of this equation shows that it not only depends on the 2-RDM but also on the 3- and 4-RDMs, which causes it to be indeterminate. Besides this difficulty, which will be discussed at length later on, one may ask whether the solutions of this equation coincide with those of the Schrödinger equation. Indeed, the derivation given above for the 2-CSE shows that the Schrödinger equation implies the 2-CSE. But does the inverse relation hold? The answer to this question was given by Nakatsuji, who showed that the p-CSE for $p \geq 2$ is equivalent to the Schrödinger equation [16] by stating and proving the following theorem.

Theorem [Nakatsuji] If the RDMs are N-representable, then the p-CSE is satisfied by the p-, $(p+1)$-, and $(p+2)$-RDM *if and only if* the N-electron density matrix (N-DM), preimage of these matrices, satisfies the Schrödinger equation.

An elegant proof, in second quantization, that the 2-CSE implies the Schrödinger equation, was given by Mazziotti and is as follows.

Proof. The Schrödinger equation is satisfied *if and only if* the well-known dispersion relation [47]

$$\langle \Psi | \hat{H}^2 | \Psi \rangle - \langle \Psi | \hat{H} | \Psi \rangle^2 = 0 \tag{33}$$

is satisfied

Let us now consider the 2-CSE as given by Eq. (31). It can be written

$$\frac{1}{4} \sum_{i,j,k,l} K_{i,j;k,l} \langle \Psi | a_i^\dagger a_j^\dagger a_l a_k a_r^\dagger a_s^\dagger a_u a_t | \Psi \rangle = E_\Psi \, ^2D^\Psi_{r,s;t,u} \tag{34}$$

and multiplying both sides of this equation by the element $K_{r,s;t,u}$ and adding over repeated indices, one has

$$\langle \Psi | \left(\frac{1}{2} \sum_{i,j,k,l} K_{i,j;k,l} a_i^\dagger a_j^\dagger a_l a_k \right) \left(\frac{1}{2} \sum_{r,s,t,u} K_{r,s;t,u} a_r^\dagger a_s^\dagger a_u a_t \right) | \Psi \rangle \\ = E_\Psi \left(\sum_{r,s,t,u} K^2_{r,s;t,u} D^\Psi_{r,s;t,u} \right) \tag{35}$$

which is the dispersion relation and what had to be proved. Since for $p > 2$ the p-CSE implies the 2-CSE, the demonstration is also valid for these higher-order equations. ∎

It should be noted that what has been demonstrated for the p-CSE does not hold for the 1-CSE because the Hamiltonian includes two-electron terms. In fact, it is easy to see that the 1-CSE is satisfied not only by the RDMs corresponding to the FCI solution but also by the 1-, 2-, and 3-RDMs corresponding to a Hartree–Fock solution.

An important consequence of the equivalence of the 2-CSE and higher-order CSEs with the Schrödinger equation is that the CSEs may be applied to the study not only of the ground-state but also of excited states.

B. The Role of the Spin

Until now the electron spin has not been explicitly taken into account. In this section we analyze the differences introduced in the 2-CSE when the spin properties are considered.

In the absence of spin interactions, the Hamiltonian may be written

$$\hat{H} = \sum_{r<s,k<l} K^{\alpha\alpha}_{r,s;k,l} a^\dagger_{r_\alpha} a^\dagger_{s_\alpha} a_{l_\alpha} a_{k_\alpha} + \sum_{u,v,m,n} K^{\alpha\beta}_{u,v;m,n} a^\dagger_{u_\alpha} a^\dagger_{v_\beta} a_{n_\beta} a_{m_\alpha} \\ + \sum_{r<s,k<l} K^{\beta\beta}_{r,s;k,l} a^\dagger_{r_\beta} a^\dagger_{s_\beta} a_{l_\beta} a_{k_\beta} \tag{36}$$

where

$$K^{\alpha\alpha}_{r,s;k,l} = K_{r,s;k,l} - K_{r,s;l,k} \tag{37}$$

$$K^{\alpha\beta}_{u,v;m,n} = K_{u,v;m,n} \tag{38}$$

$$K^{\beta\beta}_{r,s;k,l} = K_{r,s;k,l} - K_{r,s;l,k} \tag{39}$$

In this representation the 2-CSE is formed by the following three equations:

$$E_\Psi \, {}^2D^{\Psi\,\alpha\alpha}_{i,j;p,q} = \begin{cases} {}^2D^{\Psi\,\alpha\alpha}_{i,j;r,s} K^{\alpha\alpha}_{r,s;p,q} \\ - {}^3D^{\Psi\,\alpha\alpha\alpha}_{i,j,m;q,r,s} K^{\alpha\alpha}_{r,s;p,m} + {}^3D^{\Psi\,\alpha\alpha\alpha}_{i,j,m;p,r,s} K^{\alpha\alpha}_{r,s;q,m} \\ + {}^3D^{\Psi\,\alpha\alpha\beta}_{i,j,m;p,u,v} K^{\alpha\beta}_{u,v;q,m} - {}^3D^{\Psi\,\alpha\alpha\beta}_{i,j,m;q,u,v} K^{\alpha\beta}_{u,v;p,m} \\ + {}^4D^{\Psi\,\alpha\alpha\alpha\alpha}_{i,j,k,l;p,q,r,s} K^{\alpha\alpha}_{r,s;k,l} + {}^4D^{\Psi\,\alpha\alpha\beta\beta}_{i,j,k,l;p,q,r,s} K^{\beta\beta}_{r,s;k,l} \\ + {}^4D^{\Psi\,\alpha\alpha\alpha\beta}_{i,j,m,n;p,q,u,v} K^{\alpha\beta}_{u,v;m,n} \end{cases} \tag{40}$$

$(i<j, p<q)$

$$E_\Psi \, {}^2D^{\Psi\,\alpha\beta}_{i,j;p,q} = \begin{cases} {}^2D^{\Psi\,\alpha\beta}_{i,j;u,v} K^{\alpha\beta}_{u,v;p,q} \\ - {}^3D^{\Psi\,\alpha\alpha\beta}_{m,i,j;r,s,q} K^{\alpha\alpha}_{r,s;p,m} + {}^3D^{\Psi\,\alpha\beta\beta}_{i,j,m;p,r,s} K^{\beta\beta}_{r,s;q,m} \\ - {}^3D^{\Psi\,\alpha\alpha\beta}_{m,i,j;p,u,v} K^{\alpha\beta}_{u,v;m,q} - {}^3D^{\Psi\,\alpha\beta\beta}_{i,j,n;u,v,q} K^{\alpha\beta}_{u,v;p,n} \\ + {}^4D^{\Psi\,\alpha\alpha\alpha\beta}_{k,l,i,j;r,s,p,q} K^{\alpha\alpha}_{r,s;k,l} + {}^4D^{\Psi\,\alpha\beta\beta\beta}_{i,j,k,l;p,q,r,s} K^{\beta\beta}_{r,s;k,l} \\ + {}^4D^{\Psi\,\alpha\alpha\beta\beta}_{i,m,j,n;p,u,q,v} K^{\alpha\beta}_{u,v;m,n} \end{cases} \tag{41}$$

$$E_\Psi \, {}^2D^{\Psi\,\beta\beta}_{i,j;p,q} = \begin{cases} {}^2D^{\Psi\,\beta\beta}_{i,j;r,s} K^{\beta\beta}_{r,s;p,q} \\ -{}^3D^{\Psi\,\beta\beta\beta}_{m,i,j;q,r,s} K^{\beta\beta}_{r,s;p,m} + {}^3D^{\Psi\,\beta\beta\beta}_{m,i,j;p,r,s} K^{\beta\beta}_{r,s;q,m} \\ -{}^3D^{\Psi\,\alpha\beta\beta}_{m,i,j;u,v,p} K^{\alpha\beta}_{u,v;m,q} + {}^3D^{\Psi\,\alpha\beta\beta}_{m,i,j;u,v,q} K^{\alpha\beta}_{u,v;m,p} \\ +{}^4D^{\Psi\,\beta\beta\beta\beta}_{i,j,k,l;p,q,r,s} K^{\beta\beta}_{r,s;k,l} + {}^4D^{\Psi\,\alpha\alpha\beta\beta}_{k,l,i,j;r,s,p,q} K^{\alpha\alpha}_{r,s;k,l} \\ +{}^4D^{\Psi\,\alpha\beta\beta\beta}_{m,n,i,j;u,v,p,q} K^{\alpha\beta}_{u,v;m,n} \end{cases} \quad (42)$$

$(i<j, p<q)$

with the restriction $r < s$ and $k < l$. An implicit sum over repeated indices has been assumed.

In these equations, the different 2-RDM blocks are defined [46]:

$$ {}^2D^{\Psi\,\sigma\sigma'}_{i,j;k,l} \equiv {}^2D^\Psi_{i_\sigma j_{\sigma'};k_\sigma l_{\sigma'}} = \langle \Psi | a^\dagger_{i_\sigma} a^\dagger_{j_{\sigma'}} a_{l_{\sigma'}} a_{k_\sigma} | \Psi \rangle \quad (43)$$

These three block equations are only partially independent. Thus one cannot solve them separately because there are 3- and 4-RDM spin blocks appearing in more than one of the partial equations. Although an exact solution of these block equations would give the same energy when solving each of them, in practice this is not so. In consequence, one obtains different values of the energy for each of the block equations during the iterative process. Obviously, these differences should disappear or at least be negligible at convergence.

1. The Spin Contracted Equation

It is evident that a 2-RDM that corresponds to a Hamiltonian eigenstate also corresponds to a pure-spin state. However, when one is working with an approximated RDM, it is important that this RDM should correspond to a spin eigenstate.

Let us therefore consider, by analogy with the Hamiltonian (Eq. (31)) the compact form of the *contracted spin equation* (CSpE) [46]:

$$\langle \Psi | \hat{S}^2 a^\dagger_i a^\dagger_j a_l a_k | \Psi \rangle = S(S+1) \langle \Psi | a^\dagger_i a^\dagger_j a_l a_k | \Psi \rangle \quad (44)$$

Replacing into this equation the $\hat{S}^2$ by its second quantization expression

$$\hat{S}^2 = \hat{S}^2_z - \hat{S}_z + \hat{S}_+ \hat{S}_- = -\sum_{i,j} a^\dagger_{i_\alpha} a^\dagger_{j_\beta} a_{i_\beta} a_{j_\alpha}$$
$$+ \left(\frac{1}{2}\sum_i (a^\dagger_{i_\alpha} a_{i_\alpha} - a^\dagger_{i_\beta} a_{i_\beta})\right)^2 \quad (45)$$
$$+ \frac{1}{2}\sum_i (a^\dagger_{i_\alpha} a_{i_\alpha} - a^\dagger_{i_\beta} a_{i_\beta})$$

and transforming the string of one-electron operators into its normal form, one obtains

$$^2D^{\Psi\,\alpha\alpha}_{p,q;s,r} = C\left(\sum_i(-{}^3D^{\Psi\,\alpha\alpha\beta}_{i,q,p;s,r,i} + {}^3D^{\Psi\,\alpha\alpha\beta}_{i,p,q;s,r,i}) - \sum_{i,j} {}^4D^{\Psi\,\alpha\alpha\alpha\beta}_{p,q,i,j;s,r,j,i}\right) \qquad (46)$$

$$^2D^{\Psi\,\alpha\beta}_{p,q;s,r} = C\left(\begin{array}{l}-{}^2D^{\Psi\,\alpha\beta}_{q,p;s,r} + \sum_i(-{}^3D^{\Psi\,\alpha\alpha\beta}_{p,q,i;s,i,r} + {}^3D^{\Psi\,\alpha\beta\beta}_{i,p,q;s,r,i}) \\ -\sum_{i,j} {}^3D^{\Psi\,\alpha\alpha\beta\beta}_{i,p,j,q;s,j,r,i}\end{array}\right) \qquad (47)$$

$$^2D^{\Psi\,\beta\beta}_{p,q;s,r} = C\left(\sum_i(-{}^3D^{\Psi\,\alpha\beta\beta}_{p,i,q;i,s,r} + {}^3D^{\Psi\,\alpha\beta\beta}_{q,i,p;i,s,r}) - \sum_{i,j} {}^4D^{\Psi\,\alpha\beta\beta\beta}_{jpqi;i,s,r,j}\right) \qquad (48)$$

where

$$C = \frac{1}{N/2 + (N_\alpha - N_\beta)^2/4 - S(S+1)} \qquad (49)$$

By analogy with Nakatsuji, Alcoba [48] demonstrated the following theorem.

Theorem [Alcoba] Assuming that the RDMs are N-representable, then the 2-CSpE is satisfied by the 2-, 3-, 4-RDMs *if and only if* the N-DM, preimage of these matrices, satisfies the spin equation

$$\langle\Lambda|\hat{S}^2|\Psi\rangle\langle\Psi|\Omega\rangle = S(S+1)\,{}^N D^\Psi_{\Lambda;\Omega} \qquad (50)$$

The demonstration of this theorem follows in a parallel form that of the CSE.

In order to satisfy both the CSpE and the CSE, Valdemoro et al. [46] replace the 2-RDM spin blocks appearing on the right-hand side (rhs) of Eqs. (40), (41), and (42) by the corresponding expression in Eqs. (46), (47), and (48). The $\alpha\alpha$ spin block of the resulting equations has the following form:

$$E_\Psi\,{}^2D^{\Psi\,\alpha\alpha}_{i,j;p,q} = \begin{cases} -K^{\alpha\alpha}_{r,s;p,m}\,{}^3D^{\Psi\,\alpha\alpha\alpha}_{i,j,m;q,r,s} + K^{\alpha\alpha}_{r,s;q,m}\,{}^3D^{\Psi\,\alpha\alpha\alpha}_{i,j,m;p,r,s} \\ +K^{\alpha\beta}_{u,v;q,m}\,{}^3D^{\Psi\,\alpha\alpha\beta}_{i,j,m;p,u,v} - K^{\alpha\beta}_{u,v;p,m}\,{}^3D^{\Psi\,\alpha\alpha\beta}_{i,j,m;q,u,v} \\ +K^{\alpha\alpha}_{r,s;k,l}\,{}^4D^{\Psi\,\alpha\alpha\alpha\alpha}_{i,j,k,l;p,q,r,s} + K^{\beta\beta}_{r,s;k,l}\,{}^4D^{\Psi\,\alpha\alpha\beta\beta}_{i,j,k,l;p,q,r,s} \\ +K^{\alpha\beta}_{u,v;m,n}\,{}^4D^{\Psi\,\alpha\alpha\alpha\beta}_{i,j,m,n;p,q,u,v} \\ +C\,K^{\alpha\alpha}_{r,s;p,q}\,\{-{}^3D^{\Psi\,\alpha\alpha\beta}_{v,j,i;r,s,v} + {}^3D^{\Psi\,\alpha\alpha\beta}_{v,i,j;r,s,v} \\ -{}^4D^{\Psi\,\alpha\alpha\alpha\beta}_{i,j,v,t;r,s,t,v}\} \end{cases} \qquad (51)$$

$(i<j, p<q)$

The other spin blocks, $\alpha\beta$ and $\beta\beta$, have a similar structure.

The main difference between the spin-adapted 2-CSE and the nonadapted one is that in the rhs of the spin-adapted 2-CSE, the 2-RDM only appears in the $\alpha\beta$ block.

2. The Singlet Case

The studies on the spin properties of the 2-RDM and of the second-order correlation matrix [38, 49, 50] have shown that for singlet states the $\alpha\beta$ block of the 2-RDM completely determines the other two spin blocks of the 2-RDM. In consequence, in these cases, the iterative solution of the 2-CSE may be carried out by working only with the $\alpha\beta$ block of the 2-CSE, and the $\alpha\alpha$ and the $\beta\beta$ blocks of the 2-RDM are determined in terms of the $\alpha\beta$ one.

This simplification can reduce significantly the computational effort but it has its drawbacks. Thus, in our experience, the $\sigma\sigma$ blocks of the 2-CSE converge better than the $\alpha\beta$ block. This is probably due to a lower efficiency of the construction algorithms for the higher matrices involved in the $\alpha\beta$ block. The reason for the better performance of the algorithms involved in the $\sigma\sigma$ blocks is that, as will be seen later on, the electron exchange plays a large role in them, and the correlation effects are relatively less dominant. No calculations have as yet been carried out on singlet states by using just the $\alpha\beta$ block of the 2-CSE but it is possible that for large systems the computational reduction may be advantageous, even at the cost of losing some accuracy.

C. Iterative Solution of the Contracted Schrödinger Equation

The dependence of the 2-CSE on the 3- and 4-RDMs renders indeterminate this equation [18, 52]. This is the reason why the interest in this equation was initially lost. However, after the encouraging results obtained in the construction of the 2-RDM in terms of the 1-RDM [19], the possibility arose of constructing good approximations of the high-order RDMs in terms of the lower-order ones. This permitted one to remove the indeterminacy and opened the way to build an iterative method for solving the 2-CSE.

Let us represent the 2-CSE in a symbolic form.

$$^2\underline{\mathcal{M}}^\Psi \equiv \text{function } (\underline{K}, {}^2\underline{D}^\Psi, {}^3\underline{D}^\Psi, {}^4\underline{D}^\Psi) = E_\Psi \, {}^2\underline{D}^\Psi \qquad (52)$$

In a schematic way the different steps of the basic procedure proposed by Colmenero and Valdemoro in 1994 [53] are as follows:

- Let us choose a reasonable and N-representable 2-RDM and 1-RDM as initial probes, and let us use them to approximate first the 3-RDM and then the 4-RDM.
- Let us correct as finely as possible the N-representability defects of the 4-RDM and then contract it in order to obtain a new set of 1-, 2-, and 3-RDMs.

- Let us replace all these matrices into the expression of $\mathcal{M}$.
- Let us take the trace of both sides of the CSE equation, and one has

$$E' = \frac{\text{Tr}(^2\underline{\mathcal{M}})}{\binom{N}{2}} \tag{53}$$

- Using this energy value, one can obtain a new 2-RDM,

$$^2\underline{D}' = \frac{^2\underline{\mathcal{M}}}{E'} \tag{54}$$

This process is repeated until convergence. Although the above steps describe in a schematic way a process that will be later analyzed in more detail, it is convenient at this stage to comment on the second of these steps. As we saw, the equivalence between the Schrödinger equation and the 2-CSE exists only when the RDMs appearing in the equation are N-representable. Although the initial 2-RDM is chosen to be N-representable, or closely so, the approximation algorithms for the higher-order matrices only preserve some of the necessary conditions. Thus the 4-RDM must be *purified* as much as possible. Also, there must be consistency among the 4-, 3- and 2-RDMs; which is why the 4-RDM must be contracted in order to obtain a new set of the lower-order matrices. In this way, the N-representability corrections carried out on the 4-RDM are transmitted not only to the lower-order matrices but also to the partial traces. Thus, denoting as $^2\mathcal{M}^i$ the part of $^2\mathcal{M}$ depending on the i-RDM, one has

$$\text{Tr}(^2\underline{\mathcal{M}}^{(2)}) = E_\Psi \tag{55}$$

$$\text{Tr}(^2\underline{\mathcal{M}}^{(3)}) = 2(N-2)\, E_\Psi \tag{56}$$

$$\text{Tr}(^2\underline{\mathcal{M}}^{(4)}) = \frac{(N-3)(N-2)}{2}\, E_\Psi \tag{57}$$

and these traces continue to hold for the approximated equation.

In order to avoid keeping in computer memory all the 4-RDM elements, the contraction of the 4-RDM in order to get a consistent 3-RDM is simulated by another algorithm. Thus the only 4-RDM elements to be stored are the diagonal elements. All the elements are only calculated once and entered in all the places where they appear.

IV. THE REDUCED DENSITY MATRICES CONSTRUCTION ALGORITHMS

As has been mentioned earlier, the indeterminacy of the 2-CSE may be removed by approximating the 3- and 4-RDMs in terms of the 1- and 2-RDMs. These approximation algorithms and the higher-order corrections proposed by several authors are described in this section. Comparative results obtained with the different methods and concerning the 3-RDM and the 4-RDM are discussed at the end of the section.

In order to better analyze the interplay of the different terms appearing in the algorithms, we start by considering the construction of the 2-RDM, which was at the origin of the development [19].

A. The 2-RDM Construction Algorithm

In Section II we saw that, according to the second-order fermion relation, Eq. (16), the difference between the 2-HRDM and the 2-RDM, was a functional of the 1-RDM, which involved also Kronecker deltas. When replacing in that equation the Kronecker deltas in terms of the 1-RDM and the 1-HRDM, Eq. (14), one obtains

$$\left. \begin{array}{c} 2!^2 \bar{D}^{\Psi}_{i,j;p,q} \\ - \\ 2!^2 D^{\Psi}_{i,j;p,q} \end{array} \right\} = \left\{ \begin{array}{c} {}^1\bar{D}^{\Psi}_{i;p} \, {}^1\bar{D}^{\Psi}_{j;q} - {}^1\bar{D}^{\Psi}_{i;q} \, {}^1\bar{D}^{\Psi}_{j;p} \\ - \\ {}^1D^{\Psi}_{i;p} \, {}^1D^{\Psi}_{j;q} - {}^1D^{\Psi}_{i;q} \, {}^1D^{\Psi}_{j;p} \end{array} \right. \tag{58}$$

which may be written in a more compact form as

$$\left. \begin{array}{c} 2!\,^2 \bar{D}^{\Psi}_{i,j;p,q} \\ - \\ 2!^2 D^{\Psi}_{i,j;p,q} \end{array} \right\} = \left\{ \begin{array}{c} \sum_{\mathcal{P}} (-1)^{\mathcal{P}} \, \mathcal{P} \, {}^1\bar{D}^{\Psi}_{i;p} \, {}^1\bar{D}^{\Psi}_{j;q} \\ - \\ \sum_{\mathcal{P}} (-1)^{\mathcal{P}} \, \mathcal{P} \, {}^1D^{\Psi}_{i;p} \, {}^1D^{\Psi}_{j;q} \end{array} \right. \tag{59}$$

where $\sum_{\mathcal{P}} (-1)^{\mathcal{P}} \mathcal{P}$ antisymmetrizes the column labels of the 1-RDMs and of the 1-HRDMs.

This duality of holes and particles allows us to write

$$2!\,^2 D^{\Psi}_{i,j;p,q} = \sum_{\mathcal{P}} (-1)^{\mathcal{P}} \, \mathcal{P} \, ({}^1D^{\Psi}_{i;p} \, {}^1D^{\Psi}_{j;q}) + 2!\,^2 \Delta^{\Psi}_{i,j;p,q} \tag{60}$$

and

$$2!\,^2 \bar{D}^{\Psi}_{i,j;p,q} = \sum_{\mathcal{P}} (-1)^{\mathcal{P}} \, \mathcal{P} \, ({}^1\bar{D}^{\Psi}_{i;p} \, {}^1\bar{D}^{\Psi}_{j;q}) + 2!\,^2 \Delta^{\Psi}_{i,j;p,q} \tag{61}$$

where the matrix $^2\Delta$ is thus defined as

$$\begin{aligned} 2!\,^2 \Delta^{\Psi}_{i,j;p,q} &= 2!\,^2 D^{\Psi}_{i,j;p,q} - \sum_{\mathcal{P}} (-1)^{\mathcal{P}} \, \mathcal{P} \, ({}^1D^{\Psi}_{i;p} \, {}^1D^{\Psi}_{j;q}) \\ &= 2!\,^2 \bar{D}^{\Psi}_{i,j;p,q} - \sum_{\mathcal{P}} (-1)^{\mathcal{P}} \, \mathcal{P} \, ({}^1\bar{D}^{\Psi}_{i;p} \, {}^1\bar{D}^{\Psi}_{j;q}) \end{aligned} \tag{62}$$

This error was originally approximated by an iterative purification renormalizing procedure, focusing on rendering the 2-RDM and the 2-HRDM positive-semidefinite and correctly normalized [19].

In order to identify the structure of $^2\Delta^\Psi$, let us recall the relation linking the 2-RDM (Eq. (19)),

$$2!\,^2D^\Psi_{i,j;p,q} = {}^1D^\Psi_{i;p}\,{}^1D^\Psi_{j;q} - \delta_{j,p}\,{}^1D^\Psi_{i;q} + {}^2C^\Psi_{i,j;p,q} \tag{63}$$

and replace the Kronecker delta as before:

$$2!\,^2D^\Psi_{i,j;p,q} = {}^1D^\Psi_{i;p}\,{}^1D^\Psi_{j;q} - {}^1D^\Psi_{j;p}\,{}^1D^\Psi_{i;q} - {}^1\bar{D}^\Psi_{j;p}\,{}^1D^\Psi_{i;q} + {}^2C^\Psi_{i,j;p,q} \tag{64}$$

When comparing this equation with Eq. (60) one finds

$$2!\,^2\Delta^\Psi_{i,j;p,q} = -{}^1\bar{D}^\Psi_{j;p}\,{}^1D^\Psi_{i;q} + {}^2C^\Psi_{i,j;p,q} \tag{65}$$

which is formed by two terms describing correlation effects. The ${}^1\bar{D}^\Psi\,{}^1D^\Psi$ term may be interpreted as part of a self-repulsion term and is easily evaluated; while the CM term, which was discussed in Section II, is an unknown that can only be approximated. These two kinds of correlation terms balance each other and it is the CM term that causes the difficulties. It is interesting that the $^2\Delta^\Psi$ is common to both the 2-RDM and the 2-HRDM, which is why it is cancelled out when one takes the difference of these two matrices.

B. Higher-Order RDMs Construction Algorithms

The arguments just described for the construction of the 2-RDM were extended without difficulty to the higher-order RDMs. The algorithms for these high-order RDMs were originally reported by Colmenero et al. [20] in a spin-free basis; and Valdemoro et al. [46] obtained later on the algorithms in a spin-orbital basis. For the 3-RDM, the algorithm in a spin-orbital basis is

$$\begin{aligned}
3!\,^3D^\Psi_{i,j,k;p,q,r} = &-2\sum_{\mathcal{P}} (-1)^\mathcal{P}\,\mathcal{P}\,({}^1D^\Psi_{i;p}\,{}^1D^\Psi_{j;q}\,{}^1D^\Psi_{k;r}) \\
&+ \sum_{\mathcal{P}'} (-1)^{\mathcal{P}'}\,\mathcal{P}'\,2!\,({}^1D^\Psi_{i;p}\,{}^2D^\Psi_{j,k;q,r} \\
&+ {}^1D^\Psi_{j;q}\,{}^2D^\Psi_{i,k;p,r} + {}^1D^\Psi_{k;r}\,{}^2D^\Psi_{i,j;p,q}) \\
&+ 3!\,^3\Delta^\Psi_{i,j,k;p,q,r}
\end{aligned} \tag{66}$$

where $\sum_{\mathcal{P}} (-1)^\mathcal{P}\,\mathcal{P}$ antisymmetrizes the column indices of the three 1-RDM involved and $\sum_{\mathcal{P}'} (-1)^{\mathcal{P}'}\,\mathcal{P}'$ antisymmetrizes the column index of the 1-RDM with the column indices of the 2-RDM.

THEORY AND METHODOLOGY

In a similar way one obtains the 4-RDM construction algorithm [20, 46]:

$$4!\, ^4D^{\Psi}_{i,j,k,l;\, p,q,r,s} = \sum_{\mathcal{P}}(-1)^{\mathcal{P}}\,\mathcal{P}\,3!\,(^1D^{\Psi}_{i;p}\,^3D^{\Psi}_{j,k,l;q,r,s} +\,^1D^{\Psi}_{j;q}\,^3D^{\Psi}_{i,k,l;p,r,s}$$
$$+\,^3D^{\Psi}_{i,j,l;p,q,s}\,^1D^{\Psi}_{k;r} +\,^3D^{\Psi}_{i,j,k;p,q,r}\,^1D^{\Psi}_{l;s})$$
$$+\,3\sum_{\mathcal{P}'}(-1)^{\mathcal{P}'}\,\mathcal{P}'\,(^1D^{\Psi}_{i;p}\,^1D^{\Psi}_{j;q}\,^1D^{\Psi}_{k;r}\,^1D^{\Psi}_{l;s})$$
$$-\sum_{\mathcal{P}''}(-1)^{\mathcal{P}''}\,\mathcal{P}''\,2!\,(^1D^{\Psi}_{i;p}\,^1D^{\Psi}_{j;q}\,^2D^{\Psi}_{k,l;r,s} \qquad (67)$$
$$+\,^1D^{\Psi}_{i;p}\,^2D^{\Psi}_{j,l;q,s}\,^1D^{\Psi}_{k;r} +\,^1D^{\Psi}_{i;p}\,^2D^{\Psi}_{j,k;q,r}\,^1D^{\Psi}_{l;s}$$
$$+\,^2D^{\Psi}_{i,l;p,s}\,^1D^{\Psi}_{j;q}\,^1D^{\Psi}_{k;r} +\,^2D^{\Psi}_{i,k;p,r}\,^1D^{\Psi}_{j;q}\,^1D^{\Psi}_{l;s}$$
$$+\,^2D^{\Psi}_{i,j;p,q}\,^1D^{\Psi}_{k;r}\,^1D^{\Psi}_{l;s})$$
$$+\,4!\,^4\Delta^{\Psi}_{i,j,k,l;p,q,r,s}$$

In what follows this set of algorithms, based on the separation of particles and holes, will be referred to as *VCP*.

In 2001 Valdemoro et al. [54] proposed a generalization of the *VCP* basic approach. It exploits the fact that the 2-RDM has more information than the 1-RDM; and instead of replacing one Kronecker delta in terms of 1-RDM and 1-HRDM, one can replace functions of Kronecker deltas in terms of higher-order RDMs and HRDMs. In this way, one partly avoids the cancelation of the correction terms pointed out by Mazziotti [45].

Let us therefore rewrite Eq. (16):

$$\left\{ \begin{array}{l} 2!\,^2\bar{D}^{\Psi}_{i,j;k,l} - 2!\,^2D^{\Psi}_{i,j;k,l} \\ +\delta_{i,k}\,^1D^{\Psi}_{j;l} + \delta_{j,l}\,^1D^{\Psi}_{i;k} - \delta_{i,l}\,^1D^{\Psi}_{j;k} - \delta_{j,k}\,^1D^{\Psi}_{i;l} \end{array} \right\} = \delta_{i,k}\,\delta_{j,l} - \delta_{i,l}\,\delta_{j,k} \qquad (68)$$

This equation expresses an antisymmetrized product of two Kronecker deltas in terms of RDMS and HRDMs. By combining it with the expression of the simple Kronecker delta previously used (Eq. (14)), one can replace the antisymmetrized products of three/four Kronecker deltas, which appear when taking the expectation values of the anticommutator/commutator of three/four annihilators with three/four creator operators. With the help of the symbolic system Mathematica [55], and by separating as in the *VCP* approach the particles from the holes part, one obtains

$$4!\,^4D^{\Psi}_{i,j,k,l;\,p,q,r,s} = \sum_{\mathcal{P}}(-1)^{\mathcal{P}}\,\mathcal{P}\,3!\,(^1D^{\Psi}_{i;p}\,^3D^{\Psi}_{j,k,l;q,r,s} +\,^1D^{\Psi}_{j;q}\,^3D^{\Psi}_{i,k,l;p,r,s}$$
$$+\,^1D^{\Psi}_{k;r}\,^3D^{\Psi}_{i,j,l;p,q,s} +\,^3D^{\Psi}_{ijk;\,p,q,r}\,^1D^{\Psi}_{l;s})$$
$$-\sum_{\mathcal{P}'}(-1)^{\mathcal{P}'}\,\mathcal{P}'\,2!\,2!\,(^2D^{\Psi}_{i,j;p,q}\,^2D^{\Psi}_{k,l;r,s}$$
$$+\,^2D^{\Psi}_{i,l;p,s}\,^2D^{\Psi}_{j,k;q,r} +\,^2D^{\Psi}_{i,k;p,r}\,^2D^{\Psi}_{j,l;q,s}) + 4!\,^4\Delta^{\Psi}_{i,j,k,l;p,q,r,s}$$

This *generalized particles–holes separating approach* generates an algorithm (GP-H) that emphasizes the role of the 2-RDM—the variable of the 2-CSE—and it is computationally more economical [54].

C. Other Approaches

Nakatsuji and Yasuda [56, 57] derived the 3- and 4-RDM expansions, in analogy with the Green function perturbation expansion. In their treatment the $^2\Delta$ error played the role of the perturbation term. The algorithm that they obtained for the 3-RDM was analogous to the *VCP* one, but the $^3\Delta$ matrix was decomposed into two terms: one where two $^2\Delta$ elements are coupled and a higher-order one. Neither of these two terms can be evaluated exactly; thus, in a sense, the difference with the *VCP* is just formal. However, the structure of the linked term suggested a procedure to approximate the $^3\Delta$ error, as will be seen later on.

In the 4-RDM case the Nakatsuji–Yasuda algorithm adds a new term to the *VCP* one (Eq. (67)). This new term is formed by an antisymmetrized product of two $^2\Delta$ elements. These authors' algorithm may thus be expressed as

$$4!\, ^4D^{\Psi}_{i,j,k,l;p,q,r,s} = \sum_{\mathcal{P}} (-1)^{\mathcal{P}}\, \mathcal{P}\, 3!\, (^1D^{\Psi}_{i;p}\, ^3D^{\Psi}_{j,k,l;q,r,s} + ^1D^{\Psi}_{j;q}\, ^3D^{\Psi}_{i,k,l;p,r,s}$$
$$+ ^3D^{\Psi}_{i,j,l;p,q,s}\, ^1D^{\Psi}_{k;r} + ^3D^{\Psi}_{i,j,k;p,q,r}\, ^1D^{\Psi}_{l;s})$$
$$+ 3\sum_{\mathcal{P}'} (-1)^{\mathcal{P}'}\, \mathcal{P}'\, (^1D^{\Psi}_{i;p}\, ^1D^{\Psi}_{j;q}\, ^1D^{\Psi}_{k;r}\, ^1D^{\Psi}_{l;s})$$
$$- \sum_{\mathcal{P}''} (-1)^{\mathcal{P}''}\, \mathcal{P}''\, 2!\, (^1D^{\Psi}_{i;p}\, ^1D^{\Psi}_{j;q}\, ^2D^{\Psi}_{k,l;r,s}$$
$$+ ^1D^{\Psi}_{i;p}\, ^2D^{\Psi}_{j,l;q,s}\, ^1D^{\Psi}_{k;r} + ^1D^{\Psi}_{i;p}\, ^2D^{\Psi}_{j,k;q,r}\, ^1D^{\Psi}_{l;s}$$
$$+ ^2D^{\Psi}_{i,l;p,s}\, ^1D^{\Psi}_{j;q}\, ^1D^{\Psi}_{k;r} + ^2D^{\Psi}_{i,k;p,r}\, ^1D^{\Psi}_{j;q}\, ^1D^{\Psi}_{l;s}$$
$$+ ^2D^{\Psi}_{i,j;p,q}\, ^1D^{\Psi}_{k;r}\, ^1D^{\Psi}_{l;s})$$
$$+ \sum_{\mathcal{P}'''} (-1)^{\mathcal{P}'''}\, \mathcal{P}'''\, 2!\, 2!\, (^2\Delta^{\Psi}_{i,j;p,q}\, ^2\Delta^{\Psi}_{k,l;r,s}$$
$$+ ^2\Delta^{\Psi}_{i,l;p,s}\, ^2\Delta^{\Psi}_{j,k;q,r} + ^2\Delta^{\Psi}_{i,k;p,r}\, ^2\Delta^{\Psi}_{j,l;q,s})$$
$$+ 4!\, ^4\Delta^{\Psi}_{i,j,k,l;p,q,r,s} \tag{70}$$

Following Martin and Schwinger [58], Mazziotti [59] derived in 1998 a generating function for constructing a p-RDM. By differentiating this functional with respect to the p Schwinger variables, and taking the limit, he obtained a Taylor series whose coefficients were the different RDMs. By analogy with Kubo cumulant expansion [60], Mazziotti identified the p-RDM with the p-order moment of this expansion; and he identified the connected part of his RDM expansion (what is called here the $^p\Delta$) with the corresponding cumulant. The algorithm thus obtained [45, 59, 61–63], which Mazziotti expressed using the Grassmann products notation [23], is the same as that of Nakatsuji and Yasuda. His developments therefore confirm the form of the previous algorithms. Moreover, Mazziotti's approach permits one to carry out the error analysis within the framework of the cumulant theory. Since all the terms of the Nakatsuji–Yasuda expansion for the 4-RDM also appear in Mazziotti's, this algorithm is denoted *NYM* in what follows.

It is interesting to note that GP-H (derived by Valdemoro et al. [54]) contains in an implicit form the correction terms proposed by Nakatsuji and Yasuda and by Mazziotti (except for a sign).

There are several other studies of cumulant expansions of the RDMs. Thus Kutzelnigg and Mukherjee also published in 1999 [64] an RDM expansion that is similar to Mazziotti's. An extended study of this cumulant approach was given by Ziesche [65]. Also, a particularly interesting analysis of the cumulant expansions was given by Harris [66], who proposed a systematic way for obtaining the different terms of the expansion.

An interesting generalization of the construction algorithms was proposed by Herbert and Harriman [67]. In this algorithm, each of the Grassmann products appearing were multiplied by a parameter. The set of these unknown parameters was evaluated by fitting the results of the algorithm with those of a CI calculation.

From the beginning it was clear that the different RDMs involved in the 2-CSE had to be consistent with each other. These matrices had also to be as closely positive semi-definite as possible and had to have a correct trace. The renormalization procedure used by Valdemoro et al. [54] renders the diagonal elements of the 4-RDM positive and makes sure that the trace is the adequate one. Then it contracts this diagonal to obtain the diagonals of the lower-order RDMs. When contracting the 4-RDM algorithm (Eq. (67)), one obtains the algorithm with which the off-diagonal elements of the 3-RDM are directly calculated, by entering as data the 1-, 2- and 3-RDMs previously obtained. In this way one can avoid keeping the 4-RDM in computer memory. Although this procedure has been subsequently refined, the basic idea has remained invariant and in what follows will be refered to as normalization procedure. Mazziotti's approach is conceptually similar, the main difference being that he seeks the consistency among the 3- and 4-RDMs in a different manner. He realized that by using a basis set of natural orbitals the expression obtained for the 3-RDM by contracting the 4-RDM no longer depends on more 3-RDM elements than the one being evaluated. In this way he gets an uncoupled system of equations that may be solved directly. Mazziotti's approach is laborious because a basis transformation of the reduced Hamiltonian matrix must be carried out at each iteration; but it gave excellent results [45, 62, 63]. In fact, the calculations carried out with the various approaches just seen, both for the ground and excited states of a series of atoms and molecules, have generally yielded good results [45, 68–70].

D. A Unifying Algorithm

The approach that will now be described expresses in a single algorithm the VCP, the GP-H, and the NYM expansions for the 4-RDM.

Alcoba had the idea to combine the two algorithms obtained by separating the particles and holes expressions—the GP-H algorithm (Eq. (69)) and the VCP one (Eq. (67))—and he obtained [48]

$$
\begin{aligned}
4!\,^4D^{\Psi}_{i,j,k,l;p,q,r,s} = & \sum_{\mathcal{P}}(-1)^{\mathcal{P}}\mathcal{P}\,3!\,(^1D^{\Psi}_{i;p}\,^3D^{\Psi}_{j,k,l;q,r,s} +\,^1D^{\Psi}_{j;q}\,^3D^{\Psi}_{i,k,l;p,r,s} \\
& +\,^1D^{\Psi}_{k;r}\,^3D^{\Psi}_{i,j,l;p,q,s} +\,^3D^{\Psi}_{i,j,k;p,q,r}\,^1D^{\Psi}_{l;s}) \\
& -\sum_{\mathcal{P}'}(-1)^{\mathcal{P}'}\mathcal{P}'\,2!\,2!\,(^2D^{\Psi}_{i,j;p,q}\,^2D^{\Psi}_{k,l;r,s} \\
& +\,^2D^{\Psi}_{i,l;p,s}\,^2D^{\Psi}_{j,k;q,r} +\,^2D^{\Psi}_{i,k;p,r}\,^2D^{\Psi}_{j,l;q,s}) \\
& +\xi\sum_{\mathcal{P}''}(-1)^{\mathcal{P}''}\mathcal{P}''\,2!\,2!\,(^2\Delta^{\Psi}_{i,j;p,q}\,^2\Delta^{\Psi}_{k,l;r,s} \\
& +\,^2\Delta^{\Psi}_{il;ps}\,^2\Delta^{\Psi}_{j,k;q,r} +\,^2\Delta^{\Psi}_{i,k;p,r}\,^2\Delta^{\Psi}_{j,l;q,s}) \\
& +4!\,^4\Delta^{\Psi}_{i,j,k,l;p,q,r,s}
\end{aligned}
$$
(71)

where ξ is an arbitrary parameter that reproduces the *VCP* algorithm for $\xi = 1$ and the *NYM* for $\xi = 2$. Also, for $\xi = 0$ it reproduces the GP-H [48, 54].

In general, this approach generates a parametric algorithm for a p-RDM with $p - 3$ associated arbitrary parameters leading to a whole family of approximations. In what follows this *unifying algorithm* is denoted by UA.

1. A Criterion for Selecting the Parameter Value for the 4-RDM

The structure of the parametric UA for the 4-RDM satisfies the fourth-order fermion relation (the expectation value of the commutator of four annihilator and four creator operators [26]) for any value of the parameter ξ, which is a basic and necessary N-representability condition. Also, the 4-RDM constructed in this way is symmetric for any value of ξ. On the other hand, the other N-representability conditions will be affected by this value. Hence it seems reasonable to optimize this parameter in such a way that at least one of these conditions is satisfied. Alcoba's working hypothesis [48] was the determination of the parameter value by imposing the trace condition to the 4-RDM. In order to test this working hypothesis, he constructed the 4-RDM for two states of the BeH_2 molecule in its linear form $D_{\infty h}$. The calculations were carried out with a minimal basis set formed by 14 Hartree–Fock spin orbitals belonging to three different symmetries. Thus orbitals 1, 2, and 3 are σ_g; orbitals 4 and 5 are σ_u; and orbitals 6 and 7 are degenerate π orbitals. The two states considered are the ground state, where

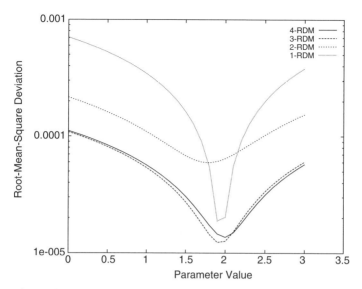

Figure 1. RMS deviation of the RDMs corresponding to the ground state of BeH$_2$ for different values of the UA parameter.

$|1\bar{1}2\bar{2}4\bar{4}\rangle$ is the dominant Slater determinant, and the singlet excited state, $|\Psi_{IV}\rangle$, which has a shared dominancy of the two Slater determinants $|1\bar{1}2\bar{2}4\bar{6}\rangle$ and $|1\bar{1}2\bar{2}6\bar{4}\rangle$. The 4-RDM was calculated for these states for different values of the parameter ξ; and then, by contraction of this matrix, the corresponding 3-, 2-, 1-RDMs were obtained. In Figs. 1 and 2 is represented the RMS deviation of these matrices with respect to the values obtained with the FCI method for each of these two states.

When calculating the trace error for different ξ values, one finds that this error vanishes for $\xi = 1.957$ in the ground state and for $\xi = 1.215$ in the excited state. These ξ values coincide with those corresponding to the minimal RMS deviations of the matrices. Therefore the selection criterion for ξ seems to be adequate.

E. Estimating the Error Matrix $^3\Delta$

A great deal of work has been dedicated by different authors [45, 56, 57, 62, 71–74] to estimate the error matrix $^3\Delta$, which was the unknown term in the construction algorithm for the 3-RDM. In fact, the experience of the different authors showed that it was crucial to have a good approximation of this matrix since the error matrix $^4\Delta$ is less determinative in order to achieve convergence in the iterative solution of the CSE.

The $^3\Delta$ matrix is Hermitian and antisymmetric with respect to the permutation of its indices. These properties significantly reduce the number of elements

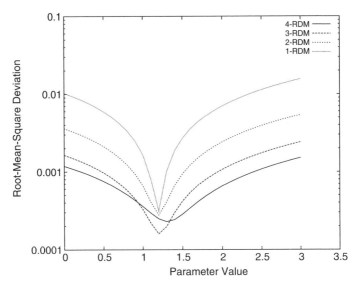

Figure 2. RMS deviation of the RDMs corresponding to the excited state of BeH$_2$ for different values of the UA parameter.

that must be calculated. This number is still further reduced in states with a singlet spin symmetry, since it has been found that in this case the blocks $^3\Delta^{\alpha\alpha\alpha}$ and $^3\Delta^{\beta\beta\beta}$ have negligible values. Moreover, for states with spin quantum number $M_s = 0$, the $^3\Delta^{\alpha\alpha\beta}$ block and the $^3\Delta^{\alpha\beta\beta}$ are equal. In consequence, the research was centered on obtaining the $^3\Delta^{\alpha\alpha\beta}$ block.

Nakatsuji and Yasuda [56, 57] focused on the term appearing in $^3\Delta$, which according to the perturbative expansion could be interpreted as a linked diagram of two $^2\Delta$ elements. In analogy to the Dyson equation, they proposed to estimate the $^3\Delta$ with a procedure whose main step, expressed in a spin-orbital basis, may be written

$$3! \, ^3\Delta^{\Psi}_{i,j,r;k,s,p} \approx \sum_{\mathcal{P}} (-1)^{\mathcal{P}} \, \mathcal{P} \sum_{l,t} 2! \, ^2\Delta^{\Psi}_{i,j;k,l} \, (^1D_{l;t} - \,^1\bar{D}_{l;t}) \, 2! \, ^2\Delta^{\Psi}_{t,r;s,p} \qquad (72)$$

where the 1-RDM and 1-HRDM appearing in this formula correspond to a Hartree–Fock calculation. In what follows this algorithm is denoted *NY*.

The results obtained with this algorithm were very good except when the indices $\{ijr\}$ corresponded to occupied orbitals in a Hartree–Fock state and $\{ksp\}$ to unoccupied ones (and vice versa); that is, when the 3-RDM element corresponded to the expectation value of a three-body elemental excitation.

In 1999 Valdemoro, Tel, and Pérez-Romero proposed a modification of the *NY* approximation, which gives slightly more accurate values for the $^3\Delta$ and which is computationally more economical [72, 73]. Moreover, the analysis of this new algorithm, which in what follows is denoted *VTP*, clarified the factors determining the $^3\Delta$ value.

Let us now see which were the observations leading to this *VTP* algorithm. When analyzing the values of the $^3\Delta$ elements obtained by decomposing the 3-RDM corresponding to a FCI calculation of the ground state of the molecule BeH$_2$, several significant features came out. Thus it was found that the error is not of the same order for all the matrix elements; and that, in fact, only some few elements of the $^3\Delta$ had nonnegligible values. Moreover, the elements having relevant values were those involving occupied and unoccupied frontier orbitals. This same observation was reported when approximating the 2-RDM [19] for the ground state of the beryllium atom. In the BeH$_2$ molecule the elements involving the highest occupied molecular orbital, *homo* (h), for each orbital symmetry, and the lowest unoccupied one, *lumo* (l), were those that showed significant $^3\Delta$ errors. Thus it was found that only the three following types of elements needed to be considered: $^3\Delta^\Psi_{h_1,l_1,\bar{h}_2;h_3,l_2,\bar{h}_4}$, $^3\Delta^\Psi_{l_1,h_1,\bar{l}_2;l_3,h_2,\bar{l}_4}$, and $^3\Delta^\Psi_{h_1,h_2,\bar{h}_3;l_1,l_2,\bar{l}_3}$. The state of reference used to define the *homo* and *lumo* orbitals is the Slater determinant, which dominates in Ψ.

The next step in this analysis was to find out whether the values obtained for these elements with the *NY* algorithm came out as the result of a sum of several factors or whether only a few terms of the sum appearing in Eq. (72) were contributing to each of the $^3\Delta$ elements value. It came out that only one of the sum terms contributed significantly; and, consequently, a new and simpler algorithm than the *NY* one could be devised. This new approximating procedure can be described as follows:

- Elements with two *homo* and one *lumo* orbitals in the column and row labels

$$3!\,^3\Delta^\Psi_{h_1,l_1,\bar{h}_2;l_2,h_3,\bar{h}_4} = -2!\,^2\Delta^\Psi_{h_1,\bar{h}_2;l_2,\bar{x}}\,2!\,^2\Delta^\Psi_{l_1,\bar{x};h_3,\bar{h}_4} \tag{73}$$

where x is the *lumo*. When, due to symmetry reasons, the product is null the index x that should be selected should be the next unoccupied frontier orbital.

- The $^3\Delta$ elements whose row and column labels have indices corresponding to two *lumo* and one *homo*

$$3!\,^3\Delta^\Psi_{l_1,h_1,\bar{l}_2;h_2,l_3,\bar{l}_4} = 2!\,^2\Delta^\Psi_{l_1,\bar{l}_2;h_2,\bar{y}}\,2!\,^2\Delta^\Psi_{h_1,\bar{y};l_3,\bar{l}_4} \tag{74}$$

where y denotes the *homo*. Again, when the product, due to symmetry reasons, is null, y should be the closest occupied orbital to the *homo*.

Due to the antisymmtry property of the $^3\Delta$ matrix,

$$^3\Delta^\Psi_{h_1,l_1,\bar{h}_2;h_3,l_2,\bar{h}_4} = -^3\Delta^\Psi_{h_1,l_1,\bar{h}_2;l_2,h_3,\bar{h}_4} \tag{75}$$

thus the other elements are obtained from those considered above.

As can be seen, the essence of these rules is to replace the nonnegligible $^3\Delta$ elements by a product of two $^2\Delta$ elements corresponding to a double excitation and to a double deexcitation, respectively. In the case examined, the other products considered by Nakatsuji and Yasuda do not contribute. However, a sum over x/y should be employed when the basis set is large and several orbitals have very close energy values to the *homo* and *lumo* of each orbital symmetry. In these cases the previous formulas are replaced by

$$3!\,^3\Delta^\Psi_{h_1,l_1,\bar{h}_2;l_2,h_3,\bar{h}_4} = -\sum_x 2!\,^2\Delta^\Psi_{h_1,\bar{h}_2;l_2,\bar{x}}\, 2!\,^2\Delta^\Psi_{l_1,\bar{x};h_3,\bar{h}_4} \tag{76}$$

and

$$3!\,^3\Delta^\Psi_{l_1,h_1,\bar{l}_2;h_2,l_3,\bar{l}_4} = \sum_y 2!\,^2\Delta^\Psi_{h_1,\bar{y};l_3,\bar{l}_4}\, 2!\,^2\Delta^\Psi_{l_1,\bar{l}_2;h_2,\bar{y}} \tag{77}$$

respectively, where x/y is the set of frontier spin orbitals and their neighbors' un-occupied/occupied ones.

1. Analytical Contraction of the Unifying Algorithm

As mentioned previously, the off-diagonal elements of the 3-RDM are determined by an algorithm obtained by contracting the 4-RDM. For simplicity sake, the expression given here for the contraction of the 4-RDM (Eq. (71)) corresponds to the spin block $^3D^{\Psi\,\sigma\sigma\sigma}$,

$$3!\,^3D^{\Psi\,\sigma\sigma\sigma}_{i,j,k;p,q,r} = (A_1 + A_2 + A_3 + A_4 + A_5 + A_6)/(N_\sigma - 3) \tag{78}$$

where

$$A_1 = -3 \sum_{\mathcal{P}'} (-1)^{\mathcal{P}'}\, \mathcal{P}'\, 2!\, \{^1D^{\Psi\,\sigma}_{i;p}\,^2D^{\Psi\,\sigma\sigma}_{j,k;q,r}$$
$$+\,^1D^{\Psi\,\sigma}_{j;q}\,^2D^{\Psi\,\sigma\sigma}_{i,k;p,r} +\,^1D^{\Psi\,\sigma}_{k;r}\,^2D^{\Psi\,\sigma\sigma}_{i,j;p,q}\} \tag{79}$$

$$A_2 = \xi\, 3 \sum_{\mathcal{P}'} (-1)^{\mathcal{P}'}\, \mathcal{P}'\, 2!\, \{((^1D^{\Psi\,\sigma})^2_{i;p} -\,^1D^{\Psi\,\sigma}_{i;p})\,^2\Delta^{\Psi\,\sigma\sigma}_{j,k;q,r}$$
$$+ ((^1D^{\Psi\,\sigma})^2_{j;q} -\,^1D^{\Psi\,\sigma}_{j;q})\,^2\Delta^{\Psi\,\sigma\sigma}_{i,k;p,r} \tag{80}$$
$$+ ((^1D^{\Psi\,\sigma})^2_{k;r} -\,^1D^{\Psi\,\sigma}_{k;r})\,^2\Delta^{\Psi\,\sigma\sigma}_{i,j;p,q}\}$$

$$A_3 = -2! \, 2! \, \{ {}^2D^{\Psi\,\sigma\sigma}_{i,j;p,s} {}^2D^{\Psi\,\sigma\sigma}_{k,s;q,r} - {}^2D^{\Psi\,\sigma\sigma}_{i,j;q,s} {}^2D^{\Psi\,\sigma\sigma}_{k,s;p,r}$$
$$- {}^2D^{\Psi\,\sigma\sigma}_{i,j;r,s} {}^2D^{\Psi\,\sigma\sigma}_{k,s;p,q} - {}^2D^{\Psi\,\sigma\sigma}_{i,k;p,s} {}^2D^{\Psi\,\sigma\sigma}_{j,s;q,r}$$
$$+ {}^2D^{\Psi\,\sigma\sigma}_{i,k;q,s} {}^2D^{\Psi\,\sigma\sigma}_{j,s;p,r} - {}^2D^{\Psi\,\sigma\sigma}_{i,k;r,s} {}^2D^{\Psi\,\sigma\sigma}_{j,s;p,q} \quad (81)$$
$$+ {}^2D^{\Psi\,\sigma\sigma}_{i,s;p,q} {}^2D^{\Psi\,\sigma\sigma}_{j,k;r,s} - {}^2D^{\Psi\,\sigma\sigma}_{i,s;p,r} {}^2D^{\Psi\,\sigma\sigma}_{j,k;q,s}$$
$$+ {}^2D^{\Psi\,\sigma\sigma}_{i,s;q,r} {}^2D^{\Psi\,\sigma\sigma}_{j,k;p,s} \}$$

$$A_4 = \xi \, 2! \, 2! \, \{ {}^2\Delta^{\Psi\,\sigma\sigma}_{i,j;p,s} {}^2\Delta^{\Psi\,\sigma\sigma}_{k,s;q,r} - {}^2\Delta^{\Psi\,\sigma\sigma}_{i,j;q,s} {}^2\Delta^{\Psi\,\sigma\sigma}_{k,s;p,r}$$
$$- {}^2\Delta^{\Psi\,\sigma\sigma}_{i,j;r,s} {}^2\Delta^{\Psi\,\sigma\sigma}_{k,s;p,q} - {}^2\Delta^{\Psi\,\sigma\sigma}_{i,k;p,s} {}^2\Delta^{\Psi\,\sigma\sigma}_{j,s;q,r}$$
$$+ {}^2\Delta^{\Psi\,\sigma\sigma}_{i,k;q,s} {}^2\Delta^{\Psi\,\sigma\sigma}_{j,s;p,r} - {}^2\Delta^{\Psi\,\sigma\sigma}_{i,k;r,s} {}^2\Delta^{\Psi\,\sigma\sigma}_{j,s;p,q} \quad (82)$$
$$+ {}^2\Delta^{\Psi\,\sigma\sigma}_{i,s;p,q} {}^2\Delta^{\Psi\,\sigma\sigma}_{j,k;r,s} - {}^2\Delta^{\Psi\,\sigma\sigma}_{i,s;p,r} {}^2\Delta^{\Psi\,\sigma\sigma}_{j,k;q,s}$$
$$+ {}^2\Delta^{\Psi\,\sigma\sigma}_{i,s;q,r} {}^2\Delta^{\Psi\,\sigma\sigma}_{j,k;p,s} \}$$

$$A_5 = -3! \, \{ {}^3D^{\Psi\,\sigma\sigma\sigma}_{i,j,k;p,q,s} {}^1D^{\Psi\,\sigma}_{s;r} + {}^3D^{\Psi\,\sigma\sigma\sigma}_{i,j,s;p,q,r} {}^1D^{\Psi\,\sigma}_{k;s}$$
$$- {}^3D^{\Psi\,\sigma\sigma\sigma}_{i,j,k;p,r,s} {}^1D^{\Psi\,\sigma}_{s;q} - {}^3D^{\Psi\,\sigma\sigma\sigma}_{i,k,s;p,q,r} {}^1D^{\Psi\,\sigma}_{j;s} \quad (83)$$
$$+ {}^3D^{\Psi\,\sigma\sigma\sigma}_{i,j,k;q,r,s} {}^1D^{\Psi\,\sigma}_{s;p} + {}^3D^{\Psi\,\sigma\sigma\sigma}_{j,k,s;p,q,r} {}^1D^{\Psi\,\sigma}_{i;s} \}$$

$$A_6 = N_\sigma \, 3! \, {}^3D^{\Psi\,\sigma\sigma\sigma}_{i,j,k;p,q,r} \quad (84)$$

and N_σ is the number of electrons with spin σ. An implicit sum over repeated indices is assumed.

Following Mazziotti's reasonings [45,62,63], Alcoba has considered Eq. (78) as a system of equations where the variables are the 3-RDM elements, which can be decoupled when the off-diagonal 1-RDM elements are null, that is, when the matrices are represented in a natural basis of orbitals.

The *NYM* algorithm is therefore a particular case for $\xi = 2$ of Alcoba's parametric expression. It must be noted, that as happens with the *NY* and the *VTP*, the UA does not correct the error in the elements whose indices correspond to a three-body elemental excitation.

The values, obtained with the different methods just decribed, of the more significant elements of ${}^3\Delta$ are given in Table I. The test system has been the BeH$_2$ molecule in its linear form D$_{\infty h}$. The state Ψ considered is the ground state, and the basis set used is the same as previously. As can be seen, the results have been very satisfactory in all the approaches. However, a detailed analysis of these values indicates that the algorithm giving the best approximation for ${}^3\Delta$ is the *VTP* followed by Alcoba's UA for the value $\xi = 1.957$. Hence the 3-RDM obtained with Eq. (66) together with the *VTP* estimations for ${}^3\Delta$ yield very good results for those states whose orbitals have occupations close to 0 or 1, except for those few elements involving three-body frontier excitations.

TABLE I
Most Significant Elements of the Matrix $^3\Delta^{\alpha\alpha\beta}$

Matrix Element	$^3\Delta$	Algorithm				UA ($\xi = 1.957$)
		NY	VTP		Mazziotti	
			Value	x or y		
233; 233	−0.00300	−0.00284	−0.00309	2	−0.00333	−0.00316
253; 343	0.00255	0.00242	0.00258	2	0.00269	0.00255
266; 266	−0.00317	−0.00304	−0.00310	2	−0.00308	−0.00293
266; 277	−0.00320	−0.00306	−0.00310	2	−0.00310	−0.00295
233; 453	−0.00340	−0.00327	−0.00344	4	−0.00349	−0.00332
232; 232	0.00300	0.00283	0.00309	3	0.00334	0.00318
344; 344	0.00670	0.00630	0.00675	3	0.00700	0.00665
234; 454	0.00341	0.00328	0.00344	3	0.00359	0.00341
232; 344	−0.00455	−0.00437	−0.00457	3	−0.00486	−0.00462
342; 454	−0.00384	−0.00364	−0.00382	3	−0.00392	−0.00373
254; 452	0.00194	0.00185	0.00194	3	0.00199	0.00190
234; 234	0.00217	0.00194	0.00216	5	0.00223	0.00211
262; 262	0.00306	0.00295	0.00310	6	0.00348	0.00330

F. The N-Representability Problem: Introducing Bounds Correction

Until now the focus has been on the construction algorithms for the 3- and 4-RDMs and the estimation of the Δ errors. However, the question of how to impose that the RDMs involved as well as the high-order G-matrices be positive must not be overlooked. This condition is not easy to impose in a rigorous way for such large matrices. The renormalization procedure of Valdemoro et al. [54], which was computationally economical but only approximate, acted only on the diagonal elements.

For the 3- and 4-G matrices, the approach can be to apply adequate bounds to these matrices' diagonal elements. This question was studied thoroughly by several authors [26, 29, 54, 75–77] and there are a large number of inequalities derived from the D, Q, and GN-representability conditions. In practice, one must select a small number of these inequalities with the criterion that they should be as restrictive as possible. In the calculations of Valdemoro et al. [54] the selected conditions for the fourth-order G-matrix lead to the three following inequalities:

1. A bound for the 4-RDM diagonal elements involving off-diagonal 3-RDM elements which presents nonlinear terms:

$$4!\,^4D^\Psi_{k,l,i,j;k,l,i,j} \geq -(^1D^\Psi_{i;j})^2\,^1D^\Psi_{k;k} - 2!\,^2D^\Psi_{k,i;k,i}$$
$$+ 2!\,^2D^\Psi_{k,l;k,l}\,(^1D^\Psi_{i;j})^2 + 3!\,^3D^\Psi_{k,l,i;k,l,i}$$
$$+ 3!\,^3D^\Psi_{k,i,j;k,i,j} + 2\cdot 2!\,^2D^\Psi_{i,k;j,k}\,^1D^\Psi_{i;j} \qquad (85)$$
$$+ (2!\,^2D^\Psi_{j,l;k,i} +\,^1D^\Psi_{i;j}\,^1D^\Psi_{k;l})^2$$
$$- 3!\,2\,^3D^\Psi_{i,k,l;j,k,l}\,^1D^\Psi_{i;j}$$

2. A condition for the 4-RDM diagonal elements involving the square of an off-diagonal element of the 2-RDM:

$$4!\,^4D^\Psi_{k,l,i,j;k,l,i,j} \geq -2!\,^2D^\Psi_{k,l;k,l} + 3!\,^3D^\Psi_{k,l,j;k,l,j} + 3!\,^3D^\Psi_{k,l,i;k,l,i} + (2!\,^2D^\Psi_{k,l;i,j})^2 \tag{86}$$

3. An upper bound for the 4-RDM diagonal elements involving a 3-RDM off-diagonal element:

$$4!\,^4D^\Psi_{k,l,i,j;k,l,i,j} \leq 2!\,^2D^\Psi_{k,l;k,l}\,(^1D^\Psi_{i;j})^2 + 3!\,^3D^\Psi_{k,l,i;k,l,i} - 2\cdot 3!\,^3D^\Psi_{i,k,l;j,k,l}\,^1D^\Psi_{i;j} \tag{87}$$

Other families of inequalities were looked into [26, 73] but they were not used in the calculations reported here.

G. Some Comparative Results

The results obtained with the different approaches described in this section are shown in Tables II–V. As before, the test probe is the linear BeH$_2$ molecule.

The 4-RDMs for the ground state and first excited singlet state were calculated with the UA (Eq. (71)) for different values of the ξ parameter. The first three columns correspond, respectively, to the GP-H, the VCP, and to the NYM algorithms. The different RDM elements were analyzed in order to find out if they satisfied the bounds inequalities. If this were not the case, these elements' values would be given the values corresponding to the upper or lower bound which had been unsatisfied. In order to examine the performance of the different algorithms and of the bounds without the influence of the quality of the data used, the 1-, 2- and 3-RDMs employed to calculate the 4-RDM were obtained in an FCI treatment. The different results were then compared with the FCI

TABLE II
Ground-State: Error of the $^4D^{\Psi\,\alpha\alpha\beta}_{i,j,k,l;i,j,k,l}$ Elements with the UA and Imposing N-Representability Bounds

Indices				Algorithm Error			Algorithm Error + Bounds (B)		
i	j	k	l	$\xi=0$	$\xi=1$	$\xi=2$	$\xi=0+B$	$\xi=1+B$	$\xi=2+B$
2	3	4	2	−0.0001847	−0.0000743	0.0000012	−0.0000001	−0.0000001	0.0000012
2	3	4	4	−0.0001985	−0.0001171	−0.0000357	0.0000000	0.0000000	0.0000000
2	4	5	2	−0.0001153	−0.0000675	−0.0000198	−0.0000006	−0.0000006	−0.0000006
2	4	6	4	−0.0001123	−0.0000851	−0.0000579	0.0000000	0.0000000	0.0000000
3	4	6	6	−0.0001114	−0.0000829	−0.0000543	0.0000000	0.0000000	0.0000000

TABLE III
Ground State: Error of the $^4D^{\Psi\,\alpha\alpha\beta\beta}_{i,j,k,l;i,j,k,l}$ Elements with the UA and Imposing N-Representability Bounds

Indices				Algorithm Error			Algorithm Error + Bounds (B)		
i	j	k	l	$\zeta=0$	$\zeta=1$	$\zeta=2$	$\zeta=0+\text{B}$	$\zeta=1+\text{B}$	$\zeta=2+\text{B}$
2	3	2	4	−0.0001827	−0.0000707	0.0000015	−0.0000222	−0.0000222	0.0000001
2	3	3	4	0.0000012	−0.0000495	−0.0001801	0.0000001	−0.0000495	−0.0001801
2	3	4	5	0.0017676	0.0017676	−0.0000661	0.0000991	0.0000991	0.0000000
2	4	2	4	0.0002840	0.0001361	−0.0000117	0.0000219	0.0000219	−0.0000117
2	4	4	6	−0.0001123	−0.0000851	−0.0000579	0.0000000	0.0000000	0.0000000
2	5	2	5	0.0013544	0.0007780	−0.0000123	0.0000571	0.0000571	−0.0000032
2	5	3	4	0.0021823	0.0017539	−0.0000488	0.0000897	0.0000897	−0.0000029
3	4	3	4	0.0031281	0.0031281	−0.0004268	0.0001789	0.0001789	−0.0000063
3	6	4	6	−0.0001127	−0.0000829	−0.0000530	0.0000000	0.0000000	0.0000000
4	5	4	5	0.0007380	0.0007380	−0.0000207	0.0000733	0.0000733	0.0000000
4	6	4	6	0.0001660	0.0001381	0.0001102	0.0000000	0.0000000	0.0000000

4-RDM. The results for each spin block of the 4-RDM are presented in Tables II–V. It should be mentioned that, before carrying the comparison with the FCI 4-RDM, the renormalization procedure previously mentioned was applied.

The results of the three last columns correspond to calculations with the same algorithm after application of the bound conditions given by the inequalities (85), (87), and (86), denoted as B in the Tables II–V.

As can be seen, while for the ground state the *NYM* algorithm corresponding to the parameter value $\zeta = 2$ gives the most satisfactory results, this is not the case for the first singlet excited state, where the errors in some of the 4-RDM elements are nonnegligible. This applies also to the algorithm given by Eq. (69), which is obtained with the parameter value $\zeta = 0$. However, it should

TABLE IV
Excited State: Some Significant Errors of the $^4D^{\Psi\,\alpha\alpha\alpha\beta}_{i,j,k,l;i,j,k,l}$ Elements with the UA and Imposing N-Representability Bounds

Indices				Algorithm Error			Algorithm Error + Bounds (B)		
i	j	k	l	$\zeta=0$	$\zeta=1$	$\zeta=2$	$\zeta=0+\text{B}$	$\zeta=1+\text{B}$	$\zeta=2+\text{B}$
2	3	4	6	−0.0151279	−0.0052737	0.0000007	0.0000000	0.0000000	0.0000002
2	3	6	4	−0.0145106	−0.0048784	0.0000030	0.0000000	0.0000000	0.0000005
2	3	6	6	0.0000000	0.0000000	−0.0053446	0.0000000	0.0000000	0.0000000
2	4	6	2	−0.0063085	−0.0033265	−0.0003444	−0.0000002	−0.0000002	−0.0000002
2	4	6	3	0.0000015	−0.0019269	−0.0080394	0.0000015	−0.0000005	−0.0000005
2	4	6	4	−0.0011146	0.0000005	0.0000005	−0.0000015	0.0000005	0.0000005
3	4	6	2	0.0000001	−0.0021817	−0.0082845	0.0000001	0.0000000	0.0000000
3	4	6	3	−0.0047042	−0.0026740	−0.0006437	0.0000000	0.0000000	0.0000000

TABLE V
Excited State: Some Significant Errors of the $^4D^{\Psi\,\alpha\alpha\beta\beta}_{i,j,k,l;i,j,k,l}$ Elements with the UA and Imposing N-Representability Bounds

Indices				Algorithm Error			Algorithm Error + Bounds (B)		
i	j	k	l	$\xi=0$	$\xi=1$	$\xi=2$	$\xi=0+\text{B}$	$\xi=1+\text{B}$	$\xi=2+\text{B}$
2	3	2	3	0.0000012	0.0000012	−0.0018960	0.0000012	0.0000012	−0.0000002
2	3	4	6	0.0145106	0.0048784	−0.0047539	0.0000000	0.0000000	−0.0000005
2	4	2	4	−0.0041440	−0.0019392	0.0000007	−0.0000060	−0.0000060	0.0000007
2	4	2	6	0.0063085	0.0033265	0.0003444	0.0000002	0.0000002	0.0000002
2	4	3	6	−0.0041857	0.0019269	0.0033128	−0.0000015	0.0000005	0.0000005
2	4	4	6	0.0000061	−0.0017806	−0.0046758	0.0000000	−0.0000005	−0.0000005
2	6	2	4	0.0063085	0.0033265	0.0003444	0.0000002	0.0000002	0.0000002
2	6	2	6	−0.0057041	−0.0030819	−0.0004596	0.0000000	0.0000000	0.0000000
2	6	3	4	−0.0039210	0.0021817	0.0064365	−0.0000001	0.0000000	0.0000000
2	6	4	6	−0.0014689	−0.0027143	−0.0039598	0.0000000	0.0000000	0.0000000

be noted that although the original *VCP* algorithm obtained with the parameter value $\xi = 1$ is not good either, its errors for the excited state are lower than those obtained with the other algorithms.

When considering the effect due to imposing bounds, the results for the 4-RDM become excellent independently of the parameter value selected. Moreover, these good results occur both in the ground and in the excited state.

Let us now see what the situation is for the 3-RDM. As has been described, the 3-RDM is obtained by contraction from the 4-RDM. On the other hand, in order to construct the 4-RDM one needs the 3-RDM. According to the results reported in Table I, the estimation of the $^3\Delta$ for the ground state is satisfactory. Also, the UA for $\xi = 0$ together with the bounds inequalities

$$3!\,^3D^{\Psi}_{i,p,j;i,p,j} \leq 2!\,^2D^{\Psi}_{i,p;i,p} - 2\,2!\,^2D^{\Psi}_{i,p,j,p}\,^1D^{\Psi}_{i;j} + (^1D^{\Psi}_{i;j})^2\,^1D^{\Psi}_{p;p} \qquad (88)$$

and

$$3!\,^3D^{\Psi}_{i,k,l;i,k,l}\,(2\,^1D^{\Psi}_{i;i} - 1) \geq (^1D^{\Psi}_{i;i})^2\,(\,2!\,^2D^{\Psi}_{k,l;k,l} -\,^1D^{\Psi}_{k;k}) \\ + 2!\,^2D^{\Psi}_{i,k;i,k}\,(2\,^1D^{\Psi}_{i;i} - 1) \\ + (\,2!\,^2D^{\Psi}_{i,l;i,k} -\,^1D^{\Psi}_{i;i}\,^1D^{\Psi}_{l;k})^2 \qquad (89)$$

give very good results for the ground state [48]. However, the results obtained for the first singlet excited state of the BeH$_2$ molecule were not good [48] since imposing bounds did not sufficiently improve the situation. From the study carried out by Alcoba, the situation with respect to the 3-RDM may be summarized as follows:

1. For those states having orbital occupations close to 0 or 1, the UA with $\xi = 0$ together with the bounds constitutes an efficient method for

constructing the 3-RDM. Alternatively, the UA with $\xi = 1$ together with the $^3\Delta$ corrections also produces good results, except for the off-diagonal elements describing three-frontier electron excitations.

2. For those states having orbital occupations clearly different from 0 or 1— that is, when the orbital occupations play simultaneously the role of particles and holes—the problem of obtaining a good 3-RDM still remains open. Consequently, the 4-RDM is also affected, notwithstanding the fact that its algorithm is efficient. However, in this case the bounds correct very well the errors in the diagonal elements.

In spite of these satisfactory results, it should be noted that being able to obtain very good results for the most significant elements when using FCI input data does not guarantee obtaining good results throughout the iterative process. Thus, in practice, when the input data for obtaining the bounds were not the FCI ones the results of the iterative process were not as good as when the bounds were not imposed. Another reason for these disapointing applicative results may be the appearance of inconsistencies among the different spin blocks of the RDMs. At present, when rather satisfying purification procedures have been devised for the 2-RDM, this analysis should be repeated, since the input data have been improved. In this way, one may either confirm or discard one of the hypothetical reasons for the bad performance in the practice of the application of bounds.

V. FACTORS AFFECTING THE CONVERGENCE OF THE ITERATIVE SOLUTION OF THE CSE

A. Influence of the Algorithms on the 2-CSE Convergence

As has been shown, the UA with $\xi = 2$ for the 4-RDM—which corresponds to the inclusion of the correction term introduced by Nakatsuji and Yasuda and by Mazziotti—gives rather accurate results. The importance that the value of this parameter has on the convergence of the iterative process is now examined.

As mentioned, the optimal value of the parameter is $\xi = 1.957$, but for simplicity's sake, the curves shown in Fig. 3 correspond to the integer values $\xi = 0$, $\xi = 1$, and $\xi = 2$.

As may be observed, the curve corresponding to the value $\xi = 2$, which is very close to the optimized parameter value, reaches a minimal value of the energy very close to the FCI one.

Besides the influence of the parameter ξ on the energy, it is important to analyze how the different values of the parameter ξ affect the 1-RDM, since this matrix is directly connected with the electronic density. Thus the elements of the 1-RDM for each value of ξ obtained at the minimum of each curve are shown

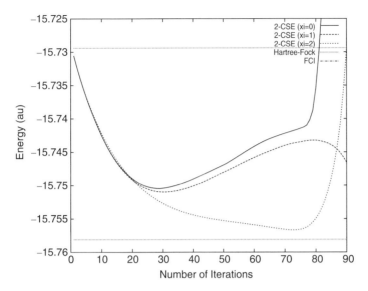

Figure 3. Values of the energy obtained for each iteration in the self-consistent solution of the 2-CSE for different values of ξ in the ground state of the BeH$_2$ molecule.

in Table VI. Since the matrix is symmetric, only the superior half of the matrix is given. Also, since the ground state is a singlet, only the α spin block is reported.

These results show that all three values of ξ give good results but it is the value $\xi = 2$ that performs best.

B. Influence of the N- and S-Representability of the 2-RDM on the Convergence of the 2-CSE Iterative Process

Initially, in order to partially correct the N-representability defects of the 2-RDM obtained at each iteration, this matrix was diagonalized and rendered positive by applying to the eigenvalues the same operations described previously, when correcting the 4-RDM diagonal elements. When carrying out these operations, it was observed that the appearance of the divergence was retarded but in many cases it was not prevented. Mazziotti [78] realized that further purifying the 2-RDM substantially improved the convergence of the 2-CSE iterative process. He applied a 2-RDM purification procedure based on Coleman's unitarily invariant decomposition of a second-order matrix. He then imposed the D and Q N-representability conditions and obtained very satisfactory results. By focusing not only on imposing the D- and Q-conditions but also the G-condition as well as the S-representability conditions, Alcoba et al. [38, 51] proposed two very complete procedures that permitted one to obtain smoothly convergent processes, leading to highly accurate solutions. The theoretical developments leading to

TABLE VI
Ground-state 1-RDM Elements

1.00	0.00	0.00	0.00	0.00	0.00	0.00
9.99×10^{-1}	-1.74×10^{-4}	1.28×10^{-4}	0.00	0.00	0.00	0.00
9.99×10^{-1}	-1.24×10^{-4}	1.31×10^{-4}	0.00	0.00	0.00	0.00
9.99×10^{-1}	-1.30×10^{-4}	1.99×10^{-4}	0.00	0.00	0.00	0.00
9.99×10^{-1}	-2.05×10^{-4}	1.99×10^{-4}	0.00	0.00	0.00	0.00
	1.00	0.00	0.00	0.00	0.00	0.00
	9.95×10^{-1}	7.73×10^{-4}	0.00	0.00	0.00	0.00
	9.95×10^{-1}	8.97×10^{-4}	0.00	0.00	0.00	0.00
	9.89×10^{-1}	4.01×10^{-3}	0.00	0.00	0.00	0.00
	9.85×10^{-1}	1.27×10^{-2}	0.00	0.00	0.00	0.00
		0.00	0.00	0.00	0.00	0.00
		4.98×10^{-3}	0.00	0.00	0.00	0.00
		4.96×10^{-3}	0.00	0.00	0.00	0.00
		1.09×10^{-2}	0.00	0.00	0.00	0.00
		1.43×10^{-2}	0.00	0.00	0.00	0.00
			1.00	0.00	0.00	0.00
			9.94×10^{-1}	-6.50×10^{-4}	0.00	0.00
			9.94×10^{-1}	-5.71×10^{-4}	0.00	0.00
			9.90×10^{-1}	-3.17×10^{-4}	0.00	0.00
			9.87×10^{-1}	-5.57×10^{-3}	0.00	0.00
Hartree–Fock				0.00	0.00	0.00
2-CSE ($\xi = 0$)				3.16×10^{-3}	0.00	0.00
2-CSE ($\xi = 1$)				3.09×10^{-3}	0.00	0.00
2-CSE ($\xi = 2$)				5.66×10^{-3}	0.00	0.00
FCI				6.20×10^{-3}	0.00	0.00
					0.00	0.00
					1.29×10^{-3}	0.00
					1.12×10^{-3}	0.00
					2.32×10^{-3}	0.00
					3.30×10^{-3}	0.00
						0.00
						1.29×10^{-3}
						1.12×10^{-3}
						2.32×10^{-3}
						3.30×10^{-3}

the purification procedures are not trivial and constitute in themselves a line of study. This is the reason why a whole chapter of this book is dedicated to this very important area, which is not further considered here. Another factor that strongly enhances the convergence of the iterative process is the shifting of the origin of energy. This matter will also be examined in the same chapter as the 2-RDM purification procedures.

VI. AN EXACT FORMAL SOLUTION TO THE CONTRACTED SCHRÖDINGER EQUATION'S INDETERMINACY

In previous sections the construction algorithms aimed to obtain good approximations of the high-order RDMs appearing in the 2-CSE in terms of the lower-order ones. Here, the approach is to obtain a set of equations, equivalent to the CSEs, which are not formally indeterminate. The most important equation of this family is a self-contained equation represented in the four-electron space. The cost of removing the indeterminacy of the CSEs is an increase in the size of the problem. The formal aspects of this treatment are very enlightening and, although at present the operativity of this approach is limited, the results obtained for a family of four electron compounds are excellent. These new equations, which are called *modified contracted Schrödinger equations* (MCSEs) are combinations of lower-order CSEs [79, 80]. They involve explicitly high-order correlation matrices (CMs), which, as will be seen, play an important theoretical role.

A. Decomposition of High-Order RDMs and a Basic Cancellation Relation

We have seen that the second-order CM is one of the terms resulting from the decomposition of the 2-RDM. Similarly, when decomposing the 3- and 4-RDMs the third- and fourth-order CMs are obtained. Thus, for instance, one of the possible decompositions of the 3-RDM is

$$3!\,^3D^\Psi_{i,k,m;j,l,n} = -\,2!\,^2D^\Psi_{i,k;j,n}\,\delta_{m,l} + 2!\,^2D^\Psi_{i,k;l,n}\,\delta_{j,m}$$
$$+ 2!\,^2D^\Psi_{i,k;j,l}\,^1D^\Psi_{m;n} + {}^{(3;2,1)}\mathcal{C}^\Psi_{i,k,m;j,l,n} \qquad (90)$$

where the elements of the third-order CM, $^{(3;2,1)}\mathcal{C}$, have the following structure:

$$^{(3;2,1)}\mathcal{C}^\Psi_{i,k,m;j,l,n} = 2!\sum_{\Psi'\neq\Psi} {}^2D^{\Psi\Psi'}_{i,k;j,l}\,{}^1D^{\Psi'\Psi}_{m;n} \qquad (91)$$

In order to refer to the *p*-order correlation matrices with $p > 2$, the notation needs to become more precise. Thus a left superscript has to be introduced. As can be seen from the definition, Eq. (91), the first index of this left superscript denotes the global order of the matrix, and the following indices denote the order of the different TRDMs involved. The right subscripts, denoting the element labels, coincide with those of the 2-RDM from which they derive.

At the basis of the theoretical developments leading to the MCSEs lie a set of relations reported by Tel et al. [81], which link CMs of different orders among themselves. These C-relations establish a set of necessary conditions that the CMs must satisfy when they correspond to a Hamiltonian eigenstate. The most important of these relations, which is essential in the derivation of the MCSEs, can be expressed as

$$^{(p;2,x,y,\ldots)}\mathcal{O}^{\Psi}_{v_1,\ldots,v_x,t_1,\ldots,t_y,\ldots;w_1,\ldots,w_x,z_1,\ldots,z_y,\ldots}$$
$$\equiv \frac{1}{2}\sum_{i,j,k,l} K_{i,j;k,l} \,^{(p;2,x,y,\ldots)}\mathcal{C}^{\Psi}_{i,j,v_1,\ldots,v_x,t_1,\ldots,t_y,\ldots;k,l,w_1,\ldots,w_x,z_1,\ldots,z_y,\ldots} = 0 \qquad (92)$$

where $p = 2 + x + y + \cdots$. Note that the labeling of the symbol $^{(p;2,x,y,\ldots)}\mathcal{O}$ coincides with that of the CMs involved in the sum.

Alcoba [80] reported four theorems showing that the cancellation of these types of terms is a sufficient condition to guarantee that these matrices correspond to eigenstates of the system. In particular, his first theorem states the following.

Theorem [Alcoba]   Assuming that the four-electron CM, $^{(4;2,2)}\mathcal{C}^{\Psi}$, can be derived by decomposition of an N-representable 4-RDM, then

$$^{(4;2,2)}\mathcal{O}^{\Psi} = 0 \qquad (93)$$

will be satisfied by this 4-CM *if and only if* the N-DM, preimage of the 4-RDM, satisfies the Schrödinger equation.

The first part of the demonstration of this theorem is contained in Eq. (92). The second part of the demonstration is as follows. Using the *unit operator* it is easy to see that

$$0 = {}^{(4;2,2)}\mathcal{O}^{\Psi} = E_{\psi}\, 2!\, 2\underline{D}^2 - 2!\,^2\mathcal{M}^{\Psi} \qquad (94)$$

This equation is equivalent to the 2-CSE, which by Nakatsuji's theorem implies that the Schrödinger equation is satisfied.

B. Derivation of the Fourth-Order MCSE

As will now be shown, the 4-CSE can be transformed into a self-contained equation, the 4-MCSE. Let us therefore consider the 4-CSE in the following compact form:

$$E_{\Psi}\, 4!\,^4 D^{\Psi}_{i,j,k,l;p,q,r,s} = \langle \Psi | \hat{H}\, a_i^{\dagger}\, a_j^{\dagger}\, a_k^{\dagger}\, a_l^{\dagger}\, a_s\, a_r\, a_q\, a_p | \Psi \rangle$$
$$\equiv 4!\,^4\mathcal{M}^{\Psi}_{i,j,k,l;p,q,r,s} \qquad (95)$$

Remembering that

$$^2\mathcal{M} = \langle \Psi | \hat{H}\, a_i^{\dagger}\, a_j^{\dagger}\, a_s\, a_r | \Psi \rangle \qquad (96)$$

one proceeds to change the order of the fermion operators in expression (95) having in mind that one wishes to have a term with the ordering $a_i^{\dagger}\, a_j^{\dagger}\, a_q\, a_p$

$a_k^\dagger a_l^\dagger a_s a_r$. Then, inserting the unit operator at the convenient places, as in the RDMs' decompositions (after the $\hat{H}$ and between an annihilator and a creator operator), one obtains

$$\begin{aligned}
E_\Psi\, 4!\, {}^4D^\Psi_{i,j,k,l;p,q,r,s} &= (\delta_{q,l}\,\delta_{k,p} - \delta_{p,l}\,\delta_{k,q})\, 2!\, {}^2\mathcal{M}^\Psi_{i,j;r,s} \\
&+ (\delta_{q,l}\,\delta_{k,s} - \delta_{k,q}\, {}^1D^\Psi_{l;s})\, 2!\, {}^2\mathcal{M}^\Psi_{i,j;p,r} \\
&+ (\delta_{l,r}\,\delta_{k,q} - \delta_{q,l}\, {}^1D^\Psi_{k;r})\, 2!\, {}^2\mathcal{M}^\Psi_{i,j;p,s} \\
&- (\delta_{l,p}\,\delta_{k,s} - \delta_{k,p}\, {}^1D^\Psi_{l;s})\, 2!\, {}^2\mathcal{M}^\Psi_{i,j;q,r} \\
&- (\delta_{k,p}\,\delta_{l,r} - \delta_{l,p}\, {}^1D^\Psi_{k;r})\, 2!\, {}^2\mathcal{M}^\Psi_{i,j;q,s} \\
&+ 2!\, {}^2\mathcal{M}^\Psi_{i,j;p,q}\, 2!\, {}^2D^\Psi_{k,l;r,s} \\
&- \delta_{q,l}\, (E_\Psi\, {}^{(3;2,1)}\mathcal{C}^\Psi_{i,j,k;p,s,r} + {}^{(5;2,2,1)}\mathcal{O}^\Psi_{i,j,k;p,s,r}) \\
&- \delta_{k,q}\, (E_\Psi\, {}^{(3;2,1)}\mathcal{C}^\Psi_{i,j,l;p,r,s} + {}^{(5;2,2,1)}\mathcal{O}^\Psi_{i,j,l;p,r,s}) \\
&+ \delta_{l,p}\, (E_\Psi\, {}^{(3;2,1)}\mathcal{C}^\Psi_{i,j,k;q,s,r} + {}^{(5;2,2,1)}\mathcal{O}^\Psi_{i,j,k;q,s,r}) \\
&+ \delta_{k,p}\, (E_\Psi\, {}^{(3;2,1)}\mathcal{C}^\Psi_{i,j,l;q,r,s} + {}^{(5;2,2,1)}\mathcal{O}^\Psi_{i,j,l;q,r,s}) \\
&+ E_\Psi\, {}^{(4;2,2)}\mathcal{C}^\Psi_{i,j,k,l;p,q,r,s} + {}^{(6;2,2,2)}\mathcal{O}^\Psi_{i,j,k,l;p,q,r,s} \\
&= 4!\, {}^4\mathcal{M}^\Psi_{i,j,k,l;p,q,r,s}
\end{aligned} \tag{97}$$

The important feature of this hierarchy equation is that the dependence of the 4-CSE upon the 5-CSE and 6-CSE has been replaced by the fifth- and sixth-order $\mathcal{O}$ cancellation terms, Eq. (92). These high-order terms can therefore be omitted and it follows that

$$\begin{aligned}
E_\Psi\, 4!\, {}^4D^\Psi_{i,j,k,l;p,q,r,s} &= (\delta_{q,l}\,\delta_{k,p} - \delta_{p,l}\,\delta_{k,q})\, 2!\, {}^2\mathcal{M}^\Psi_{i,j;r,s} \\
&+ (\delta_{q,l}\,\delta_{k,s} - \delta_{k,q}\, {}^1D^\Psi_{l;s})\, 2!\, {}^2\mathcal{M}^\Psi_{i,j;p,r} \\
&+ (\delta_{l,r}\,\delta_{k,q} - \delta_{q,l}\, {}^1D^\Psi_{k;r})\, 2!\, {}^2\mathcal{M}^\Psi_{i,j;p,s} \\
&- (\delta_{l,p}\,\delta_{k,s} - \delta_{k,p}\, {}^1D^\Psi_{l;s})\, 2!\, {}^2\mathcal{M}^\Psi_{i,j;q,r} \\
&- (\delta_{k,p}\,\delta_{l,r} - \delta_{l,p}\, {}^1D^\Psi_{k;r})\, 2!\, {}^2\mathcal{M}^\Psi_{i,j;q,s} \\
&+ 2!\, {}^2\mathcal{M}^\Psi_{i,j;p,q}\, 2!\, {}^2D^\Psi_{k,l;r,s} \\
&- \delta_{q,l}\, E_\Psi\, {}^{(3;2,1)}\mathcal{C}^\Psi_{i,j,k;p,s,r} \\
&- \delta_{k,q}\, E_\Psi\, {}^{(3;2,1)}\mathcal{C}^\Psi_{i,j,l;p,r,s} \\
&+ \delta_{l,p}\, E_\Psi\, {}^{(3;2,1)}\mathcal{C}^\Psi_{i,j,k;q,s,r} \\
&+ \delta_{k,p}\, E_\Psi\, {}^{(3;2,1)}\mathcal{C}^\Psi_{i,j,l;q,r,s} \\
&+ E_\Psi\, {}^{(4;2,2)}\mathcal{C}^\Psi_{i,j,k,l;p,q,r,s}
\end{aligned} \tag{98}$$

which is the 4-MCSE.

The iterative procedure for solving the 4-MCSE follows a very similar general scheme to that of the 2-CSE. Thus one starts with an intial set of 3-, 2-, 1-RDMs, a ^4C matrix, and the energy E that corresponds to this set of matrices. These initial data must correspond to a reasonable approximation of the eigenstate under study. This set of matrices is replaced in the 4-MCSE and, after symmetrizing the resulting matrix, $^4\tilde{\mathcal{M}}$, its trace is divided by that of the 4-RDM, $\binom{N}{4}$, which gives a new energy E'. Then a new ^4D is obtained by dividing $^4\tilde{\mathcal{M}}$ by E'. All the lower-order RDMs are obtained by contraction of this 4-RDM and then, by decomposing this same 4-RDM, the ^4C is evaluated. All these operations are straightforward and no approximated algorithm is needed. With this new set of data, a new iteration is started.

Alcoba demonstrated an important theorem concerning this equation.

Theorem[Alcoba]   Assuming that the matrices $^{(3;2,1)}\underline{C}^\Psi$ and $^{(4;2,1,1)}\underline{C}^\Psi$ can be obtained, respectively, by decomposing the set of the N-representable 3- and 4-RDMs, then the MCSE Eq. (98) will be satisfied by this set of RDMs *if and only if* the density matrix $^N D^\Psi$, preimage of the 3- and 4-RDMs, satisfies the Schrödinger equation.

When the initial 4-RDM has been obtained from an FCI treatment, then the iterative procedure described above can proceed indefinitely without variation. That is, the solution *is a fixed point*, as is to be expected in view of Alcoba's theorem.

Therefore the 4-MCSE is not only determinate but, when solved, its solution is exact. As already mentioned, the price one has to pay is the fact of working in a four-electron space; and the difficulty, as in the 2-CSE case, is that the matrices involved must be N-representable. Indeed, in order to ensure the convergence of the iterative process, the 4-RDM should be purified at each iteration, since the need for its N-representability is crucial. In practice, the optimizing procedure used is to antisymmetrize the $^4\tilde{\mathcal{M}}$ at each iteration. This operation would not be needed if all the matrices were N-representable; but, if they are not, this condition is not satisfied. In order to impose that the 4-RDM, from which all the lower-order matrices are obtained, be positive semidefinite, the procedure followed by Alcoba has been to diagonalize this matrix and to apply to the eigenvalues the same purification as that applied to the diagonal elements in the 2-CSE case, by forcing the trace to also have a correct value.

C. Some Significant Results

A set of calculations on the beryllium atom and its isoelectronic series have been carried out [48]. The starting basis set used was Clementi's double-zeta [82]. This basis was then transformed into the Hartree–Fock one, and the initial RDMs corresponded to Slater determinants built with this basis. Note that in these four-electron sytems the 4-RDM coincides with the N-electron N-DM, which simplifies the calculation of the FCI 4-RDM with which the MCSE calculations are compared. Moreover, in the N-DM case, the only and sufficient conditions that must be satisfied by this matrix are to be Hermitian, positive semidefinite, antisymmetric, and having a correct trace. These were the conditions imposed in the calculations reported here. The positive semidefiniteness of the 4-RDM together with the trace were imposed after each iteration by acting on the eigenvalues of the 4-RDM by applying a similar procedure to the one used for the 4-RDM diagonal in the 2-CSE iterative solution.

In Fig. 4 the values of the energy at each iteration are shown for the beryllium atom; and the root mean square deviations of the 2-RDM and the 4-RDM with respect to the FCI values are shown in Fig. 5.

As can be observed there is a smooth and complete convergence toward both the exact energy (FCI) and the exact RDMs.

The general performance of the method for the ions of the beryllium isoelectronic series is very similar to that for the beryllium atom; therefore only the most significant values for the series are reported in Table VII. The results obtained for the singlet excited state corresponding to the dominance of the $|1\bar{1}2\bar{3}\rangle$ and $|1\bar{1}3\bar{2}\rangle$ Slater determinants are given in Fig. 6. As can be observed, the convergence value is close to the FCI value, although the accuracy is not as good as in the ground state. Also, when continuing with the iterations, the

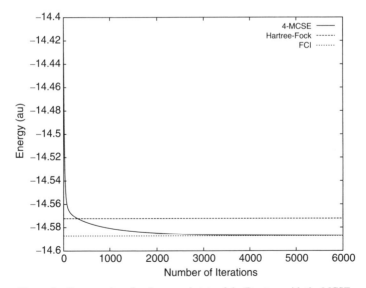

Figure 4. Energy values for the ground state of the Be atom with the MCSE.

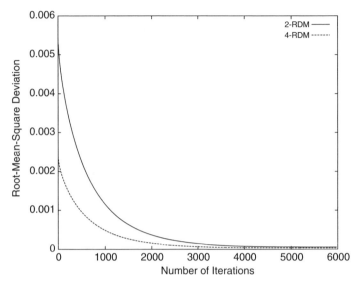

Figure 5. Root mean square deviations of the 2-RDM and the 4-RDM obtained for the ground state of the Be atom with the MCSE.

process falls into the ground state. It should be noted that when the regulating device previously mentioned was applied, the number of iterations needed to attain convergence was reduced to 10% of the number needed without this regulating device. When performing similar calculations for the BeH_2 molecule,

TABLE VII
Results Obtained with the Iterative 4-MCSE Method for the Ground State of Some Ions of the Beryllium Isoelectronic Series

System	Method	Energy (a.u.)	Deviation of ^2D	Deviation of ^4D	Iteration
B^+	Hartree–Fock	−24.23383	3.815×10^{-3}	1.691×10^{-3}	
	full CI	−24.24840	0.0	0.0	
	4-MCSE	−24.24811	8.217×10^{-6}	1.370×10^{-5}	6000
C^{2+}	Hartree–Fock	−36.40072	2.804×10^{-3}	1.245×10^{-3}	
	full CI	−36.41489	0.0	0.0	
	4-MCSE	−36.41466	3.409×10^{-6}	8.670×10^{-6}	6000
N^{3+}	Hartree–Fock	−51.06981	2.222×10^{-3}	9.869×10^{-4}	
	full CI	−51.08378	0.0	0.0	
	4-MCSE	−51.08359	1.698×10^{-6}	5.872×10^{-6}	6000
O^{4+}	Hartree–Fock	−68.23818	1.849×10^{-3}	8.213×10^{-4}	
	full CI	−68.25196	0.0	0.0	
	4-MCSE	−68.25179	9.634×10^{-7}	4.195×10^{-6}	6000

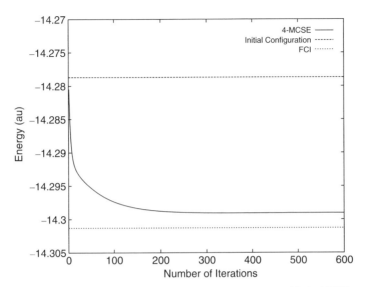

Figure 6. Energy values for the excited state of the Be atom with the MCSE.

where the positive semidefiniteness is no longer a sufficient N-representability condition, the process was not smooth. Thus, after two initial oscillations, the curve converged toward a minimum and then rapidly diverged. It must be noted that, when examining the 1-RDM at the minimum of the curve, the matrix was found to be extremely close to the FCI one. This imperfect performance must be due to the fact that in this case the N-representability conditions imposed on the 4-RDM were no longer sufficient. Therefore, for systems with more than four electrons, a strict purification procedure for the 4-RDM at each iteration must be applied. This renders the method rather expensive, which reduces its usefulness, particularly when the size of the system imposes the use of a large basis set.

VII. SOME FINAL REMARKS

The outlook given in this chapter on the theory of the second-order contracted Schrödinger equation and on its methodology has been aimed mostly at convincing the reader that this theory is not difficult to understand and that its methodology is now ready to be applied. That is, in the author's opinion, this methodology can be considered as accurate and probably more economical than the best standard quantum chemical computational methods for the study of states where the occupation number of spin orbitals is close to one or zero, in particular, when the refinements involving the 2-RDM purification and the device for accelerating convergence described by Alcoba in this book are used.

The most important question that remains open concerns the search for an appropriate approach to the study of states having some spin orbitals with occupation numbers close to 0.5. The structure of the algorithms is what causes the difficulty, since the leading terms of the 2-RDM involve products of the type $0.5 \times 0.5 = 0.25$, which is much too small. In consequence, the value of the unknown $^2\Delta$ elements is of the same order and can no longer be considered a small error. This is the case of many excited states for which the results have not been satisfactory up to now.

Since this chapter was centered on the CSE, only the most relevant aspects of the MCSE theory and practice have been treated here. It is nevertheless to be hoped that the brevity of this exposition has been sufficient to show the importance of this theory. Indeed, by considering the properties of the cancellation terms [81] jointly with Alcoba's theorems and the structure of the 4-MCSE, it can be concluded that *an N-body eigenproblem is just a four-electron one*. Moreover, the basic variable of this equation is the 2-CSE and through it the 2-RDM. Because the FCI 4-RDM determines a fixed point in the iterative process, the results obtained in the calculations reported here and in some other unpublished ones confirm the exactness of the 4-MCSE solution.

It is true that one must work in a four- or in a three-electron space; however, the reward is tantalizing: to get an exact, not approximate solution. The difficulty is of course the high computational cost of introducing all the known N-representability conditions. The question whether one could relax the N-representability conditions to be imposed while keeping the procedure convergent is still open.

Although the excited states can also be studied with the MCSE method, the results are not as clear as those for the ground state. The difficulty in this case is that while the iterative process seems to stabilize itself close to the value of the excited state energy, this value finally falls toward the ground-state energy.

Although for brevity's sake the 3-MCSE has not been considered here, it may be convenient to mention it in these final comments. This equation, which depends on the 1-CSE, does not have a unique solution. Indeed, this equation is satisfied not only by the FCI 3-RDM but also by the Hartree–Fock one. Alcoba [48] performed a series of calculations with the 3-MCSE for the beryllium isoelectronic series. Alcoba took as initial data a set of RDMs that corresponded to a state that had already some correlation and whose energy was below the Hartree–Fock's one. The results of these calculations showed that there was a smooth although very slow convergence toward the exact solution. For larger systems the situation will probably be similar to the 4-MCSE one and a strict purification should be applied to the 3-RDM. Also, in this case, enhancement of the convergence rate is necessary. Since the purification of the 3-RDM should be more economical than that of the 4-RDM, this approach seems to be more attractive. However, one must have as initial data a correlated and closely N-representable 3-RDM. Moreover, the question whether the iterative process of the 3-MCSE would be sufficiently accelerated when applying the regulating device is still an open question.

Acknowledgments

The author wishes to thank D. R. Alcoba for his helpful comments and for his generous permission to use the material of his thesis and in particular for allowing her to report here some of his unpublished results. The author also thanks Prof. L. M. Tel and Dr. E. Pérez-Romero for their helpful discussions. The author acknowledges financial support from the Ministerio de Ciencia y Tecnologia under project BFM2003-05133.

References

1. K. Husimi, Some formal properties of the density matrix. *Proc. Phys. Math. Soc. Japan* **22**, 264 (1940).
2. P. O. Lowdin, "Quantum theory of many particle systems. I. Physical interpretation by means of density matrices, natural spin orbitals, and convergence problems in the method of configuration interaction. *Phys. Rev.* **97**, 1474 (1955).
3. J. E. Mayer, Electron correlation. *Phys. Rev.* **100**, 1579 (1995).
4. R. U. Ayres, Variational approach to the many-body problem. *Phys. Rev.* **111**, 1453 (1958).
5. C. A. Coulson, Present state of molecular structure calculations. *Rev. Mod. Phys.* **32**, 175 (1960).
6. R. M. Erdahl (ed.), *Reduced Density Operators with Applications to Physical and Chemical Systems II*, Queen's University, Kingston, Ontario, 1974.
7. E. R. Davidson, *Reduced Density Matrices in Quantum Chemistry*, Academic Press, New York, 1976.
8. *Density Matrices and Density Functionals* (R. Erdahl and V. Smith, eds.), Proceedings of the A. J. Coleman Symposium, Kingston, Ontario, 1985, Reidel, Dordrecht, 1987.
9. R. G. Parr and W. Yang, *Theory of Atoms and Molecules*, Oxford University Press, New York, 1989.
10. *Many-Electron Densities and Reduced Density Matrices* (J. Cioslowski, ed.) Kluwer, Norwell, MA, 2000.
11. *Density Matrices and Density Functionals* (R. Erdahl and V. Smith, eds.), Proceedings of the A. J. Coleman Symposium, Kingston, Ontario, 1985, Reidel, Dordrecht, 1987.
12. A. J. Coleman and V. I. Yukalov, *Reduced Density Matrices: Coulson's Challenge*, Springer-Verlag, New York, 2000.
13. S. Cho, *Sci. Rep. Gumma Univ.* **11**, 1 (1962).
14. L. Cohen and C. Frishberg, Hierarchy equations for reduced density matrices. *Phys. Rev. A* **13**, 927 (1976).
15. L. Cohen and C. Frishberg, Hartree–Fock density matrix equation. *J. Chem. Phys.* **65**, 4234 (1976).
16. H. Nakatsuji, Equation for direct determination of the density matrix. *Phys. Rev. A* **14**, 41 (1976).

17. C. Valdemoro, Theory and practice of the spin adapted reduced Hamiltonian, in *Density Matrices and Density Functionals* (R. Erdahl and V. Smith, eds.), Proceedings of the A. J. Coleman Symposium, Kingston, Ontario, 1985, Reidel, Dordrecht, 1987.
18. J. E. Harriman, Limitation on the density-equation approach to many-electron problems. *Phys. Rev. A* **19**, 1893 (1979).
19. C. Valdemoro, Approximating the second-order reduced density matrix in terms of the first-order one. *Phys. Rev. A* **45**, 4462 (1992).
20. F. Colmenero, C. Perez del Valle, and C. Valdemoro, Approximating q-order reduced density matrices in terms of the lower-order ones. 1. General relations. *Phys. Rev. A* **47**, 971 (1993).
21. F. Colmenero and C. Valdemoro, Approximating q-order reduced density matrices in terms of the lower-order ones. 2. Applications. *Phys. Rev. A* **47**, 979 (1993).
22. F. Bopp, Ableitung der Bindungsenergie von N-teilchen-systemen aus 2-teilchen-dichtematrizen. *Z. Phys.* **156**, 1421 (1959).
23. A. J. Coleman and I. Absar, Reduced Hamiltonian orbitals. III. Unitarily invariant decomposition of Hermitian Operators. *Int. J. Quantum Chem.* **18**, 1279 (1980).
24. C. Valdemoro, Spin-adapted reduced Hamiltonian. I: Elementary excitations. *Phys. Rev. A* **31**, 2114 (1985).
25. C. Valdemoro, Spin-adapted reduced Hamiltonian. II: Total energy and reduced-density matrices, *Phys. Rev. A* **31**, 2123 (1985).
26. C. Valdemoro, L. M. Tel, and E. Pérez-Romero, N-representability problem within the framework of the contracted Schrödinger equation. *Phys. Rev. A* **61**, 032507 (2000).
27. A. J. Coleman, The structure of fermion density matrices. *Rev. Mod. Phys.* **35**, 668 (1963).
28. R. M. Erdahl, Representability conditions, in *Density Matrices and Density Functionals* (R. Erdahl and V. Smith, eds.), Proceedings of the A. J. Coleman Symposium, Kingston, Ontario, 1985, Reidel,Dordrecht, 1987.
29. C. Garrod and J. K. Percus, Reduction of the N-particle variational problem. *J. Math. Phys.* **5**, 1756 (1964).
30. A. J. Coleman, N-representability circunvented, in *Energy, Structure and Reactivity—Boulder Conference* (D.W Smith and W. McRae, eds.), John Wiley & Sons, Hoboken, NJ, 1973.
31. J. K. Percus, Role of model systems in few-body reduction of N-fermion problem. *Int. J. Quantum Chem.* **13**, 89 (1978).
32. R. M. Erdahl, Reduction of the N-particle variational problem. *Int. J. Quantum Chem.* **13**, 697 (1978).
33. M. Rosina, B. Golli, and R. M. Erdahl, A lower bound to the ground state energy of a boson system with fermion source, in *Density Matrices and Density Functionals* (R. Erdahl and V. Smith, eds.), Proceedings of the A. J. Coleman Symposium, Kingston, Ontario, 1985, Reidel, Dordrecht, 1987.
34. C. Valdemoro, M. P. de Lara-Castells, E. Pérez-Romero, and L. M. Tel, *Second European Workshop on Quantum Systems in Chemistry and Physics*, Oxford, 1997.
35. C. Valdemoro, M. P. de Lara-Castells, E. Pérez-Romero, and L. M. Tel, The first order contracted density equations: correlation effects. *Adv. Quantum Chem.* **31**, 37 (1999).
36. C. Valdemoro, L. M. Tel, D. R. Alcoba, E. Pérez-Romero and F. J. Casquero, Some basic properties of the correlation matrices. *Int. J. Quantum. Chem.* **90**, 1555 (2002).
37. J. E. Harriman, Grassmann products, cumulants, and two-electron reduced density matrices. *Phys. Rev. A* **65**, 052507 (2002).

38. D. R. Alcoba and C. Valdemoro, Spin structure and properties of the correlation matrices corresponding to pure spin states: controlling the S-representability of these matrices. *Int. J. Quantum Chem.* **102**, 629 (2005).
39. C. Garrod and M. Rosina, Particle–hole matrix: its connection with the symmetries and collective features of the ground-state. *J. Math. Phys.* **10**, 1855 (1969).
40. A. Beste, K. Runge, and R. Barlett, Ensuring N-representability: Coleman's algorithm. *Chem. Phys. Lett.* **355**, 263 (2002).
41. H. Kummer, N-representability problem for reduced density matrices. *J. Math. Phys.* **8**, 2063 (1967).
42. C. Valdemoro, An alternative approach to the calculation of the Hamiltonian matrix elements. *An. Fís. A* **79**, 98 (1983).
43. C. Valdemoro, A new method for the calculation of the electronic properties of atoms and molecules. *An. Fís. A* **79**, 106 (1983).
44. H. Schlosser, Hierarchy equations for reduced transition operators. *Phys. Rev. A* **15**, 1349 (1977).
45. D. A. Mazziotti, Contracted Schrödinger equation: determining quantum energies and two-particle density matrices without wavefunctions. *Phys. Rev. A* **57**, 4219 (1998).
46. C.Valdemoro, L. M. Tel, and E. Pérez-Romero, The contracted Schrödinger equation: some results. *Adv. Quantum Chem.* **28**, 33 (1997).
47. J. E. Harriman, Geometry of density-matrices. 5. Eigenstates. *Phys. Rev. A* **30**, 19 (1984).
48. D. R. Alcoba, *Estudio de las cuestiones cruciales en la solucion iterativa de la Ecuacion de Schrödinger Contraida.* Ph.D. Thesis, Universidad Autonoma de Madrid, 2004.
49. R. McWeeny and Y. Mizuno, Density matrix in many-electron quantum mechanics. 2. Separation of space and spin variables—spin-coupling problems. *Proc. Roy. Soc. A* **259**, 554 (1961).
50. C. Valdemoro, D. R. Alcoba, and L. M.Tel, Recent developments in the contracted Schrödinger equation method: controlling the N-representability of the second-order reduced density matrix. *Int. J. Quantum Chem* **93**, 212 (2003).
51. D. R. Alcoba, F. J. Casquero, L. M. Tel, E. Pérez-Romero, and C. Valdemoro, Convergence enhancement in the iterative solution of the second-order contracted Schrödinger equation. *Int. J. Quantum Chem.* **102**, 620 (2005).
52. K. Yasuda, Uniqueness of a solution of the contracted Schödinger equation. *Phys. Rev. A* **65**, 052121 (2002).
53. F. Colmenero and C. Valdemoro, Self-consistent approximate solution of the second-order contracted Schrödinger equation. *Int. J. Quantum Chem.* **51**, 369 (1994).
54. C. Valdemoro, D. R. Alcoba, L.M. Tel, and E. Pérez-Romero, Imposing bounds on the high-order reduced density matrices elements. *Int. J. Quantum Chem.* **85**, 214, (2001).
55. *The Mathematica Book*, 3rd ed, Wolfram Media, Cambridge University Press, Cambridge, 1996.
56. H. Nakatsuji and K. Yasuda, Direct determination of the quantum-mechanical density matrix using the density equation. *Phys. Rev. Lett.* **76**, 1039 (1996).
57. K. Yasuda and H. Nakatsuji, Direct determination of the quantum-mechanical density matrix using the density equation. 2, *Phys. Rev. A* **56**, 2648 (1997).
58. P. C. Martin and J. Schwinger, Theory of many particles systems. I, *Phys. Rev.* **115**, 1342 (1959).
59. D. A. Mazziotti, Approximate solution for electron correlation through the use of Schwinger probes. *Chem. Phys. Lett.* **289**, 419 (1998).
60. R. Kubo, Generalized cumulant expansion method. *J. Phys. Soc. Japan* **17**, 1100 (1962).

61. D. A. Mazziotti, 3,5-Contracted Schrödinger equation: determining quantum energies and reduced density matrices without wavefunctions. *Int. J. Quantum Chem.* **70**, 557 (1998).
62. D. A. Mazziotti, Pursuit of N-representability for the contracted Schrödinger equation through density-matrix reconstruction. *Phys. Rev. A* **60**, 3618 (1999).
63. D. A. Mazziotti, Complete reconstruction of reduced density matrices. *Chem. Phys. Lett.* **326**, 212 (2000).
64. W. Kutzelnigg and D. Mukherjee, Cumulant expansion of the reduced density matrices. *J. Chem. Phys.* **110**, 2800 (1999).
65. P. Ziesche, Cumulant expansions of reduced density matrices, reduced density matrices and Green functions, in *Many-Electron Densities and Reduced Density Matrices* (J. Cioslowski, ed.), Kluwer, Norwell, MA, 2000.
66. F. Harris Cumulant-based approximations to reduced density matrices. *Int. J. Quantum Chem.* **90**, 105 (2002).
67. J. M. Herbert and J. E. Harriman, Contraction relations for Grassman products of reduced density matrices and implications for density matrix reconstruction. *Phys. Rev. A* **65**, 022511 (2002).
68. M. Ehara, M. Nakata, H. Kou, K. Yasuda, and H. Nakatsuji, Direct determination of the density matrix using the density equation: potential energy curves of HF, CH_4, BH_3, and H_2O. *Chem. Phys. Lett.*, **305**, 483 (1999).
69. M. Nakata, M. Ehara, K. Yasuda, and H. Nakatsuji, Direct determination of second-order density matrix: open-shell system and excited state. *J. Chem. Phys.* **112**, 8772 (2000).
70. M. Nooijen, M. Wladyslawski, and A. Hazra, Cumulant approach to the direct calculation of reduced density matrices: a critical analysis. *J. Chem. Phys.* **118**, 4832, (2003).
71. C. Valdemoro, Electron correlation and reduced density matrices, in *Topics in Current Chemistry* (P. R. Surjan, ed.), Springer-Verlag, New York, 1999.
72. A. Hernandez-Laguna, J. Maruani, R. McWeeny, and S. Wilson (eds.), *Quantum Systems in Chemistry and Physics*, Vol. 1, Kluwer, Dordrecht, The Netherlands, 2000.
73. C. Valdemoro, L. M. Tel, and E. Pérez-Romero, Critical questions concerning iterative solution of the contracted Schrödinger equation, in *Many-Electron Densities and Reduced Density Matrices* (J. Cioslowski, ed.), Kluwer, Norwell, MA, 2000.
74. C. Valdemoro, L. M. Tel, E. Pérez-Romero, and A. Torre, The iterative solution of the contracted Schrödinger equation: a new quantum chemical method. *J. Mol. Struct. (Theochem).* **537**, 1 (2001).
75. D. A. Mazziotti and R. M. Erdahl, Uncertainty relations and reduced density matrices: mapping many-body quantum mechanics onto four particles. *Phys. Rev. A* **63**, 042113 (2001).
76. F. Weinhold and E. B. Wilson, Reduced density matrices of atoms and molecules. II. On the N-representability problem. *J. Chem. Phys.* **47**, 2298 (1967).
77. E. R. Davidson, Linear inequalities for density matrices III. *Int. J. Quantum Chem.* **90**, 1555 (2002).
78. D. A. Mazziotti, Purification of correlated reduced density matrices. *Phys. Rev. E* **65**, 026704 (2001).
79. D. R. Alcoba and C. Valdemoro, Family of modified-contracted Schrödinger equations. *Phys. Rev. A* **64**, 062105 (2001).
80. D. R. Alcoba, Equivalence theorems between the solutions of the fourth-order modified contracted Schrödinger equation and those of the Schrödinger equation. *Phys. Rev. A* **65**, 032519 (2002).
81. L. M. Tel, E. Pérez-Romero, C. Valdemoro, and F. J. Casquero, Cancellation of high-order electron correlation effects corresponding to eigenstates. *Int. J. Quantum Chem.* **82**, 131 (2001).
82. E. Clementi and C. Roetti, *Atom. Data Nucl. Data Tables* **14**, 177 (1974).

CHAPTER 8

CONTRACTED SCHRÖDINGER EQUATION

DAVID A. MAZZIOTTI

*Department of Chemistry and The James Franck Institute,
The University of Chicago, Chicago, IL 60637 USA*

CONTENTS

I. Introduction
II. Contracted Schrödinger Equation
   A. Derivation in Second Quantization
   B. Nakatsuji's Theorem
III. Reconstruction of the 3- and 4-RDMS
   A. Rosina's Theorem
   B. Particle–Hole Duality
   C. Cumulants
   D. Approximation of the Cumulant 3-RDM
   E. Cumulant Structure of the CSE
IV. Purification of the 2-RDM
   A. N-Representability of the 1-RDM
   B. N-Representability of the 2-RDM
      1. Unitary and Cumulant Decompositions of the 2-RDM
      2. Positivity Conditions on the 2-RDM
      3. Spin Blocks of the 2-RDM
V. Self-Consistent Iteration
VI. Algorithm for Solving the CSE
VII. Applications
VIII. A Look Ahead
Acknowledgments
Appendix: Grassmann Products
References

I. INTRODUCTION

Knowledge of the 2-particle reduced density matrix (2-RDM) allows one to calculate the energy and other observables for atomic and molecular systems with

---

*Reduced-Density-Matrix Mechanics: With Application to Many-Electron Atoms and Molecules,
A Special Volume of Advances in Chemical Physics, Volume 134*, edited by David A. Mazziotti.
Series editor Stuart A. Rice. Copyright © 2007 John Wiley & Sons, Inc.

an arbitrary number N of electrons. For a quantum system, fully characterized by a single N-particle wavefunction, the N-particle density matrix ^{N}D is the kernel of the wavefunction's projection operator. By integrating the density matrix ^{N}D over $N-2$ particles, we obtain the 2-RDM, which contains enough information to calculate the expectation values for any operator with only two particle interactions like the electronic Hamiltonian [1–4]. Calculation of the 2-RDM without the many-electron wavefunction is challenging because not every two-particle density matrix derives from an N-particle density matrix. Restricting the 2-RDM to represent an N-particle density matrix requires nontrivial constraints known as N-representability conditions [3, 5–13]. A new approach to the direct calculation of the 2-RDM was developed in the 1990s through a projection of the Schrödinger equation onto the space of two particles known as the contracted Schrödinger equation (CSE) (or density equation) [14–20, 22–37].

Nakatsuji [37] in 1976 first proved that with the assumption of N-representability [3] a 2-RDM and a 4-RDM will satisfy the CSE if and only if they correspond to an N-particle wavefunction that satisfies the corresponding Schrödinger equation. Just as the Schrödinger equation describes the relationship between the N-particle Hamiltonian and its wavefunction (or density matrix ^{N}D), the CSE connects the two-particle reduced Hamiltonian and the 2-RDM. However, because the CSE depends on not only the 2-RDM but also the 3- and 4-RDMs, it cannot be solved for the 2-RDM without additional constraints. Two additional types of constraints are required: (i) formulas for building the 3- and 4-RDMs from the 2-RDM by a process known as *reconstruction*, and (ii) constraints on the N-representability of the 2-RDM, which are applied in a process known as *purification*.

Employing the particle–hole duality, Valdemoro derived formulas for reconstructing the 3- and the 4-RDMs from the 2-RDM to remove the indeterminacy of the CSE [14, 15, 20, 38]. Yasuda and Nakatsuji [19] added an additional term to each of these formulas by considering the decoupling diagrams for Green's functions. The author systematized these reconstruction functions in the CSE by applying the theory of cumulants to the RDMs [21, 22, 24–26, 39, 40]. After presenting Rosina's theorem, which justifies the reconstruction of the higher RDMs from the 2-RDM, we derive in Sections III.B and III.C the reconstruction formulas for the 3- and 4-RDMs from the perspectives of the particle–hole duality [14, 15, 20, 38] and cumulant theory [21, 22, 24–26, 39, 40], respectively. In Section III.D the cumulant expansion for the 3-RDM is improved by two different corrections.

In addition to reconstruction within the CSE, it is important to constrain the 2-RDM to remain approximately N-representable. The process of correcting a 2-RDM to satisfy N-representability constraints is known as *purification*. In the context of an iterative solution of the CSE, early algorithms by Valdemoro checked that the 2-RDM satisfies a number of fundamental inequalities such as the nonnegativity of the diagonal elements [17]. The author developed a more general *purification* algorithm that corrects the 2-RDM so that two N-representability constraints known as the D- and Q-conditions are satisfied [28]. Alcoba and Valdemoro [34] recently extended the author's algorithm to include explicitly another N-representability constraint known as the G-condition. Purification of the 2-RDM is described in detail in Section IV.

The ingredients of (i) CSE, (ii) reconstruction of the 3- and the 4-RDMs, and (iii) purification of the 2-RDM are combined in an iterative algorithm for solving the CSE in Section VI. Applications of the CSE algorithm to a variety of atoms and molecules from Refs. [28, 29] are presented. Purification of the 2-RDM is seen to be critical for an accurate solution of the CSE [28, 29].

II. CONTRACTED SCHRÖDINGER EQUATION

A quantum system of N fermions may be characterized by the Schrödinger equation (SE)

$$H|\psi_n\rangle = E_n|\psi_n\rangle \tag{1}$$

in which the wavefunction ψ_n depends on the coordinates for the N particles. Beginning with the SE, we will obtain Valdemoro's form of the contracted Schrödinger equation (CSE) in second quantization [16, 42–44]. The derivation emphasizes the use of test functions for performing the projection (or contraction) of the SE onto the lower particle space [20]. By Nakatsuji's theorem [37] there is a one-to-one mapping between N-representable RDM solutions of the CSE and wavefunction solutions of the SE. In 1998 the author proved Nakatsuji's theorem [37] for the second-quantized CSE [20].

A. Derivation in Second Quantization

Within second quantization [41] the Hamiltonian operator may be expressed as

$$H = \frac{1}{2} \sum_{p,q;s,t} {}^2K^{p,q}_{s,t} a^\dagger_p a^\dagger_q a_t a_s \tag{2}$$

where the elements of the two-particle reduced Hamiltonian ${}^2K$ are given by

$${}^2K^{p,q}_{s,t} = {}^2V^{p,q}_{s,t} + \frac{1}{N-1}(\delta_{q,t}\epsilon_{p,s} + \delta_{p,s}\epsilon_{q,t}) \tag{3}$$

The repulsion between electrons 1 and 2 is represented by

$${}^2V^{p,q}_{s,t} = \langle \phi_p(1)\phi_q(2)|\frac{1}{r_{12}}|\phi_s(1)\phi_t(2)\rangle \tag{4}$$

while one-electron portions of the Hamiltonian are included in the matrix ϵ:

$$\epsilon_{p,s} = -\langle \phi_p(1)| \frac{\nabla_1^2}{2} + \sum_l \frac{Z_l}{r_{1,l}} |\phi_s(1)\rangle \tag{5}$$

Because the N-particle Hamiltonian (H) contains only two-electron excitations, the expectation value of H yields a formula for the energy involving just the 2-RDM,

$$E = \sum_{p,q;s,t} {}^2K_{s,t}^{p,q} \, {}^2D_{s,t}^{p,q} = \text{Tr}({}^2K\,{}^2D) \tag{6}$$

where

$$^2D_{s,t}^{p,q} = \frac{1}{2!} \langle \psi | a_p^\dagger a_q^\dagger a_t a_s | \psi \rangle \tag{7}$$

In general, the p-RDM in second quantization is defined as

$$^pD_{j_1,j_2,\ldots,j_p}^{i_1,i_2,\ldots,i_p} = \frac{1}{p!} \langle \psi | a_{i_1}^\dagger a_{i_2}^\dagger \cdots a_{i_p}^\dagger a_{j_p} a_{j_{p-1}} \cdots a_{j_1} | \psi \rangle \tag{8}$$

and the normalization is $N!/(p!(N-p)!)$. Variation of ^2D to produce the lowest energy will generate the ground-state energy of the reduced Hamiltonian ^2K, which will usually be much lower than the energy of the many-particle Hamiltonian H. To obtain the correct energy E of the N-electron Hamiltonian H, we must impose additional N-representability constraints on the 2-RDM to ensure that it is derivable from an antisymmetric N-particle wavefunction ψ through the integration of its associated density matrix.

To derive the CSE rather than the expectation value, we define functions $\langle \Phi_{k,l}^{i,j} |$ to test the two-electron space

$$\langle \Phi_{k,l}^{i,j} | = \langle \psi | a_i^\dagger a_j^\dagger a_l a_k \tag{9}$$

Taking the inner product of the test functions with the SE produces

$$\langle \psi | a_i^\dagger a_j^\dagger a_l a_k H \psi \rangle = E \langle \psi | a_i^\dagger a_j^\dagger a_l a_k | \psi \rangle = 2E \, {}^2D_{k,l}^{i,j} \tag{10}$$

If we substitute for the Hamiltonian operator in Eq. (2), we obtain the relation

$$\sum_{p,q,s,t} {}^2K_{s,t}^{p,q} \langle \psi | a_i^\dagger a_j^\dagger a_l a_k a_p^\dagger a_q^\dagger a_t a_s | \psi \rangle = 4E \, {}^2D_{k,l}^{i,j} \tag{11}$$

Rearranging creation and annihilation operators on the left-hand side to produce RDMs, we generate Valdemoro's 2,4-CSE [16]:

$$(^2\mathbf{K}^2\mathbf{D})^{i,j}_{k,l} + 3\sum_{p,q,t}\left(^2K^{p,q}_{i,t}{}^3D^{p,q,j}_{k,t,l} + {}^2K^{p,q}_{j,t}{}^3D^{p,q,i}_{l,t,k}\right)$$
$$+ 6\sum_{p,q,s,t}\left(^2K^{p,q}_{s,t}{}^4D^{p,q,i,j}_{s,t,k,l}\right) = E\ ^2D^{i,j}_{k,l} \qquad (12)$$

Evaluation of the first term in the above equation involves multiplying matrices $^2\mathbf{K}$ and $^2\mathbf{D}$ and then selecting the element of the resulting matrix, specified by the indices. We have derived the 2,4-CSE through test functions rather than the generalized matrix contraction mapping [16, 42–44]. A 1,3-CSE may also be produced by replacing the doubly excited test functions in Eq. (9) with test functions formed by single excitations of the ground-state wavefunction. Similarly, a 3,5-CSE and a 4,6-CSE may be created with test functions using triple and quadruple excitations, respectively. Since the 2,4-CSE is the focus of this chapter, we simply refer to it as the CSE.

B. Nakatsuji's Theorem

While early work [16, 19] on the CSE assumed that Nakatsuji's theorem [37], proved in 1976 for the integrodifferential form of the CSE, remains valid for the *second-quantized* CSE, the author presented the first formal proof in 1998 [20]. Nakatsuji's theorem is the following: if we assume that the density matrices are pure N-representable, then the CSE may be satisfied by ^2D and ^4D if and only if the preimage density matrix $^N D$ satisfies the Schrödinger equation (SE). The above derivation clearly proves that the SE implies the CSE. We only need to prove that the CSE implies the SE. The SE equation can be satisfied if and only if

$$\langle\psi|H^2|\psi\rangle - \langle\psi|H|\psi\rangle^2 = 0 \qquad (13)$$

known as the dispersion condition [45]. Multiplying both sides of the CSE in Eq. (11) by the reduced Hamiltonian elements $^2K^{i,j}_{k,l}$ and summing over the remaining indices produces

$$\langle\psi|\left(\frac{1}{2}\sum_{i,j;k,l}{}^2K^{i,j}_{k,l}a^\dagger_i a^\dagger_j a_l a_k\right)\left(\frac{1}{2}\sum_{p,q;s,t}{}^2K^{p,q}_{s,t}a^\dagger_p a^\dagger_q a_t a_s\right)|\psi\rangle$$
$$= E\left(\sum_{i,j;k,l}{}^2K^{i,j}_{k,l}{}^2D^{i,j}_{k,l}\right) \qquad (14)$$

By Eq. (6) the sum on the right-hand side of the above equation is equal to the energy E, and from Eq. (2) we realize that the sums on the left-hand side are just Hamiltonian operators in the second-quantized notation. Hence, when the 2-RDM corresponds to an N-particle wavefunction ψ, Eq. (12) implies Eq. (13), and the proof of Nakatsuji's theorem is accomplished. Because the Hamiltonian is defined in second quantization, the proof of Nakatsuji's theorem is also valid when the one-particle basis set is incomplete. Recall that the SE with a second-quantized Hamiltonian corresponds to a Hamiltonian eigenvalue equation with the given one-particle basis. Unlike the SE, the CSE only requires the 2- and 4-RDMs in the given one-particle basis rather than the full N-particle wavefunction. While Nakatsuji's theorem holds for the 2,4-CSE, it is not valid for the 1,3-CSE. This foreshadows the advantage of reconstructing from the 2-RDM instead of the 1-RDM, which we will discuss in the context of Rosina's theorem.

III. RECONSTRUCTION OF THE 3- AND 4-RDMs

The CSE allows us to recast N-representability as a reconstruction problem. If we knew how to build from the 2-RDM to the 4-RDM, the CSE in Eq. (12) furnishes us with enough equations to solve iteratively for the 2-RDM. Two approaches for reconstruction have been explored in previous work on the CSE: (i) the explicit representation of the 3- and 4-RDMs as functionals of the 2-RDM [17, 18, 20, 21, 29], and (ii) the construction of a family of higher 4-RDMs from the 2-RDM by imposing ensemble representability conditions [20]. After justifying reconstruction from the 2-RDM by Rosina's theorem, we develop in Sections III.B and III.C the functional approach to the CSE from two different perspectives—the particle–hole duality and the theory of cumulants.

A. Rosina's Theorem

Proving that the ground-state 2-RDM contains enough information to generate the higher RDMs provides theoretical justification for reconstruction functionals for the 3- and 4-RDMs in terms of the 2-RDM. Early work [14] on the CSE appealed to the well-known theorem of Hohenberg–Kohn (HK), which demonstrates that the 1-density and the particle number N are theoretically sufficient to determine the ground-state energies and wavefunctions for atoms and molecules [46, 47]. If the 1-density is enough to generate the wavefunction, it may seem that the 1-RDM or 2-RDM must be more than sufficient to build a unique series of higher RDMs leading to the wavefunction. However, as we will show, this argument neglects an implicit assumption in the HK theorem. The proof that the ground-state 1-density determines the ground and excited wavefunctions depends on a theoretic construction of the Hamiltonian from the 1-density [48]. For electronic structure problems the particle number N alone completely determines the form of the kinetic energy and electron repulsion terms within the Hamiltonian while the unknown one-particle part of the potential is specified through the given 1-density. When we construct the higher RDMs from lower RDMs or densities, neither the Hamiltonian nor any specific information about electronic systems appears in the reconstruction formulas. In addition to a knowledge of the particle number N and the 1-density, however, the theorem of Hohenberg and Kohn implicitly assumes a knowledge of the kinetic and repulsion terms within the Hamiltonian. Without more explicit knowledge of the Hamiltonian in a reconstruction functional, the 1-density cannot determine the wavefunction as it is not difficult to illustrate. Consider the 1-density from a wavefunction that is not a Slater determinant. Both Gilbert [49] and Harriman [50], however, have shown that every 1-density may be represented by an N-particle Slater wavefunction. Hence the 1-density clearly corresponds to at least two N-representable wavefunctions—one Slater wavefunction and one non-Slater wavefunction. Furthermore, convex combinations of these pure density matrices yield an infinite family of ensemble N-representable density matrices, which contract to the correct 1-density.

Although the 1-density alone is not sufficient to determine the ground-state wavefunction for an *unknown* Hamiltonian with two-particle interactions, the 2-RDM is enough to build the wavefunction, and the proof of this lies not in the HK theorem but in an important, less famous result, originally discussed by Rosina. Let us consider the 2-RDM $^2D(\psi)$ for the antisymmetric nondegenerate ground state of an N-particle Hamiltonian H with two-particle interactions. By $D^2(\psi)$ we indicate the 2-RDM from the contraction of a pure density matrix formed with ψ. The 2-RDM determines the energy of the eigenstate ψ by Eq. (6). If $^2D(\psi)$ may be obtained from two antisymmetric wavefunctions ψ, the ground state will be degenerate since by Eq. (6) they must have the same energy. Because this contradicts the assumption that the ground state is nondegenerate, we have that $^2D(\psi)$ has only one pure N-representable preimage, $^ND(\psi)$. Furthermore, because all of the other states of the system have higher energies, minimizing over the larger class of N-ensemble representable matrices will always produce the pure density matrix, corresponding to the ground state. For this reason the ^2D of the ground state also has only one preimage in the larger family of ensemble density matrices. Hence we have the reconstruction theorem, originally proved by Rosina at the 1967 conference on reduced density matrices at Queen's University [51] and developed by the author in the context of the RDM reconstruction problem for the CSE in 1998.

Theorem 1 *The 2-RDM for the antisymmetric, nondegenerate ground state of an unspecified N-particle Hamiltonian H with two-particle interactions has a unique preimage in the set of N-ensemble representable density matrices ND.*

B. Particle–Hole Duality

Many-body problems in quantum mechanics are usually described by the number of particles N in the system and the probabilities of finding those particles at different locations in space. If the rank of the one-particle basis is a finite number r, an equally valid description of the system may be given by specifying the number of holes $r - N$ in the system and the probabilities of finding these holes at different locations in space. This possibility for an equivalent representation of the system by particles or holes is known as the particle–hole duality. By using the fermion anticommutation relation

$$a_j a_i^\dagger + a_i^\dagger a_j = \delta_j^i \tag{15}$$

to rearrange the creation and annihilation operators in the definition Eq. (2) of the Hamiltonian such that all of the annihilators appear to the left of the creators, we generate a hole representation of the Hamiltonian $\bar{H}$ whose expectation value with the $(r-N)$-hole density matrix $^{(r-N)}\bar{D}$ produces the energy E

$$E = \text{Tr}(\bar{H}^{(r-N)}\bar{D}) \tag{16}$$
$$= \text{Tr}(^2\bar{K}^2\bar{D}) \tag{17}$$

As shown in the second line, like the expression for the energy E as a function of the 2-RDM, the energy E may also be expressed as a linear functional of the two-hole reduced density matrix $^2\bar{D}$ (2-HRDM) and the two-hole reduced Hamiltonian $^2\bar{K}$. Direct minimization of the energy to determine the 2-HRDM would require $(r-N)$-representability conditions. The definition for the p-hole reduced density matrices in second quantization is given by

$$^p\bar{D}_{i_1,i_2,\ldots,i_p}^{j_1,j_2,\ldots,j_p} = \frac{1}{p!} \langle \psi | a_{j_1} a_{j_2} \cdots a_{j_p} a_{i_1}^\dagger a_{i_2}^\dagger \cdots a_{i_p}^\dagger | \psi \rangle \tag{18}$$

Normalization of the p-HRDM in second quantization is $(r-N)!/(p!(r-N-p)!)$.

Because the hole and particle perspectives offer equivalent physical descriptions, the p-RDMs and p-HRDMs are related by a linear mapping [52, 53]. Thus if one of them is known, the other one is easily determined. The same linear mapping relates the p-particle and p-hole reduced Hamiltonian matrices (^2K and $^2\bar{K}$). An explicit form for the mapping may readily be determined by using the fermion anticommutation relation to convert the p-HRDM in Eq. (18) to the corresponding p-RDM. For $p = 1$ the result is simply

$$^1\bar{D}_i^j = {}^1\delta_j^i - {}^1D_j^i \tag{19}$$

which is equivalent to taking the expectation of the fermion anticommutation relation. Similarly, for $p = 2$ we obtain the relation

$$^2\bar{D}_{i_1,i_2}^{j_1,j_2} = (\delta_{j_1}^{i_1}\delta_{j_2}^{i_2} - \delta_{j_2}^{i_2}\delta_{j_1}^{i_1})/2 - {}^1D_{j_1}^{i_1}\delta_{j_2}^{i_2} + {}^1D_{j_2}^{i_2}\delta_{j_1}^{i_1} + {}^2D_{j_1,j_2}^{i_1,i_2} \qquad (20)$$

which contains a sum of three different kinds of terms that have (i) one 2-RDM, (ii) one 1-RDM multiplying one δ, and (iii) two δ's. This expression represents the commutation relation for a composite particle consisting of two fermions. By anticommuting the creation and annihilation operators, we can generate analogous expressions for composite particles consisting of more than two fermions.

Before introducing the general expression, we express Eq. (20) more concisely through the antisymmetric wedge product $\wedge$ from Grassmann algebra [54]. The wedge product between two matrices ^pD and ^qD involving p and q particles produces an antisymmetric matrix involving $p + q$ particles defined by

$$^pD \wedge {}^qD = \hat{A}_N {}^pD \otimes {}^qD \hat{A}_N \qquad (21)$$

where the $\hat{A}_N$ is the N-particle antisymmetrization operator and $\otimes$ is the tensor product. More details about evaluating wedge products are given in the Appendix. For the 2-HRDM as a functional of RDMs we obtain

$$^2\bar{D}_{i_1,i_2}^{j_1,j_2} = {}^2I_{j_1,j_2}^{i_1,i_2} - 2 {}^1D_{j_1}^{i_1} \wedge {}^1I_{j_2}^{i_2} + {}^2D_{j_1,j_2}^{i_1,i_2} \qquad (22)$$

where ^1I is the identity matrix

$$^1I_{j_1}^{i_1} = \delta_{j_1}^{i_1} \qquad (23)$$

and

$$^2I_{j_1,j_2}^{i_1,i_2} = {}^1I_{j_1}^{i_1} \wedge {}^1I_{j_2}^{i_2} \qquad (24)$$

In general, the linear relation between the p-HRDM and p-RDM may be expressed as

$$^p\bar{D} = {}^pI + \sum_{n=1}^{p-1} (-1)^n \binom{p}{n} {}^nD \wedge {}^{(p-n)}I + (-1)^p {}^pD \qquad (25)$$

Indices for the RDMs are not shown for notational clarity. The p-RDM as a functional of the p-HRDM may easily be obtained by switching $^p\bar{D}$ and ^pD in the above equation.

Valdemoro and co-workers [14] realized that these particle–hole relations could be written in the following form:

$$^p\bar{D} + (-1)^{p+1}\,^pD = f(^{p-1}\bar{D}) + (-1)^{p+1} f(^{p-1}D) \tag{26}$$

where $f(^{p-1}\bar{D})$ is a functional of the $(p-1)$-HRDM and lower HRDMs and $f(^{p-1}D)$ has the same functional form as $f(^{p-1}\bar{D})$ with the HRDMs replaced with the corresponding RDMs. With the appropriate f functional for each p, the relation in Eq. (26) is *exact* and *equivalent* to Eq. (25). Valdemoro and co-workers then obtain functionals for the p-RDM and p-HRDM by assuming that

$$^pD \approx\,^pD_{\text{Vald}} = f(^{p-1}D) \tag{27}$$

and

$$^p\bar{D} \approx\,^p\bar{D}_{\text{Vald}} = f(^{p-1}\bar{D}) \tag{28}$$

These formulas are approximate because some of the terms for the particle and hole RDMs cancel in relation (26). Rearranging Eq. (25) for each p as originally described by Valdemoro will produce the functionals f. We have found an easier method [20, 22] for extracting the functionals f, which, however, does not show the equivalence between Eqs. (25) and (26). Since Valdemoro's method appears in the literature [14], we explain our technique, which generates f from Eq. (25) through the following two substitutions: (i) replace ^1I with ^1D, which is equivalent to assuming that $^1\bar{D} = 0$ in Eq. (19), and (ii) set $^p\bar{D} = 0$. The technique works because it assumes a separation of particles and holes by setting all of the hole matrices in the expression to zero to produce f. For p from 2 to 5 the resulting RDM functionals are represented by the portions of the functionals in Table I, which are not underlined [20–22]. The right superscripts p in Table I indicate that an RDM is wedge with itself p times; for example, $^1D^2$ represents $^1D \wedge\,^1D$. The underlined corrections will be determined below through an extension of the particle–hole arguments and later through cumulant expansions.

TABLE I
Approximate Reconstruction Functionals for the p-RDMs in Terms of Lower RDMs Where Corrections to Valdemoro's Functionals Are Underlined

$$^2D \approx\,^1D^2$$
$$^3D \approx 3\,^2D \wedge\,^1D - 2\,^1D^3$$
$$^4D \approx 4\,^3D \wedge\,^1D - 6\,^2D \wedge\,^1D^2 + 3\,^1D^4 + \underline{3^2\Delta \wedge\,^2\Delta}$$
$$^5D \approx 5\,^4D \wedge\,^1D - 10\,^3D \wedge\,^1D^2 + 10\,^2D \wedge\,^1D^3 - 4\,^1D^5 + \underline{10\,^3\Delta \wedge\,^2\Delta}$$

Corrections for the 4-RDM and 5-RDM functionals may be obtained by searching for some terms involving the wedge products of lower RDMs, which cancel with the corresponding corrections for the HRDM functionals [20]. Consider the matrices $^2\Delta$ and $^3\Delta$ describing the errors in Valdemoro's reconstruction functionals for the 2- and 3-RDMs as well as the matrices $^2\bar{\Delta}$ and $^3\bar{\Delta}$ describing the errors in Valdemoro's reconstruction functionals for the 2- and 3-HRDMs

$$^2\Delta = {}^2D - {}^2D_{\text{Vald}} \tag{29}$$
$$= {}^2\bar{D} - {}^2\bar{D}_{\text{Vald}} \tag{30}$$
$$= {}^2\bar{\Delta} \tag{31}$$

and

$$^3\Delta = {}^3D - {}^3D_{\text{Vald}} \tag{32}$$
$$= -({}^3\bar{D} - {}^3\bar{D}_{\text{Vald}}) \tag{33}$$
$$= -{}^3\bar{\Delta} \tag{34}$$

An appropriate correction for the 4-RDM and 4-HRDM functionals is

$$^4D_{\text{corr}} = k_4 {}^2\Delta \wedge {}^2\Delta \tag{35}$$
$$= k_4 {}^2\bar{\Delta} \wedge {}^2\bar{\Delta} \tag{36}$$
$$= {}^4\bar{D}_{\text{corr}} \tag{37}$$

because this term has the same functional form for particles and holes and yet, since they are equal, they cancel in the commutation relation (26). The proportionality factor k_4 is equal to the number of distinct ways of distributing the four particles in two groups of two particles. The possibilities are $\{12\}\{34\}$, $\{13\}\{24\}$, and $\{14\}\{23\}$; hence $k_4 = 3$. The 5-RDM and 5-HRDM functionals have the following corrections:

$$^5D_{\text{corr}} = k_5 {}^3\Delta \wedge {}^2\Delta \tag{38}$$
$$= -(k_5 {}^3\bar{\Delta} \wedge {}^2\bar{\Delta}) \tag{39}$$
$$= -{}^5\bar{D}_{\text{corr}} \tag{40}$$

Again this term has the same functional form for particles and holes. Note that for odd p the corrections must have opposite signs to cancel in the anticommutation relation (26). As with k_4, the proportionality factor k_5 is equal to the number of distinct ways of distributing the five particles between a group of three particles and a group of two particles; thus $k_5 = 10$.

C. Cumulants

The reconstruction functionals, derived in the previous section through the particle–hole duality, may also be produced through the theory of cumulants [21, 22, 24, 26, 39, 55–57]. We begin by constructing a functional whose derivatives with respect to probe variables generate the reduced density matrices in second quantization. Because we require that additional derivatives increase the number of second quantization operators, we are led to the following exponential form:

$$G(J) = \langle \psi | O(\exp(\sum_k J_k a_k^\dagger + J_k^* a_k)) | \psi \rangle \tag{41}$$

where J_k and its conjugate J_k^* are Schwinger probe variables. For fermions these Schwinger probes have the property that they anticommute, $\{J_k, J_l\} = 0$. Differentiation of $G(J)$ with respect to the probes leads to the accumulation of creation and annihilation operators before the exponential. Because the annihilation and creation operators do not commute, we need to impose a specific ordering for these operators, which appear before the exponential after differentiation. Since we wish to form functionals for RDMs, we define that the creation operators should always appear to the left of the annihilation operators independent of the order in which we differentiate with respect to the probes. If we wished to produce the corresponding HRDM functionals, we would order the annihilators to the left of the creators. We represent this ordering convention through the ordering operator O in the definition of $G(J)$. This ordering process is analogous to the time ordering of the creation and annihilation operators, which appears in the theory of Green's functions [58].

The general relation between the differentiation of $G(J)$ with respect to the Schwinger probes and the RDMs may be characterized as

$$^p D^{i_1,i_2,\ldots,i_p}_{j_1,j_2,\ldots,j_p} = \lim_{J \to 0} \frac{1}{p!} \frac{\partial^p G}{\partial J_{i_p} \ldots \partial J_{i_2} \partial J_{i_1} \partial J^*_{j_1} \ldots \partial J^*_{j_{p-1}} \partial J^*_{j_p}} \tag{42}$$

$$= \frac{1}{p!} \langle \psi | a^\dagger_{i_1} a^\dagger_{i_2} \ldots a^\dagger_{i_p} a_{j_p} a_{j_{p-1}} \ldots a_{j_1} | \psi \rangle \tag{43}$$

The coefficients of the multivariable Taylor series expansion of $G(J)$ about the point where the Schwinger probes vanish are elements of the RDMs. Thus $G(J)$ is known as the *generating functional* for RDMs. Mathematically, the RDMs of the functional $G(J)$ are known as the *moments*. The moment-generating functional $G(J)$ may be used to define another functional $W(J)$, known as the cumulant-generating functional, by the relation

$$G(J) = \exp(W(J)) \tag{44}$$

Just as the moments are formed from $G(J)$ as in Eq. (43), the *cumulants* $^p\Delta$ are produced from $W(J)$ by

$$^p\Delta^{i_1,i_2,...,i_p}_{j_1,j_2,...,j_p} = \lim_{J \to 0} \frac{1}{p!} \frac{\partial^p W}{\partial J_{i_p}...\partial J_{i_2} \partial J_{i_1} \partial J^*_{j_1}...\partial J^*_{j_{p-1}} \partial J^*_{j_p}} \tag{45}$$

and the cumulants are defined as the coefficients of the multivariable Taylor series expansion of $W(J)$ about the point where the Schwinger probes vanish. The introduction of another generating functional $W(J)$ in Eq. (44) may seem unnecessary. The set of cumulants $^p\Delta$ for p ranging from 1 to q contains the same information as the set of moments ^pD for the same range of p, but the information is distributed differently. This different distribution of information will allow us to determine the reconstruction functionals for building higher RDMs from lower RDMs.

As explained by Kubo [55], cumulants have the special property that they vanish if and only if one of their particles is statistically independent of the rest. Thus for a mean field approximation (Hartree–Fock) where each of the N particles is treated independently, all cumulants except $^1\Delta$ vanish. Another way of interpreting this property of cumulants is to say that the p-particle cumulant $^p\Delta$ represents the part of the p-RDM that cannot be written as a simple wedge product of lower RDMs. The formula for $^3D_{\text{Vald}}$ from Table I accounts for situations where two of the particles are close enough to interact while the remaining particle is sufficiently separated in space for us to assume that it is statistically independent of the others. Therefore, approximating the 3-RDM as a functional of the lower RDMs is equivalent to assuming that $^3\Delta$ vanishes. Similarly, the remaining functionals in Table I, which express the given p-RDM as a functional of lower RDMs, do not accurately represent configurations in which all p particles are close enough to be simultaneously influenced by pairwise interactions. They assume that $^p\Delta$ vanishes. By analogy with the convention for Green's functions in quantum field theory [58], we define the *unconnected* p-RDM as the part of the p-RDM that can be written as wedge products of lower RDMs while the *connected* (or cumulant) p-RDM is the remaining portion of the RDM that cannot be expressed as antisymmetrized products of lower RDMs. Hence the connected RDMs are just the cumulants.

We may express the p-RDM in terms of the connected q-RDMs for q between 1 and p by differentiating Eq. (44) with respect to the Schwinger probes as in Eq. (42) and taking the limit as the probes approach zero. The derivatives of the generating functional $G(J)$ produce the p-RDM while differentiation of $\exp(W)$ on the right side produces products of elements from the connected RDMs according to Eq. (45). Because the formula for elements of the p-RDM must treat the permutation of the upper and lower indices antisymmetrically, the products between elements of connected RDMs may be replaced with wedge

TABLE II

Reconstruction Functionals for the p-RDMs in Terms of the Connected p-RDM and Lower Connected RDMs, Where Corrections Beyond Valdemoro's Approximation Are Underlined

$$^1D = {}^1\Delta$$
$$^2D = {}^1\Delta \wedge {}^1\Delta + \underline{{}^2\Delta}$$
$$^3D = {}^1\Delta^3 + 3{}^2\Delta \wedge {}^1\Delta + \underline{{}^3\Delta}$$
$$^4D = {}^1\Delta^4 + 6{}^2\Delta \wedge {}^1\Delta^2 + 4{}^3\Delta \wedge {}^1\Delta + \underline{3{}^2\Delta^2} + \underline{{}^4\Delta}$$

products. As before, this allows us to write the formulas concisely through the wedge products of Grassmann algebra. The results for the p-RDMs through $p = 4$ are summarized in Table II. These functionals for the p-RDMs are exact, but they include the connected p-RDM. An approximation for the p-RDM in terms of lower RDMs may be achieved by setting the connected portion $^p\Delta$ to zero. In this way we recover the functionals for the p-RDMs in Table I with corrections. Thus, through the particle–hole duality, we were able to generate the unconnected portion of the p-RDM exactly. Again, the terms missing in Valdemoro's approximation are denoted by an underline. In general, any terms involving only $^q\Delta$, where $q > 1$, will cancel with the corresponding p-HRDM correction and not appear in Valdemoro's approximation.

The reconstruction functionals may be understood as substantially renormalized many-body perturbation expansions. When exact lower RDMs are employed in the functionals, contributions from *all* orders of perturbation theory are contained in the reconstructed RDMs. As mentioned previously, the reconstruction exactly accounts for configurations in which at least one particle is statistically isolated from the others. Since we know the unconnected p-RDM exactly, all of the error arises from our imprecise knowledge of the connected p-RDM. The connected nature of the connected p-RDM will allow us to estimate the size of its error. For a Hamiltonian with no more than two-particle interactions, the connected p-RDM will have its first nonvanishing term in the $(p - 1)$ order of many-body perturbation theory (MBPT) with a Hartree–Fock reference. This assertion may be understood by noticing that the minimum number of pairwise potentials V required to connect p particles completely is $(p - 1)$. It follows from this that as the number of particles p in the reconstructed RDM increases, the accuracy of the functional approximation improves. The reconstruction formula in Table I for the 2-RDM is equivalent to the Hartree–Fock approximation since it assumes that the two particles are statistically independent. Correlation corrections first appear in the 3-RDM functional, which with $^3\Delta = 0$ is correct through first order of MBPT, and the 4-RDM functional with $^4\Delta = 0$ is correct through second order of MBPT.

Because the reconstruction of the 3-RDM with $^3\Delta = 0$ has a second-order error, the evaluation of the CSE with the unconnected 3- and 4-RDM cumulant expansions has a second-order error. To correct the CSE through second order, we require a second-order estimate for the connected 3-RDM. In the next section we examine two approximations for the connected 3-RDM.

D. Approximation of the Cumulant 3-RDM

Rosina's theorem states that for an *unspecified* Hamiltonian with no more than two-particle interactions the ground-state 2-RDM alone has sufficient information to build the higher RDMs and the exact wavefunction [20, 51]. Cumulants allow us to divide the reconstruction functional into two parts: (i) an *unconnected* part that may be written as antisymmetrized products of the lower RDMs, and (ii) a *connected* part that cannot be expressed as products of the lower RDMs. As shown in the previous section, cumulant theory alone generates all of the unconnected terms in RDM reconstruction, but cumulants do not directly indicate how to compute the connected portions of the 3- and 4-RDMs from the 2-RDM. In this section we discuss a systematic approximation of the connected (or cumulant) 3-RDM [24, 26].

The theory of cumulants allows us to partition an RDM into contributions that scale differently with the number N of particles. Because all of the particles are connected by interactions, the cumulant RDMs $^p\Delta$ scale linearly with the number N of particles. The unconnected terms in the p-RDM reconstruction formulas scale between N^2 and N^p according to the number of connected RDMs in the wedge product. For example, the term $^1\Delta^p$ scales as N^p since all p particles are statistically independent of each other. By examining the scaling of terms with N in the contraction of higher reconstruction functionals, we may derive an important set of relations for the connected RDMs.

In the contraction of any wedge product the position of the upper and lower indices generates two types of terms in Grassmann algebra [26]: (i) *pure* contraction terms where the upper and lower contraction indices appear on the same component of the wedge product, and (ii) *transvection* terms, where the upper and lower contraction indices appear on different components of the wedge product. To illustrate, we consider the contraction of the wedge product between $^3\Delta$ and $^1\Delta$:

$$\hat{P}(^3\Delta \wedge {}^1\Delta) + \hat{T}(^3\Delta \wedge {}^1\Delta) \tag{46}$$

where the $\hat{P}$ and the $\hat{T}$ operators denote the pure contraction and transvection terms, respectively. By having contraction indices on different connected RDMs, the transvection sum joins the two terms to produce a completely connected piece that scales linearly with N. In contrast, in the pure case where the indices are on the same RDM, the resulting unconnected expression scales as N^2. Therefore the contraction of unconnected functionals may yield connected terms through transvection.

The 3- and 4-RDMs are related by the linear contraction mapping

$$\frac{N-3}{4}{}^3D = \hat{L}_4^3({}^4D) \tag{47}$$

Only the connected RDMs ${}^3\Delta$ and ${}^4\Delta$ scale linearly with N in the reconstruction formulas for the 3- and 4-RDMs. However, the contraction of the 4-RDM reconstruction formula in Table I generates by transvection additional terms that scale linearly with N. *Without approximation* the terms that scale linearly with N on both sides of Eq. (47) may be set equal. These terms must be equal to preserve the validity of Eq. (47) for any integer value of N. In this manner we obtain a relation that reveals which terms of the 4-RDM reconstruction functional are mapped to the connected 3-RDM [26]:

$$-\frac{3}{4}{}^3\Delta = 4\hat{T}({}^3\Delta \wedge {}^1\Delta) + 3\hat{T}({}^2\Delta \wedge {}^2\Delta) + \hat{L}_4^3({}^4\Delta) \tag{48}$$

Equation (48) is an exact system of equations [20, 24, 26] relating the elements of ${}^3\Delta$ to the elements of ${}^1\Delta$, ${}^2\Delta$, and ${}^4\Delta$. Because ${}^4\Delta$ vanishes until third order, Eq. (48) suggests that a second-order approximation of ${}^3\Delta$ may be obtained from solving this system of equations with ${}^4\Delta = 0$.

Additional insight may be obtained by writing the system of equations in the natural-orbital basis set, that is, the basis set that diagonalizes the 1-RDM. In this basis set the two terms with the connected 3-RDM may be collected to obtain the formula for the elements of the connected (or cumulant) 3-RDM [26],

$$n_{q,s,t}^{i,j,k}\,{}^3\Delta_{q,s,t}^{i,j,k} = -\frac{1}{6}\sum_l \hat{A}({}^2\Delta_{q,s}^{i,l}\,{}^2\Delta_{l,t}^{j,k}) + \hat{L}_4^3({}^4\Delta)_{q,s,t}^{i,j,k} \tag{49}$$

where

$$n_{q,s,t}^{i,j,k} = {}^1D_i^i + {}^1D_j^j + {}^1D_k^k + {}^1D_q^q + {}^1D_s^s + {}^1D_t^t - 3 \tag{50}$$

and the operator $\hat{A}$ performs all distinct antisymmetric permutations of the indices excluding the summation index l. The formula in Eq. (49) is exact. By setting ${}^4\Delta = 0$, we obtain an approximation for the matrix elements of ${}^3\Delta$ which is correct through second order of perturbation theory except when three of the six indices are occupied in the zero-order (Hartree–Fock) wavefunction.

The two types of cumulant 3-RDM elements that cannot easily be constructed from elements of the 2-RDM are [24, 26]

$$\{{}^3\Delta_{o,o,o}^{x,x,x}\} \quad \text{and} \quad \{{}^3\Delta_{x,o,o}^{x,x,o}\} \tag{51}$$

where o and x denote occupied and virtual orbitals at zero order, respectively. For these classes of elements the sum of the six occupation numbers minus the scalar three in $n_{q,s,t}^{i,j,k}$ vanishes until first order of RDM perturbation theory, and hence the connected 4-RDM divided by $n_{q,s,t}^{i,j,k}$ has a nonvanishing second-order contribution. Because the errors in the cumulant reconstruction formulas, including this correction for the connected 3-RDM, are invariant under unitary transformations of the one-particle basis set, the same part of the 3-RDM (after unitary transformation to another one-particle basis set) cannot be determined by the system of equations in Eq. (48). For these classes of 3-RDM elements we assume that the connected 3-RDM is zero. In the CSE these elements may not be too important. For example, the first class of three-particle excitations do not affect the 2-RDM from the CSE until third order while the elements of the second class have been shown in calculations to be numerically small.

Nakatsuji and Yasuda have proposed a different second-order formula for the connected 3-RDM, which they derived from Feynman perturbation theory [19, 24]. Their correction may be written

$$^3\Delta_{q,s,t}^{i,j,k} \approx \frac{1}{6} \sum_l s_l \hat{A}(^2\Delta_{q,s}^{i,l}\,^2\Delta_{l,t}^{j,k}) \tag{52}$$

where s_l equals 1 if l is occupied in the Hartree–Fock reference and -1 if l is not occupied. Computational experience shows that the Nakatsuji–Yasuda correction also does not improve the classes of cumulant 3-RDM elements in Eq. (51), and in fact, like the correction in Eq. (49), it often makes these elements worse than assuming that they are zero. Hence in the CSE we assume that these elements are zero.

To complete our discussion of the formulas for the connected 3-RDM, we mention that the system of equations in Eq. (48) and the formula in Eq. (49), which is often called the Mazziotti correction [24, 26, 29] to distinguish it from the Nakatsuji–Yasuda correction [19, 24] for the 3-RDM, can also be derived from a contracted Schrödinger equation for the number operator:

$$\langle \psi | a_i^\dagger a_j^\dagger a_k^\dagger a_t a_s a_q \hat{N} | \psi \rangle = N(3!\,^3D_{q,s,t}^{i,j,k}) \tag{53}$$

where in second quantization the number operator is given by

$$\hat{N} = \sum_l a_l^\dagger a_l \tag{54}$$

The right-hand side of the CSE does not scale linearly with N because N times the 3-RDM scales between N^2 and N^4. Taking the part of the equation that scales linearly with N, we obtain

$$\langle \psi | a_i^\dagger a_j^\dagger a_k^\dagger a_t^\dagger a_s^\dagger a_q^\dagger \hat{N} | \psi \rangle_C = 0 \tag{55}$$

in which the notation $\langle \rangle_C$ indicates the connected part of the expectation value. Writing Eq. (55) in terms of the cumulant parts of the 1-, 2-, 3-, and 4-RDMs yields precisely the system of equations in Eq. (48), and upon unitary transformation to a natural-orbital basis set we can obtain the formula in Eq. (49). The only difference in the two derivations is the placement of the number operator. In second quantization the contraction in Eq. (47) is

$$\frac{1}{4!} \langle \psi | a_i^\dagger a_j^\dagger a_k^\dagger \hat{N} a_t a_s a_q | \psi \rangle = \frac{N-3}{4} {}^3 D_{q,s,t}^{i,j,k} \tag{56}$$

Taking the connected part of both sides yields

$$\frac{1}{4!} \langle \psi | a_i^\dagger a_j^\dagger a_k^\dagger \hat{N} a_t a_s a_q | \psi \rangle_C = -\frac{3}{4} {}^3 \Delta_{q,s,t}^{i,j,k} \tag{57}$$

Upon simplification it is not difficult to show that Eq. (57) is equivalent to Eqs. (48) and (55). Therefore the position of the number operator does not affect the relation that we have derived for the cumulant 3-RDM. More general relations for the cumulant p-RDM may similarly be derived by contracting the $(p+1)$-RDM to the p-RDM.

E. Cumulant Structure of the CSE

Cumulant theory offers a systematic approach to reconstructing the 3- and 4-RDMs within the CSE from the 2-RDM, but it also provides insight into the structure of the CSE. Let us define ${}^1C$ as the connected part of the left-hand side of the 1,3-CSE,

$$ {}^1C_k^i = \langle \psi | a_i^\dagger a_k \hat{H} | \psi \rangle_C \tag{58}$$

and ${}^2C$ as the connected part of the left-hand side of the CSE,

$$ {}^2C_{k,l}^{i,j} = \frac{1}{2} \langle \psi | a_i^\dagger a_j^\dagger a_l a_k \hat{H} | \psi \rangle_C \tag{59}$$

As in the previous section, by connected we mean all terms that scale linearly with N. Wedge products of cumulant RDMs can scale linearly if and only if they are connected by the indices of a matrix that scales linearly with N (*transvection*). In the previous section we only considered the indices of the one-particle identity matrix in the contraction (or number) operator. In the CSE we have the two-particle reduced Hamiltonian matrix, which is defined in Eqs. (2) and (3). Even though the one-electron part of ${}^2K$ scales as N^2, the division by $N-1$ in Eq. (3) causes it to scale linearly with N. Hence, from our definition of connected, which only requires the matrix to scale linearly with N, the transvection terms involving ^2K will be connected. Alternatively, one could define the Hamiltonian operator as

$$H = \sum_{p,s} \epsilon_s^p a_p^\dagger a_s + \frac{1}{2} \sum_{p,q;s,t} {}^2V_{s,t}^{p,q} a_p^\dagger a_q^\dagger a_t a_s \tag{60}$$

and then the transvection terms would be defined with respect to the linear-in-N-scaling and local matrices ϵ and ^2V.

It is not difficult to show that the CSE has the following structure in terms of its connected parts ^1C and ^2C and the 1- and 2-RDMs:

$$E\,{}^2D + {}^1D \wedge {}^1C + {}^2C = E\,{}^2D \tag{61}$$

For any choice for the 2-RDM the first unconnected term on the left-hand side of the CSE precisely cancels with the right-hand side. This part of the CSE, therefore, does not contain any information about the 2-RDM, and the CSE is satisfied if and only if

$$^1D \wedge {}^1C + {}^2C = 0 \tag{62}$$

These two terms, $^1D \wedge {}^1C$ and ^2C, however, scale as N^2 and N, respectively. Hence they cannot cancel each other, and we have the result that the CSE is satisfied if and only if the connected 1,3-CSE and the connected CSE vanish:

$$^1C = 0 \tag{63}$$

$$^2C = 0 \tag{64}$$

The connected structure of the CSE has also been explored by Yasuda [23] using Grassmann algebra, by Kutzelnigg and Mukherjee [27] using a cumulant version of second-quantized operators, and by Herbert and Harriman [30] using a diagrammatic technique.

IV. PURIFICATION OF THE 2-RDM

The concept of purification is well known in the linear-scaling literature for one-particle theories like Hartree–Fock and density functional theory, where it denotes the iterative process by which an arbitrary one-particle density matrix is projected onto an idempotent 1-RDM [2, 59–61]. An RDM is said to be *pure N-representable* if it arises from the integration of an N-particle density matrix $\Psi\Psi^*$, where Ψ (the preimage) is an N-particle wavefunction [3–5]. Any idempotent 1-RDM is N-representable with a unique Slater-determinant preimage. Within the linear-scaling literature the 1-RDM may be directly computed with unconstrained optimization, where iterative purification imposes the N-representability conditions [59–61]. Recently, we have shown that these methods for computing the 1-RDM directly are related to the solution of the 1,2-CSE (1,2-CSE is the contraction of the Schrödinger equation onto the one-particle space) [62].

While purification for noninteracting 1-RDMs was first pioneered by McWeeny in the late 1950s [2], the concept was not extended to correlated density matrices until 2002 [29]. We define *purification* of *correlated* RDMs as the iterative process by which an arbitrary p-particle density matrix is projected onto a p-RDM that obeys several necessary conditions for N-representability [28]. Note that the word *necessary* is used since the full set of N-representability conditions for the p-RDM ($p > 1$) is not known. Although there is a considerable literature on minimizing the energy with respect to a 2-RDM that is constrained by N-representability conditions [4, 63–74], the literature on correcting a 2-RDM that is not N-representable is not large [17, 28, 29, 34, 35, 51, 53, 75]. The need for such techniques is suggested by the iterative nature of the CSE. The extension of purification to the 2-RDM plays a role in the solution of the 2,4-CSE, which is analogous to the role of 1-RDM purification in the solution of the 1,2-CSE [28, 29].

A. N-Representability of the 1-RDM

Some of the most important N-representability conditions on the 2-RDM arise from its relationship with the 1-RDM. A 2-RDM must contract to a 1-RDM that is N-representable,

$$^1D = \left(\frac{2}{N-1}\right)\hat{L}_2^1(^2D) \tag{65}$$

where the operator $\hat{L}_2^1$ denotes the contraction operator that maps the 2-RDM to the 1-RDM. The factor of $(N-1)/2$ arises from the normalization of the 1-RDM and 2-RDM to N and $N(N-1)/2$, respectively. The N-representability conditions for the 1-RDM arise from the *particle–hole duality* [7, 53, 63]. The expectation value of the anticommutation relation for fermions in Eq. (15) yields the relation between the elements of the 1-RDM $^1D_j^i$ and the elements of the one-hole RDM $^1\bar{D}_j^i$,

$$^1\bar{D}_j^i + {}^1D_j^i = {}^1I_j^i \tag{66}$$

where ^1I is the identity matrix. Any 1-RDM is ensemble N-representable if and only if it is Hermitian with trace N and both the 1-RDM and its one-hole RDM are positive semidefinite [3, 4, 7, 53, 63], which is denoted by

$$^1D \geq 0 \tag{67}$$

and

$$^1\bar{D} \geq 0 \tag{68}$$

A matrix is *positive semidefinite* if and only if all of its eigenvalues are nonnegative. Because the 1-RDM and the one-hole RDM share the same eigenvectors, these two positivity restrictions are equivalent to constraining the occupation numbers of the 1-RDM to lie between zero and one [3].

Purification of a trial 2-RDM with the 1-RDM conditions may be accomplished by contracting the 2-RDM as in Eq. (65) and checking that the eigenvalues of the 1-RDM lie between zero and one. If the eigenvalues fall outside this interval, neither the 1-RDM nor the 2-RDM can be N-representable. Any method for adjusting the 1-RDM occupation numbers must preserve the trace of the 1-RDM, which is the number N of particles. We have employed the following algorithm to effect this purification: (i) set all of the negative 1-RDM eigenvalues to zero; (ii) correct the trace by decreasing the occupation number for the highest occupied orbital; (iii) set all 1-RDM eigenvalues greater than one to one; and (iv) correct the trace by increasing the occupation number for the lowest unoccupied orbital. We decrease the highest occupied orbital and increase the lowest unoccupied orbital since these changes are unlikely to produce occupation numbers outside the zero-to-one interval. This is only one reasonable approach to ensuring that the occupation numbers of the 1-RDM are N-representable; many variations on this simple strategy may also be employed. Once the 1-RDM has been adjusted to be N-representable, we need a method for modifying the 2-RDM so that it contracts by Eq. (65) to the updated 1-RDM.

B. N-Representability of the 2-RDM

The appropriate modification of the 2-RDM may be accomplished by combining N-representability constraints, known as positivity conditions, with both the *unitary* and the *cumulant* decompositions of the 2-RDM.

1. Unitary and Cumulant Decompositions of the 2-RDM

Any two-particle Hermitian matrix ^2A may be decomposed into three components that exist in different subspaces of the unitary group. These components reveal the structure of the matrix with respect to the contraction operation [4, 76–80],

$$^2A = {}^2A_0 + {}^2A_1 + {}^2A_2 \qquad (69)$$

where

$$^2A_0 = \frac{2\,\mathrm{Tr}(^2A)}{r(r-1)}\,{}^2I \qquad (70)$$

$$^2A_1 = \frac{4}{r-2}\,{}^1A \wedge {}^1I - \frac{4\,\mathrm{Tr}(^2A)}{r(r-2)}\,{}^2I \qquad (71)$$

and

$$^2A_2 = {}^2A - \frac{4}{r-2}{}^1A \wedge {}^1I + \frac{2\,\text{Tr}(^2A)}{(r-1)(r-2)}{}^2I \qquad (72)$$

The one-particle matrix ^1A is the contraction of the two-particle matrix ^2A,

$$^1A = \hat{L}_2^1(^2A) \qquad (73)$$

the symbol r denotes the rank of the one-particle basis set, and

$$^2I = {}^1I \wedge {}^1I \qquad (74)$$

The zeroth component $^2A_0$ contains the trace information for ^2A,

$$\hat{L}_2^0(^2A_0) = \text{Tr}(^2A_0) = \text{Tr}(^2A) \qquad (75)$$

and the first component $^2A_1$ contains the one-particle information for ^2A except for the trace:

$$\hat{L}_2^1(^2A_0 + {}^2A_1) = {}^1A \qquad (76)$$

The two-particle component of ^2A carries information that vanishes upon contraction,

$$\hat{L}_2^1(^2A_2) = 0 \qquad (77)$$

where 0 in this equation represents the zero matrix.

The unitary decomposition may be applied to any Hermitian, antisymmetric two-particle matrix including the 2-RDM, the two-hole RDM, and the two-particle reduced Hamiltonian. The decomposition is also readily generalized to treat p-particle matrices [80–82]. The trial 2-RDM to be purified may be written

$$^2D = {}^2D_0 + {}^2D_1 + {}^2D_2 \qquad (78)$$

Note that if $^2A = {}^2D$ in Eqs. (70), (71), and (72), then from Eqs. (65) and (76) we have that

$$^1A = \left(\frac{N-1}{2}\right){}^1D \qquad (79)$$

Using Eq. (71) and the adjusted 1-RDM from the last section, we can construct a modified one-particle portion of the 2-RDM $^2D_1^a$. Then the appropriate 2-RDM that contracts to the adjusted 1-RDM is readily expressed as

$$^2D^a = {}^2D_0 + {}^2D_1^a + {}^2D_2 \qquad (80)$$

Both the trace and one-particle subspaces of the 2-RDM are now N-representable. Does the 1-RDM tell us anything about the two-particle component of the 2-RDM, which vanishes when it is contracted to the one-particle space? Before examining additional N-representability conditions, we address this question.

As discussed in Sections III. B and III. C the unitary decomposition is not the only approach for expressing an RDM in terms of lower RDMs. The cumulant decomposition (or expansion) of the 2-RDM [21, 22, 24, 26, 39, 40] is

$$^2D = {}^1D \wedge {}^1D + {}^2\Delta \qquad (81)$$

The portion of the 2-RDM that may be expressed as wedge products of lower RDMs is said to be *unconnected*. The unconnected portion of the 2-RDM contains an important portion of the two-particle component from the unitary decomposition $^2D_2$, and similarly, the trace and one-particle unitary components contain an important portion of the connected 2-RDM $^2\Delta$, which corrects the contraction. Both decompositions may be synthesized by examining the unitary decomposition of the connected 2-RDM,

$$^2\Delta = {}^2\Delta_0 + {}^2\Delta_1 + {}^2\Delta_2 \qquad (82)$$

The trace and the one-particle components of the connected 2-RDM are completely determined by the 1-RDM. Hence, it is *the two-particle unitary subspace of the connected 2-RDM that may require further purification*.

Similarly, the cumulant decomposition for the two-hole RDM is

$$^2\bar{D} = {}^1\bar{D} \wedge {}^1\bar{D} + {}^2\bar{\Delta} \qquad (83)$$

With the anticommutation relation for fermions in Eq. (15) and the second-quantized definitions, it has been shown that the connected portions of the two-particle and two-hole RDMs are equal [14, 20, 38]:

$$^2\bar{\Delta} = {}^2\Delta \qquad (84)$$

It follows that

$$^2\bar{\Delta}_2 = {}^2\Delta_2 \qquad (85)$$

Therefore we have the important fact that *for a fixed 1-RDM any correction to the 2-RDM will also be a correction to the two-hole RDM*. In the next section we use this fact in purifying the 2-RDM to satisfy two N-representability restrictions.

2. Positivity Conditions on the 2-RDM

Two significant N-representability conditions on the 2-RDM are that both the two-particle and the two-hole RDMs must be positive semidefinite:

$$^2D \geq 0 \tag{86}$$

and

$$^2\bar{D} \geq 0 \tag{87}$$

In the N-representability literature these positivity conditions are known as the D- and the Q-conditions [5, 7, 63]. The two-particle RDM and the two-hole RDM are linearly related via the particle–hole duality,

$$^2\bar{D} = {}^2I - 2\,{}^1D \wedge {}^1I + {}^2D \tag{88}$$

If the trial 2-RDM does not obey the D-condition, then it has a set of eigenvectors $\{v_i\}$ whose associated eigenvalues are negative. Hence we can construct a set of two-particle matrices $\{{}^2O_i\}$

$$^2O_i = v_i v_i^\dagger \tag{89}$$

for which

$$\mathrm{Tr}({}^2O_i\,{}^2D) < 0 \tag{90}$$

Each member of the set $\{{}^2O_i\}$ is said to *expose* the 2-RDM [4, 52]. Similarly, if the trial 2-RDM does not obey the Q-condition, then the two-hole RDM has a set of eigenvectors $\{\bar{v}_i\}$ whose associated eigenvalues are negative. The bar in $\bar{v}_i$ simply distinguishes the eigenvectors of the two-hole RDM from those of the 2-RDM; it does not denote the adjoint. A set of two-hole matrices $\{{}^2\bar{O}_i\}$ may be generated

$$^2\bar{O}_i = \bar{v}_i \bar{v}_i^\dagger \tag{91}$$

for which

$$\mathrm{Tr}({}^2\bar{O}_i\,{}^2\bar{D}) < 0 \tag{92}$$

As with the D-condition, each member of the set $\{{}^2\bar{O}_i\}$ is said to expose the two-hole RDM.

The 2-RDM may be made positive semidefinite if each of the negative eigenvalues is set to zero, but this alters not only the positivity but also the contraction of the 2-RDM to the 1-RDM and even the 2-RDM trace. How can we modify the 2-RDM to prevent it from being exposed by the set $\{^2O_i\}$ and yet maintain contraction to the N-representable 1-RDM? Again we can employ the unitary decomposition. For a matrix ^2O_i the decomposition is

$$^2O_i = {}^2O_{i;0} + {}^2O_{i;1} + {}^2O_{i;2} \tag{93}$$

Zeroing the 2-RDM eigenvalue associated with v_i is equivalent to adding an appropriate amount of ^2O_i to the 2-RDM. However, this also changes the trace and the underlying 1-RDM because ^2O_i contains the zeroth and the first components of the unitary decomposition. We can modify the two-particle component only by adding just $^2O_{i;2}$ rather than ^2O_i. The adjusted 2-RDM may then be expressed as

$$^2D_a = {}^2D + \sum_i \alpha_i {}^2O_{i;2} \tag{94}$$

where the set of coefficients is determined from the system of linear equations

$$\text{Tr}(^2O_i {}^2D_a) = 0 \quad \forall i \tag{95}$$

Although the adjusted 2-RDM is not exposed by any of the matrices in the set $\{^2O_i\}$, in general there will be new eigenvectors with negative eigenvalues. However, these negative eigenvalues are in general smaller than those of the unadjusted 2-RDM. Hence, by repeating this process *iteratively*, the 2-RDM may be purified so that the D-condition is satisfied *without* modifying the contraction.

Analogously, the two-hole matrices in the set $\{^2\bar{O}_i\}$ may be decomposed:

$$^2\bar{O}_i = {}^2\bar{O}_{i;0} + {}^2\bar{O}_{i;1} + {}^2\bar{O}_{i;2} \tag{96}$$

To impose only the Q-condition, we have an adjusted two-hole RDM

$$^2\bar{D}_a = {}^2\bar{D} + \sum_i \beta_i {}^2\bar{O}_{i;2} \tag{97}$$

whose coefficients are determined from the system of linear equations

$$\text{Tr}(^2\bar{O}_i {}^2\bar{D}_a) = 0 \quad \forall i \tag{98}$$

One possibility for imposing both the D- and the Q-conditions is to update the 2-RDM via Eqs. (94) and (95), convert the 2-RDM to the two-hole RDM, update the two-hole RDM via Eqs. (97) and (98), and then to repeat this process until convergence. However, this alternating approach does not usually show good convergence since the 2-RDM changes often damage Q-positivity and the two-hole RDM changes often adversely affect D-positivity. A better approach would be to impose both the D- and the Q-updates *simultaneously*.

A simultaneous purification with respect to both the D- and the Q-conditions may be achieved by using the fact that for a fixed 1-RDM any correction to the 2-RDM will also be a correction to the two-hole RDM and vice versa. This suggests that we write the adjusted 2-RDM as

$$^2D_a = {}^2D + \sum_i \alpha_i {}^2O_{i;2} + \sum_i \beta_i {}^2\bar{O}_{i;2} \qquad (99)$$

where the expansion coefficients are determined by solving the linear equations in both Eqs. (95) and (98) simultaneously. Note that the linear mapping between the 2-RDM and the two-hole RDM must be employed in Eq. (98). The resulting adjusted 2-RDM will not be exposed by either the operators $\{^2O_i\}$ or, in its two-hole form, the operators $\{^2\bar{O}_i\}$. Repeated application of this purification produces a 2-RDM that satisfies, to a specified tolerance, the D- and the Q-conditions.

3. Spin Blocks of the 2-RDM

The RDMs for atoms and molecules have a special structure from the spin of the electrons. To each spatial orbital, we associate a spin of either α or β. Because the two spins are orthogonal upon integration of the N-particle density matrix, only RDM blocks where the net spin of the upper indices equals the net spin of the lower indices do not vanish. Hence a p-RDM is block diagonal with $(p+1)$ nonzero blocks. Specifically, the 1-RDM has two nonzero blocks, an α-block and a β-block:

$$^1D^{\alpha,i}_{\alpha,j} \neq 0 \qquad ^1D^{\beta,i}_{\beta,j} \neq 0 \qquad (100)$$

and the 2-RDM has three nonzero blocks, an α/α-block, an α/β-block, and a β/β-block:

$$^2D^{\alpha,i;\alpha,k}_{\alpha,j;\alpha,l} \neq 0 \qquad ^2D^{\alpha,i;\beta,k}_{\alpha,j;\beta,l} \neq 0 \qquad ^2D^{\beta,i;\beta,k}_{\beta,j;\beta,l} \neq 0 \qquad (101)$$

The spin structure enhances computational efficiency since each of the blocks may be purified separately.

For the remainder of this section we treat closed-shell atoms and molecules, where the α- and the β-spins are indistinguishable. Because the α- and the β-blocks of the 1-RDM are equal, we need only purify the eigenvalues for one of these blocks.

As in Section IV.A. the eigenvalues of the 1-RDM must lie in the interval $[0, 1]$ with the trace of each block equal to $N/2$. Similarly, with the α/α- and the β/β-blocks of the 2-RDM being equal, only one of these blocks requires purification. The purification of either block is the same as in Section IV.B.2 with the normalization being $N(N/2 - 1)/4$. The unitary decomposition ensures that the α/α-block of the 2-RDM contracts to the α-component of the 1-RDM. The purification of Section IV.B.2, however, cannot be directly applied to the α/β-block of the 2-RDM since the spatial orbitals are not antisymmetric; for example, the element with upper indices $\alpha, i; \beta, i$ is not necessarily zero. One possibility is to apply the purification to the entire 2-RDM. While this procedure ensures that the whole 2-RDM contracts correctly to the 1-RDM, it does not generally produce a 2-RDM whose individual spin blocks contract correctly. Usually the overall 1-RDM is correct only because the α/α-spin block has a contraction error that cancels with the contraction error from the α/β-spin block.

A better strategy is to introduce a modified unitary decomposition for the α/β-block. An appropriate decomposition is

$$^2D_0^{\alpha,\beta} = \frac{\text{Tr}(^2D^{\alpha,\beta})}{r_s^2}{}^1I_\alpha^\alpha{}^1I_\beta^\beta \tag{102}$$

$$^2D_1^{\alpha,\beta} = \frac{1}{r_s}\left(^1D^\alpha - \frac{\text{Tr}(^1D^\alpha_\alpha)}{r_s}{}^1I^\alpha_\alpha\right){}^1I^\beta_\beta + \frac{1}{r_s}{}^1I^\alpha_\alpha\left(^1D^\beta_\beta - \frac{\text{Tr}(^1D^\beta_\beta)}{r_s}{}^1I^\beta_\beta\right) \tag{103}$$

and

$$^2D_2^{\alpha,\beta} = {}^2D^{\alpha,\beta} - {}^2D_1^{\alpha,\beta} - {}^2D_0^{\alpha,\beta} \tag{104}$$

where r_s denotes the number of spatial orbitals, which equals half the number r of spin orbitals. Like the unitary decomposition for antisymmetric matrices in Section IV.B.1, the zeroth component $^2D_0^{\alpha,\beta}$ contains the trace information,

$$\hat{L}_2^0(^2D_0^{\alpha,\beta}) = \text{Tr}(^2D_0^{\alpha,\beta}) = \text{Tr}(^2D^{\alpha,\beta}) \tag{105}$$

and the first component $^2D_1^{\alpha,\beta}$ contains the one-particle information except for the trace,

$$\hat{L}_2^1(^2D_0^{\alpha,\beta} + {}^2D_1^{\alpha,\beta}) = \frac{N}{2}{}^1D^\alpha_\alpha \tag{106}$$

The two-particle component of $^2D^{\alpha,\beta}$ carries information that vanishes upon contraction,

$$\hat{L}_2^1(^2D_2^{\alpha,\beta}) = 0 \tag{107}$$

where the 0 represents the zero matrix. The purification process for the 2-RDM's α/β-block remains the same as described in Section IV.B.2 except that the decomposition in Eqs. (102)–(104) is employed.

V. SELF-CONSISTENT ITERATION

A fundamental approach to computing the ground-state wavefunction and its energy for an N-electron system is the *power method* [20, 83]. In the power method a series of trial wavefunctions $|\Phi_n\rangle$ are generated by repeated application of the Hamiltonian

$$|\Phi_{n+1}\rangle = H|\Phi_n\rangle \tag{108}$$

The Hamiltonian gradually filters the ground-state wavefunction from the trial wavefunction. To understand this filtering process, we expand the initial trial wavefunction in the exact wavefunctions of the Hamiltonian $|\Psi_i\rangle$. With n iterations of the power method, we have

$$|\Phi_{n+1}\rangle = H^n|\Phi_1\rangle \tag{109}$$
$$= E_1^n c_1|\Psi_1\rangle + E_2^n c_2|\Psi_2\rangle + \cdots + E_i^n c_i|\Psi_i\rangle \tag{110}$$

As long as the $|E_1|$ is greater than $|E_i|$ for any $i \neq 1$, the power method upon normalization will converge to $|\Psi_1\rangle$. The rate of convergence depends on the ratio of the energy with second largest magnitude to the energy with the largest magnitude, that is, $|E_2|/|E_1|$.

The power method for the wavefunction may be adapted to a power method for the N-particle density matrix:

$$^N D_{n+1} = \tfrac{1}{2}(H\,{}^N D_n + {}^N D_n H) \tag{111}$$

If $|\Delta\rangle = |\Phi_{n+1}\rangle - |\Phi_n\rangle$, then the wavefunction update in Eq. (108) corresponds to the following density-matrix update,

$$^N D_{n+1} = |\Phi_n + \Delta\rangle\langle\Phi_n + \Delta| \tag{112}$$
$$= {}^N D_n + |\Delta\rangle\langle\Phi_n| + |\Phi_n\rangle\langle\Delta| + |\Delta\rangle\langle\Delta| \tag{113}$$

while the density-matrix update in Eq. (111) is

$$^N D_{n+1} = {}^N D_n + \tfrac{1}{2}|\Delta\rangle\langle\Phi_n| + \tfrac{1}{2}|\Phi_n\rangle\langle\Delta| \tag{114}$$

The two updates differ only by a factor of one-half before the first-order change from Δ and the second-order change. Unlike the wavefunction power method, the N-particle density matrices from each iteration in Eq. (111) are not exactly positive semidefinite until convergence.

A *contracted* power method for the 2-RDM may be developed by projecting the N-particle power method onto the space of two particles

$$^2D^{i,j;k,l}_{n+1} = \frac{1}{4E}\left(\langle\Phi_n|a_i^\dagger a_j^\dagger a_l a_k H|\Phi_n\rangle + \langle\Phi_n|Ha_i^\dagger a_j^\dagger a_l a_k|\Phi_n\rangle\right) \tag{115}$$

where E is the energy associated with the nth trial 2-RDM. From the CSE we can write Eq. (115) as

$$^2D^{i,j;k,l}_{n+1} = \frac{1}{4E}\sum_{m,n,p,q}(^4G^{i,j,k,l}_{m,n,p,q} + {}^4G^{m,n,p,q}_{i,j,k,l})^2K^{m,n}_{p,q} \tag{116}$$

where the generalized G-matrix ^4G is expressible in terms of the 3- and the 4-RDMs as follows:

$$^4G^{i,j,k,l}_{m,n,p,q} = 4!\,{}^4D^{i,j,m,n}_{k,l,p,q} + 3!(^3D^{i,j,m}_{l,p,q}\delta^n_k - {}^3D^{i,j,m}_{k,p,q}\delta^n_l) \tag{117}$$
$$+ 3!(^3D^{i,j,n}_{k,p,q}\delta^m_l - {}^3D^{i,j,n}_{l,p,q}\delta^m_k) + 2^2D^{i,j}_{p,q}(\delta^m_k\delta^n_l - \delta^m_l\delta^n_k) \tag{118}$$

The 2-RDM is automatically antisymmetric, but it may require an adjustment of the trace to correct the normalization. The functionals in Table I from cumulant theory allow us to approximate the 3- and the 4-RDMs from the 2-RDM and, hence, to iterate with the contracted power method. Because of the approximate reconstruction the contracted power method does not yield energies that are strictly above the exact energy. As in the full power method the updated 2-RDM in Eq. (116) moves toward the eigenstate whose eigenvalue has the largest magnitude.

VI. ALGORITHM FOR SOLVING THE CSE

Here we synthesize the concepts of the last four sections, (i) CSE, (ii) reconstruction, (iii) purification, and (iv) a contracted power method, to obtain an iterative algorithm for the direct calculation of the 2-RDM.

CSE Algorithm

1. Select an initial 2-RDM such as the Hartree–Fock 2-RDM.
2. Reconstruct the 3- and 4-RDMs from the 2-RDM.

3. Evaluate the 2-RDM update in Eq. (116).
4. Purify the new 2-RDM.
5. Repeat steps 2, 3, and 4 until convergence.

In practice, the reconstruction of the 3- and 4-RDMs is performed while the CSE is being evaluated in step 3. With fast summation the scaling of the CSE algorithm is r^6 with the connected 3-RDM set to zero and r^7 with connected 3-RDM corrections.

VII. APPLICATIONS

With the CSE both the N-particle energy and the 2-RDM may be computed for quantum systems of fermions. In this section we illustrate the contracted power method for several molecular systems. Each of the molecules in Table III is treated in its equilibrium geometry [84], where the integrals are computed with PC Gamess [85], an implementation of the quantum chemistry package GAMESS (USA) [86]. The molecules in Tables III to V are represented with Slater-type orbitals expanded in six Gaussian functions while the molecules in Table VI are treated in a split-valence double-zeta basis set [87]. Spin orbitals are employed, and none of the core orbitals is frozen. The wavefunction methods and their abbreviations are: (i) Hartree–Fock (HF), (ii) second- and third-order many-body perturbation theory (MP2 and MP3, respectively), and (iii) full configuration interaction (FCI). The contracted power method with purification is applied with three different choices for reconstructing the RDMs in the CSE: (i) the first-order (or unconnected) formula (U), (ii) the second-order (Nakatsuji–Yasuda) formula (N), and (iii) the second-order (Mazziotti) formula (M).

For all molecules in Table III the U energies are better than those obtained from MP2. The U method yields 99.4% and 101.6% of the correlation energy

TABLE III
Molecular Energies from the CSE (STO Basis)

			% of Correlation Energy				
	Energy		ψ Methods			CSE Methods	
Molecule	HF	FCI	MP2	MP3	U	N	M
BeH$_2$	−15.7233	−15.7590	65.2	86.8	107.4	105.8	106.4
CH$_4$	−39.7144	−39.7926	74.2	91.3	118.6	105.9	103.6
CO	−112.3033	−112.4430	92.1	89.1	99.4	85.1	83.8
H$_2$O	−75.6788	−75.7289	71.4	90.9	110.9	99.7	93.3
N$_2$	−108.5418	−108.7005	97.9	94.4	101.6	87.5	86.2
NH$_3$	−55.7200	−55.7890	76.0	91.8	119.7	100.4	99.3

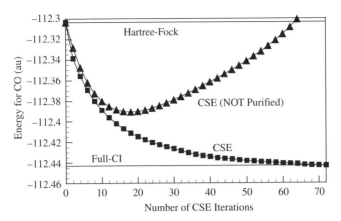

Figure 1. The energy for the molecule CO is given as a function of the number of contracted power iterations. With first-order (U) reconstruction the CSE obtains the correlation energy within 1%.

(CE) for CO and N_2, while MP3 produces only 89.1% and 94.4% of the CE for these molecules. Figure 1 presents the energy for CO as a function of the number of contracted power iterations. All of the molecules except CO and N_2 have even better energies with the second-order methods. For the molecules BeH_2, CH_4, H_2O, and NH_3, both the M and the N methods generate more of the CE than either MP2 or MP3. With MP2 the percentage of CE ranges from a low of 65.2% for BeH_2 to a high of 76% for NH_3; in contrast, the ranges for M and N are from 93.3% (H_2O) to 99.3% (NH_3) and from 94.1% (CH_4) to 99.7% (H_2O). Even for CO and N_2 the CEs of the N method differ from those of MP3 by only 0.006 au and 0.011 au while the absolute values of the CEs are 0.140 au and 0.159 au, respectively. The N and M methods are rather similar in accuracy with the methods differing in the percentage of CE by as little as 0.6% (BeH_2) and by a maximum of 6.4% (H_2O). For the molecules in Table III the energies from N are a little better than those from M except for CH_4. Both the ability of the first-order U to yield better energies than MP2 and the ability of the second-order methods (M and N) to improve the energies of MP3 reflect the perturbative renormalization within the CSE.

For each molecule the errors in the α/α- and the α/β-blocks of the 2-RDM are reported in Table IV. The 2-RDM errors are measured through a least-squares norm, which is defined by

$$\text{Error} = (\text{Tr}[(^2D_{\text{exact}} - {}^2D_{\text{approx}})^2])^{1/2} \qquad (119)$$

where the 2-RDMs are normalized to unity. Except for CO the α/β-block of the Hartree–Fock 2-RDM has more error than the α/α-block. Both the N and the M

TABLE IV
2-RDM Error from the CSE

Molecule	$\|\|^2D^{\alpha,\beta}_{CSE} - {}^2D^{\alpha,\beta}_{FCI}\|\|$			
	HF	U	N	M
BeH$_2$	0.0151	0.00567	0.00401	0.00363
CH$_4$	0.00655	0.00240	0.00143	0.00146
CO	0.00520	0.00202	0.00273	0.00277
H$_2$O	0.00519	0.00228	0.00100	0.00116
N$_2$	0.00515	0.00182	0.00247	0.00250
NH$_3$	0.00633	0.00211	0.00113	0.00131

methods improve the α/β-block of the Hartree–Fock 2-RDM by half an order of magnitude. As with the energies in Table III, both the N and the M methods yield better 2-RDMs than U except for CO and N$_2$. Hence other two-particle properties may be expected to mirror the energetic accuracy for CO and N$_2$ offered by U.

In Table V we check the N-representability of the CSE 2-RDMs through three well-known positivity conditions, the D-, the Q-, and the G-conditions [4, 5, 63]. The D- and the Q-conditions are given in Eqs. (86) and (87), while the G-condition states that the following matrix (known as the G-matrix)

$$^2G^{i,j}_{k,l} = \langle \Psi | a_i^\dagger a_j a_l^\dagger a_k | \Psi \rangle \tag{120}$$

must be positive semidefinite. These conditions are necessary but not sufficient for the 2-RDM to be N-representable. In Table V the D-, the Q- and the G-matrices are normalized to unity. Note that the energetic similarity between the N and the M methods is further reflected in positivity errors, which are quite close when compared with the positivity errors from the U approximation. We

TABLE V
Positivity of the CSE 2-RDMs

	Positivity Error Measured by the Lowest Negative Eigenvalue					
	D-Matrix		Q-Matrix		G-Matrix	
Molecule	U	M	U	M	U	M
BeH$_2$	−2.5e−5	−3.1e−5	−7.4e−6	−9.4e−6	−1.2e−4	−1.3e−4
CH$_4$	−1.9e−7	−1.3e−7	−5.7e−6	5.1e−7	−8.0e−5	−1.1e−4
CO	−6.7e−8	−7.1e−14	−1.4e−5	−5.9e−7	−5.9e−5	−1.4e−4
H$_2$O	−1.5e−5	−1.2e−13	−6.0e−5	−1.7e−6	−2.2e−4	−1.3e−4
N$_2$	−3.3e−7	−4.3e−14	−1.0e−5	−6.3e−7	−6.7e−5	−1.4e−4
NH$_3$	−8.7e−7	−3.4e−8	−2.5e−5	−1.5e−6	−8.1e−5	−8.9e−5

TABLE VI
Molecular Energies from the CSE (Double-Zeta Basis)

| | Energy | | % of Correlation Energy | | | | |
| | | | ψ Methods | | CSE Methods | | |
Molecule	HF	FCI	MP2	MP3	U	N	M
BeH_2	−15.7602	−15.8008	69.5	89.0	71.9	90.5	88.8
BH	−25.1134	−25.1740	59.7	81.3	51.3	73.5	71.1
CH_4	−39.8951	−40.0182	83.3	94.0	79.7	92.5	90.6
CO	−112.6848	−112.8934	97.3	89.4	81.3	84.7	81.1
HF	−100.0219	−100.1464	96.6	94.9	103.2	90.6	90.2
H_2O	−76.0091	−76.1419	93.5	94.4	91.8	95.5	92.8
N_2	−108.8781	−109.1043	100.9	90.6	80.7	84.9	81.1
NH_3	−55.9648	−56.0988	89.4	94.2	83.2	94.0	90.8

expect the 2-RDM to satisfy the D- and the Q-conditions since they are enforced at each iteration of the contracted power method through purification. However, it is quite important that within the framework of the CSE and reconstruction the enforcement of the D- and Q-conditions also causes the G-condition to be satisfied within 10^{-4}. As shown in Fig. 1, without purification the CSE obtains only half of the correlation energy before diverging.

In Table VI we apply the CSE algorithm to several molecules in a double-zeta basis set [87]. The first-order reconstruction U within the CSE yields energies that are similar to those from MP2 for BeH_2, BH, CH_4, HF, H_2O, and NH_3. Again, the first-order reconstruction captures second-order accuracy in the energies. The second-order N and M methods improve on the energies of the U method except for HF, CO, and N_2. Both N and M are significantly better than MP2 for BeH_2, BH, and CH_4; for example, with BeH_2 MP2 yields 69.5% while N and M give 90.5% and 88.8%, respectively. For NH_3 and H_2O the M and the N methods produce more than 90% of the CE, which is similar to the accuracy of MP2 and MP3. Even for N_2, CO, and HF the percentage of CE from N trails the percentages from MP3 by only 5.7%, 4.7%, and 4.3%. The energies from the N and the M methods are quite similar with the energies from N being slightly more accurate for the double-zeta molecules considered; the percentages of CE from the two methods differ by as little as 0.4% for HF and only by as much as 3.8% for N_2.

VIII. A LOOK AHEAD

An algorithm for solving the CSE has been developed with emphasis on three key features: (i) reconstruction of the 3- and 4-RDMs, (ii) purification of the

2-RDM, and (iii) optimization by self-consistent iteration. The CSE method allows the direct calculation of the electronic energies and 2-RDMs without calculation or storage of the many-electron wavefunction. While significant progress has been made in the last decade in using the CSE to compute the 2-RDM, there remain many open questions and directions for improving both the efficiency and accuracy of the calculations. We believe that important progress will be made in the three key areas of reconstruction, purification, and optimization, where purification will benefit from the recent advances in the variational optimization of the 2-RDM via semidefinite programming [63, 66–74], which is discussed in the next part of the book. A related, interesting area for future research is the anti-Hermitian part of the CSE [27, 31, 63], which may be written with a commutator as

$$\langle \psi | [a_p^\dagger a_q^\dagger a_t a_s, \hat{H}] | \psi \rangle = 0 \tag{121}$$

Importantly, the anti-Hermitian CSE may be evaluated through second order of a renormalized perturbation theory even when the cumulant 3-RDM is neglected in the reconstruction. The anti-Hermitian part of the CSE [27, 31, 63] is the stationary condition for two-body unitary transformations of the N-particle wavefunction [31, 32], and hence the two-body unitary transformations may easily be evaluated with the anti-Hermitian CSE and RDM reconstruction without the many-electron Schrödinger equation. The contracted Schrödinger equation in conjunction with the concepts of reconstruction and purification provides a new, important approach to computing the 2-RDM directly without the many-electron wavefunction.

Acknowledgments

The author expresses his appreciation to Dudley Herschbach, Herschel Rabitz, John Coleman, and Alexander Mazziotti for their support and encouragement. The author thanks the NSF, the Henry-Camille Dreyfus Foundation, the Alfred P. Sloan Foundation, and the David-Lucile Packard Foundation for their support.

APPENDIX: GRASSMANN PRODUCTS

The Grassmann (or wedge) product [20, 54, 78] of a q-RDM with a $(p - q)$-RDM may be expressed as

$$^q D \wedge {}^{(p-q)} D = A_N {}^q D \otimes {}^{(p-q)} D A_N \tag{A1}$$

where A_N is the N-particle antisymmetrization operator and $\otimes$ is the tensor product. To utilize this formula in a calculation, we must understand the technique for evaluating wedge products of matrices. Let us first consider the wedge product C of two one-particle matrices, ^1A and ^1B,

$$^2C = {}^1A \wedge {}^1B \tag{A2}$$

The elements $c_{k,l}^{i,j}$ of the matrix ^2C may be obtained from a_k^i and b_l^j by summing the distinct products arising from all antisymmetric permutations of the upper indices and all antisymmetric permutations of the lower indices. With the wedge product of one-particle matrices, there are only four distinct possibilities:

$$c_{k,l}^{i,j} = a_k^i \wedge b_l^j = \tfrac{1}{4}(a_k^i b_l^j - a_k^j b_l^i - a_l^i b_k^j + a_l^j b_k^i) \tag{A3}$$

More generally, we can write the elements of the wedge product as

$$a_{j_1,j_2,\ldots,j_p}^{i_1,i_2,\ldots,i_p} \wedge b_{j_{p+1},\ldots,j_N}^{i_{p+1},\ldots,i_N} = \left(\frac{1}{N!}\right)^2 \sum_{\pi,\sigma} \epsilon(\pi)\epsilon(\sigma) \pi\sigma a_{j_1,j_2,\ldots,j_p}^{i_1,i_2,\ldots,i_p} b_{j_{p+1},\ldots,j_N}^{i_{p+1},\ldots,i_N} \tag{A4}$$

in which π represents all permutations of the upper indices and σ represents all permutations of the lower indices, while the function $\epsilon(\pi)$ returns $+1$ for an even number of transpositions and -1 for an odd number of transpositions. Since both the upper and the lower indices have $N!$ permutations, there are $(N!)^2$ terms in the sum. Hence normalization requires division by $(N!)^2$. If, however, the elements $a_{j_1,j_2,\ldots,j_p}^{i_1,i_2,\ldots,i_p}$ and $b_{j_{p+1},\ldots,j_N}^{i_{p+1},\ldots,i_N}$ are already antisymmetric in their upper and lower indices, only $(N!/(p!q!))^2$ of the above terms will be distinct. This allows us to decrease the number of numerical operations required for computing the wedge product. For wedge products between matrices with the same number of upper and lower indices, we have an important commutation relation

$$^pA_{j_1,j_2,\ldots,j_p}^{i_1,i_2,\ldots,i_p} \wedge {}^qB_{j_{p+1},\ldots,j_N}^{i_{p+1},\ldots,i_N} = {}^qB_{j_1,\ldots,j_q}^{i_1,\ldots,i_q} \wedge {}^pA_{j_{q+1},\ldots,j_N}^{i_{q+1},\ldots,i_N} \tag{A5}$$

or without the indices

$$^pA \wedge {}^qB = {}^qB \wedge {}^pA \tag{A6}$$

If the sum $(p+q)$ is odd, exchanging the p upper indices with q upper indices will produce a minus sign, but this will be cancelled by another minus sign produced by exchanging the lower indices. In many cases it will be easier and clearer to write the wedge products as in the second form, Eq. (A6), without specifying a particular element through indices.

References

1. D. Ter Haar, Theory and applications of the density matrix. *Rep. Prog. Phys.* **24**, 304 (1961).
2. R. McWeeny, Some recent advances in density matrix theory. *Rev. Mod. Phys.* **32**, 335 (1960).
3. A. J. Coleman, Structure of fermion density matrices. *Rev. Mod. Phys.*, **35**, 668 (1963).
4. A. J. Coleman and V. I. Yukalov. *Reduced Density Matrices: Coulson's Challenge,* Springer-Verlag, New York, 2000.
5. C. Garrod and J. Percus, Reduction of N-particle variational problem. *J. Math. Phys.* **5**, 1756 (1964).
6. H. Kummer, N-representability problem for reduced density matrices. *J. Math. Phys.* **8**, 2063 (1967).
7. A. J. Coleman, Necessary conditions for N-representability of reduced density matrices. *J. Math. Phys.* **13**, 214 (1972).
8. A. J. Coleman, *Rept. Math. Phys.* **4**, 113 (1973).
9. E. R. Davidson, *Reduced Density Matrices in Quantum Chemistry*, Academic Press, New York, 1976.
10. A. J. Coleman, Convex structure of electrons. *Int. J. Quantum Chem.* **11**, 907 (1977).
11. A. J. Coleman, Reduced density operators and N-particle problem. *Int. J. Quantum Chem.* **13**, 67 (1978).
12. J. E. Harriman, Geometry of density matrices. 1. Defintions, N-matrices and 1-matrices. *Phys. Rev. A* **17**, 1249 (1978).
13. J. E. Harriman, Geometry of density matrices. 2. Reduced density matrices and N-representability. *Phys. Rev. A* **17**, 1257 (1978).
14. F. Colmenero, C. Perez del Valle, and C. Valdemoro, Approximating q-order reduced density-matrices in terms of the lower-order ones. 1. General relations. *Phys. Rev. A* **47**, 971 (1993).
15. F. Colmenero and C. Valdemoro, Approximating q-order reduced density-matrices in terms of the lower-order ones. 2. Applications. *Phys. Rev. A* **47**, 979 (1993).
16. F. Colmenero and C. Valdemoro, Self-consistent approximate solution of the 2nd-order contracted Schrödinger equation. *Int. J. Quantum Chem.* **51**, 369 (1994).
17. C. Valdemoro, L. M. Tel, and E. Pérez-Romero, The contracted Schrödinger equation: some results. *Adv. Quantum Chem.* **28**, 33 (1997).
18. H. Nakatsuji and K. Yasuda, Direct determination of the quantum-mechanical density matrix using the density equation. *Phys. Rev. Lett.* **76**, 1039 (1996).
19. K. Yasuda and H. Nakatsuji, Direct determination of the quantum-mechanical density matrix using the density equation. 2. *Phys. Rev. A* **56**, 2648 (1997).
20. D. A. Mazziotti, Contracted Schrödinger equation: determining quantum energies and two-particle density matrices without wavefunctions. *Phys. Rev. A* **57**, 4219 (1998).
21. D. A. Mazziotti, Approximate solution for electron correlation through the use of Schwinger probes. *Chem. Phys. Lett.* **289**, 419 (1998).
22. D. A. Mazziotti, 3,5-Contracted Schrödinger equation: determining quantum energies and reduced density matrices without wavefunctions. *Int. J. Quantum Chem.* **70**, 557 (1998).
23. K. Yasuda, Direct determination of the quantum-mechanical density matrix: Parquet theory. *Phys. Rev. A* **59**, 4133 (1999).
24. D. A. Mazziotti, Pursuit of N-representability for the contracted Schrödinger equation through density-matrix reconstruction. *Phys. Rev. A* **60**, 3618 (1999).

25. D. A. Mazziotti, Comparison of contracted Schrödinger and coupled-cluster theories. *Phys. Rev. A* **60**, 4396 (1999).
26. D. A. Mazziotti, Complete reconstruction of reduced density matrices. *Chem. Phys. Lett.* **326**, 212 (2000).
27. W. Kutzelnigg and D. Mukherjee, Irreducible Brillouin conditions and contracted Schrödinger equations for n-electron systems. I. The equations satisfied by the density cumulants. *J. Chem. Phys.* **114**, 2047 (2001).
28. D. A. Mazziotti, Correlated purification of reduced density matrices. *Phys. Rev. E* **65**, 026704 (2002).
29. D. A. Mazziotti, A variational method for solving the contracted Schrödinger equation through a projection of the N-particle power method onto the two-particle space. *J. Chem. Phys.* **116**, 1239 (2002).
30. J. M. Herbert and J. E. Harriman, Contraction relations for Grassmann products of reduced density matrices and implications for density matrix reconstruction. *J. Chem. Phys.* **117**, 7464 (2002).
31. W. Kutzelnigg and D. Mukherjee, Irreducible Brillouin conditions and contracted Schrödinger equations for n-electron systems. IV. Perturbative analysis. *J. Chem. Phys.* **120**, 7350 (2004).
32. D. A. Mazziotti, Exactness of wave functions from two-body exponential transformations in many-body quantum theory. *Phys. Rev. A* **69**, 012507 (2004).
33. M. D. Benayoun and A. Y. Lu, Invariance of the cumulant expansion under 1-particle unitary transformations in reduced density matrix theory. *Chem. Phys. Lett.* **387**, 485 (2004).
34. D. R. Alcoba and C. Valdemoro, Spin structure and properties of the correlation matrices corresponding to pure spin states: controlling the S-representability of these matrices. *Int. J. Quantum Chem.* **102**, 629 (2005).
35. D. R. Alcoba, F. J. Casquero, L. M. Tel, E. Pérez-Romero, and C. Valdemoro, Convergence enhancement in the iterative solution of the second-order contracted Schrödinger equation. *Int. J. Quantum Chem.* **102**, 620 (2005).
36. L. Cohen and C. Frishberg, Hierarchy equations for reduced density matrices. *Phys. Rev. A* **13**, 927 (1976).
37. H. Nakatsuji, Equation for direct determination of density matrix. *Phys. Rev. A* **14**, 41 (1976).
38. C. Valdemoro, Approximating the second-order reduced density matrix in terms of the first-order one. *Phys. Rev. A* **45**, 4462 (1992).
39. W. Kutzelnigg and D. Mukherjee, Cumulant expansion of the reduced density matrices. *J. Chem. Phys.* **110**, 2800 (1999).
40. D. A. Mazziotti, in *Many-Electron Densities and Density Matrices*, (J. Cioslowski, ed.), Kluwer, Boston, 2000.
41. P. R. Surján. *Second Quantized Approach to Quantum Chemistry: An Elementary Introduction*, Springer-Verlag, New York, 1989.
42. C. Valdemoro, in *Density Matrices and Density Functionals* (R. Erdahl and V. Smith, eds.), Proceedings of the A. J. Coleman Symposium, Kingston, Ontario, 1985, Reidel, Dordrecht, 1987.
43. C. Valdemoro, Spin-adapted reduced Hamiltonian. 1. Elementary excitations. *Phys. Rev. A* **31**, 2114 (1985).
44. C. Valdemoro, A. Torre, and L. Lain, in *Structure, Interaction and Reactivity* (S. Fraga, ed.), Elsevier, Amsterdam, 1993.
45. J. E. Harriman, Geometry of density matrices. 5. Eigenstates. *Phys. Rev. A* **30**, 19 (1984).
46. P. Hohenberg and W. Kohn, Inhomogeneous electron gas. *Phys. Rev. B* **136**, 864 (1964).
47. P. C. Hohenberg, W. Kohn, and L. J. Sham, *Adv. Quantum Chem.* **21**, 7 (1990).

48. M. Levy. *Adv. Quantum Chem.* **21**, 69 (1990).
49. T. L. Gilbert, Hohenberg–Kohn theorem for nonlocal potentials. *Phys. Rev. B* **12**, 2111 (1975).
50. J. E. Harriman, Orthonormal orbitals for the representation of an arbitrary density. *Phys. Rev. A* **24**, 680 (1981).
51. M. Rosina, in *Reduced Density Operators with Application to Physical and Chemical Systems* (A. J. Coleman and R. M. Erdahl, eds.), Queen's Papers in Pure and Applied Mathematics No. 11, Queen's University, Kingston, Ontario, 1968.
52. R. M. Erdahl, Convex structure of the set of N-representable reduced 2-matrices. *J. Math. Phys.* **13**, 1608 (1972).
53. M. B. Ruskai, N-representability problem: particle–hole equivalence. *J. Math. Phys.* **11**, 3218 (1970).
54. W. Slebodziński. *Exterior Forms and Their Applications*, Polish Scientific Publishers, Warsaw, 1970.
55. R. Kubo, Generalized cumulant expansion method. *J. Phys. Soc. Japan* **17**, 1100 (1962).
56. C. W. Gardiner. *Handbook of Stochastic Methods for Physics, Chemistry and the Natural Sciences*, Springer-Verlag, New York, 1983.
57. K. Kladko and P. Fulde, On the properties of cumulant expansions. *Int. J. Quantum Chem.* **66**, 377 (1998).
58. J. W. Negele and H. Orland, *Quantum Many-Particle Systems*, Addison-Wesley Publishing, New York, 1988.
59. X.-P. Li, R. W. Nunes, and D. Vanderbilt, Density-matrix electronic-structure method linear system-size scaling. *Phys. Rev. B* **47**, 10891 (1993).
60. M. S. Daw, Model for energetics of solids based on the density matrix. *Phys. Rev. B* **47**, 10895 (1993).
61. S. Goedecker, Linear scaling of electronic structure methods. *Rev. Mod. Phys.* **71**, 1085, (1999).
62. D. A. Mazziotti, Linear scaling and the 1,2-contracted Schrödinger equation. *J. Chem. Phys.* **115**, 8305 (2001).
63. D. A. Mazziotti and R. M. Erdahl, Uncertainty relations and reduced density matrices: mapping many-body quantum mechanics onto four particles. *Phys. Rev. A* **63**, 042113 (2001).
64. C. Garrod, V. Mihailović, and M. Rosina, Variational approach to 2-body density matrix. *J. Math. Phys.* **10**, 1855 (1975).
65. R. M. Erdahl, Two algorithms for the lower bound method of reduced density matrix theory. *Rep. Math. Phys.* **15**, 147 (1979).
66. M. Nakata, H. Nakatsuji, M. Ehara, M. Fukuda, K. Nakata, and K. Fujisawa, Variational calculations of fermion second-order reduced density matrices by semidefinite programming algorithm. *J. Chem. Phys.* **114**, 8282 (2001).
67. D. A. Mazziotti, Variational minimization of atomic and molecular ground-state energies via the two-particle reduced density matrix. *Phys. Rev. A* **65**, 062511 (2002).
68. M. Nakata, M. Ehara, and H. Nakatsuji, Density matrix variational theory: application to the potential energy surfaces and strongly correlated systems, *J. Chem. Phys.* **116**, 5432 (2002).
69. D. A. Mazziotti, Solution of the 1,3-contracted Schrödinger equation through positivity conditions on the 2-particle reduced density matrix. *Phys. Rev. A* **66**, 062503 (2002).
70. Z. Zhao, B. J. Braams, H. Fukuda, M. L. Overton, and J. K. Percus, The reduced density matrix method for electronic structure calculations and the role of three-index representability conditions. *J. Chem. Phys.* **120**, 2095 (2004).

71. D. A. Mazziotti, Realization of quantum chemistry without wavefunctions through first-order semidefinite programming. *Phys. Rev. Lett.* **93**, 213001 (2004).
72. D. A. Mazziotti, First-order semidefinite programming for the direct determination of two-electron reduced density matrices with application to many-electron atoms and molecules. *J. Chem. Phys.* **121**, 10957 (2004).
73. G. Gidofalvi and D. A. Mazziotti, Application of variational reduced-density-matrix theory to organic molecules. *J. Chem. Phys.* **122**, 094107 (2005).
74. J. R. Hammond and D. A. Mazziotti, Variational reduced-density-matrix calculations on radicals: an alternative approach to open-shell *ab initio* quantum chemistry. *Phys. Rev. A* **73**, 012509 (2006).
75. C. Valdemoro, L. M. Tel, and E. Pérez-Romero, N-representability problem within the framework of the contracted Schrödinger equation. *Phys. Rev. A* **61**, 032507 (2000).
76. A. J. Coleman, in *Reduced Density Operators with Application to Physical and Chemical Systems II* (R. M. Erdahl, ed.), Queen's Papers in Pure and Applied Mathematics No. 40, Queen's University, Kingston, Ontario, 1974.
77. A. J. Coleman and I. Absar, One-electron orbitals intrinsic to the reduced Hamiltonian. *Chem. Phys. Lett.* **39**, 609 (1976).
78. A. J. Coleman and I. Absar, Reduced Hamiltonian orbitals. 3. Unitarily-invariant decomposition of Hermitian operators. *Int. J. Quantum Chem.* **18**, 1279 (1980).
79. J. E. Harriman, Geometry of density matrices. 2. Reduced density matrices and N-representability. *Phys. Rev. A* **17**, 1257 (1978).
80. M. E. Casida and J. E. Harriman, Geometry of density matrices. 6. Superoperators and unitary invariance. *Int. J. Quantum Chem.* **30**, 161 (1986).
81. A. C. Tang and H. Guo, Characteristic operators and unitarily-invariant decomposition of Hermitian operators. *Int. J. Quantum Chem.* **23**, 217 (1983).
82. C. C. Sun, X. Q. Li, and A. C. Tang, Lie algebra and unitarily-invariant decomposition of Hermitian operators. *Int. J. Quantum Chem.* **25**, 653 (1984).
83. G. Strang. *Linear Algebra with Applications*, 3rd ed., Academic Press, New York, 1988.
84. *Handbook of Chemistry and Physics*, 79th ed., Chemical Rubber Company, Boca Raton, FL, 1998.
85. Alex A. Granovsky, www http://classic.chem.msu.su/gran/gamess/index.html.
86. M. W. Schmidt et al., General atomic and molecular electronic-structure system. *J. Comput. Chem.* **14**, 1347 (1993).
87. T. H. Dunning, Jr. and P. J. Hay, in *Methods of Electronic Structure Theory* (H. F. Schaefer III, ed.), Plenum Press, New York, 1977.
88. M. Ehara, M. Nakata, H. Kou, K. Yasuda, and H. Nakatsuji, Direct determination of the density matrix using the density equation: potential energy curves of HF, CH_4, BH_3, NH_3, and H_2O. *Chem. Phys. Lett.* **305**, 483 (1999).

CHAPTER 9

PURIFICATION OF CORRELATED REDUCED DENSITY MATRICES: REVIEW AND APPLICATIONS

D. R. ALCOBA

Departamento de Física, Facultad de Ciencias Exactas y Naturales, Universidad de Buenos Aires, Ciudad Universitaria, 1428, Buenos Aires, Argentina

CONTENTS

I. Introduction
II. General Notation, Basic Definitions, and Theoretical Background
   A. General Notation and Basic Definitions
   B. The Energy, the Reduced Density Matrices, and the N-Representability Problem
   C. The G-Matrices and the S-Representability Problem
      1. Spin Structure of the Second-Order G-Matrices
      2. Spin Properties of the Second-Order G-Matrices
III. Purification Procedures Based on Unitary Decompositions of Second-Order Reduced Density Matrices
   A. The Mazziotti Purification Procedure
      1. Unitary Decomposition of Antisymmetric Second-Order Matrices
      2. The Purification Procedure
   B. Improving the Mazziotti Purification Procedure
      1. Unitary Decomposition of Arbitrary Second-Order Matrices
      2. Improved Version of the Mazziotti Purification Procedure
   C. Test Calculations and Results
IV. Purification Procedures Based on the Correlation Matrix Decomposition of Second-Order Reduced Density Matrices
   A. The Pure Two-Body Correlation Matrix within the 2-RDM Formalism
      1. Basic Properties of the Pure Two-Body Correlation Matrices
      2. Spin Structure and Properties of the Pure Two-Body Correlation Matrices in the Singlet State
   B. The Alcoba–Valdemoro Purification Procedure
   C. Test Calculations and Results

Reduced-Density-Matrix Mechanics: With Application to Many-Electron Atoms and Molecules, A Special Volume of Advances in Chemical Physics, Volume 134, edited by David A. Mazziotti. Series editor Stuart A. Rice. Copyright © 2007 John Wiley & Sons, Inc.

V. Purification Within the Framework of the Second-Order Contracted Schrödinger Equation
   A. The Second-Order Contracted Schrödinger Equation and Its Iterative Solution
      1. The Second-Order Contracted Schrödinger Equation
      2. The Regulated Iterative Self-Consistent Solution
   B. Analysis of the Effects of the N- and S-Purification on the Regulated CSE Iterative Process
   C. Application of the Regulated CSE-NS Method to the Study of Potential Energy Curves
VI. Discussion and Final Remarks
Acknowledgments
References

I. INTRODUCTION

The possibility of describing a many-body system of N electrons through the use of the second-order reduced density matrix (2-RDM) [1–6] was proposed by Husimi [7], Löwdin [8], Mayer [9], McWeeny [10], Ayres [11], Coulson [12], and Coleman [13] in the 1950s. Such a description implies a simpler approach to the many-body problem, where explicit calculation of the N-electron wavefunction is abandoned in favor of a direct computation of the 2-RDM [1–6]. The difficulty of this approach lies in that even after many years of research, the set of conditions which such mathematical–physical objects must fulfill in order to ascertain that they can be derived from a well-behaved pure spin N-electron wavefunction, the so-called N- and S-representability conditions [1–6, 13, 14], cannot yet be claimed to be complete. However, the search of stringent N-and S-representability conditions has been intense and fruitful [1–6, 15–20]. Thus, although an exact procedure for determining directly an N- and S-representable 2-RDM has not been found, many mathematical properties of these matrices are now known and several variational [21–34] and nonvariational [15, 35–69] methods for approximating RDMs and for employing them have been developed.

Recently, two different general strategies for correcting the representability defects of an approximated 2-RDM have emerged [70–72]. The common idea underlying these approaches is to try to "purify" a trial 2-RDM while keeping the corresponding 1-RDM fixed, which should be N- and S-representable. Since the 1-RDM derives by contraction from the 2-RDM, the purification may be achieved by acting on the 2-RDM pure two-body component with a vanishing contraction into the one-body space. In other words, the aim is to render N- and S-representable a pure two-body matrix, which is the matrix responsible for the N- and S-representability defects. One of these purification strategies, proposed by Mazziotti [70], uses the unitarily invariant decomposition of the 2-RDM reported in 1974 by Coleman [73] and extended by other authors [74–77]. Apart from other mathematically significant properties, this decomposition guarantees that there is no contribution of the 2-RDM pure two-body term to the 2-RDM contraction into the one-body space. The second purification strategy, initially developed by Valdemoro, Alcoba, and Tel [71] and subsequently extended by Alcoba and Valdemoro [72], is based on the decomposition of the 2-RDM into two matrices: a matrix that only depends on the 1-RDM, and a pure two-body correlation matrix [6, 18, 60, 64, 71, 72, 78–87] whose contribution to the 2-RDM contraction vanishes.

Although these purification strategies present a broad field of application, the original aim was to combine them with the iterative method for solving the second-order contracted Schrödinger equation [1, 6, 45, 47–52], so that the convergence and stability of its iterative solution would be both accelerated and enhanced [87]. This question has recently been studied and the results that confirm this hypothesis were reported in Refs. [70, 88]. Indeed, the iterative process is improved and its stability is achieved. This showed not only the effectiveness of the purification strategies on several atomic and molecular systems but also their critical role within this methodology.

The aim of this chapter is to review, both from theoretical and practical points of view, the work done in this direction in the last years.

II. GENERAL NOTATION, BASIC DEFINITIONS, AND THEORETICAL BACKGROUND

A. General Notation and Basic Definitions

In what follows a two-particle interacting system having a fixed and well-defined number of particles N will be considered. It will also be considered that the one-electron space is spanned by a finite basis set of $2K$ orthonormal spin orbitals. Under these conditions the 1-RDM and 2-RDM elements are defined in second quantization language as

$$^1D_{i_\sigma;k_\sigma} = \langle\Psi|\hat{a}^\dagger_{i_\sigma}\,\hat{a}_{k_\sigma}|\Psi\rangle \qquad (1)$$

and

$$2!\,^2D_{i_\sigma j_{\sigma'};k_\sigma l_{\sigma'}} = \langle\Psi|\hat{a}^\dagger_{i_\sigma}\,\hat{a}^\dagger_{j_{\sigma'}}\,\hat{a}_{l_{\sigma'}}\,\hat{a}_{k_\sigma}|\Psi\rangle \qquad (2)$$

respectively. In these expressions, Ψ represents the N-electron state whose observables are being investigated; the indices i,j,k,l represent orthonormal orbitals; and σ,σ' represent the spin functions (α or β).

Similarly, the elements of the first- and second-order hole reduced density matrices (HRDMs) are defined, respectively, as

$$^1Q_{i_\sigma;k_\sigma} = \langle\Psi|\hat{a}_{k_\sigma}\,\hat{a}^\dagger_{i_\sigma}|\Psi\rangle \qquad (3)$$

and

$$2!\,^2Q_{i_\sigma j_{\sigma'};k_\sigma l_{\sigma'}} = \langle\Psi|\hat{a}_{l_{\sigma'}}\,\hat{a}_{k_\sigma}\,\hat{a}^\dagger_{i_\sigma}\,\hat{a}^\dagger_{j_{\sigma'}}|\Psi\rangle \qquad (4)$$

B. The Energy, the Reduced Density Matrices, and the N-Representability Problem

In this formalism the spin-independent many-body Hamiltonian may be written

$$\hat{H} = \frac{1}{2} \sum_{\sigma,\sigma'} \sum_{i,j,k,l} {}^2K_{i_\sigma j_{\sigma'};k_\sigma l_{\sigma'}} \, \hat{a}^\dagger_{i_\sigma} \hat{a}^\dagger_{j_{\sigma'}} \hat{a}_{l_{\sigma'}} \hat{a}_{k_\sigma} \tag{5}$$

where ${}^2K$ is the reduced Hamiltonian matrix [37,89]

$$ {}^2K_{i_\sigma j_{\sigma'};k_\sigma l_{\sigma'}} = \frac{1}{N-1} (\epsilon_{i_\sigma;k_\sigma} \delta_{j,l} + \epsilon_{j_{\sigma'};l_{\sigma'}} \delta_{i,k}) + \langle i_\sigma j_{\sigma'} | k_\sigma l_{\sigma'} \rangle \tag{6}$$

Here the symbol ϵ represents the one-electron integral matrix and $\langle i_\sigma j_{\sigma'} | k_\sigma l_{\sigma'} \rangle$ is the usual two-electron repulsion integral in the Condon and Shortley notation.

Thus the second-quantized expression of the energy of the state Ψ in terms of the 2-RDM is

$$E = \langle \Psi | \hat{H} | \Psi \rangle = \sum_{\sigma,\sigma'} \sum_{i,j,k,l} {}^2K_{i_\sigma j_{\sigma'};k_\sigma l_{\sigma'}} \, {}^2D_{k_\sigma l_{\sigma'};i_\sigma j_{\sigma'}} \tag{7}$$

This equation implies that if the 2-RDM corresponding to a given state is known, the energy and, in fact, all its other observables can be obtained. That is so because the Hamiltonian only has one- and two-electron operators and therefore the many-body problem may, in principle, be considered an effective two-body problem. This is only possible if the 2-RDM can be obtained directly, without previous knowledge of the N-electron wavefunction, which is a difficult mathematical–physical problem. One therefore needs to know the necessary and sufficient conditions that a 2-RDM—and, in general, a p-RDM—must satisfy in order to ensure that there exists an N-electron wavefunction from which it may be derived. To determine such conditions constitutes the N-representability problem defined by Coleman [13] in 1963, which is at the origin of a wide literature [1–6]. This problem has been solved for the 1-RDM and 1-HRDM by Coleman [13], who reported the set of necessary and sufficient ensemble N-representability conditions for these matrices. Thus the 1-RDM and 1-HRDM must be Hermitian, must be positive semidefinite, and must be normalized as

$$\sum_i {}^1D_{i_\sigma;i_\sigma} = N_\sigma \tag{8}$$

$$\sum_i {}^1Q_{i_\sigma;i_\sigma} = (K - N_\sigma) \tag{9}$$

where N_σ denotes the number of electrons with σ spin function. Also, these matrices must be linked as follows:

$$^1D_{i_\sigma;k_\sigma} + {}^1Q_{i_\sigma;k_\sigma} = \delta_{i,k} \tag{10}$$

Although a formal solution of the N-representability problem for the 2-RDM and 2-HRDM (and higher-order matrices) was reported [1], this solution is not feasible, at least in a practical sense [90]. Hence, in the case of the 2-RDM and 2-HRDM, only a set of necessary N-representability conditions is known. Thus these latter matrices must be Hermitian, Positive semidefinite (D- and Q-conditions [16, 17, 91]), and antisymmetric under permutation of indices within a given row/column. These second-order matrices must contract into the first-order ones according to the following relations:

$$2! \sum_j {}^2D_{i_\sigma j_{\sigma'};k_\sigma j_{\sigma'}} = (N_{\sigma'} - \delta_{\sigma,\sigma'}) \, {}^1D_{i_\sigma;k_\sigma} \tag{11}$$

$$2! \sum_j {}^2Q_{i_\sigma j_{\sigma'};k_\sigma j_{\sigma'}} = (K - N_{\sigma'} - \delta_{\sigma,\sigma'}) \, {}^1Q_{i_\sigma;k_\sigma} \tag{12}$$

and must be normalized as [1, 2]

$$2! \sum_{i,j} {}^2D_{i_\sigma j_{\sigma'};i_\sigma j_{\sigma'}} = N_\sigma (N_{\sigma'} - \delta_{\sigma,\sigma'}) \tag{13}$$

$$2! \sum_{i,j} {}^2Q_{i_\sigma j_{\sigma'};k_\sigma j_{\sigma'}} = (K - N_\sigma)(K - N_{\sigma'} - \delta_{\sigma,\sigma'}) \tag{14}$$

Also, these matrices must be related to each other through the second-order fermion relation [15],

$$2! \, {}^2Q_{i_\sigma j_{\sigma'};k_\sigma l_{\sigma'}} = 2! \, {}^2D_{i_\sigma j_{\sigma'};k_\sigma l_{\sigma'}} + \delta_{i,k}\delta_{j,l} - {}^1D_{j_{\sigma'};l_{\sigma'}}\delta_{i,k} - {}^1D_{i_\sigma;k_\sigma}\delta_{j,l}$$
$$- \delta_{\sigma,\sigma'}(\delta_{i,l}\delta_{j,k} - {}^1D_{j_\sigma;k_\sigma}\delta_{i,l} - {}^1D_{i_\sigma;l_\sigma}\delta_{j,k}) \tag{15}$$

whose semidefinite positiveness expresses the previously mentioned Q-condition [16, 17, 91].

Equation (15) implies that the 2-RDM and 2-HRDM matrices contain the same information. Indeed, these matrices are two of the three different matrix representations of the 2-RDM on the two-body space, the third one being the second-order G-matrix (2-G) [16]. This matrix, which may be written [24, 25]

$$^2G_{i_\sigma j_{\sigma'};k_{\sigma''} l_{\sigma'''}} = 2! \, {}^2D_{i_\sigma l_{\sigma'''};j_{\sigma'} k_{\sigma''}} - \delta_{\sigma,\sigma'}\delta_{\sigma'',\sigma'''} \, {}^1D_{i_\sigma;j_\sigma} \, {}^1D_{k_{\sigma''};l_{\sigma''}} + \delta_{\sigma,\sigma''}\delta_{\sigma',\sigma'''}\delta_{j,l} \, {}^1D_{i_\sigma;k_\sigma} \tag{16}$$

or equivalently [15]

$$
\begin{aligned}
{}^2G_{i_\sigma j_{\sigma'};k_{\sigma''} l_{\sigma'''}} &= \sum_{\Psi' \neq \Psi} \langle \Psi | \hat{a}^\dagger_{i_\sigma} \hat{a}_{j_{\sigma'}} | \Psi' \rangle \langle \Psi' | \hat{a}^\dagger_{l_{\sigma'''}} \hat{a}_{k_{\sigma''}} | \Psi \rangle \\
&\equiv \sum_{\Psi' \neq \Psi} {}^1D^{\Psi\Psi'}_{i_\sigma;j_{\sigma'}} {}^1D^{\Psi'\Psi}_{l_{\sigma'''};k_{\sigma''}}
\end{aligned}
\qquad (17)
$$

where ${}^1D^{\Psi\Psi'}$ is the First-order transition reduced density matrix (1-TRDM), must be Hermitian, must be positive semidefinite (G-condition [16, 17, 24, 25, 27, 91]), and must contract into the one-body space according to [71, 79, 83]

$$
\sum_j {}^2G_{i_\sigma j_{\sigma'};k_\sigma j_{\sigma'}} = (K - N_{\sigma'})\, {}^1D_{i_\sigma;k_\sigma} + \delta_{\sigma,\sigma'}({}^1D - {}^1D^2)_{i_\sigma;k_\sigma} \qquad (18)
$$

$$
\sum_j {}^2G_{j_\sigma i_{\sigma'};j_\sigma k_{\sigma'}} = N_\sigma(\delta_{i,k} - {}^1D_{i_{\sigma'};k_{\sigma'}}) + \delta_{\sigma,\sigma'}({}^1D - {}^1D^2)_{i_{\sigma'};k_{\sigma'}} \qquad (19)
$$

where ${}^1D^2$ represents the square of the 1-RDM. Also, this matrix must be normalized as

$$
\sum_{i,j} {}^2G_{i_\sigma j_{\sigma'};i_\sigma j_{\sigma'}} = N_\sigma\, (K - N_{\sigma'} + \delta_{\sigma,\sigma'}) - \delta_{\sigma,\sigma'} \sum_i ({}^1D^2)_{i_\sigma;i_\sigma} \qquad (20)
$$

Thus it is essential to constrain these three different matrix representations of the 2-RDM to accomplish all the basic properties just reported when considering the N-representability of the 2-RDM.

C. The G-Matrices and the S-Representability Problem

As an extension of the N-representability problem, Valdemoro and co-workers introduced the S-representability problem [14], that is, the incomplete knowledge of the set of necessary and sufficient conditions that a p-RDM must fulfil in order to ensure that it derives from an N-electron wavefunction having *well-defined* spin quantum numbers:

$$
\hat{S}^2 | \Psi_{S,M} \rangle = S(S+1) | \Psi_{S,M} \rangle \qquad (21)
$$

$$
\hat{S}_z | \Psi_{S,M} \rangle = M | \Psi_{S,M} \rangle \qquad (22)
$$

These S-representability conditions, by extension of the N-representability terminology, have recently been analyzed in depth by reconsidering the spin structure of the 2-G matrices [72].

1. Spin Structure of the Second-Order G-Matrices

As shown in Eq. (17), the 1-TRDMs are the basic building elements in the 2-G matrices. Since the 1-TRDMs only connect two states whose spin numbers differ at most in one unit, the structure of the 2-G matrices may be rewritten in terms of separate spin components characterized by the spin quantum number S' of the states $|\Psi'\rangle$ appearing in the 1-TRDMs [72]. Thus

$$^2G_{i_\sigma j_{\sigma'};k_{\sigma''}l_{\sigma'''}} = \sum_{\Psi'_{S,M'} \neq \Psi_{S,M}} {}^1D^{\Psi_{S,M}\Psi'_{S,M'}}_{i_\sigma j_{\sigma'}}\ {}^1D^{\Psi'_{S,M'}\Psi_{S,M}}_{l_{\sigma'''};k_{\sigma''}}$$
$$+ \sum_{\Psi'_{S+1,M'}} {}^1D^{\Psi_{S,M}\Psi'_{S+1,M'}}_{i_\sigma j_{\sigma'}}\ {}^1D^{\Psi'_{S+1,M'}\Psi_{S,M}}_{l_{\sigma'''};k_{\sigma''}} + \sum_{\Psi'_{S-1,M'}} {}^1D^{\Psi_{S,M}\Psi'_{S-1,M'}}_{i_\sigma j_{\sigma'}}\ {}^1D^{\Psi'_{S-1,M'}\Psi_{S,M}}_{l_{\sigma'''};k_{\sigma''}}$$

(23)

with

$$M' = \begin{cases} M-1 & \text{for}\quad \sigma = \sigma'' = \alpha, \sigma' = \sigma''' = \beta \\ M & \text{for}\quad \sigma = \sigma' = \sigma'' = \sigma''' \text{ or } \sigma = \sigma', \sigma'' = \sigma''' \\ M+1 & \text{for}\quad \sigma = \sigma'' = \beta, \sigma' = \sigma''' = \alpha \end{cases} \quad (24)$$

In what follows, the compact notation that will be used for each of these spin components is

$$\{S',M'\}^2G_{i_\sigma j_{\sigma'};k_{\sigma''}l_{\sigma'''}} \equiv \sum_{\Psi'_{S',M'} \neq \Psi_{S,M}} {}^1D^{\Psi_{S,M}\Psi'_{S',M'}}_{i_\sigma j_{\sigma'}}\ {}^1D^{\Psi'_{S',M'}\Psi_{S,M}}_{l_{\sigma'''};k_{\sigma''}} \quad (25)$$

where S' may take the values S, $S+1$, and $S-1$.

2. Spin Properties of the Second-Order G-Matrices

From their definition, it follows that each of the spin components of the 2-G matrix are positive semidefinite. The semidefinite positiveness of these matrices constitutes a much more exacting set of conditions than the well-known single N-representability G-condition, since the former conditions imply the latter one but not conversely.

Furthermore, properties of the spin components of the 2-G can be obtained by reconsidering the spin properties of the 1-TRDMs. Thus the different spin-blocks of the 1-TRDMs can be related among themselves through the action of the operator $\hat{S}^2$ on pure spin states. One therefore has

$$S(S+1)\ {}^1D^{\Psi_{S,M}\Psi'_{S',M'}}_{i_\sigma j_{\sigma'}} = \langle \Psi_{S,M}|\hat{S}^2\ \hat{a}^\dagger_{i_\sigma}\ \hat{a}_{j_{\sigma'}}|\Psi'_{S',M'}\rangle \quad (26)$$

By moving the $\hat{S}^2$ operator to the right on the right-hand side (rhs) of Eq. (26), a set of equations linking the different spin-blocks of the 1-TRDMs is obtained. These equations lead to a set of relations linking different elements of the spin components of the 2-G matrices. The resulting relations can be classified as follows:

(a) Case $S = S'$; $M = 0$.

$$\{S,0\}^2 G_{i_\beta j_\beta; k_\beta l_\beta} = \{S,0\}^2 G_{i_\alpha j_\alpha; k_\beta l_\beta}$$
$$= \{S,0\}^2 G_{i_\beta j_\beta; k_\alpha l_\alpha} = \{S,0\}^2 G_{i_\alpha j_\alpha; k_\alpha l_\alpha} \qquad (27)$$
$$\{S,1\}^2 G_{i_\beta j_\alpha; k_\beta l_\alpha} = \{S,-1\}^2 G_{i_\alpha j_\beta; k_\alpha l_\beta} \quad (S \neq 0) \qquad (28)$$

(b) Case $S = S'$; $M \neq 0$.

$$\frac{4M^2}{(S-M)(S+M+1)} \{S,M+1\}^2 G_{i_\beta j_\alpha; k_\beta l_\alpha}$$
$$= \frac{4M^2}{(S-M+1)(S+M)} \{S,M-1\}^2 G_{i_\alpha j_\beta; k_\alpha l_\beta}$$
$$= \{S,M\}^2 G_{i_\alpha j_\alpha; k_\alpha l_\alpha} + \{S,M\}^2 G_{i_\beta j_\beta; k_\beta l_\beta}$$
$$- \{S,M\}^2 G_{i_\alpha j_\alpha; k_\beta l_\beta} - \{S,M\}^2 G_{i_\beta j_\beta; k_\alpha l_\alpha} \qquad (29)$$

(c) Case $S \neq S'$.

$$\frac{(S(S+1) - S'(S'+1) + 2M)^2}{4(S'-M)(S'+M+1)} \{S',M-1\}^2 G_{i_\alpha j_\beta; k_\alpha l_\beta}$$
$$= \frac{(S(S+1) - S'(S'+1) - 2M)^2}{4(S'-M+1)(S'+M)} \{S',M+1\}^2 G_{i_\beta j_\alpha; k_\beta l_\alpha}$$
$$= \{S',M\}^2 G_{i_\beta j_\beta; k_\beta l_\beta} = -\{S',M\}^2 G_{i_\alpha j_\alpha; k_\beta l_\beta}$$
$$= -\{S',M\}^2 G_{i_\beta j_\beta; k_\alpha l_\alpha} = \{S',M\}^2 G_{i_\alpha j_\alpha; k_\alpha l_\alpha} \qquad (30)$$

Similarly, application of the properties of the spin-shifting operators, $\hat{S}_\pm$, allows one to obtain the relations connecting the 1-TRDMs corresponding to different multiplet states. Thus, by considering the action of the spin-shifting operator $\hat{S}_+$ on pure spin states,

$$\sqrt{(S-M)(S+M+1)} \, ^1D_{i_\sigma j_{\sigma'}}^{\Psi_{S,M} \Psi'_{S',M'}} = \langle \Psi_{S,M+1} | \hat{S}_+ \hat{a}^\dagger_{i_\sigma} \hat{a}_{j_{\sigma'}} | \Psi'_{S',M'} \rangle \qquad (31)$$

and by moving the $\hat{S}_+$ operator to the right on the rhs of Eq. (31), a set of recurrence relations among the 1-TRDMs corresponding to different multiplet states is obtained. The resulting equations lead to setting up the interconnections among the different spin components of the 2-G matrices corresponding to different states of a given multiplet. These connections can be summarized as follows:

(a) Case $S' = S$.

$$\{S,M'\}^2 G_{i_\sigma j_{\sigma'};k_{\sigma''} l_{\sigma'''}} = \sum_{\gamma,\gamma'} \theta(\gamma,\gamma',\sigma,\sigma',\sigma'',\sigma''') \, \{S,S\}^2 G^{(\Psi_{S,S})}_{i_\gamma j_\gamma; k_{\gamma'} l_{\gamma'}} \tag{32}$$

where

$$\theta(\gamma,\gamma',\sigma,\sigma',\sigma'',\sigma''') \equiv \delta_{\sigma,\sigma'}\delta_{\sigma'',\sigma'''}(\mu_+\delta_{\gamma,\sigma} + \mu_-(1-\delta_{\gamma,\sigma}))$$
$$\times (\mu_+\delta_{\gamma',\sigma''} + \mu_-(1-\delta_{\gamma',\sigma''}))$$
$$+ \delta_{\sigma,\sigma''}\delta_{\sigma',\sigma'''}(1-\delta_{\sigma,\sigma'})(2\delta_{\gamma,\gamma'}-1)$$
$$\times (v_-^2\delta_{\sigma,\alpha} + v_+^2\delta_{\sigma,\beta}) \tag{33}$$

and

$$\mu_\pm \equiv \frac{1 \pm M/S}{2} \tag{34}$$

$$v_\pm \equiv \frac{\sqrt{(S \mp M)(S \pm M + 1)}}{2S} \tag{35}$$

(b) Case $S' = S + 1$.

$$\{S+1,M'\}^2 G_{i_\sigma j_{\sigma'};k_{\sigma''} l_{\sigma'''}} = \theta'(\sigma,\sigma',\sigma'',\sigma''') \, \{S+1,S+1\}^2 G^{(\Psi_{S,S})}_{i_\beta j_\alpha; k_\beta l_\alpha} \tag{36}$$

where

$$\theta'(\sigma,\sigma',\sigma'',\sigma''') \equiv \delta_{\sigma,\sigma'}\delta_{\sigma'',\sigma'''}(2\delta_{\sigma,\sigma'}-1)\xi^2$$
$$+ \delta_{\sigma,\sigma''}\delta_{\sigma',\sigma'''}(1-\delta_{\sigma,\sigma'})$$
$$\times (\xi_-^2\delta_{\sigma,\alpha} + \xi_+^2\delta_{\sigma,\beta}) \tag{37}$$

and

$$\xi \equiv -\sqrt{\frac{(S-M+1)(S+M+1)}{(2S+1)(2S+2)}} \tag{38}$$

$$\xi_\pm \equiv -\sqrt{\frac{(S \pm M+1)(S \pm M+2)}{(2S+1)(2S+2)}} \tag{39}$$

(c) Case $S' = S - 1$.

$$\{S-1,M'\}{}^2G_{i_\sigma j_{\sigma'};k_{\sigma''}l_{\sigma'''}} = \theta''(\sigma,\sigma',\sigma'',\sigma''')\,\{S-1,S-1\}{}^2G^{(\Psi_{S,S})}_{i_\alpha j_\beta;k_\alpha l_\beta} \tag{40}$$

where

$$\begin{aligned}\theta''(\sigma,\sigma',\sigma'',\sigma''') &\equiv \delta_{\sigma,\sigma'}\delta_{\sigma'',\sigma'''}(2\delta_{\sigma,\sigma''} - 1)\xi'^2 \\ &+ \delta_{\sigma,\sigma''}\delta_{\sigma',\sigma'''}(1 - \delta_{\sigma,\sigma'}) \\ &\times (\xi'^2_{-}\delta_{\sigma,\alpha} + \xi'^2_{+}\delta_{\sigma,\beta})\end{aligned} \tag{41}$$

and

$$\xi' \equiv -\sqrt{\frac{(S-M)(S+M)}{2S(2S-1)}} \tag{42}$$

$$\xi'_{\pm} \equiv -\sqrt{\frac{(S \mp M - 1)(S \mp M)}{2S(2S-1)}} \tag{43}$$

The above relations, which are represented in a spin-orbital basis, are especially relevant; they are analytical results that describe all the conditions that a 2-G corresponding to any pure spin state must satisfy; hence they constitute a complete set of S-representability conditions. Their generality implies a general usefulness within the framework of any RDM methodology.

III. PURIFICATION PROCEDURES BASED ON UNITARY DECOMPOSITIONS OF SECOND-ORDER REDUCED DENSITY MATRICES

A. The Mazziotti Purification Procedure

The description of the two-electron correlation effects within the 2-RDM formalism is not unique. Indeed, different approaches have been reported in the literature [53, 61, 73–83, 86, 89, 92–96] and different aspects of this exciting subject have been analyzed. At the basis of all these descriptions of the two-body correlation effects lies a decomposition of the 2-RDM. However, as mentioned in Section I, any purification strategy that maintains the 1-RDM fixed requires that the contribution of the pure two-body term arising from the considered 2-RDM decomposition should vanish upon contraction into the one-body space. This requirement, a very strong one, significantly reduces the list of possible descriptions of the correlation term. Thus one of the only two approaches leading to a 2-RDM decomposition satisfying this requirement is the unitarily invariant partition of antisymmetric second-order matrices reported by Coleman [73]. This partitioning, which has been applied extensively in quantum chemistry in order to decompose and analyze the structure of several second-order matrices of physical interest [70, 86, 89, 97–99], is at the basis of an iterative purification procedure recently proposed by Mazziotti [70]. The method that he proposed, hereafter called the MZ purification procedure, aims at guaranteeing the positive semidefiniteness of the 2-RDM and of the 2-HRDM (D- and Q-conditions) while retaining the original 1-RDM. Before reviewing the procedure, Coleman's decomposition and its properties will be addressed.

1. Unitary Decomposition of Antisymmetric Second-Order Matrices

In 1974 Coleman [73] proposed to decompose any Hermitian antisymmetric second-order matrix ^2A as

$$^2A_{ij;kl} = {}^2_0A_{ij;kl} + {}^2_1A_{ij;kl} + {}^2_2A_{ij;kl} \tag{44}$$

where

$$^2_0A_{ij;kl} = \frac{A(\delta_{i,k}\delta_{j,l} - \delta_{i,l}\delta_{j,k})}{K(K-1)} \tag{45}$$

$$^2_1A_{ij;kl} = \frac{{}^1P_{i;k}\,\delta_{j,l} + {}^1P_{j;l}\,\delta_{i,k} - {}^1P_{i;l}\,\delta_{j,k} - {}^1P_{j;k}\,\delta_{i,l}}{K-2} - \frac{2A(\delta_{i,k}\,\delta_{j,l} - \delta_{i,l}\delta_{j,k})}{K(K-2)} \tag{46}$$

$$^2_2A_{ij;kl} = {}^2A_{ij;kl} - {}^2_0A_{ij;kl} - {}^2_1A_{ij;kl} \tag{47}$$

with

$$A = \sum_{m,n} {}^2A_{mn;mn} \tag{48}$$

$$^1P_{i;j} = \sum_m {}^2A_{im;jm} \tag{49}$$

If the second-order Hermitian matrix follows the transformation rule for a (2,2) tensor, then this decomposition is the only possible manner of expressing these matrices as a sum of simpler parts so that the decomposition remains invariant under unitary tranformations of the basis [73].

The three parts of this decomposition reveal the structure of the matrix with respect to the contraction operations. These parts have been called the 0-, 1- and 2-body part of the second-order matrix ^2A, respectively. Following the notation introduced in Ref. [73], each of these parts have been identified by a left-lower index.

While the 0-body part of this decomposition contains the 0-body information for ^2A,

$$\sum_{m,n} {}_0^2 A_{mn;mn} = A \tag{50}$$

the 1-body part contains the 1-body information for ^2A,

$$\sum_m ({}_0^2 A_{im;jm} + {}_1^2 A_{im;jm}) = {}^1P_{i;j} \tag{51}$$

and the 2-body part contains information that vanishes upon contraction,

$$\sum_m {}_2^2 A_{im;jm} = 0 \tag{52}$$

2. The Purification Procedure

As has been mentioned, the MZ purification procedure is based on Coleman's unitary decomposition of an antisymmetric Hermitian second-order matrix described earlier. When applied to *singlet* states of atoms and molecules, the computational cost of this purification procedure is reduced, since the 2-RDM (and thus the 1-RDM obtained by contraction) presents only two different spin-blocks, the $\alpha\alpha$- and $\alpha\beta$-blocks (and only one spin-block for the 1-RDM). For the remaining part of this section only this type of state will be treated.

According to this unitarily invariant decomposition, the different spin-blocks of the trial 2-RDM, which must be corrected, are decomposed as follows:

$$^2D_{i_\sigma j_{\sigma'};k_\sigma l_{\sigma'}} = {}_0^2 D_{i_\sigma j_{\sigma'};k_\sigma l_{\sigma'}} + {}_1^2 D_{i_\sigma j_{\sigma'};k_\sigma l_{\sigma'}} + {}_2^2 D_{i_\sigma j_{\sigma'};k_\sigma l_{\sigma'}} \tag{53}$$

with

$$_0^2 D_{i_\alpha j_\alpha;k_\alpha l_\alpha} = \frac{(\sum_{p,q} {}^2 D_{p_\alpha q_\alpha;p_\alpha q_\alpha})(\delta_{i,k}\delta_{j,l} - \delta_{i,l}\delta_{j,k})}{K(K-1)} \tag{54}$$

$$_1^2 D_{i_\alpha j_\alpha;k_\alpha l_\alpha} = \frac{(N/2-1)({}^1D_{i_\alpha;k_\alpha}\delta_{j,l} + {}^1D_{j_\alpha;l_\alpha}\delta_{i,k})}{2(K-2)}$$

$$- \frac{(N/2-1)({}^1D_{i_\alpha;l_\alpha}\delta_{j,k} + {}^1D_{j_\alpha;k_\alpha}\delta_{i,l})}{2(K-2)}$$

$$- \frac{(N/2-1)(\sum_p {}^1 D_{p_\alpha;p_\alpha})(\delta_{i,k}\delta_{j,l} - \delta_{i,l}\delta_{j,k})}{K(K-2)} \tag{55}$$

$$_2^2 D_{i_\alpha j_\alpha;k_\alpha l_\alpha} = {}^2D_{i_\alpha j_\alpha;k_\alpha l_\alpha} - {}_0^2 D_{i_\alpha j_\alpha;k_\alpha l_\alpha} - {}_1^2 D_{i_\alpha j_\alpha;k_\alpha l_\alpha} \tag{56}$$

where the α-block of the 1-RDM is obtained from the contraction of the 2-RDM into the 1-body space, Eq. (11).

For the $\alpha\beta$-block of the 2-RDM the decomposition was generated ad hoc [70]. This is because this block is not antisymmetric under permutation of the orbital indices within the row or column subsets of indices; and thus the unitary decomposition reported by Coleman cannot be applied. Hence the ad hoc decomposition is given here by

$$_0^2 D_{i_\alpha j_\beta;k_\alpha l_\beta} = \frac{\left(\sum_{p,q} {}^2 D_{p_\alpha q_\beta;p_\alpha q_\beta}\right) \delta_{i,k}\delta_{j,l}}{K^2} \tag{57}$$

$$_1^2 D_{i_\alpha j_\beta;k_\alpha l_\beta} = \frac{(N/2)({}^1 D_{i_\alpha;k_\alpha}\,\delta_{j,l} + {}^1 D_{j_\beta;l_\beta}\,\delta_{i,k})}{K}$$

$$- \frac{(N/2)\left(\sum_p ({}^1 D_{p_\alpha;p_\alpha} + {}^1 D_{p_\beta;p_\beta})\right)\delta_{i,k}\delta_{j,l}}{K^2} \tag{58}$$

$$_2^2 D_{i_\alpha j_\beta;k_\alpha l_\beta} = {}^2 D_{i_\alpha j_\beta;k_\alpha l_\beta} - {}_0^2 D_{i_\alpha j_\beta;k_\alpha l_\beta} - {}_1^2 D_{i_\alpha j_\beta;k_\alpha l_\beta} \tag{59}$$

In accordance with Coleman, the 0-particle part of this decomposition, Eqs. (54) and (57), contains the 0-body information for the 2-RDM,

$$\sum_{p,q} {}_0^2 D_{p_\sigma q_{\sigma'};p_\sigma q_{\sigma'}} = \sum_{p,q} {}^2 D_{p_\sigma q_{\sigma'};p_\sigma q_{\sigma'}} \tag{60}$$

the 1-particle part, Eqs. (55) and (58), contains the 1-body information for the 2-RDM,

$$2! \sum_m ({}_0^2 D_{i_\sigma m_{\sigma'};j_\sigma m_{\sigma'}} + {}_1^2 D_{i_\sigma m_{\sigma'};j_\sigma m_{\sigma'}}) = \left(\frac{N}{2} - \delta_{\sigma,\sigma'}\right) {}^1 D_{i_\sigma;j_\sigma} \tag{61}$$

and the 2-particle part, Eqs. (56) and (59), contains information that vanishes upon contractions into the 1-body space,

$$\sum_m {}_2^2 D_{i_\sigma m_{\sigma'};j_\sigma m_{\sigma'}} = 0 \tag{62}$$

By decomposing the trial 2-RDM in this way, it is possible to act upon the N-representability defects of this matrix.

1. When the 1-RDM—obtained from the contraction of the 2-RDM into the 1-body space—is not N-representable, this matrix is corrected by employing any one of the methods described in Refs. [70, 71]. With this new 1-RDM, ${}^1\tilde{D}$, the 1-particle part of the 2-RDM is recalculated. Thus

$$^2\tilde{D}_{i_\sigma j_{\sigma'};k_\sigma l_{\sigma'}} = {}_0^2 D_{i_\sigma j_{\sigma'};k_\sigma l_{\sigma'}} + {}_1^2\tilde{D}_{i_\sigma j_{\sigma'};k_\sigma l_{\sigma'}} + {}_2^2 D_{i_\sigma j_{\sigma'};k_\sigma l_{\sigma'}} \tag{63}$$

By doing so, the updated 2-RDM, $^2\tilde{D}$, presents N-representable contractions into the 1-body space

$$2! \sum_m {}^2\tilde{D}_{i_\sigma m_{\sigma'};j_\sigma m_{\sigma'}} = \left(\frac{N}{2} - \delta_{\sigma,\sigma'}\right) {}^1\tilde{D}_{i_\sigma;j_\sigma} \qquad (64)$$

It also presents correct contractions into the 0-body space, since in this purification procedure it is assumed that the trace condition Eq. (13) is fulfilled by the initial 2-RDM.

2. In order to impose the D and Q N-representability conditions on the 2-RDM and its 2-HRDM, these two matrices are diagonalized. From the eigenvectors $\{|x_{p\,\sigma\sigma';\sigma\sigma'}\rangle\}$ corresponding to the negative eigenvalues $\{x_{p\,\sigma\sigma';\sigma\sigma'}\}$ of the $\sigma\sigma'$-block of the 2-RDM, and the eigenvectors $\{|\bar{x}_{q\,\sigma\sigma';\sigma\sigma'}\rangle\}$ corresponding to the negative eigenvalues $\{\bar{x}_{q\,\sigma\sigma';\sigma\sigma'}\}$ of the $\sigma\sigma'$-block of its 2-HRDM, a set of second-order matrices is constructed,

$$^2X_{p\,i_\sigma j_{\sigma'};k_\sigma l_{\sigma'}} = \langle i_\sigma j_{\sigma'}|x_{p\,\sigma\sigma';\sigma\sigma'}\rangle\langle x_{p\,\sigma\sigma';\sigma\sigma'}|k_\sigma l_{\sigma'}\rangle \qquad (65)$$

$$^2\bar{X}_{q\,i_\sigma j_{\sigma'};k_\sigma l_{\sigma'}} = \langle i_\sigma j_{\sigma'}|\bar{x}_{q\,\sigma\sigma';\sigma\sigma'}\rangle\langle \bar{x}_{q\,\sigma\sigma';\sigma\sigma'}|k_\sigma l_{\sigma'}\rangle \qquad (66)$$

where $|i_\sigma j_{\sigma'}\rangle$ and $|k_\sigma l_{\sigma'}\rangle$ are two-electron Slater determinants.

The 2-RDM is corrected by adding a correcting-matrix $^2\Gamma_{\sigma\sigma';\sigma\sigma'}$,

$$^2\tilde{\tilde{D}}_{i_\sigma j_{\sigma'};k_\sigma l_{\sigma'}} = {}^2\tilde{D}_{i_\sigma j_{\sigma'};k_\sigma l_{\sigma'}} + {}^2\Gamma_{i_\sigma j_{\sigma'};k_\sigma l_{\sigma'}} \qquad (67)$$

which is given by

$$^2\Gamma_{i_\sigma j_{\sigma'};k_\sigma l_{\sigma'}} = \sum_p \gamma_{p\,\sigma\sigma';\sigma\sigma'} {}^2_2 X_{p\,i_\sigma j_{\sigma'};k_\sigma l_{\sigma'}} + \sum_q \epsilon_{q\,\sigma\sigma';\sigma\sigma'} {}^2_2\bar{X}_{q\,i_\sigma j_{\sigma'};k_\sigma l_{\sigma'}} \qquad (68)$$

where $^2_2X_{p\,\sigma\sigma';\sigma\sigma'}$ and $^2_2\bar{X}_{q\,\sigma\sigma';\sigma\sigma'}$ are the two-particle and two-hole parts of the matrices Eqs. (65) and (66), respectively. The parameters $\gamma_{p\,\sigma\sigma';\sigma\sigma'}$ and $\epsilon_{q\,\sigma\sigma';\sigma\sigma'}$ of $^2\Gamma_{\sigma\sigma';\sigma\sigma'}$ are chosen to satisfy the linear system

$$x_{p\,\sigma\sigma';\sigma\sigma'} + \sum_{i,j,k,l} {}^2X_{p\,i_\sigma j_{\sigma'};k_\sigma l_{\sigma'}} {}^2\Gamma_{k_\sigma l_{\sigma'};i_\sigma j_{\sigma'}} = 0 \quad (\forall\,p) \qquad (69)$$

$$\bar{x}_{q\,\sigma\sigma';\sigma\sigma'} + \sum_{i,j,k,l} {}^2\bar{X}_{q\,i_\sigma j_{\sigma'};k_\sigma l_{\sigma'}} {}^2\Gamma_{k_\sigma l_{\sigma'};i_\sigma j_{\sigma'}} = 0 \quad (\forall\,q) \qquad (70)$$

The addition of the correcting-matrix $^2\Gamma_{\sigma\sigma';\sigma\sigma'}$ does not modify the contractions of the 2-RDM.

This step is repeated until the positivity of the 2-RDM and its 2-HRDM is satisfied up to a specified tolerance.

This procedure has been applied to several trial 2-RDMs corresponding to different molecular systems [67], thus obtaining very accurate energies and closely N-representable 2-RDMs. Unfortunately, the S-representability of the resulting 2-RDMs has not been analyzed.

Thus let us consider some particular relations that must be satisfied by the 2-RDM spin-blocks corresponding to a singlet state. It is well known that in this case the $\alpha\alpha$- and $\alpha\beta$-blocks of the 2-RDM are related as follows [100]:

$$^2D_{i_\alpha j_\alpha;k_\alpha l_\alpha} = {}^2D_{i_\alpha j_\beta;k_\alpha l_\beta} - {}^2D_{i_\alpha j_\beta;l_\alpha k_\beta} \tag{71}$$

This relation imposes severe conditions on the $\alpha\beta$-block of the 2-RDM. Thus this spin-block must satisfy the relations

$$\sum_m {}^2D_{i_\alpha m_\beta;m_\alpha j_\beta} = \sum_m \langle\Psi|\hat{a}^\dagger_{i_\alpha} \hat{a}^\dagger_{m_\beta} \hat{a}_{j_\beta} \hat{a}_{m_\alpha}|\Psi\rangle$$

$$= \langle\Psi|\hat{a}^\dagger_{i_\alpha} \hat{a}_{j_\alpha}|\Psi\rangle - \sum_m \langle\Psi|\hat{a}^\dagger_{i_\alpha} \hat{a}_{j_\beta} \hat{a}^\dagger_{m_\beta} \hat{a}_{m_\alpha}|\Psi\rangle$$

$$= {}^1D_{i_\alpha;j_\alpha} - \langle\Psi|\hat{a}^\dagger_{i_\alpha} \hat{a}_{j_\beta} \hat{S}_-|\Psi\rangle = {}^1D_{i_\alpha;j_\alpha} \tag{72}$$

and

$$\sum_m {}^2D_{m_\alpha i_\beta;j_\alpha m_\beta} = {}^1D_{i_\beta;j_\beta} \tag{73}$$

which follow from the fact that $\hat{S}_+|\Psi_{0,0}\rangle = \hat{S}_-|\Psi_{0,0}\rangle = 0$. In order to keep track of the first-order matrices, we will denote them as ${}^1D'_{i_\alpha;j_\alpha}$ and ${}^1D'_{i_\beta;j_\beta}$.

Moreover, these conditions imply that

$$\sum_{m,n} {}^2D_{m_\alpha n_\beta;n_\alpha m_\beta} = \frac{N}{2} \tag{74}$$

which is directly related to the expectation value of the $\hat{S}^2$ operator, $\langle\hat{S}^2\rangle$; that is [14, 101],

$$\langle\hat{S}^2\rangle = S(S+1) = \langle\hat{S}^2_z\rangle + \frac{N}{2} - \sum_{m,n} {}^2D_{m_\alpha n_\beta;n_\alpha m_\beta} \tag{75}$$

Taking into account the results just mentioned, let us now reconsider the two steps of the procedure for the $\alpha\beta$-block of the 2-RDM, ${}^2D_{\alpha\beta;\alpha\beta}$. In the first step, the $\alpha\beta$-block of the 2-RDM is recalculated, thus yielding an updated $\alpha\beta$-block of the 2-RDM $^2\tilde{D}_{\alpha\beta;\alpha\beta}$, which presents N-representable contractions given by Eq. (11). However, this is not the case for the contractions given by Eqs. (72) and (73). For example, one of these latter contractions of the updated $\alpha\beta$-block of the 2-RDM is given by

$$\sum_m {}^2\tilde{D}_{i_\alpha m_\beta; m_\alpha j_\beta} = \sum_m ({}^2_0\tilde{D}_{i_\alpha m_\beta; m_\alpha j_\beta} + {}^2_1\tilde{D}_{i_\alpha m_\beta; m_\alpha j_\beta} + {}^2_2\tilde{D}_{i_\alpha m_\beta; m_\alpha j_\beta})$$

$$= {}^1D'_{i_\alpha;j_\alpha} + \frac{(N/2)({}^1\tilde{D}_{i_\alpha;j_\alpha} + {}^1\tilde{D}_{i_\beta;j_\beta} - {}^1D_{i_\alpha;j_\alpha} - {}^1D_{i_\beta;j_\beta})}{K} \quad (76)$$

where it has been assumed that the initial $^2D_{\alpha\beta;\alpha\beta}$ satisfies the condition Eq. (72). Thus it follows that the updated $\alpha\beta$-block of the 2-RDM violates the condition Eq. (72). A similar reasoning can be followed for the second step, where the addition of the correcting-matrix $^2\Gamma_{\alpha\beta;\alpha\beta}$ to $^2D_{\alpha\beta;\alpha\beta}$ also produces errors in the contractions given by Eqs. (72) and (73). Moreover, the expectation value of the $\hat{S}^2$ operator will also present a deviation from its original value, since the contraction Eq. (74) will also be affected. Consequently, both steps of the MZ purification procedure introduce S-representability defects in the 2-RDM when correcting the N-representability defects of this matrix. Thus this procedure yields purified 2-RDMs that do not correspond to pure spin wavefunctions.

B. Improving the Mazziotti Purification Procedure

The S-representability defects of the MZ purification procedure can be corrected by generalizing Coleman's unitarily invariant decomposition. Thus a new procedure—based on a different generalized unitarily invariant decomposition of the 2-RDM recently reported in Ref. [77]—will now be described here.

1. Unitary Decomposition of Arbitrary Second-Order Matrices

Recently, a unitarily invariant decomposition of Hermitian second-order matrices of *arbitrary* symmetry under permutation of the indices within the row or column subsets of indices has been reported by Alcoba [77]. This decomposition, which generalizes that of Coleman, also presents three components that are mutually orthogonal with respect to the trace scalar product [77]:

$$^2A_{ij;kl} = {}^2_0A_{ij;kl} + {}^2_1A_{ij;kl} + {}^2_2A_{ij;kl} \quad (77)$$

where

$$^2_0A_{ij;kl} = \frac{(KA - A')\delta_{i,k}\delta_{j,l} + (KA' - A)\delta_{i,l}\delta_{j,k}}{K(K^2 - 1)} \quad (78)$$

$$^2_1A_{ij;kl} = \frac{(4A' - 2KA)\delta_{i,k}\delta_{j,l} + (4A - 2KA')\delta_{i,l}\delta_{j,k}}{K(K^2 - 4)}$$

$$+ \frac{2(^1P_{j;l}\delta_{i,k} + {}^1P'_{i;k}\delta_{j,l} + {}^1R_{j;k}\delta_{i,l} + {}^1R'_{i;l}\delta_{j,k})}{K(K^2 - 4)}$$

$$+ \frac{(K^2 - 2)(^1P_{i;k}\delta_{j,l} + {}^1P'_{j;l}\delta_{i,k} + {}^1R_{i;l}\delta_{j,k} + {}^1R'_{j;k}\delta_{i,l})}{K(K^2 - 4)}$$

$$+ \frac{{}^1P_{i;l}\delta_{j,k} + {}^1P_{j;k}\delta_{i,l} + {}^1P'_{i;l}\delta_{j,k} + {}^1P'_{j;k}\delta_{i,l}}{4 - K^2}$$

$$+ \frac{{}^1R_{i;k}\delta_{j,l} + {}^1R_{j;l}\delta_{i,k} + {}^1R'_{i;k}\delta_{j,l} + {}^1R'_{j;l}\delta_{i,k}}{4 - K^2} \tag{79}$$

$$^2_2A_{ij;kl} = {}^2A_{ij;kl} - {}^2_0A_{ij;kl} - {}^2_1A_{ij;kl} \tag{80}$$

with

$$A = \sum_{m,n} {}^2A_{mn;mn} \tag{81}$$

$$A' = \sum_{m,n} {}^2A_{mn;nm} \tag{82}$$

$$^1P_{i;j} = \sum_m {}^2A_{im;jm} \tag{83}$$

$$^1P'_{i;j} = \sum_m {}^2A_{mi;mj} \tag{84}$$

$$^1R_{i;j} \sum_m {}^2A_{im;mj} \tag{85}$$

$$^1R'_{i;j} \sum_m {}^2A_{mi;jm} \tag{86}$$

It must be noted that, due to the *arbitrary* symmetry under permutation of indices of this second-order matrix, a larger set of contractions into the 0- and 1-body space must be taken into account.

If the second-order Hermitian matrix follows the transformation rule for a (2, 2) tensor, then this decomposition is the only possible manner of expressing these matrices as a sum of simpler parts such that the decomposition remains invariant under unitary tranformations of the basis [77].

While the 0-part of this decomposition contains the 0-body information for ^2A,

$$\sum_{m,n} {}_0^2 A_{mn;mn} = A \tag{87}$$

$$\sum_{m,n} {}_0^2 A_{mn;nm} = A' \tag{88}$$

the 1-part contains the 1-body information for ^2A,

$$\sum_m ({}_0^2 A_{im;jm} + {}_1^2 A_{im;jm}) = {}^1P_{i;j} \tag{89}$$

$$\sum_m ({}_0^2 A_{mi;mj} + {}_1^2 A_{mi;mj}) = {}^1P'_{i;j} \tag{90}$$

$$\sum_m ({}_0^2 A_{im;mj} + {}_1^2 A_{im;mj}) = {}^1R_{i;j} \tag{91}$$

$$\sum_m ({}_0^2 A_{mi;jm} + {}_1^2 A_{mi;jm}) = {}^1R'_{i;j} \tag{92}$$

and the 2-part contains information that vanishes upon contraction,

$$\sum_m {}_2^2 A_{im;jm} = \sum_m {}_2^2 A_{mi;mj} = \sum_m {}_2^2 A_{im;mj} = \sum_m {}_2^2 A_{mi;jm} = 0 \tag{93}$$

2. Improved Version of the Mazziotti Purification Procedure

The independence with respect to the type of permutation-symmetry of the decomposition just reported allows one to treat the different spin-blocks of the 2-RDM on an equal footing. Moreover, this decomposition leads to a partitioning of these blocks into three orthogonal parts, which reveal the structure of these blocks with respect to *all* contraction operations.

If this decomposition is applied to the $\alpha\alpha$-block of the 2-RDM, then it reduces itself to that given by Eqs. (54)–(56). However, if the $\alpha\beta$-block of the 2-RDM is considered, then this decomposition becomes

$$^2D_{i_\alpha j_\beta;k_\alpha l_\beta} = {}_0^2 D_{i_\alpha j_\beta,k_\alpha l_\beta} + {}_1^2 D_{i_\alpha j_\beta;k_\alpha l_\beta} + {}_2^2 D_{i_\alpha j_\beta;k_\alpha l_\beta} \tag{94}$$

with

$$
{}_0^2 D_{i_\alpha j_\beta,k_\alpha l_\beta} = \frac{\sum_{p,q} (K\, {}^2D_{p_\alpha q_\beta;p_\alpha q_\beta} - {}^2D_{p_\alpha q_\beta;q_\alpha p_\beta})\, \delta_{i,k}\delta_{j,l}}{K(K^2-1)}
$$
$$
+ \frac{\sum_{p,q}(K\, {}^2D_{p_\alpha q_\beta;q_\alpha p_\beta} - {}^2D_{p_\alpha q_\beta;p_\alpha q_\beta})\, \delta_{i,l}\delta_{j,k}}{K(K^2-1)} \tag{95}
$$

$$\begin{aligned}
{}_1^2 D_{i_\alpha j_\beta, k_\alpha l_\beta} =\ & \frac{\sum_p 2\,({}^1 D'_{p_\alpha;p_\alpha} + {}^1 D'_{p_\beta;p_\beta})\,\delta_{i,k}\delta_{j,l}}{K(K^2-4)} \\
& - \frac{\sum_p K\frac{N}{2}\,({}^1 D_{p_\alpha;p_\alpha} + {}^1 D_{p_\beta;p_\beta})\,\delta_{i,k}\delta_{j,l}}{K(K^2-4)} \\
& + \frac{\sum_p N\,({}^1 D_{p_\alpha;p_\alpha} + {}^1 D_{p_\beta;p_\beta})\,\delta_{i,l}\delta_{j,k}}{K(K^2-4)} \\
& - \frac{\sum_p K\,({}^1 D'_{p_\alpha;p_\alpha} + {}^1 D'_{p_\beta;p_\beta})\,\delta_{i,l}\delta_{j,k}}{K(K^2-4)} \\
& + \frac{N\,({}^1 D_{j_\alpha;l_\alpha}\delta_{i,k} + {}^1 D_{i_\beta;k_\beta}\delta_{j,l})}{K(K^2-4)} \\
& + \frac{2\,({}^1 D'_{j_\alpha;k_\alpha}\delta_{i,l} + {}^1 D'_{i_\beta;l_\beta}\delta_{j,k})}{K(K^2-4)} \\
& + \frac{\frac{(K^2-2)N}{2}\,({}^1 D_{i_\alpha;k_\alpha}\delta_{j,l} + {}^1 D_{j_\beta;l_\beta}\delta_{i,k})}{K(K^2-4)} \\
& + \frac{(K^2-2)\,({}^1 D'_{i_\alpha;l_\alpha}\delta_{j,k} + {}^1 D'_{j_\beta;k_\beta}\delta_{i,l})}{K(K^2-4)} \\
& + \frac{\frac{N}{2}\,({}^1 D_{i_\alpha;l_\alpha}\delta_{j,k} + {}^1 D_{j_\alpha;k_\alpha}\delta_{i,l})}{4-K^2} \\
& + \frac{\frac{N}{2}\,({}^1 D_{i_\beta;l_\beta}\delta_{j,k} + {}^1 D_{j_\beta;k_\beta}\delta_{i,l})}{4-K^2} \\
& + \frac{{}^1 D'_{j_\alpha;l_\alpha}\delta_{i,k} + {}^1 D'_{i_\alpha;k_\alpha}\delta_{j,l}}{4-K^2} \\
& + \frac{{}^1 D'_{j_\beta;l_\beta}\delta_{i,k} + {}^1 D'_{i_\beta;k_\beta}\delta_{j,l}}{4-K^2}
\end{aligned} \tag{96}$$

$${}_2^2 D_{i_\alpha j_\beta;k_\alpha l_\beta} = {}^2 D_{i_\alpha j_\beta;k_\alpha l_\beta} - {}_0^2 D_{i_\alpha j_\beta;k_\alpha l_\beta} - {}_1^2 D_{i_\alpha j_\beta;k_\alpha l_\beta} \tag{97}$$

where the matrices ${}^1 D_{\alpha;\alpha}$, ${}^1 D_{\beta;\beta}$, ${}^1 D'_{\alpha;\alpha}$, and ${}^1 D'_{\beta;\beta}$ appearing in Eq. (96) are obtained from the different contractions of the 2-RDM into the 1-body space, Eqs. (11), (72), and (73), respectively.

The different parts of this new decomposition reveal the structure of the ${}^2 D_{\alpha\beta;\alpha\beta}$ with respect to all contracting operations into the 0- and 1-body space. Thus, while the first part of this new decomposition, Eq. (95), contains the information of the two different contractions of $^2D_{\alpha\beta;\alpha\beta}$ into the 0-body space,

$$\sum_{m,n} {}^2_0D_{m_\alpha n_\beta;m_\alpha n_\beta} = \sum_{m,n} {}^2D_{m_\alpha n_\beta;m_\alpha n_\beta} \tag{98}$$

$$\sum_{m,n} {}^2_0D_{m_\alpha n_\beta;n_\alpha m_\beta} = \sum_{m,n} {}^2D_{m_\alpha n_\beta;n_\alpha m_\beta} \tag{99}$$

the second one, Eq. (96), contains the information of the four possible contractions of $^2D_{\alpha\beta;\alpha\beta}$ into the 1-body space,

$$\sum_m \left({}^2_0D_{i_\alpha m_\beta;j_\alpha m_\beta} + {}^2_1D_{i_\alpha m_\beta;j_\alpha m_\beta} \right) = \frac{N}{2} {}^1D_{i_\alpha;j_\alpha} \tag{100}$$

$$\sum_m \left({}^2_0D_{m_\alpha i_\beta;m_\alpha j_\beta} + {}^2_1D_{m_\alpha i_\beta;m_\alpha j_\beta} \right) = \frac{N}{2} {}^1D_{i_\beta;j_\beta} \tag{101}$$

$$\sum_m \left({}^2_0D_{i_\alpha m_\beta;m_\alpha j_\beta} + {}^2_1D_{i_\alpha m_\beta;m_\alpha j_\beta} \right) = {}^1D'_{i_\alpha;j_\alpha} \tag{102}$$

$$\sum_m \left({}^2_0D_{m_\alpha i_\beta;j_\alpha m_\beta} + {}^2_1D_{m_\alpha i_\beta;j_\alpha m_\beta} \right) = {}^1D'_{i_\beta;j_\beta} \tag{103}$$

and the third one, Eq. (97), contains information that vanishes upon contractions,

$$\sum_m {}^2_2D_{i_\alpha m_\beta;j_\alpha m_\beta} = \sum_m {}^2_2D_{m_\alpha i_\beta;m_\alpha j_\beta} = \sum_m {}^2_2D_{i_\alpha m_\beta;m_\alpha j_\beta} = \sum_m {}^2_2D_{m_\alpha i_\beta;j_\alpha m_\beta} = 0 \tag{104}$$

Moreover, each of these parts are related to those of the $^2D_{\alpha\alpha;\alpha\alpha}$, Eqs. (54)–(56), as follows:

$$^2_pD_{i_\alpha j_\alpha;k_\alpha l_\alpha} = {}^2_pD_{i_\alpha j_\beta;k_\alpha l_\beta} - {}^2_pD_{i_\alpha j_\beta;l_\alpha k_\beta} \quad (p = 0, 1, 2) \tag{105}$$

which is a consequence of the condition Eq. (71).

Thus we propose to use the new decomposition given by Eqs. (95)–(97) instead of that given by Eqs. (57)–(59) for correcting the $\alpha\beta$-block of an approximated 2-RDM. This leads to a new iterative procedure, hereafter called the I-MZ purification procedure, which can be summarized as follows:

1. While the $\alpha\alpha$-block of the 2-RDM to be corrected is decomposed by following Eqs. (54)–(56), the $\alpha\beta$-block of this matrix is decomposed by following Eqs. (95)–(97).
2. When the contractions of the 2-RDM into the 0-body space do not satisfy Eqs. (13) and/or (74), these contractions are corrected by modifying the

0-part of the decompositions of the different spin-blocks of this matrix. Thus the 2-RDM is recalculated as follows:

$$^2\tilde{D}_{i_\sigma j_{\sigma'};k_\sigma l_{\sigma'}} = {}^2_0\tilde{D}_{i_\sigma j_{\sigma'};k_\sigma l_{\sigma'}} + {}^2_1 D_{i_\sigma j_{\sigma'};k_\sigma l_{\sigma'}} + {}^2_2 D_{i_\sigma j_{\sigma'};k_\sigma l_{\sigma'}} \quad (106)$$

As mentioned earlier, in the MZ purification procedure it is assumed that the trace condition Eq. (13) is satisfied by the initial 2-RDM; but this is not generally the case when considering approximated 2-RDMs.

3. The $\alpha\alpha$-block of the 2-RDM is further corrected by following the two steps of the MZ purification procedure since the decomposition for this block remains unchanged.

4. When the matrices $^1D_{\alpha;\alpha}$, $^1D_{\beta;\beta}$, $^1D'_{\alpha;\alpha}$, and $^1D'_{\beta;\beta}$—obtained from the different contractions of $^2D_{\alpha\beta;\alpha\beta}$—are not N-representable, these matrices undergo the corresponding correction. With these new matrices the one-particle part of $^2D_{\alpha\beta;\alpha\beta}$ is recalculated, thus yielding a new 2-RDM:

$$^2\tilde{D}_{i_\alpha j_\beta;k_\alpha l_\beta} = {}^2_0\tilde{D}_{i_\alpha j_\beta;k_\alpha l_\beta} + {}^2_1\tilde{D}_{i_\alpha j_\beta;k_\alpha m_\beta} + {}^2_2 D_{i_\alpha j_\beta;k_\alpha l_\beta} \quad (107)$$

5. The positivity of the $\alpha\beta$-block of the resulting 2-RDM and its 2-HRDM is imposed as in the MZ purification procedure, but the new decomposition is used when calculating the correcting-matrix $^2\Gamma_{\alpha\beta;\alpha\beta}$.

With this procedure no S-representability defects are introduced into the trial 2-RDM. Thus, for instance, in the fourth step of this procedure the $\alpha\beta$-block of the updated 2-RDM will present the following contractions into the 1-body space:

$$\sum_m {}^2\tilde{D}_{i_\alpha m_\beta;j_\alpha m_\beta} = \sum_m ({}^2_0 D_{i_\alpha m_\beta;j_\alpha m_\beta} + {}^2_1\tilde{D}_{i_\alpha m_\beta;j_\alpha m_\beta} + {}^2_2 D_{i_\alpha m_\beta;j_\alpha m_\beta})$$

$$= \frac{N}{2}{}^1\tilde{D}_{i_\alpha j_\alpha} \quad (108)$$

$$\sum_m {}^2\tilde{D}_{i_\alpha m_\beta;m_\alpha j_\beta} = {}^1\tilde{D}'_{i_\alpha j_\alpha} \quad (109)$$

and similarly for the other contractions. The matrices $^1\tilde{D}_{\alpha;\alpha}$, $^1\tilde{D}_{\beta;\beta}$, $^1\tilde{D}'_{\alpha;\alpha}$, and $^1\tilde{D}'_{\beta;\beta}$ stand for the corrected N-representable first-order matrices. Note that the most difficult case has been assumed; that is, neither the condition Eq. (11) nor the conditions Eqs. (72) and (73) are satisfied by the trial $^2D_{\alpha\beta;\alpha\beta}$, which implies that the matrices $^1D_{\alpha;\alpha}$, $^1D_{\beta;\beta}$, $^1D'_{\alpha;\alpha}$, and $^1D'_{\beta;\beta}$ are not N-representable. A similar reasoning can be followed for the last step of the new procedure, where the addition of the new correcting-matrix will not affect the different contractions of the 2-RDM into the 0- and 1-body space. Thus this new procedure permits one to correct *both* the N- and S-representability defects of an approximated 2-RDM.

It must be emphasized that the $\alpha\beta$-block of the 2-RDM may also be decomposed into two subblocks, the singlet and the triplet one. This clearly enhances the computational efficiency of the purification of the 2-RDM, since these subblocks may be corrected separately. On the other hand, since the singlet and triplet subblocks are symmetric and antisymmetric, respectively, under permutation of the orbital indices within the row or column subsets of labels, it would be possible to use the unitarily invariant decomposition of Coleman [73] and that of Sun et al. [102] to correct both the N- and S-representability defects of these subblocks. However, this would be formally equivalent to use the decomposition given by Eqs. (94)–(97), since this latter decomposition implicitly presents the former ones as particular cases [77].

C. Test Calculations and Results

In order to analyze the performance of the I-MZ purification procedure and to compare it with the MZ one, a set of calculations have been carried out. The probes selected have been the beryllium atom, the isoelectronic ions B^+, C^{2+}, N^{3+}, and O^{4+}, and the Li_2 (Li—Li bond length of 2.75 au) and linear BeH_2 (Be—H bond length of 2.54 au) molecules. The basis sets used were formed by Hartree–Fock molecular orbitals built out of minimal Slater orbital basis sets. The states studied were the ground states, which present a dominant closed-shell Slater determinant configuration.

In order to get significant results, the initial data must be formed by a set of clearly non-N-representable second-order matrices, which would generate upon contraction a closely ensemble N-representable 1-RDM. It therefore seemed reasonable to choose as initial data the approximate 2-RDMs built by application of the *independent pair model* within the framework of the *spin-adapted reduced Hamiltonian* (SRH) theory [37–45]. This choice is adequate because these matrices, which are positive semidefinite, Hermitian, and antisymmetric with respect to the permutation of two row/column indices, are not N-representable, since the 2-HRDMs derived from them are not positive semidefinite. Moreover, the 1-RDMs derived from these 2-RDMs, although positive semidefinite, are neither ensemble N-representable nor S-representable. That is, the correction of the N- and S-representability defects of these sets of matrices (approximated 2-RDM, 2-HRDM, and 1-RDM) is a suitable test for the two purification procedures. Attention has been focused only on correcting the N- and S-representability of the $\alpha\beta$-block of these matrices, since the I-MZ purification procedure deals with a different decomposition of this block.

Since the performance of the procedures was found to be very similar in all the cases studied, attention here will mainly be focused on the beryllium atom.

Although the contractions of the approximated 2-RDM into the 0-body space were those given by Eqs. (13) and (74), respectively, none of the different contractions of this matrix into the 1-body space was N-representable. Thus the MZ

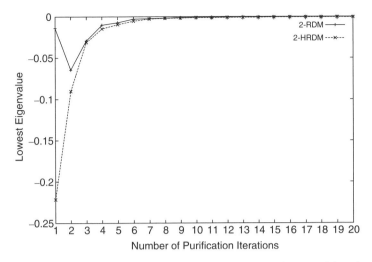

Figure 1. Lowest eigenvalue of the 2-RDM and the 2-HRDM matrices at each iteration of the I-MZ purification procedure for the ground state of the beryllium atom.

and the I-MZ purification procedures were applied. The correction of these N-representability defects was carried out by using the method described in Ref. [71]. Although the updated $\alpha\beta$-block of the 2-RDM obtained with the MZ purification procedure presented N-representable contractions given by Eq. (11), those given by Eqs. (72) and (73) were not N-representable. In consequence, the updated 2-RDM was not S-representable. Thus the updated $\alpha\beta$-block of the 2-RDM obtained with the I-MZ procedure has been used as data when correcting the positivity defects of the $^2D_{\alpha\beta;\alpha\beta}$ and the $^2Q_{\alpha\beta;\alpha\beta}$.

In Figs. 1 and 2, the values of the lower eigenvalue of each of the two matrices $^2D_{\alpha\beta;\alpha\beta}$ and $^2Q_{\alpha\beta;\alpha\beta}$ are plotted on the ordinate; and the iteration numbers appear on the abscissa. The results obtained when employing both the I-MZ purification procedure and the MZ purification procedure are given in Figs. 1 and 2, respectively. As can be seen, the convergence toward positive matrices is attained in both cases. Moreover, the procedures show similar convergence rates.

In order to study to what extent the contractions given by Eqs. (72) and (73) of the $^2D_{\alpha\beta;\alpha\beta}$ reproduce the initial N-representable 1-RDMs when correcting the positivity defects of the $^2D_{\alpha\beta;\alpha\beta}$ and the $^2Q_{\alpha\beta;\alpha\beta}$ by using the MZ purification procedure, its root mean square (RMS) deviation with respect to the initial N-representable 1-RDMs has been calculated at each iteration. The results, which are reported in Fig. 3, show that this type of contraction deviates from the initial ones. Therefore it is clearly seen that the MZ purification procedure also introduces S-representability defects when imposing these N-representability

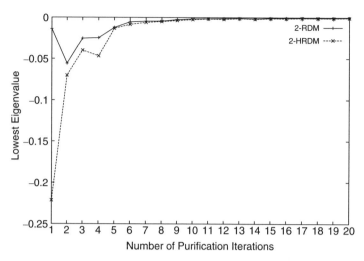

Figure 2. Lowest eigenvalue of the 2-RDM and the 2-HRDM matrices at each iteration of the MZ purification procedure for the ground state of the beryllium atom.

conditions. Furthermore, when iterating the MZ purification procedure, the expectation value of the $\hat{S}^2$ operator—calculated by using Eq. (75)—presented a deviation from its initial value $\langle \hat{S}^2 \rangle = 0$. Thus in Fig. 4 this value is plotted at each iteration.

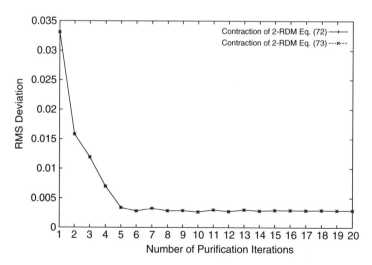

Figure 3. RMS deviation of the contractions given by Eqs. (72) and (73) from the initial ones at each iteration of the MZ purification procedure for the ground state of the beryllium atom.

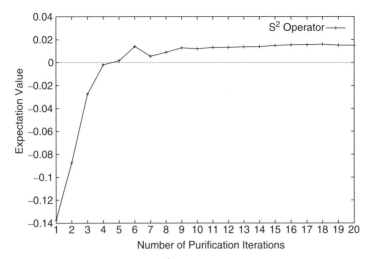

Figure 4. Expectation value of the $\hat{S}^2$ operator at each iteration of the MZ purification procedure for the ground state of the beryllium atom.

As mentioned earlier, similar behaviors to that of the berillyum atom were found when studying the B^+, C^{2+}, N^{3+}, O^{4+}, Li_2, and BeH_2 cases. The results obtained in these calculations are reported in Tables I and II, where the performances of the purification procedures are compared.

Finally, in order to illustrate the role of the I-MZ purification procedure in improving the approximated 2-RDMs obtained by application of the independent pair model within the framework of the SRH theory, all the different spin-blocks of these matrices were purified. The energy of both the initial (non-purified) and updated (purified) RDMs was calculated. These energies and those corresponding to a full configuration interaction (full CI) calculation are reported in Table III. As can be appreciated from this table, the nonpurified energies of all the test systems lie below the full CI ones while the purified ones lie above and very close to the full CI ones.

On the whole, these results show that the I-MZ purification procedure is very suitable for acting upon *both* the *N*- and *S*-representability defects of an approximated 2-RDM.

IV. PURIFICATION PROCEDURES BASED ON THE CORRELATION MATRIX DECOMPOSITION OF SECOND-ORDER REDUCED DENSITY MATRICES

A very different purification strategy was initiated by Valdemoro et al. [71] in 2003 and extended by Alcoba and Valdemoro [72] in 2005. Thus these authors

TABLE I
Performances of the MZ and the I-MZ Purification Procedures When Applied to Approximated 2-RDMs of Several Test Systems in Their Ground State

System	Purification Procedure	Iteration	Lowest Eigenvalue $^2D_{\alpha\beta;\alpha\beta}$	$^2Q_{\alpha\beta;\alpha\beta}$
Be	MZ	1	-1.386×10^{-2}	-2.216×10^{-1}
		20	-3.366×10^{-4}	-6.463×10^{-4}
	I-MZ	1	-1.386×10^{-2}	-2.216×10^{-1}
		20	-2.285×10^{-4}	-5.778×10^{-4}
B$^+$	MZ	1	-7.468×10^{-3}	-1.534×10^{-1}
		20	-2.187×10^{-4}	-3.371×10^{-4}
	I-MZ	1	-7.468×10^{-3}	-1.534×10^{-1}
		20	-2.148×10^{-4}	-2.799×10^{-4}
C^{2+}	MZ	1	-6.083×10^{-3}	-1.163×10^{-1}
		20	-1.825×10^{-4}	-2.850×10^{-4}
	I-MZ	1	-6.083×10^{-3}	-1.163×10^{-1}
		20	-9.543×10^{-5}	-1.694×10^{-4}
N^{3+}	MZ	1	-5.252×10^{-3}	-9.298×10^{-2}
		20	-1.878×10^{-4}	-2.276×10^{-4}
	I-MZ	1	-5.252×10^{-3}	-9.298×10^{-2}
		20	-8.222×10^{-5}	-1.127×10^{-4}
O^{4+}	MZ	1	-4.679×10^{-3}	-7.771×10^{-2}
		20	-4.835×10^{-5}	-9.018×10^{-5}
	I-MZ	1	-4.679×10^{-3}	-7.771×10^{-2}
		20	-4.001×10^{-5}	-8.876×10^{-5}
Li$_2$	MZ	1	-4.533×10^{-2}	-4.026×10^{-1}
		20	-1.808×10^{-4}	-3.717×10^{-3}
	I-MZ	1	-4.533×10^{-2}	-4.026×10^{-1}
		20	-5.101×10^{-7}	-5.115×10^{-4}
BeH$_2$	MZ	1	-1.770×10^{-2}	-2.080×10^{-1}
		20	-3.489×10^{-3}	-7.321×10^{-3}
	I-MZ	1	-1.770×10^{-2}	-2.080×10^{-1}
		20	-2.266×10^{-3}	-1.683×10^{-3}

developed an alternative iterative purification procedure, hereafter called the AV purification procedure, aimed at rendering positive semidefinite the three second-order matrices 2-RDM, 2-HRDM, and 2-G while ensuring that they reproduce by contraction the same ensemble N- and S-representable 1-RDM. This procedure also ensures that this 2-RDM is S-representable in the singlet case. What renders it possible is that this purification procedure is focused on correcting the N- and S-representability defects of the pure two-body correlation matrix [6, 18, 60, 64, 71, 72, 78–87]. As will be shown, this latter matrix describes the pure two-body correlation effects present not only in the 2-RDM but also in the 2-HRDM. Furthermore, this matrix is intimately related to the 2-G matrix; and the 1-RDM can be deduced from their different contractions.

TABLE II
Performances of the MZ and the I-MZ Purification Procedures When
Applied to Approximated 2-RDMs of Several Test Systems in Their Ground State

System	Purification Procedure	Iteration	RMS Deviation $^1D'_{\alpha;\alpha}$	$^1D'_{\beta;\beta}$	$\langle \hat{S}^2 \rangle$
Be	MZ	1	3.311×10^{-2}	3.311×10^{-2}	-1.375×10^{-1}
		20	2.876×10^{-3}	2.876×10^{-3}	1.495×10^{-2}
	I-MZ	1	0	0	0
		20	0	0	0
B$^+$	MZ	1	1.931×10^{-2}	1.931×10^{-2}	-7.277×10^{-2}
		20	1.161×10^{-3}	1.161×10^{-3}	5.975×10^{-3}
	I-MZ	1	0	0	0
		20	0	0	0
C^{2+}	MZ	1	1.407×10^{-2}	1.407×10^{-2}	-5.859×10^{-2}
		20	7.495×10^{-4}	7.495×10^{-4}	3.759×10^{-3}
	I-MZ	1	0	0	0
		20	0	0	0
N^{3+}	MZ	1	1.276×10^{-2}	1.276×10^{-2}	-6.606×10^{-2}
		20	2.995×10^{-4}	2.995×10^{-4}	1.334×10^{-3}
	I-MZ	1	0	0	0
		20	0	0	0
O^{4+}	MZ	1	1.057×10^{-2}	1.057×10^{-2}	-5.591×10^{-2}
		20	2.765×10^{-4}	2.765×10^{-4}	1.444×10^{-3}
	I-MZ	1	0	0	0
		20	0	0	0
Li$_2$	MZ	1	1.324×10^{-1}	1.324×10^{-1}	-1.333×10^{-1}
		20	1.250×10^{-3}	1.250×10^{-3}	-7.553×10^{-3}
	I-MZ	1	0	0	0
		20	0	0	0
BeH$_2$	MZ	1	2.120×10^{-2}	2.120×10^{-2}	-1.879×10^{-1}
		20	4.966×10^{-3}	4.966×10^{-3}	5.646×10^{-2}
	I-MZ	1	0	0	0
		20	0	0	0

That is, all the information about the three important matrices 2-RDM, 2-HRDM, and 2-G is contained and available in the pure two-body correlation matrix. Moreover, the spin properties of both the pure two-body correlation matrix and the 2-G matrices play a central role in this purification procedure.

A. The Pure Two-Body Correlation Matrix Within the 2-RDM Formalism

Reconsider a 2-RDM element

$$2!\,^2D_{i_\sigma j_{\sigma'};k_\sigma l_{\sigma'}} = \langle \Psi | \hat{a}^\dagger_{i_\sigma} \hat{a}^\dagger_{j_{\sigma'}} \hat{a}_{l_{\sigma'}} \hat{a}_{k_\sigma} | \Psi \rangle \tag{110}$$

TABLE III
Calculated Energies from Non purified and Purified Approximated 2-RDMs Corresponding to Several Test Systems in their Ground State When Applying the I-MZ Purification Procedure[a]

	Energy (au)		
System	Nonpurified 2-RDM	Purified 2-RDM	Full CI
Be	−14.669	−14.583	−14.587
B^+	−24.358	−24.247	−24.248
C^{2+}	−36.525	−36.413	−36.414
N^{3+}	−51.195	−51.083	−51.085
O^{4+}	−68.364	−68.251	−68.252
Li_2	−14.997	−14.836	−14.847
BeH_2	−15.914	−15.741	−15.764

[a] Full CI energies are quoted as references.

and reorder the operators of the rhs term in such a way that one has a string of alternating creator and annihilator operators. One of the possible reorderings is

$$\langle \Psi | \hat{a}^\dagger_{i_\sigma} \hat{a}^\dagger_{j_{\sigma'}} \hat{a}_{l_{\sigma'}} \hat{a}_{k_\sigma} | \Psi \rangle = \delta_{j,l} \langle \Psi | \hat{a}^\dagger_{i_\sigma} \hat{a}_{k_\sigma} | \Psi \rangle - \langle \Psi | \hat{a}^\dagger_{i_\sigma} \hat{a}_{l_{\sigma'}} \hat{a}^\dagger_{j_{\sigma'}} \hat{a}_{k_\sigma} | \Psi \rangle \quad (111)$$

Inserting the unity operator

$$\hat{I} = |\Psi\rangle\langle\Psi| + \sum_{\Psi' \neq \Psi} |\Psi'\rangle\langle\Psi'| \quad (112)$$

between the operators $\hat{a}_{l_{\sigma'}}$ and $\hat{a}^\dagger_{j_{\sigma'}}$ in the last term of Eq. (111), it follows that the 2-RDM may be decomposed as

$$2! \, ^2D_{i_\sigma j_{\sigma'};k_\sigma l_{\sigma'}} = {}^2\mathcal{A}_{i_\sigma j_{\sigma'};k_\sigma l_{\sigma'}} + {}^2\mathcal{C}_{i_\sigma j_{\sigma'};k_\sigma l_{\sigma'}} \quad (113)$$

where the matrices $^2\mathcal{A}$ and $^2\mathcal{C}$ have, as in any tensor decomposition, the same row and column indices as the 2-RDM from which they derive. The structure of these matrices, is respectively,

$$^2\mathcal{A}_{i_\sigma j_{\sigma'};k_\sigma l_{\sigma'}} = {}^1D_{i_\sigma;k_\sigma} \, ^1D_{j_{\sigma'};l_{\sigma'}} - \delta_{\sigma,\sigma'} \, ^1D_{j_\sigma;k_\sigma} \, ^1D_{i_\sigma;l_\sigma} + \delta_{\sigma,\sigma'} \, ^1D_{i_\sigma;l_\sigma} \, ^1Q_{j_\sigma;k_\sigma} \quad (114)$$

and

$$^2\mathcal{C}_{i_\sigma j_{\sigma'};k_\sigma l_{\sigma'}} = \sum_{\Psi' \neq \Psi} \langle \Psi | \hat{a}^\dagger_{i_\sigma} \hat{a}_{k_\sigma} | \Psi' \rangle \langle \Psi' | \hat{a}^\dagger_{j_{\sigma'}} \hat{a}_{l_{\sigma'}} | \Psi \rangle$$

$$= \sum_{\Psi' \neq \Psi} {}^1 D^{\Psi\Psi'}_{i_\sigma;k_\sigma} \, ^1 D^{\Psi'\Psi}_{j_{\sigma'};l_{\sigma'}} \quad (115)$$

According to Eq. (114), the $^2\mathcal{A}$ matrix *depends only* on the 1-RDM; it is a sum of three terms: while the first term describes a classical system of independent particles, the second and third terms may be connected with the electron self-repulsion. These self-repulsion terms, represent respectively, the exchange and that part of the correlation effects that may be described in terms of the 1-HRDM [79]. On the other hand, the ^2C matrix—first reported in 1997 [78] and subsequently thoroughly studied [79–83, 86]—describes a dynamic correlation mechanism where two electrons undergo virtual excitations and deexcitations in order to avoid each other [79], causing the polarization effects of the electron cloud. This matrix was called pure two-body correlation matrix (2-CM) because, as shown in Eq. (115), it cannot be factorized into one-body RDMs/HRDMs; and, as will be shown, it does not contribute to the contraction of the 2-RDM into the one-electron space.

1. Basic Properties of the Pure Two-Body Correlation Matrices

A very important property of the 2-CM is that [15, 83] the decomposition of the 2-HRDM yields a two-body hole correlation matrix that coincides with the 2-CM. Thus

$$2!\,^2Q_{i_\sigma j_{\sigma'};k_\sigma l_{\sigma'}} = {}^2\bar{\mathcal{A}}_{i_\sigma j_{\sigma'};k_\sigma l_{\sigma'}} + {}^2C_{i_\sigma j_{\sigma'};k_\sigma l_{\sigma'}} \tag{116}$$

where

$$^2\bar{\mathcal{A}}_{i_\sigma j_{\sigma'};k_\sigma l_{\sigma'}} = {}^1Q_{i_\sigma;k_\sigma}\,{}^1Q_{j_{\sigma'};l_{\sigma'}} - \delta_{\sigma,\sigma'}\,\delta_{i,l}\,{}^1Q_{j_\sigma;k_\sigma} \tag{117}$$

Contrary to the 2-RDM and 2-HRDM, the 2-CM is neither positive semidefinite nor antisymmetric. Indeed, the permutation of an index leads to the following relation:

$$^2C_{i_\sigma j_{\sigma'};l_{\sigma'}k_\sigma} = {}^1D_{i_\sigma;k_\sigma}\,{}^1Q_{j_{\sigma'};l_{\sigma'}} + \delta_{\sigma,\sigma'}\,{}^1D_{i_\sigma;l_\sigma}\,{}^1Q_{j_\sigma;k_\sigma} - {}^2C_{i_\sigma j_{\sigma'};k_\sigma l_{\sigma'}} \tag{118}$$

Since the 2-CMs are not antisymmetric, four different contracting operations can be performed on them [71, 79, 83]. Two of these contractions, which can be considered "natural" because they derive from the contraction of relation Eq. (113), are

$$\sum_j {}^2C_{i_\sigma j_{\sigma'};k_\sigma j_{\sigma'}} = 0 \tag{119}$$

and

$$\sum_j {}^2C_{j_\sigma i_{\sigma'};j_\sigma k_{\sigma'}} = 0 \tag{120}$$

It is because of these two relations that no contribution of the ^2C appears in the contractions of Eq. (113).

The other two less obvious contractions are

$$\sum_j {}^2C_{i_\sigma j_{\sigma'};j_{\sigma'} k_\sigma} = (K - N_{\sigma'})\, {}^1D_{i_\sigma;k_\sigma} + \delta_{\sigma,\sigma'} \left({}^1D - {}^1D^2\right)_{i_\sigma;k_\sigma} \quad (121)$$

$$\sum_j {}^2C_{j_\sigma i_{\sigma'};k_{\sigma'} j_\sigma} = N_\sigma \left(\delta_{i,k} - {}^1D_{i_{\sigma'};k_{\sigma'}}\right) + \delta_{\sigma,\sigma'} \left({}^1D - {}^1D^2\right)_{i_{\sigma'};k_{\sigma'}} \quad (122)$$

These last two contractions (Eqs. (121) and (122)) are also very important since they lead to the 1-RDM and therefore the corresponding 2-RDM and 2-HRDM [24, 25, 83]. Hence it follows that the 2-RDM N-representability problem may be studied equivalently by focusing on the N-representability conditions for the 2-CM matrix [71, 83]. Thus the set of relations given above constitutes a set of N-representability conditions—strongly exacting and necessary conditions—not only for the 2-CM matrix but also for the 2-RDM as well as for the 2-HRDM.

Another important property of the 2-CM matrices is that they are closely related with the positive semidefinite 2-G matrix [15, 83]. The interrelations between these two matrices are

$$^2C_{i_\sigma j_{\sigma'};k_\sigma l_{\sigma'}} = {}^2G_{i_\sigma k_\sigma;l_{\sigma'} j_{\sigma'}} \quad (123)$$

and,

$$^2C_{i_\sigma j_{\sigma'};l_{\sigma'} k_\sigma} = {}^2G_{i_\sigma l_{\sigma'};k_\sigma j_{\sigma'}} \quad (124)$$

That is, both the 2-CM and the 2-G matrix have common elements, but a given element occupies different positions in each matrix. In other words, while the labels of the row/column of the 2-CM refer, as in the 2-RDM, to *two particles/two holes*, the labels of the row/column of the 2-G matrix refer to particle–hole/hole–particle. Thus, although both the 2-CM and the 2-G matrices describe similar types of correlation effects, only the 2-CM describes *pure* two-body correlation effects. This is because the 2-CM "natural" tensorial contractions vanish, and thus there is no contribution to the natural contraction of the 2-RDM into the one-body space; whereas the 2-G "natural" tensorial contractions are functionals of the 1-RDM.

2. Spin Structure and Properties of the Pure Two-Body Correlation Matrices in the Singlet Case

Because the 2-CM and 2-G matrices are directly related through Eqs. (123) and (124), the 2-CM may also be decomposed into a sum of spin components. Thus one may write

$$^2C_{i_\sigma j_{\sigma'};k_{\sigma''} l_{\sigma'''}} = {}_{\{S,M'\}}{}^2C_{i_\sigma j_{\sigma'};k_{\sigma''} l_{\sigma'''}} + {}_{\{S+1,M'\}}{}^2C_{i_\sigma j_{\sigma'};k_{\sigma''} l_{\sigma'''}} + {}_{\{S-1,M'\}}{}^2C_{i_\sigma j_{\sigma'};k_{\sigma''} l_{\sigma'''}} \quad (125)$$

where

$$\{S',M'\}{}^2C_{i_\sigma j_{\sigma'};k_{\sigma''}l_{\sigma'''}} \equiv \sum_{\Psi'_{S',M'} \neq \Psi_{S,M}} {}^1D_{i_\sigma;k_{\sigma''}}^{\Psi_{S,M}\Psi'_{S',M'}} {}^1D_{j_{\sigma'};l_{\sigma'''}}^{\Psi'_{S',M'}\Psi_{S,M}} \quad (126)$$

and S' may take the values S, $S+1$, and $S-1$. Furthermore, each of these spin components are interrelated with those of the 2-G matrix according to

$$\{S',M'\}{}^2C_{i_\sigma j_{\sigma'};k_{\sigma''}l_{\sigma'''}} = \{S',M'\}{}^2G_{i_\sigma k_{\sigma''};l_{\sigma'''}j_{\sigma'}} \quad (127)$$

Hence the general relations addressed in Section II are also valid for the 2-CM. Thus the different spin components of a 2-CM are also related when this matrix corresponds to a pure spin state with spin quantum numbers S and M; and similarly for the 2-CMs corresponding to different states of a given multiplet.

In the particular case of singlet states, it can be shown that all the spin-blocks of the 2-CM are proportional to ${}^2C_{\alpha\beta;\alpha\beta}$. Thus only this spin block is needed to determine the two-body correlation matrix.

According to Eq. (125), this 2-CM spin-block may be decomposed as

$$^2C_{i_\alpha j_\beta;k_\alpha l_\beta} = {}_{\{0,0\}}{}^2C_{i_\alpha j_\beta;k_\alpha l_\beta} + {}_{\{1,0\}}{}^2C_{i_\alpha j_\beta;k_\alpha l_\beta} \quad (128)$$

In order to be able to correct each of these 2-CM spin components separately, it is necessary to obtain a relation where a component is given in terms of the 2-CM and eventually also of the 1-RDM. These relations, which have recently been reported in Ref. [72], are

$$_{\{0,0\}}{}^2C_{i_\alpha j_\beta;k_\alpha l_\beta} = {}^2C_{i_\alpha j_\beta;k_\alpha l_\beta} + \tfrac{1}{2}\left({}^1D_{i_\alpha;l_\alpha}\,{}^1Q_{j_\beta;k_\beta} - {}^2C_{i_\alpha j_\beta;l_\alpha k_\beta}\right) \quad (129)$$

and

$$_{\{1,0\}}{}^2C_{i_\alpha j_\beta;k_\alpha l_\beta} = \tfrac{1}{2}\left({}^2C_{i_\alpha j_\beta;l_\alpha k_\beta} - {}^1D_{i_\alpha;l_\alpha}\,{}^1Q_{j_\beta;k_\beta}\right) \quad (130)$$

Conversely, it follows that

$$^2C_{i_\alpha j_\beta;k_\alpha l_\beta} = \tfrac{2}{3}\left(2\,{}_{\{0,0\}}{}^2C_{i_\alpha j_\beta;k_\alpha l_\beta} + {}_{\{0,0\}}{}^2C_{i_\alpha j_\beta;l_\alpha k_\beta}\right)$$
$$- \tfrac{1}{3}\left({}^1D_{i_\alpha;k_\alpha}\,{}^1Q_{j_\beta;l_\beta} + 2\,{}^1D_{i_\alpha;l_\alpha}\,{}^1Q_{j_\beta;k_\beta}\right) \quad (131)$$

and

$$^2C_{i_\alpha j_\beta;k_\alpha l_\beta} = \tfrac{1}{2}\left({}^1D_{i_\alpha;k_\alpha}\,{}^1Q_{j_\beta;l_\beta} + {}^1Q_{i_\beta;k_\beta}\,{}^1D_{j_\alpha;l_\alpha}\right)$$
$$+ {}_{\{1,0\}}{}^2C_{i_\alpha j_\beta;l_\alpha k_\beta} + {}_{\{1,0\}}{}^2C_{j_\alpha i_\beta;k_\alpha l_\beta} \quad (132)$$

These relations show that each of the spin components presents a one-to-one correspondence with the $^2C_{\alpha\beta;\alpha\beta}$ spin-block, and therefore with the entire correlation matrix. This is because the 1-RDM—and, consequently, the 1-HRDM—appearing in Eqs. (131) and (132) can be obtained from the different contractions of the spin components. Thus, while it follows from Eqs. (118), (119), (120), (129), and (130) that

$$\sum_j {}_{\{0,0\}}{}^2C_{i_\alpha j_\beta;k_\alpha j_\beta} = \sum_j {}_{\{0,0\}}{}^2C_{j_\alpha i_\beta;j_\alpha k_\beta} = 0 \tag{133}$$

$$\sum_j {}_{\{1,0\}}{}^2C_{i_\alpha j_\beta;k_\alpha j_\beta} = \sum_j {}_{\{1,0\}}{}^2C_{j_\alpha i_\beta;j_\alpha k_\beta} = 0 \tag{134}$$

Eqs. (118), (121), (122), (129), and (130) lead to

$$\sum_j {}_{\{0,0\}}{}^2C_{i_\alpha j_\beta;j_\alpha k_\beta} = \frac{(2K-N)}{4}{}^1D_{i_\alpha;k_\alpha} + ({}^1D - {}^1D^2)_{i_\alpha;k_\alpha} \tag{135}$$

$$\sum_j {}_{\{0,0\}}{}^2C_{j_\alpha i_\beta;k_\alpha j_\beta} = \frac{N}{4}(\delta_{i,k} - {}^1D_{i_\beta;k_\beta}) + ({}^1D - {}^1D^2)_{i_\beta;k_\beta} \tag{136}$$

$$\sum_j {}_{\{1,0\}}{}^2C_{i_\alpha j_\beta;j_\alpha k_\beta} = -\frac{(2K-N)}{4}{}^1D_{i_\alpha;k_\alpha} \tag{137}$$

$$\sum_j {}_{\{1,0\}}{}^2C_{j_\alpha i_\beta;k_\alpha j_\beta} = -\frac{N}{4}(\delta_{i,k} - {}^1D_{i_\beta;k_\beta}) \tag{138}$$

and therefore

$$\sum_{i,j} {}_{\{0,0\}}{}^2C_{i_\alpha j_\beta;j_\alpha i_\beta} = \frac{N}{8}(2K-N+4) - \sum_i ({}^1D^2)_{i_\alpha;i_\alpha} \tag{139}$$

$$\sum_{i,j} {}_{\{1,0\}}{}^2C_{i_\alpha j_\beta;j_\alpha i_\beta} = -\frac{N}{8}(2K-N) \tag{140}$$

Equations (135)–(138) describe a set of unusual contractions, since the sum labels run over orbitals, which multiply different spin functions. In fact, they are exact only in the singlet case, where $\hat{S}_+|\Psi_{0,0}\rangle = \hat{S}_-|\Psi_{0,0}\rangle = 0$, ${}^1D_{\alpha;\alpha} = {}^1D_{\beta;\beta}$, and $N_\alpha = N_\beta = N/2$.

In order to solve Eqs. (135) and (136), one may follow the method reported in Refs. [24, 25]. Thus each of these equations has two solutions but one of the solutions can be ignored, since it does not correspond to an ensemble N-representable 1-RDM.

When the 2-CM is exact, all the 1-RDMs obtained from Eqs. (135)–(138) coincide; however, in practice one can only hope that the differences among these matrices are small. These latter properties constitute important S-representability conditions in the singlet case and are at the center of the N- and S-representability purification procedure, which will now be described. In what follows we will identify $^1D^p$, $^1D^q$, $^1D^r$, and $^1D^s$ with the solutions of Eqs. (135), (136), (137), and (138), respectively while keeping the symbol ^1D for the initial 1-RDM, which remains fixed throughout the iterations of the AV purification procedure.

B. The Alcoba–Valdemoro Purification Procedure

In order to be as precise as possible, the different steps of the AV N- and S-representability purification procedure will be described in the same order as they appear in the flowchart of the computer code.

1. *Initial Data.*
   (a) The initial data are the trial $^2D^{(0)}_{\alpha\beta;\alpha\beta}$ and the corresponding 1-RDM, ^1D, which has previously been rendered ensemble N-representable. This is achieved as follows: the 1-RDM matrices are diagonalized. Then the negative eigenvalues are made equal to zero, and the matrix is renormalized iteratively while keeping all the eigenvalues equal or less than one. The subroutine that performs this task is very efficient and obtains an N-representable 1-RDM extremely close to the initial one.
   (b) From the initial 2-RDM spin-block, the corresponding $^2C^{(0)}_{\alpha\beta;\alpha\beta}$ is obtained by applying formula (113).

2. *Imposing the D/Q N-Representability Condition on the 2-RDM/2-HRDM.*
   (a) The $^2Q^{(0)}_{\alpha\beta;\alpha\beta}$ is now formed using relation (116).
   (b) The 2-RDM and the 2-HRDM are diagonalized. Let us call $\{|x_p\rangle\}/\{x_p\}$ the eigenvectors/eigenvalues of the $^2D^{(0)}_{\alpha\beta;\alpha\beta}$ spin-block. Similarly, let us call $\{|\bar{x}_q\rangle\}/\{\bar{x}_q\}$ the eigenvectors/eigenvalues of the $^2Q^{(0)}_{\alpha\beta;\alpha\beta}$ spin-block. The new 2-RDM and 2-HRDM are then reconstructed as

$$^2D^{(1)}_{i_\alpha j_\beta;k_\alpha l_\beta} = \sum_p{}' x_p \langle i_\alpha j_\beta|x_p\rangle\langle x_p|k_\alpha l_\beta\rangle \tag{141}$$

and

$$^2Q^{(1)}_{i_\alpha j_\beta;k_\alpha l_\beta} = \sum_q{}' \bar{x}_q \langle i_\alpha j_\beta|\bar{x}_q\rangle\langle \bar{x}_q|k_\alpha l_\beta\rangle \tag{142}$$

where $|i_\alpha j_\beta\rangle$ and $|k_\alpha l_\beta\rangle$ are two-electron Slater determinants. The symbol Σ' indicates that only the positive eigenvalues are considered.

The $^2D^{(1)}_{\alpha\beta;\alpha\beta}$ and $^2Q^{(1)}_{\alpha\beta;\alpha\beta}$ are then multiplied by a factor in order to renormalize their trace to the value $(N_\alpha N_\beta)$ and $(K - N_\alpha)(K - N_\beta)$, respectively.

3. *Implications of the D- and Q-Conditions on the 2-CM.* The errors of the 2-RDM and 2-HRDM are

$$^2\Gamma_{i_\alpha j_\beta;k_\alpha l_\beta} = {}^2D^{(1)}_{i_\alpha j_\beta;k_\alpha l_\beta} - {}^2D^{(0)}_{i_\alpha j_\beta;k_\alpha l_\beta} \tag{143}$$

$$^2\bar{\Gamma}_{i_\alpha j_\beta;k_\alpha l_\beta} = {}^2Q^{(1)}_{i_\alpha j_\beta;k_\alpha l_\beta} - {}^2Q^{(0)}_{i_\alpha j_\beta;k_\alpha l_\beta} \tag{144}$$

Since the 1-RDM is kept fixed, these errors should necessarily correspond to those of the $^2C^{(0)}_{\alpha\beta;\alpha\beta}$. Thus the new approximation for the 2-CM is obtained as follows:

$$^2C^{(1)}_{i_\alpha j_\beta;k_\alpha l_\beta} = {}^2C^{(0)}_{i_\alpha j_\beta;k_\alpha l_\beta} + \tfrac{1}{2}\left({}^2\Gamma_{i_\alpha j_\beta;k_\alpha l_\beta} + {}^2\bar{\Gamma}_{i_\alpha j_\beta;k_\alpha l_\beta}\right) \tag{145}$$

4. *Imposing S-Representability Conditions on the 2-CM.* In order to impose the S-representability conditions on the 2-CM, we proceed by correcting its spin components in a sequential manner.

   (a) The $_{\{0,0\}}{}^2C$ spin component.
      (i) Due to Eq. (129), the $_{\{0,0\}}{}^2C^{(1)}_{\alpha\beta;\alpha\beta}$ is obtained.

$$_{\{0,0\}}{}^2C^{(1)}_{i_\alpha j_\beta;k_\alpha l_\beta} = \tfrac{1}{2}({}^1D_{i_\alpha;l_\alpha}{}^1Q_{j_\beta;k_\beta} - {}^2C^{(1)}_{i_\alpha j_\beta;l_\alpha k_\beta}) + {}^2C^{(1)}_{i_\alpha j_\beta;k_\alpha l_\beta} \tag{146}$$

      (ii) The $_{\{0,0\}}{}^2G^{(1)}_{\alpha\alpha;\beta\beta}$ matrix is obtained according to the relation

$$_{\{0,0\}}{}^2G_{i_\alpha k_\alpha;l_\beta j_\beta} = {}_{\{0,0\}}{}^2C_{i_\alpha j_\beta;k_\alpha l_\beta} \tag{147}$$

   This matrix is rendered positive by carrying out the same operations described in step 2 for the 2-RDM. The renormalization factor in this case is chosen so as to yield the trace given in Eq. (139).
      (iii) From the resulting $_{\{0,0\}}{}^2G^{(2)}_{\alpha\alpha;\beta\beta}$ a new $_{\{0,0\}}{}^2C^{(2)}_{\alpha\beta;\alpha\beta}$ is obtained.
      (iv) From Eqs. (135) and (136) the corresponding $^1D^p$ and $^1D^q$ are obtained.
      (v) Equation (131) leads to a new $^2C^{(2)}$

$$^2C^{(2)}_{i_\alpha j_\beta;k_\alpha l_\beta} = \tfrac{2}{3}\left(2_{\{0,0\}}{}^2C^{(2)}_{i_\alpha j_\beta;k_\alpha l_\beta} + {}_{\{0,0\}}{}^2C^{(2)}_{i_\alpha j_\beta;l_\alpha k_\beta}\right) \\ - \tfrac{1}{3}\left({}^1D^p_{i_\alpha;k_\alpha}{}^1Q^q_{j_\beta;l_\beta} + 2\,{}^1D^p_{i_\alpha;l_\alpha}{}^1Q^q_{j_\beta;k_\beta}\right) \tag{148}$$

Note that only at convergence does this relation coincide with Eq. (131).

(b) The $_{\{1,0\}}{}^2\mathrm{C}$ spin component. The key equations in this part of the procedure are Eqs. (130), (137), and (138), while the underlying reasoning linking them is similar to that applied in steps (a) (i)–(iv). Note that in this case the $_{\{1,0\}}{}^2\mathrm{G}^{(2)}_{\alpha\alpha;\beta\beta}$, which is obtained according to the relation

$$_{\{1,0\}}{}^2\mathrm{G}_{i_\alpha k_\alpha;l_\beta j_\beta} = {}_{\{1,0\}}{}^2\mathrm{C}_{i_\alpha j_\beta;k_\alpha l_\beta}, \tag{149}$$

must be rendered *negative semi-definite* [72] and must be renormalized to present the trace given by Eq. (140). Moreover, the 1-RDMs obtained from the contractions of this 2-CM spin component are ${}^1\mathrm{D}^r$ and ${}^1\mathrm{D}^s$. Equation (132) leads to a new ${}^2\mathrm{C}^{(2)}$:

$${}^2\mathrm{C}^{(2)}_{i_\alpha j_\beta;k_\alpha l_\beta} = \tfrac{1}{2}({}^1\mathrm{D}^r_{i_\alpha;k_\alpha}\,{}^1\mathrm{Q}^s_{j_\beta;l_\beta} + {}^1\mathrm{Q}^s_{i_\beta;k_\beta}\,{}^1\mathrm{D}^r_{j_\alpha;l_\alpha}) + {}_{\{1,0\}}{}^2\mathrm{C}^{(2)}_{i_\alpha j_\beta;l_\alpha k_\beta} + {}_{\{1,0\}}{}^2\mathrm{C}^{(2)}_{j_\alpha i_\beta;k_\alpha l_\beta} \tag{150}$$

Note that at convergency ${}^1\mathrm{D}^r = {}^1\mathrm{D}^s = {}^1\mathrm{D}$.

5. *Obtaining the Final RDMs at Each Iteration.* The RDMs constituting the initial data for the next iteration are recalculated by aplying Eqs. (113) and (116).

6. *Consistency Tests.* A set of consistency tests are carried out at the end of each iteration. These tests check the extent of coincidence of the five 1-RDMs ${}^1\mathrm{D}$, ${}^1\mathrm{D}^p$, ${}^1\mathrm{D}^q$, ${}^1\mathrm{D}^r$, and ${}^1\mathrm{D}^s$, as well as the convergence toward the value zero of the RMS deviations of the natural contractions of the spin components of the 2-CM.

The results obtained in the calculations of the ground states of several test molecules are reported in the next section.

C. Test Calculations and Results

In order to analyze the performance of this purification procedure and to compare it with those reported in the previous section, the same atomic and molecular systems in their ground state were selected as test systems. Again, the basis sets used were formed by Hartree–Fock molecular orbitals built out of minimal Slater orbital basis sets and the initial data were chosen to be the approximate 2-RDMs built by application of the independent pair model within the framework of the SRH theory.

Since the results obtained for each of the test systems studied show a very similar performance, only the results obtained for the beryllium atom will be reported in detail.

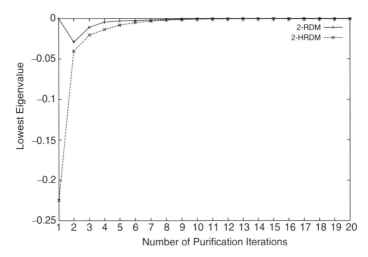

Figure 5. Lowest eigenvalue of the 2-RDM and 2-HRDM at each iteration of the AV purification procedure for the ground state of the beryllium atom.

The rate of convergence toward positivity may easily be appreciated by plotting the lowest eigenvalue of each of the different matrices at each iteration. Thus the simultaneous convergence toward positivity of the 2-RDM and 2-HRDM is shown in Fig. 5. As can be seen, convergence in both curves is smooth and rapid. After twenty iterations the negativity of these matrices is negligible (the lowest eigenvalues of the 2-RDM and 2-HRDM are -0.00023 and -0.00044, respectively).

Figure 6 shows how the S-representability is attained. Thus it can be seen from this figure that the $_{\{0,0\}}{}^2G_{\alpha\alpha;\beta\beta}/_{\{1,0\}}{}^2G_{\alpha\alpha;\beta\beta}$ spin-block converges very satisfactorily on a positive/negative semidefinite matrix. After twenty iterations the lowest/highest eigenvalue of these two matrices is -0.00010 and 0.00022, respectively. As was mentioned in Section II, these conditions are much more exacting than the well-known G-condition.

A very important reliability consistency test of the procedure is shown in Fig. 7, where it can be seen that the convergence of the four initially very different matrices $^1D^p$, $^1D^q$, $^1D^r$, $^1D^s$ toward the ^1D, which is kept fixed throughout the iterations, is clearly excellent; indeed, after twenty iterations the RMS deviation of these matrices from ^1D are 0.00008, 0.00008, 0.00005, and 0.00010, respectively. Another set of tests confirming the consistency of the results is provided by Fig. 8. In this figure the RMS deviations from zero of the natural contractions of the spin components of the 2-CM—and hence of the 2-CM—are shown. Clearly, these deviations converge toward zero very rapidly; at iteration 20 the RMS deviations of the natural-left and natural-right contractions of

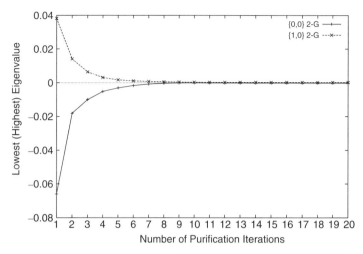

Figure 6. Lowest eigenvalue of the $_{\{0,0\}}{}^2G_{\alpha\alpha;\beta\beta}$ and highest eigenvalue of the $_{\{1,0\}}{}^2G_{\alpha\alpha;\beta\beta}$ at each iteration of the AV purification procedure for the ground state of the beryllium atom.

$_{\{0,0\}}{}^2C_{\alpha\beta;\alpha\beta}$ are 0.00015 and 0.00015, respectively, while those of $_{\{1,0\}}{}^2C_{\alpha\beta;\alpha\beta}$ are 0.00013 and 0.00013, respectively.

As mentioned earlier, similar behaviors to that of the beryllium atom case were found when applying the procedure to the approximated 2-RDMs of the B^+, C^{2+}, N^{3+}, O^{4+}, Li_2, and BeH_2 cases. The results obtained in these

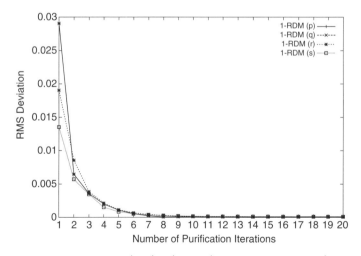

Figure 7. RMS deviation of the $^1D^p$, $^1D^q$, $^1D^r$, and $^1D^s$ from the fixed 1-RDM ^1D at each iteration of the AV purification procedure for the ground state of the beryllium atom.

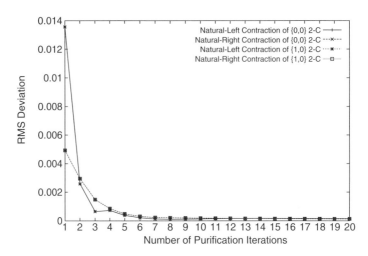

Figure 8. RMS deviation of the natural-left and natural-right contractions of the ${}_{\{0,0\}}{}^2C_{\alpha\beta;\alpha\beta}$ and ${}_{\{1,0\}}{}^2C_{\alpha\beta;\alpha\beta}$ from the null vector at each iteration of the AV purification procedure for the ground state of the beryllium atom.

calculations are reported in Tables IV–VI. These results show that the D- and Q-conditions as well as the spin G-conditions are efficiently imposed in all the cases studied. Moreover, while part of the set of S-representability conditions is also explicitly imposed in the procedure, the rest of these conditions are only indirectly taken into account. The test calculations at the end of the

TABLE IV
Performances of the AV Purification Procedure When Applied to Approximated 2-RDMs of Several Test Systems in Their Ground State

System	Iteration	Lowest (Highest) Eigenvalue			
		${}^2D_{\alpha\beta;\alpha\beta}$	${}^2Q_{\alpha\beta;\alpha\beta}$	${}_{\{0,0\}}{}^2G_{\alpha\alpha;\beta\beta}$	${}_{\{1,0\}}{}^2G_{\alpha\alpha;\beta\beta}$
Be	1	0	-2.247×10^{-1}	-6.604×10^{-2}	3.820×10^{-2}
	20	-2.296×10^{-4}	-4.380×10^{-4}	-1.031×10^{-4}	2.229×10^{-4}
B^+	1	0	-1.540×10^{-1}	-4.772×10^{-2}	2.778×10^{-2}
	20	-1.058×10^{-4}	-3.068×10^{-4}	-6.615×10^{-5}	1.442×10^{-4}
C^{2+}	1	0	-1.167×10^{-1}	-3.729×10^{-2}	2.127×10^{-2}
	20	-5.765×10^{-5}	-1.874×10^{-4}	-3.994×10^{-5}	8.482×10^{-5}
N^{3+}	1	0	-9.322×10^{-2}	-3.050×10^{-2}	1.712×10^{-2}
	20	-3.420×10^{-5}	-1.236×10^{-4}	-2.573×10^{-5}	5.385×10^{-5}
O^{4+}	1	0	-7.795×10^{-2}	-2.589×10^{-2}	1.434×10^{-2}
	20	-2.269×10^{-5}	-8.652×10^{-5}	-1.758×10^{-5}	3.669×10^{-5}
Li_2	1	0	-4.089×10^{-1}	-9.866×10^{-2}	5.194×10^{-2}
	20	4.129×10^{-4}	-1.483×10^{-4}	-3.581×10^{-5}	5.958×10^{-5}
BeH_2	1	0	-2.147×10^{-1}	-7.884×10^{-2}	4.986×10^{-2}
	20	-2.433×10^{-3}	-1.041×10^{-3}	-4.683×10^{-4}	2.800×10^{-4}

TABLE V
Performances of the AV Purification Procedure When Applied to
Approximated 2-RDMs of Several Test Systems in Their Ground State

		RMS Deviation			
System	Iteration	$^1D^p$	$^1D^q$	$^1D^r$	$^1D^s$
Be	1	2.906×10^{-2}	2.906×10^{-2}	1.903×10^{-2}	1.353×10^{-2}
	20	8.087×10^{-5}	8.087×10^{-5}	5.198×10^{-5}	9.996×10^{-5}
B^+	1	2.178×10^{-2}	2.178×10^{-2}	1.404×10^{-2}	1.034×10^{-2}
	20	3.678×10^{-5}	3.678×10^{-5}	4.208×10^{-5}	6.240×10^{-5}
C^{2+}	1	1.729×10^{-2}	1.729×10^{-2}	1.112×10^{-2}	8.350×10^{-3}
	20	1.971×10^{-5}	1.971×10^{-5}	2.690×10^{-5}	3.707×10^{-5}
N^{3+}	1	1.423×10^{-2}	1.423×10^{-2}	9.145×10^{-3}	6.933×10^{-3}
	20	1.142×10^{-5}	1.142×10^{-5}	1.830×10^{-5}	2.393×10^{-5}
O^{4+}	1	1.212×10^{-2}	1.212×10^{-2}	7.793×10^{-3}	5.942×10^{-3}
	20	7.478×10^{-6}	7.478×10^{-6}	1.295×10^{-5}	1.658×10^{-5}
Li_2	1	3.991×10^{-2}	3.991×10^{-2}	5.382×10^{-2}	1.803×10^{-2}
	20	7.065×10^{-5}	7.065×10^{-5}	7.171×10^{-5}	4.116×10^{-5}
BeH_2	1	2.095×10^{-2}	2.774×10^{-2}	5.772×10^{-3}	5.269×10^{-3}
	20	2.205×10^{-4}	3.288×10^{-4}	5.909×10^{-5}	1.216×10^{-4}

procedure iterations yield highly satisfying results, since not only the S-representability conditions imposed are satisifed but also the spin conditions not imposed are nevertheless satisfied by the resulting 2-CM. These results show that after twenty iterations the convergence towards the set of all the

TABLE VI
Performances of the AV Purification Procedure When Applied to
Approximated 2-RDMs of Several Test Systems in Their Ground State

		RMS Deviation of Natural Contractions			
		$_{\{0,0\}}{}^2C_{\alpha\beta;\alpha\beta}$		$_{\{1,0\}}{}^2C_{\alpha\beta;\alpha\beta}$	
System	Iteration	Left	Right	Left	Right
Be	1	1.357×10^{-2}	1.357×10^{-2}	4.928×10^{-3}	4.928×10^{-3}
	20	1.521×10^{-4}	1.521×10^{-4}	1.309×10^{-4}	1.309×10^{-4}
B^+	1	9.617×10^{-3}	9.617×10^{-3}	3.756×10^{-3}	3.756×10^{-3}
	20	8.480×10^{-5}	8.480×10^{-5}	7.627×10^{-5}	7.627×10^{-5}
C^{2+}	1	7.352×10^{-3}	7.352×10^{-3}	3.062×10^{-3}	3.062×10^{-3}
	20	4.930×10^{-5}	4.930×10^{-5}	4.457×10^{-5}	4.457×10^{-5}
N^{3+}	1	5.947×10^{-3}	5.947×10^{-3}	2.565×10^{-3}	2.565×10^{-3}
	20	3.152×10^{-5}	3.152×10^{-5}	2.830×10^{-5}	2.830×10^{-5}
O^{4+}	1	5.018×10^{-3}	5.018×10^{-3}	2.217×10^{-3}	2.217×10^{-3}
	20	2.187×10^{-5}	2.187×10^{-5}	1.949×10^{-5}	1.949×10^{-5}
Li_2	1	2.650×10^{-2}	2.650×10^{-2}	4.603×10^{-3}	4.603×10^{-3}
	20	6.093×10^{-5}	6.093×10^{-5}	5.290×10^{-5}	5.290×10^{-5}
BeH_2	1	1.602×10^{-2}	1.602×10^{-2}	2.312×10^{-3}	2.312×10^{-3}
	20	9.434×10^{-5}	9.434×10^{-5}	2.446×10^{-5}	2.446×10^{-5}

TABLE VII
Study of the Spin G-Conditions in the I-MZ and the AV Purification Procedures When Applied to Approximated 2-RDMs of Li_2 and BeH_2 Systems in Their Ground States

System	Iteration	I-MZ Purification $\{0,0\}{}^2G_{\alpha\alpha;\beta\beta}$	I-MZ Purification $\{1,0\}{}^2G_{\alpha\alpha;\beta\beta}$	AV Purification $\{0,0\}{}^2G_{\alpha\alpha;\beta\beta}$	AV Purification $\{1,0\}{}^2G_{\alpha\alpha;\beta\beta}$
Li_2	1	-1.598×10^{-1}	1.277×10^{-1}	-9.866×10^{-2}	5.194×10^{-2}
	20	-4.286×10^{-4}	2.823×10^{-4}	-3.581×10^{-5}	5.958×10^{-5}
BeH_2	1	-9.880×10^{-2}	8.057×10^{-2}	-7.884×10^{-2}	4.986×10^{-2}
	20	-5.256×10^{-3}	2.240×10^{-3}	-4.683×10^{-4}	2.800×10^{-4}

properties, which, to our knowledge, characterize a well-behaved 2-RDM, is achieved in a consistent and simultaneous way.

All these results have been compared with those obtained by applying the I-MZ purification procedure (Tables I and II). Although the results are very similar from a global point of view, it has been found that in certain cases (e.g., BeH_2 molecule) the latter procedure yields 2-RDMs and 2-HRDMs that oscillate markedly before converging toward positive matrices. Another important difference betweeen the results concerns the spin G-conditions. Thus, although these conditions are not imposed in the I-MZ procedure, the negativity/positivity of the spin components of the 2-G matrix are corrected as effectively as the 2-RDMs and 2-HRDMs. This may be due to the fact that the different spin-blocks of these matrices are forced to contract correctly; and therefore an indirect action on the spin components of the 2-G matrix may occur. On the other hand, the correction of the negativity of the spin components of the 2-G matrix in the AV purification procedure is carried out still more effectively than the negativity correction of the 2-RDM and 2-HRDM. This is illustrated for Li_2 and BeH_2 molecules in Table VII. The conditions imposed on the spin components of the 2-G and their contractions are essential for the N- and S-representability of this matrix and, hence, for those of the 2-RDM and 2-HRDM.

V. PURIFICATION WITHIN THE FRAMEWORK OF THE SECOND-ORDER CONTRACTED SCHRÖDINGER EQUATION

Among the several 2-RDM-oriented methods that have been developed for the study of chemical systems, one of the most recent and promising techniques is based on the iterative solution of the second-order contracted Schrödinger equation (2-CSE) [1, 6, 15, 18, 36, 45–60, 62–65, 68, 70, 79–85, 103–111]. The 2-CSE was initially derived in 1976 in first quantization in the works of Cho [103], Cohen and Frishberg [104, 105], and Nakatsuji [106] and later on deduced in second quantization by Valdemoro [45] through the contraction of the matrix representation of the Schrödinger equation into the two-electron space. This equation was shown to be equivalent to the Schrödinger equation (by the necessary and sufficient condition) within the N-representable space of RDMs [106]. The drawback of this very attractive equation is that its solution is indeterminate [63, 107]. This is due to the fact that, although in an average way, RDMs of orders higher than 2 (the 3-RDM and the 4-RDM) appear in it.

The indeterminacy of the 2-CSE had caused this equation to be overlooked for many years. In 1992 Valdemoro proposed a method to approximate the 2-RDM in terms of the 1-RDM [108], which was extended in order to approximate the 3- and 4-RDMs in terms of the lower-order matrices [46, 47] and in 1994 Colmenero and Valdemoro [48] applied these approximate constructing algorithms to avoid the indeterminacy problem and solve iteratively the 2-CSE. Since then, the study of improved constructing algorithms as well as alternative strategies in the iterative procedure have been proposed [1, 6, 15, 18, 36, 49–57, 59, 60, 62–65, 68, 70, 71, 79–85, 109–111]. Also, good results of several calculations have been reported [6, 49–52, 55, 56, 68, 70, 88, 109, 111].

One of the critical points that affect the success of the iterative solution of the 2-CSE is that of the possible N- and S-representability defects of the 2-RDM. Thus the N-representability of the initial trial 2-RDM as well as its S-representability are partially lost during the iterative process [87]. As pointed out in Refs. [15, 70, 88, 111], the divergence that appears at a certain point of the iterative process can be prevented by correcting the small representability defects of the resulting 2-RDM as the iterations proceed. Thus the 2-CSE iterative process has been linked with several iterative purification procedures [50, 70, 88, 111]. In the method that we apply here, we couple the iterative 2-CSE process with the AV purification procedure and with a regulating convergence device recently reported [88]. These two implementations render this method highly efficient. In what follows this method will be referred to as the regulated CSE-NS method.

A. The Second-Order Contracted Schrödinger Equation and Its Iterative Solution

Since most of the different steps of the method have previously been published [48, 50, 60, 82], a detailed study will only be given of the new *regulating* device, which was introduced in Ref. [88].

1. The Second-Order Contracted Schrödinger Equation

When contracting the matrix form of the Schrödinger equation into the two-electron space and transforming into normal form the resulting equation, one obtains the 2-CSE [45, 50]. In the spin-orbital representation, the 2-CSE splits into the three following coupled equations (summation over all possible values of common indices is implicit with the restrictions $r < s$ and $k < l$):

$$E\,{}^2D_{i_\alpha j_\alpha;p_\alpha q_\alpha}(i<j,p<q) = \begin{Bmatrix} {}^2K_{r_\alpha s_\alpha;p_\alpha q_\alpha}\,{}^2D_{i_\alpha j_\alpha;r_\alpha s_\alpha} \\ -{}^2K_{r_\alpha s_\alpha;p_\alpha m_\alpha}\,{}^3D_{i_\alpha j_\alpha m_\alpha;q_\alpha r_\alpha s_\alpha} \\ +{}^2K_{r_\alpha s_\alpha;q_\alpha m_\alpha}\,{}^3D_{i_\alpha j_\alpha m_\alpha;p_\alpha r_\alpha s_\alpha} \\ +{}^2K_{u_\alpha v_\beta;q_\alpha m_\beta}\,{}^3D_{i_\alpha j_\alpha m_\beta;p_\alpha u_\alpha v_\beta} \\ -{}^2K_{u_\alpha v_\beta;p_\alpha m_\beta}\,{}^3D_{i_\alpha j_\alpha m_\beta;q_\alpha u_\alpha v_\beta} \\ +{}^2K_{r_\alpha s_\alpha;k_\alpha l_\alpha}\,{}^4D_{i_\alpha j_\alpha k_\alpha l_\alpha;p_\alpha q_\alpha r_\alpha s_\alpha} \\ +{}^2K_{r_\beta s_\beta;k_\beta l_\beta}\,{}^4D_{i_\alpha j_\alpha k_\beta l_\beta;p_\alpha q_\alpha r_\beta s_\beta} \\ +{}^2K_{u_\alpha v_\beta;m_\alpha n_\beta}\,{}^4D_{i_\alpha j_\alpha m_\alpha n_\beta;p_\alpha q_\alpha u_\alpha v_\beta} \end{Bmatrix} \quad (151)$$

$$E\,{}^2D_{i_\alpha j_\beta;p_\alpha q_\beta} = \begin{Bmatrix} {}^2K_{u_\alpha v_\beta;p_\alpha q_\beta}\,{}^2D_{i_\alpha j_\beta;u_\alpha v_\beta} \\ -{}^2K_{r_\alpha s_\alpha;p_\alpha m_\alpha}\,{}^3D_{m_\alpha i_\alpha j_\beta;r_\alpha s_\alpha q_\beta} \\ +{}^2K_{r_\beta s_\beta;q_\beta m_\beta}\,{}^3D_{i_\alpha j_\beta m_\beta;p_\alpha r_\beta s_\beta} \\ -{}^2K_{u_\alpha v_\beta;m_\alpha q_\beta}\,{}^3D_{m_\alpha i_\alpha j_\beta;p_\alpha u_\alpha v_\beta} \\ -{}^2K_{u_\alpha v_\beta;p_\alpha n_\beta}\,{}^3D_{i_\alpha j_\beta n_\beta;u_\alpha v_\beta q_\beta} \\ +{}^2K_{r_\alpha s_\alpha;k_\alpha l_\alpha}\,{}^4D_{k_\alpha l_\alpha i_\alpha j_\beta;r_\alpha s_\alpha p_\alpha q_\beta} \\ +{}^2K_{r_\beta s_\beta;k_\beta l_\beta}\,{}^4D_{i_\alpha j_\beta k_\beta l_\beta;p_\alpha q_\beta r_\beta s_\beta} \\ +{}^2K_{u_\alpha v_\beta;m_\alpha n_\beta}\,{}^4D_{i_\alpha m_\alpha j_\beta n_\beta;p_\alpha u_\alpha q_\beta v_\beta} \end{Bmatrix} \quad (152)$$

The relation for the $M = -1$ spin-block follows directly from Eq. (151) by exchanging the spin functions.

As can be seen, the 2-CSE depends not only on the 2-RDM but also on the 3- and 4-RDMs. This fact lies at the root of the indeterminacy of this equation [63, 107]. As already mentioned, in the method proposed by Colmenero and Valdemoro [46–48] and in those further proposed by Nakatsuji and Yasuda [49, 51] and by Mazziotti [52, 111], a set of algorithms for approximating the higher-order RDMs in terms of the lower-order ones [46, 47, 108] allows this equation to be solved iteratively until converging to a self-consistent solution. In the approach considered in this work, the spin-adapted 2-CSE has been used. This equation is obtained by coupling the 2-CSE with the second-order contracted spin equation [50].

2. *The Regulated Iterative Self-Consistent Solution.*

Initially, the main features of the iterative self-consistent solution of the 2-CSE were the following [48]:

1. Starting from an initial set of 1- and 2-RDMs, the 3- and 4-RDMs are approximated by applying the constructing algorithms [18, 46–52, 60, 81].

The algorithms that will be used here are those reported in Ref. [18], which emphasize the role played by the 2-RDM—the basic variable of the 2-CSE—while the original ones emphasized the role played by the 1-RDM. Another advantage of these algorithms is that they are more economical than the original ones, since they have fewer terms.

2. Once the 3- and 4-RDMs are evaluated, these trial matrices, jointly with the 2-RDM obtained by contracting them, are replaced in the rhs of the spin-adapted 2-CSE coupled equations, which become a matrix represented in the two-electron space, $^2\mathcal{R}_{\sigma\sigma';\sigma\sigma'}$, such that

$$E\,^2D_{i_\sigma j_{\sigma'};p_\sigma q_{\sigma'}} = {}^2\mathcal{R}_{i_\sigma j_{\sigma'};p_\sigma q_{\sigma'}} \tag{153}$$

3. Then an intermediate energy and a new 2-RDM are obtained as follows:

$$E = \frac{1}{\binom{N}{2}} \sum_{\sigma \leq \sigma'} \sum_{i,j} {}^2\mathcal{R}_{i_\sigma j_{\sigma'};i_\sigma j_{\sigma'}} \tag{154}$$

$$^2D_{i_\sigma j_{\sigma'};p_\sigma q_{\sigma'}} = \frac{1}{E}\,^2\mathcal{R}_{i_\sigma j_{\sigma'};p_\sigma q_{\sigma'}} \tag{155}$$

4. After contracting the new 2-RDM to obtain the corresponding 1-RDM, a new iteration starts by calculating new trials for the 3- and 4-RDMs.
5. The procedure is repeated until convergence is obtained.

In order to regulate the convergence of the iterative process by damping or accelerating its rate according to convenience, the basic procedure has been further refined [88]. This is achieved by substituting the numerical reduced Hamiltonian matrix by a new one

$$^2K_{s i_\sigma j_{\sigma'};k_\sigma l_{\sigma'}} = {}^2K_{i_\sigma j_{\sigma'};k_\sigma l_{\sigma'}} - \frac{a}{\binom{N}{2}} \delta_{i,k}\,\delta_{j,l} \tag{156}$$

where the quantity a is an amount of energy. The net result of this replacement is that the electron distributions of the stationary states are not modified, but the energy levels are shifted to new values

$$E_s = E - a \tag{157}$$

This energy shifting does not modify the RDMs, which constitute the solution at convergence, that is, the full CI RDMs. On the other hand, the value of a influences the rate of convergence of the process. Denoting the 2-RDM entering the iterative process as $^2\mathrm{D}^{(0)}$ and calling $^2\mathrm{D}^{(1)}$ its outcome, the 2-CSE is now expressed as

$$E_s\,^2\mathrm{D}^{(1)}_{s\,i_\sigma j_{\sigma'};p_\sigma q_{\sigma'}} = {}^2\mathcal{R}_{s\,i_\sigma j_{\sigma'};p_\sigma q_{\sigma'}}$$
$$= {}^2\mathcal{R}_{i_\sigma j_{\sigma'};p_\sigma q_{\sigma'}} - a\,{}^2\mathrm{D}^{(0)}_{i_\sigma j_{\sigma'};p_\sigma q_{\sigma'}} \quad (158)$$

because the only action of the Kronecker deltas appearing in relation (156) is to contract the 3- and 4-RDMs into the 2-RDM. The outcome of the shifted equation, Eq. (158), is related to the nonshifted one as follows:

$$^2\mathrm{D}^{(1)}_{s\,i_\sigma j_{\sigma'};p_\sigma q_{\sigma'}} = \frac{E}{E-a}\,{}^2\mathrm{D}^{(1)}_{i_\sigma j_{\sigma'};p_\sigma q_{\sigma'}} - \frac{a}{E-a}\,{}^2\mathrm{D}^{(0)}_{i_\sigma j_{\sigma'};p_\sigma q_{\sigma'}} \quad (159)$$

Thus, when analyzing Eq. (159) in detail, one observes the following:

- For $a/E > 1$ the process does not converge to the solution aimed at but to a different one since the direction of the searching flow is reversed.
- For $0 < a/E < 1$ the convergence of the process is accelerated and what is more it converges toward a more accurate value. This acceleration is necessary in the cases where the convergence is smooth but slow. In general, as the energy value improves, the 1- and 2-RDMs become closer to the full CI ones. This is so because with a rapid convergence fewer errors are accumulated. However, there is a limit to the regulating parameter. Thus, if the regime becomes too rapid, the RDMs of different orders do not optimize to a similar extent. In general, the optimum accelerating parameter lies close to, though above, the (negative) full CI energy value. In fact, a good value of this parameter is given by the mean energy of a bielectronic configuration times the number of geminals appearing in an N-electron Slater determinant:

$$a = \frac{\binom{N}{2}}{\binom{2K}{2}} \sum_{\sigma \leq \sigma'} \sum_{i,j} {}^2\mathrm{K}_{i_\sigma j_{\sigma'};i_\sigma j_{\sigma'}} \quad (160)$$

- For $a/E < 0$ the convergence of the process is damped. This damping is necessary in the cases where large oscillations appear at the beginning of the iterative process. By introducing the damping, the modifications in the RDMs, from one iteration to the next one, are sufficiently small to allow the easy correction of the deviations of the N- and S-representability, thus avoiding the appearance of oscillations.

This regulating device, which acts, according to its sign, as either a damping or an accelerating agent, greatly fosters the convergence of the iterative process [88].

B. Analysis of the Effects of the N- and S-Purification on the Regulated CSE Iterative Process

The regulated CSE iterative process has been linked with several iterative purification procedures [50, 70, 88, 111]. In the analysis carried out here the performance of the regulated CSE is compared with the one obtained by coupling the AV purification procedure to the regulated CSE—the regulated CSE-NS iterative process.

In order to carry out this analysis, the Li_2 (Li—Li bond length of 5.50 au) and the linear BeH_2 (Be—H bond length of 2.54 au) molecules, each in its ground state, were calculated. The basis set is formed by Hartree–Fock molecular orbitals built out of a minimal basis set of Slater orbitals. In all the calculations reported in this section, the integrals were evaluated with an adapted version of the program SMILES [112, 113] and the initial RDMs are those corresponding to a Hartree–Fock calculation.

The reference calculation in the analysis of the Li_2 molecule consists of performing 160 iterations of the regulated CSE process with a regulating parameter of $a = -11.5$ au, while that of the BeH_2 molecule consists of performing 40 iterations of the regulated CSE process with a regulating parameter of $a = -13$ au. The other calculations consist of carrying out 15, 35, and 55 iterations of the AV N- and S-purification process at each iteration of the reference calculations.

As an illustrative example, the energy curves showing the convergence of the overall process for both the regulated CSE process without purification and the regulated CSE process with 15 purification iterations of the calculations of the Li_2 molecule are shown in Fig. 9. The two horizontal lines correspond to the Hartree–Fock and the full CI energy values. As can be noticed, both the accuracy and the convergence rate are remarkably improved when purification is carried out.

The results of the complete set of calculations are given in Table VIII, where we report the values of the largest negative eigenvalues of the 2-RDM, the 2-HRDM, and the 2-G matrix, as well as the energy error with respect to the full CI energy value. As can be seen, there is a clear and general improvement when the N- and S-purification procedure is added to the CSE process. The initial N- and S-representability defects, which manifest themselves either by producing a premature minimum in the energy followed by divergence or by giving a total energy below the full CI one, are corrected. Consequently, the energy error after purification becomes smaller and positive, and the absolute values of the negative eigenvalues of the three matrices, which must be positive

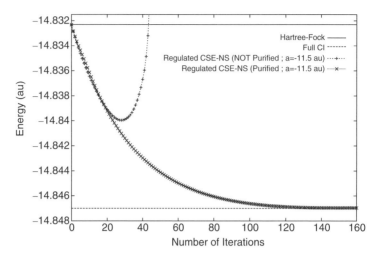

Figure 9. The ground-state electronic energy for Li_2 is shown as a function of the regulated CSE-NS iterations both *with* and *without* purification. The regulated CSE-NS with purification captures 99.12% of the correlation energy, while the regulated CSE-NS without purification achieves only 51.61% of the correlation energy before diverging.

semidefinite, diminish in one, two, or three orders of magnitude according to the system and quantity involved. As can also be seen in Table VIII, this effect is enhanced if the number of iterations of the N- and S-representability purification procedure is augmented.

C. Application of the Regulated CSE-NS Method to the Study of Potential Energy Curves

In order to judge the performance of the regulated CSE-NS method, the calculation of the potential energy curves (PECs) of small molecules such as Li_2 and BeH_2 is very illuminating.

The results obtained for the stretching PEC of the ground state of the Li_2 molecule are given in Fig. 10. In this figure we compare the results of the Hartree–Fock, of the regulated CSE-NS, and of the full CI calculations within a range of the Li—Li bond length of 4.6–8.0 au. As can be seen, the results are excellent since the regulated CSE-NS curve closely reproduces the full CI one. Indeed, the correlation energy obtained in this calculation is within the range of 97.45–99.21%.

The symmetric-stretching PEC of the ground state of the BeH_2 molecule within the range of the Be—H bond length of 2.34–2.94 au was also calculated using a minimal basis set, and the results are shown in Fig. 11. Here again, we

TABLE VIII
Representability Defects of the 2-RDM, 2-HRDM, and 2-G and Energy Error Obtained When Solving the Regulated CSE with and Without the N-, S-Purification Procedure for Li_2 and BeH_2 Molecules

System	Quantity	Representability Defect		
		Without Purification		
Li_2				
	2-RDM	-1.90×10^{-4}		
	2-HRDM	-3.86×10^{-3}		
	2-G	-1.73×10^{-3}		
	Energy error (au)	$+0.0070$		
BeH_2				
	2-RDM	$-1.79\ 10^{-4}$		
	2-HRDM	$-6.89\ 10^{-4}$		
	2-G	$-1.32\ 10^{-3}$		
	Energy error (au)	-0.0018		
		With Purification		
		(15 Iterations)	(35 Iterations)	(55 Iterations)
Li_2				
	2-RDM	7.04×10^{-6}	2.91×10^{-7}	3.56×10^{-8}
	2-HRDM	-3.08×10^{-5}	-2.05×10^{-5}	-1.62×10^{-5}
	2-G	-7.55×10^{-6}	-5.30×10^{-6}	-4.22×10^{-6}
	Energy error (au)	$+0.0001$	$+0.0001$	$+0.0001$
BeH_2				
	2-RDM	$-1.06\ 10^{-5}$	$-6.42\ 10^{-6}$	$-4.31\ 10^{-6}$
	2-HRDM	$-1.99\ 10^{-5}$	$-1.10\ 10^{-5}$	$-3.92\ 10^{-6}$
	2-G	$-2.02\ 10^{-5}$	$-1.35\ 10^{-5}$	$-1.22\ 10^{-5}$
	Energy error (au)	$+0.0006$	$+0.0006$	$+0.0006$

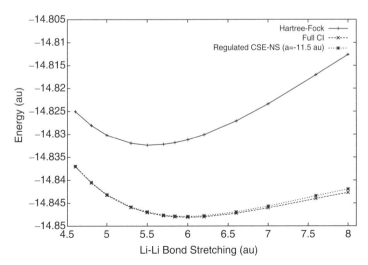

Figure 10. Stretching potential energy curve of Li_2 calculated by the Hartree–Fock, regulated CSE-NS, and full CI methods.

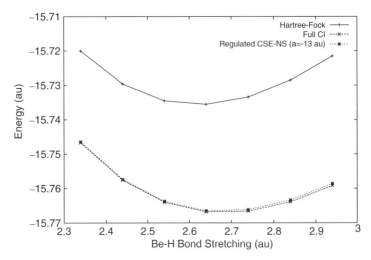

Figure 11. Symmetric-stretching potential energy curve of BeH$_2$ calculated by the Hartree–Fock, regulated CSE-NS, and full CI methods.

compare the Hartree–Fock, the regulated CSE-NS, and the full CI calculation. The correlation energy obtained in this calculation is within the range of 98.43–99.20%.

Although in both cases the results are very good, it must be noted that as the interatomic distance augments, the PECs obtained by the regulated CSE-NS slightly separate from those of the full CI. This is due to the fact that in those zones of the PECs the dominance of a single configuration is not so clear and the constructing algorithms thus become less efficient. Similar behaviors have been reported by Ehara et al. [56] and by Mazziotti [111] when studying other compounds.

The calculations reported here reproduce with a high precision and in very few iterations the full CI results. These results, as well as others reported by Nakatsuji and co-workers [6, 49, 51, 56, 109], Mazziotti [70, 111], and Alcoba and co-workers [88], render us confident that the moment has arrived when this methodology can be competitively applied with the standard quantum chemistry methods in the study of electronic states that have a clear dominance of a single configuration.

VI. DISCUSSION AND FINAL REMARKS

From the begining of the development of the RDM theory the need to render N- and S-representable a 2-RDM obtained by an approximative method was patent. The development first of the spin-adapted reduced Hamiltonian methodology and, more recently, that of the second-order contracted Schrödinger equation rendered the solution of this problem urgent. The purification strategies described in this chapter constitute a conclusive solution to this problem. From the results reported here it is clear that these purification procedures, in particular the AV one, achieve a nearly perfect purification of the 2-RDM corresponding to a singlet state. Moreover, the good performance of this procedure is accomplished in only a few iterations, which enhances considerably the applicability of the procedure. Clearly, other spin symmetries have now to be considered in order to complete the 2-RDM study.

Another extension of this theoretical study is the consideration of both an economical and an effective purification strategy for the 4-RDM. The need for such a purification scheme is motivated by the need to have an N- and S-representable 4-RDM if one wishes to solve the fourth-order modified contracted Schrödinger equation [62, 64, 87]. There have already been several attemps to purify both the 3-RDM and 4-RDM [18, 34, 52]. In particular, a set of inequalities that bound the diagonal and off-diagonal elements of these high-order matrices have been reported [18]. However, the results obtained with this approach within the framework of the fourth-order modified contracted Schrödinger equation (and the second-order contracted Schrödinger equation) were not fully satisfactory because the different spin-blocks of the matrices did not appear to be properly balanced [87, 114].

Let us finally comment on the effectiveness of the coupling of the purification procedure with the regulating device implemented within the iterative solution of the second-order contracted Schrödinger equation. It has been shown here that these two implementations highly improve the performance of the iterative method. In fact, the regulating device, which acts, according to its sign, as either a damping or an accelerating agent, greatly fosters the convergence of a purification process.

In view of all the results presented here it can be concluded that the coupling of any RDM-oriented method with the purification procedure augments its applicability in a significant way. In particular, the coupling of the purification procedure and regulating device with the iterative solution of the 2-CSE renders this approach not only reliable but also highly effective.

Acknowledgments

The author wishes to thank Prof. C. Valdemoro, Dr. E. Pérez-Romero, Prof. A. Torre, Prof. L. M. Tel, Prof. R. C. Bochicchio, and Prof. L. Lain for their valuable suggestions and helpful comments. The author acknowledges the Department of Physics, Facultad de Ciencias Exactas y Naturales, Universidad de Buenos Aires, for facilities provided during the course of this work.

References

1. A. J. Coleman and V. I. Yukalov, *Reduced Density Matrices: Coulson's Challenge*, Springer-Verlag, New York, 2000.
2. E. R. Davidson, *Reduced Density Matrices in Quantum Chemistry*, Academic Press, New York, 1976.

3. *Reduced Density Matrices with Applications to Physical and Chemical Systems* (A. J. Coleman and R. M. Erdahl, eds.), Queen's Papers on Pure and Applied Mathematics No. 11, Queen's University, Kingston, Ontario, 1968.
4. *Reduced Density Matrices with Applications to Physical and Chemical Systems II* (R. M. Erdahl, ed.), Queen's Papers on Pure and Applied Mathematics No. 40, Queen's University, Kingston, Ontario, 1974.
5. *Density Matrices and Density Functionals* (R. M. Erdahl and V. Smith, eds.), Proceedings of the A. J. Coleman Symposium, Kingston, Ontario, 1985, Reidel, Dordrecht, 1987.
6. *Many-Electron Densities and Reduced Density Matrices* (J. Cioslowski, ed.), Kluwer, Dordrecht, The Netherlands, 2000.
7. K. Husimi, Some formal properties of the density matrix. *Proc. Soc. Japan* **22**, 264 (1940).
8. P. O. Löwdin, Quantum theory of many particle systems. I. Physical interpretation by means of density matrices, natural spin orbitals, and convergence problems in the method of configuration interaction. *Phys. Rev.* **97**, 1474 (1955).
9. J. E. Mayer, Electron correlation. *Phys. Rev.* **100**, 1579 (1955).
10. R. McWeeny, The density matrix in self-consistent field theory. I. Iterative construction of the density matrix. *Proc. R. Soc. A* **235**, 496 (1956).
11. R. U. Ayres, Variational approach to the many-body problem. *Phys. Rev.* **111**, 1453 (1958).
12. C. A. Coulson, Present state of molecular structure calculations. *Rev. Mod. Phys.* **32**, 170 (1960).
13. A. J. Coleman, Structure of fermion density matrices. *Rev. Mod. Phys.* **35**, 668 (1963).
14. E. Pérez-Romero, L. M. Tel, and C. Valdemoro, Traces of spin-adapted reduced density matrices. *Int. J. Quantum Chem.* **61**, 55 (1997).
15. C. Valdemoro, L. M. Tel, and E. Pérez-Romero, N-representability problem within the framework of the contracted Schrödinger equation. *Phys. Rev. A* **61**, 032507 (2000).
16. C. Garrod and J. K. Percus, Reduction of N-particle variational problem. *J. Math. Phys.* **5**, 1756 (1964).
17. R. M. Erdahl, Representability. *Int. J. Quantum Chem.* **13**, 697 (1978).
18. C. Valdemoro, D. R. Alcoba, L. M. Tel, and E. Pérez-Romero, Imposing bounds on the high-order reduced density matrices elements. *Int. J. Quantum Chem.* **85**, 214 (2001).
19. E. R. Davidson, Linear inequalities for density matrices: III. *Int. J. Quantum Chem.* **90**, 1555 (2002).
20. R. M. Erdahl and B. Jin, The lower bound method for reduced density matrices. *J. Mol. Struct. (Theochem.)* **527**, 207 (2000).
21. L. J. Kijewski and J. K. Percus, Lower-bound method for atomic calculations. *Phys. Rev. A* **2**, 1659 (1970).
22. C. Garrod, M. V. Mihailovic, and M. Rosina, Variational approach to 2-body density matrix. *J. Math. Phys.* **16**, 868 (1975).
23. M. Rosina and C. Garrod, Variational calculation of reduced density matrices. *J. Comp. Phys.* **18**, 300 (1975).
24. C. Garrod and M. Rosina, Particle–hole matrix—its connection with symmetries and collective features of ground state. *J. Math. Phys* **10**, 1855 (1975).
25. M. V. Mihailovic and M. Rosina, Excitations as ground state variational parameters. *Nucl. Phys. A* **130**, 386 (1969).
26. C. Garrod and M. A. Fusco, Density matrix variational caculation for atomic-Be. *Int. J. Quantum Chem.* **10**, 495 (1976).
27. M. V. Mihailović and M. Rosina, Variational approach to density matrix for light-nuclei. *Nucl. Phys. A* **237**, 221 (1975).

28. R. M. Erdahl, Two algorithms for the lower bound method of reduced density matrix theory. *Rep. Math. Phys.* **15**, 147 (1979).
29. M. Nakata, H. Nakatsuji, M. Ehara, M. Fukuda, K. Nakata, and K. Fujisawa, Variational calculations of fermion second-order reduced density matrices by semidefinite programming algorithm. *J. Chem. Phys.* **114**, 8282 (2001).
30. D. A. Mazziotti, Variational minimization of atomic and molecular ground-state energies via the two-particle reduced density matrix. *Phys. Rev. A* **65**, 062511 (2002).
31. G. Gidofalvi and D. A. Mazziotti, Boson correlation energies via variational minimization with the two-particle reduced density matrix: exact N-representability conditions for harmonic interactions. *Phys. Rev. A* **69**, 042511 (2004).
32. T. Juhasz and D. A. Mazziotti, Perturbation theory corrections to the two-particle reduced density matrix variational method. *J. Chem. Phys.* **121**, 1201 (2004).
33. Z. Zhao, B. J. Braams, H. Fukuda, M. L. Overton, and J. K. Percus, The reduced density matrix method for electronic structure calculations and the role of three-index representability conditions. *J. Chem. Phys.* **120**, 2095 (2004).
34. D. A. Mazziotti, Realization of quantum chemistry without wave functions through first-order semidefinite programming. *Phys. Rev. Lett.* **93**, 213001 (2004).
35. M. Rosina, Application of 2-body density matrix of ground-state for calculations of some excited-states. *Int. J. Quantum Chem.* **13**, 737 (1978).
36. D. A. Mazziotti and R. M. Erdahl, Uncertainty relations and reduced density matrices: mapping many-body quantum mechanics onto four particles. *Phys. Rev. A* **63**, 042113 (2001).
37. C. Valdemoro, Spin-adapted reduced Hamiltonians. 1. Elementary excitations. *Phys. Rev. A* **31**, 2114 (1985).
38. C. Valdemoro, in *Computational Chemistry Structure, Interactions and Reactivity, Part A* (S. Fraga, ed.), Elsevier, Amsterdam, 1992; and references therein.
39. C. Valdemoro, Spin-adapted reduced Hamiltonians. 2. Total energy and reduced density-matrices. *Phys. Rev. A* **31**, 2123 (1985).
40. J. Karwowski, W. Duch, and C. Valdemoro, Matrix-elements of a spin-adapted reduced Hamiltonian. *Phys. Rev. A* **33**, 2254 (1986).
41. C. Valdemoro, L. Lain, F. Beitia, A. Ortiz de Zarate, and F. Castaño, Direct approximation to the reduced density-matrices: calculation of the isoelectronic sequence of beryllium up to argon. *Phys. Rev. A* **33**, 1525 (1986).
42. A. Torre, L. Lain, and J. Millan, Analysis of several methods in the direct approximation of reduced density-matrices. *Nuovo Cimento B* **108**, 491 (1993).
43. C. Valdemoro, M. P. de Lara-Castells, R. Bochicchio, and E. Perez-Romero, A relevant space within the spin-adapted reduced Hamiltonian theory. I. Study of the BH molecule. *Int. J. Quantum Chem.* **65**, 97 (1997).
44. C. Valdemoro, M. P. de Lara-Castells, R. Bochicchio, and E. Pérez-Romero, A relevant space within the spin-adapted reduced Hamiltonian theory. II. Study of the π-cloud in benzene and naphthalene. *Int. J. Quantum Chem.* **65**, 107 (1997).
45. C. Valdemoro, in *Density Matrices and Density Functionals* (R. M. Erdahl and V. Smith, eds.), Proceedings of the A. J. Coleman Symposium, Kingston, Ontario, 1985, Reidel, Dordrecht, 1987.
46. F. Colmenero, C. Pérez del Valle, and C. Valdemoro, Approximating q-order reduced density-matrices in terms of lower-order ones. 1. General relations. *Phys. Rev. A* **47**, 971 (1993).
47. F. Colmenero and C. Valdemoro, Approximating q-order reduced density-matrices in terms of lower-order ones. 1. Applications. *Phys. Rev. A* **47**, 979 (1993).

48. F. Colmenero and C. Valdemoro, Self-consistent approximate solution of the second-order contracted Schrödinger equation. *Int. J. Quantum Chem.* **51**, 369 (1994).
49. H. Nakatsuji and K. Yasuda, Direct determination of the quantum-mechanical density matrix using the density equation. *Phys. Rev. Lett.* **76**, 1039 (1996).
50. C. Valdemoro, L. M. Tel, and E. Pérez-Romero, The contracted Schrödinger equation: some results. *Adv. Quantum Chem.* **28**, 33 (1997).
51. K. Yasuda and H. Nakatsuji, Direct determination of the quantum-mechanical density matrix using the density equation. 2. *Phys. Rev. A* **56**, 2648 (1997).
52. D. A. Mazziotti, Contracted Schrödinger equation: determining quantum energies and two-particle density matrices without wavefunctions. *Phys. Rev. A* **57**, 4219 (1998).
53. D. A. Mazziotti, Approximate solution for electron correlation through the use of Schwinger probes. *Chem. Phys. Lett.* **289**, 419 (1998).
54. D. A. Mazziotti, 3,5-Contracted Schrödinger equation: determining quantum energies and reduced density matrices without wavefunctions. *Int. J. Quantum Chem.* **70**, 557 (1998).
55. K. Yasuda, Direct determination of the quantum-mechanical density matrix: Parquet theory. *Phys. Rev. A* **59**, 4133 (1999).
56. M. Ehara, M. Nakata, H. Kou, K. Yasuda, and H. Nakatsuji, Direct determination of the density matrix using the density equation: potential energy curves of HF, CH_4, BH_3, NH_3, and H_2O. *Chem. Phys. Lett.* **305**, 483 (1999).
57. D. A. Mazziotti, Pursuit of N-representability for the contracted Schrödinger equation through density-matrix reconstruction. *Phys. Rev. A* **60**, 3618 (1999).
58. D. A. Mazziotti, Comparison of contracted Schrödinger and coupled-cluster theories. *Phys. Rev. A* **60**, 4396 (1999).
59. D. A. Mazziotti, Complete reconstruction of reduced density matrices. *Chem. Phys. Lett.* **326**, 212 (2000).
60. C. Valdemoro, L. M. Tel, E. Pérez-Romero, and A. Torre, The iterative solution of the contracted Schrödinger equation: a new quantum chemical method. *J. Mol. Struct. (Theochem.)* **537**, 1 (2001).
61. W. Kutzelnigg and D. Mukherjee, Direct determination of the cumulants of the reduced density matrices. *J. Chem. Phys.* **110**, 2800 (1999).
62. D. R. Alcoba and C. Valdemoro, Family of modified-contracted Schrödinger equations. *Phys. Rev. A* **64**, 062105 (2001).
63. K. Yasuda, Uniqueness of the solution of the contracted Schrödinger equation. *Phys. Rev. A* **65**, 052121 (2002).
64. D. R. Alcoba, Equivalence theorems between the solutions of the fourth-order modified contracted Schrödinger equation and those of the Schrödinger equation. *Phys. Rev. A* **65**, 032519 (2002).
65. J. M. Herbert and J. E. Harriman, Contraction relations for Grassmann products of reduced density matrices and implications for density matrix reconstruction. *Phys. Rev. A* **65**, 022511 (2002).
66. J. M. Herbert and J. E. Harriman, Extensivity and the contracted Schrödinger equation. *J. Chem. Phys.* **117**, 7464 (2002).
67. D. A. Mazziotti, Solution of the 1,3-contracted Schrödinger equation through positivity conditions on the two-particle reduced density matrix. *Phys. Rev. A* **66**, 062503 (2002).
68. M. Nooijen, M. Wladyslawski, and A. Hazra, Cumulant approach to the direct calculation of reduced density matrices: a critical analysis. *J. Chem. Phys.* **118**, 4832 (2003).

69. D. A. Mazziotti, Extraction of electronic excited states from the ground-state two-particle reduced density matrix. *Phys. Rev. A* **68**, 052501 (2003).
70. D. A. Mazziotti, Purification of correlated reduced density matrices. *Phys. Rev. E* **65**, 026704 (2002).
71. C. Valdemoro, D. R. Alcoba, and L. M. Tel, Recent developments in the contracted Schrödinger equation method: controlling the N-representability of the second-order reduced density matrix. *Int. J. Quantum Chem.* **93**, 212 (2003).
72. D. R. Alcoba and C. Valdemoro, Spin structure and properties of the correlation matrices corresponding to pure spin states: controlling the S-representability of these matrices. *Int. J. Quantum Chem.* **102**, 629 (2005).
73. A. J. Coleman, in *Reduced Density Matrices with Applications to Physical and Chemical Systems II* (R. M. Erdahl, ed.), Queen's Papers on Pure and Applied Mathematics No. 40, Queen's University, Kingston, Ontario, 1974.
74. J. E. Harriman, Geometry of density matrices. 2. Reduced density matrices and N-representability. *Phys. Rev. A* **17**, 1257 (1978).
75. J. E. Harriman, Geometry of density matrices. 3. Spin components. *Int. J. Quantum Chem.* **15**, 611 (1979).
76. M. E. Casida and J. E. Harriman, Geometry of density matrices. 6. Superoperators and unitary invariance. *Int. J. Quantum Chem.* **30**, 161 (1986).
77. D. R. Alcoba, Unitarily invariant decomposition of arbitrary Hermitian matrices of physical interest. *Int. J. Quantum Chem.* **97**, 776 (2004).
78. C. Valdemoro, M. P. de Lara-Castells, E. Pérez-Romero, and L. M. Tel, Second European Workshop on Quantum Systems in Chemistry and Physics, Oxford, 1997.
79. C. Valdemoro, M. P. de Lara-Castells, E. Pérez-Romero, and L. M. Tel, The first order contracted density equations: correlation effects. *Adv. Quantum Chem.* **31**, 37 (1999).
80. C. Valdemoro, in *Topics in Current Chemistry: Correlation and Localization* (P. R. Surjan, ed.), Springer-Verlag, New York, 1999.
81. C. Valdemoro, L. M. Tel, and E. Pérez-Romero, in *Quantum Systems in Chemistry and Physics I* (A. Hernández-Laguna, J. Maruani, R. McWeeny, and S. Wilson, eds.) Kluwer, Dordrecht, The Netherlands, 2000.
82. C. Valdemoro, L. M. Tel, and E. Pérez-Romero, in *Many-Electron Densities and Reduced Density Matrices* (J. Cioslowski, ed.), Kluwer, Dordrecht, The Netherlands, 2000.
83. C. Valdemoro, L. M. Tel, D. R. Alcoba, E. Pérez-Romero, and F. J. Casquero, Some basic properties of the correlation matrices. *Int. J. Quantum Chem.* **90**, 1555 (2002).
84. L. M. Tel, E. Pérez-Romero, C. Valdemoro, and F. J. Casquero, Cancellation of high-order electron correlation effects corresponding to eigenstates. *Int. J. Quantum Chem.* **82**, 131 (2001).
85. L. M. Tel, E. Pérez-Romero, C. Valdemoro, and F. J. Casquero, Compact forms of reduced density matrices. *Phys. Rev. A* **67**, 052504 (2003).
86. J. E. Harriman, Grassmann products, cumulants, and two-electron reduced density matrices. *Phys. Rev. A* **65**, 052507 (2002).
87. D. R. Alcoba, *Estudio de las cuestiones cruciales en la solución iterativa de la Ecuación de Schrödinger Contraída*, Ph.D. thesis, Universidad Autónoma de Madrid, 2004.
88. D. R. Alcoba, F. J. Casquero, L. M. Tel, E. Pérez-Romero, and C. Valdemoro, Convergence enhancement in the iterative solution of the second-order contracted Schrödinger equation. *Int. J. Quantum Chem.* **102**, 620 (2005).

89. A. J. Coleman and I. Absar, Reduced Hamiltonian orbitals. 3. Unitarily-invariant decomposition of Hermitian operators. *Int. J. Quantum Chem.* **18**, 1279 (1980).
90. A. Beste, K. Runge, and R. Barlett, Ensuring N-representability: Coleman's algorithm. *Chem. Phys. Lett.* **355**, 263, (2002).
91. A. J. Coleman, Necessary conditions for N-representability of reduced density matrices. *J. Math. Phys.* **13**, 214 (1972).
92. R. Kubo, Generalized cumulant expansion method. *J. Phys. Soc. Japan* **17**, 1100 (1962).
93. T. Schork and P. Fulde, Calculating excitation-energies with the help of cumulants. *Int. J. Quantum Chem.* **51**, 113 (1994).
94. K. Kladko and P. Fulde, On the properties of cumulant expansions. *Int. J. Quantum Chem.* **66**, 377 (1998).
95. W. Kutzelnigg and D. Mukherjee, Cumulant expansion of the reduced density matrices. *J. Chem. Phys.* **110**, 2800 (1999).
96. F. E. Harris, Cumulant-based approximations to reduced density matrices. *Int. J. Quantum Chem.* **90**, 105 (2002).
97. I. Absar and A. J. Coleman, One-electron orbitals intrinsic to reduced Hamiltonian. *Chem. Phys. Lett.* **39**, 609 (1976).
98. I. Absar and A. J. Coleman, Reduced Hamiltonian orbitals. I. New approach to many-electron problem. *Int. J. Quantum Chem.* **10**, 319 (1976).
99. I. Absar, Reduced Hamiltonian orbitals. II. Optimal orbital basis sets for the many-electron problem. *Int. J. Quantum Chem.* **13**, 777 (1978).
100. R. McWeeny and Y. Mizuno, Density matrix in many-electron quantum mechanics. 2. Separation of space and spin variables—spin coupling problems. *Proc. R. Soc. London A* **259**, 554 (1961).
101. P. R. Surjan, *Second Quantized Approach to Quantum Chemistry: An Elementary Introduction*, Springer-Verlag, Berlin, 1989.
102. C. C. Sun, X. Q. Li, and A. C. Tang, On the unitarily-invariant decomposition of Hermitian operators. *Int. J. Quantum Chem.* **25**, 1045 (1984).
103. S. Cho, *Sci. Rep. Gumma Univ.* **11**, 1 (1962).
104. L. Cohen and C. Frishberg, Hierarchy equations for reduced density matrices. *Phys. Rev. A* **13**, 927 (1976).
105. L. Cohen and C. Frishberg, Hartree–Fock density matrix equation. *J. Chem. Phys.* **65**, 4234 (1976).
106. H. Nakatsuji, Equation for direct determination of density matrix. *Phys. Rev. A* **14**, 41 (1976).
107. J. E. Harriman, Limitation on the density-equation approach to many-electron problem. *Phys. Rev. A* **19**, 1893 (1979).
108. C. Valdemoro, Approximating the second-order reduced density-matrix in terms of the first-order one. *Phys. Rev. A* **45**, 4462 (1992).
109. M. Nakata, M. Ehara, K. Yasuda, and H. Nakatsuji, Direct determination of second-order density matrix using density equation: open-shell system and excited state. *J. Chem. Phys.* **112**, 8772 (2000).
110. M. Nooijen, Can the eigenstates of a many-body Hamiltonian be represented exactly using a general two-body cluster expansion? *Phys. Rev. Lett.* **84**, 2108 (2000).
111. D. A. Mazziotti, Variational method for solving the contracted Schrödinger equation through a projection of the N-particle power method onto the two-particle space. *J. Chem. Phys.* **116**, 1239 (2002).

112. J. Fernández Rico, R. López, A. Aguado, I. Ema, and G. Ramírez, Reference program for molecular calculations with Slater-type orbitals. *J. Comput. Chem.* **19**, 1284 (1998).
113. J. Fernández Rico, R. López, A. Aguado, I. Ema, and G. Ramírez, New program for molecular calculations with Slater-type orbitals. *Int. J. Quantum Chem.* **81**, 148 (2000).
114. D. R. Alcoba, unpublished results.

CHAPTER 10

CUMULANTS, EXTENSIVITY, AND THE CONNECTED FORMULATION OF THE CONTRACTED SCHRÖDINGER EQUATION

JOHN M. HERBERT[*]

Department of Chemistry, The Ohio State University, Columbus, OH 43210 USA

JOHN E. HARRIMAN

Department of Chemistry, University of Wisconsin, Madison, WI 53706 USA

CONTENTS

I. Introduction
II. Reduced Eigenvalue Equations
III. Reduced Density Matrix Cumulants
   A. Additive Versus Multiplicative Separability
   B. Cumulant Formalism
   C. Extensivity
   D. Independence of the Cumulants
IV. Diagrammatic Representations
V. The Connected Equations
   A. Cancellation of Unconnected Terms
   B. Discussion of the Connected Equations
   C. Reconstruction and Solution of the Reduced Equations
References

---

[*]E-mail: jherbert@chemistry.ohio-state.edu

Reduced-Density-Matrix Mechanics: With Application to Many-Electron Atoms and Molecules,
A Special Volume of Advances in Chemical Physics, Volume 134, edited by David A. Mazziotti.
Series editor Stuart A. Rice. Copyright © 2007 John Wiley & Sons, Inc.

I. INTRODUCTION

In recent years, several groups [1–16] have explored the possibility of circumventing the wavefunction and the electronic Schrödinger equation in quantum chemical calculations, instead solving the so-called contracted Schrödinger equation (CSE) [16–19] for the two-electron reduced density matrix (2-RDM). Within the set of N-representable [20–24] RDMs, the CSE is an equivalent [16–18] formulation of the N-electron, clamped-nuclei Schrödinger equation, and couples the 2-, 3-, and 4-RDM elements via a linear equation that does not involve the electronic wavefunction explicitly. Direct calculation of the 2-RDM, and thereby electronic properties, is accomplished using approximate *reconstruction functionals* [7–9, 15, 25–27], by means of which the 3- and 4-RDMs are expressed in terms of the 2-RDM, leading to closed nonlinear equations for matrix elements of the latter.

Much of the recent literature on RDM reconstruction functionals is couched in terms of cumulant decompositions [13, 27–38]. Insofar as the p-RDM represents a quantum mechanical probability distribution for p-electron subsystems of an N-electron supersystem, the RDM cumulant formalism bears much similarity to the cumulant formalism of classical statistical mechanics, as formalized long ago by by Kubo [39]. (Quantum mechanics introduces important differences, however, as we shall discuss.) Within the cumulant formalism, the p-RDM is decomposed into "connected" and "unconnected" contributions, with the latter obtained in a known way from the lower-order q-RDMs, $q < p$. The connected part defines the pth-order RDM cumulant (p-RDMC). In contrast to the p-RDM, the p-RDMC is an *extensive* quantity, meaning that it is additively separable in the case of a composite system composed of noninteracting subsystems. (The p-RDM is multiplicatively separable in such cases [28, 32]). The implication is that the RDMCs, and the connected equations that they satisfy, behave correctly in the limit of noninteracting subsystems *by construction*, whereas a 2-RDM obtained by approximate solution of the CSE may fail to preserve extensivity, or in other words may not be *size-consistent* [40, 42].

In this work, we derive—via explicit cancellation of unconnected terms in the CSE—a pair of simultaneous, connected equations that together determine the 1- and 2-RDMCs, which in turn determine the 2-RDM in a simple way. Because the cancellation of unconnected terms is exact, we have in a sense done nothing; the connected equations are equivalent to the CSE and, given N-representability boundary conditions, they are also equivalent to the electronic Schrödinger equation. The important difference is that the connected equations for the cumulants automatically yield a size-consistent 2-RDM, even when solved approximately, because every term in these equations is manifestly extensive.

The derivation of the connected equations that is presented here is an expanded version of the one we published previously [43]. Our derivation utilizes a diagram technique and a "first-quantized" formalism, in which the CSE is expressed in terms of position-space kernels and Hilbert-space operators. Equations that couple the RDMCs have also been published in second quantization, by Kutzelnigg and Mukherjee [29, 31] and by Nooijen et al. [44], but the derivation presented here has the conceptual advantage that it explicitly demonstrates the cancellation of all unconnected terms, and furthermore does not require the introduction of a basis set (as is tacitly assumed in second quantization). Our derivation thus proves that the final, connected equations are equivalent to the CSE as well as to the ordinary electronic Schrödinger equation. Moreover, our derivation clarifies several important differences between the connected and the unconnected equations. As explained in Section V.B, the connected CSE is in fact a pair of *implicit* equations for the 1- and 2-RDMCs, whereas the original CSE is an *explicit* equation for the RDMs. In addition, the electronic energy—an explicit parameter in the CSE—is absent from this equation's connected analogues. Formally speaking, the connected equations that we ultimately obtain are equivalent to the "irreducible" CSEs introduced by Kutzelnigg and Mukherjee [29, 31], who derived connected equations starting from the fermion anticommutation relations, in a manner that does not rely on the original CSE at all.

The remainder of this chapter is organized as follows. Section II introduces the CSE as a special case of a more general class of *reduced eigenvalue equations*, and Section III formally defines the RDMCs. In the interest of motivating our derivation of connected CSEs, we include in Section III a survey of the quantum-mechanical cumulant formalism and the basic properties of the RDMCs, focusing especially on their additive separability for noninteracting subsystems. In Section IV, we develop a diagram technique to facilitate formal manipulation of terms that appear in the CSE. These diagrams also clarify the relationship between the CSE and older, Green's function methods in many-body theory, a connection that is examined in Section V. In that section we also present the main result of this work, a derivation of the connected form of the CSE, along with a discussion of procedures for solving the connected equations.

II. REDUCED EIGENVALUE EQUATIONS

Employing the abbreviated notation "j" $\equiv \mathbf{x}_j$ for the composite space/spin coordinates of the jth electron, let

$$\hat{W}_{(1,\ldots,N)} = \sum_{j=1}^{N} \hat{h}_{(j)} + \sum_{j<k}^{N} \hat{g}_{(j,k)} \qquad (1)$$

be a symmetric operator on the N-electron Hilbert space. This implies that $\hat{g}_{(j,k)} = \hat{g}_{(k,j)}$, which reflects the indistinguishability of electrons. We wish to consider RDM analogues of the N-electron eigenvalue equation

$$\hat{W}\Psi = w\Psi \tag{2}$$

Let the eigenvalue w be fixed and assume that Ψ is nondegenerate and unit-normalized. The restriction to nondegenerate eigenstates will be relaxed in Section V, but for now we consider only pure-state density matrices. The N-electron density matrix for the pure state Ψ is

$$D_{N(1,\ldots,N;1',\ldots,N')} \stackrel{\text{def}}{=} \Psi_{(1,\ldots,N)} \Psi^*_{(1',\ldots,N')} \tag{3}$$

For $p < q \leq N$, we define a *partial trace operator*

$$\text{tr}_{p+1,\ldots,q} \stackrel{\text{def}}{=} \int d\mathbf{x}_{p+1} \cdots d\mathbf{x}_q \, d\mathbf{x}'_{p+1} \cdots d\mathbf{x}'_q \, \delta(\mathbf{x}_{p+1} - \mathbf{x}'_{p+1}) \cdots \delta(\mathbf{x}_q - \mathbf{x}'_q) \tag{4}$$

that generates the p-RDM from the q-RDM,

$$D_p = \left[\frac{q!(N-q)!}{p!(N-p)!}\right] \text{tr}_{p+1,\ldots,q} D_q \tag{5}$$

and furthermore establishes the normalization

$$\text{tr}\, D_p \equiv \text{tr}_{1,\ldots,p} D_p = \frac{N}{p} \tag{6}$$

This is the convention that is most convenient for calculating expectation values, since in this case $\langle \Psi | \hat{W} | \Psi \rangle = \text{tr}(\hat{W}_2 D_2)$, where

$$\hat{W}_{2(1,2)} \stackrel{\text{def}}{=} \hat{g}_{(1,2)} + \frac{\hat{h}_{(1)} + \hat{h}_{(2)}}{N-1} \tag{7}$$

is the two-electron *reduced operator* corresponding to $\hat{W}$ [45, 46].

Most of this chapter utilizes the first-quantized formulation of the RDMs introduced above. However, some concepts related to separability and extensivity are more easily discussed in second quantization, and the second-quantized formalism is therefore employed in Section III. Introducing an orthonormal spin-orbital basis $|\phi_j\rangle = \hat{a}_j^\dagger |0\rangle$, the elements of the p-RDM are expressed directly in second quantization as

$$\mathbf{D}_{i_1,\ldots,i_p;j_1,\ldots,j_p} = \frac{1}{p!} \langle \Psi | \hat{a}_{i_1}^\dagger \cdots \hat{a}_{i_p}^\dagger \hat{a}_{j_p} \cdots \hat{a}_{j_1} | \Psi \rangle \tag{8}$$

We denote the tensor of such elements as $\boldsymbol{D}_p$, which is the tensor representation of the kernel D_p in a basis of p-electron direct products of the spin orbitals $\{|\phi_j\rangle\}$ [46]. The convention introduced in Eq. (8), that the number of indices implicitly specifies the tensor rank, is followed wherever tensors are used in this chapter.

From the N-electron Hilbert-space eigenvalue equation, Eq. (2), follows a hierarchy of p-electron *reduced* eigenvalue equations [13, 17, 18, 47] for $1 \leq p \leq N-2$. The pth equation of this hierarchy couples D_p, D_{p+1}, and D_{p+2} and can be expressed as

$$\Omega_{p(1,\ldots,p;1',\ldots,p')} \equiv 0 \tag{9}$$

in which Ω_p is the p-electron kernel [11]

$$\Omega_{p(1,\ldots,p;1',\ldots,p')} \stackrel{\text{def}}{=} \left[\sum_{j=1}^{p} \hat{h}_{(j)} + (1-\delta_{p,1})\sum_{j<k}^{p}\hat{g}_{(j,k)} - w\right]D_p$$
$$+ (p+1)\text{tr}_{p+1}\left\{\left[\hat{h}_{(p+1)} + \sum_{j=1}^{p}\hat{g}_{(j,p+1)}\right]D_{p+1}\right\} \tag{10}$$
$$+ \binom{p+2}{2}\text{tr}_{p+1,p+2}\{\hat{g}_{(p+1,p+2)}D_{p+2}\}$$

Here $D_n = D_{n(1,\ldots,n;1',\ldots,n')}$. The quantity Ω_p is called the pth-order *energy density matrix*.

Following Kutzelnigg and Mukherjee [29–31], we refer to Eq. (9) the pth-order CSE, or CSE(p) for brevity. (CSE(p) has also been called the $(p, p+2)$-CSE [13].) Strictly speaking, the term "CSE" implies that $\hat{W}$ is an electronic Hamiltonian, which is clearly the most important case, but the formal structure of Eqs. (9) and (10) is the same for any $\hat{W}$ having the form specified in Eq. (1). In the case of spin eigenstates, for example, the reduced equations for $\hat{W} = \hat{S}^2$ may be useful as boundary conditions to enforce while solving CSE(p) [3].

The remarkable fact, first demonstrated by Nakatsuji [18], is that for each $p \geq 2$, CSE(p) is equivalent (in a necessary and sufficient sense) to the original Hilbert-space eigenvalue equation, Eq. (2), provided that CSE(p) is solved subject to boundary conditions (N-representability conditions) appropriate for the $(p+2)$-RDM. CSE(p), in other words, is a closed equation for the $(p+2)$-RDM (which determines the $(p+1)$- and p-RDMs by partial trace) and has a unique N-representable solution D_{p+2} for each electronic state, including excited states. Without N-representability constraints, however, this equation has many spurious solutions [48, 49]. CSE(2) is the most tractable reduced equation that is still equivalent to the original Hilbert-space equation, and ultimately it is CSE(2) that we wish to solve. Importantly, we do not wish to solve CSE(2) for the 4-RDM, as this quantity is an eight-index tensor subject to four-electron N-representability conditions. Rather, we wish to solve CSE(2) in terms of the 2-RDM, via reconstruction of the 3- and 4-RDMs.

III. REDUCED DENSITY MATRIX CUMULANTS

In this section we introduce the p-RDMC, Δ_p, which encapsulates the part of the p-RDM that is additively separable in the limit of noninteracting subsystems. Although the RDMCs have been discussed at length in the literature [27–38], this section provides an introduction and summary of the most important points. In this section we use the second-quantized formulation of the RDMs (see Eq. (8)), as separability properties are most easily introduced using this formalism.

A. Additive Versus Multiplicative Separability

Although the RDMs provide a compact and appealing description of electronic structure, this description is unsatisfactory in at least one respect, namely, expectation values calculated from RDMs are not *manifestly* extensive, so do not necessarily become additively separable in the limit of noninteracting subsystems. This basic flaw ultimately arises because the RDMs are multiplicatively separable rather than additively separable [28–32].

To illustrate this point, consider a composite system composed of two non-interacting subsystems, one with p electrons (subsystem A) and the other with $q = N - p$ electrons (subsystem B). This would be the case, for example, in the limit that a diatomic molecule A—B is stretched to infinite bond distance. Because subsystems A and B are noninteracting, there must exist disjoint sets $\mathcal{B}_A$ and $\mathcal{B}_B$ of orthonormal spin orbitals, one set associated with each subsystem, such that the composite system's Hamiltonian matrix can be written as a direct sum,

$$\mathbf{H} = \mathbf{H}_A \oplus \mathbf{H}_B \tag{11}$$

where $\mathbf{H}_X (X \in \{A, B\})$ consists of matrix elements between determinants Φ_X constructed exclusively from spin orbitals in $\mathcal{B}_X$. Thus $\langle \Phi_A | \hat{H} | \Phi_B \rangle = 0$.

Let Ψ_X be an eigenfunction of $\mathbf{H}_X$, normalized to unity. Then the wavefunction for the composite system is

$$\Psi_{(1,\ldots,N)} = \frac{1}{\sqrt{N!}} \hat{P}_N (\Psi_{A(1,\ldots,p)} \Psi_{B(p+1,\ldots,N)}) \tag{12}$$

in which the operator $\hat{P}_N$ antisymmetrizes the product function $\Psi_A \Psi_B$ by generating all $N!$ signed permutations of the coordinates $\mathbf{x}_1, \ldots, \mathbf{x}_N$. In Dirac notation,

$|\Psi\rangle = |\Psi_A \Psi_B\rangle$, and one says that Ψ is *multiplicatively separable* in the two subsystems, recognizing that in quantum mechanics $|\Psi\rangle$ is separable only up to an overall antisymmetrization (or a symmetrization, in the case of bosons) that renders all coordinates equivalent. The separation of the wavefunction in Eq. (12) is equivalent, in a necessary and sufficient sense, to the block structure of the Hamiltonian in Eq. (11) [32, 50–52].

Because subsystems A and B do not interact, it must be that Ψ_A consists of a determinantal expansion in functions Φ_A taken solely from the set $\mathcal{B}_A$, and similarly Ψ_B uses only those spin orbitals in $\mathcal{B}_B$. It follows that Ψ_A and Ψ_B are strongly orthogonal [53]. Two antisymmetric functions $f(x_1, \ldots, x_p)$ and $g(y_1, \ldots, y_q)$ are said to be *strongly orthogonal* if

$$\int dz\, f^*(x_1,\ldots,x_{p-1},z)\, g(y_1,\ldots,y_{q-1},z) \equiv 0 \qquad (13)$$

Note that the integral above is nominally a function of $p + q - 2$ coordinates. Furthermore, because the functions of interest are antisymmetric, it does not matter which coordinates are chosen for the dummy integration variable z.

Consider the RDMs obtained from the separable wavefunction in Eq. (12). Since Ψ_A and Ψ_B are strongly orthogonal, it follows from Eq. (8) that $\langle \Psi_A \Psi_B | \hat{a}_i^\dagger \hat{a}_j | \Psi_A \Psi_B \rangle = 0$ unless ϕ_i and ϕ_j are associated with the same subsystem. Thus the 1-RDM separates into subsystem 1-RDMs,

$$D_1(\mathbf{x}; \mathbf{x}') = D_1^A(\mathbf{x}; \mathbf{x}') + D_1^B(\mathbf{x}; \mathbf{x}') \qquad (14)$$

The case $p = 1$ is the unique example for which D_p is additively separable. This is equivalent to the statement that D_1 equals its own cumulant (see Section III.B).

To obtain D_2, we need to evaluate matrix elements $\langle \Psi_A \Psi_B | \hat{a}_i^\dagger \hat{a}_j^\dagger \hat{a}_l \hat{a}_k | \Psi_A \Psi_B \rangle$. For reasons that will become clear, let us introduce the quantity

$$\Lambda_{ij;kl} \stackrel{\text{def}}{=} \mathbf{D}_{ij;kl} - \frac{1}{2}(\mathbf{D}_{i;k}\mathbf{D}_{j;l} - \mathbf{D}_{i;l}\mathbf{D}_{j;k}) \qquad (15)$$

The interesting scenario is when two of the four indices in this equation refer to subsystem A and the other two refer to subsystem B. Suppose, for definiteness, that $\phi_i, \phi_j \in \mathcal{B}_A$ and $\phi_k, \phi_l \in \mathcal{B}_B$. Then the strong orthogonality of Ψ_A and Ψ_B implies that $\mathbf{D}_{ij;kl} = 0$. More interesting is the case when $\phi_i, \phi_k \in \mathcal{B}_A$ and $\phi_j, \phi_l \in \mathcal{B}_B$. In this case $\mathbf{D}_{ij;kl}$ is generally nonzero; hence the 2-RDM mixes indices from different non-interacting subsystems, and thus fails to be additively separable. What about $\Lambda_{ij;kl}$? According to Eq. (14), $\mathbf{D}_{i;l} = 0$ since i and l refer to different subsystems, and therefore $\Lambda_{ij;kl} = \mathbf{D}_{ij;kl} - \frac{1}{2}\mathbf{D}_{i;k}\mathbf{D}_{j;l}$. The 2-RDM part of this expression can be simplified using the anticommutation relations, noting that $i \neq l$ and $j \neq k$. The result is

$$\begin{aligned}\mathbf{D}_{ij;kl} &\equiv \frac{1}{2}\langle \Psi_A \Psi_B | \hat{a}_i^\dagger \hat{a}_k \hat{a}_j^\dagger \hat{a}_l | \Psi_A \Psi_B \rangle \\ &= \frac{1}{2} \langle \Psi_A | \hat{a}_i^\dagger \hat{a}_k | \Psi_A \rangle \langle \Psi_B | \hat{a}_j^\dagger \hat{a}_l | \Psi_B \rangle\end{aligned} \qquad (16)$$

which is a product of 1-RDM elements from different subsystems. It follows that $\Lambda_{ij;kl} = 0$ for the case in question, and since $\Lambda_{ij;kl}$ as defined in Eq. (15) is antisymmetric, this quantity must in fact be zero unless all four indices refer to the same subsystem. Thus, unlike $\mathbf{D}_{ij;kl}$, the quantity $\Lambda_{ij;kl}$ is additively separable in the two noninteracting subsystems A and B,

$$\Delta_2(\mathbf{x}_1, \mathbf{x}_2; \mathbf{x}_1', \mathbf{x}_2') = \Delta_2^A(\mathbf{x}_1, \mathbf{x}_2; \mathbf{x}_1', \mathbf{x}_2') + \Delta_2^B(\mathbf{x}_1, \mathbf{x}_2; \mathbf{x}_1', \mathbf{x}_2') \qquad (17)$$

Δ_2 is precisely the 2-RDMC, and from Eq. (15) we note that expectation values for the composite A + B system can be computed using either D_2 alone, or $D_1 \equiv \Delta_1$ together with Δ_2. From the standpoint of *exact* quantum mechanics, either method yields exactly the same expectation value and, in particular, both methods respect the extensivity of the electronic energy. If D_2 is calculated by means of *approximate* quantum mechanics, however, one cannot generally expect that extensivity will be preserved, since exchange terms mingle the coordinates on different subsystems, and exact cancellation cannot be anticipated unless built in from the start. Methods that respect this separability by construction are said to be size-consistent [40–42].

In careful usage, extensivity is actually a more general concept than size consistency [42]. The former term implies a complete absence of unconnected terms in one's working equations, while size-consistency merely indicates that the energy is additively separable for noninteracting subsystems, a necessary consequence of extensivity. Methods that violate extensivity will yield per-particle correlation energies that tend to zero in the limit of an infinite system [42]. Hence the conventional wisdom is that use of manifestly extensive methods (coupled-cluster theory being the canonical example) is crucial for "large" systems containing subunits so distant as to be essentially noninteracting. It is not entirely clear how large one can go before this becomes a problem, though the effective range of the spin-traced 1-RDM may provide an indication. Computational studies suggest that for linear alkanes (i.e., one-dimensional insulators) the effective range $|\mathbf{r} - \mathbf{r}'|$ over which $D_1(\mathbf{r}; \mathbf{r}')$ is nonnegligible is about 15–20 carbon atoms [54], depending on drop tolerances, and we may judge that for larger systems extensivity violations may have important consequences. Lack of size-consistency is also a concern when breaking bonds, dissociating clusters, or comparing correlation energies between systems with different numbers of electrons.

In the present context, the way to ensure extensivity is to reformulate the CSE so that the RDMCs and not the RDMs are the basic variables. One can always recover the RDMs from the cumulants, but only the cumulants satisfy connected equations that do not admit the possibility of mixing noninteracting subsystems. Connected equations are derived in Section V. Before introducing that material, we first provide a general formulation of the p-RDMC for arbitrary p.

B. Cumulant Formalism

Following Ziesche [35, 55], in order to develop the theory of cumulants for non-commuting creation and annihilation operators (as opposed to classical variables), we introduce field operators $f(\mathbf{x})$ and $f^\dagger(\mathbf{x})$ satisfying the anticommutation relations for a Grassmann field,

$$[f(\mathbf{x}), f(\mathbf{x}')]_+ = 0 \quad (18a)$$

$$[f(\mathbf{x}), f^\dagger(\mathbf{x}')]_+ = 0 \quad (18b)$$

These field operators are sometimes termed *probe variables* because they function as dummy placeholders in the formal differentiations that follow but do not appear in the final expressions for the cumulants, which are obtained formally in the limit that $f, f^\dagger \to 0$.

First, we define a functional $\mathcal{G}[f, f^\dagger]$ whose derivatives generate the RDMs. In terms of the usual field operators $\hat{\psi}(\mathbf{x})$ and $\hat{\psi}^\dagger(\mathbf{x}')$,

$$\hat{\psi}(\mathbf{x}) = \sum_k \phi k(\mathbf{x}) \, \hat{a}_k \quad (19)$$

the RDM generating functional is [55]

$$\mathcal{G}[f, f^\dagger] = \left\langle \Psi \left| \mathcal{N} \exp\left(\int d\mathbf{x} \, [f(\mathbf{x}) \hat{\psi}^\dagger(\mathbf{x}) + f^\dagger(\mathbf{x}) \hat{\psi}(\mathbf{x})] \right) \right| \Psi \right\rangle \quad (20)$$

This is an analogue of the classical moments-generating functional discussed by Kubo [39]. Upon expanding the exponential as a power series, the operator $\mathcal{N}$ acts to place each term in so-called normal order, in which all creation operators $\hat{\psi}^\dagger$ are to the left of all annihilation operators $\hat{\psi}$. By virtue of this ordering (and *only* by virtue of this ordering),

$$\mathcal{G}[f, f^\dagger] = \left\langle \exp\left(\int d\mathbf{x} f(\mathbf{x}) \hat{\psi}^\dagger(\mathbf{x}) \right) \exp\left(\int d\mathbf{x}' f^\dagger(\mathbf{x}') \hat{\psi}(\mathbf{x}') \right) \right\rangle$$
$$= 1 + \mathcal{F}[f, f^\dagger] \quad (21)$$

where $\langle \cdots \rangle = \langle \Psi | \cdots | \Psi \rangle$ and

$$\mathcal{F}[f, f^\dagger] = \sum_{p=1}^{\infty} \frac{1}{p!^2} \left\langle \int d\mathbf{x}_1 \cdots d\mathbf{x}_p \, d\mathbf{x}'_1 \cdots d\mathbf{x}'_p \, f(\mathbf{x}_1) \hat{\psi}^\dagger(\mathbf{x}_1) \cdots f(\mathbf{x}_p) \hat{\psi}^\dagger(\mathbf{x}_p) \right. \\ \left. \times f^\dagger(\mathbf{x}'_p) \hat{\psi}(\mathbf{x}'_p) \cdots f^\dagger(\mathbf{x}'_1) \hat{\psi}(\mathbf{x}'_1) \right\rangle \quad (22)$$

The expectation value in Eq. (21) eliminates terms that do not conserve particle number; hence the two exponentials in Eq. (21) yield only a single summation in Eq. (22). The factor of $1/p!^2$ ensures that $\mathrm{tr}\, D_p = \binom{N}{p}$.

Formally, the logarithm $\ln \mathcal{G}$ provides a generating functional for the cumulants. That is, a formal expression for the p-RDMC is

$$\Delta_{p(1,\ldots,p;1',\ldots,p')} = \frac{1}{p!} \left(\lim_{f, f^\dagger \to 0} \frac{\delta^{2p}}{\delta f(1)\, \delta f^\dagger(1') \cdots \delta f(p)\, \delta f^\dagger(p')} \ln \mathcal{G} \right) \quad (23)$$

(The normalization of the cumulants is more complicated than that of the RDMs, but some specific examples are given in Section III.C.) Although ostensibly tedious, the above definition of Δ_p is operationally easy to use. In a formal expansion of $\ln \mathcal{G} = \ln(1 + \mathcal{F})$, the functional derivatives in Eq. (23) serve to select all terms consisting of exactly p creation operators $\hat{\psi}^\dagger$ and exactly p annihilation operators $\hat{\psi}$, while at the same time eliminating the integrals and replacing the dummy integration variables with particle coordinates $\mathbf{x}_1, \ldots, \mathbf{x}_p$ and $\mathbf{x}'_1, \ldots, \mathbf{x}'_p$.

As introduced above, the functional $\ln \mathcal{G}[f, f^\dagger]$ generates the cumulants as position-space kernels. As an alternative, Mazziotti [13, 33] has introduced a generating functional for the expansion coefficients $\Delta_{i_1,\ldots,i_p; j_1,\ldots,j_p}$ of Δ_p in a basis $\{\phi_k\}$ of orthonormal spin orbitals. Mazziotti's formalism can be obtained from the expressions above by expanding the Grassmann fields f and $f^\dagger$ in this basis,

$$f(\mathbf{x}) = \sum_k J_k\, \phi_k(\mathbf{x}) \quad (24)$$

The J_k are the probe variables in this formulation (which Mazziotti [33] terms "Schwinger probes"). We mention also Kutzelnigg and Mukherjee's treatment of RDMCs [28], which utilizes an antisymmetrized logarithm function, along with some special creation and annihilation operators, to generate the elements $\Delta_{i_1,\ldots,i_p; j_1,\ldots,j_p}$.

Using either Eq. (23) or Mazziotti's adaptation of it, one may derive exact expression for the RDMs in terms of their cumulants. The first few such expressions are

$$D_1 = \Delta_1 \tag{25a}$$

$$D_2 = \Delta_1^{\wedge 2} + \Delta_2 \tag{25b}$$

$$D_3 = \Delta_1^{\wedge 3} + 3\Delta_2 \wedge \Delta_1 + \Delta_3 \tag{25c}$$

$$D_4 = \Delta_1^{\wedge 4} + 6\Delta_2 \wedge \Delta_1^{\wedge 2} + 3\Delta_2^{\wedge 2} + 4\Delta_3 \wedge \Delta_1 + \Delta_4 \tag{25d}$$

$$D_5 = \Delta_1^{\wedge 5} + 10\Delta_2 \wedge \Delta_1^{\wedge 3} + 10\Delta_3 \wedge \Delta_1^{\wedge 2} + 5\Delta_4 \wedge \Delta_1 \tag{25e}$$
$$+ 15\,\Delta_1 \wedge \Delta_2^{\wedge 2} + 10\,\Delta_2 \wedge \Delta_3$$

Here "$\wedge$" denotes an antisymmetrized product (*Grassmann product* [47–56])

$$(\Delta_p \wedge \Delta_q)_{(1,\ldots,p+q;1',\ldots,(p+q)')} = \frac{1}{(p+q)!^2}\,\hat{P}_{p+q}\,\hat{P}'_{p+q}$$
$$\times \left[\Delta_{p(1,\ldots,p;1',\ldots,p')} \times \Delta_{q(p+1,\ldots,p+q;(p+1)',\ldots,(p+q)')}\right] \tag{26}$$

where $\hat{P}'_{p+q}$ and $\hat{P}_{p+q}$ indicate sums over signed permutations of the primed and unprimed coordinates, respectively (cf. Eq. (12)). "Wedge" exponents appearing in Eqs. (25a)–(25e) are defined according to

$$\Delta_p^{\wedge n} \stackrel{\text{def}}{=} \underbrace{\Delta_p \wedge \Delta_p \wedge \cdots \wedge \Delta_p}_{n \text{ factors}} \tag{27}$$

and should not be confused with matrix products such as Δ_1^n, the matrix product of n copies of Δ_1.

The decomposition of D_2 in Eq. (25b) is sometimes called the Levy–Lieb partition of the 2-RDM [57,58]. Formulas essentially equivalent to Eqs. (25a)–(25e) were known long ago, in the context of time-dependent Green's functions [59–61], but this formalism was rediscovered in the present context by Mazziotti [33].

Implicit in Eqs. (25a)–(25e) are definitions of the cumulants in terms of the RDMs, for example,

$$\Delta_2 = D_2 - D_1 \wedge D_1 \tag{28a}$$

$$\Delta_3 = D_3 - 2D_1^{\wedge 3} - 3D_1 \wedge D_2 \tag{28b}$$

$$\Delta_4 = D_4 + 13D_1 \wedge D_3 + 6D_1^{\wedge 2} \wedge D_2 - 4D_1 \wedge D_3 \tag{28c}$$

These equations *define* the RDMCs in terms of the RDMs and do not depend on the validity of perturbative expansions of the RDMs, although insofar as perturbation theory is applicable, Δ_p is precisely the sum of connected diagrams in the expansion of D_p.

The cumulant formulas in Eqs. (28a)–(28c) can be generated easily using a convenient mnemonic introduced by Harris [62]. To obtain the cumulant decomposition of D_{p+1} from that of D_p, one sums—for each term in the D_p—all possible ways in which the particle number can be increased by one.

Particle number can be increased either by replacing Δ_n with Δ_{n+1}, or by incorporating an additional Grassmann product with Δ_1. As an example, consider generating D_3 (Eq. (25c)) from D_2 (Eq. (25b)). Given the first term in Eq. (25b), $\Delta_1 \wedge \Delta_1$, one can increase particle number in three ways, and from these one obtains $\Delta_2 \wedge \Delta_1 + \Delta_1 \wedge \Delta_2 + \Delta_1 \wedge \Delta_1 \wedge \Delta_1$. The second term in Eq. (25b), Δ_2, affords $\Delta_3 + \Delta_2 \wedge \Delta_1$ upon increase in particle number. Together, these terms afford Eq. (25c).

This mnemonic emphasizes the combinatorial nature of the cumulants. For example, the term $3D_2 \wedge D_1$ in D_3 carries a coefficient that reflects the fact that there are three ways to obtain a three-particle distribution from one- and two-particle distributions, namely, $D_1 \wedge D_1 \wedge D_1$, $D_1 \wedge D_2$, and $D_2 \wedge D_1$. In contrast, the term $D_1^{\wedge 3}$ in D_3 has a coefficient of unity because there is only one way to combine one-particle distributions.

The combinatorial point of view is reminiscent of the classical cumulant formalism developed by Kubo [39], and indeed the structure of Eqs. (25) and (28) is essentially the same as the equations that define the classical cumulants, up to the use of an antisymmetrized product in the present context. In further analogy to the classical cumulants, the p-RDMC is identically zero if simultaneous p-electron correlations are negligible. In that case, the p-RDM is precisely an antisymmetrized product of lower-order RDMs.

C. Extensivity

For a multiplicatively separable wavefunction like the one in Eq. (12), the matrix elements of Δ_p vanish unless all indices correspond to the same subsystem [28, 32]. Using the notation introduced previously, this means that $\Delta_{j_1,\ldots,j_p;k_1,\ldots,k_p} = 0$ unless $\phi_m \in \mathcal{B}_A$ for each index m or else $\phi_m \in \mathcal{B}_B$ for each m. This is the essential difference that allows for an extensive formulation of quantum mechanics in terms of the RDMCs but not in terms of the RDMs. From the standpoint of extensivity, the basic problem with the RDMs is the manner in which the exchange terms in their unconnected parts mix the coordinates corresponding to noninteracting subsystems. Such exchange terms are identified by the presence of a Grassmann product. Examining the cumulant decompositions of the RDMs in Eqs. (25a)–(25e), it is evident that any term containing a

Grassmann product scales asymptotically ($N \to \infty$) like N^n, for some $n > 1$. For example, the Grassmann product

$$[\Delta_1 \wedge \Delta_1]_{(1,2;1',2')} = \frac{1}{2}\left[\Delta_{1(1;1')}\Delta_{1(2;2')} - \Delta_{1(1;2')}\Delta_{1(2;1')}\right] \quad (29)$$

appearing as part of D_2 has a trace given by

$$\text{tr}(\Delta_1 \wedge \Delta_1) = N^2 - \text{tr}(\Delta_1^2) \quad (30)$$

As $N \to \infty$, $\text{tr}(\Delta_1 \wedge \Delta_1) \sim N^2$. One says that $\Delta_1 \wedge \Delta_1$ scales like N^2.

One convenient consequence of binomial normalization for the RDMs (Eq. (6)) is that when this convention is followed, extensive quantities such as Δ_p have traces proportional to N, while nonextensive quantities possess traces that scale as some higher power of N (e.g., $\text{tr}D_p \sim N^p$). Let us define a set of quantities

$$\tau_k \stackrel{\text{def}}{=} \frac{\text{tr}(\Delta_1^k)}{N} \quad (31)$$

that satisfy the property

$$1 = \tau_1 \geq \tau_2 \geq \tau_3 \geq \cdots \geq 0 \quad (32)$$

which follows from the fact that all eigenvalues of Δ_1 lie in the interval $[0, 1]$. Equation (32) is valid even for extended systems, where $N \to \infty$. In fact, without loss of generality one may assume that $\tau_k > 0$ for each k, since the N-electron wavefunction can always be expanded in terms of natural spin orbitals having strictly positive occupation numbers [24]. The limiting case in which $\tau_k = 1$ for all k is obtained if and only if the two-electron interaction $\hat{g} \equiv 0$. In this case, the wavefunction is a single determinant, D_1 is idempotent, and [45]

$$D_p = D_1^{\wedge p} \quad \text{(single determinant)} \quad (33)$$

This form of D_p implies that $\Delta_p \equiv 0$ for each $p > 1$, a reflection of the fact that an independent-electron wavefunction consists of one-electron subsystems coupled only by exchange.

Traces of the RDMCs can be expressed conveniently in terms of the τ_k. For example,

$$\text{tr}\Delta_2 = \frac{1}{2}N(\tau_2 - 1) \quad (34)$$

and

$$\text{tr}\,\Delta_3 = \frac{1}{3}N(1 - 3\tau_2 + 2\tau_3) \tag{35}$$

Given the inequalities in Eq. (32), these trace expressions make it clear that $\text{tr}\Delta_2 \sim N$ and $\text{tr}\Delta_3 \sim N$, even as $N \to \infty$, and furthermore they demonstrate that the normalization of the p-RDMC depends on the system in question. (In particular, the traces depend on how far D_1 deviates from idempotency.) A few absolute bounds can be derived, such as

$$-\frac{1}{2}N \leq \text{tr}\Delta_2 \leq 0 \tag{36}$$

These inequalities do not exclude the possibility that Δ_2 has both positive and negative eigenvalues, which is generally the case. Traces of Δ_2 and Δ_3 have been examined for some model problems by Kutzelnigg and Mukherjee [28].

Partial traces of cumulants are also extensive, unlike those of the RDMs themselves. Starting from Eq. (25c), for example, one may show that

$$\text{tr}_3\Delta_3 = -\frac{2}{3}\Delta_2 + \frac{1}{6}\left[\hat{P}'_2(\Delta_1\Delta_2) + \hat{P}_2(\Delta_2\Delta_1)\right] \tag{37}$$

where the matrix products are defined, for example, as

$$(\boldsymbol{\Delta_1\Delta_2})_{i_1,i_2\,;\,j_1,j_2} = \sum_k \Delta_{i_1;k}\Delta_{i_2,k\,;\,j_1,j_2} \tag{38}$$

One may verify directly that $\text{tr}(\Delta_1\Delta_2) = N(\tau_3 - \tau_2)/2$ and therefore $\text{tr}(\Delta_1\Delta_2) \sim N$.

A word about notation is in order, regarding Eq. (37). Previously (cf. Eq. (26)), $\hat{P}'_n$ and $\hat{P}_n$ were defined to act upon primed and unprimed coordinates of n-electron kernels. Where tensors are involved, such as in Eq. (37), $\hat{P}'_n$ represents signed permutations over the row indices, (i.e., the first set of indices) and $\hat{P}_n$ denotes signed permutations over column indices. Thus, for example, when $\hat{P}'_2$ acts on $\Delta_1\Delta_2$ in Eq. (37), this operation antisymmetrizes the indices i_1 and i_2 appearing in Eq. (38). The column indices (j_1 and j_2) of this product are already antisymmetric, having inherited this property from Δ_2.

As noted earlier, $\text{tr}\Delta_n \sim N$ when binomial normalization is used for the RDMs, while nonextensive terms have traces that scale as higher powers of N. This is certainly a convenient means to recognize terms that are not extensive, but in some sense this trick overlooks the physical picture behind extensivity, which does not depend on any particular normalization convention. Similarly, insofar as perturbation theory is applicable, the fact that the RDMCs scale as N can be viewed as a consequence of the linked-cluster theorem [63, 64], but the deeper concept of extensivity does not depend on the validity of perturbation theory. Mathematically, extensivity is a statement about connectivity in the sense of matrix products, as in Eq. (38). In Section V, we introduce a nonperturbative diagram notation that emphasizes connectivity and extensivity, and demonstrates that D_n (as opposed to Δ_n) contains unconnected products, up to and including the product of n unconnected one-electron diagrams.

Thus far we have discussed connectivity and extensivity in terms of the RDMs and RDMCs, but our ultimate goal is to apply these concepts to CSE(2). Replacing the RDMs in Ω_2 with their cumulant decompositions elucidates the unconnected terms in CSE(2). Consider, as an example, the following term in $\Omega_{2(1,2;1',2')}$:

$$\hat{h}_{(1)} D_{2(1,2;1',2')} = \hat{h}_{(1)} \left[\Delta_{2(1,2;1',2')} + \frac{1}{2} \Delta_{1(1;1')} \Delta_{1(2;2')} - \frac{1}{2} \Delta_{1(1;2')} \Delta_{1(2;1')} \right] \quad (39)$$

(This is the first term on the right-hand side of Eq. (10), for the case $p = 2$.) The first term on the right-hand side in Eq. (39) is obviously connected, and we may deduce that the second term is unconnected because its trace equals $N^2 \langle \hat{h} \rangle / 2$. The third term, which constitutes a *transvection* [27, 62] of Δ_1 with itself, is actually connected, but differs from the second term by a coordinate permutation. If the second term is removed from CSE(2), then the third term ought to be removed as well, for otherwise we destroy the antisymmetry of Ω_2. This example illustrates the complexity of formulating an extensive version of CSE(2). It is not enough to eliminate unconnected terms; one must eliminate their exchange counterparts as well.

D. Independence of the Cumulants

Before deriving equations that determine the RDMCs, we ought to clarify precisely which are the RDMCs of interest. It is clear, from Eqs. (25a) and (25b), that Δ_1 and Δ_2 contain the same information as D_2 and can therefore be used to calculate expectation values $\langle \hat{W} \rangle$, where $\hat{W}$ is any symmetric two-electron operator of the form given in Eq. (1). Whereas the 2-RDM contains all of the information available from the 1-RDM, and affords the value of $\langle \hat{W} \rangle$ with no additional information, the 2-RDMC in general does not determine the 1-RDM [43, 65], so both Δ_1 and Δ_2 must be determined independently in order to calculate $\langle \hat{W} \rangle$. More generally, $\Delta_1, \ldots, \Delta_n$ are all independent quantities, whereas the RDMs $D_1, \ldots, D_n$ are related by the partial trace operation. The n-RDM determines all of the lower-order RDMs and lower-order RDMCs, but

Δ_n alone is insufficient to specify *any* of the other cumulants, or any RDMs at all (save for the trivial $n = 1$ case).

A simple proof that Δ_1 and Δ_2 are independent proceeds as follows. First, observe that

$$\text{tr}_2 \Delta_2 = \frac{1}{2}(\Delta_1^2 - \Delta_1) \tag{40}$$

from which it follows that Δ_1 and $\text{tr}_2 \Delta_2$ share a common set of eigenvectors, namely, the natural spin orbitals. Let $\{n_k\}$ be the natural occupation numbers (eigenvalues of Δ_1), and for each n_k, let e_k be the eigenvalue of $\text{tr}_2 \Delta_2$ associated with the same eigenvector. These two eigenvalues are related according to

$$e_k = \frac{1}{2} n_k(n_k - 1) \tag{41}$$

or in other words,

$$n_k = \frac{1}{2}(1 \pm \sqrt{1 + 8 e_k}) \tag{42}$$

Thus n_k is a double-valued function of e_k, as depicted in Fig. 1. Strictly speaking, then, the eigenvalues of $\text{tr}_2 \Delta_2$ do not determine those of Δ_1, and consequently Δ_1 cannot be determined from Δ_2 alone.

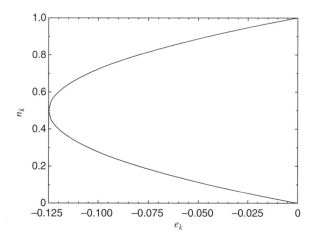

Figure 1. An eigenvalue n_k of Δ_1 as a (double-valued) function of the corresponding eigenvalue e_k of $\text{tr}_2 \Delta_2$.

That being said, in reality each eigenvalue of Δ_1 will likely be near either 0 or 1, except in certain open-shell systems with significant multideterminant character. Excluding such cases, it may be possible that, given Δ_2 (and thus the e_k), one can choose, for each k, one of the two solutions n_k in Eq. (42), based on whether the kth natural spin orbital is expected to be strongly or weakly occupied. (This could be determined by its expansion in Hartree–Fock orbitals.) Suppose that either $n_k = \varepsilon$ or $n_k = 1 - \varepsilon$, where ε is small. Upon calculating e_k corresponding to each, and substituting this back into Eq. (42), one obtains in either case a choice between solutions $n_k = 1 - \varepsilon + O(\varepsilon^2)$ and $n_k = \varepsilon + O(\varepsilon^2)$. As long as $\varepsilon^2 \ll \varepsilon$, and assuming that one can ascertain which natural spin orbitals are strongly occupied, Δ_2 effectively does determine Δ_1. In such cases, $\langle \hat{W} \rangle$ can be determined from Δ_2 alone.

IV. DIAGRAMMATIC REPRESENTATIONS

As outlined earlier, our task is to eliminate from CSE(2) both the unconnected terms *and* their exchange counterparts. These are readily identified using diagrammatic representations of Ω_1 and Ω_2 that we introduce in this section. The diagrams are not strictly necessary, but are quite convenient and (in the authors' opinion) easier to check for mistakes than lengthy algebraic formulas. In addition, certain reconstruction functionals for the 3- and 4-RDMs have been derived using diagrammatic many-body perturbation theory [7, 8, 11], and a diagrammatic representation for CSE(2) clarifies the role of this equation in improving approximate reconstruction functionals (see Section V.C). Our diagram conventions are conceived with this purpose in mind and are unrelated to the CSE diagrams introduced by Mukherjee and Kutzelnigg [30, 31].

The basic diagram elements representing $D_1 \equiv \Delta_1, \hat{g}, \hat{h}$, and Δ_p (for $p \geq 2$) are illustrated in Fig. 2. Recall that CSE(p) is given by the equation $\Omega_p \equiv 0$, where Ω_p is the p-electron kernel defined in Eq. (10). The terms in this kernel consist of $\hat{h}$ and $\hat{g}$ acting on RDMs, followed in some cases by a trace over one or two coordinate indices. Upon replacing the RDMs with their cumulant

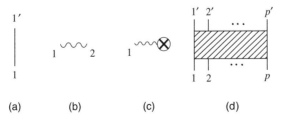

Figure 2. Basic diagram elements used in this work: (a) $D_{1(1;1')}$; (b) $\hat{g}_{(1,2)}$; (c) $\hat{h}_{(1)}$; and (d) $\Delta_{p(1,\ldots,p;1',\ldots,p')}$, for $p \geq 2$.

expansions (Eqs. (28a)–(28c)), we construct a diagrammatic representation of each term by connecting operator diagrams to cumulant diagrams, at the coordinates on which the operators act. For instance,

$$\begin{array}{c} \text{(diagram)} \end{array} = \hat{g}(2,3)\,\Delta_1(1;2')\,\Delta_2(2,3;1',3') \qquad (43)$$

is obtained by attaching a $\hat{g}$ diagram at the lower endpoints of a $\Delta_{2(2,3;1',3')}$ diagram, since according to Fig. 2(d) these endpoints represent coordinates $\mathbf{x}_2$ and $\mathbf{x}_3$. A factor of $\Delta_{1(1;2')}$ is present, as indicated, but because $\hat{g}_{(2,3)}$ operates on neither $\mathbf{x}_1$ nor $\mathbf{x}'_2$, this part of the diagram is not connected to the rest. Inspection of either the diagram or the algebra in Eq. (43) reveals that this term is unconnected.

A trace over coordinate $\mathbf{x}_n$ is indicated by connecting the line labeled n to the line labeled n'. The labels n and n' are then deleted, since these coordinates become a single dummy integration variable. Diagrammatically, this creates a loop in the case that both $\mathbf{x}_n$ and $\mathbf{x}'_n$ are arguments of the same cumulant. As an example, we apply tr_3 to Eq. (43) to obtain

$$\begin{array}{c} \text{(diagram)} \end{array} = \Delta_1(1;2')\,\mathrm{tr}_3\left\{\hat{g}(2,3)\,\Delta_2(2,3;1',3')\right\} \qquad (44)$$

If, on the other hand, $\mathbf{x}_n$ and $\mathbf{x}'_n$ are arguments of different cumulants, then a trace over $\mathbf{x}_n$ serves to connect two cumulant diagrams:

$$\begin{array}{c} \text{(diagram)} \end{array} = \Delta_1(1;2')\,\mathrm{tr}_{3,4}\left\{\hat{g}(3,4)\,\Delta_1(2;3')\,\Delta_2(3,4;1',4')\right\} \qquad (45)$$

Note carefully the subtle difference between this diagram and the previous one, at the position labeled "2." These two examples illustrate that *internal* operator vertices (those not appearing at the endpoint of a cumulant line) are each associated with a coordinate integration, whereas a vertex that appears at the endpoint of a cumulant line does not imply an integration. Thus, in Eq. (45), both ends of the $\hat{g}$ diagram are internal vertices, reflecting the fact that both arguments of $\hat{g}$ are integration variables. In Eq. (44), only one argument of $\hat{g}$ is an integration variable and thus the $\hat{g}$ diagram has one internal and one external vertex, the latter associated with $\mathbf{x}_2$.

In close analogy to diagrammatic perturbation theory (although our diagrams are not perturbative), we have transformed the problem of generating terms in Ω_p into a problem of generating topologically distinct diagrams, which makes it relatively easy to incorporate symmetries such as $\hat{g}_{(j,k)} = \hat{g}_{(k,j)}$ that reduce the number of terms in Ω_p. The nontrivial terms in Ω_1 and Ω_2 that involve only the one-electron cumulant are

$$6\,\mathrm{tr}_3\left\{\hat{h}(3)\,\Delta_1^{\wedge 3}(1,2,3;1',2',3')\right\} = \hat{P}_2'\left\{\begin{array}{c}1'\ 2'\\ \diagram \\ 1\ 2\end{array} - \hat{P}_2\begin{array}{c}1'\ 2'\\ \diagram \\ 1\ 2\end{array}\right\} \tag{46}$$

$$6\,\mathrm{tr}_3\left\{[\hat{g}(1,3) + \hat{g}(2,3)]\Delta_1^{\wedge 3}(1,2,3;1',2',3')\right\}$$
$$= \hat{P}_2\hat{P}_2'\left\{\begin{array}{c}1'\ 2'\\ \diagram \\ 1\ 2\end{array} - \begin{array}{c}1'\ 2'\\ \diagram \\ 1\end{array}2 - \begin{array}{c}1'\ 2'\\ \diagram \\ 2\end{array}\right\} \tag{47}$$

$$6\,\mathrm{tr}_{2,3}\left\{\hat{g}(2,3)\,\Delta_1^{\wedge 3}(1,2,3;1',2',3')\right\} = 2\left\{\begin{array}{c}1'\\ \diagram \\ 1\end{array} - \begin{array}{c}1'\\ \diagram \\ 1\end{array} + (\diagram - \diagram)\begin{array}{c}1'\\ |\\ 1\end{array}\right\} \tag{48}$$

and

$$24\,\mathrm{tr}_{3,4}\left\{\hat{g}(3,4)\,\Delta_1^{\wedge 4}(1,2,3,4;1',2',3',4')\right\}$$
$$= \hat{P}_2\left\{2\begin{array}{c}1'\ 2'\\ \diagram \\ 1\ 2\end{array} + (\diagram - \diagram)\begin{array}{c}1'\ 2'\\ \diagram \\ 1\ 2\end{array} + 2\hat{P}_2'\left(\begin{array}{c}1'\ 2'\\ \diagram \\ 1\ 2\end{array} - \begin{array}{c}1'\ 2'\\ \diagram \\ 1\ 2\end{array}\right)\right\} \tag{49}$$

These expressions are highly compact compared to brute-force expansions of the Grassmann products $\Delta_1^{\wedge 3}$ and $\Delta_1^{\wedge 4}$. For example, $\Delta_1^{\wedge 4}$ ostensibly contains $4!^2 = 576$ terms, as compared to the 14 terms that appear in Eq. (49) if one writes out all permutations.

Certain diagrams in the expressions above have no coordinate dependence and are related to the eigenvalue w in Eq. (2). Let us decompose $w = w_1 + w_2$ into a one-electron contribution

$$w_1 = N\langle \hat{h} \rangle \tag{50}$$

and a two-electron contribution

$$w_2 = \binom{N}{2}\langle \hat{g} \rangle \tag{51}$$

with $\langle \hat{h} \rangle = \text{tr}(\hat{h} D_1)$ and $\langle \hat{g} \rangle = \text{tr}(\hat{g} D_2)$. These equations are expressed diagrammatically as

$$w_1 = \text{tr}\left\{\hat{h}_{(1)} \Delta_1(1;1')\right\} = \text{○⤳⊗} \tag{52}$$

and

$$w_2 = \text{tr}\left\{\hat{g}_{(1,2)} D_2(1,2;1',2')\right\} = \tfrac{1}{2}\text{○⤳○} - \tfrac{1}{2}\text{⊖} + \text{(▨)} \tag{53}$$

When w is the electronic energy, the first two terms on the right-hand side of Eq. (53) are the Coulomb and exchange energies, respectively, while the third term defines what we term the *cumulant correlation energy*. The cumulant decomposition of D_2 thus provides universal, extensive definitions for the exchange and correlation energies, and these definitions do not depend on any independent-electron (Hartree–Fock or Kohn–Sham) reference state. According to this definition, the (exact) Coulomb and exchange energies are available from the (exact) 1-RDM, while the cumulant correlation energy requires the 2-RDM.

For expressions involving higher-order cumulants, one can utilize the antisymmetry of Δ_p to reduce the number of terms. For example, the identity

$$\text{[diagram]} = -\text{[diagram]} \tag{54}$$

is obtained by exchanging the lines entering the top of Δ_2, which corresponds to a permutation of the primed coordinates in Δ_2. After gaining some facility with the diagrams, one can write down the remaining terms in Ω_2:

$$9\,\text{tr}_3\left\{\hat{h}_{(3)}\left(\Delta_1 \wedge \Delta_2\right)(1,2,3;1',2',3')\right\}$$

$$= \hat{P}_2 \hat{P}'_2 \text{[diagram]} + \text{[diagram]} - \hat{P}'_2 \text{[diagram]} - \hat{P}_2 \text{[diagram]} \tag{55}$$

$$9\,\text{tr}_3\left\{\hat{g}_{(1,3)}\left(\Delta_1 \wedge \Delta_2\right)(1,2,3;1',2',3')\right\}$$

$$= \hat{P}'_2 \left\{\text{[diagram]} + \text{[diagram]} - \text{[diagram]}\right\} + \text{[diagram]} - \text{[diagram]} - \text{[diagram]} \tag{56}$$

$$16\,\mathrm{tr}_{3,4}\left\{\hat{g}(3,4)\left(\Delta_1\wedge\Delta_3\right)(1,2,3,4;1',2',3',4')\right\}$$

$$=\hat{P}_2\hat{P}_2'\;[\text{diagram}] - 2\hat{P}_2\;[\text{diagram}] - 2\hat{P}_2'\;[\text{diagram}]$$

$$+\;2\;[\text{diagram}] - 2\;[\text{diagram}] \tag{57}$$

$$72\,\mathrm{tr}_{3,4}\left\{\hat{g}(3,4)\left(\Delta_1^{\wedge 2}\wedge\Delta_2\right)(1,2,3,4;1',2',3',4')\right\}$$

$$=(\text{diagram}-\text{diagram})\,[\text{diagram}] + \hat{P}_2'\left\{[\text{diagram}]+2\,[\text{diagram}]-2\,[\text{diagram}]\right\}$$

$$+\,\hat{P}_2\left\{[\text{diagram}]+2\,[\text{diagram}]-2\,[\text{diagram}]+[\text{diagram}]\right\}$$

$$+\,2\hat{P}_2\hat{P}_2'\left\{-[\text{diagram}]-[\text{diagram}]+[\text{diagram}]\right.$$
$$\left.-[\text{diagram}]-[\text{diagram}]-[\text{diagram}]\right\} \tag{58}$$

and

$$18\,\mathrm{tr}_{3,4}\left\{\hat{g}(3,4)\,\Delta_2^{\wedge 2}(1,2,3,4;1',2',3',4')\right\}$$

$$=2\hat{P}_2\left\{[\text{diagram}]-[\text{diagram}]-[\text{diagram}]\right\}$$

$$-\,2\hat{P}_2'\,[\text{diagram}] + [\text{diagram}] + [\text{diagram}] \tag{59}$$

V. THE CONNECTED EQUATIONS

Recall that the equation CSE(p) is written as $\Omega_p(1,\ldots,p;1',\ldots,p') \equiv 0$, where Ω_p is the pth-order energy density matrix. Yasuda [11] has introduced a generating functional for the energy density matrices and used this functional to demonstrate that $\Omega_p = \Omega_p^C + \Omega_p^U$ can be decomposed into a connected part Ω_p^C and an unconnected part Ω_p^U. The diagrammatic technique introduced in the previous section brings this to the forefront, and in this section we use diagrammatic representations of Ω_1 and Ω_2 to formulate connected versions of CSE(1) and CSE(2). Whereas CSE(1) is necessarily satisfied if CSE(2) is satisfied (since the former is merely a partial trace of the latter [47]), the connected versions of CSE(1) and CSE(2) are independent conditions on the 1- and 2-RDMCs, which must be satisfied simultaneously.

A. Cancellation of Unconnected Terms

Clearly Ω_1, as defined in Eq. (10), contains unconnected terms, including, for example,

$$w\,D_1(1;1') = \left(\text{diagrams}\right)\Big|_1^{1'} \tag{60}$$

but these terms cancel exactly and $\Omega_1^U \equiv 0$. Since an approximate solution of CSE(1) may not lead to exact cancellation of the unconnected terms, instead of solving the equation $\Omega_1 \equiv 0$, one ought to solve the manifestly extensive equation

$$\Omega_1^C \equiv 0 \tag{61}$$

The connected part of Ω_1 is found to be

$$\Omega_1^C(1;1') = \{\text{diagrams}\} \tag{62}$$

Since the unconnected terms cancel exactly, Eq. (61) is equivalent, in a necessary and sufficient sense, to CSE(1). Following Kutzelnigg and Mukherjee [29–31] we refer to Eq. (61) as the first-order *irreducible* contracted Schrödinger equation, ICSE(1). To obtain an equation that is equivalent, within a finite basis set, to our ICSE(p), one must solve the Kutzelnigg–Mukherjee version [30] of ICSE(p) simultaneously with its adjoint equation, whereas our version of ICSE(p) is equal to its own adjoint, thus ensuring that its solution Δ_p is self-adjoint.

Neither CSE(1) nor ICSE(1) is equivalent to the original Hilbert-space eigenvalue equation; for that we need CSE(2). The unconnected part of Ω_2 is [11]

$$\Omega_2^U = D_1 \wedge \Omega_1 = D_1 \wedge \Omega_1^C \tag{63}$$

This relationship can be verified directly using the expressions in the previous section. Thus, if D_1 satisfies CSE(1)—a necessary condition if D_2 is to satisfy CSE(2)—then $\Omega_2^U = 0$ and we obtain the extensive equation

$$\Omega_2^C \equiv 0 \tag{64}$$

which we call ICSE(2). Carrying out the cancellation is relatively easy using diagrams, and one obtains

$$\Omega_2^C(1,2;1',2') = \tag{65}$$

Including permutations, this expression for Ω_2^C contains 68 terms, a significant reduction as compared to the unsimplified Grassmann products.

Equations (62) and (65), expressed in diagrammatic notation, are the only forms of ICSE(1) and ICSE(2) that appear in our original publication of the connected equations [43], although a short time later a connected, algebraic version of CSE(2) was published by Nooijen and co-workers [44]. Here, we translate our diagrammatic version equations into algebraic ones, using the diagram rules introduced in Section IV. For pedagogical purposes, and owing to the complexity of the result, we break up Eqs. (62) and (65) line-by-line and present each line as a separate algebraic expression. In addition, certain obvious factorizations are bypassed in the algebraic formulation that follows, in order that diagrams on the left-hand side of the equality match up with algebraic expressions on the right-hand side term-by-term and in the same order. This facilitates comparison between the diagrammatic and the algebraic equations.

The algebraic form of Ω_1^C is contained in the equations

$$= \hat{h}(1)\,\Delta_1(1;1') + \mathrm{tr}_2\left\{\hat{g}(1,2)\left[\Delta_1(1;1')\,\Delta_1(2;2') - \Delta_1(1;2')\,\Delta_1(2;1')\right]\right\}$$

$$- \mathrm{tr}_2\left[\hat{h}(2)\,\Delta_1(1;2')\,\Delta_1(2;1')\right]$$

$$- \mathrm{tr}_{2,3}\left\{\hat{g}(2,3)\left[\Delta_1(1;2')\,\Delta_1(2;1')\,\Delta_1(3;3') - \Delta_1(1;2')\,\Delta_1(2;3')\,\Delta_1(3;1')\right]\right\} \quad (66a)$$

$$= 3\,\mathrm{tr}_{2,3}\left[\hat{g}(2,3)\,\Delta_3(1,2,3;1',2',3')\right] + 2\,\mathrm{tr}_2\left[\hat{h}(2)\,\Delta_2(1,2;1',2')\right]$$

$$+ 2\,\mathrm{tr}_{2,3}\left[\hat{g}(2,3)\,\Delta_2(1,2;1',2')\,\Delta_1(3;3')\right] \quad (66b)$$

and

$$= -2\,\mathrm{tr}_{2,3}\left[\hat{g}(2,3)\,\Delta_2(1,2;1',3')\,\Delta_1(3;2')\right] + 2\,\mathrm{tr}_2\left[\hat{g}(1,2)\,\Delta_2(1,2;1',2')\right]$$

$$- 2\,\mathrm{tr}_{2,3}\left[\hat{g}(2,3)\,\Delta_2(2,3;1',2')\,\Delta_1(1;2')\right] - 2\,\mathrm{tr}_{2,3}\left[\hat{g}(2,3)\,\Delta_2(1,3;2',3')\,\Delta_1(2;1')\right] \quad (66c)$$

The kernel $\Omega^C_{2\,(1;1')}$ is equal to the sum of the terms given in Eqs. (66a)–(66c). The various terms in ICSE(2) are

$$= \left[\hat{h}(1) + \hat{h}(2) + \hat{g}(1,2)\right] \Delta_2(1,2;1',2') + \text{tr}_{3,4}\left[\hat{g}(3,4)\,\Delta_2(1,2;3',4')\,\Delta_2(3,4;1',2')\right]$$
$$+ 3\hat{P}_2\left\{\text{tr}_3\left[\hat{g}(2,3)\,\Delta_3(1,2,3;1',2',3')\right] + \text{tr}_{3,4}\left[\hat{g}(3,4)\,\Delta_3(1,3,4;1',2',4')\,\Delta_1(2;3')\right]\right\} \quad (67\text{a})$$

$$= 3\hat{P}'_2\,\text{tr}_{3,4}\left[\hat{g}(3,4)\,\Delta_3(1,2,3;1',4',3')\,\Delta_1(4;2')\right] + 3\,\text{tr}_3\left[\hat{h}(3)\,\Delta_3(1,2,3;1',2',3')\right]$$
$$+ 3\,\text{tr}_{3,4}\left[\hat{g}(3,4)\,\Delta_3(1,2,3;1',2',3')\,\Delta_1(4;4')\right] \quad (67\text{b})$$
$$- 3\,\text{tr}_{3,4}\left[\hat{g}(3,4)\,\Delta_3(1,2,3;1',2',4')\,\Delta_1(4;3')\right]$$
$$+ 6\,\text{tr}_{3,4}\left[\hat{g}(3,4)\,\Delta_3(1,2,3,4;1',2',3',4')\right]$$

$$= \hat{P}'_2\left\{\tfrac{1}{2}\,\hat{g}(1,2)\,\Delta_1(1;1')\,\Delta_1(2;2') - 2\,\text{tr}_{3,4}\left[\hat{g}(3,4)\,\Delta_2(1,2;1',3')\,\Delta_2(3,4;2',4')\right]\right.$$
$$+ \tfrac{1}{2}\,\text{tr}_{3,4}\left[\hat{g}(3,4)\,\Delta_2(1,2;3',4')\,\Delta_1(3;1')\,\Delta_1(4;2')\right]$$
$$\left. + \text{tr}_{3,4}\left[\hat{g}(3,4)\,\Delta_2(1,2;1',3')\left(\Delta_1(3;4')\,\Delta_1(4;2') - \Delta_1(3;2')\,\Delta_1(4;4')\right)\right]\right\} \quad (67\text{c})$$

$$= \hat{P}'_2\hat{P}_2\left\{\text{tr}_3\left[\hat{g}(1,3)\,\Delta_2(2,3;2',3')\,\Delta_1(1;2')\right] - \text{tr}_3\left[\hat{g}(1,3)\,\Delta_2(1,2;3',2')\,\Delta_1(3;1')\right]\right.$$
$$- \text{tr}_{3,4}\left[\hat{g}(3,4)\,\Delta_2(1,3;1',3')\,\Delta_1(2;4')\,\Delta_1(4;2')\right] \quad (67\text{d})$$
$$+ \text{tr}_{3,4}\left[\hat{g}(3,4)\,\Delta_2(1,3;1',4')\,\Delta_1(2;3')\,\Delta_1(4;2')\right]$$
$$\left. - \tfrac{1}{2}\,\text{tr}_3\left[\Delta_1(1;1')\,\Delta_1(2;3')\,\Delta_1(3;2')\right]\right\}$$

$$\hat{P}_2\left\{\frac{1}{2}\;\diagup\!\!\!\!\diagup + \diagup\!\!\!\!\diagup - \diagup\!\!\!\!\diagup - \diagup\!\!\!\!\diagup - \diagup\!\!\!\!\diagup + \diagup\!\!\!\!\diagup - \diagup\!\!\!\!\diagup\right\}$$

$$= \hat{P}_2\left\{\tfrac{1}{2}\,\mathrm{tr}_{3,4}\left[\hat{g}(3,4)\,\Delta_2(1,2;3',4')\,\Delta_1(3;1')\,\Delta_1(4;2')\right]\right.$$
$$+ \mathrm{tr}_{3,4}\left[\hat{g}(3,4)\,\Delta_2(1,3;1',2')\bigl(\Delta_1(4;3')\,\Delta_1(2;4') - \Delta_1(2;3')\,\Delta_1(4;4')\bigr)\right]$$
$$- \mathrm{tr}_3\left[\hat{h}(3)\,\Delta_2(1,3;1',2')\,\Delta_1(2;3')\right] - \mathrm{tr}_3\left[\hat{h}(3)\,\Delta_2(1,2;1',3')\,\Delta_1(3;2')\right]$$
$$\left.- \mathrm{tr}_3\left[\hat{g}(2,3)\,\Delta_2(1,3;1',2')\,\Delta_1(2;3')\right]\right\} \tag{67e}$$

and

$$\hat{P}_2\left\{-\diagup\!\!\!\!\diagup + \tfrac{1}{2}\,\diagup\!\!\!\!\diagup + 2\,\diagup\!\!\!\!\diagup - 2\,\diagup\!\!\!\!\diagup - 2\,\diagup\!\!\!\!\diagup\right\}$$

$$= \hat{P}_2\left\{-\mathrm{tr}_3\left[\hat{g}(1,3)\,\Delta_2(1,3;1',2')\,\Delta_1(2;3')\right]\right.$$
$$+ \tfrac{1}{2}\mathrm{tr}_{3,4}\left[\hat{g}(3,4)\,\Delta_1(1;3')\,\Delta_1(2;4')\,\Delta_1(3;1')\,\Delta_1(4;2')\right]$$
$$+ 2\,\mathrm{tr}_{3,4}\left[\hat{g}(3,4)\,\Delta_2(1,3;1',3')\,\Delta_2(4,2;4',2')\right]$$
$$- 2\,\mathrm{tr}_{3,4}\left[\hat{g}(3,4)\,\Delta_2(1,3;1',2')\,\Delta_2(2,4;3',4')\right]$$
$$\left.- 2\,\mathrm{tr}_{3,4}\left[\hat{g}(3,4)\,\Delta_2(1,3;1',4')\,\Delta_2(4,2;3',2')\right]\right\} \tag{67f}$$

The kernel $\Omega_2^C(1,2;1',2')$ is equal to the sum of the terms given in Eqs. (67a)–(67f).

B. Discussion of the Connected Equations

Perhaps the most striking feature of ICSE(1) and ICSE(2) is the absence of the eigenvalue w in these equations. In hindsight its disappearance should not be surprising, since w appears in Ω_p as the product wD_p. The observable w scales as N, as does the connected part of D_p; hence no part of wD_p exhibits correct scaling, and this entire term must cancel with some other part of CSE(p). (This is analogous to the fact that the coupled-cluster amplitude equations, which are extensive, contain the cluster amplitudes but not the electronic energy.) Certainly, w is specified *implicitly* in ICSE(1) and ICSE(2), insofar as the cumulants Δ_1 and Δ_2 together determine D_2 and thus also determine $w = \text{tr}(\hat{W}_2 D_2)$. The absence of w in ICSE(p) has important consequences, to which we shall return later in this section.

In deriving ICSE(1) and ICSE(2) from the corresponding CSEs, we have merely identified and removed terms that cancel exactly; as such, these two connected equations, when solved simultaneously, are entirely equivalent to CSE(2) and thus equivalent to the original Hilbert-space eigenvalue equation (Schrödinger equation), provided that appropriate N-representability constraints are enforced. Since necessary and sufficient N-representability constraints are not known, one must in practice contend with an infinite number of spurious solutions to these equations. Recent calculations [7, 9, 34, 66] in which CSE(2) is solved starting from an N-representable (actually, Hartree–Fock) 2-RDM indicate that, for ground states, the solution usually converges to a 2-RDM that is nearly consistent with the necessary P-, Q-, and G-conditions [24, 66, 67] for N-representability. (These conditions demand that the two-particle density matrix, the two-hole density matrix, and the particle–hole density matrix, respectively, be positive semi–definite, and by "nearly consistent" we mean that any negative eigenvalues are small in magnitude.)

Even given a hypothetical set of necessary and sufficient N-representability constraints, however, the solution of CSE(2) is only unique provided that the eigenvalue w is specified and fixed. Because w does not appear in the ICSEs, a unique solution of ICSE(1) and ICSE(2) is obtained only by simultaneous solution of these equations subject not only to N-representability constraints but also subject to the constraint that $w = \text{tr}(\hat{W}_2 D_2)$ remains fixed. For auxiliary constraint equations, such as the reduced eigenvalue equation for the operator $\hat{S}^2$, one would know the target expectation value $\langle \hat{S}^2 \rangle$ in advance and could therefore constrain $\langle \hat{S}^2 \rangle = \text{tr}(\hat{S}_2^2 D_2)$. In the basic equations of our theory, however, $\hat{W}$ is an electronic Hamiltonian and such a constraint would require us to know the electronic energy in advance. Foregoing the energy constraint, ICSE(1) and ICSE(2) possess N-representable solutions corresponding to the ground state, the excited states, and also all superposition states that can be formed from degenerate eigenfunctions of $\hat{W}$. This is again analogous to coupled-cluster theory, whose connected working equations do not contain the electronic energy explicitly, and have solutions corresponding to both ground and excited electronic states [68]; the ground-state solution is selected by means of the initial guess. Compared to CSE(2), the absence of the electronic energy in the ICSEs is not a serious disadvantage, since in the former case the energy is not known a priori, and therefore w appearing in CSE(2) must be iteratively updated during the course of achieving a self-consistent solution.

Before discussing further how ICSE(1) and ICSE(2) can be solved, let us first discuss the solution of CSE(2). For $w \neq 0$, CSE(2) may be written

$$D_2 = w^{-1} F_w[D_2, D_3, D_4] \qquad (68)$$

where the functional $F_w \equiv \Omega_2 + wD_2$ (cf. Eq. (10)). Assuming that one posesses approximate reconstruction functionals $D_3[D_2]$ and $D_4[D_2]$, Eq. (68) can be solved for D_2 by one of two means. The first option is to substitute the reconstruction functionals directly into F_w, effectively making F_w a functional of D_2 only. Upon expanding Eq. (68) in a finite basis set, this leads to a closed set of nonlinear equations for the tensor elements of D_2, and these equations can be solved, for example, by a Newton–Raphson procedure [7, 9, 11]. Alternatively, Eq. (68) can be solved by self-consistent iteration, employing the reconstruction functionals at each iteration to generate updated 3- and 4-RDMs from the current 2-RDM, and using the current 2-RDM to estimate w. Several algorithms for carrying out this iteration scheme have been described [1, 2, 6, 14].

It does not appear that the ICSEs can be solved by self-consistent iteration, however. In Eq. (68), CSE(2) is expressed in a form that affords the 2-RDM as an *explicit* functional of the 2-, 3-, and 4-RDMs, but no analogous formulation of ICSE(1) or ICSE(2) is possible, since the 1- and 2-RDMCs appearing in these equations are always acted upon by $\hat{h}$ or $\hat{g}$ (cf. Eqs. (66) and (67)). Thus the ICSEs are *implicit* equations for the cumulants.

Using cumulant reconstruction functionals $\Delta_3[\Delta_1, \Delta_2]$ and $\Delta_4[\Delta_1, \Delta_2]$, one can certainly derive closed, nonlinear equations for the elements of Δ_1 and Δ_2, which could be solved using an iterative procedure that does not exploit the reconstruction functionals at each iteration. Of the RDM reconstruction functionals derived to date, several [7, 8, 11] utilize the cumulant decompositions in Eqs. (25c) and (25d) to obtain the unconnected portions of D_3 and D_4 exactly (in terms of the lower-order RDMs), then use many-body perturbation theory to estimate the connected parts Δ_3 and Δ_4 in terms of Δ_1 and Δ_2, the latter essentially serving as a renormalized pair interaction. Reconstruction functionals of this type are equally useful in solving ICSE(1) and ICSE(2), but the reconstruction functionals introduced by Valdemoro and co-workers [25, 26] cannot be used to solve the ICSEs because they contain no connected terms in D_3 or D_4 (and thus no contributions to Δ_3 or Δ_4).

C. Reconstruction and Solution of the Reduced Equations

Next, we present some observations concerning the connection between the reconstruction process and the iterative solution of either CSE(p) or ICSE(p). The perturbative reconstruction functionals mentioned earlier each constitute a finite-order ladder-type approximation to the 3- and 4-RDMCs [46, 69]; examples of the lowest-order corrections of this type are shown in Fig. 3. The hatched squares in these diagrams can be thought of as arising from the 2-RDM, which serves as an effective pair interaction for a form of many-body perturbation theory. Ordinarily, ladder-type perturbation expansions neglect three-electron (and higher) correlations, even when extended to infinite order in the effective pair interaction [46, 69], but iterative solution of the CSEs (or ICSEs) helps to

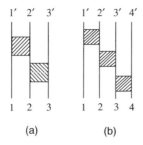

Figure 3. Lowest order connected corrections to (a) Δ_3 and (b) Δ_4, within a renormalized ladder-type approximation.

build these correlations back into the cumulants. This becomes clear upon examination of the diagrammatic representations of these equations, together with diagrammatic representations of the reconstruction functionals.

In Fig. 4(a) we show a typical diagram in the expansion of Δ_3 that cannot be incorporated into any ladder-type diagram because it involves simultaneous correlation between three particles [69]. As it appears in CSE(2) and ICSE(2), however, Δ_3 is always traced over coordinate $\mathbf{x}_3$, and in Fig. 4(b) we show the effect of tr_3 on the diagram in Fig. 4(a). Diagram 4(b) *is* included in the partial trace of a *third*-order ladder-type diagram, namely, the one shown in Fig. 4(c). Thus the presence of tr_3 in the two-particle equations allows one to incorporate three- and higher-body effects that would not otherwise be present in a ladder approximation for the three- and four-electron cumulants.

Actually three-particle correlations such as that in Fig. 4(a) are introduced by the CSEs and ICSEs, even within a second-order ladder approximation. To understand why, consider the diagram in Fig. 4(d), which represents one of the terms in Ω_2^C. Within a second-order ladder approximation to Δ_3, diagram 4(b) is included within diagram 4(d). Thus three- and higher-body effects are incorporated into the cumulants Δ_3 and Δ_4 by the CSEs or ICSEs, even when

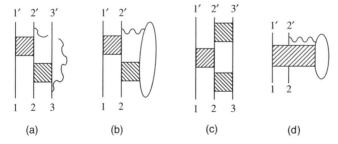

Figure 4. Diagrams illustrating the connection between reconstruction and solution of the CSEs or ICSEs. See the text for an explanation.

these effects are absent from approximate reconstruction functionals. In effect, solution of these equations corresponds to a partial summation of the perturbation series for D_2, in the case of CSE(2), or Δ_1 and Δ_2, in the case that ICSE(1) and ICSE(2) are solved simultaneously. The connection between reconstruction and solution of coupled Green's function equations of motion, which are time-dependent hierarchies analogous to the CSE(p) hierarchy, has received some attention [70, 71], though a more thorough exploration of this connection would be welcome.

References

1. F. Colmenero and C. Valdemoro, *Int. J. Quantum Chem.* **51**, 369 (1994).
2. C. Valdemoro, Reduced density matrix versus wave function: recent developments, in *Strategies and Applications in Quantum Chemistry*, (Y. Ellinger and M. Defranceschi, eds.), Kluwer, Dordrecht, 1996, p. 55.
3. C. Valdemoro, L. M. Tel, and E. Pérez-Romero, *Adv. Quantum Chem.* **28**, 33 (1997).
4. C. Valdemoro, D. R. Alcoba, and L. M. Tel, *Int. J. Quantum Chem.* **93**, 212 (2003).
5. C. Valdemoro, L. M. Tel, and E. Pérez-Romero, Critical questions concerning iterative solution of the contracted Schrödinger equation, in *Many-Electron Densities and Reduced Density Matrices*, (J. Cioslowski, ed.), Plenum, New York, 2000.
6. C. Valdemoro, L. M. Tel, E. Pérez-Romero, and A. Torre, *J. Mol. Struct. (Theochem.)* **537**, 1 (2001); erratum: **574**, 255 (2001).
7. K. Yasuda and H. Nakatsuji, *Phys. Rev. A* **56**, 2648 (1997).
8. H. Nakatsuji and K. Yasuda, *Phys. Rev. Lett.* **76**, 1039 (1996).
9. M. Nakata, M. Ehara, K. Yasuda, and H. Nakatsuji, *J. Chem. Phys.* **112**, 8772 (2000).
10. M. Ehara, M. Nakata, H. Kou, K. Yasuda, and H. Nakatsuji, *Chem. Phys. Lett.* **305**, 483 (1999).
11. K. Yasuda, *Phys. Rev. A* **59**, 4133 (1999).
12. H. Nakatsuji, Density equation theory in chemical physics, in *Many-Electron Densities and Reduced Density Matrices*, (J. Cioslowski, ed.), Plenum, New York, 2000, p. 85.
13. D. A. Mazziotti, *Int. J. Quantum Chem.* **70**, 557 (1998).
14. D. A. Mazziotti, *J. Chem. Phys.* **116**, 1239 (2002).
15. D. A. Mazziotti, *Phys. Rev. A* **60**, 3618 (1999).
16. D. A. Mazziotti, *Phys. Rev. A* **57**, 4219 (1998).
17. L. Cohen and C. Frishberg, *Phys. Rev. A* **13**, 927 (1976).
18. H. Nakatsuji, *Phys. Rev. A* **14**, 41 (1976).
19. H. Nakatsuji, *Theor. Chem. Acc.* **102**, 97 (1999).
20. A. J. Coleman, *Rev. Mod. Phys.* **35**, 668 (1963).
21. H. Kummer, *J. Math. Phys.* **8**, 2063 (1967).
22. R. M. Erdahl, *Int. J. Quantum Chem.* **13**, 697 (1978).
23. E. R. Davidson, *Reduced Density Matrices in Quantum Chemistry*, Academic, New York, 1976.
24. A. J. Coleman and V. I. Yukalov, *Reduced Density Matrices: Coulson's Challenge*, Springer, Berlin, 2000.
25. F. Colmenero, C. Pérez del Valle, and C. Valdemoro, *Phys. Rev. A* **47**, 971 (1993).
26. F. Colmenero and C. Valdemoro, *Phys. Rev. A* **47**, 979 (1993).

27. D. A. Mazziotti, *Chem. Phys. Lett.* **326**, 212 (2000).
28. W. Kutzelnigg and D. Mukherjee, *J. Chem. Phys.* **110**, 2800 (1999).
29. W. Kutzelnigg and D. Mukherjee, *Chem. Phys. Lett.* **317**, 567 (2000).
30. D. Mukherjee and W. Kutzelnigg, *J. Chem. Phys.* **114**, 2047 (2001); erratum: **114**, 8226 (2001).
31. W. Kutzelnigg and D. Mukherjee, *J. Chem. Phys.* **116**, 4787 (2002).
32. W. Kutzelnigg, *Int. J. Quantum Chem.* **95**, 404 (2003).
33. D. A. Mazziotti, *Chem. Phys. Lett.* **289**, 419 (1998).
34. D. A. Mazziotti, Cumulants and the contracted Schrödinger equation, in *Many-Electron Densities and Reduced Density Matrices*, (J. Cioslowski, ed.), Plenum, New York, 2000, p. 139.
35. P. Ziesche, Cumulant expansions of reduced densities, reduced density matrices, and Green's functions, in *Many-Electron Densities and Reduced Density Matrices*, (J. Cioslowski, ed.), Plenum, New York, 2000, p. 33.
36. P. Ziesche and F. Tasnádi, *Int. J. Quantum Chem.* **100**, 495 (2004).
37. L. Lain, A. Torre, and R. Bochicchio, *J. Chem. Phys.* **117**, 5497 (2002).
38. J. E. Harriman, *Phys. Rev. A* **65**, 052507 (2002).
39. R. Kubo, *J. Phys. Soc. (Japan)* **17**, 1100 (1962).
40. R. J. Bartlett and G. D. Purvis, *Int. J. Quantum Chem.* **14**, 561 (1978).
41. R. J. Bartlett and G. D. Pervis III, *Phys. Scr.* **21**, 255 (1980).
42. R. J. Bartlett, Coupled-cluster theory: an overview of recent developments, in *Modern Electronic Structure Theory, Part II*, (D. R. Yarkony, ed.), World Scientific, Rivers Edge, NJ, 1995, p. 1047.
43. J. M. Herbert and J. E. Harriman, *J. Chem. Phys.* **117**, 7464 (2002).
44. M. Nooijen, M. Wladyslawski, and A. Hazra, *J. Chem. Phys.* **118**, 4832 (2003).
45. P.-O. Löwdin, *Phys. Rev.* **97**, 1490 (1955).
46. J. M. Herbert, *Reconstructive Approaches to One- and Two-Electron Density Matrix Theory*, Ph.D. thesis, University of Wisconsin, Madison, WI, 2003.
47. J. M. Herbert and J. E. Harriman, *Phys. Rev. A* **65**, 022511 (2002).
48. J. E. Harriman, *Phys. Rev. A* **19**, 1893 (1979).
49. K. Yasuda, *Phys. Rev. A* **65**, 052121 (2002).
50. W. Kutzelnigg, *J. Chem. Phys.* **77**, 3081 (1982).
51. W. Kutzelnigg and S. Koch, *J. Chem. Phys.* **79**, 4315 (1983).
52. W. Kutzelnigg, Quantum chemistry in Fock space, in *Aspects of Many-Body Effects in Molecules and Extended Systems* (D. Mukherjee, ed.), Volume 50 of *Lecture Notes in Chemistry*, Springer-Verlag, Berlin, 1989, p. 35.
53. T. Arai, *J. Chem. Phys.* **33**, 95 (1960).
54. P. E. Maslen, C. Oshsenfeld, C. A. White, and M. Head-Gordon, *J. Phys. Chem. A* **102**, 2215 (1998).
55. P. Ziesche, *Solid State Commun.* **82**, 597 (1992).
56. A. J. Coleman and I. Absar, *Int. J. Quantum Chem.* **18**, 1279 (1980).
57. E. H. Lieb, *Phys. Rev. Lett.* **46**, 457 (1981); erratum: **47**, 69 (1981).
58. M. Levy, Correlation energy functionals of one-matrices and Hartree–Fock densities, in *Density Matrices and Density Functionals* (R. Erdahl and V. H. Smith, Jr., ed.), Reidel, Dordrecht, The Netherlands, 1987, p. 479.
59. C. de Dominicis and P. C. Martin, *J. Math. Phys.* **5**, 14 (1964); *ibid.* **5**, 31 (1964).

60. S. Weinberg, *Phys. Rev.* **133**, B232 (1964).
61. S. Weinberg, *The Quantum Theory of Fields, Vol. I: Foundations*, Cambridge University Press, Cambridge, 1995.
62. F. Harris, *Int. J. Quantum Chem.* **90**, 105 (2002).
63. R. D. Mattuck, *A Guide to Feynman Diagrams in the Many-Body Problem*, 2nd ed., McGraw-Hill, New York, 1976.
64. F. E. Harris, H. J. Monkhorst, and D. L. Freeman, *Algebraic and Diagrammatic Methods in Many-Fermion Theory*, Oxford University Press, New York, 1992.
65. M. Nooijen, *J. Chem. Phys.* **111**, 8356 (1999).
66. D. A. Mazziotti and R. M. Erdahl, *Phys. Rev. A* **63**, 042113 (2001).
67. C. Garrod and J. Percus, *J. Math. Phys.* **5**, 1756 (1964).
68. P. Piecuch and K. Kowalski, In search of the relationship between multiple solutions characterizing coupled-cluster theories, in *Computational Chemistry: Reviews of Current Trends*, Vol. 5 (J. Leszczynski, ed.), World Scientific, New York, 2000, p. 1.
69. J. M. Herbert, *Phys. Rev. A* **66**, 052502 (2002).
70. R. D. Mattuck and A. Theumann, *Adv. Phys.* **20**, 721 (1971).
71. S.-J. Wang, W. Zuo, and W. Cassing, *Nucl. Phys. A* **573**, 245 (1994).

CHAPTER 11

GENERALIZED NORMAL ORDERING, IRREDUCIBLE BRILLOUIN CONDITIONS, AND CONTRACTED SCHRÖDINGER EQUATIONS

WERNER KUTZELNIGG

Lehrstuhl für Theoretische Chemie Ruhr-Universität Bochum, D-44780 Bochum, Germany

DEBASHIS MUKHERJEE

Department of Physical Chemistry, Indian Association for the Cultivation of Science, Calcutta 700 032, India and Jawaharlal Centre for Advanced Scientific Research, Bangalore, India

CONTENTS

I. Introduction
II. Many-Body Theory in Fock-Space Formulation
   A. Excitation Operators
   B. k-Particle Density Matrices
   C. Spin-Free Excitation Operators and k-Particle Density Matrices
   D. Cumulants of the k-Particle Density Matrices
   E. Properties of Density Cumulants
   F. Bounds and Other Inequalities
   G. Density Cumulants for Degenerate States
   H. Intrinsic Definition of the Correlation Energy
III. Generalized Normal Ordering
   A. Particle–Hole Formalism
   B. Particle–Hole Formalism in an Arbitrary Basis
   C. Normal Ordering with Respect to Arbitrary Reference Function
   D. Generalized Wick Theorem
   E. Diagrammatic Representation
   F. Hamiltonion in Generalized Normal Order

Reduced-Density-Matrix Mechanics: With Application to Many-Electron Atoms and Molecules, A Special Volume of Advances in Chemical Physics, Volume 134, edited by David A. Mazziotti. Series editor Stuart A. Rice. Copyright © 2007 John Wiley & Sons, Inc.

IV. Stationarity Conditions for the Energy
   A. Contracted Schrödinger Equations
   B. k-particle Brillouin Conditions
   C. Irreducible Contracted Schrödinger Equations
   D. Irreducible Brillouin Conditions
V. Toward the Solution of the IBC_k and CSE_k
   A. General Considerations
   B. One-Particle Approximation
   C. Two-Particle Approximation
   D. Perturbative Analysis
VI. Conclusions
Acknowledgments
References

I. INTRODUCTION

A product of annihilation and creation operators is said to be in *normal order* if all creation operators are left of all annihilation operators. Any Fock-space operator can, according to Wick's theorem [1], be expanded into a sum of normal-ordered operators. If one wants to privilege a particular single Slater determinant *reference function* Φ, one can redefine normal ordering with respect to Φ as a *physical vacuum*, introducing *hole creation* and *hole annihilation* operators. It is, however, possible [2, 3], though not yet very popular, to generalize the concept of normal ordering with respect to arbitrary reference functions Ψ, and also to formulate a generalization of Wick's theorem for these. In this generalized formulation *contractions* appear, which involve the *density cumulants* λ_k corresponding to Ψ. For a single Slater determinant reference function, this reduces to the traditional particle–hole formalism with only one-particle or one-hole contractions.

Formulating conditions for the energy to be stationary with respect to variations of the wavefunction Ψ in this generalized normal ordering, one is led to the *irreducible Brillouin conditions* and *irreducible contracted Schrödinger equations*, which are conditions on the one-particle density matrix and the k-particle cumulants λ_k, and which differ from their traditional counterparts (even after *reconstruction* [4]) in being strictly separable (size consistent) and describable in terms of *connected diagrams* only.

In terms of these conditions, a k-particle hierarchy of approximations can be defined, with Hartree–Fock as the one-particle approximation for closed-shell states. Unfortunately, the stationarity conditions do not determine the λ_k fully, and for their construction additional information is required, which essentially guarantees n-representability. Nevertheless, the k-particle hierarchy based on the irreducible stationarity conditions opens a promising way for the solution of the n-electron problem.

II. MANY-BODY THEORY IN FOCK-SPACE FORMULATION

A. Excitation Operators

In our formalism [5–9] *excitation operators* play a central role. Let an *orthonormal basis* $\{\psi_p\}$ of *spin orbitals* be given. This basis has usually a finite dimension d, but it should be chosen such that in the limit $d \to \infty$ it becomes complete (in the so-called first Sobolev space [10]). We start from *creation* and *annihilation operators* for the ψ_p in the usual way, but we use a tensor notation, in which *subscripts* refer to annihilation and *superscripts* to creation:

$$a_p; \quad a^q = a_q^\dagger \tag{1}$$

These operators satisfy the anticommutation relations

$$[a_p, a_q]_+ = 0; \quad [a^p, a^q]_+ = 0; \quad [a_p, a^q]_+ = \delta_p^q \tag{2}$$

The Kronecker delta is written here in a tensor notation. One can define excitation operators as *normal products* (or products in *normal order*) of the same number of creation and annihilation operators (*normal order* in the original sense means that all creation operators have to be on the left of all annihilation operators).

$$a_q^p = a^p a_q \tag{3}$$

$$a_{rs}^{pq} = a^q a^p a_r a_s \tag{4}$$

$$a_{stu}^{pqr} = a^r a^q a^p a_s a_t a_u; \ldots \tag{5}$$

These operators are *particle-number conserving*, that is, action of any excitation operator on an n-electron wavefunction (with n arbitrary) leads again to an n-electron wavefunction (or deletes it).

In order to define excitation operators, one need not start from the creation and annihilation operators; one can instead simply require that action of, for example, a_q^p on a Slater determinant Φ with ψ_q occupied and (for $p \neq q$) ψ_p unoccupied replaces ψ_q by ψ_p. Otherwise it annihilates Φ.

Any *particle-number conserving operator* can be expanded in the a_q^p, a_{rs}^{pq}, and so on. Let a Hamiltonian defined in *configuration space* be given:

$$H_n = H(1, 2, \ldots, n) = \sum_{k=1}^n h(k) + \sum_{k<l=1}^n g(k,l); \quad g(k,l) = \frac{1}{r_{kl}} \tag{6}$$

If we define the matrix elements of the one-electron and two-electron parts of H

$$h_q^p = \langle \psi_q | h | \psi_p \rangle \tag{7}$$

$$g_{rs}^{pq} = \langle \psi_r(1)\psi_r(2) | g(1,2) | \psi_p(1)\psi_q(2) \rangle \tag{8}$$

then the *Fock-space Hamiltonian H*, corresponding to H_n, is

$$H = h_q^p a_p^q + \tfrac{1}{2} g_{rs}^{pq} a_{pq}^{rs} \tag{9}$$

where the Einstein summation convention over repeated indices has been implied [6].

We keep in mind the option to extend the one-electron basis, in terms of which H is defined, to the limit where it becomes complete [10].

The Fock-space Hamiltonian H is equivalent to the configuration-space Hamiltonian H_n insofar as both have the same matrix elements between n-electron Slater determinants. The main difference is that H has eigenstates of arbitrary particle number n: it is, in a way, the direct sum of all H_n. Another difference, of course, is that H_n is defined independently of a basis and hence does not depend on the dimension of the latter. One can also define a basis-independent Fock-space Hamiltonian H, in terms of *field operators* [11], but this is not very convenient for our purposes.

Any *product* of two or more excitation operators can be written as a *sum* of excitation operators, for example,

$$a_q^p a_s^r = a_{qs}^{pr} + \delta_q^r a_s^p \tag{10}$$

$$a_q^p a_{tu}^{rs} = a_{qtu}^{prs} + \delta_q^r a_{tu}^{ps} + \delta_q^s a_{tu}^{rp} \tag{11}$$

(The Kronecker δ in the tensor notation has an obvious meaning.) Each product gives rise to the *normal product* (maximum excitation rank), that is, the first term on the right-hand side (rhs) of Eqs. (10), (11), and so on and all possible *contractions*. The latter involve an upper right and a lower left label. In the contractions the original pairing of upper and lower labels must be kept as much as possible. The relations (10) and (11) are generalizations of *Wick's theorem* [1], which was originally formulated for general products of creation and annihilation operators (a_p and a^q) rather than for products of excitation operators. Note that in our formulation of the generalized Wick's theorem there is *no sign rule* to be observed (unlike in the traditional Wick theorem). Of course Eqs. (10) and (11) follow directly from the anticommutation relations, but they can alternatively be derived from the indicated "direct" definition of the excitation operators.

B. *k*-Particle Density Matrices

Consider a state described by the wavefunction Ψ, normalized to unity. Then we define the *k*-particle *(reduced) density matrices* as expectation values of the excitation operators:

$$\boldsymbol{\gamma} = \boldsymbol{\gamma}_1: \gamma_q^p = \langle \Psi | a_q^p | \Psi \rangle \tag{12}$$

$$\boldsymbol{\gamma}_2: \gamma_{rs}^{pq} = \langle \Psi | a_{rs}^{pq} | \Psi \rangle \tag{13}$$

$$\boldsymbol{\gamma}_3: \gamma_{stu}^{pqr} = \langle \Psi | a_{stu}^{pqr} | \Psi \rangle \tag{14}$$

We use boldface letters like $\boldsymbol{\gamma}_2$ for the respective full matrices. The matrix elements are defined in terms of *spin-orbital* labels. For the normalization one easily gets

$$\mathrm{Tr}\, \boldsymbol{\gamma}_1 = \gamma_p^p = n \tag{15}$$

$$\mathrm{Tr}\, \boldsymbol{\gamma}_2 = \gamma_{pq}^{pq} = n(n-1) \tag{16}$$

$$\mathrm{Tr}\, \boldsymbol{\gamma}_k = n!/(n-k)! \tag{17}$$

These definitions are easily generalized from a pure state, described by Ψ, to *ensemble states*, described by a system density matrix $\mathcal{P}$, for which an expectation value is

$$\langle A \rangle = \mathrm{Tr}\{A\mathcal{P}\} = \sum_p c_p \langle \Psi_p | A | \Psi_p \rangle; \quad \mathcal{P} = \sum_p c_p | \Psi_p \rangle \langle \Psi_p |$$

$$\sum_p c_p = 1; \quad \text{for all } c_p > 0 \tag{18}$$

It is convenient to use a one-electron basis of *natural spin orbitals* (NSOs) in terms of which $\boldsymbol{\gamma}$ is diagonal.

$$\gamma_q^p = n_p \delta_q^p; \quad 0 \leq n_p \leq 1 \tag{19}$$

One calls n_p the *occupation number* of the pth NSO.

A proof of the bounds for the occupation numbers will be given in Section II. F.

Expectation values of (particle-number conserving) operators are easily expressed in terms of the density matrices. For example, for the energy

$$E = \langle \Psi | H | \Psi \rangle = h_q^p \gamma_p^q + \tfrac{1}{2} g_{rs}^{pq} \gamma_{pq}^{rs} \tag{20}$$

C. Spin-Free Excitation Operators and k-Particle Density Matrices

Most Hamiltonians of physical interest are spin-free. Then the matrix elements in Eq. (9) depend only on the space part of the spin orbitals and vanish for different spin by integration over the spin part. Then it is recommended to eliminate the spin and to deal with spin-free operators only. We start with a basis of spin-free orbitals φ_P, from which we construct the spin orbitals $\varphi_P \alpha$ and $\varphi_P \beta$. All excitation operators carry orbital labels (capital letters) and spin labels

(Greek letters). We define spin-free excitation operators carrying only orbital labels, by summation over spin

$$E_Q^P = a_{Q\alpha}^{P\alpha} + a_{Q\beta}^{P\beta} \tag{21}$$

$$E_{RS}^{PQ} = a_{R\alpha S\alpha}^{P\alpha Q\alpha} + a_{R\alpha S\beta}^{P\alpha Q\beta} + a_{R\beta S\alpha}^{P\beta Q\alpha} + a_{R\beta S\beta}^{P\beta Q\beta}; \ldots \tag{22}$$

It would be in the spirit of a systematic formulation with lowercase letters for spin-orbital labels and capital letters for labels of spin-free orbitals, to choose the symbols $A_Q^P \ldots$ rather than $E_Q^P \ldots$. We use $E_Q^P \ldots$ nevertheless for the spin-free counterparts of the a_q^p, in particular, since the symbol E_Q^P has some tradition [13] (though *not* in a tensor notation). The E_Q^P are often called *generators of the unitary group* U(n) [13].

The Fock-space Hamiltonian, Eq. (9), then becomes

$$H = h_Q^P E_P^Q + \tfrac{1}{2} g_{RS}^{PQ} E_{PQ}^{RS} \tag{23}$$

with matrix elements over spin-free orbitals in analogy to Eq. (8).

Spin-free density matrices are obtained as

$$\Gamma_1 : \Gamma_Q^P = \gamma_{Q\alpha}^{P\alpha} + \gamma_{Q\beta}^{P\beta} = \langle \Psi | E_Q^P | \Psi \rangle \tag{24}$$

$$\Gamma_2 : \Gamma_{RS}^{PQ} = \gamma_{R\alpha S\alpha}^{P\alpha Q\alpha} + \gamma_{R\alpha S\beta}^{P\alpha Q\beta} + \gamma_{R\beta S\alpha}^{P\beta Q\alpha} + \gamma_{R\beta S\beta}^{P\beta Q\beta} = \langle \Psi | E_{RS}^{PQ} | \Psi \rangle \tag{25}$$

The spin-free one-particle density matrix $\Gamma_1 = \Gamma$ is diagonal in the basis of the (spin-free) *natural orbitals* (NOs)

$$\Gamma_Q^P = n_P \delta_Q^P \tag{26}$$

The occupation numbers n_P lie between 0 and 2:

$$0 \leq n_P \leq 2 \tag{27}$$

The spin-free two-particle excitation operators E_{RS}^{PQ} and density matrices Γ_{RS}^{PQ} are symmetric with respect to simultaneous exchange of the upper and lower indices, but neither symmetric nor antisymmetric with respect to exchange of either upper or lower indices separately:

$$E_{RS}^{PQ} = E_{SR}^{QP}; \quad E_{RS}^{QP} \neq E_{RS}^{PQ}; \quad E_{RS}^{QP} \neq -E_{RS}^{PQ} \tag{28}$$

$$\Gamma_{RS}^{PQ} = \Gamma_{SR}^{QP}; \quad \Gamma_{RS}^{QP} \neq \Gamma_{RS}^{PQ}; \quad \Gamma_{RS}^{QP} \neq -\Gamma_{RS}^{PQ} \tag{29}$$

Note that E_{RS}^{PQ} and Γ_{RS}^{PQ} usually do not vanish for $P = Q$ and/or $R = S$. Two electrons can be in the *same* spin-free orbital. One can define operators and density matrices that are symmetry adapted to the separate exchange of upper or lower labels [8].

$$(E^+)^{PQ}_{RS} = \tfrac{1}{2}(E^{PQ}_{RS} + E^{QP}_{RS}); \quad (E^-)^{PQ}_{RS} = \tfrac{1}{2}(E^{PQ}_{RS} - E^{QP}_{RS}) \qquad (30)$$

$$(\Gamma^+)^{PQ}_{RS} = \tfrac{1}{2}(\Gamma^{PQ}_{RS} + \Gamma^{QP}_{RS}); \quad (\Gamma^-)^{PQ}_{RS} = \tfrac{1}{2}(\Gamma^{PQ}_{RS} - \Gamma^{QP}_{RS}) \qquad (31)$$

One finds [14] that (for eigenstates of $\hat{S}^2$ with eigenvalue $S(S+1)$)

$$\mathrm{Tr}(\Gamma^+) = (\Gamma^+)^{PQ}_{PQ} = \tfrac{1}{4}n(n+2) - S(S+1) \qquad (32)$$

$$\mathrm{Tr}(\Gamma^-) = (\Gamma^-)^{PQ}_{PQ} = \tfrac{3}{4}n(n+2) + S(S+1) \qquad (33)$$

The eigenstates of Γ^+ are symmetric spin-free two-electron functions corresponding to *singlet pairs*; those of Γ^- are antisymmetric spin-free two-electron functions corresponding to *triplet pairs*. $\mathrm{Tr}(\Gamma^+)$ is the probability that two electrons are coupled to a singlet pair, while $\mathrm{Tr}(\Gamma^-)$ is the probability that two electrons are coupled to a triplet pair.

For spin-free k-particle excitation operators and density matrices, linear combinations that transform as *irrep* of the symmetric group $\mathcal{S}_k$ can be defined in an analogous way [15].

The expectation value of the Hamiltonian in Eq. (23) becomes

$$E = h^P_Q \Gamma^Q_P + \tfrac{1}{2} g^{PQ}_{RS} \Gamma^{RS}_{PQ} = h^P_Q \Gamma^Q_P + \tfrac{1}{2}(g^+)^{PQ}_{RS}(\Gamma^+)^{RS}_{PQ} + \tfrac{1}{2}(g^-)^{PQ}_{RS}(\Gamma^-)^{RS}_{PQ} \qquad (34)$$

$$(g^+)^{PQ}_{RS} = g^{PQ}_{RS} + g^{QP}_{RS}; \quad (g^-)^{PQ}_{RS} = g^{PQ}_{RS} - g^{QP}_{RS} \qquad (35)$$

There is also a spin-free Wick theorem [3, 5, 16],

$$E^P_Q E^R_S = E^{PR}_{QS} + \delta^R_Q E^P_S \qquad (36)$$

$$E^P_Q E^{RS}_{TU} = E^{PRS}_{QTU} + \delta^R_Q E^{PS}_{TU} + \delta^S_Q E^{RP}_{TU} \qquad (37)$$

that formally agrees with its spin-dependent counterpart, Eqs. (10) and (11).

D. Cumulants of the k-Particle Density Matrices

The k-particle density matrices, in particular, γ_1 and γ_2, are extremely useful quantities. They are much simpler than the wavefunction but contain all relevant information. Yet, except γ_1, they have one important drawback. They are not additively separable (extensive). For a supersystem AB consisting of two noninteracting subsystems A and B, we have

$$\gamma_k(AB) \neq \gamma_k(A) + \gamma_k(A); \quad \text{for } k \geq 2 \tag{38}$$

The equality sign will hold for the cumulants, which we introduce now.

The *cumulant* λ_2 — with matrix elements λ_{rs}^{pq} — of the two-particle density matrix γ_2 — with matrix elements γ_{rs}^{pq} — is the *difference* between γ_2 and *what one expects for independent particles* that obey Fermi statistics [71]:

$$\lambda_2 : \lambda_{rs}^{pq} = \gamma_{rs}^{pq} - \gamma_r^p \gamma_s^q + \gamma_s^p \gamma_r^q \tag{39}$$

Cumulants of any order can be defined via a *generating function* [17, 18]. Consider the expectation value of the exponential of an arbitrary one-particle operator $\tilde{k}$:

$$A = \langle \Phi | : \exp \hat{k} : | \Phi \rangle = \langle \Phi | 1 + \hat{k} + \tfrac{1}{2} : \hat{k}^2 : + \cdots | \Phi \rangle$$
$$\hat{k} = k_q^p a_p^q \tag{40}$$

Double dots (: $\cdots$:) mean normal products (with respect to the genuine vacuum), for example, $a_q^p a_s^r := a_{qs}^{pr}$. We get

$$A = 1 + k_q^p \gamma_p^q + \tfrac{1}{2} k_q^p k_s^r \gamma_{pr}^{qs} + \cdots$$
$$= 1 + k_q^p \gamma_p^q + \frac{1}{2} \sum_{p<r} (k_q^p k_s^r - k_s^p k_q^r) \gamma_{pr}^{qs} + \cdots \tag{41}$$

and realize that γ_p^q is the coefficient of k_q^p, γ_{pr}^{qs} is the coefficient of $\tfrac{1}{2}(k_q^p k_s^r - k_s^p k_q^r)$, and so on. Let us now define the *antisymmetrized logarithm* of an expression like A, in terms of the Taylor expansion of $\ln(1+x)$, but with products of γ-factors replaced by the corresponding antisymmetrized products: for example,

$$\gamma_p^q \gamma_r^s \to \det\{\gamma_p^q \gamma_r^s\} = \gamma_p^q \gamma_r^s - \gamma_r^q \gamma_p^s \tag{42}$$

$$\gamma_p^q \gamma_r^s \gamma_t^u \to \det\{\gamma_p^q \gamma_r^s \gamma_t^u\} \tag{43}$$

$$\gamma_p^q \gamma_{rt}^{su} \to \sum (-1)^P \gamma_p^q \gamma_{rt}^{su} = \gamma_p^q \gamma_{rt}^{su} + \gamma_r^s \gamma_{pt}^{qu} + \gamma_t^u \gamma_{pr}^{qs}$$
$$- \gamma_r^q \gamma_{pt}^{su} - \gamma_t^q \gamma_{rp}^{su} - \gamma_t^s \gamma_{pr}^{qu} - \gamma_p^s \gamma_{pr}^{qu} - \gamma_p^u \gamma_{tr}^{qs} - \gamma_r^u \gamma_{pt}^{qs} \tag{44}$$

In the last expression the sum goes over all nontrivial partitions of the lower labels, and of the upper labels, with a sign factor $(-1)^P$ depending on the parity

P of the partition. The determinant $\det\{\cdots\}$ is a special case for the partition into single elements, that is, a permutation. Then

$$B = \ln_a A = k_q^p \gamma_p^q + \tfrac{1}{4}(k_q^p k_s^r - k_s^p k_q^r)\{\gamma_{pr}^{qs} - \gamma_p^q \gamma_r^s + \gamma_r^q \gamma_p^s\} + \cdots$$
$$= k_q^p \gamma_p^q + \tfrac{1}{4}(k_q^p k_s^r - k_s^p k_q^r)\lambda_{pr}^{qs} + \cdots \tag{45}$$

Now γ_p^q is the coefficient of k_q^p, and λ_{pr}^{qs} that of $\tfrac{1}{2}(k_q^p k_s^r - k_s^p k_q^r)$, and so on. So B is the generating function for the λ_k in the same sense as A is the generating function for the γ_k. To take first the expectation value of an exponential and then the logarithm is a standard way to arrive at cumulants [19].

For the higher-order cumulants λ_3 and λ_4 we get

$$\lambda_3 : \lambda_{qsu}^{prt} = \gamma_{qsu}^{prt} - \sum(-1)^P \gamma_q^p \lambda_{su}^{rt} - \det\{\gamma_q^p \gamma_s^r \gamma_u^t\} \tag{46}$$

$$\lambda_4 : \lambda_{qsuw}^{prtv} = \gamma_{qsuw}^{prtv} - \sum(-1)^P \gamma_q^p \lambda_{suw}^{rtv} - \frac{1}{2}\Sigma(-1)^P \lambda_{qs}^{pr} \lambda_{uw}^{tv}$$
$$- \sum(-1)^P \gamma_q^p \gamma_s^r \lambda_{uw}^{tv} - \det\{\gamma_q^p \gamma_s^r \gamma_u^t \gamma_w^v\} \tag{47}$$

with the sums going over all partitions of lower labels, as explained after Eq. (44). A more compact formulation is in terms of the Grassmann (or wedge) products

$$\lambda_2 = \gamma_2 - \gamma_1 \wedge \gamma_1 \tag{48}$$
$$\lambda_3 = \gamma_3 - \gamma_1 \wedge \lambda_2 - \gamma_1 \wedge \gamma_1 \wedge \gamma_1 \tag{49}$$
$$\lambda_4 = \gamma_4 - \gamma_1 \wedge \lambda_3 - \lambda_2 \wedge \lambda_2 - \gamma_1 \wedge \gamma_1 \wedge \lambda_2 - \gamma_1 \wedge \gamma_1 \wedge \gamma_1 \wedge \gamma_1 \tag{50}$$

The two-particle cumulant is a *correlation increment*. It describes *Coulomb correlation*, since the Fermi correlation is already contained in the description in terms of γ_1 only. In terms of the cumulants, the energy expectation value can be written

$$E = \tfrac{1}{2}(h_q^p + f_q^p)\gamma_p^q + \tfrac{1}{2}\bar{g}_{rs}^{pq}\lambda_{pq}^{rs} \tag{51}$$

$$f_q^p = h_q^p + \bar{g}_{qs}^{pr}\gamma_r^s; \quad \bar{g}_{qs}^{pr} = g_{qs}^{pr} - g_{sq}^{pr} \tag{52}$$

Here the *generalized Fock operator* f with matrix elements f_q^p appears for the first time. It looks familiar and resembles the Fock operator of Hartree–Fock theory. However, now the γ_r^s are matrix elements of the *exact* one-particle density matrix

γ_1, which, unlike in Hartree–Fock theory, is *not* assumed to be *idempotent* (corresponding to a single Slater determinant).

E. Properties of Density Cumulants

The cumulants of the density matrices, for short *density cumulants*, have a few nice properties [17]. We note in particular the *Hermiticity*, for example,

$$\lambda_{rs}^{pq} = (\lambda_{pq}^{rs})^* \tag{53}$$

and the *antisymmetry*,

$$\lambda_{rs}^{pq} = -\lambda_{rs}^{qp} = -\lambda_{sr}^{pq} = \lambda_{sr}^{qp} \tag{54}$$

Density cumulants are *separable* in the following sense.

For $\Psi = \mathcal{A}\{\Psi_A(1,2,\ldots,n_A)\Psi_B(n_A+1,\ldots,n_A+n_B)\}$ with Ψ_A and Ψ_B *strongly orthogonal*, and $\mathcal{A}$ the *antisymmetrizer*, $\lambda_{rs}^{pq} = 0$ unless *all labels* refer either to subsystem A or B.

$$\lambda_{rs}^{pq} = (\lambda_A)_{rs}^{pq} + (\lambda_B)_{rs}^{pq} \tag{55}$$

The cumulants have the important property (and this holds for the cumulants λ_k of arbitrary particle rank) of being *additively separable*. For a noninteracting supersystem AB we have

$$\lambda_k(AB) = \lambda_k(A) + \gamma_k(B); \quad \text{for all } k \tag{56}$$

One further sees easily that (in an NSO basis) matrix elements of a cumulant are nonvanishing *only* if *all* its labels refer to *partially occupied (active)* NSOs with occupation number *different from 0 or 1*.

$$\lambda_{rs}^{pq} = 0, \quad \text{in an NSO basis if any } n_p = 0 \text{ or } = 1 \tag{57}$$

There are *trace relations* [20] like

$$\mathrm{Tr}(\lambda_1) = \sum_k n_k = n \tag{58}$$

$$\mathrm{Tr}(\lambda_2) = \mathrm{Tr}(\gamma_1^2 - \gamma_1) = \sum_k (n_k^2 - n_k) = O(n) \tag{59}$$

$$\mathrm{Tr}(\lambda_3) = \mathrm{Tr}(-4\gamma_1^3 + \gamma_1^2 - 2\gamma_1) \tag{60}$$

$$= 6\sum_k n_k(n_k - \tfrac{1}{2})(n_k - 1) = O(n) \tag{61}$$

and *partial trace relations*

$$\lambda_{qr}^{pr} = -\gamma_q^p + \gamma_r^p \gamma_q^r = (\gamma^2 - \gamma)_q^p \tag{62}$$

$$\lambda_{qst}^{prt} = 2\lambda_{qs}^{pr} - \gamma_t^p \lambda_{sq}^{rt} - \gamma_t^r \lambda_{qs}^{pt} - \gamma_q^t \lambda_{ts}^{pr} - \gamma_s^t \lambda_{qt}^{pr} \tag{63}$$

$$\lambda_{qrst}^{prst} = (-2\gamma^3 + 4\gamma^2 - 2\gamma)_q^p - \gamma_t^r \lambda_{rq}^{tp} - \gamma_r^t \lambda_{qt}^{pr} \tag{64}$$

One notes that the traces of all λ_k are of $O(n)$, while the traces of the corresponding γ_k are of $O(n^k)$.

Particularly noteworthy is the *particle–hole symmetry*. Let us define (one- and two-) *hole density matrices* [17]

$$\eta_q^p = \langle \Phi | a_q a^p | \Phi \rangle = \delta_q^p - \gamma_q^p \tag{65}$$

$$\eta_{rs}^{pq} = \langle \Phi | a_s a_r a^p a^q | \Phi \rangle$$
$$= \delta_r^p \delta_s^q - \delta_s^p \delta_r^q - \delta_r^p \gamma_s^q - \gamma_r^p \delta_s^q + \delta_s^p \gamma_r^q + \gamma_s^p \delta_r^q + \gamma_{rs}^{pq} \tag{66}$$

$$\eta_q^p = \delta_q^p (1 - n_p) \quad \text{in an NSO basis} \tag{67}$$

These η_m matrices have the *same* irreducible components (*cumulants*) as the corresponding γ_m matrices, just with γ_q^p replaced by η_q^p and with some sign changes, for example,

$$\eta_{rs}^{pq} = \lambda_{rs}^{pq} + \eta_r^p \eta_s^q - \eta_s^p \eta_r^q \tag{68}$$

$$\eta_{stu}^{pqr} = -\lambda_{stu}^{pqr} + \eta_s^p q r_{tu} + \cdots + \eta_s^p \eta_t^q \eta_u^r + \cdots \tag{69}$$

One can further define a *particle–hole density matrix*, which also has the same two-particle cumulant

$$\beta_{r,s}^{p,q} = \langle \Phi | (a_r^p - \gamma_r^p)(a_s^q - \gamma_s^q) | \Phi \rangle = \eta_r^q \gamma_s^p + \lambda_{rs}^{pq} \tag{70}$$

The special cases of APSG (antisymmetrized product of strongly orthogonal geminals) and AGP (antisymmetrized geminal power) functions have been analyzed [17]. Among other things, conditions were found for vanishing of the three-particle cumulant λ_3 [17].

F. Bounds and Other Inequalities

Some important inequalities for the matrix elements of γ and λ_2 can be derived. Actually all γ_k with $k < n$ must be *nonnegative*.

Let us express γ_1 in terms of its NSOs. Then

$$n_p = \gamma_p^p = \langle \Psi | a^p a_p \Psi \rangle = \langle a_p \Psi | a_p \Psi \rangle \geq 0 \tag{71}$$

(This is formulated here for a pure state, but the generalization to an ensemble state is straightforward.) Also the one-hole density matrix must be nonnegative [21]:

$$1 - n_p = \eta_p^p = \langle \Psi | a_p a^p \Psi \rangle = \langle a^p \Psi | a^p \Psi \rangle \geq 0 \tag{72}$$

hence

$$0 \leq n_p \leq 1 \tag{73}$$

This is necessary for pure-state n-representability. Together with

$$\sum_p n_p = n \tag{74}$$

it is necessary and sufficient for ensemble n-representability.

For $k = 2$ we have the nonnegativity of the diagonal elements of γ_2 [21, 22]:

$$\gamma_{pq}^{pq} = \langle a_p a_q \Psi | a_p a_q \Psi \rangle \geq 0 \tag{75}$$

Similar inequalities hold for the two-hole and particle–hole density matrices [21, 22]:

$$\eta_{pq}^{pq} = \langle a^p a^q \Psi | a^p a^q \Psi \rangle \geq 0 \tag{76}$$

$$\beta_{p,q}^{q,p} = \langle \tilde{a}_q^p \Psi | \tilde{a}_q^p \Psi \rangle \geq 0 \tag{77}$$

In the literature on n-representability [21, 22], the nonnegativity of γ_2, η_2, and β_2 is referred to as D-, Q-, and G-conditions respectively [22]. Also known are a B- and a C-condition [21, 22], but these are implied by the G-condition [22, 23].

From the nonnegativity of γ_2, η_2, and β_2, some important inequalities for the diagonal matrix elements of λ_2 can be derived:

$$0 \leq \gamma_{pq}^{pq} = \lambda_{pq}^{pq} + \gamma_p^p \gamma_q^q - \gamma_q^p \gamma_p^q \tag{78}$$

$$0 \leq \eta_{pq}^{pq} = \lambda_{pq}^{pq} + \eta_p^p \eta_q^q - \eta_q^p \eta_p^q \tag{79}$$

$$0 \leq \beta_{q,p}^{p,q} = \lambda_{pq}^{qp} + \eta_q^q \gamma_p^p = -\lambda_{pq}^{pq} + \eta_q^q \gamma_p^p \tag{80}$$

$$0 \leq \beta_{q,p}^{q,p} = \lambda_{pq}^{qp} + \eta_p^p \gamma_q^q = -\lambda_{pq}^{pq} + \eta_p^p \gamma_q^q \tag{81}$$

Especially in an NSO basis we obtain

$$\lambda_{pq}^{pq} \geq \max\{-n_p n_q, -(1-n_p)(1-n_q)\} \geq -\tfrac{1}{4} \tag{82}$$

$$\lambda_{pq}^{pq} \leq \min\{n_p(1-n_q), n_q(1-n_p)\} \leq +\tfrac{1}{4} \tag{83}$$

There are *Cauchy–Schwarz-type* relations between the diagonal and nondiagonal elements of γ_2, η_2, and β_2:

$$\gamma^{pq}_{rs}\gamma^{rs}_{pq} = \langle a_p a_q \Psi | a_r a_s \Psi \rangle \langle a_r a_s \Psi | a_p a_q \Psi \rangle$$
$$\leq \langle a_p a_p \Psi | a_p a_q \Psi \rangle \langle a_r a_s \Psi | a_r a_s \Psi \rangle = \gamma^{pq}_{pq}\gamma^{rs}_{rs} \quad (84)$$

$$\eta^{pq}_{rs}\eta^{rs}_{pq} = \langle a^p a^q \Psi | a^r a^s \Psi \rangle \langle a^r a^s \Psi | a^p a^q \Psi \rangle$$
$$\leq \langle a^p a^q \Psi | a^p a^q \Psi \rangle \langle a^r a^s \Psi | a^r a^s \Psi \rangle = \eta^{pq}_{pq}\eta^{rs}_{rs} \quad (85)$$

$$\beta^{p,r}_{q,s}\beta^{s,q}_{r,p} = \langle \tilde{a}^p_q \Psi | \tilde{a}^r_s \Psi \rangle \langle \tilde{a}^r_s \Psi | \tilde{a}^q_p \Psi \rangle$$
$$\leq \langle \tilde{a}^p_q \Psi | \tilde{a}^p_q \Psi \rangle \langle \tilde{a}^r_s \Psi | \tilde{a}^r_s \Psi \rangle = \beta^{p,q}_{q,p}\beta^{s,r}_{r,s} \quad (86)$$

Somewhat weaker conditions are obtained by summing the Cauchy–Schwarz inequalities over $\{r, s\}$, for example,

$$\gamma^{pq}_{rs}\gamma^{rs}_{pq} \leq n(n-1)\gamma^{pq}_{pq} \quad (87)$$

In Eq. (87) the equality sign holds for a *pure state of a two-electron system*. In terms of Møller–Plesset perturbation theory (see Section V. D) one gets, independently of n [24, 25],

$$\lambda^{ij}_{ij} = -\tfrac{1}{2}\lambda^{ab}_{ij}\lambda^{ij}_{ab} + O(\mu^3) \quad (88)$$
$$\lambda^{ab}_{ab} = \tfrac{1}{2}\lambda^{ij}_{ab}\lambda^{ab}_{ij} + O(\mu^3) \quad (89)$$

where μ is the perturbation parameter.

There is an interesting *pairing relation* between the eigenvalues $v^{(p)}_k$ of the p-particle density matrix γ_p (with eigenfunctions, i.e., natural p-states, $\chi^{(p)}_k$) and those $v^{(n-p)}_k$ of γ_{n-p}:

$$(n-p)!v^{(p)}_k = p!v^{(n-p)}_k \quad (90)$$

This is closely related to the possibility of writing the n-electron wavefunction Ψ as

$$\Psi(1, 2, \ldots, n) = \sum_k c_k \chi^{(p)}_k(1, 2, \ldots, p)\chi^{(n-p)}_k(p+1, \ldots, n) \quad (91)$$

$$|c_k|^2 = \frac{(n-p)!}{n!}v^{(p)}_k = \frac{p!}{n!}v^{(n-p)}_k \quad (92)$$

where $\chi_k^{(p)}$ and $\chi_k^{(n-p)}$ are *conjugate* natural p- and $n-p$ states. This is a special case [26] of a theorem derived a century ago by Erhard Schmidt [27]. From Eq. (92) and the Cauchy–Schwarz inequality one gets

$$
\begin{aligned}
c_k &= \langle \Psi | \chi_k^{(p)}(1,\ldots,p) \chi_k^{(n-p)}(p+1,\ldots,n) \rangle \\
&= \langle \mathcal{A}\Psi | \chi_k^{(p)}(1,2,\ldots,p) \chi_k^{(n-p)}(p+1,\ldots,n) \rangle \\
&= \langle \Psi | \mathcal{A}\chi_k^{(p)}(1,2,\ldots,p) \chi_k^{(n-p)}(p+1,\ldots,n) \rangle
\end{aligned}
\tag{93}
$$

$$
\begin{aligned}
|c_k|^2 &\leq \langle \chi_k^{(p)}(1,\ldots,p) \chi_k^{(n-p)}(p+1,\ldots,n) | \\
&\quad \times \mathcal{A}\chi_k^{(p)}(1,\ldots,p) \chi_k^{(n-p)}(p+1,\ldots,n) \rangle
\end{aligned}
\tag{94}
$$

where $\mathcal{A}$ is the *idempotent* antisymmetrizer

$$
\mathcal{A} = \frac{1}{n!} \sum_p \epsilon_p P = \mathcal{A}^2
\tag{95}
$$

in which the sum goes over all permutations of n elements and where ϵ_p is the parity of the permutation. Using a lemma, based on combinatorial arguments, probably first found by Sasaki [28],

$$
\langle f(1,\ldots,p)g(p+1,\ldots,n) | \mathcal{A} f(1,\ldots,p)g(p+1,\ldots,n) \rangle \leq \frac{1}{n-p+1}
\tag{96}
$$

one gets an upper bound for the eigenvalues of γ_k,

$$
v_k^{(p)} \leq \frac{n!}{(n-p+1)!} = \frac{\text{Tr } \gamma_p}{n-p-1}
\tag{97}
$$

especially

$$
n_k = v_k^{(1)} \leq 1 \tag{98}
$$

$$
v_k^{(2)} \leq n \tag{99}
$$

$$
v_k^{(3)} \leq n(n-1) \tag{100}
$$

The upper bound n for the eigenvalues $v_k^{(2)}$ of γ_2 is rather large compared to the value $v_k^{(2)} = 2$ for the nonzero eigenvalues of γ_2 for a *single Slater determinant*. However, one can construct a wavefunction, namely, an *antisymmetrized geminal power of extreme type* for which such a large eigenvalue is actually realized. Such wavefunctions play an important role in the theory of superconductivity [21]. For a *well-closed shell state* a *large eigenvalue* of γ_2 is rather unlikely. We cannot go into details here [22].

G. Density Cumulants for Degenerate States

For degenerate states a problem arises with the definition of cumulants. We consider here only *spin degeneracy*. Spatial degeneracy can be discussed on similar lines. For $S \neq 0$ there are $(2S+1)$ different M_S-values for one S. The n-particle density matrix $\rho(M_S) = |\Psi_{M_S}\rangle\langle\Psi_{M_S}|$ of a single one of these states does not transform as an *irreducible representation* (*irrep*) of the spin rotation group SU_2. However, the $(2S+1)^2$-dimensional set of n-particle density matrices and transition density matrices

$$\rho(M_S, M'_S) = |\Psi_{M_S}\rangle\langle\Psi_{M'_S}| \tag{101}$$

span a $(2S+1)^2$-dimensional *reducible* representation of SU_2. One can construct the irreducible tensor components ρ_σ with $\sigma = (0, 1, \ldots, 2S)$ as linear combinations of the $\rho(M_S, M'_S)$. Especially the (normalized) totally symmetric component is given as

$$\rho_0 = (2S+1)^{-1/2} \sum_{M_S=-S}^{S} |\Psi_{M_S}\rangle\langle\Psi_{M_S}| \tag{102}$$

Except for the normalization factor this is equal to the spin-averaged n-particle *ensemble state*. The spin-free density matrices (24), (25), and so on defined previously (in Section II.C) correspond to such an ensemble averaging.

Examples for non-totally-symmetric components in the decomposition of density matrix into *irreducible tensor components* are the one-particle *spin density* matrices:

$$(D^0)_S^P = 1/(\sqrt{2}) \, \langle\Psi|a_{S\alpha}^{P\alpha} - a_{S\beta}^{P\beta}|\Psi\rangle$$
$$(D^+)_S^P = \langle\Psi|a_{S\beta}^{P\alpha}|\Psi\rangle$$
$$(D^-)_S^P = \langle\Psi|a_{S\alpha}^{P\beta}|\Psi\rangle \tag{103}$$

which transform as a vector, that is, an irreducible tensor operator of rank 1. For a singlet state all these components vanish.

If one tries to apply the definition (39) *naively* to degenerate states, one is faced with the problem that, for example, γ_{rs}^{pq} and $\gamma_r^p \gamma_s^q$ have a different transformation behavior with respect to the spin rotation group SU_2; hence λ_{rs}^{pq} would have no acceptable transformation behavior at all. The way out of this dilemma is to define the *irreducible tensor components* of λ_{rs}^{pq} in terms of those of γ_{rs}^{pq} and γ_r^p.

The easiest and, in many respects, the most satisfactory way is to consider only the totally symmetric tensor components (i.e., the spin-free density matrices) and to define the spin-free cumulants in terms of these [17, 30]. This corresponds to replacing the considered state by an M_S-averaged ensemble.

We arrive then at the definitions

$$\Lambda^{PQ}_{RS} = \Gamma^{PQ}_{RS} - \Gamma^{P}_{R}\Gamma^{Q}_{S} + \tfrac{1}{2}\Gamma^{P}_{S}\Gamma^{Q}_{R} \tag{104}$$

$$\Lambda^{PQR}_{STU} = \Gamma^{PQR}_{STU} - \Gamma^{P}_{S}\Lambda^{QR}_{TU} - \Gamma^{Q}_{T}\Lambda^{PR}_{SU} - \Gamma^{R}_{U}\Lambda^{PQ}_{ST}$$
$$+ \tfrac{1}{2}\Gamma^{P}_{T}\Lambda^{QR}_{SU} + \tfrac{1}{2}\Gamma^{P}_{U}\Lambda^{QR}_{TS} + \tfrac{1}{2}\Gamma^{Q}_{S}\Lambda^{PR}_{TU} + \tfrac{1}{2}\Gamma^{Q}_{U}\Lambda^{PR}_{ST}$$
$$+ \tfrac{1}{2}\Gamma^{R}_{S}\Lambda^{PQ}_{UT} + \tfrac{1}{2}\Gamma^{R}_{T}\Lambda^{PQ}_{SU} - \Gamma^{P}_{S}\Gamma^{Q}_{T}\Gamma^{R}_{U}$$
$$- \tfrac{1}{4}\Gamma^{P}_{T}\Gamma^{Q}_{U}\Gamma^{R}_{S} - \tfrac{1}{4}\Gamma^{P}_{U}\Gamma^{Q}_{S}\Gamma^{R}_{T} + \tfrac{1}{2}\Gamma^{P}_{T}\Gamma^{Q}_{S}\Gamma^{R}_{U}$$
$$+ \tfrac{1}{2}\Gamma^{P}_{U}\Gamma^{Q}_{T}\Gamma^{R}_{S} + \tfrac{1}{2}\Gamma^{P}_{S}\Gamma^{Q}_{U}\Gamma^{R}_{T} \tag{105}$$

For the symmetry-adapted two-particle cumulants we get

$$(\Lambda^{+})^{PQ}_{RS} = (\Gamma^{+})^{PQ}_{RS} - \tfrac{1}{4}\Gamma^{P}_{R}\Gamma^{Q}_{S} - \tfrac{1}{4}\Gamma^{P}_{S}\Gamma^{Q}_{R} \tag{106}$$

$$(\Lambda^{-})^{PQ}_{RS} = (\Gamma^{-})^{PQ}_{RS} - \tfrac{3}{4}\Gamma^{P}_{R}\Gamma^{Q}_{S} + \tfrac{3}{4}\Gamma^{P}_{S}\Gamma^{Q}_{R} \tag{107}$$

The definitions given in this section are valid both for degenerate and nondegenerate states.

H. Intrinsic Definition of the Correlation Energy

Let us start from the energy expression (34). We write it as a sum of a one-electron energy E_1 and an electron interaction energy E_2.

$$E = E_1 + E_2 \tag{108}$$
$$E_1 = h^{P}_{Q}\Gamma^{Q}_{P} \tag{109}$$
$$E_2 = \tfrac{1}{2}\Gamma^{PQ}_{RS} g^{RS}_{PQ} = \int \frac{\varrho^{(2)}(1,2)}{2r_{12}} d\tau_1 d\tau_2 \tag{110}$$

The electron interaction energy E_2 can be further decomposed into the following contributions:

$$E_{\text{Coul}} = \int \frac{\varrho(1)\varrho(2)}{2r_{12}} d\tau_1 d\tau_2 \tag{111}$$

$$E_x = -\int \frac{\Gamma(1,2)\Gamma(2,1)}{4r_{12}} d\tau_1 d\tau_2 \tag{112}$$

$$E_{\text{corr}} = \int \frac{\lambda^{(2)}(1,2)}{2r_{12}} d\tau_1 d\tau_2 \tag{113}$$

with the cumulant $\lambda^{(2)}(1,2)$ in configuration space defined in analogy to its counterpart λ_2 in Fock space. This decomposition differs from that common in both ab initio quantum chemistry and density functional theory (DFT). It has the advantage of being *intrinsic*, that is, referring to a single wavefunction, and not depending on any approximation.

Actually the sum $E_1 + E_{\text{Coul}} + E_x$ is of the form of a Hartree–Fock energy, except that $\frac{1}{2}\Gamma$ is not necessarily idempotent. Since this sum is minimized for an idempotent $\frac{1}{2}\Gamma$, the value of this sum is slightly above the Hartree–Fock energy. So the correlation energy defined by Eq. (113) will usually be somewhat larger in absolute value than that defined by Löwdin [31]. However, one will not expect a very large difference, provided that we are in a closed-shell situation, where $\frac{1}{2}\Gamma$ is not *very far* from idempotent. The sum $E_1 + E_{\text{Coul}} + E_x$ will, unlike the Hartree–Fock energy, not satisfy a virial theorem. The correlation energy in Eq. (113) is purely a two-electron potential energy term and does not, unlike in the traditional definition, consist of two-electron and one-electron contributions, the latter even split into a kinetic and a potential part.

In density functional theory (DFT) one uses the same definition (111) for the Coulomb energy (including self-interaction) as we propose here. However, both E_1 and the exchange energy (including self-exchange) are defined differently. These are not evaluated in terms of the exact $\frac{1}{2}\Gamma$, but in terms of that corresponding to the *Kohn–Sham determinant*, which is idempotent. The errors due to this replacement are then absorbed into the correlation energy, which can therefore not be identified with that defined here.

III. GENERALIZED NORMAL ORDERING

A. Particle–Hole Formalism

We introduce the generalized normal ordering in various steps, starting with the traditional particle–hole formalism.

The particle–hole formalism has been introduced as a simplification of many-body perturbation theory for closed-shell states, for which a single Slater determinant Φ dominates and is hence privileged. One uses the labels $i, j, k, \ldots$ for spin orbitals *occupied* in Φ and $a, b, c, \ldots$ for spin orbitals unoccupied (*virtual*) in Φ.

Then one redefines the annihilation operator a_i for an *occupied* spin orbital as the *hole creation operator* $b_i^\dagger$, and the creation operator $a_i^\dagger$ for an *occupied* spin orbital as the *hole annihilation operator* b_i. The fermion operators for the *virtual* spin orbitals remain unchanged.

The essential step is to define a new *normal ordering* with respect to Φ regarded as a *physical vacuum*. Now a product of a and b operators is said to be in *normal order* with respect to Φ, if all a_a and b_i are right of all $a_a^\dagger$ and $b_i^\dagger$.

This convention is in conflict with our tensor notation, that we do not want to abandon [8]. However, what really matters is only the change of the definition of normal ordering, and this can easily be formulated in our language as well.

In fact, it turns out that the excitation operators *in normal order in the particle–hole sense* can be written as linear combinations of operators *in (the original) normal order with respect to the genuine vacuum*. We put a *tilde* on operators in the *new* normal ordering. We get then

$$\tilde{a}_q^p = a_q^p - \delta_q^p n_p \tag{114}$$

$$\tilde{a}_{rs}^{pq} = a_{rs}^{pq} - \delta_r^p n_p a_s^q - \delta_s^q n_q a_r^p + \delta_s^p n_p a_r^q + \delta_r^q n_q a_s^p + \delta_{rs}^{pq} n_p n_q \tag{115}$$

$$n_p = 1 \text{ for } \psi_p \text{ occupied in } \Phi; \qquad n_p = 0 \text{ for } \psi_p \text{ unoccupied in } \Phi \tag{116}$$

$$\delta_{rs}^{pq} = \delta_r^p \delta_s^q - \delta_s^p \delta_r^q \tag{117}$$

We note in particular that

$$\langle \Phi | \tilde{a}_q^p | \Phi \rangle = 0; \qquad \langle \Phi | \tilde{a}_{rs}^{pq} | \Phi \rangle = 0 \tag{118}$$

that is, expectation values of $\tilde{a}$ operators with respect to Φ vanish, as do the expectation values of a operators with respect to the *genuine vacuum* $|0\rangle$,

$$\langle 0 | a_q^p | 0 \rangle = 0; \qquad \langle 0 | a_{rs}^{pq} | 0 \rangle = 0 \tag{119}$$

For products of $\tilde{a}$ operators one finds again a *generalized Wick theorem* like Eqs. (10) and (11), but in addition to *particle contractions* we also have *hole contractions* and *combined contractions*, even *full contractions*, that is, simple numbers:

$$\tilde{a}_q^p \tilde{a}_s^r = \tilde{a}_{qs}^{pr} + \delta_q^r (1 - n_q) \tilde{a}_s^p - \delta_s^p n_p \tilde{a}_q^r + \delta_q^r \delta_s^p n_p (1 - n_q) \tag{120}$$

$$\tilde{a}_q^p \tilde{a}_{tu}^{rs} = \tilde{a}_{qtu}^{prs} + \delta_q^r (1 - n_q) \tilde{a}_{tu}^{ps} + \delta_q^s (1 - n_q) \tilde{a}_{tu}^{rp} - \delta_t^p n_p \tilde{a}_{qu}^{rs}$$
$$- \delta_u^p n_p \tilde{a}_{tq}^{rs} + \delta_q^r (1 - n_q) \delta_t^p n_p \tilde{a}_u^s - \delta_q^r (1 - n_q) \delta_u^p n_u a_t^s$$
$$- \delta_q^s (1 - n_p) \delta_t^p n_p \tilde{a}_u^r + \delta_q^s (1 - n_q) \delta_u^p n_p \tilde{a}_t^r \tag{121}$$

While *particle contractions* connect an upper right with a lower left label (and are associated with a factor $(1 - n_p)$), *hole contractions* go from upper left to lower right, with a factor $-n_p$. *Closed loops* introduce another factor -1. A graphical interpretation is possible, in agreement with that for the conventional particle–hole picture. We postpone this to the more general case of an arbitrary reference function (Section III.E).

Expectation values of products of $\tilde{a}_q^p$ operators with respect to Φ are simply the sum of all *full contractions*, for example,

$$\langle \Phi | \tilde{a}_q^p \tilde{a}_s^r | \Phi \rangle = \delta_q^r \delta_s^p n_p (1 - n_q) \tag{122}$$

$$\langle \Phi | \tilde{a}_a^i \tilde{a}_i^a | \Phi \rangle = 1 \tag{123}$$

This result represents the most important advantage of the particle–hole formalism. Many-body perturbation theory (MBPT) consists mainly in the evaluation of expectation values (with respect to the *physical vacuum*) of products of excitation operators. This is easily done by means of Wick's theorem in the particle–hole formalism.

B. Particle–Hole Formalism in an Arbitrary Basis

The results of the last section, which are essentially a reformulation of the traditional particle–hole formalism for excitation operators, were first presented in 1984 [8]. At that time it was not realized that only two very small steps are necessary to generalize this formalism to *arbitrary reference states*. Only after Mukherjee approached the formulation of a generalized normal ordering on a rather different route [2], did it come to our attention how easy this generalization actually is, when one starts from the results of the last section.

As in the previous section we consider a *single Slater determinant* reference function Φ with the spin orbitals ψ_i occupied. However, we express our excitation operators in a completely arbitrary basis of spin orbitals ψ_p, which is no longer the *direct sum* of occupied and unoccupied spin orbitals. Then the following replacements must be made [3]:

$$\delta_q^p n_p \Rightarrow \langle \Phi | a_q^p | \Phi \rangle = \gamma_q^p \tag{124}$$

$$n_p n_q \delta_{rs}^{pq} \Rightarrow \langle \Phi | a_{rs}^{pq} | \Phi \rangle = \gamma_{rs}^{pq} \tag{125}$$

In the original basis $\{\psi_i, \psi_a\}$, in which γ_q^p is diagonal, we retrieve, of course, the results of the last section, but in a general basis we get the $\tilde{a}$ operators in the following form:

$$\tilde{a}_q^p = a_q^p - \gamma_q^p \tag{126}$$

$$\tilde{a}_{rs}^{pq} = a_{rs}^{pq} - \gamma_r^p a_s^q - \gamma_s^q a_r^p + \gamma_s^p a_r^q + \gamma_r^q a_s^p + \gamma_{rs}^{pq} \tag{127}$$

$$\begin{aligned}\tilde{a}_{stu}^{pqr} = {} & a_{stu}^{pqr} - \gamma_s^p a_{tu}^{qr} + \gamma_s^q a_{tu}^{pr} + \gamma_s^r a_{tu}^{qp} + \gamma_t^p a_{su}^{qr} \\ & - \gamma_t^q a_{su}^{pr} + \gamma_t^r a_{su}^{pq} + \gamma_u^p a_{ts}^{qr} + \gamma_u^q a_{st}^{pr} + \gamma_u^r a_{st}^{pq} \\ & + \gamma_{st}^{pq} a_u^r + \gamma_{su}^{pr} a_t^q + \gamma_{tu}^{qr} a_s^p - \gamma_{su}^{pq} a_t^r - \gamma_{ut}^{pq} a_s^r \\ & - \gamma_{st}^{pr} a_u^q - \gamma_{tu}^{pr} a_s^q - \gamma_{ts}^{qr} a_u^p - \gamma_{su}^{qr} a_t^p - \gamma_{stu}^{pqr} \end{aligned} \tag{128}$$

It is easily checked that

$$\langle\Phi|\tilde{a}_q^p|\Phi\rangle = \langle\Phi|a_q^p|\Phi\rangle - \gamma_q^p = 0 \qquad (129)$$

$$\langle\Phi|\tilde{a}_{rs}^{pq}|\Phi\rangle = \gamma_{rs}^{pq} - 2\gamma_r^p\gamma_s^q + 2\gamma_s^p\gamma_r^q + \gamma_{rs}^{pq} = 0;\ldots \qquad (130)$$

that is, *expectation values* of $\tilde{a}$ operators with respect to Φ *vanish*, provided that

$$\gamma_{rs}^{pq} = \gamma_r^p\gamma_s^q - \gamma_s^p\gamma_r^q \qquad (131)$$

which is, of course, the case for a single Slater determinant reference state.

C. Normal Ordering with Respect to Arbitrary Reference Function

In order to generalize the concept of *normal ordering* such that it is valid with respect to *any arbitrary reference function* Ψ, we start from the following guiding principles:

(i) Normal order operators $\tilde{a}$ should be expressible as linear combinations of the operators a in normal order with respect to the genuine vacuum.

$$\tilde{a}_{rs}^{pq} = \alpha a_{rs}^{pq} + \beta a_r^p + \gamma a_s^p + \delta a_r^q + \varepsilon a_s^q + \eta \qquad (132)$$

On the r-h-s there should be excitation operators of the same and of lower particle ranks.

(ii) Expectation values of normal order operators $\tilde{a}$ with respect to the reference function Ψ should vanish.

$$\langle\Psi|\tilde{a}_{rs}^{pq}|\Psi\rangle = 0 \qquad (133)$$

(iii) The known results, valid when Ψ is single Slater determinant, must be recovered.

One meets these requirements, if one rewrites the definitions (126) – (128) in a recursive way.

$$\tilde{a}_q^p = a_q^p - \gamma_q^p \qquad (134)$$

$$\begin{aligned}\tilde{a}_{rs}^{pq} &= a_{rs}^{pq} - \gamma_r^p(\tilde{a}_s^q + \gamma_s^q) - \gamma_s^q(\tilde{a}_r^p + \gamma_r^p) \\ &\quad + \gamma_s^p(\tilde{a}_r^q + \gamma_r^q) + \gamma_r^q(\tilde{a}_s^p + \gamma_s^p) + \gamma_{rs}^{pq} \\ &= a_{rs}^{pq} - \gamma_r^p\tilde{a}_s^q - \gamma_s^q\tilde{a}_r^p + \gamma_s^p\tilde{a}_r^q + \gamma_r^q\tilde{a}_s^p - \gamma_{rs}^{pq}\end{aligned} \qquad (135)$$

$$\begin{aligned}\tilde{a}_{stu}^{pqr} &= a_{stu}^{pqr} - \gamma_s^p\tilde{a}_{tu}^{qr} + \gamma_s^q\tilde{a}_{tu}^{pr} + \gamma_s^r\tilde{a}_{tu}^{qp} + \gamma_t^p\tilde{a}_{su}^{qr} \\ &\quad - \gamma_t^q\tilde{a}_{su}^{pr} + \gamma_t^r\tilde{a}_{su}^{pq} + \gamma_u^p\tilde{a}_{ts}^{qr} + \gamma_u^q\tilde{a}_{st}^{pr} - \gamma_u^r\tilde{a}_{st}^{pq} \\ &\quad - \gamma_{st}^{pq}\tilde{a}_u^r - \gamma_{su}^{pr}\tilde{a}_t^q - \gamma_{tu}^{qr}\tilde{a}_s^p + \gamma_{su}^{pq}\tilde{a}_t^r + \gamma_{ut}^{pq}\tilde{a}_s^r \\ &\quad + \gamma_{st}^{pr}\tilde{a}_u^q + \gamma_{tu}^{pr}\tilde{a}_s^q + \gamma_{ts}^{qr}\tilde{a}_u^p + \gamma_{su}^{qr}\tilde{a}_t^p - \gamma_{stu}^{pqr}\end{aligned} \qquad (136)$$

This is, of course, just a trivial reformulation for Φ, a single Slater determinant. However, *only* in this special case are we free to use the identity (131) to rewrite Eqs. (134)–(136) to get back Eqs. (126)–(128). Otherwise Eqs. (134)–(136) are not identical to Eqs. (126)–(128). Fortunately, it turns out that Eqs. (134)–(136) are the searched-for generalizations.

In fact, let us evaluate the expectation values of the operators (134) – (136) with respect to an arbitrary Ψ! For Eq. (134) we get

$$\langle \Psi | \tilde{a}_q^p | \Psi \rangle = \langle \Psi | a_q^p | \Psi \rangle - \gamma_q^p = 0 \tag{137}$$

provided that γ_q^p is consistent with Ψ. We now use Eqs. (137) and (135) to get

$$\langle \Psi | \tilde{a}_{rs}^{pq} | \Psi \rangle = \langle \Psi | a_{rs}^{pq} | \Psi \rangle - \gamma_{rs}^{pq} = 0 \tag{138}$$

provided that γ_{rs}^{pq} is consistent with Ψ. So we find recursively that all these expectation values vanish, provided that the γ-elements that arise in the definitions (134) – (136) are those corresponding to Ψ. Equations (134)–(136) hence provide the searched-for generalization of normal ordering to an arbitrary reference function. A formulation for any particle rank is possible if we introduce the short-hand notation

$$\Sigma(-1)^P A_{q_1 q_2 \cdots q_k}^{p_1 p_2 \cdots p_k} B_{q_{k+1} \cdots q_n}^{p_{k+1} \cdots p_n} \tag{139}$$

for the sum over *all partitions* of both the n upper labels p_i and the n lower labels q_i into respective *subsets* of k and $n - k$ labels, keeping the *original pairing* of the p_i and q_i as much as possible, with appropriate *sign factors*, depending on the parity P of the partition, for example, a factor (-1) for each permutation of a pair $(p_j, q_j; p_l, q_l)$ to $(p_j, q_l; p_l, q_j)$. An example is

$$\begin{aligned}\Sigma(-1)^P \gamma_q^p a_{su}^{rt} = {} & \gamma_q^p a_{su}^{rt} + \gamma_s^r a_{qu}^{pt} + \gamma_u^t a_{qs}^{pr} - \gamma_s^p a_{qu}^{rt} - \gamma_u^p a_{sq}^{rt} - \gamma_q^r a_{su}^{pt} \\ & - \gamma_u^r a_{qs}^{pt} - \gamma_q^t a_{su}^{rp} - \gamma_u^t a_{qs}^{pr}\end{aligned} \tag{140}$$

Then the excitation operators in generalized normal ordering are in a compact form:

$$\tilde{a}_q^p = a_q^p - \gamma_q^p \tag{141}$$

$$\tilde{a}_{qs}^{pr} = a_{qs}^{pr} - \sum(-1)^P \gamma_q^p \tilde{a}_s^r - \gamma_{qs}^{pr} \tag{142}$$

$$\tilde{a}_{qsu}^{prt} = a_{qsu}^{prt} - \sum(-1)^P \gamma_q^p \tilde{a}_{su}^{rt} - \sum(-1)^P \gamma_{qs}^{pr} \tilde{a}_u^t - \gamma_{qsu}^{prt} \tag{143}$$

$$\begin{aligned}\tilde{a}_{qsuw}^{prtv} = {} & a_{qsuw}^{prtv} - \sum(-1)^P \gamma_q^p \tilde{a}_{suw}^{rtv} - \sum(-1)^P \gamma_{qs}^{pr} \tilde{a}_{uw}^{tv} \\ & - \sum(-1)^P \gamma_{qsu}^{prt} \tilde{a}_w^v - \gamma_{qsuw}^{prtv}\end{aligned} \tag{144}$$

Note that these expressions as such are never needed explicitly. What one does need are the *contraction rules* (i.e., the generalized Wick theorem) derived from these expressions. These will be given in the next section.

D. Generalized Wick Theorem

For the formulation of the generalized Wick theorem corresponding to the generalized normal ordering, we need the matrix element η_q^p, Eq. (65), of the *one-hole density matrix* and the *cumulants* λ_k, Eqs. (39)–(47), of the k-particle density matrices.

For products of operators $\tilde{a}$ in the generalized normal order we then get

$$\tilde{a}_q^p \tilde{a}_s^r = \tilde{a}_{qs}^{pr} + \eta_q^r \tilde{a}_s^p - \gamma_s^p \tilde{a}_q^r + \gamma_s^p \eta_q^r + \lambda_{qs}^{pr} \tag{145}$$

$$\tilde{a}_q^p \tilde{a}_{tu}^{rs} = \tilde{a}_{qtu}^{prs} + \eta_q^r \tilde{a}_{tu}^{ps} + \eta_q^s \tilde{a}_{ut}^{pr} - \gamma_t^p \tilde{a}_{qu}^{rs} - \gamma_u^p \tilde{a}_{tq}^{rs}$$
$$+ \{\eta_q^r \gamma_t^p + \lambda_{qt}^{pr}\} \tilde{a}_u^s - \{\eta_q^r \gamma_u^p + \lambda_{qu}^{pr}\} \tilde{a}_t^s$$
$$- \{\eta_q^s \gamma_t^p + \lambda_{qt}^{ps}\} \tilde{a}_u^r + \{\eta_q^s \gamma_u^p + \lambda_{qu}^{ps}\} \tilde{a}_t^r$$
$$- \lambda_{tq}^{rs} \tilde{a}_u^p - \lambda_{qu}^{rs} \tilde{a}_t^p - \lambda_{ut}^{pr} \tilde{a}_q^s - \lambda_{tu}^{ps} \tilde{a}_q^r$$
$$+ \eta_q^r \lambda_{tu}^{ps} + \eta_q^s \lambda_{ut}^{pr} - \gamma_t^p \lambda_{qu}^{rs} - \gamma_u^p \lambda_{tq}^{rs} + \lambda_{qtu}^{prs} \tag{146}$$

The elements of the cumulants λ_{qs}^{pr} and λ_{qsu}^{prt} vanish if Φ is a single Slater determinant. Then we retrieve the known result, Eqs. (120) and (121).

For a more general reference function there are not only *particle and hole contractions*, but also *contractions that involve cumulants*. Again it holds that the expectation value of a product of $\tilde{a}$ operators with respect to the reference function Ψ is equal to the sum of all full contractions:

$$\langle \Psi | \tilde{a}_q^p \tilde{a}_s^r | \Psi \rangle = \gamma_s^p \eta_q^r + \lambda_{qs}^{pr} \tag{147}$$

Although the theory has been formulated in terms of *excitation operators* only, the extension to arbitrary Fock-space operators is straightforward.

E. Diagrammatic Representation

The Fock-space expressions are conveniently illustrated by diagrams. A matrix element, say, $\bar{g}_{rs}^{pq}$, is represented by a *vertex*—in this case by a dot (·)—with two *ingoing lines* (with arrows *toward* the vertex) corresponding to the lower labels (rs), and two *outgoing lines* (with arrows *leaving* the vertex) corresponding to the upper labels (pq). A line between two vertices (carrying a spin-orbital label) means a *contraction*. For a line connecting two vertices *directly* (i.e., not involving a λ element) an upgoing (particle) line is associated with a factor $(1 - n_p)$, while a downgoing (hole) line is associated

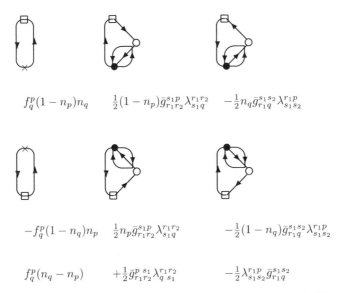

Figure 1. Diagrammatic representation of the terms that contribute to the irreducible one-particle Brillouin conditions, Eq. (167), with their algebraic equivalents. The last line represents the commutators (i.e., the difference of the above values).

with a factor $-n_p$. In contractions involving a λ element, the latter is represented by an open circle.

The diagrams that we use represent contractions of operators. Unlike in the diagrams of traditional MBPT no energy denominators are implied. Such denominators arise in the special context of perturbation theory and must then be indicated explicitly [5–7].

As an example we give in Fig. 1 the diagrammatic illustration of the terms that contribute to the IBC$_1$ of Eq. (167), together with the corresponding algebraic expressions. (Commutators are split into differences of two products.) On Fig. 2 the diagrams for the IBC$_2$ of Eq. (169) are given without their algebraic counterparts.

Note that a dot (·) always means a matrix element of the antisymmetrized electron interaction $\bar{g}$, a cross (×) a matrix element of the one-particle operator f, while an open square (□) collects the *free labels* in any of these contractions. If the reference function is a single Slater determinant, all cumulants λ vanish; one is then left with particle and hole contractions, like in traditional MBPT in the particle–hole picture.

The diagrams just discussed—with antisymmetrized vertices at spin-orbital level — are of *Hugenholtz* type [32]. One can alternatively define [30] diagrams of *Goldstone* type [33] with spin-free vertices.

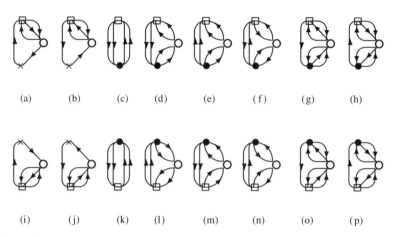

Figure 2. Diagrammatic representation of the terms that contribute to the irreducible two-particle Brillouin conditions, Eq. (169).

In these diagrams *self-contractions* are forbidden; that is, one line cannot *enter and leave the same vertex*. The lines entering or leaving a λ vertex cannot all connect to the same vertex.

Contraction rules: A contraction by a single line (not involving a λ vertex) implies a factor n_p (in terms of NSOs) for a single-hole contraction (downgoing line), and a factor $1 - n_p$ for a single-particle contraction (upgoing line). More than one multiple contraction or combinations of single and multiple contractions are allowed.

Sign rule: Write the diagram in Goldstone form with line vertices (instead of point vertices), to recognize which ingoing line corresponds to—or is paired with—which outgoing line. Another factor (-1) arises for (a) any down going single contraction, (b) any closed loop, including those in partial self-contractions, (c) any lambda matrix element of odd particle rank (e.g. λ^{pqr}_{stu}), and (d) any odd permutation of the original pairing of external lines.

F. Hamiltonian in Generalized Normal Order

The Hamiltonian in *normal order* with respect to its own exact eigenfunction (of full-CI type) is

$$H = E + f^p_q \tilde{a}^q_p + \tfrac{1}{2} g^{pq}_{rs} \tilde{a}^{rs}_{pq} \tag{148}$$

The energy (to be evaluated from Eq. (51)) appears as a constant term, consistent with the fact that the expectation values of the last two terms with respect to Ψ automatically vanish. The f^p_q are the matrix elements of the generalized Fock operator, Eq. (52).

IV. STATIONARITY CONDITIONS FOR THE ENERGY

A. Contracted Schrödinger Equations

Before we come to the *irreducible* stationarity conditions for the energy—our main concern—let us have a short look at the traditional stationarity conditions.

We start from the Hamiltonian (in normal order with respect to the genuine vacuum)

$$H = h^p_q a^q_p + \tfrac{1}{2} g^{pq}_{rs} a^{rs}_{pq} \qquad (149)$$

and we ask for the condition that the energy expectation value

$$E = \langle \Psi | H | \Psi \rangle = h^p_q \gamma^q_p + \tfrac{1}{2} g^{pq}_{rs} \gamma^{rs}_{pq} \qquad (150)$$

is stationary with respect to arbitrary variations of Ψ

$$\Psi \to (1 + X)\Psi = \Psi + \delta\Psi \qquad (151)$$

We get

$$\delta E = \langle \Psi | (H - E) X | \Psi \rangle = 0 \qquad (152)$$

where E plays the role of a Lagrange multiplier. We can now consider that X is a one-particle, two-particle, and so on excitation operator, and so get the hierarchy of k-particle contractred Schrödinger equations. The one-particle CSE reads

$$0 = \langle \Psi | a^p_q (H - E) | \Psi \rangle = h^r_s \gamma^{qs}_{pr} + h^r_p \gamma^q_r + \tfrac{1}{2} \bar{g}^{rs}_{pt} \gamma^{qt}_{rs} + \tfrac{1}{2} \bar{g}^{rs}_{tu} \gamma^{qtu}_{prs} \qquad (153)$$

while the two-particle CSE is

$$0 = \langle \Psi | a^{pq}_{rs} (H - E) | \Psi \rangle \qquad (154)$$

(For the explicit expression see Ref. [20].) This is a hierarchy of equations, first proposed independently by Nakatsuji [35] and Cohen and Frishberg [36]. In Eq. (153) γ_1 is expressed through γ_2 and γ_3, in Eq. (154) γ_2 is expressed through γ_3 and γ_4, and so on. This hierarchy is of no direct practical use, because there is no justification for a truncation of the γ_k at some k. Valdemoro [4] found an ingenious way to approximate the γ_k for higher particle rank k in terms of those of lower k and was so able to achieve a truncation of the hierarchy. This *reconstruction* was later refined by Nakatsuji and Yasuda [37], and analyzed by Mazziotti [18], who showed that the reconstruction can be rationalized if one intermediately expresses the γ_k in terms of the cumulants λ_k, and truncates the λ_k at some k, which is justified for the λ_k, but not the γ_k. The method based on this reconstruction has been called *contracted Schrödinger equations* by Valdemoro [4], the *density equation* by Nakatsuji and Yasuda [37], and *VNM-method* (for Valdemoro, Nakatsuji, Mazziotti) by Coleman and Yukalov [22]. It is described elsewhere in this book and we cannot go into details.

We could, of course, start from the CSE_k, express the γ_k in terms of the λ_k, and so arrive at a hierarchy of equations for the λ_k that can be truncated at some particle rank k. We prefer, however, to derive such a hierarchy directly, bypassing that in terms of the γ_k, as will be done in Section IV.C.

B. k-Particle Brillouin Conditions

Let us now consider the stationarity of the energy with respect to *unitary variations*

$$\Psi \to e^Z \Psi; \quad Z = -Z^\dagger \tag{155}$$

$$\{Z\} = \{a_q^p, a_{rs}^{pq}, a_{stu}^{pqr}, \ldots\} \tag{156}$$

The stationarity conditions are the k-particle Brillouin condition BC_k

$$\langle \Psi | [H, Z] | \Psi \rangle = 0 \tag{157}$$

$$\langle \Psi | [H, a_q^p] | \Psi \rangle = 0 \tag{158}$$

$$\langle \Psi | [H, a_{rs}^{pq}] | \Psi \rangle = 0, \ldots \tag{159}$$

While the (one-particle) Brillouin condition BC_1 has been known for a long time, and has played a central role in Hartree–Fock theory and in MC-SCF theory, the generalizations for higher particle rank were only proposed in 1979 [38], although a time-dependent formulation by Thouless [39] from 1961 can be regarded as a precursor.

The BC_1 (Eq. (158)) and BC_2 (Eq. (159)) are explicitly

$$h_s^p \gamma_q^s - \gamma_r^p h_q^r + \tfrac{1}{2} \bar{g}_{tu}^{pr} \gamma_{qr}^{tu} - \tfrac{1}{2} \gamma_{rs}^{pu} \bar{g}_{qu}^{rs} = 0 \tag{160}$$

$$h_u^p \gamma_{rs}^{uq} + h_u^q \gamma_{rs}^{pu} - \gamma_{ts}^{pq} h_r^t - \gamma_{rt}^{pq} h_s^t$$
$$+ \tfrac{1}{2} \bar{g}_{vw}^{pq} \gamma_{rs}^{vw} - \tfrac{1}{2} \gamma_{tu}^{pq} \bar{g}_{rs}^{tu} + \tfrac{1}{2} \bar{g}_{vw}^{pt} \gamma_{rts}^{vwq}$$
$$+ \tfrac{1}{2} \bar{g}_{vw}^{qu} \gamma_{sur}^{vwp} - \tfrac{1}{2} \bar{g}_{rw}^{tu} \gamma_{uts}^{wpq} - \tfrac{1}{2} \bar{g}_{rs}^{tu} \gamma_{tru}^{vpq} = 0 \tag{161}$$

with $\bar{g}$ the antisymmetrized electron interaction. As for the CSE_k, there is no obvious truncation of the hierarchy of k-particle equations, so like the latter they are of no direct use.

While the CSE_k expresses γ_k in terms of γ_{k+1} and γ_{k+2}, the BC_k only needs γ_{k+1}. The BC_k are obviously simpler than the CSE_k. On the other hand, the BC_k can be derived order by order from the CSE_k, while the converse is not true. Only if one goes up to $k = n$, do the two sets of stationarity conditions become equivalent.

A reconstruction in analogy to that for the CSE_k has, to the authors' knowledge, never been tried. It is, however, straightforward to express the γ_k through the λ_k and so get a hierarchy for the latter. Again we prefer the direct way, to be described in Section IV.C.

C. Irreducible Contracted Schrödinger Equations

Generally, the irreducible counterparts $ICSE_k$ of the CSE_k are obtained (consider also the Hermitian conjugates!) if one replaces the excitation operators by those in normal order with respect to Ψ:

$$\langle \Psi | \tilde{a}_q^p (H - E) | \Psi \rangle = 0 \tag{162}$$

$$\langle \Psi | \tilde{a}_{rs}^{pq} (H - E) | \Psi \rangle = 0, \ldots \tag{163}$$

The Lagrange multiplier E cancels with the constant part of H in Eq. (148) such that the $ICSE_1$ in Eq. (162) becomes

$$\begin{aligned} 0 &= f_r^s \langle \Psi | \tilde{a}_q^p \tilde{a}_s^r | \Psi \rangle + \tfrac{1}{2} g_{r_1 r_2}^{s_1 s_2} \langle \Psi | \tilde{a}_q^p \tilde{a}_{s_1 s_2}^{r_1 r_2} | \Psi \rangle \\ &= f_r^s \{ \lambda_{qs}^{pr} + \gamma_s^p \eta_q^r \} \\ &\quad + \tfrac{1}{2} \bar{g}_{r_1 r_2}^{s_1 s_2} \left\{ \tfrac{1}{2} \lambda_{q s_1 s_2}^{p r_1 r_2} - \gamma_{s_1}^p \lambda_{q s_2}^{r_1 r_2} + \eta_q^{r_1} \lambda_{s_1 s_2}^{p r_2} \right\} \end{aligned} \tag{164}$$

The result in an NSO basis is

$$\begin{aligned} 0 &= f_r^s \lambda_{qs}^{pr} + f_q^p (1 - n_q) n_p + \tfrac{1}{4} \bar{g}_{r_1 r_2}^{s_1 s_2} \lambda_{p r_1 r_2}^{q s_1 s_2} \\ &\quad - \tfrac{1}{2} n_p \bar{g}_{r_1 r_2}^{s_1 p} \lambda_{s_1 q}^{r_1 r_2} + \tfrac{1}{2} (1 - n_q) \bar{g}_{r_1 q}^{s_1 s_2} \lambda_{s_1 s_2}^{r_1 p} \end{aligned} \tag{165}$$

Let us point out, without going into detail, that the generalized normal ordering can, of course, also be formulated in a *spin-free form* [30].

The $ICSE_k$ of this section are essentially equivalent to the CSE_k of section IV.A, provided that one expresses the latter in terms of γ and λ_k for $k \geq 2$ rather than the γ_k. There are, however, two subtle differences.

The CSE$_2$ imply the CSE$_1$ (or generally the CSE$_k$ all CSE$_l$ with $l < k$). The ICSE$_k$ for different k, however, are *linearly independent*. The CSE$_2$ turns out to be a linear combination of the ICSE$_2$ and the ICSE$_1$. This is why we refer to the ICSE$_k$ as *irreducible* conditions. One may alternatively call these conditions *separable*. For a supersystem consisting of two noninteracting subsystems, one need only consider the ICSE$_k$ for the subsystems. In a diagrammatic representation, the ICSE$_k$ only contain connected diagrams, while the CSE$_k$ contain disconnected diagrams as well.

D. Irreducible Brillouin Conditions

The condition for E to be stationary with respect to unitary one-particle transformations in generalized normal order is

$$\langle \Psi | [H, \tilde{a}^p_q] | \Psi \rangle = 0 \tag{166}$$

This is nothing but the Brillouin condition of MC-SCF theory. Explicitly, in an NSO basis it reads

$$f^p_q (n_q - n_p) + \tfrac{1}{2} \bar{g}^{p\,s_1}_{r_1 r_2} \lambda^{r_1 r_2}_{q\,s_1} - \tfrac{1}{2} \lambda^{r_1 p}_{s_1 s_2} \bar{g}^{s_1 s_2}_{r_1 q} = 0 \tag{167}$$

In the ordinary Hartree–Fock case (with Ψ a single Slater determinant) the λ^{rs}_{pq} vanish. A graphical representation is given in Fig. 1.

One can formulate a two-particle analogue of the Brillouin condition, the IBC$_2$ ("I" stands for irreducible, which essentially means connected. For details see Refs. [20, 29].)

$$\langle \Psi | [H, \tilde{a}^{pq}_{rs}] | \Psi \rangle = 0, \ldots \tag{168}$$

Its explicit form is

$$\begin{aligned}
& f^{p_2}_r \lambda^{p_1 r}_{q_1 q_2} + f^{p_1}_r \lambda^{r\,p_2}_{q_1 q_2} - \lambda^{p_1 p_2}_{s\,q_2} f^s_{q_1} - \lambda^{p_1 p_2}_{q_1 s} f^s_{q_2} \\
& + \bar{g}^{p_1 p_2}_{q_1 q_2} \left\{ (1-n_{p_1})(1-n_{p_2}) n_{q_1} n_{q_2} - (1-n_{q_1})(1-n_{q_2}) n_{p_1} n_{p_2} \right\} \\
& + \tfrac{1}{2} \bar{g}^{p_1 p_2}_{r_1 r_2} \lambda^{r_1 r_2}_{q_1 q_2} (1 - n_{p_1} - n_{p_2}) - \tfrac{1}{2} \bar{g}^{s_1 s_2}_{q_1 q_2} \lambda^{p_1 p_2}_{s_1 s_2} (1 - n_{q_1} - n_{q_2}) \\
& + (n_{p_2} - n_{q_1}) \bar{g}^{p_2 s_1}_{r_1 q_1} \lambda^{p_1 r_1}_{s_1 q_2} - (n_{p_2} - n_{q_2}) \bar{g}^{p_2 s_1}_{r_1 q_2} \lambda^{p_1 r_1}_{s_1 q_1} \\
& - (n_{p_1} - n_{q_1}) \bar{g}^{p_1 s_1}_{r_1 q_1} \lambda^{p_2 r_1}_{s_1 q_2} + (n_{p_1} - n_{q_2}) \bar{g}^{p_1 s_1}_{r_1 q_2} \lambda^{p_2 r_1}_{s_1 q_1} \\
& + \tfrac{1}{2} \bar{g}^{p_2 s_2}_{r_1 r_2} \lambda^{r_1 r_2 p_1}_{q_2 s_2 q_1} + \tfrac{1}{2} \bar{g}^{p_1 s_2}_{r_1 r_2} \lambda^{r_1 r_2 p_2}_{q_1 s_2 q_2} \\
& - \tfrac{1}{2} \bar{g}^{s_1 s_2}_{r_1 q_1} \lambda^{p_1 p_2 r_1}_{s_2 q_2 s_1} - \tfrac{1}{2} \bar{g}^{s_1 s_2}_{r_1 q_2} \lambda^{p_1 p_2 r_1}_{q_1 s_2 s_1} = 0
\end{aligned} \tag{169}$$

A graphical representation is given in Fig. 2.

The relation of the IBC_k to the BC_k is the same as that of the $ICSE_k$ to the CSE_k. *Irreducible* has essentially the same meaning. Further more, the IBC_k are related to the $ICSE_k$, in the same way as the BC_k are related to the CSE_k; that is, the $ICSE_k$ imply the IBC_k, and not conversely.

V. TOWARD THE SOLUTION OF THE IBC_k AND CSE_k

A. General Considerations

1. We have already discussed the relations between the four stationarity conditions. In view of their separability, the two irreducible conditions are the right choice in the spirit of a many-body theory in terms of connected diagrams.

2. For any of the four conditions, a k-particle hierarchy is possible, with the k-particle approximation defined by the neglect of all λ_l with $l > k$.

3. Any of the four conditions has an *infinity of solutions*. Actually, the energy is stationary for any eigenstate of the Hamiltonian, so one has to specify in which state one is interested. This will usually be done at the iteration start. Moreover, the stationarity conditions do not discriminate between pure states and ensemble states. The stationarity conditions are even independent of the particle statistics. One must hence explicitly take care that one describes an n-fermion state. The hope that by means of the CSE_k or one of the other sets of conditions the n-representability problem is automatically circumvented has, unfortunately, been premature.

4. It is straightforward to solve the hierarchy of equations iteratively. However, the simplest iteration schemes converge only *linearly* (i.e., poorly). One should better consider quadratic iteration schemes.

5. As we shall see, the stationarity conditions determine essentially the *non-diagonal* elements of γ and the λ_k, while the diagonal elements are determined by the specification of the considered state and the n-representability.

6. There is a price to pay for the separability, or equivalently for the presence of only connected diagrams. Somewhat like in traditional many-body theory, one must be ready to accept so-called EPV (exclusion-principle violating) cumulants. Typical EPV cumulants are nonvanishing λ_k for $k > n$, while $\gamma_k = 0$ for $k > n$.

7. This means that the two-particle approximation in terms of the IBC_k or the $ICSE_k$ is *not exact* for a genuine two-electron system. For this a theory based on the CSE_2 is the right choice, but this not recommended for $n > 2$.

B. One-Particle Approximation

Let us consider the IBC$_1$, Eq. (166), and assume that $\lambda_2 = 0$. This is one possibility to define the one-particle approximation. We get

$$[\boldsymbol{\gamma}, \mathbf{f}] = 0 \tag{170}$$

that is, $\boldsymbol{\gamma}$ and $\mathbf{f}$ commute and have common eigenfunctions. In terms of the eigenstates of $\mathbf{f}$ with eigenvalues ε_p, we have

$$\gamma_q^p(\varepsilon_q - \varepsilon_p) = 0; \quad \gamma_q^p = 0 \quad \text{for } \varepsilon_p \neq \varepsilon_q \tag{171}$$

The stationarity condition (170) does not give any information on the diagonal elements γ_p^p, not even on the γ_q^p with $\varepsilon_p = \varepsilon_q$. We must get this information from another source. Fortunately, for an n-electron state, vanishing of λ_2 is only compatible with $\lambda_k = 0$ for $k > 2$ and with idempotency of $\boldsymbol{\gamma}$,

$$\boldsymbol{\gamma}^2 = \boldsymbol{\gamma}; \quad n_a = 0 \text{ or } n_i = 1 \tag{172}$$

Occupied spin orbitals are labeled as ψ_i, unoccupied (virtual) spin orbitals as ψ_a. We are thus led automatically to closed-shell Hartree–Fock theory with

$$f_q^p = h_q^p + g_{qs}^{pr}\gamma_r^s = h_q^p + g_{qi}^{pi} \tag{173}$$

If we start from the ICSE$_k$ rather than the IBC$_k$, the result is the same.

Let us now consider a generalized one-particle approximation. We no longer require that $\lambda_2 = 0$, but only that $\lambda_3 = 0$. Then we can use the following partial trace relations, which hold for $\lambda_3 = 0$ in an NSO basis [25]:

$$0 = \sum_t \lambda_{qst}^{prt} = \lambda_{qs}^{pr}\{n_p + n_q + n_r + n_s - 2\} \tag{174}$$

$$0 = \sum_{rt} \lambda_{qrt}^{prt} = (2n_p^3 - 4n_p^2 + 2n_p)\delta_q^p + 2\sum_r n_r \lambda_{qr}^{pr} \tag{175}$$

$$0 = \sum_{prt} \lambda_{prt}^{prt} = 4\sum_q n_q(n_q - \tfrac{1}{2})(n_q - 1) \tag{176}$$

Obviously, λ_{qs}^{pr} vanishes, unless $n_p + n_q + n_r + n_s = 2$. Sufficient though not necessary for Eq. (176) is that the n_p are equal to $0, \tfrac{1}{2}$, or 1; that is, now open-shell states with fractional NSO occupation numbers are also possible. This implies via Eq. (174) that the only nonvanishing elements of λ_2 are those

λ_{zu}^{xy}, where all labels x, y, z, u refer to an NSO with occupation number $\frac{1}{2}$. These are determined by spin coupling [30] rather than by stationarity conditions. For closed-shell states there is no difference with respect to the standard hierarchy.

C. Two-Particle Approximation

Let us now take the IBC$_2$, Eq. (169) in a basis in which both γ and $\mathbf{f}$ are diagonal and neglect λ_3.

$$\{\varepsilon_{p_1} + \varepsilon_{p_2} - \varepsilon_{q_1} - \varepsilon_{q_2}\} \lambda_{q_1 q_2}^{p_1 p_2} = F(\mathbf{g}, \lambda_2) \tag{177}$$

$$F(\mathbf{g}, \lambda_2) = -\bar{g}_{q_1 q_2}^{p_1 p_2} \Big\{ (1 - n_{p_1})(1 - n_{p_2}) n_{q_1} n_{q_2} + (1 - n_{q_1})(1 - n_{q_2}) n_{p_1} n_{p_2} \Big\}$$
$$- \tfrac{1}{2} \bar{g}_{r_1 r_2}^{p_1 p_2} \lambda_{q_1 q_2}^{r_1 r_2} (1 - n_{p_1} - n_{p_2}) + \tfrac{1}{2} \bar{g}_{q_1 q_2}^{s_1 s_2} \lambda_{s_1 s_2}^{p_1 p_2} (1 - n_{q_1} - n_{q_2})$$
$$- (n_{p_2} - n_{q_1}) \bar{g}_{r_1 q_1}^{p_2 s_1} \lambda_{s_1 q_2}^{p_1 r_1} + (n_{p_2} - n_{q_2}) \bar{g}_{r_1 q_2}^{p_2 s_1} \lambda_{s_1 q_1}^{p_1 r_1}$$
$$+ (n_{p_1} - n_{q_1}) \bar{g}_{r_1 q_1}^{p_1 s_1} \lambda_{s_1 q_2}^{p_2 r_1} - (n_{p_1} - n_{q_2}) \bar{g}_{r_1 q_2}^{p_1 s_1} \lambda_{s_1 q_1}^{p_2 r_1} \tag{178}$$

We limit ourselves to the closed-shell case. We try to solve this system iteratively, inserting $n_i = 1$ and $n_a = 0$ and neglecting λ_2 on the rhs in the first iteration. Then we obtain in the first iteration

$$\lambda_{ab}^{ij} = \{\varepsilon_i + \varepsilon_j - \varepsilon_a - \varepsilon_b\}^{-1} \bar{g}_{ab}^{ij} \tag{179}$$

$$\lambda_{ij}^{ab} = \{\varepsilon_i + \varepsilon_j - \varepsilon_a - \varepsilon_b\}^{-1} \bar{g}_{ij}^{ab} \tag{180}$$

while all other elements of λ_2 vanish. Even in higher iterations we only get an information for those elements of λ_2, for which $\varepsilon_i + \varepsilon_j \neq \varepsilon_a + \varepsilon_b$. Before we continue the iterations, we have to update γ. In principle, this is possible either from the partial trace relations (Eq. (62)) or from the IBC$_1$ (Eq. (166)). In either case, we need the diagonal elements of λ_2, especially λ_{ij}^{ij} and λ_{ab}^{ab}, which are obviously undetermined by the IBC$_2$. As in the one-particle approximation, we must try to get another source of information. This is more difficult than in the one-particle approximation.

There are three possibilities that are more easily understood in a perturbative analysis, which we discuss in the next section.

1. We try to impose exact n-representability conditions.
2. We update γ and the λ_k by means of a unitary transformation and regard the latter as unknown. This has the big advantage that a unitary transformation preserves n-representability. So it only matters to have an n-representable start.
3. We use the ICSE$_k$ rather than the IBC$_k$. In fact, we do get information on the diagonal elements of λ_2 from the ICSE$_k$. However, this requires

knowledge of the diagonal elements of λ_2, so the problem is shifted to the next particle rank. We shall see from a perturbative analysis that in order to get the energy correct to second order (or in a revised form to third order) in a perturbation parameter μ, one has to go up to the *three-particle* approximation. This makes methods based on the ICSE$_k$ by far inferior to traditional coupled-cluster methods. If one wants to compete with the latter, one cannot but choose the IBC$_k$ combined with at least approximative n-representability.

D. Perturbative Analysis

We expand the Hamiltonian and the IBC$_k$ in terms of a perturbation parameter μ in the spirit of Møller–Plesset perturbation theory [34]. Details are found in Ref. [25]. We need not worry about the particle rank to which we have to go, since this is fully controlled by the perturbation expansion. We limit ourselves to a closed-shell state, such that the zeroth order is simply closed-shell Hartree–Fock.

$$\text{IBC}_1^{(0)} : (f_0)_s^p (\gamma_0)_q^s - (f_0)_q^r (\gamma_0)_r^p = 0 \tag{181}$$

$$(f_0)_q^p = (f_0)_q^p + \bar{g}_{qs}^{pr}(\gamma_0)_r^s \tag{182}$$

$$E_0 = \tfrac{1}{2}\{h_q^p + (f_0)_q^p\}(\gamma_0)_p^q \tag{183}$$

To zeroth order λ_2 and all λ_k of higher particle rank vanish, while γ is idempotent and has eigenvalues 0 or 1. There is no first-order contribution to γ or E. The only first-order contribution is in λ_2:

$$(\lambda_1)_{ab}^{ij} = \{\varepsilon_i^{(0)} + \varepsilon_j^{(0)} - \varepsilon_a^{(0)} - \varepsilon_b^{(0)}\}^{-1}\bar{g}_{ab}^{ij} \tag{184}$$

$$(\lambda_1)_{ij}^{ab} = \{\varepsilon_i^{(0)} + \varepsilon_j^{(0)} - \varepsilon_a^{(0)} - \varepsilon_b^{(0)}\}^{-1}\bar{g}_{ij}^{ab} \tag{185}$$

In order not to overcharge the notation, we indicate the order in perturbation theory (PT) for a matrix element by a subscript (that we otherwise reserve for the particle rank), if the particle rank is obvious from the labels. For matrices like λ_2 we indicate the order of PT by a subscript such as $\lambda_2^{(1)}$.

The second-order energy is

$$\begin{aligned}E_2 &= \tfrac{1}{2}\{h_q^p + (f_0)_q^p\}(\gamma_2)_p^q + \tfrac{1}{2}(f_2)_q^p(\gamma_0)_p^q + \tfrac{1}{2}g_{rs}^{pq}(\lambda_1)_{pq}^{rs} \\ &= (f_0)_q^p(\gamma_2)_p^q - \tfrac{1}{2}\bar{g}_{qs}^{pr}(\gamma_0)_r^s(\gamma_2)_p^q + \tfrac{1}{2}\bar{g}_{qs}^{pr}(\gamma_2)_r^s(\gamma_0)_p^q + \tfrac{1}{2}g_{rs}^{pq}(\lambda_1)_{pq}^{rs} \\ &= \varepsilon_p^{(0)}(\gamma_2)_p^p + \tfrac{1}{4}\bar{g}_{rs}^{pq}(\lambda_1)_{pq}^{rs}\end{aligned} \tag{186}$$

The contribution involving $\lambda_2^{(1)}$ is easily evaluated as

$$\tfrac{1}{2} \bar{g}^{pq}_{rs} (\lambda_1)^{rs}_{pq} = \tfrac{1}{4} \bar{g}^{pq}_{rs} (\lambda_1)^{rs}_{pq} = \tfrac{1}{2} \{\varepsilon_i^{(0)} + \varepsilon_j^{(0)} - \varepsilon_a^{(0)} - \varepsilon_b^{(0)}\}^{-1} \bar{g}^{ij}_{ab} \bar{g}^{ab}_{ij} \qquad (187)$$

This will turn out to be equal to $2E_2$, that is, half of Eq. (187) will be compensated by the first term in the last expression in Eq. (186). To evaluate this, we need $\gamma^{(2)}$, and here we are faced with the problem that the $\mathrm{IBC}_1^{(2)}$ *does not give information on the diagonal elements* of $\gamma^{(2)}$.

Let us now discuss three routes for the construction of these diagonal elements.

The n-particle density matrix of an n-particle state is *pure-state n-representable* if—for unit trace—it is idempotent. Since we normalize γ_n as

$$\mathrm{Tr}(\gamma_n) = n! \qquad (188)$$

the pure-state n-representability condition is

$$(\gamma_n)^2 = n!\gamma_n \qquad (189)$$

For the perturbation expansion in powers of μ we get

$$(\gamma_n^{(0)})^2 = n!\gamma_n^{(0)} \qquad (190)$$
$$\gamma_n^{(0)}\gamma_n^{(1)} + \gamma_n^{(1)}\gamma_n^{(0)} - n!\gamma_n^{(1)} = 0 \qquad (191)$$
$$\gamma_n^{(0)}\gamma_n^{(2)} + \gamma_n^{(2)}\gamma_n^{(0)} - n!\gamma_n^{(2)} = -(\gamma_n^{(1)})^2 \qquad (192)$$

In a slightly tedious but elementary way [25], we finally arrive at

$$(\gamma_2)^a_a = \frac{1}{2}\sum_{i,j,b}(\lambda_1)^{ij}_{ab}(\lambda_1)^{ab}_{ij} \qquad (193)$$

$$(\gamma_2)^i_i = -\frac{1}{2}\sum_{a,b,j}(\lambda_1)^{ij}_{ab}(\lambda_1)^{ab}_{ij} \qquad (194)$$

That is, we can express the second-order corrections $\gamma^{(2)}$ to γ through the first-order corrections $\lambda_2^{(1)}$ to λ_2. This allows us to evaluate E_2. One can also evaluate the second-order corrections $\lambda_k^{(2)}$ to the diagonal elements of λ_2 and λ_3 in terms of the first-order corrections of λ_2 and show that there are no second-order corrections to λ_k for $k \geq 4$.

$$(\lambda_2)^{ab}_{ab} = \frac{1}{2} \sum_{i,j} (\lambda_1)^{ij}_{ab} (\lambda_1)^{ab}_{ij} \tag{195}$$

$$(\lambda_2)^{ij}_{ij} = -\frac{1}{2} \sum_{a,b} (\lambda_1)^{ij}_{ab} (\lambda_1)^{ab}_{ij} \tag{196}$$

$$(\lambda_2)^{ia}_{ia} = -\sum_{j,b} (\lambda_1)^{ij}_{ab} (\lambda_1)^{ab}_{ij} \tag{197}$$

$$(\lambda_2)^{iab}_{iab} = -\sum_{j} (\lambda_1)^{ij}_{ab} (\lambda_1)^{ab}_{ij} \tag{198}$$

$$(\lambda_2)^{ija}_{ija} = \sum_{b} (\lambda_1)^{ab}_{ij} (\lambda_1)^{ij}_{ab} \tag{199}$$

We do get information on the second-order corrections to γ from the perturbation expansion of the CSE_k [25]. However, in order to evaluate $(\gamma_2)^p_p$ we need to know $(\lambda_2)^{pr}_{pr}$, which is not yet known, when one needs it. One can construct $(\lambda_2)^{pr}_{pr}$ in terms of $(\lambda_2)^{prs}_{prs}$, that is, from the three-particle approximation, and one might expect that this continues *ad infinitum*. Fortunately, $(\lambda_2)^{prst}_{prst}$ vanishes, and one does not have to go beyond the three-particle approximation, in order to finally get $(\gamma_2)^p_p$ and from it E_2, but this is bad enough, and a disqualification of $ICSE_k$-based methods.

In the method based on the unitary transformation, we start by writing the "exact" wavefunction Ψ in terms of the "reference function" Φ and a unitary transformation operator e^σ in Fock space:

$$\Psi = e^\sigma \Phi \tag{200}$$

The cluster expansion of σ (which is additively separable and hence extensive) is

$$\sigma = \sigma_1 + \sigma_2 + \cdots = \sigma^p_q \tilde{a}^q_p + \tfrac{1}{4} \sigma^{pq}_{rs} \tilde{a}^{rs}_{pq} + \cdots \tag{201}$$

The tilde now indicates *normal ordering with respect to the reference function* Φ.

The energy and the Brillouin conditions BC_k become

$$\begin{aligned} E &= \langle \Phi | e^{-\sigma} H e^\sigma | \Phi \rangle \\ &= \langle \Phi | H + [H, \sigma] + \tfrac{1}{2}[[H, \sigma], \sigma] + \cdots | \Phi \rangle \end{aligned} \tag{202}$$

$$\begin{aligned} 0 &= \langle \Phi | e^{-\sigma} [H, \tilde{X}_k] e^\sigma | \Phi \rangle \\ &= \langle \Phi | [H, \tilde{X}_k] + [[H, \tilde{X}_k], \sigma] + \tfrac{1}{2}[[[H, \tilde{X}_k], \sigma], \sigma] + \cdots \Phi \rangle \end{aligned} \tag{203}$$

with both H and $\tilde{X}_k$ (a k-particle excitation operator) written in normal order with respect to Φ.

If one could solve Eq. (203) exactly for σ and insert this into Eq. (202), one would, of course, get the exact energy—provided that the reference function is n-representable (e.g., is a normalized Slater determinant). The unitary transformation *preserves the n-representability*. Equation (203) is an infinite-order nonlinear set of equations and not easy to solve. However, the perturbation expansion terminates at any finite order. We have [6, 12]

$$H = H_0 + \mu V; \quad E = E_0 + \mu E_1 + \mu^2 E_2 + \cdots \quad (204)$$

$$\sigma = \mu \sigma^{(1)} + \mu^2 \sigma^{(2)} + \cdots \quad (205)$$

$$\sigma_k = \mu \sigma_k^{(1)} + \mu^2 \sigma_k^{(2)} + \cdots \quad (206)$$

$$E_0 = \langle \Phi | H_0 | \Phi \rangle \quad (207)$$

$$E_1 = \langle \Phi | [H_0, \sigma^{(1)}] + V | \Phi \rangle \quad (208)$$

$$E_2 = \langle \Phi | [H_0, \sigma^{(2)}] + \tfrac{1}{2} [[H_0, \sigma^{(1)}], \sigma^{(1)}] + [V, \sigma^{(1)}] | \Phi \rangle \quad (209)$$

$$E_3 = \langle \Phi | [H_0, \sigma^{(3)}] + \tfrac{1}{2} [[H_0, \sigma^{(1)}], \sigma^{(2)}] + \tfrac{1}{2} [[H_0, \sigma^{(2)}], \sigma^{(1)}]$$
$$+ \tfrac{1}{6} [[[H_0, \sigma^{(1)}], \sigma^{(1)}], \sigma^{(1)}] + [V, \sigma^{(2)}] + \tfrac{1}{2} [[V, \sigma^{(1)}], \sigma^{(1)}] | \Phi \rangle \quad (210)$$

$$0 = \langle \Phi | [H_0, \tilde{X}_k] | \Phi \rangle \quad (211)$$

$$0 = \langle \Phi | [[H_0, \tilde{X}_k], \sigma^{(1)}] + [V, \tilde{X}_k] | \Phi \rangle \quad (212)$$

$$0 = \langle \Phi | [[H_0, \tilde{X}_k], \sigma^{(2)}] + \tfrac{1}{2} [[[H_0, \tilde{X}_k], \sigma^{(1)}], \sigma^{(1)}] + [[V, \tilde{X}_k], \sigma^{(1)}] | \Phi \rangle \quad (213)$$

If the stationarity conditions (211)–(213) are satisfied, the energy expressions are simplified to

$$E_1 = \langle \Phi | V | \Phi \rangle \quad (214)$$

$$E_2 = \tfrac{1}{2} \langle \Phi | [V, \sigma^{(1)}] | \Phi \rangle \quad (215)$$

$$E_3 = \tfrac{1}{6} \langle \Phi | [[[H_0, \sigma^{(1)}], \sigma^{(1)}], \sigma^{(1)}] | \Phi \rangle + \tfrac{1}{2} \langle \Phi | [[V, \sigma^{(1)}], \sigma^{(1)}] | \Phi \rangle \quad (216)$$

provided that the $\sigma^{(k)}$ are expressible in the basis of operators $\tilde{X}_k$ for which Eqs. (211)–(213) hold.

Equation (211) is obviously satisfied for all $\tilde{X}_k$ if Φ is a closed-shell Hartree–Fock function. We get $\sigma^{(1)}$ from Eq. (212). The matrix elements $\langle \Phi | [V, \tilde{X}_k] | \Phi \rangle$ vanish (for Φ a Slater determinant), except for $\tilde{X}_k = \tilde{a}_{pq}^{rs}$, a two-particle operator.

Hence $\sigma^{(1)}$ consists only of $\sigma_2^{(1)}$. We obtain it from

$$\langle\Phi|[[H_0,\sigma_2^{(1)}],\tilde{X}_k]\Phi\rangle = -\langle\Phi|[V,\tilde{X}_k]|\Phi\rangle \quad (217)$$

The only nonvanishing expressions are actually

$$\langle\Phi|[[H_0,\sigma_2^{(1)}],\tilde{a}_{ab}^{ij}]|\Phi\rangle = -(\sigma_1)_{ab}^{ij}(\varepsilon_i^{(0)}+\varepsilon_j^{(0)}-\varepsilon_a^{(0)}-\varepsilon_b^{(0)}) \quad (218)$$

$$\langle\Phi|[V,\tilde{a}_{ab}^{ij}]|\Phi\rangle = -\bar{g}_{ab}^{ij} \quad (219)$$

$$\langle\Phi|[[H_0,\sigma_2^{(1)}],\tilde{a}_{ij}^{ab}]|\Phi\rangle = (\sigma_1)_{ij}^{ab}(\varepsilon_i^{(0)}+\varepsilon_j^{(0)}-\varepsilon_a^{(0)}-\varepsilon_b^{(0)}) \quad (220)$$

$$\langle\Phi|[V,\tilde{a}_{ij}^{ab}]|\Phi\rangle = \bar{g}_{ij}^{ab} \quad (221)$$

hence

$$(\sigma_1)_{ab}^{ij} = -(\varepsilon_i^{(0)}+\varepsilon_j^{(0)}-\varepsilon_a^{(0)}-\varepsilon_b^{(0)})^{-1}\bar{g}_{ab}^{ij} \quad (222)$$

$$(\sigma_1)_{ij}^{ab} = (\varepsilon_i^{(0)}+\varepsilon_j^{(0)}-\varepsilon_a^{(0)}-\varepsilon_b^{(0)})^{-1}\bar{g}_{ij}^{ab} \quad (223)$$

Equation (217) only determines the nondiagonal elements of $\sigma_2^{(1)}$. However, unlike for the $\lambda_2^{(1)}$, *there is no loss of generality* [6] to impose that the diagonal elements of $\sigma_2^{(1)}$ vanish.

Knowing $\sigma_2^{(1)}$, we can easily construct the nonvanishing elements of $\lambda_2^{(1)}$ and of the diagonal part of $\gamma^{(2)}$ that we need for the evaluation of E_2.

$$(\lambda_1)_{ab}^{ij} = \langle\Phi|[\tilde{a}_{ab}^{ij},\sigma_2^{(1)}]|\Phi\rangle = (\sigma_1)_{ab}^{ij}\langle\Phi|[\tilde{a}_{ab}^{ij},\tilde{a}_{ij}^{ab}]|\Phi\rangle = (\sigma_1)_{ab}^{ij} \quad (224)$$

$$(\lambda_1)_{ij}^{ab} = \langle\Phi|[\tilde{a}_{ij}^{ab},\sigma_2^{(1)}]|\Phi\rangle = (\sigma_1)_{ij}^{ab}\langle\Phi|[\tilde{a}_{ij}^{ab},\tilde{a}_{ab}^{ij}]|\Phi\rangle = -(\sigma_1)_{ij}^{ab} \quad (225)$$

$$(\gamma_2)_i^i = \tfrac{1}{2}\langle\Phi|[[\tilde{a}_i^i,\sigma_2^{(1)}],\sigma_2^{(1)}]|\Phi\rangle = \tfrac{1}{2}(\sigma_1)_{ab}^{ij}(\sigma_1)_{ij}^{ab} \quad (226)$$

$$(\gamma_2)_a^a = \tfrac{1}{2}\langle\Phi|[[\tilde{a}_a^a,\sigma_2^{(1)}],\sigma_2^{(1)}]|\Phi\rangle = -\tfrac{1}{2}(\sigma_1)_{ab}^{ij}(\sigma_1)_{ij}^{ab} \quad (227)$$

This is consistent with Eqs. (184), (185), (193) and (194), but obtained in a much simpler way. One also reproduces easily the expressions (197)–(198) and finds that the d.e. of $\lambda_4^{(2)}$ vanish. In terms of the σ_k the two-particle-approximation (i.e., the truncation at $k=2$) appears to work better than in terms of λ_k. Taking the σ_k up to the particle rank k, we get the energy correct to E_{2k-1}.

VI. CONCLUSIONS

The concept of generalized normal ordering is very powerful, but still waiting to become a standard tool in many-body physics. It is the natural generalization of the particle–hole formalism for those states, which are not well described by a single Slater determinant reference state (i.e., for open-shell or multireference states). Using the generalized normal ordering it is possible to develop a coupled-cluster formalism starting from a multireference function, which bears a close resemblance to the single-reference coupled-cluster formalism [40, 41]. The concept of generalized normal ordering has recently been applied in quantum field theory and has been formulated in the context of Hopf algebras [42].

Generalized normal ordering is intimately linked to the cumulants λ_k of the k-particle density matrices γ_k (for short, density cumulants). The contractions in the sense of the generalized Wick theorem involve the λ_k.

If one formulates the conditions for stationarity of the energy expectation value in terms of generalized normal ordering, one is led to either the irreducible k-particle Brillouin conditions IBC$_k$ or the irreducible k-particle contracted Schrödinger equations (IBC$_k$), which are conditions to be satisfied by $\gamma = \gamma_1$ and the λ_k. One gets a hierarchy of k-particle approximations that can be truncated at any desired order, without any need for a *reconstruction*, as is required for the *reducible* counterparts.

There are also some unexpected problems, related to the fact that the stationarity conditions do not discriminate between ground and excited states, between pure states and ensemble states, and not even between fermions and bosons. The IBC$_k$ give only information about the *nondiagonal* elements of γ and the λ_k, whereas for the *diagonal* elements other sources of information must be used. These elements are essentially determined by the requirement of n-representability. This can be imposed exactly to the leading order of perturbation theory. Some information on the diagonal elements is obtained from the ICSE$_k$, though in a very expensive and hence not recommended way. The best way to take care of n-representability is probably via a unitary Fock-space transformation of the reference function, because this transformation preserves the n-representability.

Acknowledgments

This work was supported by Deutsche Forschungsgemeinschaft (DFG), Fonds der Chemie (FCh), and the Alexander-von-Humboldt Foundation (AvH).

References

1. G. C. Wick, *Phys. Rev.* **80**, 268 (1950).
2. D. Mukherjee, *Chem. Phys. Lett.* **274**, 561 (1997).
3. W. Kutzelnigg and D. Mukherjee, *J. Chem. Phys.* **107**, 432 (1997).
4. C. Valdemoro, *Phys. Rev.* **31**, 2114 (1985).
5. W. Kutzelnigg, *Chem. Phys. Lett.* **83**, 156 (1981).
6. W. Kutzelnigg, *J. Chem. Phys.* **77**, 3081 (1982).
7. W. Kutzelnigg and S. Koch, *J. Chem. Phys.* **79**, 4315 (1983).

8. W. Kutzelnigg, *J. Chem. Phys.* **80**, 822 (1984).
9. W. Kutzelnigg, Quantum chemistry in Fock space, in *Aspects of Many-Body Effects in Molecules and Extended Systems* (D. Mukherjee, ed.), *Lecture Notes in Chemistry Vol.* 50, Springer, Berlin, 1989.
10. B. Klahn and W. A. Bingel, *Theor. Chim. Acta* **44**, 9, 27 (1977).
11. D. J. Thouless, *The Quantum Mechanics of Many-Body Systems*, Academic Press, New York, 1961; P. Nozières, *Le problème à N corps*, Dunod, Paris, 1963; and *The Theory of Interacting Fermi Systems*, Benjamin, New York, 1964; I. Lindgren and J. Morrison, *Atomic Many-Body Theory*, Springer, Berlin, 1982.
12. H. Primas, *Helv. Phys. Acta* **34**, 331 (1961); *Rev. Mod. Phys.* **35**, 710 (1963).
13. *The Unitary Group* (J. Hinze, ed.), *Lecture Notes in Chemistry Vol.* 22, Springer, Berlin, 1981.
14. W. Kutzelnigg, *Z. Naturforschung* **18a**, 1058 (1963).
15. W. Kutzelnigg, *J. Chem. Phys.* **82**, 4166 (1984).
16. J. Paldus and B. Jeziorski, *Theor. Chim. Acta* **73**, 81 (1988).
17. W. Kutzelnigg and D. Mukherjee, *J. Chem. Phys.* **110**, 2800 (1999).
18. D. Mazziotti, *Chem. Phys. Lett.* **289**, 419 (1998); *Int. J. Quantum Chem.* **70**, 557 (1998); *Phys. Rev. A* **57**, 4219 (1998); *Phys. Rev. A* **60**, 4396 (1999).
19. R. Kubo, *J. Phys. Soc. Japan* **17**, 1100 (1962).
20. D. Mukherjee and W. Kutzelnigg, *J. Chem. Phys.* **114**, 2047 (2001); erratum **114**, 8226 (2001).
21. A. J. Coleman, *Rev. Mod. Phys.* **35**, 668 (1963).
22. A. J. Coleman and V. I. Yukalov, *Reduced Density Matrices, Lecture Notes in Chemistry*, Springer, Berlin, 2000.
23. H. Kummer, *Int. J. Quantum Chem.* **12**, 1033 (1977).
24. W. Kutzelnigg and D. Mukherjee, *J. Chem. Phys.* **120**, 7340 (2004).
25. W. Kutzelnigg and D. Mukherjee, *J. Chem. Phys.* **120**, 7350 (2004).
26. D. C. Carlson and J. H. Keller, *Phys. Rev.* **121**, 659 (1961).
27. E. Schmidt, *Math. Ann.* **63**, 433 (1907).
28. F. Sasaki, *Phys. Rev.* **138B**, 1338 (1965).
29. W. Kutzelnigg and D. Mukherjee, *Chem. Phys. Lett.* **317**, 567 (2000).
30. W. Kutzelnigg and D. Mukherjee, *J. Chem. Phys.* **116**, 4787 (2002).
31. P. O. Löwdin, *Adv. Chem. Phys.* **22**, 207 (1959).
32. N. M. Hugenholtz, *Physica* **23**, 481 (1957).
33. J. Goldstone, *Proc. R. Soc. London A* **239**, 267 (1957).
34. L. Møller and M. S. Plesset, *Phys. Rev.* **46**, 618 (1934).
35. H. Nakatsuji, *Phys. Rev.* **14**, 41 (1976).
36. L. Cohen and C. Frishberg, *Phys. Rev.* **13**, 927 (1976).
37. H. Nakatsuji and K. Yasuda, *Phys. Rev. Lett* **76**, 1039 (1996).
38. W. Kutzelnigg, *Chem. Phys. Lett.* **64**, 383 (1979); *Int. J. Quantum Chem.* **18**, 3 (1980).
39. D. J. Thouless, *The Quantum Mechanics of Many-Body Systems*, Academic Press, New York, 1961.
40. D. Mukherjee, in *Recent Progress in Many-Body Theories*, Vol. 4, (E. Schachinger, ed.), Plenum, New York, 1995.
41. U. S. Mahapatra, B. Datta, B. Bandyopadhyay, and D. Mukherjee *Adv. Quantum Chem.* **30**, 163 (1998).
42. C. Brouder, B.Fauser, A. Frabetti, and R. Oeckl, *J. Phys. A* **37**, 5895 (2004).

CHAPTER 12

ANTI-HERMITIAN FORMULATION OF THE CONTRACTED SCHRÖDINGER THEORY

DAVID A. MAZZIOTTI

Department of Chemistry and The James Franck Institute, The University of Chicago, Chicago, IL 60637 USA

CONTENTS

I. Introduction
II. Anti-Hermitian Contracted Schrödinger Equation
III. Reconstruction of the 3-RDM
IV. Optimization of the 2-RDM
V. Applications
VI. Some Connections and a Second Look Ahead
Acknowledgments
References

I. INTRODUCTION

Molecular electronic energies and two-electron reduced density matrices (2-RDMs) from the contracted Schrödinger equation (CSE) [1–26] can be significantly improved by solving only the *anti-Hermitian part* of the equation, also known as the Brillouin condition [21, 23]. The anti-Hermitian contracted Schrödinger equation (ACSE) [27] has two significant attributes: (i) the 2-RDMs and energies from solving the ACSE with *only a first-order* approximation to the 3-RDM are correct through second and third orders of perturbation theory, and (ii) the 2-RDMs from solving the ACSE are nearly N-representable without imposing N-representability conditions. In the first two sections of this chapter the ACSE and the reconstruction of the 3-RDM are presented, and in the third section a system of initial-value differential equations is derived whose solution coincides with the solution of the

---

*Reduced-Density-Matrix Mechanics: With Application to Many-Electron Atoms and Molecules,
A Special Volume of Advances in Chemical Physics, Volume 134*, edited by David A. Mazziotti.
Series editor Stuart A. Rice. Copyright © 2007 John Wiley & Sons, Inc.

ACSE for the 2-RDM [27]. The ACSE method is illustrated for the atom Be and a variety of molecules BH, H$_2$O, NH$_3$, HF, and N$_2$ as well as dissociation of BeH$_2$. Correlation energies are obtained within 98–100% of full configuration interaction, and the 2-RDMs closely satisfy the known N-representability conditions [28]. The ACSE energies are competitive with those from wavefunction methods like coupled-cluster singles–doubles (CCSD). Advantages of the ACSE in comparison with wavefunction methods include (i) the direct calculation of the 2-RDM and (ii) flexibility in the initial 2-RDM that allows the implicit reference wavefunction to be a Salter determinant or any approximate correlated wavefunction such as the wavefunction from a multiconfiguration self-consistent-field calculation. In the last section connections are drawn between the ACSE and other methods including canonical diagonalization [29, 30], the effective valence Hamiltonian [31, 32], unitary coupled cluster [33–35], and the variational 2-RDM method [36–39].

II. ANTI-HERMITIAN CONTRACTED SCHRÖDINGER EQUATION

The ACSE in a finite basis of spin orbitals can be expressed as

$$\langle \psi | [a_i^\dagger a_j^\dagger a_l a_k, \hat{H}] | \psi \rangle = 0 \tag{1}$$

where the brackets denote the quantum mechanical commutator, $a_i^\dagger$ and a_i are the second-quantized creation and annihilation operators, and the Hamiltonian operator $\hat{H}$ is

$$\hat{H} = \sum_{p,s} {}^1K_s^p a_p^\dagger a_s + \sum_{p,q,s,t} {}^2V_{s,t}^{p,q} a_p^\dagger a_q^\dagger a_t a_s \tag{2}$$

The reduced matrices ^1K and ^2V represent a partitioning of the Hamiltonian into one- and two-electron parts. Rearranging the second-quantized operators and using the definition of the 2- and 3-RDMs,

$$^2D_{k,l}^{i,j} = \frac{1}{2} \langle \psi | a_i^\dagger a_j^\dagger a_l a_k | \psi \rangle \tag{3}$$

$$^3D_{q,s,t}^{i,j,k} = \frac{1}{6} \langle \psi | a_i^\dagger a_j^\dagger a_k^\dagger a_t a_s a_q | \psi \rangle \tag{4}$$

we can write the ACSE in terms of the 2- and 3-RDMs only:

$$\sum_s {}^1K_s^k {}^2D_{s,l}^{i,j} - \sum_s {}^1K_s^l {}^2D_{s,k}^{i,j} \tag{5}$$

$$+ \sum_p {}^1K_j^p {}^2D_{k,l}^{p,i} - \sum_p {}^1K_i^p {}^2D_{k,l}^{p,j} \tag{6}$$

$$+ 6\sum_{p,s,t} {}^2V^{p,k}_{s,t} {}^3D^{i,j,p}_{s,t,l} - 6\sum_{p,s,t} {}^2V^{p,l}_{s,t} {}^3D^{i,j,p}_{s,t,k} \qquad (7)$$

$$+ 6\sum_{p,q,s} {}^2V^{p,q}_{s,j} {}^3D^{p,q,i}_{k,l,s} - 6\sum_{p,q,s} {}^2V^{p,q}_{s,i} {}^3D^{p,q,j}_{k,l,s} \qquad (8)$$

$$+ 2\sum_{s,t} {}^2V^{k,l}_{s,t} {}^2D^{i,j}_{s,t} + 2\sum_{p,q} {}^2V^{p,q}_{j,i} {}^2D^{p,q}_{k,l} = 0 \qquad (9)$$

Importantly, the 3-RDM appears only in terms with the perturbative part ^2V of the Hamiltonian, which is responsible for the improved accuracy of the ACSE in comparison with the CSE.

III. RECONSTRUCTION OF THE 3-RDM

Because the ACSE depends on both the 2- and the 3-RDMs, the 3-RDM must be approximated as a functional of the 2-RDM. As with the CSE method, the 3-RDM can be reconstructed from the 2-RDM by its *cumulant expansion* [8, 9, 11, 13–15, 21],

$$^3D^{i,j,k}_{q,s,t} = {}^1D^i_q \wedge {}^1D^j_s \wedge {}^1D^k_t + 3\, {}^2\Delta^{i,j}_{q,s} \wedge {}^1D^k_t + {}^3\Delta^{i,j,k}_{q,s,t} \qquad (10)$$

where

$$^2\Delta^{i,j}_{k,l} = {}^2D^{i,j}_{k,l} - {}^1D^i_k \wedge {}^1D^j_l \qquad (11)$$

and the operator $\wedge$ denotes the antisymmetric tensor product known as the Grassmann wedge product [7]. The cumulant (or connected) parts $^p\Delta$ of p-RDMs vanish unless all p particles are statistically dependent. Hence the cumulant RDMs scale linearly with the number N of particles in the system.

Neglecting the cumulant 3-RDM,

$$^3\Delta^{i,j,k}_{q,s,t} = 0 \qquad (12)$$

yields a first-order reconstruction of the 3-RDM from the 1- and 2-RDMs, which we call Valdemoro's (V) reconstruction [1]. Some important second-order contributions can be included by approximating the cumulant 3-RDM. Such approximations have been introduced by Nakatsuji and Yasuda [5] and Mazziotti [11, 15], which we will denote as NY and M. The NY reconstruction for $^3\Delta$ is

$$^3\Delta^{i,j,k}_{q,s,t} \approx \frac{1}{6}\sum_l s_l \hat{A}({}^2\Delta^{i,l}_{q,s}\, {}^2\Delta^{j,k}_{l,t}) \qquad (13)$$

where s_l equals 1 if l is occupied in the Hartree–Fock reference and -1 if l is not occupied and the operator $\hat{A}$ performs all distinct antisymmetric permutations of the indices excluding the summation index l. In a natural-orbital basis set the M reconstruction is

$$n_{q,s,t}^{i,j,k}\,{}^3\Delta_{q,s,t}^{i,j,k} \approx -\frac{1}{6}\sum_l \hat{A}({}^2\Delta_{q,s}^{i,l}\,{}^2\Delta_{l,t}^{j,k}) \qquad (14)$$

where

$$n_{q,s,t}^{i,j,k} = {}^1D_i^i + {}^1D_j^j + {}^1D_k^k + {}^1D_q^q + {}^1D_s^s + {}^1D_t^t - 3 \qquad (15)$$

Each of the reconstructions contains many contributions from higher orders of perturbation theory via the 1- and 2-RDMs and thus may be described as highly *renormalized*. The CSE requires a second-order correction of the 3-RDM functional to generate second-order 2-RDMs and energies, but the ACSE can produce second-order 2-RDMs and third-order energies from only a first-order reconstruction of the 3-RDM.

IV. OPTIMIZATION OF THE 2-RDM

By examining a sequence of infinitesimal unitary transformations applied to the wavefunction, we can derive a system of differential equations for solving the ACSE for the ground-state energy and its 2-RDM. We order these transformations by a continuous time-like variable λ. After an infinitesimal transformation over the interval ϵ, the energy at $\lambda + \epsilon$ is given by

$$\begin{aligned} E(\lambda+\epsilon) &= \langle \psi(\lambda)|e^{-\epsilon S(\lambda)}\hat{H}e^{\epsilon S(\lambda)}|\psi(\lambda)\rangle \\ &= E(\lambda) + \epsilon\langle \psi(\lambda)|[\hat{H},\hat{S}(\lambda)]|\psi(\lambda)\rangle + O(\epsilon^2) \end{aligned} \qquad (16)$$

This equation becomes a differential equation in the limit that $\epsilon \to 0$,

$$\frac{dE}{d\lambda} = \langle \Psi(\lambda)|[\hat{H},\hat{S}(\lambda)]|\Psi(\lambda)\rangle \qquad (17)$$

With the two-body operator $\hat{\Gamma}_{k,l}^{i,j} = a_i^\dagger a_j^\dagger a_l a_k$ the change in 2-RDM with λ can be written

$$\frac{d\,{}^2D_{k,l}^{i,j}}{d\lambda} = \langle \Psi(\lambda)|[\hat{\Gamma}_{k,l}^{i,j},\hat{S}(\lambda)]|\Psi(\lambda)\rangle \qquad (18)$$

If $\hat{S}(\lambda)$ is restricted to be an anti-Hermitian operator with no more than two-particle interactions, the variational degrees of freedom of $\hat{S}(\lambda)$ can be represented by the two-particle reduced matrix $^2S_{k,l}^{i,j}(\lambda)$, which we choose at each λ to minimize the energy along its gradient:

$$^2S_{k,l}^{i,j}(\lambda) = \langle\Psi(\lambda)|[\hat{\Gamma}_{k,l}^{i,j},\hat{H}]|\Psi(\lambda)\rangle \tag{19}$$

Equations (17)–(19) can then be evaluated with only the 2- and the 3-RDMs, where the right-hand side of Eq. (19) is the residual of the ACSE in Eqs. (5)–(9) and the right-hand side of Eq. (18) is the residual of the ACSE with $\hat{H}$ replaced by the anti-Hermitian $\hat{S}(\lambda)$. This system of differential equations produces energy and 2-RDM trajectories in λ that minimize the energy until the ACSE is satisfied.

V. APPLICATIONS

In this section we examine the accuracy of the ACSE method with calculations on a variety of molecules. Calculations are performed at equilibrium geometries [40] in a valence double-zeta basis set [41] with frozen cores, and electron integrals are computed with GAMESS (USA) [42]. The energy and 2-RDM of the ACSE are optimized by integrating Eqs. (17)–(19) with an extrapolated Euler's method. At $\lambda = 0$ the energy and 2-RDM are initialized to their values from a Hartree–Fock (mean-field) calculation. The evolution of the energy and the 2-RDM with λ continues until either (i) the energy or (ii) the least-squares error in the ACSE or the 1, 3-CSE ceases to decrease. Figure 1 displays the energy as a function of λ for Be where the 3-RDM in Eqs. (17)–(19) is reconstructed with the M functional. The ACSE energy converges to 0.2 mH above the full configuration interaction (FCI) energy.

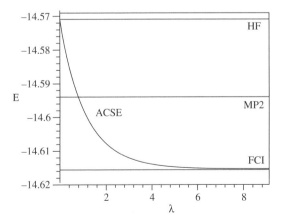

Figure 1. For Be the ACSE energy converges to 0.2 mH above the FCI energy.

TABLE I
Ground-State Energies from the ACSE with V, NY, and M 3-RDM Reconstructions Compared with the Energies from Several Wavefunction Methods, Including Hartree–Fock (HF), Second-Order Many-Body Perturbation Theory (MP2), Coupled-Cluster Singles–Doubles (CCSD), and Full Configuration Interaction (FCI), for Molecules in Valence Double-Zeta Basis Sets.[a]

		Energy Error (mH)					
		Wavefunction Methods			Anti-Hermitian CSE Methods		
System	FCI Energies (H)	HF	MP2	CCSD	V	NY	M
Be	−14.615569	44.663	21.615	0.020	−3.124	0.589	0.221
BH	−25.173472	60.076	23.910	0.775	−6.964	−1.282	0.317
H_2O	−76.141146	132.000	7.964	1.658	−3.172	1.199	0.988
NH_3	−56.303459	127.751	14.654	1.773	−2.835	1.052	0.942
HF	−100.145846	123.952	3.669	1.592	−0.825	2.137	2.064
N_2	−109.104089	225.950	−2.310	8.311	−12.640	−0.297	0.321

[a] The ACSE with N or M reconstruction produces 98–100% of the correlation energy, which markedly improves upon the 71–96% recovered by the CSE with N or M reconstruction [18]. Using the second-order N or M reconstruction is critical for consistently matching or exceeding the energy accuracy of CCSD. The difference between the ACSE and the MP2 energies highlights that each of the ACSE methods contains all second- and third-order as well as higher-order correlation effects.

For the beryllium atom and five molecules, Table I presents the errors in the ground-state correlation energies from the ACSE with the V, NY, and M 3-RDM reconstructions as well as several wavefunction methods, including Hartree–Fock (HF), second-order many-body perturbation theory (MP2), coupled-cluster singles–doubles (CCSD), and FCI. The ACSE methods are not variational, but the ACSE with V reconstruction consistently yields a lower bound on the ground-state energy. Supplementing the V reconstruction with either the NY or M estimate of the cumulant 3-RDM greatly improves the energy of each molecule with the largest improvement being observed for the triple-bonded N_2. The ACSE with NY or M reconstruction produces 98–100% of the correlation energy, which markedly improves upon the 71–96% recovered by the CSE with NY or M reconstruction [18]. Using the second-order NY or M reconstruction is critical for consistently matching or exceeding the energy accuracy of CCSD. The difference between the ACSE and the MP2 energies highlights that each of the ACSE methods contains all second- and third-order as well as higher-order correlation effects.

Earlier iterative solutions of the CSE for the 2-RDM often required that the 2-RDM be adjusted to satisfy important N-representability conditions in a process called *purification* [18, 24]. The solution of the ACSE automatically maintains the N-representability of the 2-RDM within the accuracy of the 3-RDM reconstruction. Necessary N-representability conditions require keeping the eigenvalues of three different forms of the 2-RDM, known as the ^2D, ^2Q, and ^2G

TABLE II
The 2-RDMs from Solving the ACSE Very Nearly Satisfy Known
N-Representability Conditions, Which Require the Eigenvalues of Three
Forms of the 2-RDM Matrix, Known as ^2D, ^2Q, and ^2G, to be Nonnegative.[a]

System	Method	Lowest Eigenvalue of 2-RDM Matrices		
		^2D	^2Q	^2G
BeH$_2$	V	−0.00092	−0.00011	−0.00054
	NY	−0.00002	0.00012	−0.00003
	M	−0.00009	0.00002	−0.00009
NH$_3$	V	−0.00113	−0.00022	−0.00038
	NY	−0.00000	0.00004	−0.00000
	M	−0.00001	0.00006	−0.00002

[a] In general, the NY and M reconstructions of the 3-RDM decrease the absolute value of the most negative eigenvalue by an order of magnitude.

matrices, nonnegative [36]. For H$_2$O and CH$_4$, Table II shows the lowest eigenvalues of these matrices, normalized to $N(N-1)$, $(r-N)(r-N-1)$, and $N(r-N+1)$, where r is the rank of the spin-orbital basis set. The largest negative eigenvalue for each of these matrices is three-to-five orders of magnitude smaller than the largest positive eigenvalue, which is near unity. For NH$_3$ the NY and M 3-RDM reconstructions decrease the absolute value of the most negative eigenvalue from the V reconstruction by an order of magnitude. Results similar to Table II are obtained for the other molecules in Table I.

Figure 2 shows the energy of BeH$_2$ as a function of the asymmetric stretch of one hydrogen, where the energy is computed by the ACSE with NY

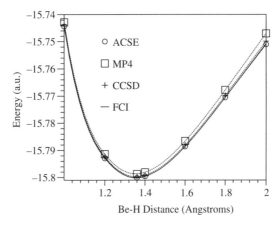

Figure 2. The asymmetric stretch in BeH$_2$ from the ACSE with NY reconstruction is compared with the stretch from MP4, CCSD, and FCI.

reconstruction, fourth-order many-body perturbation theory (MP4), coupled-cluster with single and double excitations (CCSD), and full configuration interaction (FCI). The absolute error in the energy from the ACSE ranges in the interval [0.083, 2.0] from a minimum of 0.02 mH at $R = 1.6$ Å to a maximum of 0.3 mH at $R = 2.0$ Å. The ACSE with NY reconstruction is slightly more accurate than CCSD, with CCSD having a maximum error of 0.9 mH at $R = 2.0$ Å. The significant improvement of the MP4 curve by the ACSE illustrates the physical significance of higher-order correlation effects included in the solution of the ACSE.

VI. SOME CONNECTIONS AND A SECOND LOOK AHEAD

As the applications demonstrate, the ACSE significantly enhances the accuracy of the CSE method for the direct determination of the ground-state 2-RDM and its energy without the many-electron wavefunction. The ACSE is solved by propagating a system of initial-value differential equations whose solution optimizes the 2-RDM with a series of infinitesimal unitary transformations. Unlike the CSE, the ACSE translates the first-order reconstruction of the 3-RDM (V) into molecular energies that contain all third-order and many higher-order correlation effects. The accuracy and N-representability of the energies and 2-RDMs can be enhanced further with the second-order reconstructions of Nakatsuji–Yasuda (NY) and Mazziotti (M). The ACSE with these second-order reconstructions produces energies that are competitive with the best wavefunction methods of comparable computational efficiency.

The ACSE has important connections to other approaches to electronic structure including: (i) variational methods that calculate the 2-RDM directly [36–39] and (ii) wavefunction methods that employ a two-body unitary transformation including canonical diagonalization [22, 29, 30], the effective valence Hamiltonian method [31, 32], and unitary coupled cluster [33–35]. A 2-RDM that is representable by an ensemble of N-particle states is said to be *ensemble N-representable*, while a 2-RDM that is representable by a single N-particle state is said to be *pure N-representable*. The variational method, within the accuracy of the N-representability conditions, constrains the 2-RDM to be ensemble N-representable while the ACSE, within the accuracy of 3-RDM reconstruction, constrains the 2-RDM to be pure N-representable. The ACSE and variational methods, therefore, may be viewed as complementary methods that provide approximate solutions to, respectively, the pure and ensemble N-representability problems.

Both the effective valence Hamiltonian method [31, 32] and unitary coupled cluster [33–35] employ a single two-body unitary transformation. In the effective valence Hamiltonian method [31, 32], the unitary transformation, selected by perturbation theory, is applied to the Hamiltonian to produce an effective valence Hamiltonian that can be diagonalized in a truncated Hilbert space of valence (or active) orbitals. The aim of the effective valence Hamiltonian method is to produce both ground- and excited-state wavefunctions and energies that include multireference correlation effects. In unitary coupled cluster [33, 35], the transformation, computed from the stationary condition for the ground-state energy, is applied to the Hartree–Fock wavefunction to produce a correlated approximation of the wavefunction and its energy. The unitary coupled cluster is a variant of the traditional coupled cluster method in which the exponential transformation is restricted to be unitary. In canonical diagonalization the Hamiltonian is transformed by a series of two-body unitary transformations [29]. Canonical diagonalization and the cumulant expansion from the CSE literature [8, 9, 11, 13–15] have recently been combined as discussed in the next chapter [30]. This approach to canonical diagonalization [30], which can be interpreted as a solution of the ACSE in the Heisenberg representation, is similar to the solution of the ACSE with V reconstruction.

Both canonical diagonalization and the ACSE differ significantly from the effective valence Hamiltonian method and the unitary coupled-cluster method in their use of a series of unitary transformations. The ACSE method differs from the cumulant approach to canonical diagonalization in that (i) the ACSE, formulated in the Schrödinger representation, produces both an energy and a 2-RDM, (ii) second-order formulas for the cumulant 3-RDM by Nakatsuji and Yasuda [5, 6] and Mazziotti [11, 15] are implemented in the reconstruction of the 3-RDM, and (iii) the infinitesimal unitary transformations are selected in the ACSE to minimize the energy along its gradient. Calculation of the 2-RDM is important for not only computing properties but also checking the N-representability of the ACSE solution.

Future research will (i) optimize the present implementation of the ACSE and (ii) explore the use of different initial 2-RDMs to initiate the solution of the ACSE. In an optimized form, the ACSE scales in floating-point operations as r^6 and in memory as r^4, where r is the number of spatial orbitals. While the calculations in this chapter employ an initial 2-RDM from the Hartree–Fock method, the ACSE method permits the selection of any initial 2-RDM, including a 2-RDM from a multi-reference self-consistent-field calculation. With this flexibility the ACSE method can be adapted to treat strong multireference correlation effects that are often important at nonequilibrium geometries. Building on the CSE, the ACSE yields 98–100% of the correlation energy and accurate 2-RDMs. Both the accuracy and the N-representability of the 2-RDMs are controlled by the reconstruction without any additional purification. The ACSE in conjunction with the variational 2-RDM method opens a new frontier for the accurate calculation of many-electron quantum mechanics.

Acknowledgments

The author expresses his appreciation to Dudley Herschbach, Herschel Rabitz, John Coleman, and Alexander Mazziotti for their support and encouragement. The author thanks the NSF, the Henry-Camille Dreyfus Foundation, the Alfred P. Sloan Foundation, and the David-Lucile Packard Foundation for their support.

References

1. F. Colmenero, C. Perez del Valle, and C. Valdemoro, Approximating q-order reduced density-matrices in terms of the lower-order ones. 1. General relations. *Phys. Rev. A* **47**, 971 (1993).
2. F. Colmenero and C. Valdemoro, Approximating q-order reduced density-matrices in terms of the lower-order ones. 2. Applications. *Phys. Rev. A* **47**, 979 (1993).
3. F. Colmenero and C. Valdemoro, Self-consistent approximate solution of the 2nd-order contracted Schrödinger equation. *Int. J. Quantum Chem.* **51**, 369 (1994).
4. C. Valdemoro, L. M. Tel, and E. Pérez-Romero, The contracted Schrödinger equation: some results. *Adv. Quantum Chem.* **28**, 33 (1997).
5. H. Nakatsuji and K. Yasuda, Direct determination of the quantum-mechanical density matrix using the density equation. *Phys. Rev. Lett.* **76**, 1039 (1996).
6. K. Yasuda and H. Nakatsuji, Direct determination of the quantum-mechanical density matrix using the density equation. 2. *Phys. Rev. A* **56**, 2648 (1997).
7. D. A. Mazziotti, Contracted Schrödinger equation: determining quantum energies and two-particle density matrices without wave functions. *Phys. Rev. A* **57**, 4219 (1998).
8. D. A. Mazziotti, Approximate solution for electron correlation through the use of Schwinger probes. *Chem. Phys. Lett.* **289**, 419 (1998).
9. D. A. Mazziotti, 3,5-Contracted Schrödinger equation: determining quantum energies and reduced density matrices without wave functions. *Int. J. Quantum Chem.* **70**, 557 (1998).
10. K. Yasuda, Direct determination of the quantum-mechanical density matrix: Parquet theory. *Phys. Rev. A* **59**, 4133 (1999).
11. D. A. Mazziotti, Pursuit of N-representability for the contracted Schrödinger equation through density-matrix reconstruction. *Phys. Rev. A* **60**, 3618 (1999).
12. D. A. Mazziotti, Comparison of contracted Schrödinger and coupled-cluster theories. *Phys. Rev. A* **60**, 4396 (1999).
13. W. Kutzelnigg and D. Mukherjee, Cumulant expansion of the reduced density matrices. *J. Chem. Phys.* **110**, 2800 (1999).
14. D. A. Mazziotti, Cumulants and the contracted Schrödinger equation, in *Many-Electron Densities and Density Matrices* (J. Cioslowski, ed.), Kluwer, Boston, 2000.
15. D. A. Mazziotti, Complete reconstruction of reduced density matrices. *Chem. Phys. Lett.* **326**, 212 (2000).
16. A. J. Coleman and V. I. Yukalov. *Reduced Density Matrices: Coulson's Challenge*, Springer-Verlag, New York, 2000.
17. W. Kutzelnigg and D. Mukherjee, Irreducible Brillouin conditions and contracted Schrödinger equations for n-electron systems. I. The equations satisfied by the density cumulants. *J. Chem. Phys.* **114**, 2047 (2001).
18. D. A. Mazziotti, Correlated purification of reduced density matrices. *Phys. Rev. E* **65**, 026704 (2002).

19. D. A. Mazziotti, A variational method for solving the contracted Schrödinger equation through a projection of the N-particle power method onto the two-particle space. *J. Chem. Phys.* **116**, 1239 (2002).
20. J. M. Herbert and J. E. Harriman, Contraction relations for Grassmann products of reduced density matrices and implications for density matrix reconstruction. *J. Chem. Phys.* **117**, 7464 (2002).
21. W. Kutzelnigg and D. Mukherjee, Irreducible Brillouin conditions and contracted Schrödinger equations for n-electron systems. IV. Perturbative analysis. *J. Chem. Phys.* **120**, 7350 (2004).
22. D. A. Mazziotti, Exactness of wave functions from two-body exponential transformations in many-body quantum theory. *Phys. Rev. A* **69**, 012507 (2004).
23. M. D. Benayoun, A. Y. Lu, and D. A Mzzioti, Invariance of the cumulant expansion under 1-particle unitary transformations in reduced density matrix theory. *Chem. Phys. Lett.* **387**, 485 (2004).
24. D. R. Alcoba and C. Valdemoro, Spin structure and properties of the correlation matrices corresponding to pure spin states: controlling the S-representability of these matrices. *Int. J. Quantum Chem.* **102**, 629 (2005).
25. D. R. Alcoba and F. J. Casquero, L. M. Tel, E. Pérez-Romero, and C. Valdemoro, Convergence enhancement in the iterative solution of the second-order contracted Schrödinger equation. *Int. J. Quantum Chem.* **102**, 620 (2005).
26. D. A. Mazziotti, Quantum chemistry without wave functions: two-electron reduced density matrices. *Acc. Chem. Res.* **39**, 207 (2006).
27. D. A. Mazziotti, Anti-Hermitian contracted Schrödinger equation: direct determination of the two-electron reduced density matrices of many-electron molecules. **97**, 143002 (2006).
28. D. A. Mazziotti and R. M. Erdahl, Uncertainty relations and reduced density matrices: mapping many-body quantum mechanics onto four particles. *Phys. Rev. A* **63**, 042113 (2001).
29. S. R. White, Numerical canonical transformation approach to quantum many-body problems. *J. Chem. Phys.* **117**, 7472 (2002).
30. T. Yanai and G. K. L. Chan, Canonical transformation theory for dynamic correlations in multireference problems, in *Reduced-Density-Matrix Mechanics: With Application to Many-Electron Atoms and Molecules, A Special Volume of Advances in Chemical Physics, Volume 134* (D.A. Mazziotti, ed.), Wiley, Hoboken, NJ, 2007.
31. M. G. Sheppard and K. F. Freed, Effective valence shell Hamiltonian calculations using third-order quasi-degenerate many-body perturbation theory. *J. Chem. Phys.* **75**, 4507 (1981).
32. J. E. Stevens, R. K. Chaudhuri, and K. F. Freed, Global three-dimensional potential energy surfaces of H_2S from the *ab initio* effective valence shell Hamiltonian method. *J. Chem. Phys.* **105**, 8754 (1996).
33. W. Kutzelnigg, in *Methods of Electronic Structure Theory* (H. F. Schaefer III, ed.), Plenum Press, New York, 1977.
34. M. R. Hoffmann and J. A. Simons, A unitary multiconfigurational coupled-cluster method—theory and applications. *J. Chem. Phys.* **88**, 993 (1988).
35. R. J. Bartlett, S. A. Kucharski, and J. Noga, Alternative coupled-cluster anstaze. 2. The unitary coupled-cluster method. *Chem. Phys. Lett.* **155**, 133 (1989).
36. D. A. Mazziotti, Realization of quantum chemistry without wavefunctions through first-order semidefinite programming. *Phys. Rev. Lett.* **93**, 213001 (2004).
37. G. Gidofalvi and D. A. Mazziotti, Application of variational reduced-density-matrix theory to organic molecules. *J. Chem. Phys.* **122**, 094107 (2005).

38. J. R. Hammond and D. A. Mazziotti, Variational reduced-density-matrix calculations on radicals: an alternative approach to open-shell *ab initio* quantum chemistry. *Phys. Rev. A* **73**, 012509 (2006).
39. G. Gidofalvi and D. A. Mazziotti, Variational reduced-density-matrix theory applied to the potential energy surfaces of carbon monoxide in the presence of electric fields. *J. Phys. Chem. A* **110**, 5481 (2006).
40. *Handbook of Chemistry and Physics*, 79th ed., Chemical Rubber Company, Boca Raton, FL, 1998.
41. T. H. Dunning, Jr. and P. J. Hay, in *Methods of Electronic Structure Theory* (H. F. Schaefer III, ed.), Plenum Press, New York, 1977.
42. M. W. Schmidt et al., General atomic and molecular electronic-structure system. *J. Comput. Chem.* **14**, 1347 (1993).

CHAPTER 13

CANONICAL TRANSFORMATION THEORY FOR DYNAMIC CORRELATIONS IN MULTIREFERENCE PROBLEMS

GARNET KIN-LIC CHAN AND TAKESHI YANAI

Department of Chemistry and Chemical Biology, Cornell University, Ithaca, New York 14853-1301 USA

CONTENTS

I. Introduction
II. Theory
    A. General Considerations
    B. Canonical Transformation Theory
    C. The Linearized Model
    D. Perturbative Analysis and Relation to Coupled-Cluster Theory
    E. Variation of the Reference
III. Implementation of the Linearized Canonical Transformation Theory
    A. Computational Algorithm
    B. Computational Scaling
    C. Classes of Excitations for the Exponential Operator
IV. Numerical Results
    A. Simultaneous Bond Breaking of Water Molecule with 6-31G and cc-pVDZ Basis Sets
    B. Bond Breaking of Nitrogen Molecules with 6-31G Basis Sets
    C. Comparison with MR-CISD and MR-LCCM on the Two-Configuration Reference Insertion of Be in H_2 Molecule
    D. Single- and Multireference Linearized CT for HF and BH Molecules
V. Conclusions
Acknowledgments
References

Dynamic correlation, associated with the scattering and relaxation of electrons into nonbonding degrees of freedom, must be described accurately to make quantitative predictions about electronic structure. This task can be complicated

---

Reduced-Density-Matrix Mechanics: With Application to Many-Electron Atoms and Molecules,
A Special Volume of Advances in Chemical Physics, Volume 134, edited by David A. Mazziotti.
Series editor Stuart A. Rice. Copyright © 2007 John Wiley & Sons, Inc.

by the presence of significant amounts of nondynamic correlation. We have recently developed a *canonical transformation* (CT) theory that targets dynamic correlations in bonding situations, where there is also significant nondynamic character. When combined with a suitable nondynamic correlation method, such as the *complete-active-space self-consistent field* or *density matrix renormalization group* theories, it achieves quantitative accuracies typical of equilibrium region *coupled-cluster theory* and better than that of *multireference perturbation theory*, for the complete potential energy surface ranging from equilibrium to dissociation. In this chapter, expanding on our earlier account [1], we describe the basic ideas and implementation of the canonical transformation theory and its relation to existing coupled-cluster and effective Hamiltonian theories, and we present and analyze calculations on several molecular bond-breaking reactions.

I. INTRODUCTION

In quantum chemistry, it is useful to distinguish between two kinds of electron correlation. The first is nondynamic, or strong, correlation. This is associated with the overlap of near-degenerate valence degrees of freedom and is characterized by multireference wavefunctions containing several determinants with large weights. A good treatment of nondynamic correlation is necessary to establish the correct qualitative electronic structure for a problem. The second kind of correlation is dynamic, or weak, correlation. This is associated with the scattering and relaxation of electrons into nonbonding degrees of freedom. A good treatment of dynamic correlation is necessary for fully quantitative predictions.

Many of the systems that are challenging for current ab initio quantum chemistry—for example, transition metal chemistry, molecular properties away from equilibrium geometries, open-shell molecules, and highly excited states—involve wavefunctions containing significant nondynamic character. For an accurate description of these systems, a theory must satisfy two requirements. First, it should be sufficiently flexible to encompass the very general types of wavefunctions associated with nondynamic correlation. Second, it should efficiently capture the many short-range dynamical scatterings necessary for quantitative accuracy, but whose treatment is complicated by the presence of nondynamic correlation.

In modest sized systems, we can treat the nondynamic correlation in an active space. For systems with up to 14 orbitals, the complete-active-space self-consistent field (CASSCF) theory provides a very satisfactory description [2, 3]. More recently, the ab initio density matrix renormalization group (DMRG) theory has allowed us to obtain a balanced description of nondynamic correlation for up to 40 active orbitals and more [4–13]. CASSCF and DMRG potential energy surfaces are qualitatively well behaved from equilibrium to dissociation. However, quantitative accuracy is not achieved in these theories since dynamic correlation is neglected. We understand that dynamic correlation is well captured by high-order perturbative approaches and, in particular, coupled-cluster (CC) theory [14–17]. However, coupled-cluster theory in its usual form does not describe nondynamic correlation. Thus coupled-cluster potential energy curves are quantitatively accurate near equilibrium, but not at stretched geometries. Multireference perturbation theories [18–21] present an approach to include both nondynamic and dynamic correlations, but they do not attain quantitative "chemical" accuracy—say, equal to that of coupled-cluster theory in the equilibrium region—because the perturbation theory can only practically be applied to low order. Thus the challenge remains to construct a general purpose theory with a complete, quantitative description of dynamic and nondynamic correlation that is appropriate to all bonding situations.

In the present chapter, we describe our recent progress in developing such a theory—which we call *canonical transformation* or *CT theory*—to describe dynamic correlation in situations where there can also be significant nondynamic character. As a starting point we assume that a suitable description of nondynamic correlation (e.g., through a CASSCF or DMRG calculation) can be obtained. When this is then combined with the CT theory, a complete and quantitative description of a potential energy surface from equilibrium to stretched geometries is achieved. The CT theory is size consistent and uses a unitary exponential description of dynamic correlation. The computational cost is of the same order as a *single-reference* coupled-cluster method with the same level of excitations. Our theory employs a generalized cumulant expansion to simplify the complicated expressions in the energy and amplitude equations. In this way, we avoid the difficulties of working with a complex multireference wavefunction. The work described here was first presented in Ref. [1], and this chapter is intended to provide an expanded account.

The most direct influence on the current work is the recent canonical diagonalisation theory of White [22]. This, in turn, is an independent redevelopment of the flow-renormalization group (flow-RG) of Wegner [23] and Glazek and Wilson [24]. As pointed out by Freed [25], canonical transformations are themselves a kind of renormalization, and our current theory may be viewed also from a renormalization group perspective.

We will not attempt to survey the vast existing literature for the problem of constructing dynamic correlations when nondynamic correlations are present. Instead, we refer the reader to a number of excellent recent reviews [17, 26]. The many different theories in the literature all share common elements but typically employ different approximations or adopt different points of view. It is common to stress either a wavefunction ansatz picture or an effective

Hamiltonian picture and while these pictures are essentially equivalent, we can usefully classify previous contributions along these lines. Of those that emphasize a wavefunction language, multireference coupled-cluster theories are most closely related to our current work. These have been reviewed in the article by Paldus and Li [17]. In these methods difficulties can arise from the need to handle very complicated multireference wavefunctions in the coupled-cluster equations. In the CT theory, by using a cumulant expansion, we avoid a direct manipulation of the multireference wavefunction and instead characterize the nondynamic correlation in the reference using only the one- and two-particle reduced density matrices. A recent multireference CC ansatz with a similar emphasis on simplicity, but that uses the T1 and T2 amplitudes to characterize the nondynamic correlation in the reference, is the *tailored CC theory* of Kinoshita et al. [27, 28]. Coupled-cluster theories based on the unitary operators used in the CT theory are less common. We mention in particular the work of Hoffmann and Simons [29], who formulated a unitary multireference coupled-cluster theory. Also in this context, we refer to the earlier work of Kutzelnigg [30, 31], Bartlett et al. [32, 33], and Pal and co-workers [34, 35] on single-reference unitary coupled-cluster theory. Methods that emphasize the effective Hamiltonian perspective have been reviewed in an article by Hoffmann [26]. We mention, in particular, the pioneering work on effective valence shell Hamiltonians by Freed [25], and also the work on generalized Van Vleck transformations by Kirtman [36]. These methods are both based on perturbation theory. In our CT theory, we do not use a perturbative approach. Rather, through the use of a cumulant closure, we construct effective Hamiltonians with infinite order contributions. The work of Freed has also been concerned with developing an ab initio route to the simple descriptions of electron correlation embodied in semiempirical Hamiltonians. While this is not directly addressed in the current chapter, this concern also motivates our current work.

In Section II we present the basic ideas and equations of the canonical transformation theory. We introduce two numerical models, namely, the *linearized canonical transformation* (L-CTD and L-CTSD) models, and carry out a perturbative analysis of the two. We also demonstrate that CT theory can be interpreted as a familiar generalization of Hartree–Fock theory to a two-particle mean-field theory that includes electron correlation. In Section III we describe the computational implementation of the canonical transformation theory and discuss some issues of convergence. In Section IV, we report ground-state calculations on the water and nitrogen potential energy curves, the BeH_2 insertion reaction, and hydrogen fluoride and boron hydride bond breaking. Finally, our conclusions and future directions of the theory are presented in Section V.

II. THEORY

A. General Considerations

The generic chemical problem involving both dynamic and nondynamic correlation is illustrated in Fig. 1. The orbitals are divided into two sets: the active orbitals, usually the valence orbitals, which display partial occupancies (assuming spin orbitals) very different from 0 or 1 for the state of interest, and the external orbitals, which are divided into the core (largely occupied in the target state) or virtual (largely unoccupied in the target state) orbitals. The asymmetry between

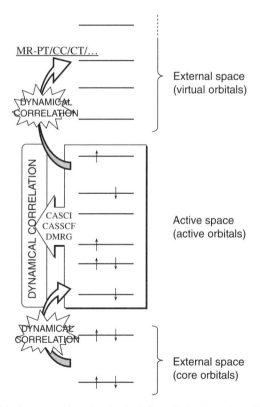

Figure 1. Multireference problems involve both dynamical and nondynamical correlation. The nondynamical correlation is accounted for by the CASCI/CASSCF/DMRG wavefunction, which is made of multiple configurations generated in the active space with a fixed number of active electrons. The dynamical correlation is recovered on top of the multiconfigurational reference by correlating the active orbitals with orbitals in the external space (i.e., core and virtual orbitals.)

the core and virtual orbitals can formally be removed by transforming to a core Fermi vacuum, where all core states are filled (although in our numerical work, we retain the distinction between core and virtuals for reasons of efficiency). The electronic Hamiltonian can be separated into active and external contributions via

$$\hat{H} = \hat{H}_{\text{act}} + \hat{H}_{\text{act-ext}} + \hat{H}_{\text{ext}} \qquad (1)$$

Nondynamic correlation is associated with active–active space correlations, while dynamic correlation is associated with correlations between the active–external and external–external spaces.

We begin by assuming that we have a reference wavefunction Ψ_0, which accounts for the nondynamic correlation and which (relative to the core Fermi vacuum) exists only in the active space. (In general, we work with a single starting state, although extensions to multiple starting states are easily considered.) An exact eigenfunction Ψ of $\hat{H}$ that incorporates the remaining dynamical correlations out of the active space can be obtained by an appropriate canonical transformation of Ψ_0,

$$\Psi = \hat{U}\Psi_0 \qquad (2)$$

A canonical transformation, which may be single particle or many particle in nature, is one that preserves the commutation relations of the particles involved. Strictly speaking, it need not be unitary (it need only be isometric; e.g., see Ref. [37]), but this distinction is less important for calculational purposes and we shall henceforth consider only unitary canonical transformations where $\hat{U}$ satisfies $\hat{U}\hat{U}^\dagger = \hat{1}$.

If Ψ_0 is itself an eigenfunction of $\hat{H}_{\text{act}}$, then no active–active rotations are necessary and $\hat{U}$ rotates only between the active–external and external–external spaces. Without loss of generality, we can write $\hat{U}$ in exponential form, namely,

$$\hat{U} = e^{\hat{A}} \qquad (3)$$

$$\hat{A} = \sum_{ae} A^a_e c^\dagger_a c_e + \sum_{e_1 e_2} A^{e_1}_{e_2} c^\dagger_{e_1} c_{e_2}$$

$$+ \sum_{a_1 a_2 e_3 e_4} A^{a_1 a_2}_{e_3 e_4} c^\dagger_{a_1} c^\dagger_{a_2} c_{e_3} c_{e_4}$$

$$+ \sum_{a_1 a_2 a_3 e_4} A^{a_1 a_2}_{a_3 e_4} c^\dagger_{a_1} c^\dagger_{a_2} c_{a_3} c_{e_4}$$

$$+ \sum_{a_1 e_2 e_3 e_4} A^{a_1 e_2}_{e_3 e_4} c^\dagger_{a_1} c^\dagger_{e_2} c_{e_3} c_{e_4}$$

$$+ \sum_{e_1 e_2 e_3 e_4} A^{e_1 e_2}_{e_3 e_4} c^\dagger_{e_1} c^\dagger_{e_2} c_{e_3} c_{e_4} + \cdots + h.c. \qquad (4)$$

where a, e denote active, external indices, respectively, and all amplitudes A are anti-Hermitian to ensure unitarity. $\hat{A}$ may be decomposed as the difference of an operator and its Hermitian conjugate, $\hat{A} = \hat{T} - T^\dagger$. The terms in Eq. (4) include not only the usual single and double excitations, but also semi-internal excitations that couple relaxation in the active space with external excitations.

Equivalently, the unitary operator can be viewed as transforming the reference Hamiltonian. This yields an effective Hamiltonian that has Ψ_0 as an eigenfunction and the exact eigenenergy E of the target state as its eigenvalue. Thus

$$\hat{\bar{H}} = e^{-\hat{A}} \hat{H} e^{\hat{A}} \tag{5}$$

$$\hat{\bar{H}} \Psi_0 = E \Psi_0 \tag{6}$$

$$E = \langle \Psi_0 | \hat{\bar{H}} | \Psi_0 \rangle \tag{7}$$

Note that in contrast to a general similarity transformation (e.g., as found in the usual coupled-cluster theory) the canonical transformation produces a Hermitian effective Hamiltonian, which is computationally very convenient. When $\hat{U}$ is expressed in exponential form, the effective Hamiltonian can be constructed termwise via the formally infinite Baker–Campbell–Hausdorff (BCH) expansion,

$$\hat{\bar{H}} = \hat{H} + [\hat{H}, \hat{A}] + \tfrac{1}{2}[[\hat{H}, \hat{A}], \hat{A}] + \cdots \tag{8}$$

The two pictures above (where $\hat{U}$ is viewed as acting on the wavefunction or acting on the Hamiltonian) are clearly mathematically equivalent. However, it is worth considering their physical equivalence in the language of canonical transformations. (A similar discussion of this issue may also be found in White [22].) In the first picture the Hamiltonian $\hat{H}$, wavefunctions Ψ_0 and Ψ, and transformation $\hat{U}$ are associated with particles defined by the operators $c_i^\dagger, c_j$; thus

$$H = \sum_{ij} t_{ij} c_i^\dagger c_j + \sum_{ijkl} v_{ijkl} c_i^\dagger c_j^\dagger c_k c_l \tag{9}$$

$$\Psi_0 = \Psi_0(c_i^\dagger, c_j^\dagger, \ldots)|-\rangle_c \tag{10}$$

$$\Psi = \Psi(c_i^\dagger, c_j^\dagger, \ldots)|-\rangle_c \tag{11}$$

$$\hat{U} = e^{\hat{A}(c_i^\dagger, c_j)} = e^{\sum_{ij} A_{ij} c_i^\dagger c_j + \sum_{ijkl} A_{ijkl} c_i^\dagger c_j^\dagger c_k c_l + \cdots} \tag{12}$$

The transformation $\hat{U}$ incorporates the additional correlations when going from the reference wavefunction Ψ_0 to the target wavefunction Ψ. In the second picture, we appear to have a different Hamiltonian $\hat{\bar{H}}$ and eigenstate Ψ_0.

However, they represent the same physical quantities as in the first picture, and only our choice of coordinates has changed. We can define a new set of creation and annihilation operators $\bar{c}_i^\dagger, \bar{c}_j$ through

$$\bar{c}_i^\dagger = e^{\hat{A}(c_i^\dagger, c_j)} c_i^\dagger e^{-\hat{A}(c_i^\dagger, c_j)} \tag{13}$$

$$\bar{c}_j = e^{\hat{A}(c_i^\dagger, c_j)} c_j e^{-\hat{A}(c_i^\dagger, c_j)} \tag{14}$$

$$c_i^\dagger = e^{-\hat{A}(\bar{c}_i^\dagger, \bar{c}_j)} \bar{c}_i^\dagger e^{\hat{A}(\bar{c}_i^\dagger, \bar{c}_j)} \tag{15}$$

$$c_j = e^{-\hat{A}(\bar{c}_i^\dagger, \bar{c}_j)} \bar{c}_j e^{\hat{A}(\bar{c}_i^\dagger, \bar{c}_j)} \tag{16}$$

Compared to the original particles, the new particles associated with $\bar{c}_i^\dagger, \bar{c}_j$ have been dressed by the dynamic correlations in $\hat{U}$. Consequently, we can represent the exact wavefunction as a simpler function of the new particles,

$$\Psi_0(\bar{c}_i^\dagger, \bar{c}_j^\dagger, \ldots)|-\rangle_{\bar{c}} = \Psi(c_i^\dagger, c_j^\dagger, \ldots)|-\rangle_c \tag{17}$$

with the equivalence between the two established by

$$\begin{aligned}\Psi_0(\bar{c}_i^\dagger, \bar{c}_j^\dagger, \ldots)|-\rangle_{\bar{c}} &= \Psi_0(e^{\hat{A}} c_i^\dagger e^{-\hat{A}}, e^{\hat{A}} c_j^\dagger e^{-\hat{A}}, \ldots)|-\rangle_{\bar{c}} \\ &= e^{\hat{A}} \Psi_0(c_i^\dagger, c_j^\dagger \ldots) e^{-\hat{A}}|-\rangle_{\bar{c}} \\ &= \Psi(c_i^\dagger, c_j^\dagger \ldots)|-\rangle_c \end{aligned} \tag{18}$$

Similarly, the Hamiltonian, in terms of the new particles, must involve different matrix elements and thus

$$\begin{aligned}\hat{\bar{H}}(\bar{c}_i^\dagger, \bar{c}_j) &= \hat{H}(c_i^\dagger, c_j) \\ &= e^{-\hat{A}(\bar{c}_i^\dagger, \bar{c}_j)} \hat{H}(\bar{c}_i^\dagger, \bar{c}_j) e^{\hat{A}(\bar{c}_i^\dagger, \bar{c}_j)} \\ &= e^{-\hat{A}(\bar{c}_i^\dagger, \bar{c}_j)} \left[\sum_{ij} t_{ij} \bar{c}_i^\dagger \bar{c}_j + \sum_{ijkl} v_{ijkl} \bar{c}_i^\dagger \bar{c}_j^\dagger \bar{c}_k^\dagger \bar{c}_l^\dagger \right] e^{\hat{A}(\bar{c}_i^\dagger, \bar{c}_j)} \\ &= \sum_{ij} \bar{t}_{ij} \bar{c}_i^\dagger \bar{c}_j + \sum_{ijkl} \bar{v}_{ijkl} \bar{c}_i^\dagger \bar{c}_j^\dagger \bar{c}_k^\dagger \bar{c}_l^\dagger + \cdots \end{aligned} \tag{19}$$

The last relation is simply our original Eq. (5), demonstrating the equivalence of the two Hamiltonians.

If $\hat{U}$ is obtained exactly, the effective Hamiltonian $\hat{\bar{H}}$ has no matrix elements between the active and external spaces—that is, the two spaces are completely decoupled—and all matrix elements containing both active and external indices are zero. Although the initial Hamiltonian only contains two-particle interactions between the active and external spaces, to achieve a full decoupling the exact $\hat{A}$ must contain contributions from single-particle to N-particle operators. This is because the decoupling of lower-particle rank terms in the Hamiltonian reintroduces matrix elements of higher-particle rank.

B. Canonical Transformation Theory

Up to this point our discussion of canonical transformations has been exact. We now proceed to the specific approximations that characterize our formulation of CT theory and discuss their relationship with approximations commonly made in other theories involving canonical (i.e., unitary) transformations.

First, we note that the determination of the exact many-particle operator $\hat{U}$ is equivalent to solving for the full interacting wavefunction Ψ. Consequently, some approximation must be made. The ansatz of Eq. (2) recalls perturbation theory, since (as contrasted with the most general variational approach) the target state Ψ is parameterized in terms of a reference Ψ_0. A perturbative construction of $\hat{U}$ is used in the effective valence shell Hamiltonian theory of Freed and the generalized Van Vleck theory of Kirtman. However, a more general way forward, which is not restricted to low order, is to determine U (and the associated amplitudes in $\hat{A}$) directly. In our CT theory, we adopt the projection technique as used in coupled-cluster theory [17]. By projecting onto excited determinants, we obtain a set of nonlinear amplitude equations, namely,

$$\langle \Psi_0 | \hat{\bar{H}} \hat{\gamma}_\alpha | \Psi_0 \rangle = 0 \qquad (20)$$

where $\hat{\gamma}_\alpha$ denotes the operator $c^\dagger_{e_2} c^\dagger_{e_1} c_{a_2} c_{a_1}$ and α denotes the indices $a_1 a_2, e_1 e_2$. From the Hermiticity of $\hat{\bar{H}}$, and the fact that (relative to the core Fermi vacuum) Ψ_0 exists only in the active space, it follows that

$$\langle \Psi_0 | [\hat{\bar{H}}, \hat{\gamma}_\alpha - \hat{\gamma}^\dagger_\alpha] | \Psi_0 \rangle = 0 \qquad (21)$$

In this form, the amplitude equations (21) have been previously studied by Kutzelnigg and named the generalized Brillouin conditions [38].

The primary questions that remain to be answered are the following: to evaluate the energy and amplitude equations (7) and (21), we need to (i) construct $\hat{\bar{H}}$ and (ii) have some means of evaluating expectation values of operators $\hat{O}$ with the reference Ψ_0.

Let us first discuss (i). The primary difficulty associated with the infinite BCH expansion comes from the fact that each term in the expansion generates operators of greater particle rank (i.e., involving a longer string of creation and annihilation operators) than the previous term. Thus it is necessary to assume some closure or truncation when constructing $\hat{\bar{H}}$. This is commonly cited as an obstacle to the adoption of unitary coupled-cluster theory, but as we shall see, a simple yet accurate closure can be found. More recently, this concern has also arisen in studies of generalized two-body exponential theories of correlation [39–42].

The approach typically taken in unitary coupled-cluster theories is to truncate either by stopping the BCH expansion at some low order (e.g., second) [29] or by keeping terms in the effective Hamiltonian such that it is correct through a given order of perturbation theory [33]. The accuracy of such a truncation is therefore tied to the accuracy of the underlying perturbation series. We start with a different nonperturbative approach used in the theory of canonical diagonalization, that is, to restrict the *form* of $\hat{\tilde{H}}$ to contain only certain classes of operators [22]. (In our numerical work, we restrict $\hat{\tilde{H}}$ to contain only one- and two-particle operators.) If we then neglect all the higher particle-rank operators, we obtain the approximation used in canonical diagonalization. In CT theory, we go one step further and account for the higher particle-rank operators appearing in $\hat{\tilde{H}}$ in an approximate way. We achieve this by using an analogue of the *cumulant decomposition* to express high particle-rank operators in terms of lower particle-rank operators and effective fields. This may be regarded as generalizing Hartree–Fock theory, where the effective Hamiltonian (the Fock operator) contains an average over the two-body interaction with a density field (see also Section II.E).

The procedure is clearest with the aid of an example. Consider the first commutator in the BCH expansion, $[\hat{H}, \hat{A}]$. Let $\hat{A}$ be the two-particle operator,

$$\sum_{a_1 a_2 e_1 e_2} A^{a_1 a_2}_{e_1 e_2} c^\dagger_{a_1} c^\dagger_{a_2} c_{e_1} c_{e_2}$$

and consider a two-particle term in the Hamiltonian, $V^{g_1 g_2}_{g_3 g_4} c^\dagger_{g_1} c^\dagger_{g_2} c_{g_3} c_{g_4}$, where g denotes a general index (i.e., a or e). Then the commutator of the two terms yields both two-particle (through double contraction) and three-particle (through single contraction) operators. We wish to decompose the new three-particle operators in terms of one- and two-particle quantities. Recall that the cumulant decomposition in statistical mechanics offers the best statistical decomposition of a high-particle-rank correlation function in terms of lower-particle-rank functions. In the context of reduced density matrices, it has been studied extensively by Valdemoro, Nakatsuji, Mazziotti, and others [43–48]. The cumulant decomposition for a three-particle density matrix element $\langle c^\dagger_i c^\dagger_j c^\dagger_k c_l c_m c_n \rangle$ is

$$\begin{aligned}\langle c^\dagger_i c^\dagger_j c^\dagger_k c_l c_m c_n \rangle &\Rightarrow 9 \langle c^\dagger_i c^\dagger_n \rangle \wedge \langle c^\dagger_j c^\dagger_k c_l c_m \rangle \\ &- 12 \langle c^\dagger_i c_n \rangle \wedge \langle c^\dagger_j c_m \rangle \wedge \langle c^\dagger_k c_l \rangle\end{aligned} \quad (22)$$

This is not quite what we need, as in the current context, we require a cumulant decomposition of a three-particle *operator*. We can construct an *operator* cumulant expansion by (i) requiring it to yield the same expectation value as the cumulant expansion for a reduced density matrix element, and (ii) keeping two-particle operators rather than single-particle operators, when the choice arises. This gives

$$c_i^\dagger c_j^\dagger c_k^\dagger c_l c_m c_n \Rightarrow 9\langle c_i^\dagger c_n^\dagger\rangle \wedge \left(c_j^\dagger c_k^\dagger c_l c_m\right) - 12\langle c_i^\dagger c_n\rangle \wedge \langle c_j^\dagger c_m\rangle \wedge \left(c_k^\dagger c_l\right)$$
$$= \langle c_i^\dagger c_n\rangle c_j^\dagger c_k^\dagger c_l c_m + \langle c_j^\dagger c_m\rangle c_k^\dagger c_i^\dagger c_n c_l + \langle c_k^\dagger c_l\rangle c_i^\dagger c_j^\dagger c_m c_n$$
$$- \langle c_i^\dagger c_l\rangle c_j^\dagger c_k^\dagger c_n c_m - \langle c_j^\dagger c_n\rangle c_k^\dagger c_i^\dagger c_m c_l - \langle c_k^\dagger c_m\rangle c_i^\dagger c_j^\dagger c_l c_n$$
$$- \langle c_i^\dagger c_m\rangle c_j^\dagger c_k^\dagger c_l c_n - \langle c_j^\dagger c_l\rangle c_k^\dagger c_i^\dagger c_n c_m - \langle c_k^\dagger c_n\rangle c_i^\dagger c_j^\dagger c_m c_l$$
$$- \tfrac{2}{3}\bigl\{\bigl(\langle c_i^\dagger c_n\rangle\langle c_j^\dagger c_m\rangle - \langle c_i^\dagger c_m\rangle\langle c_j^\dagger c_n\rangle\bigr)c_k^\dagger c_l$$
$$+ \bigl(\langle c_j^\dagger c_m\rangle\langle c_k^\dagger c_l\rangle - \langle c_j^\dagger c_l\rangle\langle c_k^\dagger c_m\rangle\bigr)c_i^\dagger c_n$$
$$+ \bigl(\langle c_k^\dagger c_l\rangle\langle c_i^\dagger c_n\rangle - \langle c_k^\dagger c_n\rangle\langle c_i^\dagger c_l\rangle\bigr)c_j^\dagger c_m \quad (23)$$
$$- \bigl(\langle c_i^\dagger c_l\rangle\langle c_j^\dagger c_m\rangle - \langle c_i^\dagger c_m\rangle\langle c_j^\dagger c_l\rangle\bigr)c_k^\dagger c_n$$
$$- \bigl(\langle c_j^\dagger c_n\rangle\langle c_k^\dagger c_l\rangle - \langle c_j^\dagger c_l\rangle\langle c_k^\dagger c_n\rangle\bigr)c_i^\dagger c_m$$
$$- \bigl(\langle c_k^\dagger c_m\rangle\langle c_i^\dagger c_n\rangle - \langle c_k^\dagger c_n\rangle\langle c_i^\dagger c_m\rangle\bigr)c_j^\dagger c_l$$
$$- \bigl(\langle c_i^\dagger c_m\rangle\langle c_j^\dagger c_n\rangle - \langle c_i^\dagger c_n\rangle\langle c_j^\dagger c_m\rangle\bigr)c_k^\dagger c_l$$
$$- \bigl(\langle c_j^\dagger c_l\rangle\langle c_k^\dagger c_m\rangle - \langle c_j^\dagger c_m\rangle\langle c_k^\dagger c_l\rangle\bigr)c_i^\dagger c_n$$
$$- \bigl(\langle c_k^\dagger c_n\rangle\langle c_i^\dagger c_l\rangle - \langle c_k^\dagger c_l\rangle\langle c_i^\dagger c_n\rangle\bigr)c_j^\dagger c_m\bigr\}$$

where $\wedge$ denotes an antisymmetrization over all indices with an associated factor $1/(P!)^2$ (P is the particle rank of the original operator) and $\langle\cdots\rangle$ denotes an average with the reference wavefunction Ψ_0 (this yields McWeeny normalization for the density matrices, i.e., $\mathrm{Tr}\langle c_i^\dagger c_j^\dagger c_m c_n\rangle = N(N-1)$, where N is the number of particles).

Unlike the density cumulant expansion, which can in principle be exact for certain states (such as Slater determinants), the operator cumulant expansion is never exact, in the sense that we cannot reproduce the full spectrum of a three-particle operator faithfully by an operator of reduced particle rank. However, if the density cumulant expansion is good for the state of interest, we expect the operator cumulant expansion to also be good for that state and also for states nearby.

With the above decomposition, the commutator $[\hat{H},\hat{A}]$ is reduced to an expression containing only terms of the form we wish to keep (i.e., one- and two-particle operators). Let us denote this approximate form of the commutator as $[\hat{H},\hat{A}]_{(1,2)}$, to indicate that the cumulant decomposition retains only one- and

TABLE I
Expansion of $\hat{\bar{H}} = \exp(-\hat{A})\hat{H}\exp(\hat{A})$ for H_2O ($R_{OH} = 1.0R_e$) with cc-pVDZ Basis Set[a]

n	$E(n)$ (E_h)	$E(n) - E(0)$	$\|[\ldots[[\hat{H},\hat{A}],\hat{A}]\ldots]/n!\|_2$
0	−76.075 858		
1	−76.408 264	−0.332 406	3.545
2	−76.238 122	−0.162 264	4.338×10^{-1}
3	−76.238 259	−0.162 401	2.272×10^{-2}
4	−76.238 185	−0.162 327	1.275×10^{-3}
5	−76.238 187	−0.162 329	6.812×10^{-5}
6	−76.238 187	−0.162 329	3.367×10^{-6}
7	−76.238 187	−0.162 329	1.164×10^{-7}
8	−76.238 187	−0.162 329	4.025×10^{-9}
9	−76.238 187	−0.162 329	1.092×10^{-10}

[a]The 2-norm of amplitudes ($A = A_1 + A_2$): $\|A_1\|_2 = 5.97 \times 10^{-2}$ and $\|A_2\|_2 = 1.48 \times 10^{-1}$.

two-particle operators. Then we can apply the procedure recursively, and thus the next commutator in the BCH expansion is approximated as

$$[[\hat{H},\hat{A}],\hat{A}] = [[\hat{H},\hat{A}]_{(1,2)},\hat{A}]_{(1,2)} \qquad (24)$$

Consequently, we can carry out the BCH expansion to arbitrarily high order without any increase in the complexity of the terms in the effective Hamiltonian. In practice, the expansion is carried out until convergence in a suitable norm of the operator coefficients is achieved, as illustrated in Table I. Rapid convergence is usually observed. Note that through the decomposition (23), the effective Hamiltonian depends on the one- and two-particle density matrices and therefore becomes state specific, much like the Fock operator in Hartree–Fock theory.

Size-consistency is a desirable feature of any approximate theory. Since we truncate in the *operator* space (as opposed to the Hilbert space of wavefunctions), the current approximation is naturally size-consistent. Consider two widely separated systems X and Y. Then we can construct two bases of creation/annihilation operators $c_X^\dagger, c_Y^\dagger$ that generate/destroy the Fock spaces of X and Y, respectively, and that commute by virtue of separation. In terms of these operators, the starting Hamiltonian is separable into components that act only on X and Y, respectively, $\hat{H} = \hat{H}_X + \hat{H}_Y$, and so too is the exponential operator $\exp(\hat{A}) = \exp(\hat{A}_X + \hat{A}_Y) = \exp(\hat{A}_X)\exp(\hat{A}_Y)$. Consequently, the effective Hamiltonian is also separable $\hat{\bar{H}} = \hat{\bar{H}}_X + \hat{\bar{H}}_Y$. The amplitude equations (21) are solvable separately in the X and Y spaces. Consequently, the total effective Hamiltonian is the sum of the corresponding effective Hamiltonians for the systems X and Y considered in isolation, the total wavefunction is a product, and the total energy is additive, as is required in a size-consistent theory.

We now discuss (ii), the evaluation of operator expectation values with the reference Ψ_0. We are interested in multireference problems, where Ψ_0 may be extremely complicated (i.e., a very long Slater determinant expansion) or a compact but complex wavefunction, such as the DMRG wavefunction. By using the cumulant decomposition, we limit the terms that appear in the effective Hamiltonian to only low-order (e.g., one- and two-particle operators), and thus we only need the one- and two-particle density matrices of the reference wavefunction to evaluate the expectation value of the energy in the energy expression (7). To solve the amplitude equations, we further require the commutator of $[\bar{H}, \hat{\gamma}_\alpha]$, which, for a two-particle effective Hamiltonian and two-particle operator $\hat{\gamma}_\alpha$, again involves the expectation value of three-particle operators. We therefore invoke the cumulant decomposition once more, and solve instead the modified amplitude equation

$$\langle \Psi_0 | [\hat{\bar{H}}_{(1,2)}, \hat{\gamma}_\alpha]_{(1,2)} | \Psi_0 \rangle = 0 \qquad (25)$$

This modified amplitude equation does not correspond to the minimization of the energy functional Eq. (7), and thus the generalized Hellmann–Feynman theorem [49] does not apply.

Consequently, with the simplifications above, all the working equations of the canonical transformation theory can be evaluated entirely in terms of a limited number of reduced density matrices (e.g., one- and two-particle density matrices) and *no explicit manipulation of the complicated reference function is required.*

C. The Linearized Model

To summarize the theory: dynamic correlations are described by the unitary operator $\exp \hat{A}$ acting on a suitable reference function, where $\hat{A}$ consists of excitation operators of the form (4). We employ a cumulant decomposition to evaluate all expressions in the energy and amplitude equations. Since we are applying the cumulant decomposition after the first commutator (the term "linear" in the amplitudes), we call this theory *linearized* canonical transformation theory, by analogy with the coupled-cluster usage of the term. The key features of the linearized CT theory are summarized and compared with other theories in Table II.

In the current work, we consider primarily two theoretical models: the linearized canonical transformation with doubles (L-CTD) and linearized canonical transformation with singles and doubles (L-CTSD) theories. These are defined by the choice of operators in $\hat{A}$. The L-CTD theory contains only two-particle

TABLE II
Features of Some Theories that Use the Language of Canonical/Unitary Transformations

Theory	Reference	How to Obtain $\hat{A}$?	Exponential Type	BCH Expansion?
EVH[a]	N.A.[b]	Perturbation theory	N.A.	Perturbation theory
UCC(n)[c]	Slater determinant	Projection equation	$e^{\hat{A}}$	Perturbation theory
CD[d]	N.A.	Minimize off-diagonals in second-quantized $\hat{H}$	$e^{\hat{A}_a} e^{\hat{A}_b} \ldots$	Truncated operator manifold
CT[e]	Multireference state (e.g., CASSCF)	Projection equation	$e^{\hat{A}}$	Truncated operator manifold, cumulant decomposition

[a]Effective valence-shell Hamiltonian [19, 25].
[b]Assumes degenerate valence space.
[c]Unitary coupled-cluster theory [33, 50].
[d]Canonical diagonalization [22].
[e]Canonical transformation (this work).

operators (including also the two-particle semi-internal excitations) and the L-CTSD theory contains both one- and two-particle operators; thus

$$\hat{A}(\text{L-CTSD}) = \hat{A}_1 + \hat{A}_2$$
$$\hat{A}(\text{L-CTD}) = \hat{A}_2$$
$$\hat{A}_1 = \sum_{ae} A_e^a c_a^\dagger c_e + \sum_{e_1 e_2} A_{e_2}^{e_1} c_{e_1}^\dagger c_{e_2}$$
$$\hat{A}_2 = \sum_{a_1 a_2 e_3 e_4} A_{e_3 e_4}^{a_1 a_2} c_{a_1}^\dagger c_{a_2}^\dagger c_{e_3} c_{e_4}$$
$$+ \sum_{a_1 a_2 a_3 e_4} A_{a_3 e_4}^{a_1 a_2} c_{a_1}^\dagger c_{a_2}^\dagger c_{a_3} c_{e_4}$$
$$+ \sum_{a_1 e_2 e_3 e_4} A_{e_3 e_4}^{a_1 e_2} c_{a_1}^\dagger c_{e_2}^\dagger c_{e_3} c_{e_4}$$
$$+ \sum_{e_1 e_2 e_3 e_4} A_{e_3 e_4}^{e_1 e_2} c_{e_1}^\dagger c_{e_2}^\dagger c_{e_3} c_{e_4} + \cdots + h.c.$$

Although L-CTD theory does not include explicit one-particle single excitations (it does include two-particle semi-internal single excitations), in most of the applications in this work it is combined with a CASSCF reference, which is already based on optimized orbitals.

D. Perturbative Analysis and Relation to Coupled-Cluster Theory

Perturbative analyses have yielded many insights into single-reference coupled-cluster theory. Although we generally are using the canonical transformation theory together with a *multireference* wavefunction, it is informative to carry out the analogous perturbative analysis for the *single-reference* canonical transformation theory, to highlight the connections with existing coupled-cluster methods. The analysis in this section follows that of Bartlett et al. [32, 33, 50].

First consider a Hartree–Fock reference function and transform to the Fermi vacuum (all occupied orbitals are in the vacuum). Then all particle density matrices are zero and the cumulant decomposition, Eq. (23), based on this reference corresponds to simply neglecting all three and higher particle-rank operators generated by commutators. This type of operator truncation is used in the canonical diagonalization theory of White [22].

Now write the Hamiltonian as

$$\hat{H} = E_{\text{HF}} + \hat{F} + \hat{W} \tag{26}$$

where $\hat{F}$ is the one-particle Fock operator and $\hat{W}$ is the two-particle fluctuation potential. From Brillouin's theorem, we recognize that $\hat{A}_2$ is first order in $\hat{W}$, while $\hat{A}_1$ is second order in $\hat{W}$. (To make contact with the analysis of unitary coupled-cluster theory in Refs. [32, 33], write $\hat{A}_1$ as $(\hat{T}_1 - \hat{T}_1^\dagger)$ and $\hat{A}_2 = (\hat{T}_2 - \hat{T}_2^\dagger)$.) Then consider the expectation value of the energy $E = \langle \exp \hat{A}^\dagger \hat{H} \exp \hat{A} \rangle$ *without* using any cumulant decomposition. Expanding in powers of the fluctuation operator, we have

$$E = E^0 + E^1 + E^2 + E^3 + E^4 + \cdots \tag{27}$$

where these are defined as

$$E^0 = \langle E_{HF} + \hat{F} \rangle \tag{28}$$

$$E^1 = \langle \hat{W} \rangle \tag{29}$$

$$E^2 = \langle [\hat{W}, \hat{A}_2] + [\hat{F}, \hat{A}_1] + [[\hat{F}, \hat{A}_2], \hat{A}_2] \rangle \tag{30}$$

$$E^3 = \langle \tfrac{1}{2}[[\hat{W}, \hat{A}_2], \hat{A}_2] + [\hat{W}, \hat{A}_1] + \tfrac{1}{2}[[\hat{F}, \hat{A}_2], \hat{A}_1] \tag{31}$$

$$+ \tfrac{1}{2}[[\hat{F}, \hat{A}_1], \hat{A}_2] \rangle \tag{31}$$

$$E^4 = \langle \tfrac{1}{6}[[[\hat{W}, \hat{A}_2], \hat{A}_2], \hat{A}_2] + \tfrac{1}{2}[[\hat{W}, \hat{A}_1], \hat{A}_2]$$

$$+ \tfrac{1}{2}[[W, \hat{A}_2], \hat{A}_1] + \frac{1}{2}[[\hat{F}, \hat{A}_1], \hat{A}_1]$$

$$+ \tfrac{1}{24}[[[[\hat{F}, \hat{A}_2], \hat{A}_2], \hat{A}_2], \hat{A}_2] \rangle \tag{32}$$

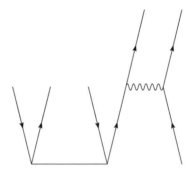

Figure 2. An example of a diagram of the three-particle operator appearing in $[\hat{W}, \hat{A}_2]$.

Now consider the effect of the cumulant decomposition on the different orders of energy contribution. First, no decomposition is involved in computing E^0, E^1. For E^2, the cumulant decomposition corresponds to

$$E^2 \Rightarrow \langle [\hat{W}, \hat{A}_2]_{(1,2)} + [\hat{F}, \hat{A}_1] + [[\hat{F}, \hat{A}_2], \hat{A}_2]_{(1,2)} \rangle \qquad (33)$$

We have used the subscript (1, 2) only when the commutator generates three-particle terms; for example, $[\hat{F}, \hat{A}_2]$ generates only two-particle terms and thus no decomposition is applied. We can illustrate the different terms diagrammatically using the coupled-cluster-type diagrams popularized by Bartlett [51]. Figure 2 illustrates a three-particle term that appears in $[\hat{W}, \hat{A}_2]$. Additional "double" lines are used to indicate contractions with a reduced density matrix. When these double lines are cut and rotated, one recovers the usual CC type diagram. (The cumulant decomposition of $[\hat{W}, \hat{A}_2]_{(1,2)}$ yields four kinds of diagrams for one- and two-particle operators, shown in Fig. 3, but for the single reference case we are considering, all these terms vanish since all particle density matrices are zero from the Fermi vacuum.) Now $\langle [\hat{F}, \hat{A}_1] \rangle$ vanishes (from Brillouin's theorem). Both $[\hat{W}, \hat{A}_2]$ and $[[\hat{F}, \hat{A}_2], \hat{A}_2]$ generate three-particle operators that are approximated in the cumulant decomposition, but these have no expectation value with the Fermi vacuum and thus do not contribute to the energy. Thus no error is made in Eq. (33) for E^2.

In the expression for E^3, we apply the cumulant decomposition twice for the double commutator $[[\hat{W}, \hat{A}_2]_{(1,2)}, \hat{A}_2]_{(1,2)}$. Once again, only the fully contracted term contributes to the energy. The only way fully contracted terms arise is from double contractions in $[\hat{W}, \hat{A}_2]$ to produce a two-particle operator, which then doubly contracts with the final $\hat{A}_2$ commutator, to contribute to the energy. Since double contractions are involved in each step, no cumulant decomposition is involved for this term. There is no contribution from the three-particle

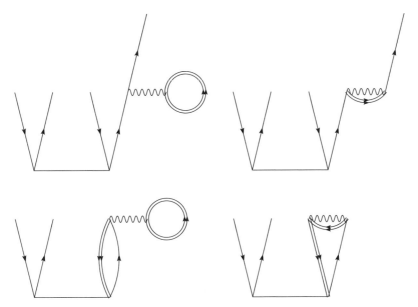

Figure 3. A diagrammatic representation of the cumulant decomposition ($[\hat{W},\hat{A}_2]_{(1,2)}$) for the three-particle operator drawn in Fig. 2. Four kinds of one- and two-particle operators are obtained. The double line is the contraction for the particle-rank reduction (closure), where the correlation is averaged with the effective field (i.e., density matrices).

operators generated by either commutator, and the cumulant-decomposition approximation is exact for E^3 (see Fig. 4).

In the expression for E^4 we find our first error from using the cumulant decomposition. Here, the three-particle operator arising from the first commutator $[\hat{W},\hat{A}_2]$,

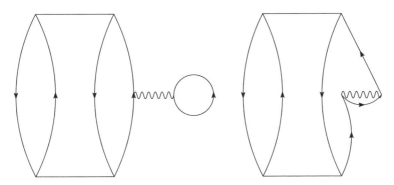

Figure 4. Two connected diagrams in the term $\langle[[\hat{W},\hat{A}_2],\hat{A}_2]\rangle$, which contribute to E_3 (Eq. (31)).

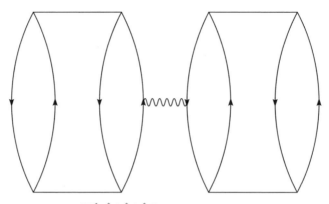

Figure 5. A diagram in $\langle[[[\hat{W},\hat{A}_2],\hat{A}_2],\hat{A}_2]\rangle$ that yields nonzero energy in E_4 (Eq. (32)) and that is missed in the cumulant decomposition in L-CTSD theory. In this diagram, the three-particle operator arising from $[\hat{W},\hat{A}_2]$ contracts successively with two other $\hat{A}_2$ terms.

which is dropped in the cumulant decomposition, can contract successively with two other $\hat{A}_2$ terms, in $[[[\hat{W},\hat{A}_2],\hat{A}_2],\hat{A}_2]$, to yield a fully contracted term and a contribution to the energy. Although the cumulant decomposition misses this contribution, it *does*, however, contain the contribution that arises from contracting the two-particle operators generated in the first commutator $[\hat{W},\hat{A}_2]$. A diagrammatic illustration of the same result is shown in Fig. 5. By a similar analysis, we find that the cumulant decomposition also provides an incomplete evaluation of $[[[[\hat{F},\hat{A}_2],\hat{A}_2],\hat{A}_2],\hat{A}_2]$, arising from intermediate three-particle operators.

In the usual coupled-cluster hierarchy, $\sum_{i=0}^{2} E_i$ is the MP2 energy functional, while $\sum_{i=0}^{3} E_i$ is the linearized coupled-cluster single–doubles (L-CCSD) energy functional. $\sum_{i=0}^{4} E_i$ is the unitary CCSD energy functional. The linearized CTSD energy is correct up to third order in perturbation theory, like the linearized CCSD theory. However, unlike linearized CCSD theory, fourth-order terms (such as $[[[\hat{W},\hat{A}_2],\hat{A}_2],\hat{A}_2]$) are not completely neglected but partly included as discussed previously. From this, we might expect the single-reference L-CTSD theory to perform intermediate between linearized CCSD and the full CCSD theory. But in fact there are an *infinite* number of additional diagrams that are included in linearized CTSD theory as compared to the usual CC and UCC(n) theories, because the energy functional does not terminate at finite order, but contains further partial contributions from E^5, E^6, and indeed to infinite order. For example, all terms involving pure orbital rotations (i.e., $\hat{A}_1$) are included to *all* orders in the energy functional. Terms involving $\hat{A}_2$, where all $\hat{A}_2$ operators are at least doubly contracted with one other operator, are also included to all orders. Examples of these additional

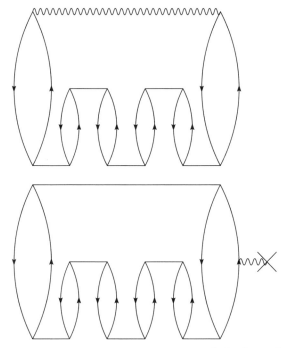

Figure 6. An example of two diagrams in $[[\ldots[[\hat{W},\hat{A}_2],\hat{A}_2]\ldots],\hat{A}_2]$ (upper) and $[[\ldots[[\hat{F},\hat{A}_2],\hat{A}_2]\ldots],\hat{A}_2]$ (lower) that appear at higher orders in linearized CTD and CTSD. The diagrams involve $\hat{A}_2$, where all $\hat{A}_2$ operators (here six $\hat{A}_2$) are at least doubly contracted with one other operator.

diagrams contained in L-CTSD, but not in the usual CC theories, are shown in Fig. 6. One might speculate that these additional diagrams would yield an improved theory, but in the general case, and certainly when we extend the discussion to cases where a multideterminantal reference wavefunction is used, the significance of these additional contributions can only be assessed numerically.

E. Variation of the Reference

A further point is of interest in the formal discussion of the canonical transformation theory. So far we have assumed that the reference function is fixed and have considered only solving for the amplitudes in the excitation operator. We may also consider optimization of the reference function itself in the presence of the excitation operator $\hat{A}$. This consideration is useful in understanding the nature of the cumulant decomposition in the canonical transformation theory.

Using the energy functional (7) and the cumulant decomposition, and making the energy stationary with respect to variations in Ψ_0^*, we find that the optimal reference Ψ_0 satisfies

$$\hat{\bar{H}}_{(1,2)}\Psi_0 + \langle\Psi_0|\frac{\delta\hat{\bar{H}}_{(1,2)}}{\delta\Psi_0^*}|\Psi_0\rangle = \hat{\bar{F}}\Psi_0 = 0 \tag{34}$$

where the second term arises because the effective Hamiltonian is state dependent through the usage of the cumulant decomposition. Thus the optimal reference function is an eigenfunction not of the effective Hamiltonian, but a correlated two-particle "Fock" operator $\hat{\bar{F}}$.

To understand this more clearly, consider a simpler model where $\hat{A}$ consists of single excitations, only single-particle operators are retained in the effective Hamiltonian, and we choose the reference function Ψ_0 to be a single determinant. Then, from a cumulant decomposition of the two-particle terms, the effective Hamiltonian becomes

$$\hat{\bar{H}}_{(1)} = \sum_{ij} T_{ij} c_i^\dagger c_j + \frac{1}{2}\sum_{ijkl} V_{ijkl}[\langle c_i^\dagger c_l\rangle c_j^\dagger c_k - \langle c_j^\dagger c_l\rangle c_i^\dagger c_k] \tag{35}$$

Note this resembles the (N-particle) Fock operator that appears in Hartree–Fock theory, but the contribution of the two-electron term is only half the normal contribution in the Fock operator. However, if we consider making the energy stationary w.r.t. variations in the reference, we must also consider the second term in Eq. (34), where we find

$$\frac{\delta\hat{\bar{H}}_{(1)}}{\delta\Psi_0^*} = \frac{1}{2}\sum_{ijkl}[V_{ijkl}(\langle c_i^\dagger c_l\rangle c_j^\dagger c_k - \langle c_j^\dagger c_l\rangle c_i^\dagger c_k)]\Psi_0 \tag{36}$$

and thus the final Fock operator $\hat{\bar{F}}$ that determines the optimal reference function is identical to the usual Hartree–Fock operator

$$\hat{\bar{F}} = \sum_{ij} T_{ij} c_i^\dagger c_j + \sum_{ijkl} V_{ijkl}[\langle c_i^\dagger c_l\rangle c_j^\dagger c_k - \langle c_j^\dagger c_l\rangle c_i^\dagger c_k] \tag{37}$$

Thus we see that Hartree–Fock theory is identical to a canonical transformation theory retaining only one-particle operators with an optimized reference, and the canonical transformation model retaining one- and two-particle operators employed in the current work, if employed with an optimized reference, is a natural extension of Hartree–Fock theory to a two-particle theory of correlation.

Finally, we note that if we retain two-particle operators in the effective Hamiltonian, but restrict $\hat{A}$ to single-particle form, we recover exactly the orbital rotation formalism of the multiconfigurational self-consistent field. Indeed, this is the way in which we obtain the CASSCF wavefunctions used in this work.

III. IMPLEMENTATION OF THE LINEARIZED CANONICAL TRANSFORMATION THEORY

A. Computational Algorithm

Considering the one- and two-particle operators in Eq. (25) separately, we obtain equations for the L-CTSD and L-CTD models:

$$R_q^p = \langle [\hat{\bar{H}}_{(1,2)}, \hat{\gamma}_q^p - \hat{\gamma}_q^{p\dagger}] \rangle = 0 \qquad \text{L-CTSD} \qquad (38)$$

$$R_{rs}^{pq} = \langle [\hat{\bar{H}}_{(1,2)}, \hat{\gamma}_{rs}^{pq} - \hat{\gamma}_{rs}^{pq\dagger}]_{(1,2)} \rangle = 0 \quad \text{L-CTD/L-CTSD} \qquad (39)$$

where $\hat{\gamma}_q^p = c_p^\dagger c_q$ and $\hat{\gamma}_{rs}^{pq} = c_p^\dagger c_q^\dagger c_s c_r$. These nonlinear equations must be solved for the amplitudes A that define the effective Hamiltonian. A sketch of our implementation is as follows:

Step 1. Given the electronic Hamiltonian $\hat{H}$ (Eq. (1)), determine the reference function in the active space (e.g., CASSCF, CASCI, or HF). Compute the one- and two-particle density matrices $\gamma_q^p = \langle c_p^\dagger c_q \rangle$ and $\gamma_{rs}^{pq} = \langle c_p^\dagger c_q^\dagger c_s c_r \rangle$ of the reference function.

Step 2. Compute the preconditioner for the amplitude equations, given by the diagonal linear terms of the amplitude equations (Eqs. (38) and (39)).

$$D_q^p = \langle [[\hat{H}, \hat{\gamma}_q^p - \hat{\gamma}_q^{p\dagger}], \hat{\gamma}_q^p - \hat{\gamma}_q^{p\dagger}] \rangle \qquad (40)$$

$$D_{rs}^{pq} = \langle [[\hat{H}, \hat{\gamma}_{rs}^{pq} - \hat{\gamma}_{rs}^{pq\dagger}]_{(1,2)}, \hat{\gamma}_{rs}^{pq} - \hat{\gamma}_{rs}^{pq\dagger}]_{(1,2)} \rangle \qquad (41)$$

Step 3. Choose the initial amplitudes A (which we set as zero).

Step 4. Compute the transformed Hamiltonian $\hat{\bar{H}}$ via the BCH equation (Eq. (8)). This is done by iterating a subroutine that computes the cumulant-decomposed commutator $\hat{\bar{H}}^{(n+1)} = [\hat{\bar{H}}^{(n)}, \hat{A}]_{(1,2)}$, which contains only one- and two-particle operators, and where $\hat{\bar{H}}^{(0)} = \hat{H}$. The full $\hat{\bar{H}}$ is then obtained as a sum of one- and two-particle operators $\hat{\bar{H}} = \sum_{n=0} \hat{\bar{H}}^{(n)}/n!$. The sum is truncated after the norm of coefficients of the nth term $\hat{\bar{H}}^{(n)}/n!$ is less than a given threshold, which we set as 10^{-9}. As illustrated in Table I, exponential convergence is observed. The energy is then computed as $E = \langle \hat{\bar{H}} \rangle$

with the density matrix elements γ_q^p and γ_{rs}^{pq} of the reference wavefunction.

Step 5. Compute the new residuals of the amplitude equations, R_q^p (Eq. (38)) and R_{rs}^{pq} (Eq. (39)).

Step 6. Correct the amplitudes by adding the preconditioned residuals.

$$A_q^p \leftarrow A_q^p + R_q^p/D_q^p \tag{42}$$

$$A_{rs}^{pq} \leftarrow A_{rs}^{pq} + R_{rs}^{pq}/D_{rs}^{pq} \tag{43}$$

Step 7. Repeat steps 4–6 until convergence.

Within the above scheme, we implemented the generalized minimal residual (GM-RES) method [52], which is a robust linear solver that ensures convergence of the iterative solution.

B. Computational Scaling

In active-space calculations, the total orbital space is usually partitioned into external core orbitals (c), active orbitals (a), and unoccupied virtual (external) orbitals (v). (There can additionally be some frozen core orbitals that remain doubly occupied throughout the calculation.)

In the iterative algorithm outlined in Section III.A the computational scalings are $n^2 n_{ca}^2 n_{av}^2$ for step 4, and $n n_{ca} n_{av}^4$ and $n n_{ca}^2 n_{av}^3$ for step 5, where $n = n_c + n_a + n_v$, $n_{ca} = n_c + n_a$, and $n_{av} = n_a + n_v$. Note that, unlike conventional multireference methods, these scalings do not depend on the number of configurations in the expansion of the reference wavefunction. In fact, the scaling is roughly $\sim O(n_a^2 n_v^4)$, which is *essentially the same as that of single-reference coupled-cluster theory.*

C. Classes of Excitations for the Exponential Operator

Iterative multireference descriptions of dynamic correlation must carefully consider the problem of convergence. There are two reasons why these calculations are more difficult than single-reference calculations of dynamic correlation. First, dynamic excitations from an active space can give rise to intruder states, which are configurations that lie *inside* the range of energies spanned by the active space. Thus these states are associated with negative or small preconditioner elements, giving rise to convergence problems. Second, there can be near-linear dependency associated with states generated by the excitation operators $\hat{A}_1$ and $\hat{A}_2$. For example, internal single-type excitations of the form $c_{o'}^\dagger c_v^\dagger c_{o'} c_o$ generated by $\hat{A}_2$ become exactly degenerate with the single excitations of the form $c_v^\dagger c_o$, created by $\hat{A}_1$, when the starting reference is a single determinant and the orbital o is occupied.

TABLE III
The Smallest Nonzero Elements of D^{pq}_{rs} Used for Preconditioning, as Defined and Used in Eqs. (41), (38), and (39), for the Different Classes of Operators[a]

Internal excitation	
ccaa	1.05×10^{-5}
caaa	2.62×10^{-6}
Semi-internal excitation	
ccav	1.43×10^{-2}
caav	2.57×10^{-3}
aaav	3.50×10^{-5}
Double excitation	
ccvv	1.25×10^{2}
cavv	2.15×10^{-1}
aavv	2.30×10^{-2}

[a]Taken from a calculation on H_2O at the equilibrium structure with a 6-31G basis set.

To improve the convergence in the iterative solution of the CT amplitude equations, we have adopted a simple scheme, where we classify the amplitude equations by the type of excitation operators involved and solve the classified sets of the amplitude equations in successive steps, which achieves some decoupling between different scales of variation. The smallest preconditioner elements for each class of excitation are shown in Table III for a calculation on the water molecule. Generally, convergence was harder to achieve near equilibrium geometries, which is consistent with the larger spread of valence orbital energies (and consequently increased importance of intruder states), and the higher linear dependency of the excitation manifold. In future work we will investigate more robust convergence algorithms such as the augmented Hessian techniques used in multiconfigurational SCF theory [53].

First number the orbitals thus: core orbitals range from 1 through the number of core orbitals N_{core}, that is, $c = 1, \ldots, N_{core}$. Active orbitals range from $N_{core} + 1$ to $N_{core} + N_{active}$, and the virtual orbitals start from $N_{core} + N_{active} + 1$. Then we divide the equations into eight classes (see also Tables IV and V): (i) internal double excitations ccaa, (ii) external double excitations (ccvv, cavv, and aavv) where the indices of ca, ca and aa are $\leq$ the number of electrons N_{elec}, (iii) external double excitations where either of the indices of ca, ca and aa is $\leq N_{elec}$ and the other is $> N_{elec}$, (iv) external double excitations where the indices of ca, ca and aa are $> N_{elec}$ (v) single excitations (ca, cv, and av), (vi) semi-internal excitations (ccav, caav, and aaav) where the indices of ca, ca and aa are $\leq N_{elec}$, (vii) semi-internal excitations where either of the indices of ca, ca and aa is $\leq N_{elec}$ and the other is $> N_{elec}$, and (viii) semi-internal excitations where the indices of ca, ca and aa are $> N_{elec}$. When solving for a given class

TABLE IV
Classification of One-Particle Excitation Operators[a]

Classified Operator	$\hat{A}_q^p = \hat{\gamma}_q^p - \hat{\gamma}_q^{p\dagger}$
ca	$\hat{\gamma}_{c_1}^{a_1} = c_{a_1}^\dagger c_{c_1}$
cv	$\hat{\gamma}_{c_1}^{v_1} = c_{v_1}^\dagger c_{c_1}$
av	$\hat{\gamma}_{a_1}^{v_1} = c_{v_1}^\dagger c_{a_1}$

[a] c, a, and v denote active core, active, and virtual (external) orbitals.

TABLE V
Classification of Two-Particle Excitation Operators[a]

Classified Operator	$\hat{A}_{rs}^{pq} = \hat{\gamma}_{rs}^{pq} - \hat{\gamma}_{rs}^{pq\dagger}$
Internal excitation	
ccaa	$\hat{\gamma}_{c_1 c_2}^{a_1 a_2} = c_{a_1}^\dagger c_{a_2}^\dagger c_{c_2} c_{c_1}$
caaa	$\hat{\gamma}_{c_1 a_2}^{a_1 a_2'} = c_{a_1}^\dagger c_{a_2'}^\dagger c_{a_2} c_{c_1}$
Semi-internal excitation	
ccav	$\hat{\gamma}_{c_1 c_2}^{v_1 a_2} = c_{v_1}^\dagger c_{a_2}^\dagger c_{c_2} c_{c_1}$
caav	$\hat{\gamma}_{c_1 a_2}^{v_1 a_2'} = c_{v_1}^\dagger c_{a_2'}^\dagger c_{a_2} c_{c_1}$
aaav	$\hat{\gamma}_{a_1 a_2}^{v_1 a_2'} = c_{v_1}^\dagger c_{a_2'}^\dagger c_{a_2} c_{a_1}$
Double excitation	
ccvv	$\hat{\gamma}_{c_1 c_2}^{v_1 v_2} = c_{v_1}^\dagger c_{v_2}^\dagger c_{c_2} c_{c_1}$
cavv	$\hat{\gamma}_{c_1 a_2}^{v_1 v_2} = c_{v_1}^\dagger c_{v_2}^\dagger c_{a_2} c_{c_1}$
aavv	$\hat{\gamma}_{a_1 a_2}^{v_1 v_2} = c_{v_1}^\dagger c_{v_2}^\dagger c_{a_2} c_{a_1}$

[a] c, a, and v denote active core, active, and virtual (external) orbitals.

of amplitudes, the amplitudes in all previous classes are also allowed to vary, while the amplitudes in all later classes are 0. The excitation operators for *caaa* are entirely neglected, and those for (i), (vi), (vii), and (viii) are partially neglected when the corresponding preconditioner element D_{rs}^{pq} (Eq. (41)) is smaller than the truncation threshold 0.5. Note that this separation scheme may break the orbital invariance of the CT theory.

IV. NUMERICAL RESULTS

A. Simultaneous Bond Breaking of Water Molecule with 6-31G and cc-pVDZ Basis Sets

As a prototype multireference application, we performed calculations of potential curves for the symmetric breaking of the water molecule H_2O in which the two O–H bonds are stretched simultaneously. We used both the 6-31G [54] and cc-pVDZ [55] basis sets. The results of L-CTD and L-CTSD calculations, together with a number of conventional methods—Hartree–Fock (HF), full configuration interaction (FCI), multireference second-order perturbation theory (MRMP) [20, 21], and coupled-cluster theory (CCSD/CCSDT)—are presented in Tables VI and VII. All multireference calculations used a CAS with six active electrons in five active orbitals (denoted by (6e, 5o)). The 1s orbital in the O atom was held frozen in all calculations. The FCI and MRMP calculations were carried out using GAMESS [56], and the CC calculations were carried out using the TCE [57] implemented in UTCHEM [58].

Figures 7 and 8 plot deviations of total energies from FCI results for the various methods. It is clear that the CASSCF/L-CTD theory performs best out of all the methods studied. (We recall that although the canonical transformation operator $\exp \hat{A}$ does not explicitly include single excitations, the main effects are already included via the orbital relaxation in the CASSCF reference.) The absolute error of the CASSCF/L-CTD theory at equilibrium—1.57 mE_h (6-31G), 2.26 mE_h (cc-pVDZ)—is slightly better than that of CCSD theory—1.66 mE_h (6-31G), 3.84 mE_h (cc-pVDZ); but unlike for the CCSD and CCSDT theories, the CASSCF/L-CTD error stays quite constant as the molecule is pulled apart while the CC theories exhibit a nonphysical turnover and a qualitatively incorrect dissociation curve. The largest error for the CASSCF/L-CTD method occurs at the intermediate bond distance of $1.8R_e$ with an error of -2.34 mE_h (6-31G), -2.42 mE_h (cc-pVDZ). Although the MRMP curve is qualitatively correct, it is not quantitatively correct especially in the equilibrium region, with an error of 6.79 mE_h (6-31G), 14.78 mE_h (cc-pVDZ). One measure of the quality of a dissociation curve is the nonparallelity error (NPE), the absolute difference between the maximum and minimum deviations from the FCI energy. For MRMP the NPE is 4 mE_h (6-31G), 9 mE_h (cc-pVDZ), whereas for CASSCF/L-CTD the NPE is 5 mE_h (6-31G), 6 mE_h (cc-pVDZ), showing that the CASSCF/L-CTD provides a quantitative description of the bond breaking with a nonparallelity error competitive with that of MRMP.

We now discuss the other CT calculations on water that are presented here. The CASSCF/L-CTSD method incorporates an additional orbital rotation on top of those contained in the CASSCF optimization, by the inclusion of one-particle operators in the excitation operator $\hat{A}$. Comparison of the CASSCF/L-CTSD with the CASSCF/L-CTD results shows that although the broad features of the potential energy curves are similar (small errors near in and far out, larger errors in the intermediate region), the absolute errors of CASSCF/L-CTSD are often larger than that of CASSCF/L-CTD. We suggest that this may arise from a lack of balance between the one-particle single excitations (which are always treated exactly) and the semi-internal single excitations, which are treated approximately both from the cumulant decomposition (Section II.B) and due to the use of the numerical cutoff in solving the amplitude equations (Section III.C). This cutoff also makes the CT calculation not orbital invariant (w.r.t. to active–active, external–external rotations), and this is probably the reason for the

TABLE VI
Total Energies (E_h) for the Simultaneous Bond Breaking of the H_2O Molecule with 6-31G Basis Sets[a]

Method Reference CAS Orbitals r_{OH} (R_e)	HF	FCI	CASSCF (6e, 5o)	L-CTSD CASSCF (6e, 5o)	L-CTSD CASSCF (6e, 5o) NOs	L-CTSD(2) CASSCF (6e, 5o) NOs	L-CTD CASSCF (6e, 5o)	L-CTD CASSCF (6e, 5o) NOs	L-CTD CASCI (6e, 5o) NOs	MRMP CASSCF (6e, 5o)	CCSD	CCSDT
1.0	−75.981 92 (139.09)	−76.121 02	−76.038 02 (83.00)	−76.120 65 (0.37)	−76.120 63 (0.38)	−76.119 91 (1.11)	−76.119 63 (1.39)	−76.119 45 (1.57)	−76.118 03 (2.99)	−76.114 23 (6.79)	−76.119 35 (1.66)	−76.120 54 (0.48)
1.4	−75.816 77 (187.77)	−76.004 54	−75.877 86 (80.06)	−76.003 77 (0.77)	−76.003 85 (0.69)	−76.003 24 (1.30)	−76.003 10 (1.44)	−76.002 70 (1.84)	−75.998 37 (6.17)	−76.000 44 (4.10)	−75.999 56 (4.98)	−76.003 47 (1.07)
1.8	−75.636 59 (262.67)	−75.899 26	−75.827 79 (71.47)	−75.905 43 (−6.17)	−75.905 22 (−5.95)	−75.904 60 (−5.33)	−75.903 24 (−3.97)	−75.901 69 (−2.42)	−75.897 38 (1.89)	−75.896 01 (3.25)	−75.889 54 (9.72)	−75.898 68 (0.59)
2.2	−75.500 43 (352.34)	−75.852 77	−75.788 37 (64.40)	−75.854 97 (−2.20)	−75.853 97 (−1.00)	−75.853 25 (−0.49)	−75.854 53 (−1.76)	−75.852 64 (0.13)	−75.857 70 (−4.93)	−75.848 54 (4.23)	−75.830 50 (22.27)	−75.858 13 (−5.36)
2.6	−75.439 89 (399.66)	−75.839 55	−75.778 09 (61.46)	−75.838 19 (1.36)	−75.839 21 (0.34)	−75.838 57 (0.98)	−75.838 52 (1.03)	−75.839 42 (0.13)	−75.848 42 (−8.87)	−75.834 47 (5.08)		
3.0	−75.415 73 (420.59)	−75.836 32	−75.775 68 (60.64)	−75.834 14 (2.18)	−75.835 95 (0.37)	−75.835 24 (1.08)	−75.835 17 (1.15)	−75.836 61 (−0.29)	−75.848 23 (−11.91)	−75.830 99 (5.33)		
3.4	−75.403 84 (431.60)	−75.835 44	−75.775 03 (60.41)	−75.832 36 (3.07)	−75.834 94 (0.50)	−75.834 30 (1.14)	−75.833 58 (1.85)	−75.835 82 (−0.39)	−75.848 36 (−12.93)	−75.830 05 (5.39)		
3.8	−75.397 02 (438.12)	−75.835 14	−75.774 80 (60.34)	−75.831 41 (3.73)	−75.834 75 (0.40)	−75.834 07 (1.08)	−75.833 63 (1.51)	−75.835 68 (−0.54)	−75.848 54 (−13.39)	−75.829 74 (5.40)		

[a]The value in parentheses is the difference from the FCI total energy in mE_h. The bond angle is fixed at $\angle HOH = 109.57°$. $R_e = 0.9929 Å$.

TABLE VII
Total Energies (E_h) for the Simultaneous Bond Breaking of the H_2O Molecule with cc-pVDZ Basis Sets[a]

r_{OH} (R_e)	Method Reference CAS Orbitals	HF	FCI	CASSCF (6e, 5o)	L-CTSD CASSCF (6e, 5o) NOs	L-CTSD(2) CASSCF (6e, 5o) NOs	L-CTD CASSCF (6e, 5o) NOs	MRMP CASSCF (6e, 5o)	CCSD	CCSDT
1.0		−76.021 67 (217.18)	−76.238 85	−76.075 86 (162.99)	−76.238 19 (0.66)	−76.238 12 (0.73)	−76.236 60 (2.26)	−76.224 07 (14.78)	−76.235 01 (3.84)	−76.238 34 (0.51)
1.4		−75.841 12 (257.89)	−76.099 02	−75.945 57 (153.45)	−76.098 80 (0.22)	−76.098 75 (0.27)	−76.097 14 (1.88)	−76.089 46 (9.56)	−76.090 46 (8.55)	−76.097 78 (1.23)
1.8		−75.651 86 (326.28)	−75.978 14	−75.840 02 (138.12)	−75.984 11 (−5.97)	−75.984 01 (−5.88)	−75.980 47 (−2.34)	−75.970 98 (7.16)	−75.960 19 (17.94)	−75.976 80 (1.34)
2.2		−75.510 38 (416.84)	−75.927 22	−75.799 46 (127.76)	−75.926 65 (0.58)	−75.926 49 (0.74)	−75.924 89 (2.34)	−75.919 86 (7.37)	−75.903 63 (23.60)	−75.938 71 (−11.48)
2.6		−75.408 77 (504.64)	−75.913 41	−75.789 38 (124.04)	−75.910 66 (2.75)	−75.910 31 (3.10)	−75.910 24 (3.17)	−75.905 47 (7.94)	−75.896 46 (16.95)	−75.943 97 (−30.56)
3.0		−75.336 38 (573.65)	−75.910 03	−75.787 02 (123.01)	−75.907 02 (3.01)	−75.906 60 (3.43)	−75.907 06 (2.97)	−75.901 90 (8.13)	−75.900 36 (9.67)	−75.951 09 (−41.06)
3.4		−75.285 62 (623.46)	−75.909 08	−75.786 37 (122.71)	−75.905 80 (3.28)	−75.905 47 (3.61)	−75.906 05 (3.03)	−75.900 91 (8.17)	−75.903 81 (5.27)	−75.955 33 (−46.25)
3.8		−75.250 43 (658.36)	−75.908 78	−75.786 17 (122.61)	−75.905 60 (3.18)	−75.905 23 (3.55)	−75.905 92 (2.86)	−75.900 60 (8.18)	−75.905 98 (2.80)	−75.957 69 (−48.91)

[a]The value in parentheses is the difference from the FCI total energy in mE_h. The bond angle is fixed at ∠HOH = 109.57°. R_e = 0.9929Å.

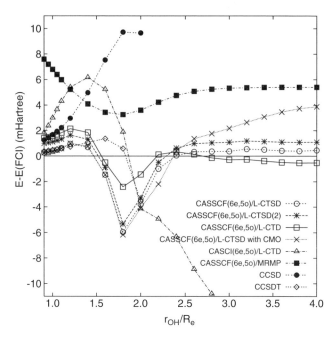

Figure 7. Energy differences $E - E(\text{FCI})$ for the simultaneous bond breaking of H_2O molecule with 6-31G basis sets.

observed differences between calculations using a CASSCF density matrix expressed in the CASSCF natural-orbital basis, and a density matrix expressed in the canonical CASSCF orbital basis. Generally, the canonical CASSCF orbital-based calculations perform less well at longer bond lengths. Finally, the importance of choosing a reasonable set of active-space orbitals is reinforced by the CASCI/L-CTD curve, which does not include any direct mechanism for orbital relaxation. The energy computed with this method breaks down in the intermediate region, past 1.8 R_e.

Table VIII lists the norms of internal, semi-internal, and external excitation amplitudes obtained in the CASSCF(6e, 5o, NO)/L-CTSD calculations with the 6-31G basis set. The percentage of retained amplitudes for internal and semi-internal excitations are also shown in the table. The norm of the external amplitudes, which primarily contribute to dynamic correlation, does not fluctuate much across the potential curve. The maximums of the internal and semi-internal amplitudes are found at the intermediate bond region. More internal and semi-internal excitation operators are retained as the bond length r_{OH} is increased.

We also measured the energy contributions from the different classes of excitation operators used in solving the amplitude equations (Section III.C).

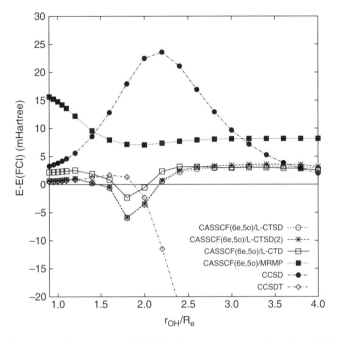

Figure 8. Energy differences $E - E(\text{FCI})$ for the simultaneous bond breaking of H_2O molecule with cc-pVDZ basis sets.

The changes in the total energy occurring during the solution of the eight classes of amplitudes are shown in Table IX. The external excitation operators give the largest contribution to the correlation energy. The contributions from steps (iii) and (iv), which correlate non-HF configurations with the external orbitals, grow larger with the longer OH bond. This reflects the importance of a

TABLE VIII
Norms and Nontruncation Ratios of Amplitudes in CASSCF(6e, 5o)/L-CTSD with NOs for H_2O Molecule with 6-31G Basis Sets

r_{OH}	Norm × 10^2			Nontruncation Ratio	
	Internal	Semi-internal	External	Internal	Semi-internal
$1.0R_e$	0.57	2.88	5.07	16%	34%
$1.4R_e$	5.76	3.31	4.84	32%	65%
$1.8R_e$	3.70	6.41	4.77	64%	79%
$2.2R_e$	3.60	3.52	4.77	64%	80%
$3.8R_e$	1.87	1.87	4.69	64%	80%

TABLE IX
Energy Changes (in millihartree) in the Eight Steps Used in Solving the CT Equations[a]

r_{OH}	(i)	(ii)	(iii)	(iv)	(v)	(vi)	(vii)	(viii)
	$ccaa$[b]	$ccvv + cavv + aavv$[c]			$\hat{A}_1^d$	$ccav + caav + aaav$[e]		
$1.0R_e$	−1.38	−43.88	−0.39	+0.01	+0.13	−37.11	+0.00	+0.00
$1.4R_e$	−2.08	−45.41	−1.91	−0.04	+0.16	−29.98	+0.05	−0.15
$1.8R_e$	−1.08	−44.82	−5.53	−0.19	+0.59	−19.61	−1.88	−4.90
$2.2R_e$	−0.40	−43.16	−9.85	−0.27	+0.80	−10.67	−0.50	−1.35
$3.8R_e$	−0.00	−41.05	−14.12	−0.29	+0.95	−5.48	+0.00	+0.04

[a]Values are from CASSCF(6e, 5o)/L-CTSD calculations with NOs for the H_2O molecule with 6-31G basis sets.
[b]Step (i) solves the equations for $ccaa$.
[c]Steps (ii)–(iv) solve the equations for $ccvv$, $cavv$, and $aavv$.
[d]Step (v) solves the equations for one-particle operators.
[e]Steps (vi)–(viii) solve the equations for $ccav$, $caav$, and $aaav$.

multiconfigurational description for dissociation. The energy contribution from semi-internal operators is significant and is largest at the equilibrium structure.

Finally, to assess the convergence of the commutator expansion in the effective Hamiltonian as the bond is stretched, we computed a "second-order" energy using the L-CTSD amplitudes, denoted CASSCF/L-CTSD(2). Here the energy expression is evaluated as $\langle \hat{H}^{(0)} + \hat{H}^{(1)} + \hat{H}^{(2)} \rangle$. As seen from Tables VI and VII and Figs. 7 and 8, the second-order energy curve faithfully follows the parent CASSCF/L-CTSD curve. This is promising for the development of hybrid CT–perturbation theories, along the lines of CC(2) theory [59, 60].

B. Bond Breaking of Nitrogen Molecules with 6-31G Basis Sets

The second application focuses on the dissociation of the triple bond in the nitrogen molecule N_2, which has been chosen by many multireference studies as a simple multireference model that is difficult to solve. For these calculations, we used the 6-31G basis set and froze the 1s orbitals. We chose two types of CAS space, (6e, 6o) and (10e, 8o). The triple bonding in N_2 is primarily formed by three valence orbitals, one σ and two π orbitals. The bond breaking is basically a chemical reaction that involves these three bonding orbitals and the corresponding higher-lying antibonding orbitals. The CAS(6e, 6o) is thus the smallest active space that allows for a qualitatively correct treatment of the bond breaking. We also look at CAS(10e, 8o), which includes the valence 2s occupied orbitals in addition to CAS(6e, 6o). This allows a correct description of the primary 2s-2p relaxation effects during bond breaking.

Table X shows the total energies as a function of bond length r_{NN} using several CT methods, as well as those using the HF, FCI, CASSCF, MRMP, CCSD, and CCSDT methods. Figure 9 plots the energy differences from the FCI results.

TABLE X

Total Energies (E_h) for the Bond Breaking of the N_2 Molecule with 6-31G Basis Sets[a]

r_{NN}	Method Reference CAS Orbitals HF	CASSCF (6e, 6o)	L-CTSD CASSCF (6e, 6o)	L-CTSD CASSCF (6e, 6o) NOs	L-CTSD(2) CASSCF (6e, 6o) NOs	L-CTD CASSCF (6e, 6o) NOs	L-CTD CASCI (6e, 6o) NOs	MRMP CASSCF (6e, 6o)
1.15 Å	−108.857 17 (248.75)	−109.017 61 (88.31)	−109.098 86 (7.06)	−109.098 86 (7.06)	−109.098 44 (7.48)	−109.099 93 (5.99)	−109.095 25 (10.67)	−109.090 51 (15.41)
1.40 Å	−108.699 62 (322.44)	−108.929 73 (92.33)	−109.014 14 (7.92)	−109.014 16 (7.90)	−109.013 66 (8.40)	−109.014 65 (7.41)	−109.002 08 (19.98)	−109.007 56 (14.50)
1.80 Å	−108.420 80 (467.23)	−108.796 15 (91.88)	−108.885 71 (2.32)	−108.885 71 (2.32)	−108.885 22 (2.81)	−108.881 90 (6.13)	−108.873 88 (14.15)	−108.878 82 (9.21)
2.20 Å	−108.216 46 (631.10)	−108.766 57 (80.99)	−108.846 20 (1.36)	−108.846 20 (1.36)	−108.845 78 (1.78)	−108.844 69 (2.87)	−108.853 23 (−5.67)	−108.840 01 (7.55)
2.60 Å	−108.076 48 (764.11)	−108.764 50 (76.09)	−108.839 61 (0.98)	−108.839 61 (0.98)	−108.838 95 (1.64)	−108.839 66 (0.93)	−108.859 87 (−19.28)	−108.832 49 (8.10)
3.00 Å	−107.982 60 (855.29)	−108.764 36 (73.53)	−108.838 73 (−0.84)	−108.838 73 (−0.84)	−108.837 95 (−0.06)	−108.839 34 (−1.45)	−108.866 29 (−28.40)	−108.830 82 (7.07)

r_{NN}	Method Reference CAS Orbitals FCI	CASSCF (10e, 8o)	L-CTSD CASSCF (10e, 8o)	L-CTSD CASSCF (10e, 8o) NOs	L-CTSD(2) CASSCF (10e, 8o) NOs	MRMP CASSCF (10e, 8o)	CCSD	CCSDT
1.15 Å	−109.105 92	−109.031 25 (74.67)	−109.102 40 (3.52)	−109.102 40 (3.52)	−109.101 85 (4.07)	−109.097 25 (8.67)	−109.094 89 (11.03)	−109.103 55 (2.37)
1.40 Å	−109.022 06	−108.944 10 (77.96)	−109.017 23 (4.83)	−109.017 24 (4.82)	−109.016 71 (5.35)	−109.013 89 (8.17)	−108.999 78 (22.28)	−109.015 75 (6.31)
1.80 Å	−108.888 03	−108.804 16 (83.87)	−108.887 24 (0.79)	−108.887 19 (0.84)	−108.886 64 (1.39)	−108.882 03 (6.00)	−108.852 93 (35.10)	−108.887 70 (0.33)
2.20 Å	−108.847 56	−108.768 65 (78.91)	−108.846 65 (0.91)	−108.846 68 (0.88)	−108.846 25 (1.31)	−108.841 24 (6.32)	−108.920 53 (−72.97)	−108.961 90 (−114.34)
2.60 Å	−108.840 59	−108.765 12 (75.47)	−108.839 34 (1.25)	−108.839 66 (0.93)	−108.838 87 (1.72)	−108.832 96 (7.63)		
3.00 Å	−108.837 89	−108.764 56 (73.33)	−108.838 15 (−0.26)	−108.838 87 (−0.98)	−108.837 62 (0.27)	−108.831 03 (6.86)		

[a]The value in parentheses is the difference from the FCI total energy in mE_h.

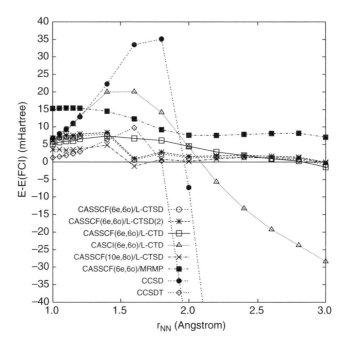

Figure 9. Energies differences ($E - E$ (FCI)) for the bond breaking curve of the N_2 molecule with 6-31G basis sets.

Comparing the different methods we see once again that the CASSCF/L-CTD method yields the most accurate description of the potential energy curve out of all the theories. The error at equilibrium (5.99 mE_h) is better than that of CCSD (11.03 mE_h) and once again this error stays roughly constant across the curve, while that of the CC-based approaches exhibit a nonphysical turnover. For comparison, the MRMP error at equilibrium is 15.41 mE_h. The nonparallelity errors for CASSCF/L-CTD and MRMP are 8.9 and 8.3 mE_h, respectively, demonstrating again that CASSCF/L-CTD yields quantitatively accurate curves with NPEs competitive with that of MRMP theory.

A source of error in the CASSCF(6e, 6o)-based methods is an incomplete treatment of the active–core relaxation. Although some effects of active–core relaxation are incorporated via the exponential operator in the CT calculations, this is incomplete due to the truncation of some operators in the *ccaa* class as explained in Section III.C. Comparing CAS(10e, 8o) with CAS(6e, 6o) shows us the effects of the truncation. At the equilibrium structure ($r_{NN} = 1.15$ Å), we observe that the L-CTSD energy with CAS(6e, 6o) is 3.5 mE_h higher than that with CAS(10e, 8o). For comparison, the MRMP energy with CAS(6e, 6o) is 6.7 mE_h higher that that with CAS(10e, 8o). Thus the truncated *ccaa* operators capture much, but not all, of the active–core relaxation. In the region far from the equilibrium structure, the discrepancy in energies between the two types of CAS disappears, as the active–core relaxation is less important.

In the nitrogen molecule, we found no significant difference in the CT energies between using NOs and canonical CAS orbitals over the potential curve, unlike in the H_2O case. However, it is still important to optimize the orbitals, as the CASCI-based L-CTD does not yield a correct potential curve. Finally, as in the water calculations, the L-CTSD(2) approximation recovers most of L-CTSD correlation energy across the potential energy curve.

C. Comparison with MR-CISD and MR-LCCM on the Two-Configuration Reference Insertion of Be in H_2 Molecule

Several authors have studied the insertion of Be in H_2 as an example of a true two-configuration multireference problem [21, 29, 61]. Laidig and Bartlett presented a multireference coupled-cluster method, which they called MR-LCCM, in which they applied a linearized form of coupled-cluster theory to a two-configuration reference [61]. Table XI shows the multireference results obtained in that work using the MR-LCCM and multireference configuration interaction (MR-CISD) methods at three structures in the C_{2v} insertion of Be into H_2. The details of multiconfigurational MR-CISD and MR-LCCM calculations are described in Ref. [61]. For comparison, the corresponding CASSCF-based linearized CT calculations with the same Gaussian basis set are also presented. Figure 10 plots the total energies.

Comparing the different calculations, we see that MR-LCCM generally overestimates the correlation energy, while MR-CISD generally underestimates the correlation energy. The CASSCF/L-CTD method yields the best energies out of all the methods, at all points, with a maximum absolute error of 1.23 mE_h. In agreement with our previous findings, the CASSCF/L-CTSD method is less accurate, with a maximum error of 4.86 mE_h. The nonparallelity errors are 2.0 mE_h (CASSCF/L-CTD), 2.2 mE_h (MR-CISD), and 3.1 mE_h (MR-LCCM). From this (admittedly small) sample of results, we can say that the CASSCF/L-CTD theory outperforms both the MR-LCCM and MR-CISD methods.

D. Single- and Multireference Linearized CT for HF and BH Molecules

We have so far examined the performance of the canonical transformation theory when paired with a suitable multireference wavefunction, such as the CASSCF wavefunction. As we have argued, because the exponential operator describes dynamic correlation, this hybrid approach is the way in which the theory is intended to be used in general bonding situations. However, we can also examine the behavior of the single-reference version of the theory (i.e., using a Hartree–Fock reference). In this way, we can compare in detail with the related

TABLE XI
BeH$_2$ Energies ($^1A_1$ State) at Three Selected Geometries (Bohr)[a]

Points	r_{Be-H_2}	r_{HH}	Method Reference CAS Orbitals	HF	FCI	CASSCF (2e, 2o)	L-CTSD CASSCF (2e, 2o) NOs	L-CTD CASSCF (2e, 2o) NOs	MR-CISD[b] CASSCF (2e, 2o)	MR-LCCM[b] CASSCF (2e, 2o)
(1)	2.5	2.78		−15.562 68[c] (60.20)	−15.622 88	−15.569 57 (53.31)	−15.622 20 (0.68)	−15.622 10 (0.78)	−15.622 04 (0.84)	−15.625 50 (−2.62)
(2)	2.75	2.55		−15.521 19[c] (81.73)	−15.602 92	−15.538 57 (64.35)	−15.598 06 (4.86)	−15.603 07 (−0.15)	−15.600 91 (2.01)	−15.605 32 (−2.40)
(3)	3.0	2.32		−15.536 47[d] (88.49)	−15.624 96	−15.558 28 (66.68)	−15.627 23 (−2.27)	−15.626 19 (−1.23)	−15.621 89 (3.01)	−15.630 46 (−5.50)

[a]The value in parentheses is the difference from the FCI total energy in mE_h.
[b]Laidig and Bartlett [61].
[c]Configuration: $1a_1^2 2a_1^2 1b_2^2$.
[d]Configuration: $1a_1^2 2a_1^2 1a_1^2$.

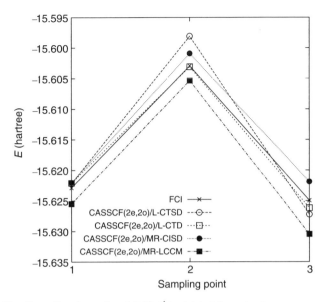

Figure 10. Plots of total energies of BeH$_2$ ($^1A_1$ state) at three structures.

single-reference coupled-cluster theory, and make contact with the perturbative analysis of the cumulant decomposition in Section II.D.

In Tables XII and XIII we present calculations using single-reference linearized (HF/CTD) and (HF/CTSD) theories on several geometries of the HF and BH molecules. Note the HF/CTD theory does not include any semi-internal two-particle excitations, since the active space is completely occupied. The results of calculations using both standard CC theories (CCSD, CCSDT, and linearized CCSD), as well as two alternative coupled-cluster theories, called the expectation value CC (XCC) method and the unitary CC (UCC) method (introduced in Refs. [33, 49, 61]), are also shown. The CC energies were obtained from Ref. [33]. CASSCF/L-CTD and CASSCF/L-CTSD calculations are presented for comparison. All calculations used the Dunning DZP basis sets and the geometries described in Ref. [33].

In both the HF molecule and BH molecules, the HF/L-CTSD and HF/L-CTD theories give energies comparable to CCSD at the equilibrium geometry. However, as the bond is stretched, they both display significantly increased errors, typical of a single-reference theory. At stretched geometries, the errors of the HF/L-CTSD method are worse than those of CCSD and comparable to those of the linearized CCSD theory.

The perturbative analysis in Section II.D showed that single-reference L-CTSD is exact through third order in the fluctuation potential, much like L-CCSD theory, and our results are consistent with this analysis. This suggests that one of the things we

TABLE XII
Energy Differences $E - E(\text{FCI})$ for the HF Molecule (Units Are mE_h)[a]

r_{HF}	Method Reference CAS Orbitals	HF	CCSD[b]	CCSDT[b]	L-CCSD	XCCSD(4)	XCC(4)[b]	UCCSD(4)	UCC(4)[b]
$1.0R_e^c$		203.88	3.01	0.26	0.81	2.65	−0.38	0.96	−4.11
$1.5R_e$		227.17	5.10	0.65	−2.45	4.62	−0.51	2.25	−3.00
$2.0R_e$		263.54	10.18	1.12	−47.34	8.60	−3.00	8.51	−2.69

r_{HF}	Method Reference CAS Orbitals	L-CTSD HF	L-CTD HF	CASSCF (2e, 2o)	L-CTSD CASSCF (2e, 2o) NOs	L-CTD CASSCF (2e, 2o) NOs	d-CT-D[d] CASSCF [2e, 2o] NOs	d-CASSCF[d] [2e, 2o] NOs	d-CT-D[d] d-CASSCF[d] [2e, 2o] NOs
$1.0R_e$		2.87	0.46	180.78	1.67	2.01	0.41	201.93	−1.18
$1.5R_e$		7.00	0.66	173.97	1.85	2.30	−8.37	197.99	−10.71
$2.0R_e$		11.93	3.59	160.12	−0.93	1.42	20.85	179.52	−21.95

[a]FCI energy: $E(1.0R_e) = -100.25097\ E_h$, $E(1.5R_e) = -100.16039\ E_h$, and $E(2.0R_e) = -100.08111\ E_h$.
[b]The numbers for CCSD, CCSDT, XCC, and UCC are from Ref. [33].
[c]$R_e = 1.733$ bohr.
[d]Natural orbitals are used. The 2-RDM is then decomposed into the 1-RDM through the cumulant expansion.

can do to improve the linearized CT results would be to treat properly some of the higher-order particle terms in E^4 (see Eq. (32)), which the present linearized theories approximate using the cumulant expansion. The importance of these high-order terms is lessened with a multireference starting point, as illustrated by the CASSCF/L-CTSD and CASSCF/L-CTD results, where even a very small active space restores the correct quantitative behavior across the entire potential energy curve, as we have found in our earlier calculations. Further information on the cumulant decomposition can be obtained by recalculating our CASSCF/L-CTD results using a more approximate expansion, where the two-particle reduced density matrices in Eq. (23) are replaced by their one-particle cumulant expansion ($\langle c_i^\dagger c_j^\dagger c_k c_l \rangle \Rightarrow \frac{1}{4}[\langle c_i^\dagger c_l \rangle \wedge \langle c_j^\dagger c_k \rangle - \langle c_k^\dagger c_l \rangle \wedge \langle c_j^\dagger c_l \rangle]$). In this way, we would be characterizing the reference state by its one-particle density matrix alone. Results from these calculations are presented in the last column of Table XII. Here we observe that the one-particle reduced density matrix provides a good characterization of the reference state at equilibrium, but, as we expect, this description becomes progressively worse as the bond is stretched.

V. CONCLUSIONS

We have proposed a canonical transformation (CT) theory to describe dynamic correlation in bonding situations where there is also significant nondynamic

TABLE XIII
Energy Differences $E - E(\text{FCI})$ for the BH Molecule (Units Are mE_h)[a]

												L-CTSD	L-CTD
	Method								L-CTSD	L-CTD	CASSCF	CASSCF	CASSCF
	Reference												
	CAS												
r_{HF}	Orbitals	HF	CCSD[b]	CCSDT[b]	L-CCSD	XCCSD(4)	XCC(4)[b]	UCCSD(4)	UCC(4)[b]	HF	HF	(4, 3o)	(4e, 3o)	(4e, 3o)
$1.0R_e$[c]		102.37	1.78	0.06	−3.86	1.99	0.65	1.92	0.57	−4.87	−2.92	95.98	−2.43	−0.21
$1.5R_e$		113.76	2.64	0.02	−6.44	2.94	0.82	2.81	0.67	−8.33	−3.38	80.42	−3.38	−1.72
$2.0R_e$		139.15	5.04	0.02	−26.74	5.60	0.91	5.01	0.17	−35.32	−11.89	77.08	−5.48	−3.62

[a]FCI energy: $E(1.0R_e) = -25.22763\,E_h$, $E(1.5R_e) = -25.17598\,E_h$ and $E(2.0R_e) = -25.12735\,E_h$.
[b]The numbers for CCSD, CCSDT, XCC, and UCC are from Ref. [33].
[c]$R_e = 2.329$ bohr.

character. By pairing this theory with a suitable multireference description of the nondynamic correlation, such as provided by the CASSCF wavefunction, we have obtained consistently quantitative descriptions for a variety of molecules over a wide range of different geometries. The best-performing method we have found is the linearized CT with doubles model with a CASSCF reference (CASSCF/L-CTD). Using this method, the accuracy obtained is comparable to or better than that of CCSD theory in the equilibrium region of a potential energy curve, but unlike in coupled-cluster theories, this accuracy persists all the way out to bond dissociation. The CASSCF/L-CTD nonparallelity errors are competitive with, or better than, those obtained with multireference perturbation theory.

In addition to the encouraging numerical results, the canonical transformation theory has a number of appealing formal features. It is based on a unitary exponential and is therefore a Hermitian theory; it is size-consistent; and it has a cost comparable to that of single-reference coupled-cluster theory. Cumulants are used in two places in the theory: to close the commutator expansion of the unitary exponential, and to decouple the complexity of the multireference wavefunction from the treatment of dynamic correlation.

There are a number of clear areas where the current theory can be improved. First, our perturbative analysis demonstrates that the linearized CT method used here is an approximation to a "full" CT theory, since it neglects certain higher-order terms, in much the same way that linearized coupled-cluster theory neglects certain terms in the full coupled-cluster equations. This may be remedied by including additional classes of operators into our effective Hamiltonian. For example, by retaining the diagonal four-particle operator $n_i n_j n_k n_l$ (where n denotes a number operator), we obtain an improved CTSD theory that is accurate through the same order of single-reference perturbation theory as CCSD theory. In addition, the numerical solution of the CT equations is currently challenging due to the presence of low-energy intruder excitations and degeneracy in the excitation manifold. Here, we note that CASSCF theories also suffered from convergence problems before the development of sophisticated second-order convergence techniques, and we can incorporate these modern algorithms into our CT calculations.

Building on the investigations in this work, we can imagine many generalizations and applications of the CT theory. Excited states are well accommodated within the theoretical framework. There will be the choice of performing single-state or state-averaged calculations. In the latter case, a single effective Hamiltonian using an average of the density matrices from several states would be constructed, much in the same spirit as state-averaged CASSCF. State-averaged CT calculations would also provide a starting point for deriving model effective Hamiltonian parameters (which need to be appropriate to multiple bonding situations), which could then be reused independently of the initial nondynamic calculation. The ability to describe dynamic correlation in a cleanly separated fashion from nondynamic correlation also opens up the possibility of reduced scaling theories of dynamic correlation out of an active-space reference. For a suitably complete active space, this would avoid problems of orbital localization in the reference, and furthermore we would expect the correlation length associated with the remaining dynamic contributions to be quite short. Finally, CT theory is naturally linked to time-dependent theories, as the time-evolution operator is itself a canonical transformation. Here, the cumulant decompositions used in CT theory will be useful in reducing the complexity of the time-dependent description of many-body electron dynamics.

Acknowledgments

We would like to acknowledge useful discussions with S. R. White. D. A. Mazziotti is thanked for the invitation to contribute this chapter. Cornell University and the Cornell Center for Materials Research are acknowledged for financial support.

References

1. T. Yanai and G. K-L. Chan. Canonical transformation theory for multi-reference problems. *J. Chem. Phys.* **124**, 194106 (2006).
2. B. O. Roos, *The Complete Active Space Self-Consistent Field Method and Its Application in Electronic Structure Calculation*, Volume 69 of *Advances in Chemical Physics*, Wiley, Chichester, 1987, p. 399.
3. K. Ruedenberg, M. W. Schmidt, M. M. Gilbert, and S. T. Elbert, Are atoms intrinsic to molecular electronic wavefunctions? *Chem. Phys.* **71**, 41 (1982).
4. S. R. White and R. L. Martin, Ab initio quantum chemistry using the density matrix renormalization group. *J. Chem. Phys.* **110**, 4127 (1998).
5. A. O. Mitrushenkov, G. Fano, F. Ortolani, R. Linguerri, and P. Palmieri, Quantum chemistry using the density matrix renormalization group. *J. Chem. Phys.* **115**, 6815 (2001).
6. A. O. Mitrushenkov, R. Linguerri, P. Palmieri, and G. Fano, Quantum chemistry using the density matrix renormalization group. II. *J. Chem. Phys.* **119**, 4148 (2003).
7. G. K.-L. Chan, An algorithm for large scale density matrix renormalization group calculations. *J. Chem. Phys.* **120**, 3172 (2004).
8. G. K.-L. Chan and M. Head-Gordon, Highly correlated calculations with a polynomial cost algorithm: a study of the density matrix renormalization group. *J. Chem. Phys.* **116**, 4462 (2002).
9. G. K.-L. Chan and M. Head-Gordon, Exact solution (within a triple-zeta, double polarization basis set) of the electronic Schrödinger equation for water. *J. Chem. Phys.* **118**, 8551 (2003).
10. G. K.-L. Chan, M. Ka'llay, and J. Gauss, State-of-the-art density matrix renormalization group and coupled cluster theory studies of the nitrogen binding curve. *J. Chem. Phys.* **121**, 6110 (2004).
11. G. K.-L. Chan and Troy Van Voorhis, Density-matrix renormalization-group algorithms with nonorthogonal orbitals and non-Hermitian operators, and applications to polyenes. *J. Chem. Phys.* **122**, 204101 (2005).
12. Ö. Legeza, J. Röder, and B. A. Hess, Controlling the accuracy of the density-matrix renormalization-group method: the dynamical block state selection approach. *Phys. Rev. B* **67**, 125114 (2003).

13. Ö. Legeza, J. Röder, and B. A. Hess, QC-DMRG study of the ionic-neutral curve crossing of LiF. *Mol. Phys.* **101**, 2019 (2003).
14. R. J. Bartlett, Many-body perturbation theory and coupled cluster theory for electron correlation in molecules. *Ann. Rev. Phys. Chem.* **32**, 359 (1981).
15. J. Cizek, On the correlation problem in atomic and molecular systems. Calculation of wavefunction components in Ursell-type expansion using quantum-field theoretical methods. *J. Chem. Phys.* **45**, 4256 (1969).
16. J. Paldus, J. Cizek, and I. Shavitt, Correlation problems in atomic and molecular systems. IV. Extended coupled-pair many-electron theory and its application to the BH_3 molecule. *Phys. Rev. A* **5**, 50 (1972).
17. J. Paldus and X. Li, *A Critical Assessment of Coupled Cluster Method in Quantum Chemistry*, Volume 110 of *Advances in Chemical Physics*, Wiley, Hoboken, NJ, 1999, p. 1.
18. K. Andersson, P.-Å. Malmqvist, B. O. Roos, A. J. Sadlej, and K. Wolinski, Second-order perturbation theory with a CASSCF reference function. *J. Phys. Chem.* **94**, 5483 (1990).
19. R. K. Chaudhuri, K. F. Freed, G. Hose, P. Piecuch, K. Kowalski, M. Woch, S. Chattopadhyay, D. Mukherjee, Z. Rolik, A. Szabados, G. Toth, and P. R. Surjan, Comparison of low-order multireference many-body perturbation theories. *J. Chem. Phys.* **122**, 134105 (2005).
20. K. Hirao, Multireference Möller–Plesset method. *Chem. Phys. Lett.* **190**, 374–380 (1992).
21. H. Nakano, Quasidegenerate perturbation theory with multiconfigurational self-consistent-field reference functions. *J. Chem. Phys.* **99**, 7983–7992 (1993).
22. S. R. White, Numerical canonical transformation approach to quantum many-body problems. *J. Chem. Phys.* **117**, 7472 (2002).
23. F. Wegner, Flow equations for Hamiltonians. *Ann. Phys. (Leipzig)* **3**, 77 (1994).
24. S. D. Glazek and K. G. Wilson, Perturbative renormalization group for Hamiltonians. *Phys. Rev. D* **49**, 4214 (1994).
25. K. F. Freed, Tests and applications of complete model space quasidegenerate many-body perturbation theory for molecules, in *Many-Body Methods in Quantum Chemistry* (U. Kaldor, ed.), Springer, Berlin, 1989, p. 1.
26. M. R. Hoffmann, Quasidegenerate perturbation theory using effective Hamiltonians, in *Modern Electronic Structure Theory, Part II*, Volume 2 of *Advanced Series in Physical Chemistry*, World Scientific, 1995, p. 1166.
27. T. Kinoshita, O. Hino, and R. J. Bartlett, Coupled-cluster method tailored by configuration interaction. *J. Chem. Phys.* **123**, 074106 (2005).
28. O. Hino, T. Kinoshita, G. K.-L. Chan, and R. J. Bartlett, Tailored coupled cluster singles and doubles method applied to calculations on molecular structure and harmonic vibrational frequencies of ozone. *J. Chem. Phys.* **124**, 114311 (2006).
29. M. R. Hoffmann and J. Simons, A unitary multiconfigurational coupled-cluster method: theory and applications. *J. Chem. Phys.* **88**, 993 (1988).
30. W. Kutzelnigg, Quantum chemistry in Fock space. I. The universal wave and energy operator. *J. Chem. Phys.* **77**, 3081 (1982).
31. W. Kutzelnigg, Quantum chemistry in Fock space. III. Particle–hole formalism. *J. Chem. Phys.* **80**, 822 (1984).
32. R. J. Bartlett, S. A. Kucharski, and J. Noga, Alternative coupled-cluster ansatz. II. The unitary coupled-cluster method. *Chem. Phys. Lett.* **155**, 133–140 (1989).
33. J. D. Watts, G. W. Trucks, and R. J. Bartlett, The unitary coupled-cluster approach and molecular properties. Applications of the UCC(4) method. *Chem. Phys. Lett.* **157**, 359–366 (1989).

34. S. Pal, A variational method to calculate static electronic properties. *Theor. Chim. Acta* **66**, 207 (1984).
35. S. Pal, M. D. Prasad, and D. Mukherjee, Use of a size consistent energy functional in many-electron theory for closed shells. *Theor. Chim. Acta* **62**, 523 (1983).
36. B. Kirtman, Simultaneous calculation of several interacting electronic states by generalized Van Vleck perturbation theory. *J. Chem. Phys.* **75**, 798 (1981).
37. Arlen Anderson, Quantum canonical transformations and integrability: beyond unitary transformations. *Phys. Lett. B* **319**, 157 (1993).
38. W. Kutzelnigg, Generalized k-particle Brillouin conditions and their use for the construction of correlated electronic wave functions. *Chem. Phys. Lett.* **64**, 383 (1979).
39. H. Nakatsuji, Structure of the exact wave function. *J. Chem. Phys.* **113**, 2949 (2000).
40. M. Nooijen, Can the eigenstates of a many-body Hamiltonian be represented exactly using a general two-body cluster expansion? *Phys. Rev. Lett.* **84**, 2108 (2000).
41. P. Piecuch, K. Kowalski, P.-D. Fan, and K. Jedziniak, Exactness of two-body cluster expansions in many-body quantum theory. *Phys. Rev. Lett.* **90**, 113001 (2003).
42. T. V. Voorhis and M. Head-Gordon, Two-body coupled cluster expansions. *J. Chem. Phys.* **115**, 5033 (2001).
43. F. Colmenero and C. Valdemoro, Approximating q-order reduced density matrices in terms of the lower-order ones. II. Applications. *Phys. Rev. A* **47**, 979 (1993).
44. F. Colmenero and C. Valdemoro, Self-consistent approximate solution of the second-order contracted Schrödinger equation. *Int. J. Quantum Chem.* **51**, 369 (1994).
45. H. Nakatsuji and K. Yasuda, Direct determination of the quantum-mechanical density matrix using the density equation. *Phys. Rev. Lett.* **76**, 1039 (1996).
46. K. Yasuda and H. Nakatsuji, Direct determination of the quantum-mechanical density matrix using the density equation. II. *Phys. Rev. A* **56**, 2648 (1997).
47. D. A. Mazziotti, Contracted Schrödinger equation: determining quantum energies and two-particle density matrices without wave functions. *Phys. Rev. A* **57**, 4219 (1998).
48. D. A. Mazziotti, Approximate solution for electron correlation through the use of Schwinger probes. *Chem. Phys. Lett.* **289**, 419 (1998).
49. R. J. Bartlett and J. Noga, The expectation value coupled-cluster method and analytical energy derivatives. *Chem. Phys. Lett.* **150**, 29–36 (1988).
50. R. J. Bartlett, S. A. Kucharski, J. Noga, J. D. Watts, and G. W. Trucks, Some consideration of alternative ansatze in coupled-cluster theory, in *Many-Body Methods in Quantum Chemistry* (U. Kaldor, ed.), Springer, Berlin, 1989, p. 125.
51. S. A. Kucharski and R. J. Bartlett, *Fifth-Order Many-Body Perturbation Theory and Its Relationship to Various Coupled-Cluster Approaches*, Volume 18 of *Advances in Quantum Chemistry* Wiley, Hoboken, NJ, 1986, p. 281.
52. Y. Saad and M. Schultz, GMRES: a generalized minimal residual algorithm for solving nonsymmetric linear systems. *SIAM J. Sci. Statist. Comput.* **7**, 856–869 (1986).
53. H.-J. Werner and W. Meyer, A quadratically convergent multiconfiguration self-consistent field method with simultaneous optimization of orbitals and CI coefficients. *J. Chem. Phys.* **73**, 2342 (1980).
54. W. J. Hehre, R. Ditchfield, and J. A. Pople, Self-consistent molecular orbital methods. XII. Further extensions of Gaussian-type basis sets for use in molecular orbital studies of organic molecules. *J. Chem. Phys.* **56**, 2257–2261 (1972).
55. T. H. Dunning, Gaussian basis sets for use in correlated molecular calculations. I. The atoms boron through neon and hydrogen. *J. Chem. Phys.* **90**, 1007–1023 (1989).

56. M. W. Schmidt, K. K. Baldridge, J. A. Boatz, S. T. Elbert, M. S. Gordon, J. J. Jensen, S. Koseki, N. Matsunaga, K. A. Nguyen, S. Su, T. L. Windus, M. Dupuis, and J. A. Montgomery, *J. Comput. Chem.* **14**, 1347 (1993).

57. So Hirata, Tensor contraction engine: abstraction and automated parallel implementation of configuration-interaction, coupled-cluster, and many-body perturbation theories. *J. Phys. Chem. A* **107**, 4940 (2003).

58. T. Yanai, H. Nakano, T. Nakajima, S. Hirata, T. Tsuneda, Y. Kawashima, Y. Nakao, M. Kamiya, H. Sekino, and K. Hirao, UTCHEM—a program for ab initio quantum chemistry, in *Lecture Notes in Computer Science*, Springer-Verlag, Berlin, 2003.

59. S. R. Gwaltney and M. Head-Gordon, A second-order perturbation correction to the coupled-cluster singles and doubles method: CCSD(2). *J. Chem. Phys.* **115**, 2014 (2001).

60. S. Hirata, M. Nooijen, I. Grabowski, and R. J. Bartlett, Perturbative corrections to coupled-cluster and equation-of-motion coupled-cluster energies: a determinantal analysis. *J. Chem. Phys.* **114**, 3919 (2001).

61. W. D. Laidig and R. J. Bartlett, A multi-reference coupled-cluster method for molecular applications. *Chem. Phys. Lett.* **104**, 424–430 (1984).

PART IV

CHAPTER 14

NATURAL ORBITAL FUNCTIONAL THEORY

MARIO PIRIS

Kimika Fakultatea, Euskal Herriko Unibertsitatea, and Donostia International Physics Center (DIPC) P.K. 1072, 20080 Donostia, Euskadi, Spain

CONTENTS

I. Introduction
II. Groundwork Toward a ^{1}D-Functional Theory
III. 1- and 2-Reduced Density Matrices
    A. Spin Structure
IV. Natural Orbital Functional (NOF)
V. Euler Equations
    A. Gilbert Nonlocal Potential
    B. Relation with the Extended Koopmans' Theorem
    C. Pernal Nonlocal Potential
VI. Cumulant of the 2-RDM
    A. Approximate Cumulant
VII. Restricted Closed-Shell NOF
    A. Two-Electron Systems
    B. N-Electron Systems
    C. Effective One-Electron Operator
VIII. Restricted Open-Shell NOF
IX. Practical NOF
X. Numerical Implementation
XI. Results
    A. Energetics
    B. Dipole Moment
    C. Equilibrium Geometries and Vibrational Frequencies
XII. Concluding Remarks
Acknowledgements
References

Reduced-Density-Matrix Mechanics: With Application to Many-Electron Atoms and Molecules,
A Special Volume of Advances in Chemical Physics, Volume 134, edited by David A. Mazziotti.
Series editor Stuart A. Rice. Copyright © 2007 John Wiley & Sons, Inc.

I. INTRODUCTION

In the last few years, the improvements in computer hardware and software have allowed the simulation of molecules and materials with an increasing number of atoms. However, the most accurate electronic structure methods based on N-particle wavefunctions, for example, the configuration interaction (CI) method or the coupled-cluster (CC) method, are computationally too expensive to be applied to large systems. There is a great need for treatments of electron correlation that scale favorably with the number of electrons.

As discussed in previous chapters of this book, since the interactions between electrons are pairwise within the Hamiltonian, the energy may be determined exactly from a knowledge of the two-particle reduced density matrix (2-RDM) $^2\mathbf{D}$. The 2-RDM carries all the relevant information if one is interested in expectation values of one- and two-particle operators. In this manner, the N-particle dependence can be avoided given that the 2-RDM is a much more economic storage of information. There remains, however, the long standing problem that not every 2-RDM is derivable from an N-particle wavefunction (the N-representability problem) [1]. As we have seen in the last few chapters, realistic variational 2-RDM calculations have recently become possible through the use of both the contracted Schrödinger equation and the optimization techniques known as semidefinite programming. Nevertheless, even the best first-order algorithms of semidefinite programming scale as r^6, where r is the rank of the one-electron basis set.

In 1964, Hohenberg and Kohn (HK) [2] demonstrated that the ground-state energy could be expressed as a functional of the one-electron density ρ only. This result led to density functional theory (DFT) [3], which has become very popular thanks to its relatively low computational cost. Practical implementations of DFT are based mainly on the formulation of Kohn–Sham (KS) [4], in which the kinetic energy is not constructed as a functional of ρ but rather from an auxiliary Slater determinant. Since the noninteracting kinetic energy differs from the many-body kinetic energy, there is a contribution from a part of the kinetic energy contained in the correlation potential. This correlation kinetic energy is the main source of problems of present-day KS functionals.

It seems that there is no N-representability problem since the conditions that ensure that a one-particle density comes from an N-particle wavefunction are well known [5]. Here, the obstacle is the construction of the functional $E[\rho]$ capable of describing a quantum mechanical N-electron system. This functional N-representability is still related to the N-representability problem of the 2-RDM. Many currently available functionals are not N-representable [6]. Consequently, the energies produced by these functionals can lie below the exact value. Even though these energy values may lie quite close to the exact ones, they do not guarantee, however, that the calculations are accurate.

Another drawback of the most popular correlation functionals is that they exhibit an incorrect behavior for N-electron atoms as the nuclear charge Z increases [7]. Recently, the accuracy of DFT has been improved by using approximations constructed to satisfy exact constraints on the exchange-correlation energy functional [8]. Despite the great success achieved for high-Z atomic ions with density functionals of this new generation, they cannot as yet adequately describe highly degenerate systems [9]. This corroborates the fact that KS functionals are better suited for describing dynamic correlation due to the short-range interelectronic repulsion than static correlation due to near degeneracy effects. The inability of approximate density functionals to account for the dispersion interactions constitutes a serious drawback. The density functional for the correlation kinetic energy remains unknown and how important it is for the dynamics of chemical reactions is an open question [9].

A direction for improving DFT lies in the development of a functional theory based on the one-particle reduced density matrix (1-RDM) $^1\mathbf{D}$ rather than on the one-electron density ρ. Like 2-RDM, the 1-RDM is a much simpler object than the N-particle wavefunction, but the ensemble N-representability conditions that have to be imposed on variations of $^1\mathbf{D}$ are well known [1]. The existence [10] and properties [11] of the total energy functional of the 1-RDM are well established. Its development may be greatly aided by imposition of multiple constraints that are more strict and abundant than their DFT counterparts [12, 13].

The major advantage of a 1-RDM formulation is that the kinetic energy is explicitly defined and does not require the construction of a functional. The unknown functional in a $^1\mathbf{D}$-based theory only needs to incorporate electron correlation. It does not rely on the concept of a fictitious noninteracting system. Consequently, the $^1\mathbf{D}$ scheme is not expected to suffer from the above mentioned limitations of KS methods. In fact, the correlation energy in 1-RDM theory scales homogeneously in contrast to the scaling properties of the correlation term in DFT [14]. Moreover, the 1-RDM completely determines the natural orbitals (NOs) and their occupation numbers (ONs). Accordingly, the $^1\mathbf{D}$ functional incorporates fractional ONs in a natural way, which should provide a correct description of both dynamical and nondynamical correlation.

II. GROUNDWORK TOWARD A ^{1}D-FUNCTIONAL THEORY

The idea of a 1-RDM functional appeared some decades ago. At the commencement, the main effort was focused on the existence of this functional. In 1974, Gilbert proved the analog of the Hohenberg–Kohn (HK) theorem for the 1-RDMs including nonlocal external potentials [15]. Berrondo and Goscinski [16] added a nonlocal external potential to the Hamiltonian and obtained a variational principle involving the 1-RDM for a local external potential by eliminating the nonlocal external source. Donnelly and Parr [17] proved that the existence was already implied in the original HK theorem. They discussed extensively the properties of this energy functional and derived the Euler equations associated with the exact ground state [17, 18].

In 1979, an elegant proof of the existence was provided by Levy [10]. He demonstrated that the universal variational functional for the electron–electron repulsion energy of an N-representable trial 1-RDM can be obtained by searching all antisymmetric wavefunctions that yield a fixed $^1\mathbf{D}$. It was shown that the functional does not require that a trial function for a variational calculation be associated with a ground state of some external potential. Thus the v-representability is not required, only N-representability. As a result, the 1-RDM functional theories of preceding works were unified. A year later, Valone [19] extended Levy's pure-state constrained search to include all ensemble representable 1-RDMs. He demonstrated that no new constraints are needed in the occupation-number variation of the energy functional. Diverse constrained-search density functionals by Lieb [20, 21] also afforded insight into this issue. He proved independently that the constrained minimizations exist.

It is well known that the exact electronic energy can also be given explicitly in terms of the spinless 1-RDM and the two-particle charge density (2-CD). This suggests an alternative viewpoint regarding $^1\mathbf{D}$-functional theory. One could employ the exact functional but with an approximate 2-CD that is built from $^1\mathbf{D}$ using a reconstruction functional $^2\mathbf{D}[^1\mathbf{D}]$. Perhaps the first explicit approximate relation between $^2\mathbf{D}$ and $^1\mathbf{D}$ containing one free parameter was that proposed by Müller in 1984 [22]. The case where the parameter was set to $\frac{1}{2}$ was discussed years later by Buijse [23], who performed self-consistent calculations for the H_2 molecule.

Zumbach and Maschke [24] discussed the $^1\mathbf{D}$ functional using ensemble search. They derived the set of self-consistent equations that include the eigenvalues of the 1-RDM, and clarified Gilbert's relation between these eigenvalues and the chemical potential. Ludeña et al. [25] considered alternatively a variational procedure with built-in pure-state N-representability conditions. The N-representability of the 1-RDM was accomplished by taking into account the conditions for the mapping of the nth-order density operator into a given $^1\mathbf{D}$. They arrived at Valone's result: the problem of obtaining a pure-state N-representable 1-RDM requires only the known ensemble constraints, if the proper functional is known.

Levy identified the unknown part of the exact universal $^1\mathbf{D}$ functional as the correlation energy $E_c[^1\mathbf{D}]$ and investigated a number of properties of $E_c[^1\mathbf{D}]$, including scaling, bounds, convexity, and asymptotic behavior [11]. He suggested approximate explicit forms for $E_c[^1\mathbf{D}]$ for computational purposes as well. Redondo presented a density-matrix formulation of several ab initio methods [26]. His generalization of the HK theorem followed closely Levy's demonstration of a 1-RDM functional. López-Sandoval and Pastor [27] investigated Levy's functional $E_c[^1\mathbf{D}]$ on lattice models.

Valdemoro [28] achieved a close approximation to the 2-RDM by using the anticommutating relation of fermion operators, or what is equivalent, the N-representability conditions. This work indicated that the development of 1-RDM functional theories should be couched in terms of explicitly antisymmetric reconstructions of the 2-RDM.

More recently, a renewed interest has appeared in the literature. An approximate exchange-correlation functional based on the diagonal elements of $^1\mathbf{D}$ and $(^1\mathbf{D})^2$ was proposed by Carlsson [29]. This functional was derived within tight-binding theory and used successfully in simple model calculations of several physical properties. Klein and Dreizler [30] developed a correlated 1-RDM theory with close connections to DFT perturbation theory. They derived formulas for first-order corrections to the Hartree–Fock (HF) 1-RDM and wrote down a formally exact expression for the correlation energy by using the linked-cluster expansion. Cioslowski and Lopez-Boada obtained by application of the hypervirial theorem an approximate functional of the HF 1-RDM in terms of three-electron integrals and an unknown screening function [31]. Their formalism incorporated dispersion effects and yielded two distinct asymptotics of the correlation energy of atoms and monoatomic ions at the limit of a large nuclear charge. Nooijen explored the possibility of using Green's functions and the extended Koopman's theorem (EKT) [32, 33] to arrive at a 1-RDM formulation [34].

The 1-RDM functional is called *natural orbital functional* (NOF) when it is based on the spectral expansion of $^1\mathbf{D}$. The first parameter-free NOF constructed and tested on real physical systems is probably the one by Goedecker and Umrigar (GU) [35]. The basic form of this functional can be traced to Müller [22] but self-interaction corrected. The GU functional considering diagonal terms coincides with the Buijse and Baerends (BB) reconstruction [23, 36]. By optimizing typically 50 NOs and ONs for a variety of atoms and ions, it was found that the GU functional yielded energies and densities that were comparable to or better than those from the generalized-gradient approximation (GGA) in the DFT. The GU functional does not suffer from dissociation problems. It satisfies the Hermiticity and particle permutation conditions but violates the nonnegativity condition for the diagonal elements of the 2-RDM. Moreover, the GU functional gives a wrong description of the ONs for the lower occupied levels and overestimates correlation effects in jellium at intermediate and high densities.

Following these first encouraging numerical results, several authors suggested new approximations to the 1-RDM functional. Holas [37] and Cioslowski and Pernal [38] proposed different generalizations of GU functional. These functionals were analyzed in detail from the perspective of the homogeneous electron gas (HEG) [39]. Csanyi and Arias (CA) [40] proposed another functional from the condition that the two-matrix is a tensor product of one-particle operators and that it satisfies the Hermiticity and particle-permutation constraints. Unfortunately, the CA functional gives almost vanishing correlation energies in contrast to the GU functional. Expressions for the second-order energy variations in the 1-RDM theory were derived, resulting in a formalism for time-independent response properties and stability conditions [41]. The derivatives of the electronic energy with respect to the number of electrons were found to be very sensitive to the used $^1\mathbf{D}$ functional.

Another route to construction of the approximate 1-RDM functional involves employment of expressions for E and $^1\mathbf{D}$ afforded by some size-consistent formalism of electronic structure theory. Mazziotti [42] proposed a geminal functional theory (GFT) where an antisymmetric two-particle function (geminal) serves as the fundamental parameter. The one-matrix-geminal relationship allowed him to define a $^1\mathbf{D}$-based theory from GFT [43]. He generalized Levy's constrained search to optimize the universal functionals with respect to 2-RDMs rather than wavefunctions.

Csanyi and Arias [40] classified the above mentioned NOFs into two types: corrected Hartree (CH) and corrected Hartree–Fock (CHF). The performance of CH and CHF approximations in molecular calculations was investigated, taking molecules H_2, Li_2, and LiH as examples [44]. A $^1\mathbf{D}$-based functional combining the properties of the CH and CHF approximations was also proposed [45]. An improved CHF-type functional leading to better results for the free-electron gas was suggested too [46]. The 2-CDs, intraculate and extraculate densities, N-representability, and variational stability obtained with reconstruction functionals $^2\mathbf{D}[^1\mathbf{D}]$ that yield these $^1\mathbf{D}$-based theories were investigated in detail [47].

The CA functional is similar to the Hartree–Fock-Bogoliubov (HFB) energy expression but differs by the sign of the post-HF correction term [48]. Existence of an HFB-type 1-RDM for systems with repulsive interactions was anticipated by other authors [49]. Different HFB-like functionals were proposed before for describing electronic structure [50–54].

Yasuda [55] obtained a correlation energy functional $E_c[^1\mathbf{D}]$ from the first- and second-order density equations together with the decoupling approximations for the 3- and 4-reduced density matrices. The Yasuda functional is capable of properly describing a high-density HEG [56] and encouraging results were reported for atoms and molecules [55]. Some shortcomings of this functional were also pointed out [57].

Piris and Otto (PO) achieved a reconstruction functional $^2\mathbf{D}[^1\mathbf{D}]$ satisfying the most general properties of the 2-RDM [58]. They kept the spin structure from Refs. [52, 53], but introduced a new spatial dependence in the correction term of the 2-RDM. Calculated values for polarizabilities [59], ionization energies, equilibrium geometries, and vibrational frequencies [60] in molecules were obtained. An extension of this functional to periodic polymers was also considered [61]. In this line of constructing the 2-RDM from formal criteria, Kollmar and Hess [62] obtained an implicit functional where the elements of an idempotent matrix were used as variational parameters. The first application of the 1-RDM functional theory to open-shell molecular systems was presented using the PO functional [60]. Recently, an open-shell formulation of the GU functional considering spin-dependent ONs was applied to the first-row atoms [63]. This approach conserves the z component, but not the total spin. A study of the partitioning of the 1-RDM according to the theory of atoms in molecules has been reported too [64].

An extension of the antisymmetrized product of strongly orthogonal geminals (APSG) theory provided a "JK-only" expression for the electron–electron repulsion energy V_{ee} of a closed-shell system [65]. An implicit NOF involving pair-excitation coefficients instead of the ONs was proposed by Kollmar and Hess [66]. They considered a size-consistent extension of a limited multiconfiguration self-consistent-field wavefunction taking into account only pair excitations (PEMCSCF). In the case of four-electron systems, an approximate expression for V_{ee} was proposed [67] using a permanent-based parameterization of coefficients in a pair-excitation configuration interaction (CI) expansion. Moreover, a particular parameterization of coefficients in a CI expansion led to an explicit functional in terms of the Coulomb and exchange integrals over NOs, and an idempotent matrix, whose diagonal elements equal the ONs [68]. The obtained functional cannot, however, be employed in practical calculations due to the necessity to carry out minimizations over a large number of possible combinations of CI coefficient signs (phase dilemma). The size and volume extensivity of such functionals has also been analyzed by applying them to the HEG [69]. A JK-only 1-RDM functional starting from the PEMCSCF method and using necessary N-representability conditions for the 2-RDM has recently been derived [70]. This functional gives a small fraction of the total correlation energy for the water molecule at the equilibrium distance, indicating that the accuracy of the JK-only functional form may be limited. Nevertheless, two new JK-only approximations that recover a reasonable fraction of the total correlation energy at the equilibrium geometries have recently been proposed.

Gritsenko et al. [71] have introduced several physical motivated repulsive corrections to the BB functional (BBC). With these corrections, they improved the quality of the BB potential energy curves for prototype few-electron molecules and with BBC3 the average error of correlation energies for atomic systems at the equilibrium geometry was only 6%. The ionization potentials produced by the GU and BBC functionals have also been investigated [72] using the EKT.

Piris [73] has recently proposed an explicit form for the cumulant [74, 75] of the 2-RDM in terms of two symmetric matrices, Δ and Λ. The suggested form of these matrices (as functions of the ONs) produces a NOF that reduces to the exact expression for the total energy in two-electron closed-shell systems [76, 77]. One can generalize it to the N-electron systems, except for the off-diagonal elements of Λ. Alternatively, the mean value theorem and the partial sum rule for matrix Λ provided a prescription for deriving a practical NOF. An assessment of this practical functional in molecules for calculating molecular properties, namely, polarizabilities [78], equilibrium bond distances, harmonic vibrational frequencies, and vertical and adiabatic ionization potentials [79], has been performed. An extension of this functional to open-shell systems has also been considered [80].

We continue this chapter with a presentation of the basic concepts and notations relevant to ^{1}D-functional theory (Section III). We then review the fundaments of the NOF theory (Section IV) and derive the corresponding Euler equations (Section V). The Gilbert [15] and Pernal [81] formulations, as well as the relation of Euler equations with the EKT, are considered here. The following sections are devoted to presenting our NOF theory. The cumulant of the 2-RDM is discussed in detail in Section VI. The spin-restricted formulations for closed and open-shells are analyzed in Sections VII and VIII, respectively. Section IX is dedicated to our further simplification in order to achieve a practical functional. In Section X, we briefly describe the implementation the NOF theory for numerical calculations. We end with some results for selected molecules (Section XI).

III. 1- AND 2-REDUCED DENSITY MATRICES

The electronic energy E for N-electron systems is an exactly and explicitly known functional of the 1- and 2-RDMs. The energy expression in spin-orbital (SO) representation is given by

$$E = \sum_{ik} {}^1D_k^i h_i^k + \sum_{ijkl} {}^2D_{kl}^{ij} \langle kl|ij\rangle \tag{1}$$

where h_i^k are the one-electron matrix elements of the core Hamiltonian,

$$h_i^k = \int d\mathbf{x}\, \phi_k^*(\mathbf{x}) \left[-\frac{1}{2}\nabla^2 - \sum_I \frac{Z_I}{|\mathbf{r} - \mathbf{r}_I|} \right] \phi_i(\mathbf{x}) \tag{2}$$

and $\langle kl|ij\rangle$ are the two-electron integrals of the Coulomb interaction,

$$\langle kl|ij\rangle = \int d\mathbf{x}_1\, d\mathbf{x}_2\, \phi_k^*(\mathbf{x}_1)\phi_l^*(\mathbf{x}_2) r_{12}^{-1} \phi_i(\mathbf{x}_1)\phi_j(\mathbf{x}_2) \tag{3}$$

Atomic units are used. Here and in the following $\mathbf{x} \equiv (\mathbf{r}, \mathbf{s})$ stands for the combined spatial and spin coordinates, $\mathbf{r}$ and $\mathbf{s}$, respectively. The SOs $\{\phi_i(\mathbf{x})\}$ constitute a complete orthonormal set of single-particle functions,

$$\langle \phi_k | \phi_i \rangle = \int d\mathbf{x}\, \phi_k^*(\mathbf{x}) \phi_i(\mathbf{x}) = \delta_i^k \tag{4}$$

with an obvious meaning of the Kronecker delta δ_i^k. Every normalizable function $\phi(\mathbf{x})$ of a single coordinate $\mathbf{x}$ may be expanded in the form

$$\phi(\mathbf{x}) = \sum_i \phi_i(\mathbf{x}) c_i, \qquad c_i = \int d\mathbf{x}\, \phi_i^*(\mathbf{x}) \phi(\mathbf{x}) \tag{5}$$

We employ Löwdin's normalization convention [82] in which the trace of the 1-RDM equals the number of electrons,

$$\mathrm{Tr}\,{}^1\mathbf{D} = \sum_i {}^1\mathbf{D}_i^i = N \tag{6}$$

and the trace of the 2-RDM gives the number of electron pairs in the system,

$$\mathrm{Tr}\,{}^2\mathbf{D} = \sum_{ik} {}^2D_{kk}^{ii} = \frac{N(N-1)}{2} = \binom{N}{2} \tag{7}$$

The 1- and 2-RDMs can be obtained in the coordinate-space representation via the expansion theorem:

$$^1D(\mathbf{x}_1'|\mathbf{x}_1) = \sum_{ik} {}^1D_k^i \phi_i^*(\mathbf{x}_1') \phi_k(\mathbf{x}_1) \tag{8}$$

$$^2D(\mathbf{x}_1', \mathbf{x}_2'|\mathbf{x}_1, \mathbf{x}_2) = \sum_{ijkl} {}^2D_{kl}^{ij} \phi_i^*(\mathbf{x}_1') \phi_j^*(\mathbf{x}_2') \phi_k(\mathbf{x}_1) \phi_l(\mathbf{x}_2) \tag{9}$$

The diagonal elements of ${}^1\mathbf{D}$ and ${}^2\mathbf{D}$ are always nonnegative, since ${}^1D(\mathbf{x}_1|\mathbf{x}_1)$ is related to the probability of finding one electron at $\mathbf{x}_1$, and ${}^2D(\mathbf{x}_1, \mathbf{x}_2|\mathbf{x}_1, \mathbf{x}_2)$ is related to the probability of finding one electron at $\mathbf{x}_1$ and another at $\mathbf{x}_2$. The diagonal elements ${}^1D_i^i$ and ${}^2D_{kk}^{ii}$ may be interpreted analogously: ${}^1D_i^i$ is related to the probability of finding one electron in the spin orbital i when all the other electrons occupy arbitrary spin orbitals, ${}^2D_{kk}^{ii}$ is related to the probability of finding one electron in the spin orbital i and another in the spin orbital k, when all other particles may occupy arbitrary spin orbitals.

As discussed in preceding chapters of this book, the 2-RDM is Hermitian:

$$^2D(\mathbf{x}_1', \mathbf{x}_2'|\mathbf{x}_1, \mathbf{x}_2) = {}^2D^*(\mathbf{x}_1, \mathbf{x}_2|\mathbf{x}_1', \mathbf{x}_2') \qquad [{}^2D_{kl}^{ij} = ({}^2D_{ij}^{kl})^*] \tag{10}$$

It is antisymmetric in each set of indices,

$$^2D(\mathbf{x}'_1, \mathbf{x}'_2 | \mathbf{x}_1, \mathbf{x}_2) = -^2D(\mathbf{x}'_1, \mathbf{x}'_2 | \mathbf{x}_2, \mathbf{x}_1) \quad (^2D^{ij}_{kl} = -^2D^{ij}_{lk}) \quad (11)$$

$$^2D(\mathbf{x}'_1, \mathbf{x}'_2 | \mathbf{x}_1, \mathbf{x}_2) = -^2D(\mathbf{x}'_2, \mathbf{x}'_1 | \mathbf{x}_1, \mathbf{x}_2) \quad (^2D^{ij}_{kl} = -^2D^{ji}_{kl}) \quad (12)$$

and symmetric with respect to particle permutation,

$$^2D(\mathbf{x}'_1, \mathbf{x}'_2 | \mathbf{x}_1, \mathbf{x}_2) = {}^2D(\mathbf{x}'_2, \mathbf{x}'_1 | \mathbf{x}_2, \mathbf{x}_1) \quad (^2D^{ij}_{kl} = {}^2D^{ji}_{lk}) \quad (13)$$

There is an important contraction relation between 1- and 2-RDMs that is in agreement with the previous normalization,

$$^1D(\mathbf{x}'_1 | \mathbf{x}_1) = \frac{2}{N-1} \int d\mathbf{x}_2\, ^2D(\mathbf{x}'_1, \mathbf{x} | \mathbf{x}_1, \mathbf{x}) \quad \left(^1D^i_k = \frac{2}{N-1} \sum_j {}^2D^{ij}_{kj}\right) \quad (14)$$

This implies that the energy functional (1) is just of the 2-RDM, because $^2\mathbf{D}$ determines $^1\mathbf{D}$. However, attempts to determine the energy by minimizing $E[^2\mathbf{D}]$ are complicated due to the lack of a simple set of necessary and sufficient conditions for ensuring that the two-matrix corresponds to an N-particle wavefunction (the N-representability problem) [1]. Nevertheless, some necessary conditions have been derived (see Chapter 9). The so called D-condition is equivalent to the requirement that the 2-RDM be positive semidefinite ($^2\mathbf{D} \geq 0$). This constraint prevents the probability distribution for finding two particles in two SOs from being anywhere negative. The G- and Q-conditions state that the electron–hole density matrix $\mathbf{G}$ and the two-hole density matrix $\mathbf{Q}$ must be positive semidefinite too. The constraint $\mathbf{G} \geq 0$ ($\mathbf{Q} \geq 0$) enforces likewise this nonnegativity for a particle and a hole (two holes).

A. Spin Structure

The N-electron Hamiltonian $\widehat{\mathcal{H}}$ corresponding to the expectation value in Eq. (1) does not contain any spin coordinates; hence both operators $\widehat{S}_z$ and $\widehat{S}^2$ commute with $\widehat{\mathcal{H}}$. Consequently, the eigenfunctions of the Hamiltonian are also eigenfunctions of these two spin operators. In particular, according to Löwdin's expressions [82], we have

$$\langle \widehat{S}_z \rangle = \int d\mathbf{x}_1\, \widehat{S}_z\, ^1D(\mathbf{r}_1 s_1 | \mathbf{r}_1 s_1) = M_S \quad (15)$$

$$\langle \widehat{S}^2 \rangle = -\frac{N(N-4)}{4} + \int d\mathbf{x}_1\, d\mathbf{x}_2\, ^2D(\mathbf{r}_1 s_1, \mathbf{r}_2 s_2 | \mathbf{r}_1 s_2, \mathbf{r}_2 s_1) = S(S+1) \quad (16)$$

where M_S and S are the spin quantum numbers describing the z component and the total spin of an N-electron eigenstate. A state with total spin S has multiplicity $(2S+1)$.

The SOs are direct products $|\phi_i\rangle = |\varphi_p\rangle \otimes |\sigma\rangle$, so the set of SOs $\{\phi_i(\mathbf{x})\}$ may be split into two subsets: $\{\varphi_p^\alpha(\mathbf{r})\alpha(\mathbf{s})\}$ and $\{\varphi_p^\beta(\mathbf{r})\beta(\mathbf{s})\}$. Given a set of $2R$ spin orbitals $\{\phi_i | i = 1, \ldots, 2R\}$, we have two sets of R orthonormal spatial functions, $\{\varphi_p^\alpha(\mathbf{r})\}$ and $\{\varphi_p^\beta(\mathbf{r})\}$, such that in general the first set is not orthogonal to the second one. Nevertheless, the original set

$$\phi_{2p-1}(\mathbf{x}) = \varphi_p^\alpha(\mathbf{r})\alpha(\mathbf{s}), \quad p = 1, \ldots, R$$
$$\phi_{2p}(\mathbf{x}) = \varphi_p^\beta(\mathbf{r})\beta(\mathbf{s}), \quad p = 1, \ldots, R \qquad (17)$$

continues being orthonormal via the orthogonality of the spin functions

$$\int d\mathbf{s}\, \alpha^*(\mathbf{s})\beta(\mathbf{s}) = \int d\mathbf{s}\, \beta^*(\mathbf{s})\alpha(\mathbf{s}) = 0 \qquad (18)$$

For $\widehat{S}_z$ eigenstates, only density matrix blocks that conserve the number of each spin type are nonvanishing. It is easily seen that two components of the 1-RDM, namely, $^1\mathbf{D}_\beta^\alpha$ and $^1\mathbf{D}_\alpha^\beta$, must vanish. One obtains

$$^1D(\mathbf{x}_1'|\mathbf{x}_1) = {}^1D_\alpha^\alpha(\mathbf{r}_1'|\mathbf{r}_1)\alpha^*(\mathbf{s}_1')\alpha(\mathbf{s}_1) + {}^1D_\beta^\beta(\mathbf{r}_1'|\mathbf{r}_1)\beta^*(\mathbf{s}_1')\beta(\mathbf{s}_1) \qquad (19)$$

From Eq. (15), considering the normalization condition for the 1-RDM (Eq. (6)), it then follows that

$$\langle \widehat{S}_z \rangle = \int d\mathbf{r}_1 Q_z(\mathbf{r}_1|\mathbf{r}_1) = \frac{N^\alpha - N^\beta}{2} = M_S \qquad (20)$$

where N^σ is the number of electrons with σ spin and $Q_z(\mathbf{r}_1|\mathbf{r}_1)$ represents the spin density [83],

$$Q_z(\mathbf{r}_1|\mathbf{r}_1) = \tfrac{1}{2}[{}^1D_\alpha^\alpha(\mathbf{r}_1|\mathbf{r}_1) - {}^1D_\beta^\beta(\mathbf{r}_1|\mathbf{r}_1)] \qquad (21)$$

On the other hand, the 2-RDM generally has 16 spin blocks. As a result of the requirement $m_s(1) + m_s(2) = m_s(1') + m_s(2')$ for $\widehat{S}_z$ eigenstates, only six spin components are nonzero. Expanding the 2-RDM by spin components, we have

$$\begin{aligned}^2D(\mathbf{x}_1',\mathbf{x}_2'|\mathbf{x}_1,\mathbf{x}_2) &= ({}^2\mathbf{D}_{\alpha\alpha}^{\alpha\alpha})\alpha^*\alpha^*\alpha\alpha + ({}^2\mathbf{D}_{\alpha\beta}^{\alpha\beta})\alpha^*\beta^*\alpha\beta + ({}^2\mathbf{D}_{\beta\alpha}^{\beta\alpha})\beta^*\alpha^*\beta\alpha \\ &\quad + ({}^2\mathbf{D}_{\beta\alpha}^{\alpha\beta})\alpha^*\beta^*\beta\alpha + ({}^2\mathbf{D}_{\alpha\beta}^{\beta\alpha})\beta^*\alpha^*\alpha\beta + ({}^2\mathbf{D}_{\beta\beta}^{\beta\beta})\beta^*\beta^*\beta\beta\end{aligned} \qquad (22)$$

In fact only three of these components are independent, for example (see Eq. (11)),

$$^2D^{\alpha\beta}_{\alpha\beta}(\mathbf{r}'_1,\mathbf{r}'_2|\mathbf{r}_1,\mathbf{r}_2) = -{}^2D^{\alpha\beta}_{\beta\alpha}(\mathbf{r}'_1,\mathbf{r}'_2|\mathbf{r}_2,\mathbf{r}_1) \quad (^2D^{p\alpha,q\beta}_{r\alpha,t\beta} = -{}^2D^{p\alpha,q\beta}_{t\beta,r\alpha}) \quad (23)$$

We may take the independent components to be $^2\mathbf{D}^{\alpha\alpha}_{\alpha\alpha}$, $^2\mathbf{D}^{\alpha\beta}_{\alpha\beta}$, $^2\mathbf{D}^{\beta\beta}_{\beta\beta}$. The parallel-spin components must be antisymmetric, but $^2\mathbf{D}^{\alpha\beta}_{\alpha\beta}$ possesses no special symmetry. Each of these two-matrix blocks must contract to the appropriate one-matrix block, namely,

$$\sum_q {}^2D^{p\sigma,q\sigma}_{r\sigma,q\sigma} = \frac{(N^\sigma - 1)}{2} {}^1D^{p\sigma}_{r\sigma} \quad (24)$$

$$\sum_q {}^2D^{p\alpha,q\beta}_{r\alpha,q\beta} = \frac{N^\beta}{2} {}^1D^{p\alpha}_{r\alpha} \quad (25)$$

It is readily demonstrated that the sum rules (24) and (25) are compatible with the Eq. (14). The traces of these two-matrix components read

$$\mathrm{Tr}\,{}^2\mathbf{D}^{\sigma\sigma}_{\sigma\sigma} = \frac{N^\sigma(N^\sigma - 1)}{2}, \quad \mathrm{Tr}\,{}^2\mathbf{D}^{\alpha\beta}_{\alpha\beta} = \frac{N^\alpha N^\beta}{2} \quad (26)$$

From Eqs. (16) and (22), taking into account the orthonormality conditions in Eq. (4) for each spin type, one obtains

$$\langle \widehat{S}^2 \rangle = -\frac{N(N-4)}{4} + \sum_{pq}({}^2D^{p\alpha,q\alpha}_{p\alpha,q\alpha} + {}^2D^{p\beta,q\beta}_{p\beta,q\beta}) - 2\sum_{pqrt} {}^2D^{p\alpha,q\beta}_{r\alpha,t\beta} S^{p\alpha}_{t\beta} S^{q\beta}_{r\alpha} \quad (27)$$

where $S^{p\sigma}_{t\sigma'} = \langle \varphi^\sigma_p | \varphi^{\sigma'}_t \rangle$ is the overlap matrix.

In this chapter, we later consider spin-polarized systems. One avenue of approach is to apply the spin unrestricted formalism, where SOs have different spatial orbitals for different spins. However, this procedure can introduce important spin contamination effects through the last term of Eq. (27) since the overlap matrix $S^{p\alpha}_{t\beta} \neq \delta^p_t$. These effects can be avoided by the use of spin-restricted theory. In this case only a single set of orbitals is used for α and β spins,

$$\varphi^\alpha_p(\mathbf{r}) = \varphi^\beta_p(\mathbf{r}) = \varphi_p(\mathbf{r}) \quad (28)$$

The orthonormality requirement ($S^{p\alpha}_{t\beta} = \delta^p_t$) leads to the expectation value of the total spin

$$\langle \widehat{S}^2 \rangle = -\frac{N(N-4)}{4} + \sum_{pq}({}^2D^{p\alpha,q\alpha}_{p\alpha,q\alpha} + {}^2D^{p\beta,q\beta}_{p\beta,q\beta} - 2\,{}^2D^{p\alpha,q\beta}_{q\alpha,p\beta}) \quad (29)$$

By combining Eq. (1) with Eqs. (19) and (22), one arrives at the energy expression

$$E = \sum_{pr} h_p^r ({}^1D_{r\alpha}^{p\alpha} + {}^1D_{r\beta}^{p\beta})$$
$$+ \sum_{pqrt} \langle rt|pq \rangle ({}^2D_{r\alpha,t\alpha}^{p\alpha,q\alpha} + {}^2D_{r\alpha,t\beta}^{p\alpha,q\beta} + {}^2D_{t\alpha,r\beta}^{q\alpha,p\beta} + {}^2D_{r\beta,t\beta}^{p\beta,q\beta}) \quad (30)$$

We must note that the two nonzero blocks ${}^2D_{r\beta,t\alpha}^{p\alpha,q\beta}$ and ${}^2D_{r\alpha,t\beta}^{p\beta,q\alpha}$ do not contribute to Eq. (30) since the corresponding two-electron matrix elements of the Coulomb interaction vanish. It is convenient now to introduce spinless density matrices [83]:

$${}^1\tilde{D}_r^p = \sum_\sigma {}^1D_{r\sigma}^{p\sigma} = {}^1D_{r\alpha}^{p\alpha} + {}^1D_{r\beta}^{p\beta} \quad (31)$$

$${}^2\tilde{D}_{rt}^{pq} = {}^2D_{r\alpha,t\alpha}^{p\alpha,q\alpha} + {}^2D_{r\alpha,t\beta}^{p\alpha,q\beta} + {}^2D_{r\beta,t\alpha}^{p\beta,q\alpha} + {}^2D_{r\beta,t\beta}^{p\beta,q\beta} \quad (32)$$

or in the coordinate-space representation

$${}^1\tilde{D}(\mathbf{r}_1'|\mathbf{r}_1) = \int d\mathbf{s}_1 \, {}^1D(\mathbf{r}_1'\mathbf{s}_1|\mathbf{r}_1\mathbf{s}_1) = \sum_\sigma {}^1D_\sigma^\sigma(\mathbf{r}_1'|\mathbf{r}_1) \quad (33)$$

$${}^2\tilde{D}(\mathbf{r}_1'\mathbf{r}_2'|\mathbf{r}_1\mathbf{r}_2) = \int d\mathbf{s}_1 d\mathbf{s}_2 \, {}^2D(\mathbf{r}_1'\mathbf{s}_1, \mathbf{r}_2'\mathbf{s}_2|\mathbf{r}_1\mathbf{s}_1, \mathbf{r}_2\mathbf{s}_2)$$
$$= \sum_{\sigma\sigma'} {}^2D_{\sigma\sigma'}^{\sigma\sigma'}(\mathbf{r}_1'\mathbf{r}_2'|\mathbf{r}_1\mathbf{r}_2) \quad (34)$$

The diagonal elements ${}^1\tilde{D}(\mathbf{r}_1|\mathbf{r}_1)$ and ${}^2\tilde{D}(\mathbf{r}_1\mathbf{r}_2|\mathbf{r}_1\mathbf{r}_2)$ are the electron density and pair density, respectively. Then Eq. (30) can be rewritten

$$E = \sum_{pr} {}^1\tilde{D}_r^p h_p^r + \sum_{pqrt} {}^2\tilde{D}_{rt}^{pq} \langle rt|pq \rangle \quad (35)$$

IV. NATURAL ORBITAL FUNCTIONAL (NOF)

Let us replace the last term in Eq. (1), which is an explicit functional of the 2-RDM, by an unknown functional of the 1-RDM:

$$E[N, {}^1\mathbf{D}] = h[N, {}^1\mathbf{D}] + V_{ee}[N, {}^1\mathbf{D}] \quad (36)$$

$h[N, {}^1\mathbf{D}]$ is the contribution from the kinetic energy and the external potential,

$$h[N, {}^1\mathbf{D}] = \sum_{ik} {}^1D_k^i h_i^k \quad (37)$$

while the electron–electron repulsion energy $V_{ee}[N,{}^1\mathbf{D}]$ constitutes a universal functional in the sense that it is invariant from one molecule to another for a given N, and hence it is independent of the external field. The constrained-search formalism [43] provides a proof by construction of the existence of this functional; in other words, it is given by the expression

$$V_{ee}[N,{}^1\mathbf{D}] = \min_{{}^2\mathbf{D}\,\in\,{}^2\mathbf{D}({}^1\mathbf{D})} U[N,{}^2\mathbf{D}] \qquad (38)$$

where

$$U[N,{}^2\mathbf{D}] = \int d\mathbf{x}_1\, d\mathbf{x}_2\,{}^2D(\mathbf{x}_1,\mathbf{x}_2|\mathbf{x}_1,\mathbf{x}_2)r_{12}^{-1} = \sum_{ijkl} {}^2D_{kl}^{ij}\langle kl|ij\rangle \qquad (39)$$

The notation ${}^2\mathbf{D}({}^1\mathbf{D})$ indicates the family of 2-RDMs that contract to the 1-RDM in agreement with Eq. (14). Restricting the 2-RDM in Eq. (38) to be pure or ensemble N-representable yields the universal functionals of Levy [10] and Valone [19], respectively.

The properties of the universal functional V_{ee} are well known [11–13]. Moreover, the exact 1-RDM functional for the two-electron closed-shell systems like H_2 or He is known too [76, 77]. However, V_{ee} is highly difficult to approximate because what we have done is just to change the variational unknown from the 2-RDM to the 1-RDM, but the 2-RDM N-representability problems remain as revealed explicitly in Eq. (38).

The 1-RDM can be diagonalized by a unitary transformation of the spin orbitals $\{\phi_i(\mathbf{x})\}$ with the eigenvectors being the natural spin orbitals (NSOs) and the eigenvalues $\{n_i\}$ representing the ONs of the latter,

$${}^1D_k^i = n_i \delta_k^i, \quad {}^1D(\mathbf{x}_1'|\mathbf{x}_1) = \sum_i n_i \phi_i^*(\mathbf{x}_1')\phi_i(\mathbf{x}_1) \qquad (40)$$

Restriction of the ONs $\{n_i\}$ to the range $0 \leq n_i \leq 1$ represents a necessary and sufficient condition for N-representability of the 1-RDM [1]. In the following, all representations used are assumed to refer to the basis of NSOs. The NO energy functional (36) reads

$$E[N,\{n_i\},\{\phi_i(\mathbf{x})\}] = \sum_i n_i h_i^i + V_{ee}[N,\{n_i\},\{\phi_i(\mathbf{x})\}] \qquad (41)$$

We may conclude that the 1-RDM and the functional N-representability problems are entirely different. The former is trivially solved since ONs sum up to the number of electrons N and lie between 0 and 1, assuring an N-representable 1-RDM. The latter refers to the conditions that guarantee the one-to-one correspondence between $E[N,{}^2\mathbf{D}]$ and $E[N,{}^1\mathbf{D}]$, which is a related problem to the N-representability of the 2-RDM. Consequently, any approximation for $V_{ee}[N,\{n_i\},\{\phi_i(\mathbf{x})\}]$ must comply at least with the known necessary conditions for the N-representability of the 2-RDM.

V. EULER EQUATIONS

Minimization of the functional (41) has to be performed under the orthonormality requirement in Eq. (4) for the NSOs, whereas the ONs conform to the N-representability conditions for ${}^1\mathbf{D}$. Bounds on the ONs are enforced by setting $n_i = \cos^2 \gamma_i$ and varying γ_i without constraints. The other two conditions may easily be taken into account by the method of Lagrange multipliers.

Associate the Lagrange multiplier μ (chemical potential) with the normalization condition in Eq. (6), the set of Hermitian–Lagrange multipliers $\{\lambda_k^i\}$ with orthonormality constraints in Eq. (4), and define the auxiliary functional Ω by the formula

$$\Omega[N,\{\gamma_i\},\{\phi_i(\mathbf{x})\}] = E - \mu\left(\sum_i \cos^2 \gamma_i - N\right) - \sum_{ik} \lambda_k^i (\langle \phi_k | \phi_i \rangle - \delta_i^k) \quad (42)$$

The functional (42) has to be stationary with respect to variations in $\{\phi_i(\mathbf{x})\}$, $\{\phi_i^*(\mathbf{x})\}$, and $\{\gamma_i\}$:

$$\delta\Omega = \sum_i \sin(2\gamma_i)\left[\mu - \frac{\partial E}{\partial n_i}\right]d\gamma_i + \sum_i \int d\mathbf{x}\,\delta\phi_i^*(\mathbf{x})\left[\frac{\delta E}{\delta\phi_i^*(\mathbf{x})} - \sum_k \lambda_i^k \phi_k(\mathbf{x})\right]$$
$$+ \sum_i \int d\mathbf{x}\left[\frac{\delta E}{\delta\phi_i(\mathbf{x})} - \sum_k \lambda_k^i \phi_k^*(\mathbf{x})\right]\delta\phi_i(\mathbf{x}) = 0$$
$$(43)$$

The partial derivative $(\partial E/\partial n_i)$ is taken holding the orbitals fixed. It satisfies the relation

$$\frac{\partial E}{\partial n_i} = h_i^i + \frac{\partial V_{ee}}{\partial n_i} = \mu \quad (44)$$

For a fixed set of occupations, the orbital Euler equations are

$$n_i \widehat{V}(\mathbf{x})\phi_i(\mathbf{x}) = \sum_k \lambda_i^k \phi_k(\mathbf{x}) \quad (45)$$

where

$$\widehat{V}(\mathbf{x}) = \widehat{h}(\mathbf{x}) + \frac{1}{n_i \phi_i(\mathbf{x})} \frac{\delta V_{ee}}{\delta \phi_i^*(\mathbf{x})} \quad (46)$$

Considering the orthonormality conditions in Eq. (4), the elements of the Langrangian λ read

$$\lambda_i^k = V_i^k n_i = h_i^k n_i + (g_k^i)^* \qquad (\lambda = \mathbf{h}^1 \mathbf{D} + \mathbf{g}^\dagger) \qquad (47)$$

where the matrix $\mathbf{g}$ possesses the elements

$$g_i^k = \int d\mathbf{x} \frac{\delta V_{ee}}{\delta \phi_k(\mathbf{x})} \phi_i(\mathbf{x}) \qquad (48)$$

A complex conjugated set of equations to Eq. (45) results for $\{\phi_i^*(\mathbf{x})\}$. Simultaneous solution of Eqs. (44)–(46) yields the optimum bound on the ground-state energy.

A. Gilbert Nonlocal Potential

By making use of the Gilbert formal identity [15],

$$\frac{\delta V_{ee}}{\delta \phi_i^*(\mathbf{x})} = n_i \phi_i(\mathbf{x}) \hat{v}_{ee}(\mathbf{x}) \qquad (49)$$

we may write the orbital Euler equations (45) in a convenient form:

$$n_i [\hat{h}(\mathbf{x}) + \hat{v}_{ee}(\mathbf{x})] \phi_i(\mathbf{x}) = n_i \hat{F}(\mathbf{x}) \phi_i(\mathbf{x}) = \sum_k \lambda_i^k \phi_k(\mathbf{x}) \qquad (50)$$

or in matrix representation

$$F_i^k n_i = \lambda_i^k, \quad F_i^k = \lambda_i^k n_i^{-1} \qquad (\mathbf{F}^1 \mathbf{D} = \lambda, \quad \mathbf{F} = \lambda (^1\mathbf{D})^{-1}) \qquad (51)$$

in which all matrices are Hermitian for a stationary state. Since the product of two Hermitian matrices is itself Hermitian if and only if they commute, it follows that the Hermiticity of λ and Eq. (51) are together equivalent to the commutation relations

$$[\mathbf{F}, {}^1\mathbf{D}] = 0, \quad [\lambda, {}^1\mathbf{D}] = 0 \qquad (52)$$

This implies that the 1-RDM and the Langrangian, λ, may be simultaneously brought to diagonal form by the same unitary transformation $\mathbf{U}$, from which it follows that

$$[\hat{h}(\mathbf{x}) + \hat{v}_{ee}(\mathbf{x})] \phi_i(\mathbf{x}) = \lambda_i n_i^{-1} \phi_i(\mathbf{x}) = \epsilon_i \phi_i(\mathbf{x}) \qquad (53)$$

The canonical orbitals thus satisfy concomitantly Eqs. (44) and (53), so we then have

$$(v_{ee})_i^i = \frac{\partial V_{ee}}{\partial n_i}, \quad \epsilon_i = \frac{\lambda_i}{n_i} = \mu \qquad (54)$$

Assuming Eq. (49), this result shows that for canonical NSOs the operator $\widehat{F}$ has an essentially degenerate eigenvalue spectrum; that is, all the NO eigenvalues are the same (μ) and are equal to minus the vertical IP [84]. Unfortunately, apart from the special case of the HF energy that may be viewed as the simplest 1-RDM functional, none of the currently known functionals (including the exact functional for the total energy in two-electron closed-shell systems) have effective potentials that satisfy the formal relation (49).

B. Relation with the Extended Koopmans' Theorem

At present, most known NOFs are defined in terms of NOs and ONs being only implicitly dependent on the 1-RDM. In this case, the energy functional still depends explicitly on the 2-RDM. As was pointed out by Donnelly [18], there is a fundamental difference between energy functionals based explicitly on 1- and on 2-RDMs. Equation (45) for the optimum orbitals is actually Löwdin's equation [82]. This equation cannot be reduced to an eigenvalue problem diagonalizing the matrix λ, although by slight manipulation the operator $\widehat{V}(\mathbf{x})$ can be transformed into a Hermitian operator with a nondegenerate spectrum of eigenvalues v. Such construction is provided by extension of Koopmans' theorem [32, 33].

The equation for the EKT for ionization potentials may be derived by expressing the wavefunction of the $(N-1)$-electron system as a simple linear combination

$$|\Psi^{N-1}\rangle = \sum_i C_i \widehat{a}_i |\Psi^N\rangle \qquad (55)$$

In Eq. (55), $\widehat{a}_i$ is the annihilation operator for an electron in orbital i, $|\Psi^N\rangle$ is the wavefunction of the N-electron system, $|\Psi^{N-1}\rangle$ is the wavefunction of the $(N-1)$-electron system, and $\{C_i\}$ are a set of coefficients to be determined. Optimizing the energy of the state Ψ^{N-1} with respect to the parameters $\{C_i\}$ and subtracting the energy of Ψ^N gives the EKT equations as a generalized eigenvalue problem,

$$\mathbf{TC} = v^1 \mathbf{DC} \qquad (56)$$

where v are the EKT ionization potentials. In Eq. (56), the metric matrix is $^1\mathbf{D}$ with the ONs $\{n_i\}$ along the diagonal and zeros in off-diagonal elements, and the transition matrix elements are given by

$$T_i^k = \langle \Psi^N | \widehat{a}_k^\dagger [\widehat{H}, \widehat{a}_i] | \Psi^N \rangle = -\left(n_i h_i^k + 2 \sum_{jlm} {}^2 D_{lm}^{kj} \langle lm | ij \rangle\right) = -V_i^k n_i \qquad (57)$$

Equation (56) can be transformed by canonical orthonormalization using $(^1\mathbf{D})^{-1/2}$. With this transformation it can be written

$$\mathbf{T'C'} = v\mathbf{C'} \tag{58}$$

It is now clear from Eqs. (47) and (57) that the diagonalization of the matrix v with the elements

$$v_i^k = -\frac{\lambda_i^k}{\sqrt{n_k n_i}} \tag{59}$$

yields ionization potentials as eigenvalues [18, 72].

C. Pernal Nonlocal Potential

Let's return to the problem of finding the optimal nonlocal potential $\widehat{v}_{ee}$ where the one-electron functions $\{\phi_i(\mathbf{x})\}$ resulting from solving the eigenproblem

$$\widehat{F}(\mathbf{x})\phi_i(\mathbf{x}) = [\widehat{h}(\mathbf{x}) + \widehat{v}_{ee}(\mathbf{x})]\phi_i(\mathbf{x}) = \epsilon_i \phi_i(\mathbf{x}) \tag{60}$$

satisfy the set of Euler equations (45).

Since the energy is real, the Langrangian λ is a Hermitian matrix at the extremum,

$$\lambda_i^k - (\lambda_k^i)^* = 0 \qquad (\lambda - \lambda^\dagger = 0) \tag{61}$$

From Eq. (61), by using Eq. (47), it follows without difficulty that

$$(n_i - n_k)h_i^k + (g_k^i)^* - g_i^k = 0 \qquad ([\mathbf{h},^1\mathbf{D}] + \mathbf{g}^\dagger - \mathbf{g} = 0) \tag{62}$$

This equation can be rewritten as a commutator between $\mathbf{F}$ and $^1\mathbf{D}$,

$$[\mathbf{F},^1\mathbf{D}] = [\mathbf{h} + v_{ee},^1\mathbf{D}] = 0 \tag{63}$$

if we define the nonlocal potential matrix v_{ee} by the commutation relation,

$$(n_i - n_k)(v_{ee})_i^k = (g_k^i)^* - g_i^k \qquad ([v_{ee},^1\mathbf{D}] = \mathbf{g}^\dagger - \mathbf{g}) \tag{64}$$

According to Eq. (63), the 1-RDM and the generalized Fockian $\mathbf{F}$ commute at the extremum; hence the NSOs are the solutions of the eigenproblem (60) with the nonlocal potential defined by the identity (64). One should note, however, that Eq. (64) does not completely define v_{ee}. In fact, the diagonal elements

$(v_{\text{ee}})_i^i$ and the elements $(v_{\text{ee}})_i^k$ corresponding to orbitals of equal ONs ($n_i = n_k$) may be arbitrary.

On the other hand, Eq. (44) must also be satisfied by the optimal set of the NOs, suggesting that the diagonal elements of $\widehat{v}_{\text{ee}}$ are defined by the formula

$$(v_{\text{ee}})_i^i = \frac{\partial V_{\text{ee}}}{\partial n_i} \tag{65}$$

This selection implies that $\epsilon_i = \mu$ for the optimal 1-RDM (compare with the Eq. (54)). The Fockian matrix elements are as follows:

$$F_i^k = h_i^k + \frac{\partial V_{\text{ee}}}{\partial n_i}\delta_i^k + \frac{1-\delta_i^k}{n_i - n_k}\int d\mathbf{x}\left[\phi_k^*(\mathbf{x})\frac{\delta V_{\text{ee}}}{\delta\phi_i^*(\mathbf{x})} - \frac{\delta V_{\text{ee}}}{\delta\phi_k(\mathbf{x})}\phi_i(\mathbf{x})\right] \tag{66}$$

In this manner, we have arrived at the Pernal nonlocal potential [81]. It can be shown, using the invariance of V_{ee} with respect to an arbitrary unitary transformation and its extremal properties [13] or by means of the first-order perturbation theory applied to the eigenequation of the 1-RDM [81], that the off-diagonal elements of $\widehat{v}_{\text{ee}}$ may also be derived via the functional derivative

$$(v_{\text{ee}})_i^k = \frac{\partial V_{\text{ee}}}{\partial^1 D_k^i} \tag{67}$$

proposed by Gilbert [15].

The other degree of freedom in defining $\widehat{v}_{\text{ee}}$ related to the degeneracy of the 1-RDM can be avoided with a proper definition of the diagonal part $(v_{\text{ee}})_i^i$ (see Ref. [81]). As a consequence, the NOs with the same ON are determined only up to the unitary transformation among them.

We conclude that the problem of finding optimal NOs turns into the iterative diagonalization of Eq. (60) with a Fockian matrix, Eq. (66). The corresponding eigenfunctions are certainly orthonormal and optimize the total energy functional, Eq. (41).

VI. CUMULANT OF THE 2-RDM

The 2-RDM formulation, Eq. (38), allows us to generalize the constrained search to approximately N-representable sets of 2-RDMs. In order to approximate the unknown functional $V_{\text{ee}}[N, {}^1\mathbf{D}]$, we use here a reconstructive functional ${}^2\mathbf{D}[{}^1\mathbf{D}]$; that is, we express the elements ${}^2 D_{kl}^{ij}$ in terms of the ${}^1 D_k^i$. We neglect any explicit dependence of ${}^2\mathbf{D}$ on the NOs themselves because the energy functional already has a strong dependence on the NOs via the one- and two-electron integrals.

The 2-RDM can be partitioned into an antisymmetrized product of the 1-RDMs, which is simply the HF approximation, and a correction $^2\Gamma$ to it,

$$^2D_{kl}^{ij} = \frac{1}{2}(^1D_k^i {}^1D_l^j - {}^1D_k^j {}^1D_l^i) + {}^2\Gamma_{kl}^{ij} = \frac{n_j n_i}{2}(\delta_k^i \delta_l^j - \delta_k^j \delta_l^i) + {}^2\Gamma_{kl}^{ij} \quad (68)$$

This decomposition of the 2-RDM is well known from the cumulant theory (see earlier chapters). $^2\Gamma$ is the cumulant matrix of the 2-RDM. Since it arises from interactions in the Hamiltonian, it might also be called the pair correlation matrix. This definition of correlation differs from the traditional one since ^1D is the one-matrix of the correlated system and not that corresponding to independent particles.

The first two terms on the right-hand side (rhs) of Eq. (68) together satisfy properties (10)–(13) of the 2-RDM. Therefore the cumulant matrix $^2\Gamma$ should satisfy these relations too. We further see that matrix elements of $^2\Gamma$ are non-vanishing only if all its labels refer to partially occupied NSOs with ON different from 0 or 1. For a single Slater determinant, the cumulant matrix vanishes.

It is important to note that the 2-RDM is not additively separable (extensive), but its cumulant matrix $^2\Gamma$ satisfies this essential property. Finally, we must note that the trace of $^2\Gamma$ is of $O(N)$; that is, it scales linearly with the size of the system, while the trace of the corresponding 2-RDM is of $O(N^2)$,

$$\text{Tr}(^2\Gamma) = \sum_i \frac{n_i^2 - n_i}{2} = O(N) \quad (69)$$

It can easily be shown from Eqs. (24) and (25), taking into account the normalization condition (6) for each spin type, that spin components of $^2\Gamma$ fulfill the following sum rules:

$$2\sum_q {}^2\Gamma_{r\sigma,q\sigma}^{p\sigma,q\sigma} = n_p^\sigma(n_p^\sigma - 1)\delta_r^p \quad (70)$$

$$\sum_q {}^2\Gamma_{r\alpha,q\beta}^{p\alpha,q\beta} = 0 \quad (71)$$

Using Eqs. (31), (32), (40), and (68), the energy, Eq. (35), reads as

$$\begin{aligned} E = &\sum_p (n_p^\alpha + n_p^\beta)h_p^p + \frac{1}{2}\sum_{pq}(n_q^\alpha + n_q^\beta)(n_p^\alpha + n_p^\beta)J_{pq} \\ &- \frac{1}{2}\sum_{pq}(n_q^\alpha n_p^\alpha + n_q^\beta n_p^\beta)K_{pq} + \sum_{pqrt} {}^2\widetilde{\Gamma}_{rt}^{pq}\langle rt|pq\rangle \end{aligned} \quad (72)$$

where $J_{pq} = \langle pq|pq \rangle$ and $K_{pq} = \langle pq|qp \rangle$ (see Eq. (3) with $\phi_i(\mathbf{x})$ replaced by $\varphi_p(\mathbf{r})$) are the usual Coulomb and exchange integrals, respectively. $^2\widetilde{\Gamma}_{rt}^{pq}$ denotes the spinless cumulant matrix,

$$^2\widetilde{\Gamma}_{rt}^{pq} = {}^2\Gamma_{r\alpha,t\alpha}^{p\alpha,q\alpha} + {}^2\Gamma_{r\alpha,t\beta}^{p\alpha,q\beta} + {}^2\Gamma_{t\alpha,r\beta}^{q\alpha,p\beta} + {}^2\Gamma_{r\beta,t\beta}^{p\beta,q\beta} \tag{73}$$

Taking into account the sum rule, Eq. (70), the expectation value of $\widehat{S}^2$ is likewise obtained from Eq. (29),

$$\langle \widehat{S}^2 \rangle = \frac{N^\alpha + N^\beta}{2} + \frac{(N^\alpha - N^\beta)^2}{4} - \sum_p n_p^\alpha n_p^\beta - \sum_{pq} 2\, {}^2\Gamma_{q\alpha,p\beta}^{p\alpha,q\beta} \tag{74}$$

A. Approximate Cumulant

A large number of choices for the cumulant $^2\Gamma$ are possible. It has a dependence of four indices and direct computation with such magnitudes is too expensive to be applied to large systems. We want to maximize the physical content of $^2\Gamma$ to a few terms. We express $^2\Gamma$ by means of two-index matrices Δ and Π,

$$\begin{aligned}
{}^2\Gamma_{r\sigma'_1, t\sigma'_2}^{p\sigma_1, q\sigma_2} = &-\frac{\Delta_{q\sigma_2}^{p\sigma_1}}{2}\left(\delta^p_r \delta^{\sigma_1}_{\sigma'_1} \delta^q_t \delta^{\sigma_2}_{\sigma'_2} - \delta^q_r \delta^{\sigma_2}_{\sigma'_1} \delta^p_t \delta^{\sigma_1}_{\sigma'_2}\right) \\
&+ \frac{m_s(1)m_s(1')}{2} \Pi_{t\sigma'_2}^{p\sigma_1} \delta^t_r \delta^{\sigma'_1}_{-\sigma'_2} \delta^q_p \delta^{\sigma_1}_{-\sigma_2}
\end{aligned} \tag{75}$$

Here m_s denotes the sign of the spin projection (it takes two values, $+1$ and -1). By taking into account the cumulant properties, Eqs. (10)–(13) with ^2D replaced by $^2\Gamma$, it can be shown that Δ must be a real symmetric matrix ($\Delta_{q\sigma_2}^{p\sigma_1} = \Delta_{p\sigma_1}^{q\sigma_2}$) with no unique diagonal elements, whereas Π is a spin-independent ($\Pi_{r\alpha}^{p\alpha} = \Pi_{r\beta}^{p\alpha} = \Pi_{r\alpha}^{p\beta} = \Pi_{r\beta}^{p\beta} = \Pi_r^p$) Hermitian matrix ($\Pi_r^p = (\Pi_p^r)^*$). We have the following spin structure for the cumulant matrix:

$$^2\Gamma_{r\sigma,t\sigma}^{p\sigma,q\sigma} = -\frac{\Delta_{q\sigma}^{p\sigma}}{2}\left(\delta^p_r \delta^q_t - \delta^q_r \delta^p_t\right) \tag{76}$$

$$^2\Gamma_{r\alpha,t\beta}^{p\alpha,q\beta} = -\frac{\Delta_{q\beta}^{p\alpha}}{2} \delta^p_r \delta^q_t + \frac{\Pi_r^p}{2} \delta^t_r \delta^q_p \tag{77}$$

The sum rule, Eq. (70), and the approximate ansatz, Eq. (76), imply the constraint

$$\sideset{}{'}\sum_q \Delta_{q\sigma}^{p\sigma} = n_p^\sigma (1 - n_p^\sigma) \tag{78}$$

where the prime indicates that the $q = p$ term is omitted. Analogously, one obtains, using Eqs. (71) and (77), the following sum rule:

$$\sum_{q}' \Delta_{q\beta}^{p\alpha} = \Pi_p^p \tag{79}$$

By combining Eq. (72) with Eqs. (76) and (77), one arrives at the energy expression

$$E = \sum_p (n_p^\alpha + n_p^\beta) h_p^p + \frac{1}{2} \sum_{pq} [(n_q^\alpha + n_q^\beta)(n_p^\alpha + n_p^\beta) - \widetilde{\Delta}_q^p] J_{pq}$$
$$- \frac{1}{2} \sum_{pq} [(n_q^\alpha n_p^\alpha + n_q^\beta n_p^\beta) - (\Delta_{q\alpha}^{p\alpha} + \Delta_{q\beta}^{p\beta})] K_{pq} + \sum_{pr} \Pi_r^p L_p^r \tag{80}$$

where $\widetilde{\Delta}_q^p$ denotes the spinless Δ matrix,

$$\widetilde{\Delta}_q^p = \Delta_{q\alpha}^{p\alpha} + \Delta_{q\beta}^{p\alpha} + \Delta_{q\alpha}^{p\beta} + \Delta_{q\beta}^{p\beta} \tag{81}$$

The new integral $L_p^r = \langle rr|pp \rangle$ arises from the correlation between particles with opposite spins and may be called the exchange and time-inversion integral [53]. In fact, one may obtain it as follows:

$$L_p^r = \int \frac{d\mathbf{x}_1 \, d\mathbf{x}_2}{r_{12}} [\varphi_r^\alpha(\mathbf{r}_1)\alpha(\mathbf{s}_1)]^* [\varphi_r^\beta(\mathbf{r}_2)\beta(\mathbf{s}_2)]^* \varphi_p^\alpha(\mathbf{r}_1)\alpha(\mathbf{s}_1)\varphi_p^\beta(\mathbf{r}_2)\beta(\mathbf{s}_2)$$
$$= \int \frac{d\mathbf{x}_1 \, d\mathbf{x}_2}{r_{12}} [\varphi_r^\alpha(\mathbf{r}_1)\alpha(\mathbf{s}_1)]^* [\widehat{I}\varphi_p^\alpha(\mathbf{r}_2)\alpha(\mathbf{s}_2)]^* \widehat{I}(2)\widehat{P}_{12}\varphi_r^\alpha(\mathbf{r}_1)\alpha(\mathbf{s}_1)\varphi_p^\alpha(\mathbf{r}_2)\alpha(\mathbf{s}_2)$$

or, equivalently,

$$L_p^r = \langle rr|pp \rangle = \langle r\alpha, r\beta|r_{12}^{-1}|p\alpha, p\beta \rangle = \langle r\alpha, p\alpha|\widehat{I}^\dagger(2)r_{12}^{-1}\widehat{I}(2)\widehat{P}_{12}|r\alpha, p\alpha \rangle \tag{82}$$

where the $\widehat{P}_{12}$ operator permutes electrons 1 and 2, and the time-inversion antiunitary operator $\widehat{I}$ changes a ket vector into a bra vector and $\alpha(\beta)$ into $\beta(\alpha)$; that is,

$$\widehat{I}|p\alpha \rangle = \langle p\beta|, \quad \widehat{I}|p\beta \rangle = -\langle p\alpha| \tag{83}$$

Note that if $\Delta_{q\sigma'}^{p\sigma} = 0$ and $\Pi_r^p = 0$, then the reconstruction proposed here yields the HF case as expected.

Similarly, combining Eq. (74) with Eqs. (76), (77), and (78), one arrives at the average total spin

$$\langle \hat{S}^2 \rangle = \frac{N^\alpha + N^\beta}{2} + \frac{(N^\alpha - N^\beta)^2}{4} + \sum_p (\Delta_{p\beta}^{p\alpha} - n_p^\alpha n_p^\beta) - \sum_p \Pi_{p\beta}^{p\alpha} \qquad (84)$$

In Eq. (84), the α- and β-dependences of $\Pi_{p\beta}^{p\alpha}$ have been retained only to emphasize that it is related to these SOs. We recall that Π is a spin-independent matrix as a consequence of the antisymmetric properties of cumulant $^2\Gamma$.

VII. RESTRICTED CLOSED-SHELL NOF

So far in this chapter we have discussed the NOF theory in terms of general set of SOs $\{\phi_i(\mathbf{x})\}$ or in terms of restricted SOs, which are constrained to have the same spatial function $\{\varphi_p(\mathbf{r})\}$ for α and β spin functions. In this section we are concerned only with closed-shell systems. Our molecules are thus allowed to have only an even number of electrons, with all electrons paired such that the spatial orbitals are doubly occupied. In this case of spin-compensated systems, the two nonzero blocks of the 1-RDM are the same ($^1\mathbf{D}_\alpha^\alpha = {}^1\mathbf{D}_\beta^\beta$); that is,

$$n_p^\alpha = n_p^\beta = n_p \qquad (85)$$

The trace of the one-matrix, Eq. (6), becomes

$$\sum_p n_p = N^\alpha = N^\beta = \frac{N}{2} \qquad (86)$$

and from Eq. (20) it follows that

$$\langle \hat{S}_z \rangle = \frac{N^\alpha - N^\beta}{2} = 0 \qquad (87)$$

For singlet states, the first and the last blocks of the 2-RDM are also equal ($^2\mathbf{D}_{\alpha\alpha}^{\alpha\alpha} = {}^2\mathbf{D}_{\beta\beta}^{\beta\beta}$), so hereafter we deal only with $^2\mathbf{D}_{\alpha\alpha}^{\alpha\alpha}$ and $^2\mathbf{D}_{\alpha\beta}^{\alpha\beta}$. We assume further that $\Delta_{q\alpha}^{p\alpha} = \Delta_{q\beta}^{p\alpha} = \Delta_{q\alpha}^{p\beta} = \Delta_{q\beta}^{p\beta} = \Delta_q^p$. For convenient purposes as we see below, we define the matrix Π in terms of a new spin-independent Hermitian matrix Λ:

$$\Pi_r^p = n_r n_p - \Delta_r^p - \Lambda_r^p \qquad (88)$$

Combining both sum rules (78) and (79) with Eq. (88) results in

$$\sum_{q}{}' \Delta_q^p = n_p(1 - n_p) \tag{89}$$

$$2\Delta_p^p + \Lambda_p^p = 2n_p^2 - n_p \tag{90}$$

Taking into account Eq. (80), the closed-shell energy can be expressed as

$$E = 2\sum_p n_p h_p^p + \sum_{pq}(n_q n_p - \Delta_q^p)(2J_{pq} - K_{pq} + L_p^q) - \sum_{pr} \Lambda_r^p L_p^r \tag{91}$$

A. Two-Electron Systems

NOF theory provides an exact energy functional for two-electron systems [76, 77]. In the weak correlation limit, the total energy is given by

$$E = 2\sum_{p=1}^{\infty} n_p h_p^p + n_1 L_1^1 - 2\sum_{p=2}^{\infty} \sqrt{n_1 n_p} L_p^1 + \sum_{p,r=2}^{\infty} \sqrt{n_r n_p} L_p^r \tag{92}$$

As can be seen from Eq. (92), the dependence of the 2-RDM on the ONs requires a distinction between spatial orbitals.

Since $^2\mathbf{D}_{\alpha\alpha}^{\alpha\alpha} = 0$ for $N = 2$, one easily deduces from Eqs. (68) and (76) that $\Delta_q^p = n_q n_p$. Consequently, it is not difficult to see from Eqs. (68), (77), and (88) that $^2\mathbf{D}_{\alpha\beta}^{\alpha\beta}$ nonzero elements have the form $^2D_{r\alpha,r\beta}^{p\alpha,p\beta} = -\Lambda_r^p/2$. Thus the total energy in Eq. (91) turns into

$$E = 2\sum_p n_p h_p^p - \sum_{pr} \Lambda_r^p L_p^r \tag{93}$$

From the requirement that for any two-electron system the expression (93) should yield Eq. (92), one has to set

$$\Lambda_r^p = -n_p \delta_r^p + (1 - \delta_r^p)[1 - 2\theta(0.5 - n_r)\theta(0.5 - n_p)]\sqrt{n_r n_p} \tag{94}$$

where $\theta(x)$ is the unit step function also known as the Heaviside function.

It is worth noting that the chosen Δ and Λ satisfy the constraints in Eqs. (78) and (90). Moreover, the expression (94) also holds true for systems where the largest occupation deviates significantly from one, indicating that it may possibly be valid for arbitrary correlation strengths [76].

B. N-Electron Systems

Let us assume the functional form in Eq. (94) for the matrix Λ in the general case of N-electron systems. This assumption and the equality in Eq. (90), affords

$$\Lambda_p^p = -n_p, \quad \Delta_p^p = n_p^2 \tag{95}$$

or, equivalently,

$$\Lambda_{p\sigma'}^{p\sigma} = -\sqrt{n_p^\sigma n_p^{\sigma'}} = -n_p, \quad \Delta_{p\sigma'}^{p\sigma} = n_p^\sigma n_p^{\sigma'} = n_p^2 \tag{96}$$

By taking into account Eq. (96), last two terms in Eq. (84) are found:

$$\sum_p (\Delta_{p\beta}^{p\alpha} - n_p^\alpha n_p^\beta) = 0, \quad \sum_p \Pi_{p\beta}^{p\alpha} = \frac{N}{2} \tag{97}$$

which leads to the exact result $\langle \hat{S}^2 \rangle = 0$ for singlet states ($2S+1 = 1$).

A very different functional form is expected for Δ in molecules with more than two electrons. Without any calculations it is clear that H$_2$ is atypical, since $^2\mathbf{D}_{\alpha\alpha}^{\alpha\alpha}$ vanishes for this system. $\Delta_q^p = n_q n_p$ ($q \neq p$), taken from the $N = 2$ case, violates the sum rule, Eq. (89), in the general case of $N > 2$. This means that the functional form of nondiagonal elements is unknown for N-electron systems, as yet. Nevertheless, some constraints can be achieved for these quantities using known necessary conditions of 2-RDM N-representability.

The above mentioned positivity conditions state that the 2-RDM $^2\mathbf{D}$, the electron–hole density matrix $\mathbf{G}$, and the two-hole density matrix $\mathbf{Q}$ must be positive semidefinite. A matrix is positive semidefinite if and only if all of its eigenvalues are nonnegative. The solution of the corresponding eigenproblems is readily carried out [73]. For $^2\mathbf{D}$, it yields the following set of eigenvalues:

$$d = \left\{ 0, n_q n_p - \Delta_q^p, \frac{1}{2}(n_q n_p - \Delta_q^p) \right\}, \quad q \neq p \tag{98}$$

There is also a single $R \times R$ block of the spin component $^2\mathbf{D}_{\alpha\beta}^{\alpha\beta}$, which has elements equal to $\Pi_r^p/2$. Accordingly, we have analytic expressions for all eigenvalues of $^2\mathbf{D}$, except those arising from the single $R \times R$ block. Consequently, our reconstructive functional satisfies the D-condition ($d \geq 0$) if $\Delta_q^p \leq n_q n_p$ and the $R \times R$ block is positive.

Considering that $\mathbf{Q}$ has the same block structure as $^2\mathbf{D}$, one obtains analogously that if one takes $\Delta_q^p \leq (1-n_q)(1-n_p)$, and the $R \times R$ block of $\mathbf{Q}_{\alpha\beta}^{\alpha\beta}$ is positive, the Q-condition is fulfilled. Consequently, the D-condition is more restrictive than the Q-condition between orbitals with ONs close to zero, whereas for Δ elements between orbitals with ONs close to one, the Q-condition is predominant.

The spin component $\mathbf{G}_{\alpha\alpha}^{\alpha\alpha}$ contains a single block $R \times R$, for which the eigenvalues have no analytic expression, and 1×1 blocks. The latter blocks have nonnegative eigenvalues if $\Delta_q^p \geq n_p(n_q - 1)$. This inequality is easy to satisfy on the domain of allowed ONs ($n_q \leq 1$) if we consider nonnegative Δ_q^p.

The opposite spin component consists entirely of 1×1 blocks $G_{p\alpha,p\beta}^{p\alpha,p\beta} = 0$, and 2×2 blocks that afford the eigenvalues

$$g_q^p = \frac{n_q + n_p}{4} + \frac{\Delta_q^p - n_q n_p}{2} \pm \frac{1}{4}\sqrt{(n_q - n_p)^2 + 4(n_q n_p - \Delta_q^p - \Lambda_q^p)^2} \quad (99)$$

In order to ensure that $g_q^p \geq 0$, expression (99) gives rise to the inequality

$$\Delta_q^p \leq n_q n_p + \frac{n_q n_p - (\Lambda_q^p)^2}{2\Lambda_q^p - n_q - n_p} = n_q n_p \quad (100)$$

To satisfy the known necessary N-representability conditions for the 2-RDM, the matrix elements of $\mathbf{\Delta}$ have to conform to the following analytic constraints:

$$\Delta_q^p \leq n_q n_p, \quad \Delta_q^p \leq (1 - n_q)(1 - n_p), \quad \Delta_q^p \geq n_p(n_q - 1), \quad q \neq p \quad (101)$$

C. Effective One-Electron Operator

The best NSOs are those that minimize the electronic energy subject to orthonormality constraints (4), and hence satisfy Löwdin's Eqs. (45) and (46). For the energy functional, Eq. (91), these equations become the spatial orbital Euler equations,

$$n_p \widehat{V}^p(\mathbf{r}_1) \varphi_p(\mathbf{r}_1) = \sum_r \lambda_p^r \varphi_r(\mathbf{r}_1) \quad (102)$$

where

$$\widehat{V}^p(\mathbf{r}_1) = \widehat{h}(\mathbf{r}_1) + \widehat{\mathcal{V}}^p(\mathbf{r}_1) \quad (103)$$

$$\widehat{\mathcal{V}}^p(\mathbf{r}_1) = \sum_q{}' \left(n_q - \frac{\Delta_q^p}{n_p}\right)[2\widehat{J}_q(\mathbf{r}_1) - \widehat{K}_q(\mathbf{r}_1) + \widehat{L}_q(\mathbf{r}_1)]$$

$$- \sum_q \frac{\Lambda_q^p}{n_p} \widehat{L}_q(\mathbf{r}_1) \quad (104)$$

$$\widehat{J}_q(\mathbf{r}_1) = \int d\mathbf{r}_2\, \varphi_q^*(\mathbf{r}_2) r_{12}^{-1} \varphi_q(\mathbf{r}_2) \quad (105)$$

$$\widehat{K}_q(\mathbf{r}_1) = \int d\mathbf{r}_2\, \varphi_q^*(\mathbf{r}_2) r_{12}^{-1} \widehat{P}_{12} \varphi_q(\mathbf{r}_2) \quad (106)$$

$$\widehat{L}_q(\mathbf{r}_1) = \int d\mathbf{r}_2\, \varphi_q(\mathbf{r}_2) r_{12}^{-1} \widehat{I}(2) \widehat{P}_{12} \varphi_q(\mathbf{r}_2) \quad (107)$$

where $\widehat{J}_q(\mathbf{r}_1)$ and $\widehat{K}_q(\mathbf{r}_1)$ are the usual Coulomb and exchange operators, respectively, whereas $\widehat{L}_q(\mathbf{r}_1)$ is the exchange and time-inversion operator [53]. Here, $\widehat{I}(2)$ becomes the complex conjugate operator. It is important to note that the operator $\widehat{V}^p(\mathbf{r}_1)$ does not fulfill the Gilbert relation, Eq. (49). Consequently, Eq. (102) cannot be reduced to an eigenvalue problem diagonalizing the matrix λ.

On the other hand, the one-electron functions $\{\varphi_p(\mathbf{r})\}$ that satisfy the set of equations (102) may be obtained from solving the eigenproblem, Eq. (60), with the Fockian matrix (see Eq. (66)),

$$F_p^r = h_p^r + (v_{ee})_p^r \qquad (108)$$

where the nonlocal potential matrix elements are

$$(v_{ee})_p^r = \frac{\partial V_{ee}}{\partial n_p}\delta_p^r + \frac{1-\delta_p^r}{n_p - n_r}\int d\mathbf{r}\, \varphi_r^*(\mathbf{r})[n_p\widehat{v}^p(\mathbf{r}) - n_r\widehat{v}^r(\mathbf{r})]\varphi_p(\mathbf{r}) \qquad (109)$$

$$\frac{\partial V_{ee}}{\partial n_p} = L_p^p + \sum_q{}'\left(n_q - \frac{\partial \Delta_q^p}{\partial n_p}\right)(2J_{pq} - K_{pq}) \qquad (110)$$

$$+ \sum_q{}'\left(n_q - \frac{\partial \Delta_q^p}{\partial n_p} - \frac{\partial \Lambda_q^p}{\partial n_p}\right)\left(\frac{L_p^q + L_q^p}{2}\right) \qquad (111)$$

Let us assume that the set of spatial orbitals $\{\varphi_p(\mathbf{r})\}$ is real; then $L_p^q = L_q^p = K_{pq}$, which allows Eqs. (104) and (110) to be further simplified, yielding

$$\widehat{v}^p(\mathbf{r}) = \widehat{J}_p(\mathbf{r}) + 2\sum_q{}'\left(n_q - \frac{\Delta_q^p}{n_p}\right)\widehat{J}_q(\mathbf{r}) - \sum_q{}'\Lambda_q^p\widehat{K}_q(\mathbf{r}) \qquad (112)$$

$$\frac{\partial V_{ee}}{\partial n_p} = J_{pp} + 2\sum_q{}'\left(n_q - \frac{\partial \Delta_q^p}{\partial n_p}\right)J_{pq} - \sum_q{}'\frac{\partial \Lambda_q^p}{\partial n_p}K_{pq} \qquad (113)$$

The closed-shell energy in Eq. (91) for real orbitals can be rewritten

$$E = \sum_p(2h_p^p + J_{pp})n_p + 2\sum_{pq}{}'(n_qn_p - \Delta_q^p)J_{pq} - \sum_{pq}{}'\Lambda_q^p K_{pq} \qquad (114)$$

VIII. RESTRICTED OPEN-SHELL NOF

We consider now situations in which a molecule has one or more unpaired electrons; hence we may have spin-polarized systems. The spatial orbitals are thus divided into two categories. Namely, those that are double occupied with two electrons of opposite spin (n_p^α and n_p^β), called closed shells (cl), and singly occupied (n_i^α or n_i^β), called open shells (op). We assume further spin-independent

ONs. All electron spins corresponding to the closed-shell part are then paired ($n_p^\alpha = n_p^\beta = n_p$) and thus they are coupled as a pure singlet. By the help of the following result (last term in Eq. (84)),

$$\sum_p \Pi_{p\beta}^{p\alpha} = \sum_p \sqrt{n_p^\alpha n_p^\beta} = \sum_p^{cl} n_p = \frac{N_{cl}}{2} \tag{115}$$

it is easy to calculate (see Eqs. (20) and (84)) the expectation values for spin operators $\hat{S}_z$ and $\hat{S}^2$; in particular,

$$\langle \hat{S}_z \rangle = \sum_p^{op} \frac{n_p^\alpha - n_p^\beta}{2} = \frac{N_{op}^\alpha - N_{op}^\beta}{2} \tag{116}$$

$$\langle \hat{S}^2 \rangle = \frac{N_{op}^\alpha}{2}\left(\frac{N_{op}^\alpha}{2} + 1\right) + \frac{N_{op}^\beta}{2}\left(\frac{N_{op}^\beta}{2} + 1\right) - \frac{1}{2} N_{op}^\alpha N_{op}^\beta \tag{117}$$

The situation here is completely analogous to that obtained in the restricted open HF theory (ROHF). The states are not eigenfunctions of $\hat{S}^2$, except when all the open-shell electrons have parallel spins ($N_{op}^\alpha = 0$ or $N_{op}^\beta = 0$). This result is a consequence of the expansion (22) used to obtain Eq. (27). Actually, the spin decomposition, Eq. (22), for $^2\mathbf{D}$ does not conserve in general the total spin S. However, we can form appropriate linear combinations of two-electron spin functions $\{\sigma_n^{(2)}(\mathbf{s}_1', \mathbf{s}_2'|\mathbf{s}_1, \mathbf{s}_2)\}$ that are simultaneously eigenfunctions of $\hat{S}_z$ and $\hat{S}^2$, and achieve a correct spin decomposition of the 2-RDM [85]:

$$^2D(\mathbf{x}_1', \mathbf{x}_2'|\mathbf{x}_1, \mathbf{x}_2) = \sum_{n=1}^{6} {}^2\mathbf{D}^n(\mathbf{r}_1', \mathbf{r}_2'|\mathbf{r}_1, \mathbf{r}_2)\sigma_n^{(2)}(\mathbf{s}_1', \mathbf{s}_2'|\mathbf{s}_1, \mathbf{s}_2) \tag{118}$$

where, for example,

$$\sigma_1^{(2)}(\mathbf{s}_1', \mathbf{s}_2'|\mathbf{s}_1, \mathbf{s}_2) = \tfrac{1}{2}[\alpha^*\alpha^*\alpha\alpha + \beta^*\beta^*\beta\beta + \alpha^*\beta^*\alpha\beta + \beta^*\alpha^*\beta\alpha] \tag{119}$$

Afterward, we have to find approximations for 2-RDM spin components $\{^2\mathbf{D}^n\}$.

Let us now focus on high-spin cases only, such as doublet, triplet, quartet... spins for one, two, three ... unpaired electrons outside the closed shells. Accordingly, singly occupied orbitals will always have the same spin ($N_{op}^\alpha = 0$ or $N_{op}^\beta = 0$) so the trace of the one-matrix, Eq. (6), becomes

$$2\sum_p^{cl} n_p + \sum_p^{op} n_p = N_{cl} + N_{op} = N \tag{120}$$

In fact, the value of N_{op} is determined by the conservation of the spin, so we have two constraints,

$$\frac{1}{2}\sum_p^{op} n_p = \frac{1}{2}N_{op} = S \qquad (121)$$

$$2\sum_p^{cl} n_p = N_{cl} = N - 2S \qquad (122)$$

where S is the quantum number describing the total spin of the N-electron high-spin coupled multiplet state.

We assume further that the ON of the open shell p is always one ($n_p = 1$). This assumption is trivial for a doublet, but it is more restrictive for higher multiplet with a corresponding underestimation of the energy. Remember that matrix elements of $^2\Gamma$ are nonvanishing only if all its labels refer to partially occupied NOs; therefore $\Delta = 0$ and $\Pi = 0$ if we consider a cumulant made up of at least one open-shell level. Since Δ and Π refer only to closed shells, we consider them spin-independent. The sum rule, Eq. (89), becomes

$$\sum_q^{cl}{}' \Delta_q^p = n_p(1 - n_p) \qquad (123)$$

and the energy expression, Eq. (80), for such a system is

$$E = \sum_p^{cl}(2h_p^p + J_{pp})n_p + 2\sum_{pq}^{cl}{}'(n_q n_p - \Delta_q^p)J_{pq} - \sum_{pq}^{cl}{}'\Delta_q^p K_{pq}$$
$$+ \sum_p^{cl}\sum_q^{op} n_p(2J_{pq} - K_{pq}) + \sum_p^{op} h_{pp} + \frac{1}{2}\sum_{pq}^{op}(J_{pq} - K_{pq}) \qquad (124)$$

where we have considered real orbitals ($L_q^p = K_{pq}$).

IX. PRACTICAL NOF

Electronic structure computations would be greatly simplified by the finding of practical NOFs. One may attempt to approximate the unknown off-diagonal elements of Δ considering the sum rule (89) and analytic constraints (101) imposed by the D-, G-, and Q-conditions. However, it is not evident how to approach Δ_q^p, for $p \neq q$, in terms of the ONs. Due to this fact, let's rewrite the energy term, which involves Δ, as

$$\sum_{pq}{}'\Delta_q^p J_{pq} = \sum_p J_p^* \sum_q{}'\Delta_q^p \qquad (125)$$

where J_p^* denotes the mean value of the Coulomb interactions J_{pq} for a given orbital p taking over all orbitals $q \neq p$. From the property shown in Eq. (89), it follows immediately that

$$\sum_{pq}{}' \Delta_q^p J_{pq} = \sum_p n_p(1-n_p) J_p^* \tag{126}$$

Inserting this expression into Eq. (114), one obtains

$$E^{\text{cl}} = \sum_p (2n_p h_p^p + n_p^2 J_{pp}) + \sum_{pq}{}' (2n_q n_p J_{pq} - \Lambda_q^p K_{pq}) \\ + \sum_p n_p(1-n_p)(J_{pp} - 2J_p^*) \tag{127}$$

A further simplification is accomplished by setting $J_p^* \approx J_{pp}/2$, which produces

$$E^{\text{cl}} = \sum_p (2n_p h_p^p + n_p^2 J_{pp}) + \sum_{pq}{}' (2n_q n_p J_{pq} - \Lambda_q^p K_{pq}) \tag{128}$$

The high-spin open-shell energy expression can be obtained in completely similar manner from Eq. (124),

$$E^{\text{op}} = \sum_p^{\text{cl}} (2n_p h_p^p + n_p^2 J_{pp}) + \sum_{pq}^{\text{cl}}{}' (2n_q n_p J_{pq} - \Lambda_q^p K_{pq}) \\ + \sum_p^{\text{cl}} \sum_q^{\text{op}} n_p(2J_{pq} - K_{pq}) + \sum_p^{\text{op}} h_{pp} + \frac{1}{2} \sum_{pq}^{\text{op}} (J_{pq} - K_{pq}) \tag{129}$$

We have thus arrived at an approximate NOF that coincides with the self-interaction-corrected Hartree functional proposed by Goedecker and Umrigar except for the choice of phases given by the sign of (Λ_q^p). Unfortunately, this NOF gives a wrong description of the ONs for the lowest occupied orbitals. In order to ensure that these ONs only are close to unity, we propose to add a new term to the functional form (94) of matrix Λ, namely,

$$\Lambda_r^p = -n_p \delta_r^p + (1-\delta_r^p)[1 - 2\theta(0.5 - n_r)\theta(0.5 - n_p)]\sqrt{n_r n_p} \\ + (1-\delta_r^p)\theta(n_r - 0.5)\theta(n_p - 0.5)\sqrt{(1-n_r)(1-n_p)} \tag{130}$$

X. NUMERICAL IMPLEMENTATION

In Section IV, we obtained the Euler equations (44)–(46), which yield the optimum bound on the ground-state energy. Usually, the simultaneous solution of these equations is established with an embedded loop algorithm. In the inner loop we look for the optimal ONs for a given set of orbitals fulfilling Eq. (44), whereas in the outer loop we optimize the NOs under the orthonormality condition for fixed ONs. We recall that bounds on the ONs are enforced by setting $n_i = \cos^2 \gamma_i$.

An important task in NOF theory is to find an efficient procedure for carrying out the orbital optimization. Direct minimization has proved [44, 47, 76] to be a costly method. First, there are directions at the energy minima with very low curvature associated with the high-energy NOs. The latter have very small ONs and hence give very small contribution to the energy. There is also a need for other starting orbitals that are closer to the optimized ones than are the HF orbitals, which are poor initial guesses in this iterative procedure and lead to lengthy orbital optimizations. Finally, and most critically, is that the NO coefficient matrix must be reorthogonalized during the course of the optimization, which is the rate-limiting step.

The one-electron equations (60) offer a new possibility for finding the optimal NSOs by iterative diagonalization of the Fockian, Eq. (66). The main advantage of this method is that the resulting orbitals are automatically orthogonal. The first calculations based on this diagonalization technique has confirmed its practical value [81].

XI. RESULTS

The study of different properties provides a measure of accuracy that can be employed in testing approximate functionals. In this section, we just quote some relevant results for selected molecules. Both the inner- and outer-loop optimizations were implemented using a sequential quadratic programming (SQP) method [86], which, computationally speaking, is a very demanding algorithm. Accordingly, we have chosen Pople medium-size basis sets [87] for the calculations and we have compared our results with the results using other methods at the same level. Among the approaches compared are the coupled-cluster technique including all single and double excitations and a perturbational estimate of the connected triple excitations [CCSD(T)], as well as the Becke-3–Lee–Yang–Parr (B3LYP) density functional [88]. The CC and B3LYP values were calculated with the GAUSSIAN system of programs [89], using the basis set keyword 5D.

A. Energetics

In Tables I and II we report the values obtained for the total energies employing the experimental geometry [90].

A survey of these tables reveals that NOF values are more like CCSD(T) calculations, which are very accurate results for the basis-set correlation energies on

TABLE I
Closed-Shell Total Energies in Hartrees (6-31G**)

Molecule	HF[a]	CCSD(T)[b]	NOF[c]	B3LYP[d]
FH	−100.009834	−100.198698	−100.178202	−100.425817
H_2O	−76.022615	−76.228954	−76.207940	−76.417892
NH_3	−56.194962	−56.399981	−56.378970	−56.556343
CO	−112.736756	−113.032821	−113.020010	−113.306694
HNO	−129.783338	−130.149630	−130.185528	−130.467334
H_2CO	−113.867947	−114.202886	−114.208902	−114.500848
HCl	−460.066040	−460.224196	−460.262033	−460.800097
PH_3	−342.452229	−342.605974	−342.635555	−343.142663
BeS	−412.103398	−412.284575	−412.310091	−412.894462
N_2O	−183.675227	−184.203202	−184.233374	−184.656028
O_3	−224.242821	−224.867269	−224.992524	−225.400708
NaCl	−621.397562	−621.546221	−621.586299	−622.556987
P_2	−681.421021	−681.667624	−681.742405	−682.683866
SO_2	−547.165010	−547.686081	−547.805959	−548.579504
Cl_2	−918.908639	−919.198123	−919.345443	−920.341830

[a]Hartree–Fock total energies.
[b]CCSD(T) total energies.
[c]Natural orbital functional total energies.
[d]B3LYP total energies.

these small molecules. The B3LYP values, as is well known, tend to be too low. We note that the percentage of the correlation energy obtained by CCSD(T) decreases as the number of electrons increases, whereas our functional keeps giving a slightly larger portion of the correlation energy (e.g., P_2, SO_2, Cl_2).

The EKT provides an alternative assessment for approximate NOFs, which is not directly related to total energy values. Table III lists the obtained vertical IPs

TABLE II
Open-Shell Total Energies in Hartrees (6-31++G**)

Molecule	HF[a]	CCSD(T)[b]	NOF[c]	B3LYP[d]
CH_2	−38.919309	−39.037939	−38.981051	−39.158525
NH	−54.957745	−55.091649	−55.017865	−55.227209
OH	−75.388827	−75.555805	−75.501346	−75.739015
MgH	−200.135519	−200.169572	−200.163233	−200.632922
SH	−398.064176	−398.195038	−398.192977	−398.745129

[a]Hartree–Fock total energies.
[b]CCSD(T) total energies.
[c]Natural orbital functional total energies.
[d]B3LYP total energies.

TABLE III
Vertical Ionization Potentials in eV (6-31G**)

Molecule	MO	KT[a]	NOF-EKT[b]	Experimental[c]
FH	π	17.06	16.82	16.19
	σ	20.23	19.99	19.90
H_2O	b_1	13.53	13.08	12.78
	a_1	15.56	15.54	14.83
	b_2	19.10	18.93	18.72
NH_3	a_1	11.44	11.05	10.80
	e	16.87	16.93	16.80
CO	σ	14.90	14.59	14.01
	π	17.22	17.28	16.85
	σ	21.67	21.58	19.78
N_2	σ_g	17.13	16.74	15.60
	π_u	16.63	17.14	16.68
	σ_u	21.11	20.96	18.78
H_2CO	b_2	11.87	11.54	10.90
	b_1	14.46	14.63	14.50
	a_1	17.58	17.44	16.10
	b_2	18.72	18.40	17.00
HCl	π	12.93	12.51	12.77[d]
	σ	17.00	16.73	16.60[d]
	σ	30.42	26.57	25.80[d]
P_2	π_u	10.14	10.39	10.65[e]
	σ_g	11.14	10.63	10.84[e]

[a] $-\varepsilon_i^{HF}$.
[b] Natural orbital functional vertical ionization potentials obtained from the extended Koopmans' theorem.
[c] Experimental vertical ionization potentials from Ref. [91].
[d] Experimental vertical ionization potential for HCl from Ref. [92].
[e] Experimental vertical ionization potential for P_2 from Ref. [93].

with our practical NOF using EKT together with Koopmans' theorem (KT) IPs and experimental values. All values were calculated at experimental geometries of neutral molecules given in Ref. [90].

KT states that the IP is given by the HF orbital energy with opposite sign ($-\epsilon_i$), calculated in the neutral system. It has long been recognized that in general there is an excellent agreement between the KT values and the experimental ones because of the fortuitous cancellation of the correlation and orbital relaxation effects. A survey of Table III reveals that KT consistently overestimates the first IPs except for P_2.

The prevailing trend is that the values decrease in moving from KT to NOF-EKT, and then from NOF-EKT to experimental data. Conversely, in the case of P_2, the first IP increases in moving from KT to NOF-EKT, and then from NOF-EKT to experimental values. For the HCl molecule, the NOF-EKT first IP is smaller than KT and experimental values. Generally, the NOF-EKT first IPs move closer to experimental data.

Table III also lists higher IPs calculated via KT and NOF-EKT methods. In general, the NOF-EKT and KT results are systematically larger than the experimental values. The behavior of higher IPs is quite similar to that for the first IP results, discussed previously. For most molecular orbitals, NOF-EKT values are smaller than KT and greater than the experimental data. There are molecular orbitals—for example, orbital π for CO and orbital b_1 for H_2CO—with NOF-EKT values greater than KT and the experimental data. The exception is P_2, for which the NOF-EKT σ_g IP is decreased. The agreement of NOF-EKT and KT IPs with the experimental values is less precise for inner valence molecular orbitals, but again the NOF-EKT IPs move closer to experimental data.

An important case is the N_2 molecule. It is well known that the KT σ_g and π_u IPs are in the wrong order for this molecule. NOF-EKT calculations on N_2 give valence shell IPs that are in correct order, and in general a numerical improvement is obtained over KT IPs. In the case of the orbital π_u for N_2, the KT IPs are closer to experimental data but due to the mentioned wrong ordering. The results are in good agreement with the corresponding experimental vertical IPs considering the small basis sets used for these calculations.

Electron affinities (EAs) are considerably more difficult to calculate than IPs. For example, EAs are much more sensitive to the basis set than the corresponding IPs. The EKT provides also the means for calculation of these magnitudes, but unfortunately the EKT-EA description is often very poor. On the other hand, vertical EAs can be calculated by the energy difference for neutral molecules (M^0) and negative ions (M^-): $E(M^0) - E(M^-)$ at near-experimental geometries of M^0. Table IV lists the obtained vertical EAs for selected open-shell molecules.

TABLE IV
Vertical Electron Affinities in eV (6-31++G**)

Molecule	S_N[a]	S_A[b]	ΔSCF[c]	ΔCCSD(T)[d]	ΔNOF[e]	Experimental[f]
CH_2	1	$\frac{1}{2}$	−3.577	−2.028	−0.091	0.652 (0.006)
NH	1	$\frac{1}{2}$	−1.439	−0.219	0.113	0.370 (0.004)
OH	$\frac{1}{2}$	0	−0.147	1.360	1.830	1.828 (0.001)
MgH	$\frac{1}{2}$	0	−0.057	0.612	0.976	1.05 (0.06)
SH	$\frac{1}{2}$	0	1.246	1.774	2.761	2.314 (0.003)

[a]S_N: total spin of the neutral molecule.
[b]S_A: total spin of the anion.
[c]ΔSCF $= E_{HF}(M^0) - E_{HF}(M^-)$.
[d]ΔCCSD(T) $= E_{CCSD(T)}(M^0) - E_{CCSD(T)}(M^-)$.
[e]ΔNOF $= E_{NOF}(M^0) - E_{NOF}(M^-)$.
[f]Experimental adiabatic electron affinities from Ref. [90]. The uncertainty is shown in parentheses.

As can be seen, generally all electron affinities predicted by ΔSCF are negative, indicating a more stable neutral system with respect to the anion. The inclusion of correlation via CCSD(T) and NOF approximates them to the available adiabatic experimental EAs, accordingly with the expected trend. The EAs tend to increase in moving from ΔCCSD(T) to ΔNOF and then from ΔNOF to the experiment. It should be noted that the NH anion is predicted to be unbound by CCSD(T), whereas the positive vertical EA value via NOF corresponds to the bound anionic state.

For OH and SH, the NOF EAs are larger than the experimental values. This trend is due to the expected underestimation of the correlation energy for open-shell states with our approach. In fact, we fix the unpaired electron in the corresponding HF higher-occupied molecular orbital (HOMO) of the neutral molecule, and then this level does not participate in the correlation. Note that for these molecules the total spin of the neutral molecule is greater than the total spin of the anion ($S_N > S_A$). The underestimation of the total energy is for neutral molecules larger than for anions and therefore the NOF vertical EAs are overestimated.

Comparisons with experimental results show that vertical NOF-EAs are better than those predicted by the CCSD(T) method within the 6-31++G** basis set.

B. Dipole Moment

The dipole moments (DMs), different from zero, obtained using HF, CCSD(T), and NOF methods are presented in Table V. For comparison, we have also included in this table the available experimental data [90].

The correlated dipole moments tend to be lower compared to HF DMs. An exception to this trend is noticed in the case of CH_2, where the CCSD(T) DM is predicted to be higher than the HF value. Important cases are the CO and N_2O molecules for which the HF approximation gives a DM in the wrong direction, whereas correlation methods approach the experimental value. The quality of the NOs is critical for the accuracy of NOF theory. For example, in the case of the CO molecule, we achieved the sign inversion of the DM after further improving the basis sets. The DMs obtained with the NOF method are in good agreement with the available experimental data considering the basis sets (6-31G**) used for these calculations.

C. Equilibrium Geometries and Vibrational Frequencies

We have employed the nongradient geometry optimization to determine the equilibrium bond distances (r_e). For each molecule, we have calculated the total energy $U(r)$ at a dense grid of bond distances r, separated from each other by 10^{-3} Å. The harmonic vibrational frequencies (ω_e) were determined from the second derivatives of the energy with respect to the nuclear positions. The

TABLE V
Dipole Moments in Debyes (6-31G**)

Molecule	HF[a]	CCSD(T)[b]	NOF[c]	Exp[d]
CH_2	0.54	0.57	0.51	—
NH	1.73	1.61	1.62	1.39[f]
OH	1.88	1.77	1.77	1.66
FH	1.98	1.87	1.84	1.82
H_2O	2.20	2.09	2.08	1.85
NH_3	1.89	1.81	1.76	1.47
CO	0.33[f]	0.07	0.03[g]	0.11
MgH	1.41	1.10	1.22	—
SH	1.07	0.97	0.91	0.76
HCl	1.48	1.37	1.30	1.08
HNO	2.02	1.63	1.61	1.67
H_2CO	2.75	2.18	2.36	2.33
PH_3	0.80	0.80	0.60	0.58
BeS	6.42	4.59	5.38	—
N_2O	0.60[f]	0.07[f]	0.03	0.17
O_3	0.78	0.49	0.50	0.53
NaCl	9.39	8.97	9.03	9.0
SO_2	2.21	1.78	1.56	1.63

[a]Hartree–Fock dipole moments.
[b]CCSD(T) dipole moments.
[c]Natural orbital functional dipole moments.
[d]Experimental dipole moments from Ref. [90].
[e]Experimental dipole moment for NH from Ref. [94].
[f]This value has an opposite sign relative to the experimental value.
[g]This value was obtained with 6-31G**.

equilibrium force constants $k_e = U''(r_e)$ were obtained from least squares fits of the energy to a second-order polynomial in the distances,

$$U(r) = U(r_e) + \tfrac{1}{2}k_e(r - r_e)^2 \tag{131}$$

Table VI lists the equilibrium geometries for the selected closed-shell diatomic molecules. It has long been recognized that the HF method gives reduced bond lengths. From Table VI we see that the correlated bond lengths are mostly longer than the experimental values. Exceptions are the bond distances predicted by CCD and NOF methods for the HCl molecule.

The values of the harmonic vibrational frequencies for the selected set of molecules are presented in Table VII. The expected trends of this property with respect to experiment are well reproduced. The HF results are systematically larger and the correlated vibrational frequencies move closer to

TABLE VI
Equilibrium Bond Lengths in Angstroms (6-31G**)

Molecule	HF[a]	CCD[b]	NOF[c]	Exp[d]
FH	0.900	0.920	0.917	0.917
CO	1.114	1.139	1.132	1.128
HCl	1.265	1.271	1.273	1.275
BeS	1.733	1.744	1.746	1.742
N_2	1.078	1.112	1.099	1.098
P_2	1.859	1.905	1.893	1.893

[a]Hartree–Fock equilibrium bond lengths.
[b]CCD equilibrium bond lengths.
[c]Natural orbital functional equilibrium bond lengths.
[d]Experimental equilibrium bond lengths from Ref. [90].

experimental data. By both correlated methods, the experimental frequencies are still overestimated, except for the HCl molecule, for which the NOF frequency is lower than the experimental value.

The performance of our practical NOF to predict equilibrium bond distances and vibrational frequencies is similar to CCD.

XII. CONCLUDING REMARKS

^{1}D-functional theory, explicitly given in terms of natural orbitals and their occupation numbers, has emerged as an alternative method to conventional approaches for considering the electronic correlation. This chapter has introduced important basic concepts for understanding the NOF formalism. We have also offered a brief characterization of almost all references concerning this theory published hitherto.

TABLE VII
Vibrational Frequencies in cm^{-1} (6-31G**)

Molecule	HF[a]	CCD[b]	NOF[c]	Exp[d]
FH	4500	4213	4216	4138
CO	2442	2251	2284	2170
HCl	3182	3089	2927	2991
BeS	1069	1045	1047	998
N_2	2761	2433	2544	2359
P_2	909	804	814	781

[a]Hartree–Fock vibrational frequencies.
[b]CCD vibrational frequencies.
[c]Natural orbital functional vibrational frequencies.
[d]Experimental vibrational frequencies from Ref. [90].

A free-parameter functional based on a new approach for the two-electron cumulant has been reviewed. This functional reduces to the exact expression for the total energy in two-electron systems and to the HF energy for idempotent 1-RDMs. Moreover, it is derived from a rigorous N-representable ansatz for the 2-RDM.

The mean value theorem provides a prescription for deriving a practical NOF that yields reasonable correlation energies for molecules. Accurate results for closed- and open-shell systems are obtained with energy expressions that only include two-index two-electron integrals. We shall improve this functional by providing a better approximation for the mean value J_i^* of the Coulomb interactions. It is highly desirable to develop our restricted open-shell formulation by appropriately expressing the 2-RDM in two-electron spin functions that are simultaneously eigenfunctions of $\widehat{S}_z$ and $\widehat{S}^2$.

The explicit form derived by Pernal for the effective nonlocal potential allows one to establish one-electron equations that may be of great value for the development of efficient computational methods in NOF theory. Although recent progress has been made, NOF theory needs to continue its assessment. Some other essential conditions such as the reproduction of the homogeneous electron gas should be utilized in the evaluation of approximate implementations.

Acknowledgments

The author is grateful to Prof. Jesus M. Ugalde, Prof. Eduardo V. Ludeña, and Dr. Xabier Lopez for helpful discussions and comments.

References

1. A. J. Coleman, *Rev. Mod. Phys.* **35**, 668 (1963).
2. P. Hohenberg and W. Kohn, *Phys. Rev. B* **136**, 846 (1964).
3. R. G. Parr and W. Yang, *Density-Functional Theory of Atoms and Molecules*, Oxford University Press, Oxford, 1989; E. S. Kryachko and E. V. Ludeña, *Energy Density Functional Theory of Many-Electron Systems*, Kluwer, Dordrecht, 1990.
4. W. Kohn and L. J. Sham, *Phys. Rev. A* **140**, 1133 (1965).
5. H. Eschrig, *The Fundamentals of Density Functional Theory*, Teubner, Stuttgart, Germany, 1996.
6. E. V. Ludeña, V. Karasiev, A. Artemiev, and D. Gómez, in *Many-Electron Densities and Reduced Density Matrices* (J. Cioslowski, ed.), Kluwer, New York, 2000, p. 209.
7. A. A. Jarzecki and E. R. Davidson, *Phys. Rev. A* **58**, 1902 (1998).
8. J. Tao, J. P. Perdew, V. N. Staroverov, and G. E. Scuseria, *Phys. Rev. Lett.* **91**, 146401(2003); J. P. Perdew, J. Tao, V. N. Staroverov and G. E. Scuseria, *J. Chem. Phys.* **120**, 6898 (2004).
9. R. C. Morrison and L. J. Bartolotti, *J. Chem. Phys.* **121**, 12151 (2004).
10. M. Levy, *Proc. Natl. Acad. Sci. U.S.A.* **76**, 6062 (1979).
11. M. Levy, in *Density Matrices and Density Functionals* (R. Erdahl and V. H. Smith, Jr., eds.), Reidel, Dordrecht, 1987, p. 479.
12. J. Cioslowski, K. Pernal, and P. Ziesche, *J. Chem. Phys.* **117**, 9560 (2002).

13. J. Cioslowski, *J. Chem. Phys.* **123**, 164106 (2005).
14. M. Levy and J. P. Perdew, *Phys. Rev. A* **32**, 2010 (1985).
15. T. L. Gilbert, *Phys. Rev. B* **12**, 2111 (1975).
16. M. Berrondo and O. Goscinski, *Int J. Quantum Chem.* **9S**, 67 (1975).
17. R. A. Donnelly and R. G. Parr, *J. Chem. Phys.* **69**, 4431 (1978).
18. R. A. Donnelly, *J. Chem. Phys.* **71**, 2874 (1979).
19. S. M. Valone, *J. Chem. Phys.* **73**, 1344 (1980).
20. E. H. Lieb, *Phys. Rev. Lett.* **46**, 457 (1981).
21. E. H. Lieb, *Int J. Quantum Chem.* **24**, 243 (1983).
22. A. M. K. Müller, *Phys. Lett. A* **105**, 446 (1984).
23. M. A. Buijse, Ph.D. thesis, Vrije Universiteit Amsterdam, Amsterdam, 1991.
24. G. Zumbach and K. Maschke, *J. Chem. Phys.* **82**, 5604 (1985).
25. T. Tung Nguyen-Dang, E. V. Ludeña, and Y. Tal, *J. Mol. Struct. (Theochem.)* **120**, 247 (1985); E. V. Ludeña and A. Sierraalta, *Phys. Rev. A* **32**, 19 (1985); E. V. Ludeña, in *Density Matrices and Density Functionals* (R. Erdahl and V. H. Smith, Jr., eds.), Reidel, Dordrecht, 1987, p. 289.
26. A. Redondo, *Phys. Rev. A* **39**, 4366 (1989).
27. R. López-Sandoval and G. M. Pastor, *Phys. Rev. B* **61**, 1764 (2000).
28. C. Valdemoro, *Phys. Rev. A* **45**, 4462 (1992).
29. A. E. Carlsson, *Phys. Rev. B* **56**, 12058 (1997); R. G. Hennig and A. E. Carlsson, *Phys. Rev. B* **63**, 115116 (2001).
30. A. Klein and R. M. Dreizler, *Phys. Rev. A* **57**, 2485 (1998).
31. J. Cioslowski, and R. Lopez-Boada, *J. Chem. Phys.* **109**, 4156 (1998); *Chem. Phys. Lett.* **307**, 445 (1999).
32. D. W. Smith and O. W. Day, *J. Chem. Phys.* **62**, 113 (1975); O. W. Day, D. W. Smith, and C. Garrot, *Int. J. Quantum Chem. Symp.* **8**, 501 (1974).
33. M. M. Morrell, R. G. Parr, and M. Levy, *J. Chem. Phys.* **62**, 549 (1975).
34. M. Nooijen, *J. Chem. Phys.* **111**, 8356 (1999).
35. S. Goedecker and C. J. Umrigar, *Phys. Rev. Lett.* **81**, 866 (1998).
36. E. J. Baerends, *Phys. Rev. Lett.* **87**, 133004 (2001); M. A. Buijse and E. J. Baerends, *Mol. Phys.* **100**, 401 (2002).
37. A. Holas, *Phys. Rev. A* **59**, 3454 (1999).
38. J. Cioslowski and K. Pernal, *J. Chem. Phys.* **111**, 3396 (1999).
39. J. Cioslowski and K. Pernal, *Phys. Rev. A* **61**, 34503 (2000); J. Cioslowski, P. Ziesche, and K. Pernal, *Phys. Rev. B* **63**, 205105 (2001).
40. G. Csanyi and T. A. Arias, *Phys. Rev. B* **61**, 7348 (2000).
41. J. Cioslowski and K. Pernal, *J. Chem. Phys.* **115**, 5784 (2001).
42. D. A. Mazziotti, *J. Chem. Phys.* **112**, 10125 (2000).
43. D. A. Mazziotti, *Chem. Phys. Lett.* **338**, 323 (2001).
44. A. J. Cohen and E. J. Baerends, *Chem. Phys. Lett.* **364**, 409 (2002).
45. V. N. Staroverov and G. E. Scuseria, *J. Chem. Phys.* **117**, 2489 (2002).
46. G. Csanyi, S. Goedecker, and T. A. Arias, *Phys. Rev. A* **65**, 32510 (2002).
47. J. M. Herbert and J. E. Harriman, *Int. J. Quantum Chem.* **90**, 355 (2002); *J. Chem. Phys.* **118**, 10835 (2003); *Chem. Phys. Lett.* **382**, 142 (2003).

48. V. N. Staroverov and G. E. Scuseria, *J. Chem. Phys.* **117**, 11107 (2002).
49. P. Durand, *Cah. Phys.* **21**, 285 (1967); O. Goscinski, *Int. J. Quantum Chem. Quantum Chem. Symp.* **16**, 591 (1982).
50. B. Lukman, J. Koller, Borštnik, and A. Ažman, *Mol. Phys.* **18**, 857 (1970).
51. W. B. England, *J. Phys. Chem.* **86**, 1204 (1982); *Int. J. Quantum Chem.* **23**, 905 (1983); *Int. J. Quantum Chem. Quantum Chem. Symp.* **17**, 357 (1983).
52. M. Piris and R. Cruz, *Int. J. Quantum Chem.* **53**, 353 (1995); M. Piris, L. A. Montero, and N. Cruz, *J. Chem. Phys.* **107**, 180 (1997); M. Piris, *J. Math. Chem.* **23**, 399 (1998); M. Piris and P. Otto, *J. Chem. Phys.* **1112**, 8187 (2000).
53. M. Piris, *J. Math. Chem.* **25**, 47 (1999).
54. B. Barbiellini and A. Bansil, *J. Phys. Chem. Solids* **62**, 2181 (2001).
55. K. Yasuda, *Phys. Rev. A* **63**, 32517 (2001); K. Yasuda, *Phys. Rev. Lett.* **88**, 53001 (2002).
56. J. Cioslowski, P. Ziesche, and K. Pernal, *J. Chem. Phys.* **115**, 8725 (2001).
57. J. Cioslowski and K. Pernal, *J. Chem. Phys.* **116**, 4802 (2001); J. Cioslowski and K. Pernal, *J. Chem. Phys.* **117**, 67 (2002).
58. M. Piris and P. Otto, *Int. J. Quantum Chem.* **94**, 317 (2003).
59. M. Piris, A. Martinez, and P. Otto, *Int. J. Quantum Chem.* **97**, 827 (2004).
60. P. Leiva and M. Piris, *J. Mol. Struct. (Theochem.)* **719**, 63 (2005).
61. M. Piris and P. Otto, *Int. J. Quantum Chem.* **102**, 90 (2005).
62. C. Kollmar and B. A. Hess, *J. Chem. Phys.* **120**, 3158 (2004).
63. N. N. Lathiotakis, N. Helbig, and E. K. U. Gross, *Phys. Rev. A* **72**, 30501 (2005).
64. D. R. Alcoba, L. Lain, A. Torre, and R. C. Bochicchio, *J. Chem. Phys.* **123**, 144113 (2005).
65. J. Cioslowski, K. Pernal, and M. Buchowiecki, *J. Chem. Phys.* **119**, 6443 (2003).
66. C. Kollmar and B. A. Hess, *J. Chem. Phys.* **119**, 4655 (2003).
67. J. Cioslowski, M. Buchowiecki, and P. Ziesche, *J. Chem. Phys.* **119**, 11570 (2003).
68. K. Pernal and J. Cioslowski, *J. Chem. Phys.* **120**, 5987 (2004).
69. J. Cioslowski and K. Pernal, *J. Chem. Phys.* **120**, 10364 (2004).
70. C. Kollmar, *J. Chem. Phys.* **121**, 11581 (2004).
71. O. Gritsenko, K. Pernal, and E. J. Baerends, *J. Chem. Phys.* **122**, 204102 (2005).
72. K. Pernal and J. Cioslowski, *Chem. Phys. Lett.* **412**, 71 (2005).
73. M. Piris, *Int. J. Quantum Chem.* **106**, 1093 (2006).
74. D. A. Mazziotti, *Chem. Phys. Lett.* **289**, 419 (1998); D. A. Mazziotti, *Int. J. Quantum Chem.* **70**, 557 (1998).
75. W. Kutzelnigg and D. Mukherjee, *J. Chem. Phys.* **110**, 2800 (1999).
76. S. Goedecker and C. J. Umrigar, in *Many-Electron Densities and Reduced Density Matrices* (J. Cioslowski, ed.), Kluwer/Plenum, New York, 2000, p. 165.
77. W. Kutzelnigg, *Theor. Chim. Acta* **1**, 327 (1963).
78. P. Leiva and M. Piris, *J. Theor. Comp. Chem.* **4**, 1165 (2005).
79. P. Leiva and M. Piris, *J. Chem. Phys.* **123**, 214102 (2005).
80. P. Leiva and M. Piris, *Int. J. Quantum Chem.* **x, x (2006).**
81. K. Pernal, *Phys. Rev. Lett.* **94**, 233002 (2005).
82. P.-O. Löwdin, *Phys. Rev.* **97**, 1474 (1955).

83. R. McWeeny, *Rev. Mod. Phys.* **32**, 335 (1960); R. McWeeny and Y. Mizuno, *Proc. R. Soc. London Ser. A* **259**, 554 (1961).
84. J. Katriel and E. R. Davidson, *Proc. Natl. Acad. Sci. USA* **77**, 4403 (1980).
85. W. A. Bingel and W. Kutzelnigg, *Adv. Quantum Chem.* **5**, 201 (1970); J. E. Harriman, *Int. J. Quantum Chem.* **12**, 1039 (1977).
86. R. Fletcher, *Practical Methods of Optimization*, 2nd ed. Wiley, Hoboken, NJ, 1987.
87. P. C. Hariharan and J. A. Pople, *Theor. Chim. Acta* **28**, 213 (1973); M. M. Francl, W. J. Petro, W. J. Hehre, J. S. Binkley, M. S. Gordon, D. J. DeFrees, and J. A. Pople, *J. Chem. Phys.* **77**, 3654 (1982).
88. I. N. Levine, *Quantum Chemistry*, Prentice Hall, Upper Saddle River, NJ, 2000.
89. M. J. Frisch, G. W. Trucks, H. B. Schlegel, P. M. W. Gill, B. G. Johnson, M. A. Robb, J. R. Cheeseman, T. Keith, G. A. Petersson, J. A. Montgomery, K. Raghavachari, M. A. Al-Laham, V. G. Zakrzewski, J. V. Ortiz, J. B. Foresman, J. Cioslowski, B. B. Stefanov, A. Nanayakkara, M. Challacombe, C. Y. Peng, P. Y. Ayala, W. Chen, M. W. Wong, J. L. Andres, E. S. Replogle, R. Gomperts, R. L. Martin, D. J. Fox, J. S. Binkley, D. J. Defrees, J. Baker, J. P. Stewart, M. Head-Gordon, C. Gonzalez, and J. A. Pople, GAUSSIAN 94 Inc., Pittsburgh, PA, 1995.
90. R. D. Johnson III (ed.), *NIST Computational Chemistry Comparison and Benchmark Database*, NIST Standard Reference Database Number 101, Release 11, May 2005 (http://srdata.nist.gov/cccbdb).
91. R. C. Morrison, *J. Chem. Phys.* **96**, 3718 (1996).
92. W. von Niessen, L. Asbrink, and G. Bieri, *J. Electron. Spectrosc. Relat. Phenom.* **26**, 173 (1982).
93. A. W. Potts, K. G. Glenn, and W. C. Price, *Discuss. Faraday Soc.* **54**, 65 (1972).
94. N. Oliphant and R. J. Bartlett, *J. Chem. Phys.* **100**, 6550 (1994).

CHAPTER 15

GEMINAL FUNCTIONAL THEORY

B. C. RINDERSPACHER

Department of Chemistry, The University of Georgia, Athens, GA 30602 USA

CONTENTS

I. Introduction
II. Strongly Orthogonal Antisymmetrized Geminal Products
    A. Singlet-Type Strongly Orthogonal Geminals
    B. Perturbation Theory on SOAGP
III. Antisymmetrized Geminal Products
IV. Formal GFT
    A. An Application of GFT
V. Unrestricted Antisymmetrized Geminal Products
VI. Conclusions
References

I. INTRODUCTION

Similar to density functional theory, geminal functional theory (GFT) endeavors to compute molecular properties from geminals, supplanting the electron density as primary variable. A geminal is a function of just two electrons. We will understand GFT for now in a general way as some form of computing the energy of a system from a geminal. The idea of using geminals is very appealing because the Hamiltonian consists of at most two-particle interactions.

II. STRONGLY ORTHOGONAL ANTISYMMETRIZED GEMINAL PRODUCTS

The earliest such attempts go back to 1953, when strongly orthogonal antisymmetrized geminal products (SOAGP) were employed [1, 2]. A strongly orthogonal geminal is such that $\int g_1(1,2)g_2(2,3)d2 = 0$, while the weaker

---

Reduced-Density-Matrix Mechanics: With Application to Many-Electron Atoms and Molecules,
A Special Volume of Advances in Chemical Physics, Volume 134, edited by David A. Mazziotti.
Series editor Stuart A. Rice. Copyright © 2007 John Wiley & Sons, Inc.

condition of orthogonality requires $\int g_1(1,2)g_2(1,2)d1d2 = 0$. Using such geminals to construct a wavefunction leads to a simple 2-RDM (see Eq. (2) and previous chapters on reduced density matrices).

$$\Psi_{\text{SOAGP}} = \bigwedge_{i=1}^{n/2} g_i, \quad \text{where} \quad (i \neq j \Rightarrow \int g_i(1,2)g_j(2,3)d2 = 0) \tag{1}$$

$$^2D(\Psi_{\text{SOAGP}}; 12, 1'2') = \sum b(ijkl) \cdot |\phi_i(1)\phi_j(2)| \cdot |\phi_k(1')\phi_l(2')|^* \tag{2}$$

where ϕ_i are the natural spin orbitals of the geminals g_k and

- $\exists n : \{i,j\} = \{k,l\} = \{2n-1, 2n\} =: \sigma \Rightarrow b(\sigma\sigma) = c_i \lambda_\sigma$.
- $\sigma \neq \tau \exists i : (\hat{\sigma}, g_i) \neq 0 \neq (\hat{\tau}, g_i) \Rightarrow b(\sigma\tau) = c_i \xi_\sigma \xi_\tau^*$.
- $\sigma \neq \tau, i \in \sigma, j \in \tau, (\hat{\tau}, g_m) \neq 0, (\hat{\sigma}, g_n) \neq 0, n \neq m \Rightarrow b(ijij) = c_{ij} \lambda_\sigma \lambda_\tau$.
- $c_i = \prod_{j \neq i} \sum_{(\hat{\sigma}, g_j) \neq 0} \lambda_\sigma / c,\ c_{ij} = \prod_{k \neq i, k \neq j} \sum_{(\hat{\sigma}, g_k) \neq 0} \lambda_\sigma / c,\ c = 5 \prod_j \sum_{(\hat{\sigma}, g_j) \neq 0} \lambda_\sigma$ and in all other cases $b = 0$.
- $\sigma = \{2i-1, 2i\} \Rightarrow \hat{\sigma} := |\phi_{2i-1}\phi_{2i}|$.
- $(g, g') = \int g(1, 2')g'(2', 2)d2', (g_i, \hat{\sigma}) \neq 0 \Rightarrow \forall j \neq i (g_j, \hat{\sigma}) = 0$.

This second-order reduced density matrix for SOAGP has a block diagonal form—one dense block for each geminal and one diagonal block for the mixing between geminals.

$$^2D_{\text{SOAGP}} = \begin{pmatrix} b(\sigma_1 \tau_1) & & & \\ & \ddots & & \\ & & \boxed{b(\sigma_l \tau_l)} & \\ & & & b(1_1 1_2 1_1 1_2) \\ & & & & \ddots \end{pmatrix} \tag{3}$$

where $i_j \in \sigma : \hat{\sigma} g_j^* \neq 0$ and $\hat{\sigma}_i g_i^* \neq 0$. The natural orbitals of this density matrix are the natural orbitals of the generating geminals. Also, the generating geminals are natural geminals of the wavefunction. Optimizing the energy for SOAGP poses the interesting problem of partitioning of the geminal spaces. There are three variables for optimization:

1. Basis functions ϕ_i.
2. Geminal coefficients ξ_i of the basis function ϕ_i.
3. Partitioning of ϕ among the $N/2$ geminals comprising the SOAGP wavefunction.

The simultaneous optimization of the latter two points can be done by solving coupled eigenvalue equations. These equations can be derived from imposing variational conditions on the energy [2]:

$$\epsilon_\mu c_{\mu j} = \sum_i c_{\mu i} H_{ij}(\mu) \tag{4}$$

$$H_{ij} = \langle \phi_{2i-1} \wedge \phi_{2i}|^2 \hat{K}|\phi_{2j-1} \wedge \phi_{2j}\rangle + \delta_{ij} \sum_{v \neq \mu} \langle \phi_{\mu 2i-1} \wedge \phi_{\mu 2i}|^2 \hat{K}|g_v\rangle \tag{5}$$

Similarly, equations can be derived for the optimization of the basis functions.

SOAGP has been applied to systems like LiH, BH [3], or NH [4]. In these calculations, it was found that the majority of the energy lowering was due to explicit correlation found in the individual geminals, while the intergeminal interactions displayed positive as well as negative character. Due to the strong orthogonality condition, the intergeminal correlation is insufficiently described. For n electrons, there are $n(n-2)/8$ intergeminal correlations, while there are only $n/2$ intrageminal interactions. As may be expected, the percentage of correlation energy retrieved decreases with the number of electrons.

A. Singlet-Type Strongly Orthogonal Geminals

Recently, an alternative scheme based on singlet-type strongly orthogonal geminals (SSG) was proposed [5]. In this scheme, the wavefunction is split into geminal subspaces depending on the number of spin-up or spin-down electrons, n_α and n_β, respectively, while the wavefunction is "filled up" with one Slater determinant.

$$\Psi_{SSG} = \mathbf{A}_n \left[\prod_{i=1}^{n_\alpha} g_i(2i-1, 2i) \right] \cdot |\phi_k(2n_\alpha + 1) \cdots \phi_{k+n_\beta-n_\alpha}(n_\alpha + n_\beta)| \tag{6}$$

$$ND_{SSG}^{(2)} = s \sum_{i=1}^{n_\alpha} c_i g_i(1,2) g_i(1'2') + g \sum_{k,l} s_{kl} |\phi_k(1)\phi_l(2)| \cdot |\phi_k(1')\phi_l(2')|$$
$$+ \sum_{k,i} c_i s_k |g_{i\sigma}|^2 |\phi_k(1) g_{ij}(2)| \cdot |\phi_k(1') g_{ij}(2')| \tag{7}$$

$$j \in \sigma : (\hat{\sigma}, g_i) \neq 0$$

where $\Delta n = n_\beta - n_\alpha$, $s_{kl} = \prod_{i \neq k, i \neq l} \|\phi_i\|^2$, $s_k = \prod_{j \neq k} \|\phi_j\|^2$, $s = \prod_j \|\phi_j\|^2$, $c_i = \prod_{j \neq i} \|g_j\|^2$, $g = \prod \|g_i\|^2$, and $N = (n_\alpha + \Delta n(\Delta n - 1)/2 + 2n_\alpha \Delta n) sg$. Hence the SSG wavefunction is the antisymmetrized product of a Slater determinant with an SOAGP wavefunction. Here the geminal subspaces are optimized by comparing the energies of a given geminal primitive $|g_{i,2j-1}g_{i,2j}|$ in geminal i and the same primitive in geminal j.

The SSG approximation has been used to study a group of diatomics from the G2/97 test set [6, 7] as well as the potential energy surface (PES) of carbon monoxide. Except for noncovalently bonded (e.g., highly polarized) molecules, SSG is superior to Hartree–Fock theory (HF) [8, 9] in describing geometry and often comparable to coupled cluster in the singles and doubles approximation (CCSD). When comparing harmonic frequencies, SSG performs even better, *generally* being comparable to CCSD.

The PES of CO for SSG is only slightly better than HF but qualitatively correct in that it is smooth and describes dissociation toward infinity. Although unrestricted MP2 [10–12] and CCSD(T) recover much more of the correlation energy than SSG, the associated curves are not as smooth as for SSG. The equilibrium distance found with SSG is extremely close to experiment ($r_e^{SSG} = 1.126$ Å, $r_e^{expt.} = 1.1283$ Å) [5].

B. Perturbation Theory on SOAGP

Encouraged by the previous results, more recently, SOAGP has been used as the reference state in PT [13, 14]. The full Hamiltonian is split into two parts, of which one, H_0, can be solved for a state (Eqs. (8) and (9)). The resulting conditions for first order (Eqs. (9)–(12)) describe an easily solved linear system of equations:

$$\hat{H} = \hat{H}_0 + \hat{W} \tag{8}$$

$$(\hat{H}_0 - E_0)\Psi^{(0)} = 0 \tag{9}$$

$$\Psi^{(i)} = \sum_i d_i \Psi_i, \quad \langle \Psi_0 | \Psi_i \rangle = 0 \tag{10}$$

$$\langle \Psi_j | \hat{W} | \Psi^{(0)} \rangle + \sum_i d_i \langle \Psi_j | (\hat{H}_0 - E_0) | \Psi_i \rangle = 0 \forall j \tag{11}$$

$$E^{(2)} = \langle \Psi_0 | \hat{W} | \Psi^{(1)} \rangle, \quad E^{(1)} = \langle \Psi_0 | \hat{W} | \Psi_0 \rangle \tag{12}$$

In order to define $\hat{H}_0$ such that Ψ_{SOAGP} is an eigenfunction, $\hat{H}_0$ was split into a sum of operators $\hat{H}_K$. These operators $\hat{H}_K$ are defined so that $\hat{H}_K g_K = E_K^0 g_K$ (Eq. (14)). Then the zeroth-order energy is merely $\sum E_K^0$ and $\hat{W} = \hat{H} - \hat{H}_0$.

$$h_{\mu\nu}^{eff} = h_{\mu\nu} + \sum_{\lambda\sigma} {}^1 D_{SOAGP} \langle \nu \wedge \sigma | \hat{r}_{12}^{-1} | \mu \wedge \lambda \rangle \tag{13}$$

$$\hat{H}_K = \sum_{\mu,\nu \text{ of } g_k} h_{\mu\nu}^{eff} \hat{a}_\mu^+ \hat{a}_\nu^- + \frac{1}{2} \sum_{\mu\nu\lambda\sigma \text{ of } g_K} \langle \mu\nu | \hat{r}_{12}^{-1} | \lambda\sigma \rangle \hat{a}_\lambda^+ \hat{a}_\mu^+ \hat{a}_\sigma^- \hat{a}_\nu^- \tag{14}$$

where, given γ, γ is of g_K, if γ is a natural orbital of g_K, and $\hat{a}_\gamma^+, \hat{a}_\gamma^-$ represent creation and annihilation operators of γ, respectively, in the second quantization formalism.

PESs for fluorine, hydrogen fluoride, and water using this ansatz showed promise [14]. In all cases the perturbative approach improved the accuracy considerably at a small increase of computational cost. Especially interesting is the possibility of linear scaling.

III. ANTISYMMETRIZED GEMINAL PRODUCTS

While SOAGP utilizes $n/2$ geminals, the possibility has also been explored using a single geminal. Given only one geminal g, a simple function for a geminal functional theory can be derived for even numbers of electrons. This *ansatz* may be generalized to systems of odd numbers of electrons by multiplying a single strongly orthogonal function to an AGP function (see Eqs. (15) and (16)):

$$\Psi_{\text{AGP}} = g^{n/2} = \mathbf{A}_n \prod_{i=1}^{n/2} g(2i-1, 2i) \qquad (15)$$

$$\Psi_{\text{GAGP}} = g^{(n-1)/2} \wedge \phi = \mathbf{A}_n g^{(n-1)/2} \phi(n) \qquad (16)$$

Since the generating geminals of an SOAGP wavefunction can be described in the same basis, it is tempting to assume that summing the geminals leads to an AGP generating geminal for the SOAGP wavefunction; such is not the case.

$$\Psi_{\text{SOAGP}} \propto \Psi_{\text{AGP}}(g) / \Leftarrow g = \sum_i g_i \qquad (17)$$

The AGP function for the summation includes contributions from g_i^N and other mixed higher-order products, which are not present in the SOAGP function.

The second-order density matrix of an AGP function has the following form [15], which is reminiscent of the SOAGP 2-RDM:

$$\rho_2(\Psi_{\text{AGP}}; 12, 1'2') = \sum b(ijkl) \cdot |\phi_i(1)\phi_j(2)| \cdot |\phi_k(1')\phi_l(2')|^* \qquad (18)$$

where ϕ_i are the natural spin orbitals of g and

- $\exists n: \{i,j\} = \{k,l\} = \{2n-1, 2n\} =: \sigma \Rightarrow b(\sigma\sigma) = 2c\lambda_\sigma a_{m-1}(\hat{\sigma})$.
- $\sigma \neq \tau \Rightarrow b(\sigma\tau) = 2c\xi_\sigma \xi_\tau^* a_{m-1}(\hat{\sigma}\hat{\tau})$.
- $\sigma \neq \tau, i \in \sigma, j \in \tau \Rightarrow b(ijij) = 2c\lambda_\sigma \lambda_\tau a_{m-2}(\hat{\sigma}\hat{\tau})$.
- $\Pi(1 + \lambda_\sigma t) = \sum a_m t^m; a_{m-1}(\hat{\sigma}) = \partial a_m/\partial \lambda_\sigma; a_{m-2}(\hat{\sigma}\hat{\tau}) = \partial^2 a_m/\partial \lambda_\sigma \partial \lambda_\tau$, $c = 1/a_m$

Again, the 2-RDM has a very simple block-diagonal structure of only two blocks, of which one is diagonal:

$$\hat{\rho}_2(\Psi_{\text{AGP}}) = \begin{pmatrix} \overline{b(\sigma\tau)} & & \\ & b(1313) & \\ & & \ddots \end{pmatrix} \qquad (19)$$

As is easily verified, the natural orbitals of the geminal and the AGP wavefunction coincide. Equation (21) implies that the first-order reduced density matrix is degenerate for every eigenvalue. The final trace normalizes to $n(n-1)$ as expected. The degeneracy of ρ_1 is intricately linked to AGP functions as, for every evenly degenerate 1-RDM, there exists an AGP function that reproduces that 1-RDM [15–17]. Any Slater determinant $|\phi_1 \cdots \phi_n|$, for example, the HF solution, can be described by a very simple geminal, for instance, $g = \sum_{i=1}^{n/2} |\phi_{2i-1}\phi_{2i}|$. Since all natural orbitals are degenerate, there is no unique geminal that represents the Slater determinant.[1]

$$(n-1)\hat{\rho}^{(1)} = \sum_\sigma |i\rangle\langle i|c\lambda_\sigma a_{m-1}(\hat{\sigma})$$

$$+ \sum_{\substack{\sigma>\tau \\ i\in\sigma \\ j\in\tau}} \sum (|i\rangle\langle i| + |j\rangle\langle j|)c\lambda_\sigma\lambda_\tau a_{m-2}(\hat{\sigma}\hat{\tau}) \qquad (20)$$

$$= \sum_{\substack{\sigma \\ i\in\sigma}} |i\rangle\langle i|c\lambda_\sigma a_{m-1}(\hat{\sigma}) + \sum_{\substack{\sigma\neq\tau \\ i\in\sigma \\ j\in\tau}} |i\rangle\langle i|c\lambda_\sigma\lambda_\tau a_{m-2}(\hat{\sigma}\hat{\tau})$$

$$= \sum_{\substack{\sigma \\ i\in\sigma}} |i\rangle\langle i|c\lambda_\sigma (2m-1)a_{m-1}(\hat{\sigma}) \qquad (21)$$

As Eqs. (18) and (2) demonstrate, the exact wavefunction is no longer explicitly needed nor is an approximate wavefunction used. Instead, the generating geminals determine the wavefunction as well as the reduced density matrices and thereby the energy of an AGP wavefunction.

IV. FORMAL GFT

So far the term GFT has been used in a suggestive way of mapping a geminal to an energy that approximates the ground-state energy of some system. Not until

---

[1] Any collection of pairs of spin orbitals will produce a geminal that gives rise to the Slater determinant.

recently was a formal description of geminal functional theory proposed [18]. The parameterization of 1-RDMs by AGP functions allows the derivation of a functional based on geminals ($R : g \to \Psi_{AGP}[g] \to {}^1D[\Psi_{AGP}[g]]$) similar to density functional theory [19, 20]. In order to devise a functional, the general (quadratic) energy functional $E(\Psi) = \langle \Psi | \hat{H} | \Psi \rangle$ is split into three contributions:

$$E[\Psi] = T({}^1D[\Psi]) + V({}^1D[\Psi]) + W({}^2D[\Psi]) \qquad (22)$$

$$T({}^1D) = \frac{n}{2} \cdot \mathrm{tr}({}^1\hat{D}\hat{p}^2) \qquad (23)$$

$$V({}^1D) = n \cdot \mathrm{tr}({}^1\hat{V}{}^1\hat{D}) \qquad (24)$$

$$W({}^2D) = \frac{n(n-1)}{2} \cdot \mathrm{tr}({}^2\hat{D}\hat{r}_{12}^{-1}) \qquad (25)$$

Due to the bijective correspondence R of geminals to evenly degenerate 1-RDMs [17, 18], the kinetic and nuclear potential terms (T and V) can be expressed as functionals of a geminal through $\tilde{T} := T \circ R$ and $\tilde{V} := V \circ R$, respectively.

The interelectronic interactions W are defined using constrained search [21, 22] over all N-representable 2-RDMs that reduce to $R(g)$. Since the set of 2-RDMs in the definition of $\tilde{W}$ contains the AGP 2-RDM of g, that set is not empty and $\tilde{W}$ is well defined. Through this construction, $\tilde{E}$ still follows the variational principle and coincides with the energy of a wavefunction Ψ', which reproduces $R(g) = {}^1D[\Psi']$ and $W({}^2D[\Psi']) = \tilde{W}[g]$. The latter is due to the completeness of $\{{}^2D : C^1({}^2D) = R(g)\}$ and the boundedness of W, where C^1 is the contraction operator of one variable. $\tilde{W}$ depends solely on the geminal and the number of electrons and thus is universal for any physical system of equal number of electrons. In other words, $\tilde{W}[g]$ for an atom or molecule of n electrons is the same for any of those atoms or molecules. This should not be confused with the geminal g corresponding to the ground state being the same for those systems.

$$\left(C^1(D(Q, x_n; Q', x'_n))\right) = \int D(Q, x_n; Q', x_n) dx_n \qquad (26)$$

$$\tilde{W}[g] = \min_{{}^2D : C^1({}^2D) = R(g)} W[{}^2D] \qquad (27)$$

$$\tilde{E}[g] = \tilde{T}[g] + \tilde{V}[g] + \tilde{W}[g] \qquad (28)$$

As in density functional theory, the functional $\tilde{W}$ is unknown but can be approximated by restricting or expanding the set of positive semidefinite matrices ${}^2D$. The SOAGP, SSG, and AGP methods represent true geminal functional theories in this sense as they restrict the set of N-representable

1- and 2-RDMs to those derivable by the respective wavefunctions. If the set is a subset of N-representable 2-RDMs, as for SOAGP, SSG, or AGP, the variational principle remains true ($\tilde{E} \geq E_{g.s.}$). Perturbation theory on SSG is a nonvariational geminal functional.

A. An Application of GFT

Revisiting the AGP method from this approach, higher-quality calculations for the AGP approximation of the ground state have recently been achieved [23]. The electron–electron correlation is approximated by Eq. (29):

$$\bar{W}_{\text{RDM/GFT}}[^1D] := \min_{g:C^1(^2D[g])=^1D} W(^2D[g]) \tag{29}$$

Based on the 1,3-contracted Schrödinger equation (CSE), the AGP energies for various systems have been recalculated with considerable success [23]. Starting from a geminal g, the 2- and 3-RDMs of the respective AGP wavefunction were used in the evaluation of the 1,3-CSE for the determination of a 1-RDM. From this 1-RDM the natural orbitals ϕ_i and occupation numbers $|\xi_i|^2$ of the next geminal $g' = \sum \xi_i |\phi_{2i-1}\phi_{2i}|$ were derived. Then $W(^2D[\Psi_{AGP}[g]])$ was optimized with respect to the phase factors of ξ_i to obtain the next geminal iterative.[2] Table I shows exemplary calculations. For four-electron systems, the electron correlation is quantitatively recovered.

Since $\tilde{W}$ is universal, the discrepancy between the exact solution and the GFT may be corrected via a factor based on a trial computation.

$$E_{\text{GFT}} = E_{\text{HF}} + (E_{\text{HF}} - E_{\text{RDM/GFT}}) \cdot \mu(n) \tag{30}$$

$$\mu(n) = \frac{E_{\text{HF,trial}} - E_{\text{exact,trial}}}{E_{\text{HF,trial}} - E_{\text{RDM/GFT,trial}}} \tag{31}$$

TABLE I
Energies and Electron Correlation Using the RDM/GFT Algorithm[a]

Molecule	n	E_{HF}	$E_{\text{RDM/GFT}}$	Full CI	Electron Correlation (%)
Be	4	−14.57091	−14.61530	−14.61557	99.3
BH	6	−25.11340	−25.15148	−25.17402	62.8
CH$_4$	10	−39.89506	−39.92536	−40.01405	25.4
HCN	14	−92.83712	−92.88268	−93.04386	22.0

[a]Data by D. A. Mazziotti originally appeared in Ref. [23].

[2]The sequence of steps has been rotated from the original to emphasize the geminal as primary variable.

TABLE II
Energies Using $\tilde{W}$ and the Scaled Extension

Molecule	n	E_{HF}	$E_{RDM/GFT}$	E_{GFT}	Full CI	Electron Correlation	RDMGFT versus GFT
LiH	4	−7.98074	−7.99913	−7.99924	−7.99929	99.8	99.1
BeH$_2$	6	−15.76024	−15.77818	−15.78879	−15.80076	70.5	44.2
HF	10	−100.02189	−100.05103	−100.13630	−100.14446	93.3	23.7
NH$_3$	10	−55.96479	−55.99806	−56.09543	−56.09474	100.5	25.6
H$_2$O	10	−76.00915	−76.03931	−76.12760	−76.13894	91.3	23.2
N$_2$	14	−108.87814	−108.93163	−109.12085	−109.09747	110.7	24.3
CO	14	−112.68484	−112.72695	−112.87590	−112.88839	93.9	20.6
C$_2$H$_2$	14	−76.79917	−76.83612	−76.96682	−76.98616	89.7	19.7

^{a}Data by D. A. Mazziotti originally appeared in Ref. [23].

With this correction all but one computed electron correlation energy fell within 10% of the exact solution. As Table II shows, the very simple scaling correction yields huge improvements on modest initial electron correlations. The use of the correction factor implies the loss of the variational principle and does not account for the use of different basis sets; for example, the Dunning double-zeta basis set for Be is different for LiH [24, 25].

V. UNRESTRICTED ANTISYMMETRIZED GEMINAL PRODUCTS

The success of AGP has retained its allure over more than five decades. Just recently, the AGP wavefunction of a set of geminals has been investigated. Applying the *aufbau* principle of Hartree–Fock theory generalizes the AGP *ansatz*. In this case, a set of geminals instead of a single geminal is used to generate the 2-RDM. In terms of GFT, $\tilde{W}$ is approximated as in Eq. (33):

$$\Psi_{UAGP} = \mathbf{A}_n \prod_{i=1}^{n/2} g_i(2i-1, 2i) \qquad (32)$$

$$\tilde{W}[g] \approx \min_{C^1(^2D_{UAGP}) = R(g)} W(^2D_{UAGP}) \qquad (33)$$

where g_i are geminals. The term "aufbau" is used in the sense that the wavefunction is built successively from nonidentical geminals. While Hartree–Fock functions are AGP functions as well as antisymmetrized products of strongly orthogonal geminals (SOAGP), this is generally not the case for SOAGP functions and therefore unrestricted AGP (UAGP) functions, as they represent a superset of SOAGP as well as AGP functions.

An interpretation of the meaning of the geminals may be derived from ordering them with respect to $\langle g_i|^2 K^2 D_{\text{aufbau}}|g_i\rangle$. In keeping with Hartree–Fock theory, the lowest energy geminals can be determined to describe the core electrons, while increasing energy terms describe increased mixing with the valence electrons and therefore explain, for instance, bonding.

Due to higher-order contributions, the most general case does not allow for a direct description of the 2-RDM, but it is possible to build the 2-RDM iteratively [26]. Of the six contributions (Eqs. (34)–(40)), F_1 and F_6 are of immediate interest. Equation (35) contains the geminal's direct contribution to the 2-RDM. The cofactor is large for a single geminal on the diagonal. ^2D_p is retained and weighed accordingly in Eq. (40). This guarantees that large contributions of previous iterations will remain large in the next iteration.

$$^2D_{p+2} = F_1 + (F_2 + F_2^*) + (F_3 + F_3^*) + F_4 + (F_5 + F_5^*) + F_6 \tag{34}$$

$$\mathbf{F_1} = \mathbf{g(12)g(1'2')} \parallel \mathbf{\Psi_p} \parallel^2 \tag{35}$$

$$F_2 = -p \int [g(12)g^*(1'3)\,^1D_p(3;2') - g(12)g^*(2'3)\,^1D_p(3;1')]d3 \tag{36}$$

$$F_3 = \binom{p}{2} \int g(12)g^*(34)\,^2D_p(34;1'2')d3\,d4 \tag{37}$$

$$F_4 = 2\binom{p}{2} \int [g(13)g^*(2'4)\,^2D_p(24;1'3) - g(23)g^*(2'4)\,^2D_p(14;1'3)]d3\,d4$$

$$+ 2\binom{p}{2} \int [g(13)g^*(1'4)\,^2D_p(24;2'3) - g(23)g^*(1'4)\,^2D_p(14;2'3)]d3\,d4$$

$$+ p(^1D_g(1,1')\,^1D_p(2;2') + \,^1D_g(2,2')\,^1D_p(1;1'))$$

$$- p(^1D_g(2,1')\,^1D_p(1;2') + \,^1D_g(1,2')\,^1D_p(2;1')) \tag{38}$$

$$F_5 = 2\binom{p}{2} \int [^1D_g(1;4)D_p^{(2)}(24;1'2') - \,^1D_g(2;4)\,^2D_p(14;1'2')]d4$$

$$- 6\binom{p}{3} \int g(13)g^*(45)D^{(3)}(245;1'2'3)d3\,d4\,d5$$

$$+ 6\binom{p}{3} \int g(23)g^*(45)D^{(3)}(145;1'2'3)d3\,d4\,d5 \tag{39}$$

$$F_6 = \binom{\mathbf{p}}{2}\,^2\mathbf{D_p(12;1'2')} \parallel \mathbf{g} \parallel^2$$

$$+ 2(p-2)\binom{p}{2} \int \,^1D_g(4;5)D_p^{(3)}(125;1'2'4)d4\,d5$$

$$+ \binom{p-2}{2}\binom{p}{2} \int g(34)g^*(56)D_p^{(4)}(1256;1'2'34)d3\,d4\,d5\,d6 \tag{40}$$

Equations (39) and (40) introduce problematic higher-order contributions. Since $g(34) = \sum_i \xi_i \alpha_i(3) \beta_i(4)$,

$$\int g(34)g(35)d3 = \sum_i |\xi_i|^2 \beta_i(4)\beta_i(5)$$

Also,

$$^2D(12;34) = \sum_i \int D^{(3)}(125;346)\beta_i(5)\beta_i(6) d5\, d6$$

Due to $\sum_i |\xi_i|^2 = 1$, $|\xi_i|^2 \leq 1/2$, and

$$^2D_p(12;1'2') = \sum_i \int \alpha_i(3) D_p^{(3)}(123;1'2'3')\alpha_i(3') d3\, d3'$$

we can conclude

$$\frac{1}{2} {}^2D_p(12;1'2') \geq \int g(34')^* g(3'4) D_p^{(3)}(123;1'2'3') d3\, d3'\, d4' \qquad (41)$$

By a similar argument,

$$\frac{1}{p} {}^2D_p(12;1'2') \geq \int g(34)^* g(3'4') D_p^{(4)}(1234;1'2'3'4') d3\, d4\, d3'\, d4' \qquad (42)$$

The UAGP *ansatz* has been applied to a variety of four-electron systems, which do not suffer from approximations to the higher-order contributions [26]. Since $\langle g_1 g_2 | \hat{A}_4 \hat{K} \hat{A}_4 | g_1' g_2' \rangle = \langle g_1 | \hat{H}'[g_2] | g_1' \rangle$ and $\langle g_1 g_2 | \hat{A}_4 | g_1' g_2' \rangle = \langle g_1 | \hat{O}[g_2] | g_1' \rangle$ are bilinear functionals of g_1 and g_1', it is possible to solve the generalized eigenvalue problem for the energy for each generating geminal while keeping the remaining geminal constant. For the majority, UAGP functions recovered in excess of 99% of the electron correlation. The overlap of the two generating geminals showed that the UAGP functions did not collapse to the already good AGP functions.

VI. CONCLUSIONS

Geminal functional theory is a very promising research area. The different varieties of antisymmetrized products are very flexible and inherently handle difficult problems, like multideterminantal molecules. The computational effort is low compared to the quality of the solutions. The perturbation theoretical approach to SSG should essentially be possible for AGP and UAGP as well. The formal definition of GFT is a flexible framework that opens up many new opportunities for exploring the nature of solutions to the Schrödinger equation.

References

1. A. C. Hurley, J. Lennard-Jones, and J. A. Pople, The molecular orbital theory of chemical valency. 16. A theory of paired-electrons in polyatomic molecules. *Proc. R. Soc. London Ser. A* **220**(1143), 446–455 (1953).
2. D. M. Silver, E. L. Mehler, and K. Ruedenberg, Electron correlation and separated pair approximation in diatomic molecules. 1. Theory. *J. Chem. Phys.* **52**(3), 1174–1180 (1970).
3. D. M. Silver, K. Ruedenberg, and E. L. Mehler, Electron correlation and separated pair approximation in diatomic molecules. 2. Lithium hydride and boron hydride. *J. Chem. Phys.* **52**(3), 1181–1205 (1970).
4. D. M. Silver, K. Ruedenberg, and E. L. Mehler, Electron correlation and separated pair approximation in diatomic molecules. 3. Imidogen. *J. Chem. Phys.* **52**(3), 1206–1227 (1970).
5. V. A. Rassolov, A geminal model chemistry. *J. Chem. Phys.* **117**(13), 5978–5987 (2002).
6. L. A. Curtiss, P. C. Redfern, K. Raghavachari, and J. A. Pople, Assessment of Gaussian-2 and density functional theories for the computation of ionization potentials and electron affinities. *J. Chem. Phys.* **109**(1), 42–55 (1998).
7. L. A. Curtiss, K. Raghavachari, P. C. Redfern, and J. A. Pople, Assessment of Gaussian-2 and density functional theories for the computation of enthalpies of formation. *J. Chem. Phys.* **106**(3), 1063–1079 (1997).
8. D. R. Hartree, *Proc. Camb. Phil. Soc.* **24**, 89–110 (1928).
9. P. M. W. Gill, Density functional theory (DFT), Hartree–Fock (HF), and the self-consistent field, in *Encyclopedia of Computational Chemistry* (P. von Ragué Schleyer, N. L. Allinger, T. Clark, J. Gasteiger, P. A. Kollman, H. F. Schaefer, and P. R. Schreiner, eds.) Wiley, Hoboken, NJ 1998, pp. 678–689.
10. D. Cremer, Møller–Plesset perturbation theory, in *Encyclopedia of Computational Chemistry* P. von Ragué Schleyer, N. L. Allinger, T. Clark, J. Gasteiger, P. A. Kollman, H. F. Schaefer, and P. R. Schreiner, eds., Wiley, Hoboken, NJ, 1998, pp. 1706–1735.
11. C. Møller and M. S. Plesset, *Phys. Rev.* **46**, 618 (1934).
12. J. Simons, An experimental chemists guide to ab initio quantum-chemistry. *J. Phys. Chem.* **95**(3), 1017–1029 (1991).
13. E. Rosta and P. R. Surjan, Interaction of chemical bonds. V. Perturbative corrections to geminal-type wave functions. *Int. J. Quantum Chem.* **80**(2), 96–104 (2000).
14. E. Rosta and P. R. Surjan, Two-body zeroth order Hamiltonians in multireference perturbation theory: the APSG reference state. *J. Chem. Phys.* **116**(3), 878–890 (2002).
15. A. J. Coleman, Structure of fermion density matrices. 2. Antisymmetrized geminal powers. *J. Math. Phys.* **6**(9), 1425–1431 (1965).
16. A. J. Coleman, The AGP model for fermion systems. *Int. J. Quantum Chem.* **63**(1), 23–30 (1997).
17. A. J. Coleman and V. I. Yukalov, *Reduced Density Matrices: Coulson's Challenge*, Springer, Berlin, 2000.
18. D. A. Mazziotti, Geminal functional theory: a synthesis of density and density matrix methods. *J. Chem. Phys.* **112**(23), 10125–10130 (2000).
19. P. Hohenberg and W. Kohn, *Phys. Rev. B* **79**, 361–376 (1964).
20. W. Kohn and L. J. Sham, Self-consistent equations including exchange and correlation effects. *Phys. Rev.* **140**(4A), 1133 (1965).

21. M. Levy, Universal variational functionals of electron-densities, 1st-order density-matrices, and natural spin-orbitals and solution of the v-representability problem. *Proc. Natl. Acad. Sci. U.S.A.* **76**(12), 6062–6065 (1979).
22. S. M. Valone, A one-to-one mapping between one-particle densities and some normal-particle ensembles. *J. Chem. Phys.* **73**(9), 4653–4655 (1980).
23. D. A. Mazziotti, Energy functional of the one-particle reduced density matrix: a geminal approach. *Chem. Phys. Lett.* **338**(4–6), 323–328 (2001).
24. T. H. Dunning, Gaussian-basis sets for use in correlated molecular calculations. 1. The atoms boron through neon and hydrogen. *J. Chem. Phys.* **90**(2), 1007–1023 (1989).
25. T. H. Dunnning, K. A. Peterson, and D. Woon, Basis sets: correlation consistent sets, in *Encyclopedia of Computational Chemistry* Wiley, Hoboken, NJ, 1998, p. 88.
26. B. C. Rinderspacher and P. R. Schreiner, An aufbau ansatz for geminal functional theory. *J. Chem. Phys.* **123**(21), 214104 (2005).

CHAPTER 16

LINEAR INEQUALITIES FOR DIAGONAL ELEMENTS OF DENSITY MATRICES

PAUL W. AYERS

Department of Chemistry, McMaster University, Hamilton, Ontario L85 4M1, Canada

ERNEST R. DAVIDSON

Department of Chemistry, University of Washington, Bagley Hall 303C/Box 351700, Seattle, WA 98195-1700 USA

CONTENTS

I. Introduction
II. Necessary and Sufficient Conditions for N-Representability
III. Linear Inequalities from the Orbital Representation
    A. The Slater Hull
    B. Derivation of the (R, R) Conditions
    C. Necessary Conditions from the Slater Hull; (R, R) Conditions
        1. $(1, 1)$ Conditions
        2. $(2, 2)$ Conditions
        3. $(3, 3)$ Conditions
    D. Necessary Conditions from the Slater Hull; (Q, R) Conditions
    E. Sufficiency of the Slater Hull Constraints for Diagonal Elements of Density Matrices
    F. Constraints on Off-Diagonal Elements from Unitary Transformation of the Slater Hull Constraints
    G. Constraints on Off-Diagonal Elements from Other Positive-Definite Hamiltonians
IV. Linear Inequalities from the Spatial Representation
V. Linking the Orbital and Spatial Representations
    A. Review of Orbital-Based Density Functional Theory
    B. Orbital-Based Q-Density Functional Theory
VI. Conclusion
References

Reduced-Density-Matrix Mechanics: With Application to Many-Electron Atoms and Molecules,
A Special Volume of Advances in Chemical Physics, Volume 134, edited by David A. Mazziotti.
Series editor Stuart A. Rice. Copyright © 2007 John Wiley & Sons, Inc.

I. INTRODUCTION

In 1994, Ziesche [1] proposed using the diagonal elements of the two-electron reduced density matrix, or electron-pair density, as the fundamental descriptor of electronic structure. The resulting pair-density functional theory is a logical refinement of density functional theory (which is based on the diagonal elements of the one-electron reduced density matrix). Extending Ziesche's idea to the diagonal elements of the Q-electron reduced density matrix, or Q-density, yields a hierarchy of Q-density functional theories that converges to the exact answer as Q approaches the total number of electrons in the system [2,3]. In this way, Q-density functional theory is a solution to the longstanding problem of how to construct a "systematically improvable" density functional theory analogous to the hierarchies of methods (CISDTQ, CCSDTQ, etc.) commonly employed in wavefunction-based ab initio quantum chemistry.

Most of the formal mathematical structure of conventional density functional theory transfers to Q-density functional theories without change [1–5]. As in density functional theory, the key for practical implementations is the variational principle,

$$E_{\text{g.s.}}[v; N] = \min_{\substack{N-\text{representable} \\ \rho_Q(\mathbf{x}_1, \ldots, \mathbf{x}_Q)}} \left(T[\rho_Q] + \int \cdots \int \rho_Q(\mathbf{x}_1, \ldots \mathbf{x}_Q) \left(\begin{array}{c} \frac{(N-k)!(Q-1)!}{(N-1)!} \left(\sum_{j=1}^{Q} v(\mathbf{x}_j) \right) \\ + \frac{(N-Q)!(Q-2)!}{(N-2)!} \left(\sum_{j_1=1}^{Q} \sum_{j_2 \neq j_1}^{Q} \frac{1}{2|\mathbf{x}_{j_1} - \mathbf{x}_{j_2}|} \right) \end{array} \right) d\mathbf{x}_1 \cdots d\mathbf{x}_Q \right)$$

(1)

Here $v(\mathbf{x})$ denotes the external potential of the molecule; for an isolated molecule in the absence of external electric fields, this is simply the potential due to nuclear–electron attraction.

Note that for $Q \geq 2$, only the kinetic energy functional needs to be approximated. Equation (1) is equally valid whether ρ_Q corresponds to a pure state or to an ensemble with a fixed number of electrons, N. In the rest of this chapter we will assume an ensemble average.

In writing Eq. (1), we have chosen to define the Q-density so that it is normalized to $N!/(N-Q)!Q!$. That is,

$$\rho_Q(\mathbf{y}_1, \ldots, \mathbf{y}_Q) = \frac{1}{Q!} \sum_i p_i \left\langle \Psi_i \left| \sum_{j_1=1}^{N} \sum_{\substack{j_2=1 \\ j_2 \neq j_1}}^{N} \cdots \sum_{\substack{j_Q=1 \\ j_Q \neq j_{Q-1}, j_{Q-2}, \ldots, j_1}}^{N} \delta(\mathbf{x}_{j_1} - \mathbf{y}_1) \cdots \delta(\mathbf{x}_{j_Q} - \mathbf{y}_Q) \right| \Psi_i \right\rangle$$

(2)

Implicit in this notation, the wavefunction is written as a function of the electronic coordinates, $x_1, x_2, \ldots, x_N$, and the bracket indicates integration over these coordinates. In this equation, the weighting coefficients, p_i, represent the probability of observing the system in the state associated with Ψ_i, and thus must satisfy the constraints

$$0 \leq p_i \leq 1$$
$$1 = \sum_{i=1}^{\infty} p_i \tag{3}$$

When determining the ground-state energy by optimizing the wavefunction, one restricts the search to wavefunctions that represent possible states of N-electron systems. In practical terms, this means that when minimizing the electronic energy with respect to the wavefunction, one considers only wavefunctions that are antisymmetric. Similarly, when one determines the ground-state energy by optimizing the Q-density, one restricts the search to Q-densities that represent N-electron systems. In practical terms, this means that when minimizing the energy with respect to the Q-density, one considers only the Q-densities that can be written in the form of Eq. (2). Such Q-densities are said to be N-representable.

- A Q-density, ρ_Q, is *N-representable* if and only if there exists some set of antisymmetric wavefunctions, $\{\Psi_i\}$, and weighting coefficients, $\{p_i\}$, for which Eq. (2) holds.
- A Q-density, ρ_Q, is *pure-state N-representable* if and only if it can be associated with a specific antisymmetric wavefunction. (In this case, one of the weights in Eq. (2) is one and all the other p_i are zero.)
- A Q-density, ρ_Q, is *non-N-representable* if and only if there is no way to satisfy Eq. (2) without violating either the requirement that $p_i \geq 0$ or the requirement that Ψ_i is antisymmetric. A non-N-representable Q-density does not represent an N-fermion system.
- A *necessary condition* for ρ_Q to be N-representable is an equation or inequality that holds for every N-representable Q-density. Although some non-N-representable Q-densities might also satisfy a necessary condition, every N-representable Q-density always satisfies a necessary condition.
- A *sufficient condition* for ρ_Q to be N-representable is an equation or inequality that guarantees any Q-density is N-representable. Although some N-representable Q-densities might violate a sufficient condition, every non-N-representable Q-density will violate a sufficient condition.

If one minimizes the energy subject to the satisfaction of some necessary, but not sufficient, conditions for N-representability, then one will include some non-N-representable Q-densities in the variational principle. Since the domain of Q-densities considered is too large, the energy obtained will be a lower bound to the full-CI energy in the basis set; the full-CI energy is an upper bound to the true energy. If one minimizes the energy subject to the satisfaction of a sufficient, but not necessary, condition for N-representability, then one will exclude some N-representable Q-densities from the search for a minimum and so the energy obtained will be an upper bound to the true energy. If one constrains the variational optimization of the *exact* energy functional with a necessary *and* sufficient condition for the N-representability of the Q-density, the exact ground-state energy will be obtained. Because all the necessary and sufficient conditions that we know are very complicated, developing approximate methods based on necessary *or* sufficient conditions is critically important. This chapter presents a family of necessary conditions for N-representability.

There are only a few studies of Q-density functional theory for $Q > 2$ [2, 3, 6]. Most studies have concentrated on the pair density, or 2-density functional theory. Excepting the fundamental work of Ziesche, early work in 2-density functional theory focused on a differential equation for the pseudo-two-electron wavefunction [7–11] defined by

$$\chi(\mathbf{x}_1, \mathbf{x}_2) = \sqrt{\frac{2}{N(N-1)}\rho_2(\mathbf{x}_1, \mathbf{x}_2)}$$

This is the natural extension of density functional methods based on $\chi(\mathbf{x}) = (1/\sqrt{N})\sqrt{\rho(\mathbf{x})}$ and it is an attractive approach to this problem partly because of how much is known about the analogous density functional method [12–16]. However, it seems difficult to deal with the restrictions N-representability imposes on $\chi(\mathbf{r}_1, \mathbf{r}_2)$ and none of these studies explicitly addresses the N-representability issue. Until recently, this seemed sensible because most researchers believed that the N-representability constraints were not very restrictive, so that the hard part of the problem was approximating the kinetic energy functional. Now, however, approximating the kinetic energy functional seems to be tractable and the most difficult aspect of pair-density theory seems to be the imposition of appropriate N-representability constraints on the variational procedure [2, 10, 17]. This is in marked contrast to density functional theory, where the N-representability conditions—the electron density must be nonnegative and normalized to the number of electrons [18, 19]—are trivial but formulating accurate kinetic energy functionals is very difficult [20, 21].

In retrospect, the importance of N-representability constraints on the pair density should have been clear from the very beginning. An N-representability condition for the diagonal elements of the Q-electron reduced density matrices

$$\Gamma_Q(\mathbf{x}_1,\ldots,\mathbf{x}_Q;\mathbf{x}'_1,\ldots,\mathbf{x}'_Q) = \frac{N!}{(N-Q)!Q!}\sum_i p_i \int \cdots$$

$$\int \Psi_i^*(\mathbf{x}'_1,\ldots,\mathbf{x}'_Q,\mathbf{x}_{Q+1},\ldots,\mathbf{x}_N)\Psi_i(\mathbf{x}_1,\ldots,\mathbf{x}_N)d\mathbf{x}_{Q+1}\cdots d\mathbf{x}_N \quad (4)$$

even appears in the early work of Garrod and Percus [22] on the N-representability problem. Weinhold and Wilson [23] derived a longer list of necessary conditions for N-representability, which were subsequently rederived, generalized, and extended [24–26]. Shortly after, Ziesche published his first work on pair-density functional theory, Davidson showed that in addition to the "obvious" requirements that $\rho_2(\mathbf{x}_1,\mathbf{x}_2) = \Gamma_2(\mathbf{x}_1,\mathbf{x}_2;\mathbf{x}_1,\mathbf{x}_2)$ be nonnegative, normalized, and symmetric with respect to exchange of coordinates, there are additional N-representability conditions associated with the eigenvalues of $\rho_2(\mathbf{x}_1,\mathbf{x}_2)$, where $\rho_2(\mathbf{x}_1,\mathbf{x}_2)$ is viewed as an integral kernel [27]. Nonetheless, until recently the N-representability problem was largely neglected in pair-density functional theory. It now seems clear, however, that if one fails to constrain the variational principle to N-representable pair densities, the energy obtained from the variational procedure could be catastrophically low [17, 28].

The importance of N-representability for pair-density functional theory was not fully appreciated probably because most research on pair-density theories has been performed by people from the density functional theory community, and there is no "N-representability problem" in conventional density functional theory. Perhaps this also explains why most work on the pair density has been performed in the "first-quantized" spatial representation ($\rho_2(\mathbf{x}_1,\mathbf{x}_2) = \Gamma_2(\mathbf{x}_1,\mathbf{x}_2;\mathbf{x}_1,\mathbf{x}_2)$) instead of the "second-quantized" orbital representation

$$\rho_{ij} = \Gamma_{ij;ij} \quad (5)$$

$$\Gamma_{ij;kl} = \iint \iint (\phi_i^*(\mathbf{x}_1)\phi_j^*(\mathbf{x}_2)\Gamma_2(\mathbf{x}_1,\mathbf{x}_2;\mathbf{x}'_1,\mathbf{x}'_2)\phi_k(\mathbf{x}'_1)\phi_l(\mathbf{x}'_2))d\mathbf{x}_1\,d\mathbf{x}'_1\,d\mathbf{x}_2\,d\mathbf{x}'_2 \quad (6)$$

favored by the density-matrix theory community. Most of the known N-representability constraints on the pair density are formulated in the orbital representation and were originally conceived as constraints on the diagonal elements of the two-electron reduced density matrix, $\Gamma_{ij;ij}$. These constraints can be very powerful because they must hold in any orbital basis; the invariance of the density matrix to unitary transformations between orbital basis sets means that constraints on the diagonal elements of the density matrix in one orbital representation imply nontrivial constraints on the off-diagonal elements of the density matrix in all other choices of the orbital basis [23–25, 29]. Indeed, all of the

N-representability conditions on the Q-density were originally intended for use in variational optimization with respect to the Q-electron reduced density matrix.

Compared to computational approaches based on the Q-electron density, approaches based on the Q-electron reduced density matrix have the advantage that the kinetic energy functional can be written in an explicit form:

$$T[\Gamma_Q] = \frac{(N-Q)!(Q-1)!}{(N-1)!} \int \cdots$$
$$\int \left(\prod_{j=1}^{Q} \delta(\mathbf{x}_j - \mathbf{x}'_j) \sum_{k=1}^{Q} -\frac{\nabla^2_{\mathbf{x}_k}}{2} \Gamma_Q(\mathbf{x}_1, \ldots, \mathbf{x}_Q; \mathbf{x}'_1, \ldots, \mathbf{x}'_Q) \right) d\mathbf{x}_1 d\mathbf{x}'_1 \cdots d\mathbf{x}_Q d\mathbf{x}'_Q \quad (7)$$

Partly for this reason, there has been a persistent interest in direct optimization of the electronic energy with respect to a reduced density matrix [30–39].

$$E_{\text{g.s.}}[v; N] = \min_{N\text{-rep.} \Gamma_Q} \text{Tr}[\hat{H}_{Q,N} \Gamma_Q] \quad (8)$$

Here, the trace notation means "operate with $\hat{H}_{Q,N}$ on Γ_Q; remove the primes from the primed variables; then integrate with respect to the remaining unprimed variables." Because Γ_Q only depends on Q-electronic coordinates, evaluating the energy using Eq. (8) requires rewriting the energy operator so that it depends only on the coordinates of Q electrons. The resulting Hamiltonian, $\hat{H}_{Q,N}$, is called the Q-electron reduced Hamiltonian for the N-electron system. For example, for an observable that can be expressed as a sum of zero-, one-, and two-body linear Hermitian operators,

$$\hat{L}_N = \hat{l}_0 + \sum_{j_1=1}^{N} \hat{l}_1(\mathbf{x}_{j_1}) + \sum_{j_1=1}^{N-1} \sum_{\substack{j_2=1 \\ j_2 \neq j_1}}^{N} \hat{l}_2(\mathbf{x}_{j_1}, \mathbf{x}_{j_2}) \quad (9)$$

the Q-body reduced operator is defined as

$$\hat{L}_{Q,N} = \frac{(N-Q)!Q!}{N!} \hat{l}_0 + \frac{(N-Q)!(Q-1)!}{(N-1)!} \sum_{j=1}^{Q} \hat{l}_1(\mathbf{x}_j)$$
$$+ \frac{(N-k)!(Q-2)!}{(N-2)!} \sum_{j_1=1}^{Q} \sum_{\substack{j_2=1 \\ j_2 \neq j_1}}^{Q} \hat{l}_2(\mathbf{x}_{j_1}, \mathbf{x}_{j_2}) \quad (10)$$

The Hamiltonians of interest in molecular electronic structure theory correspond to the choices

$$\hat{l}_1(\mathbf{x}) = -\frac{\nabla_{\mathbf{x}}^2}{2} + v(\mathbf{x}) \tag{11}$$

$$\hat{l}_2(\mathbf{x}_1, \mathbf{x}_2) = \frac{1}{2|\mathbf{x}_1 - \mathbf{x}_2|} \tag{12}$$

The zero-body operator, $\hat{l}_0$, merely shifts the zero of energy. To reduce notational complexity, we will occasionally omit the second subscript on $\hat{L}_{Q,N}$.

Constraints on the diagonal element of the density matrix can be useful in the context of the density matrix optimization problem, Eq. (8). As Weinhold and Wilson [23] stressed, the N-representability constraints on the diagonal elements of the density matrix have conceptually appealing probabilistic interpretations; this is not true for most of the other known N-representability constraints.

The similarities and differences between methods based on the electron density, electron-pair density, and reduced density matrices have recently been reviewed [5]. This chapter is not intended as a comprehensive review, but as a focused consideration of N-representability constraints that are applicable to diagonal elements of reduced density matrices. Such constraints are useful both to researchers working with the Q-density and to researchers working with Q-electron reduced density matrices, and so we shall attempt to review these constraints in a way that is accessible to both audiences. Our focus is on inequalities that arise from the Slater hull because the Slater hull provides an exhaustive list of N-representability conditions for the diagonal elements of the density matrix, $\rho_{i_1 i_2 \ldots i_Q} = \Gamma_{i_1 \ldots i_Q; i_1 \ldots i_Q}$. Although the Slater hull constraints are merely necessary, and not sufficient, for the N-representability of the Q-electron reduced density matrix, it is still an important class of constraints for density-matrix approaches.

The Slater hull constraints are not directly applicable to existing approaches to pair-density functional theory because they are formulated in the orbital representation. Toward the conclusion of this chapter, we will also address N-representability constraints that are applicable when the spatial representation of the pair density is used.

For simplicity, we shall commonly refer to the Q-electron distribution function as the Q-density and the Q-electron reduced density matrix as the Q-matrix. In position-space discussions, the diagonal elements of the Q-matrix are commonly referred to as the Q-density. In this chapter, we will also refer to the diagonal element of orbital-space representation of the Q-matrix as the Q-density.

II. NECESSARY AND SUFFICIENT CONDITIONS FOR N-REPRESENTABILITY

When implementing the variational procedure, the energy expectation value of a candidate density matrix can be written as the trace of the Q-electron reduced Hamiltonian and the Q-electron reduced density matrix, $E[\Gamma_Q] = \text{Tr}[\hat{H}_{Q,N}\Gamma_Q]$. In order to ensure that the variational principle does not give too low an energy, we remove from the variational procedure any density matrix for which $\text{Tr}[\hat{H}_{Q,N}\Gamma_Q]$ is below the true ground-state energy. If $\text{Tr}[\hat{H}_{Q,N}\Gamma_Q]$ is greater than or equal to the ground-state energy, $E_{\text{g.s.}}(\hat{H}_N)$, for every possible Q-body Hamiltonian, then Γ_Q will not cause any problems in the variational procedure and so Γ_Q is not a problematic choice as a trial density matrix. That is, the variational principle "works" if the domain of the minimization is restricted to density matrices that satisfy $\text{Tr}[\hat{H}_{Q,N}\Gamma_Q] \geq E_{\text{g.s.}}(\hat{H}_N)$ for every possible reduced Hamiltonian. In this context the "Hamiltonian" is an arbitrary Q-body Hermitian operator; it need not have any physical significance.

Though "having a variational principle that works" is all that is technically required in a useful theory, this condition is actually necessary and sufficient for the N-representability of the Q-matrix. That is,

- Γ_Q is N-representable if and only if, for every Q-body Hamiltonian, $\hat{H}_N$, $\text{Tr}[\hat{H}_{Q,N}\Gamma_Q] \geq E_{\text{g.s.}}(\hat{H}_N)$. Here, $\hat{H}_{Q,N}$ is the Q-body reduced Hamiltonian as written, for example, in Eq. (10) [22].

It is clear from the variational principle for the wavefunction that if Γ_Q is N-representable, then $\text{Tr}[\hat{H}_{Q,N}\Gamma_Q] \geq E_{\text{g.s.}}(\hat{H}_N)$ for every reduced Hamiltonian. To show that the converse is true, we need three key facts:

- The set of N-representable Γ_Q is a convex set. This follows directly from the definition, Eq. (4) [22, 40].
- The space of linear operators on Γ_Q is equal to the space of possible Q-body reduced Hamiltonians, $\hat{H}_{Q,N}$.
- Given a convex set, $\mathcal{C}$, and an element, x, that is not in the set, there exists a linear operator, l, such that $l(y) > l(x)$ for every $y \in \mathcal{C}$ [41].

Suppose that $\Gamma_Q^{\text{not}\,N\text{-rep}}$ is not N-representable. Since $\Gamma_Q^{\text{not}\,N\text{-rep}}$ is not in the convex set of N-representable density matrices, there exists a linear operator, $\hat{H}_{Q,N}$, such that

$$\text{Tr}[\hat{H}_{Q,N}\Gamma_Q^{N\text{-rep}}] > \text{Tr}[\hat{H}_{Q,N}\Gamma_Q^{\text{not}\,N\text{-rep}}] \tag{13}$$

for all N-representable Q-matrices. Thus

$$\min_{N-\text{representable}\,\Gamma_Q} \text{Tr}[\hat{H}_{Q,N}\Gamma_Q] = E_{\text{g.s.}}(\hat{H}_N) > \text{Tr}[\hat{H}_{Q,N}\Gamma_Q^{\text{not}\,N\text{-rep}}] \tag{14}$$

This indicates that if a Q-matrix is not N-representable, then there exists at least one Hamiltonian capable of diagnosing this malady. We conclude that if $\text{Tr}[\hat{H}_{Q,N}\Gamma_Q] \geq E_{\text{g.s.}}(\hat{H}_N)$ for all Hamiltonians, then Γ_Q is N-representable. In practice, it is useful to shift Hamiltonians so that the ground-state energy is zero. That is, we define Hamiltonians by

$$\hat{P}_N = \hat{H}_N - E_{\text{g.s.}}(\hat{H}_N)\hat{I}_N \qquad (15)$$

where $\hat{I}_N$ is the identity operator. This leads to a simpler statement of the N-representability conditions, namely,

- Γ_Q is N-representable if and only if, for every Q-body Hamiltonian with zero ground-state energy, $\text{Tr}[\hat{P}_{Q,N}\Gamma_Q] \geq 0$. Here $\hat{P}_{Q,N}$ is the reduced Hamiltonian corresponding to $\hat{P}_N$, as defined through Eq. (10).

The necessary and sufficient condition for N-representability we have presented is not practicable because it requires determining the ground-state energy of *every* possible Hamiltonian, including the Hamiltonian of interest. Suppose, however, that one can find the ground-state energy for a few select Hamiltonians analytically. It is necessary (but not sufficient!) that $\text{Tr}[\hat{P}_{Q,N}\Gamma_Q] \geq 0$ for all the $\hat{P}_Q$ in this set. Thinking geometrically, $\text{Tr}[\hat{P}_{Q,N}\Gamma_Q] = 0$ is a hyperplane tangent to the set of N-representable Q-matrices, and so a collection of constraints with the form of $\text{Tr}[\hat{P}_{Q,N}\Gamma_Q] = 0$ constructs a convex polyhedral "hull" whose faces are tangent to the set of N-representable densities (see Fig. 1). As the number of faces in the convex polyhedron increases—that is, as the number of Hamiltonians for which the ground-state energy is known increases—the gap between the set of N-representable reduced density matrices and the polyhedral hull containing the set decreases to zero.

III. LINEAR INEQUALITIES FROM THE ORBITAL REPRESENTATION

A. The Slater Hull

As laid out in the previous section, the quest for necessary conditions for N-representability reduces to a quest for Hamiltonians whose ground-state energy is known. As shown in Eq. (15), the Hamiltonian can then be shifted so that its ground-state energy is zero, so that a necessary condition for the N-representability of the Q-matrix can be written in terms of the reduced Hamiltonian

$$\text{Tr}[\hat{P}_{Q,N}\Gamma_Q] \geq 0 \qquad (16)$$

By imposing enough linear inequalities of this type, one constructs a polyhedral hull that bounds the set of N-representable Q-matrices.

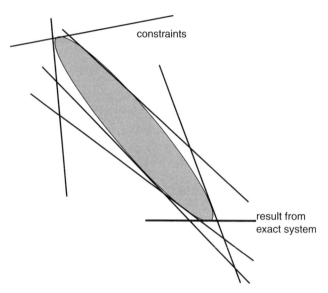

Figure 1. Pictorial representation of how constraints based on known Hamiltonians provide necessary conditions for the N-representability of density matrices. Here, the shaded region represents the convex set of N-representable density matrices.

It is easier to construct Hamiltonians whose exact solutions are known when one uses the orbital representation. Within the orbital representation, the most fundamental operator is the number operator for a spin orbital, $\hat{n}_j = a_j^+ a_j$. (a_j^+ and a_j are the operators for creating and annihilating the spin orbital $\phi_j(\mathbf{x}) = \phi_j(\mathbf{r}, \sigma)$.) When $\hat{n}_j$ operates on a Slater determinant of orbitals, $\Phi = |\phi_{i_1} \phi_{i_2} \cdots \phi_{i_N}\rangle$, the eigenvalue is one if ϕ_j is included in the Slater determinant, and zero otherwise. That is,

$$\hat{n}_j |\phi_{i_1} \phi_{i_2} \cdots \phi_{i_N}\rangle = \left(\sum_{n=1}^{N} \delta_{j,i_n} \right) |\phi_{i_1} \phi_{i_2} \cdots \phi_{i_N}\rangle \qquad (17)$$

The number operator is clearly a projector: $\hat{n}_j^2 = \hat{n}_j$. The Q-matrix can be converted to and from the orbital representation using

$$\Gamma_{i_1 \ldots i_Q; j_1 \ldots j_Q} = \iint \cdots \iint \left(\phi_{i_1}^*(\mathbf{x}_1) \cdots \phi_{i_Q}^*(\mathbf{x}_Q) \Gamma_Q(\mathbf{x}_1, \ldots \mathbf{x}_Q; \mathbf{x}'_1, \ldots, \mathbf{x}'_Q) \phi_{j_1}(\mathbf{x}'_1) \right.$$
$$\left. \cdots \phi_{j_Q}(\mathbf{x}'_Q) \right) d\mathbf{x}_1 \, d\mathbf{x}'_1 \ldots d\mathbf{x}_Q \, d\mathbf{x}'_Q \qquad (18)$$

$$\Gamma_Q(\mathbf{x}_1,\ldots,\mathbf{x}_Q;\mathbf{x}'_1,\ldots,\mathbf{x}'_Q) = \sum_{i_1\ldots i_Q=1}^{K}\sum_{j_1\ldots j_Q=1}^{K}$$
$$\times \left(\phi_{i_1}(\mathbf{x}_1)\cdots\phi_{i_Q}(\mathbf{x}_Q)\Gamma_{i_1\ldots i_Q;j_1\ldots j_Q}\phi_{j_1}^*(\mathbf{x}'_1)\cdots\phi_{j_Q}^*(\mathbf{x}'_Q)\right) \quad (19)$$

Recall that the Q-density is just the diagonal elements of the Q-matrix,

$$\rho_{i_1\ldots i_Q} = \Gamma_{i_1\ldots i_Q;i_1\ldots i_Q} \quad (20)$$
$$\rho_Q(\mathbf{x}_1,\ldots,\mathbf{x}_Q) = \Gamma_Q(\mathbf{x}_1,\ldots,\mathbf{x}_Q;\mathbf{x}_1,\ldots,\mathbf{x}_Q) \quad (21)$$

Note that there is no simple method for converting the Q-density from the spatial representation to the orbital representation and back. The Q-density in the spatial representation depends on off-diagonal elements of the Q-matrix in the orbital representation and the Q-density in the orbital representation depends on off-diagonal elements of the Q-matrix in the spatial representation.

Because the eigenfunctions of the number operators are the Slater determinants, any polynomial of number operators will also have Slater determinant eigenfunctions. Starting with a basis set of K spin orbitals, $\phi_1(\mathbf{x}), \phi_2(\mathbf{x}), \ldots, \phi_K(\mathbf{x})$, let us select a subset of R orbitals for additional scrutiny. Store the indices of this orbital subset in $\mathbf{I}_R = \{j_1, j_2, \ldots, j_R\}$.

- Any Qth-degree polynomial of the number operators for these R orbitals can be written in the form

$$\hat{L}_N^{(Q,\mathbf{I}_R)} = \sum_{i_1\in\mathbf{I}_R}\sum_{i_2\in\mathbf{I}_R}\cdots\sum_{i_Q\in\mathbf{I}_R} c_{i_1 i_2\ldots i_Q}\hat{n}_{i_1}\hat{n}_{i_2}\cdots\hat{n}_{i_Q} \quad (22)$$

To demonstrate the truth of this statement, note if any one orbital occurs more than once in Eq. (22) (i.e., if $i_q = i_r$), then using the relationship $\hat{n}_{i_q}^2 = \hat{n}_{i_q}$ will reduce the degree of this term in the polynomial. Because of this, the summation in Eq. (22) includes monomials of every degree between one (when $i_1 = i_2 = \cdots = i_Q$) and Q (when $i_1 \neq i_2 \neq \cdots \neq i_Q$). Because the eigenvalues of $\hat{L}_N^{(Q,\mathbf{I}_R)}$ should be real, we force the coefficients to be real numbers. Because each polynomial $\hat{L}_N^{(Q,\mathbf{I}_R)}$ has at most degree Q, the "energy" of this "Hamiltonian" can be evaluated using the Q-matrix. Specifically, one has

$$E[\Gamma_Q] = \text{Tr}\left[\hat{L}_{Q,N}^{(Q,\mathbf{I}_R)}\Gamma_Q\right] \quad (23)$$

To obtain an expression for the Q-electron reduced form of $\hat{L}_N^{(Q,\mathbf{I}_R)}$ that appears in this equation, we first group the monomials in $\hat{L}_N^{(Q,\mathbf{I}_R)}$ according to their degree,

$$\hat{L}_N^{(Q,\mathbf{I}_R)} = d_0 + \sum_{i_1 \in \mathbf{I}_R} d_{i_1} \hat{n}_{i_1} + \sum_{\substack{i_1 \in \mathbf{I}_R}} \sum_{\substack{i_2 \in \mathbf{I}_R \\ i_2 \neq i_1}} d_{i_1 i_2} \hat{n}_{i_1} \hat{n}_{i_2} + \cdots$$

$$+ \sum_{i_1 \in \mathbf{I}_R} \sum_{\substack{i_2 \in \mathbf{I}_R \\ i_2 \neq i_1}} \cdots \sum_{\substack{i_Q \in \mathbf{I}_R \\ i_Q \neq i_{Q-1} \neq \cdots i_1}} d_{i_1 i_2 \ldots i_Q} \hat{n}_{i_1} \hat{n}_{i_2} \cdots \hat{n}_{i_Q} \qquad (24)$$

The Q-electron reduced form of $\hat{L}_N^{(Q,\mathbf{I}_R)}$ is then

$$\hat{L}_{Q,N}^{(Q,\mathbf{I}_R)} = \frac{(N-Q)!Q!}{N!} d_0 + \frac{(N-Q)!(Q-1)!}{(N-1)!} \sum_{i_1 \in \mathbf{I}_R} d_{i_1} \hat{n}_{i_1}$$

$$+ \frac{(N-Q)!(Q-2)!}{(N-2)!} \sum_{i_1 \in \mathbf{I}_R} \sum_{\substack{i_2 \in \mathbf{I}_R \\ i_2 \neq i_1}} d_{i_1 i_2} \hat{n}_{i_1} \hat{n}_{i_2} \qquad (25)$$

$$+ \cdots + \sum_{i_1 \in \mathbf{I}_R} \sum_{\substack{i_2 \in \mathbf{I}_R \\ i_2 \neq i_1}} \cdots \sum_{\substack{i_Q \in \mathbf{I}_R \\ i_Q \neq i_{Q-1} \neq \cdots i_1}} d_{i_1 i_2 \ldots i_Q} \hat{n}_{i_1} \hat{n}_{i_2} \cdots \hat{n}_{i_Q}$$

The resulting expression is directly analogous to the expression for the reduced Hamiltonian in Eq. (10).

The eigenfunctions of $\hat{L}_N^{(Q,\mathbf{I}_R)}$ are Slater determinants and because of this it is relatively easy to determine the lowest eigenvalue of $\hat{L}_N^{(Q,\mathbf{I}_R)}$. Requiring $\mathrm{Tr}[\hat{L}_{Q,N}^{(Q,\mathbf{I}_R)} \Gamma_Q] \geq E_{\mathrm{g.s.}}(\hat{L}_N^{(Q,\mathbf{I}_R)})$ is a necessary condition for the N-representability of the Q-matrix. The (Q, R) necessary conditions for N-representability are obtained by requiring that this inequality hold for *every* polynomial of degree less than or equal to Q and for *every* possible subset of R orbitals.

- A Q-matrix, Γ_Q, is said to satisfy the *(Q,R) family of necessary conditions* for N-representability if

$$\mathrm{Tr}\left[\hat{L}_{Q,N}^{(Q,\mathbf{I}_R)} \Gamma_Q\right] \geq E_{\mathrm{g.s.}}\left(\hat{L}_N^{(Q,\mathbf{I}_R)}\right) \qquad (26)$$

for all choices of coefficients, $c_{i_1 i_2 \ldots i_Q}$, and all possible subsets of R orbitals, $\mathbf{I}_R$, in Eq. (22) [25, 26].

The (Q, R) conditions form a polyhedral hull containing the set of N-representable Γ_Q. The (Q, R) conditions either contain, or imply, every necessary condition that can be stated for the diagonal elements of the Q-matrix without using more than R distinct orbital indices. In this sense, the (Q, R) conditions are the complete set of R-orbital necessary conditions for the Q-density.

Clearly, the $(Q, R+1)$ conditions contain the (Q, R) conditions as a special case. (Just choose $c_{i_1 \ldots i_Q} = 0$ any time the orbital that is in $\mathbf{I}_{R+1}$, but not $\mathbf{I}_R$, occurs.) The most stringent necessary conditions, then, are obtained when all the orbitals in the orbital basis set are included ($R = K$). The (Q, K) family of constraints defines the Slater hull [29].

- A Q-matrix is in the *Slater hull* if it satisfies the (Q, K) constraints, where K is the number of spin orbitals in the basis set. If Γ_Q is in the Slater hull, it satisfies

$$\text{Tr}\left[\hat{L}_{Q,N}^{(Q,\mathbf{I}_K)} \Gamma_Q\right] \geq E_{\text{g.s.}}\left(\hat{L}_N^{(Q,\mathbf{I}_K)}\right) \tag{27}$$

where $\hat{L}_N^{(Q,\mathbf{I}_K)}$ is any Qth degree polynomial of the number operators in this orbital basis, $E_{\text{g.s.}}(\hat{L}_N^{(Q,\mathbf{I}_K)})$ is the smallest eigenvalue of this "Hamiltonian," and $\hat{L}_{Q,N}^{(Q,\mathbf{I}_K)}$ is the reduced Hamiltonian defined in Eq. (25).

Henceforth, we will use the statements "Γ_Q is in the Slater hull," "Γ_Q satisfies the Slater hull conditions," and "Γ_Q satisfies the (Q, K) conditions" interchangeably.

Necessary conditions based on the Slater hull have been pursued by Kummer, McRae and Davidson, Yoseloff and Kuhn, and Deza and Laurent, among others [24–26, 29, 42–44].

Even for a relatively small system, obtaining the (Q, R) conditions is computationally challenging [25]. For example, using the 2-matrix to describe the beryllium atom in a minimal basis would require the (2,10) conditions for a four-electron system. In this case, the Slater hull is a polyhedron with on the order of ten billion facets. Only small R is interesting for computational applications.

In the next section, we will show that the (R, R) necessary conditions take an especially simple form. If Γ_R satisfies the (R, R) conditions, then the Q-matrix obtained from the partial trace of Γ_R,

$$\Gamma_Q = \Gamma_{i_1\ldots i_Q;j_1\ldots j_Q} = \frac{(N-R)!R!}{(N-Q)!Q!} \sum_{i_{Q+1}=1}^{K} \sum_{i_{Q+2}=1}^{K} \cdots \sum_{i_R=1}^{K} \Gamma_{i_1\ldots i_R;j_1\ldots j_Q,i_{Q+1},\ldots i_R} \tag{28}$$

satisfies the (Q, R) conditions. This is a consequence of the fact that, in Eq. (22), polynomials of degree less than or equal to Q arise if the coefficients, $c_{i_1 i_2 \ldots i_R}$, are zero whenever the R indices have more than Q distinct values. In this way, the (Q, R) conditions arise as "special cases" implied by the (R, R) conditions. Because of the simple form of the (R, R) conditions, however, it is easier to derive the (Q, R) conditions from the (R, R) conditions than it is to derive them directly. We shall do this later, after deriving the form of the (R, R) conditions and presenting results for some special cases of interest.

The fact that every R-matrix that satisfies the (R, R) inequalities reduces to a Q-matrix that satisfies the (Q, R) conditions has clear implications for hierarchies of equations based on density matrices [45–50] and many-electron densities [2, 51]. Forcing the highest-order density matrices (or densities) that appear in these equations to satisfy the (R, R) N-representability conditions is sufficient to ensure that all of the lower-order density matrices in these equations will satisfy the (Q, R) conditions. When solving hierarchies of equations it may be better to impose N-representability conditions at the top of the hierarchy, rather than at the bottom.

B. Derivation of the (R, R) Conditions

As previously discussed, when developing necessary conditions for N-representability, it is useful to consider the subset of Hamiltonians whose ground-state energy is zero. Applying this idea to the Slater hull, the following Hamiltonians arise as important constraints:

$$\hat{P}_N^{(R,\mathbf{I}_R)}(w_{i_1}, \ldots, w_{i_R}) = \prod_{r=1}^{R} [w_{i_r} \hat{n}_{i_r} + (1 - w_{i_r})(1 - \hat{n}_{i_r})], \quad w_{i_r} = \{0, 1\} \quad (29)$$

where the weighting coefficients, w_{i_r}, are either zero or one [24–26]. For $R \leq N$ and a set of spin orbitals that contains at least $K = N + R$ functions, the ground-state energy of Eq. (29) is always zero. (Matters are complicated when $R > N$, or when the number of basis functions is less than $N + R$; in those cases there are additional constraints because it is possible that some of the Slater polynomials, $\hat{L}_N^{(R,\mathbf{I}_R)}$, that would ordinarily be negative for some choices of occupation numbers are actually positive for all subsets of N orbitals. The impossible complexity of the Slater hull conditions for large R ensures that these complications have no practical importance; hence they will be neglected.) Note that the ground state of $\hat{P}_N^{(R,\mathbf{I}_R)}(w_{i_1}, \ldots, w_{i_R})$ is highly degenerate: every Slater determinant that includes any orbital with $w_{i_r} = 0$ or omits any orbital with $w_{i_r} = 1$ is a ground-state eigenfunction of $\hat{P}_N^{(R,\mathbf{I}_R)}$.

If the expectation value of the reduced Hamiltonian, $\mathrm{Tr}\left[\hat{P}_{R,N}^{(R,\mathbf{I}_R)} \Gamma_R\right]$, is greater than or equal to zero for every Hamiltonian formed from Eq. (29), then $\mathrm{Tr}\left[\hat{L}_{R,N}^{(R,\mathbf{I}_R)} \Gamma_R\right] \geq E_{\mathrm{g.s.}}\left(\hat{L}_N^{(R,\mathbf{I}_R)}\right)$ for any polynomial of number operators depending only on the R orbitals under scrutiny, $\hat{L}_N^{(R,\mathbf{I}_R)}$. To understand why this is true, consider the set of N-electron Slater determinants that can be constructed using the subset of R orbitals contained in $\mathbf{I}_R$,

$$\Phi^{(R,\mathbf{I}_R)}(w_{i_1}, \ldots, w_{i_R}) = (w_{i_R} a_R^+)(w_{i_{R-1}} a_{R-1}^+) \cdots (w_{i_1} a_1^+) |\Phi_{\notin \mathbf{I}_R}(w_{i_1}, \ldots, w_{i_R})\rangle, \quad w_{i_r} = \{0, 1\} \quad (30)$$

$\Phi_{\notin \mathbf{I}_R}(w_{i_1}, \ldots, w_{i_R})$ represents any Slater determinant that comprises $(N - w_{i_1} - w_{i_2} - \cdots - w_{i_R})$ orbitals that are not included in the R orbitals under scrutiny. (Because R is usually chosen to be a small number and the number of one-electron basis functions is usually significantly larger than the number of electrons, there are usually many different choices for $\Phi_{\notin \mathbf{I}_R}(w_{i_1}, \ldots, w_{i_R})$.)

Consider the ensemble of states that can be constructed from these Slater determinants,

$$\Gamma_N^{(R,\mathbf{I}_R)} = \Gamma_{j_1\ldots j_N;k_1\ldots k_N}^{(R,\mathbf{I}_R)} = \sum_{\substack{w_{i_1},w_{i_2},\ldots,w_{i_R} \\ \text{choices for } \Phi_{\notin \{i_r\}}}} p_{\Phi_{\notin \mathbf{I}_R};w_{i_1},\ldots,w_{i_R}} \left| \Phi^{(R,\mathbf{I}_R)}(w_{i_1},\ldots,w_{i_R}) \right\rangle$$

$$\times \left\langle \Phi^{(R,\mathbf{I}_R)}(w_{i_1},\ldots,w_{i_R}) \right| \quad (31)$$

(Note that this ensemble includes contributions from different choices of $\Phi_{\notin \mathbf{I}_R}(w_{i_1}, \ldots, w_{i_R})$. Note also that the N-electron density matrix has been defined so that it is normalized to one.) This convex set of N-electron density matrices can be reduced to a convex set of R-electron reduced density matrices using the definition

$$\Gamma_R^{(R,\mathbf{I}_R)} \equiv \Gamma_{j_1\ldots j_R;k_1\ldots k_R}^{(R,\mathbf{I}_R)} \equiv \frac{N!}{(N-R)!R!} \sum_{j_{R+1}=1}^{K} \sum_{j_{R+2}=1}^{K} \cdots \sum_{j_N=1}^{K} \Gamma_{j_1\ldots j_N;k_1\ldots k_R j_{R+1}\ldots j_N}^{(R,\mathbf{I}_R)} \quad (32)$$

Geometrically, this set of density matrices can be pictured as an $(R+1)$-dimensional 2^R-hedron, where the vertices are defined by the pure-state Slater determinant R-matrices, where one of the $p_{\Phi_{\notin \mathbf{I}_R};w_{i_1},\ldots,w_{i_R}} = 1$. (The $R=2$ case is depicted in Fig. 2.) Each face of the 2^R-hedron lies in one of the 2^R hyperplanes defined by

$$\text{Tr}\left[\hat{P}_{R,N}^{(R,\mathbf{I}_R)}(w_{i_1},\ldots,w_{i_R})\Gamma_R\right] = 0, \quad w_{i_r} = \{0,1\} \quad (33)$$

For example, if one considers the "base" of the 2^R-hedron to include every vertex except the one in which all R of orbitals under scrutiny are occupied ($w_{i_1} = w_{i_2} = \cdots = w_{i_R} = 1$), then the base of the 2^R-hedron is contained in the hyperplane

$$\text{Tr}\left[\hat{P}_{R,N}^{(R,\mathbf{I}_R)}(1,1,\ldots,1)\Gamma_R\right] = \text{Tr}[\hat{n}_{i_1}\hat{n}_{i_2}\cdots\hat{n}_{i_R}\Gamma_R] = 0 \quad (34)$$

and the vertex that does not lie in the hyperplane satisfies

$$\text{Tr}\left[\hat{P}_N^{(R,\mathbf{I}_R)}(1,1,\ldots,1)|\Phi^{(R,\mathbf{I}_R)}(1,1\ldots,1)\rangle\langle\Phi^{(R,\mathbf{I}_R)}(1,1\ldots,1)|\right] = 1 \quad (35)$$

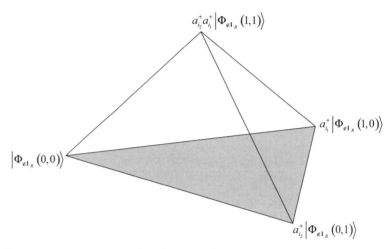

Figure 2. Pictorial representation of how one defines the Slater hull using hyperplanes. The vertices of the tetrahedron correspond to Slater determinants. The "base" of the Slater hull, which is shaded in this figure, lies in the hyperplane defined by $\text{Tr}[\hat{n}_{i_1}\hat{n}_{i_2}\Gamma_N] = 0$. Moving clockwise, starting at the front of the tetrahedron, the "sides" of the Slater hull are contained in the hyperplanes $\text{Tr}[\hat{n}_{i_1}(1-\hat{n}_{i_2})\Gamma_N] = 0$, $\text{Tr}[(1-\hat{n}_{i_1})\hat{n}_{i_2}\Gamma_N] = 0$, and $\text{Tr}[(1-\hat{n}_{i_1})(1-\hat{n}_{i_2})\Gamma_N] = 0$.

Clearly, the set of $\Gamma_R^{(R,\mathbf{I}_R)}$ defined in Eq. (32) lies on the "positive" side of all of the hyperplanes, so that

$$\text{Tr}\left[\hat{P}_{R,N}^{(R,\mathbf{I}_R)}(w_{i_1},\ldots,w_{i_R})\Gamma_R^{(R,\mathbf{I}_R)}\right] \geq 0 \qquad (36)$$

is satisfied for all 2^R possible choices of $w_{i_1},\ldots,w_{i_R} = \{0,1\}$. If one or more of the w_{i_r} is chosen to be *between* zero and one, one obtains a hyperplane that is tangent to an "edge" (if at least one of the w_{i_r} is still zero or one) or a vertex (if none of the w_{i_r} are zero or one) of the polyhedron. (Cases where any of the w_{i_r} are less than zero or greater than one correspond to Hamiltonians with negative ground-state energies unless the one-electron basis has fewer than $N+R$ orbitals.)

We are now prepared to explain why $\text{Tr}\left[\hat{L}_{R,N}^{(R,\mathbf{I}_R)}\Gamma_R\right] \geq E_{\text{g.s.}}\left(\hat{L}_N^{(R,\mathbf{I}_R)}\right)$ whenever $\text{Tr}\left[\hat{P}_{R,N}^{(R,\mathbf{I}_R)}\Gamma_R\right] \geq 0$ for all of the $\hat{P}_N^{(R,\mathbf{I}_R)}$ that one can construct based on the form in Eq. (29). Recall that the ground-state wavefunction of $\hat{L}_N^{(R,\mathbf{I}_R)}$ is a Slater determinant. This means that

$$E_{\text{g.s.}}\left(\hat{L}_N^{(R,\mathbf{I}_R)}\right) = \underbrace{\min}_{\substack{\text{Slater}\\\text{det.}}}\left\langle\Phi\left|\hat{L}_N^{(R,\mathbf{I}_R)}\right|\Phi\right\rangle \qquad (37)$$

or—since every R-matrix from a Slater determinant satisfies $\mathrm{Tr}\left[\hat{P}_{R,N}^{(R,\mathbf{I}_R)}(w_{i_1},\ldots,w_{i_R})\Gamma_R^{(R,\mathbf{I}_R)}\right] \geq 0$ (cf. Eq. (36))—one can minimize the energy with respect to the R-matrices that satisfy this constraint:

$$E_{\text{g.s.}}\left(\hat{H}_N^{(R,\mathbf{I}_R)}\right) = \min_{\left\{\Gamma_R\big|\mathrm{Tr}\left[\hat{P}_{R,N}^{(R,\mathbf{I}_R)}\Gamma_R\right]\geq 0\right\}} \mathrm{Tr}\left[\hat{L}_{R,N}^{(R,\mathbf{I}_R)}\Gamma_R\right] \qquad (38)$$

This establishes what we wished to show:

- As long as Γ_R satisfies the constraints $\mathrm{Tr}\left[\hat{P}_{R,N}^{(R,\mathbf{I}_R)}(w_{i_1},\ldots,w_{i_R})\Gamma_R\right] \geq 0$, $\mathrm{Tr}\left[\hat{L}_{R,N}^{(R,\mathbf{I}_R)}\Gamma_R\right]$ is greater than or equal to the ground-state energy for any reduced Hamiltonian, $\hat{L}_{R,N}^{(R,\mathbf{I}_R)}$, built from a polynomial of Rth degree or less from the R orbitals in $\mathbf{I}_R$, $\hat{L}_N^{(R,\mathbf{I}_R)}$.

An immediate corollary of this result is

- The (R, R) conditions for the N-representability of Γ_R are that

$$\mathrm{Tr}\left[\hat{P}_{R,N}^{(R,\mathbf{I}_R)}(w_{i_1},\ldots,w_{i_R})\Gamma_R\right] \geq 0, \quad \forall \mathbf{I}_R, w_{i_r} \in \{0,1\} \qquad (39)$$

for every possible subset of R orbitals, $\mathbf{I}_R$, and all possible choices of $w_{i_r} = \{0,1\}$. The positive semidefinite Hamiltonian, $\hat{P}_N^{(R,\mathbf{I}_R)}(w_{i_1},\ldots,w_{i_R})$, is defined in Eq. (29) and the method for constructing the reduced Hamiltonian, $\hat{P}_{R,N}^{(R,\mathbf{I}_R)}(w_{i_1},\ldots,w_{i_R})$, is based on Eq. (25).

$\mathrm{Tr}\left[\hat{P}_{R,N}^{(R,\mathbf{I}_R)}\Gamma_R\right] \geq 0$ represents the complete set of constraints on the diagonal elements of the R-matrix that can be expressed using no more than R distinct orbital indices. The (R, R) conditions are necessary, but not sufficient, for the N-representability of the R-matrix.

In fact, even the full set of (R, K) conditions is insufficient. Since the full set of Slater hull constraints is not sufficient to ensure the N-representability of the R-matrix, there exist some Γ_R that satisfy the Slater hull constraints, $\mathrm{Tr}\left[\hat{L}_{R,N}^{(R,\mathbf{I}_K)}\Gamma_R\right] \geq 0$, but give a ground-state energy below the correct energy for some Hamiltonian. We are assured, however, that (i) this problematic Hamiltonian is not an Rth-degree polynomial of number operators and (ii) the ground-state wavefunction of this Hamiltonian is not a Slater determinant.

C. Necessary Conditions from the Slater Hull; (R, R) Conditions

For each choice of w_{i_r} in Eq. (29), one obtains necessary conditions on the diagonal elements of the R-matrix (cf. Eq. (39)). We now discuss a few of the simpler (R, R) conditions. We refer the reader to the references for further details [25, 26].

1. (1, 1) Conditions

The (1,1) conditions, $\text{Tr}\left[\hat{P}_{1,N}^{(1,\mathbf{I}_1)}\Gamma_1\right] \geq 0$, can be stated in terms the reduced one-electron Hamiltonian (cf. Eq. (25)),

$$\text{Tr}[\hat{n}_i\Gamma_1] = \Gamma_{i;i} \geq 0$$
$$\text{Tr}\left[\left(\frac{1}{N} - \hat{n}_i\right)\Gamma_1\right] = 1 - \Gamma_{i;i} \geq 0 \qquad (40)$$

The above inequalities immediately yield the constraint on the possible diagonal elements of the 1-matrix, namely,

$$0 \leq \Gamma_{i;i} \leq 1 \qquad (41)$$

The 1-matrix can be diagonalized and its eigenfunctions are the natural orbitals. Equation (41) then implies that the natural orbital occupation numbers lie between zero and one, inclusive. Except for the normalization condition,

$$N = \sum_{i=1}^{K} \Gamma_{i;i} \qquad (42)$$

this is the only N-representability constraint on the 1-matrix [22, 40].

2. (2, 2) Conditions

We could obtain the (2, 2) conditions by performing the same sequence of steps we used to derive the (1, 1) conditions: construct all the N-electron $\hat{P}_N^{(2,\mathbf{I}_2)}$ Hamiltonians using Eq. (29); reduce these to the 2-electron form, $\hat{P}_{2,N}^{(2,\mathbf{I}_2)}$, using Eq. (25); then simplify inequalities of the form $\text{Tr}\left[\hat{P}_{2,N}^{(2,\mathbf{I}_2)}\Gamma_2\right] \geq 0$. Using the N-electron Hamiltonian directly, though, is less tedious and error prone, since it doesn't require multiplying out the factors in Eq. (29). That is, it is more convenient to compute the (2, 2) conditions as $\langle\Psi|\hat{P}_N^{2,\mathbf{I}_2}|\Psi\rangle$ (for a pure state) or $\text{Tr}\left[\hat{P}_N^{(2,\mathbf{I}_2)}\Gamma_N\right] \geq 0$ (for a N-electron ensemble), namely,

$$\begin{aligned}\text{Tr}[\hat{n}_i\hat{n}_j\Gamma_N] &\geq 0 \\ \text{Tr}[(1-\hat{n}_i)\hat{n}_j\Gamma_N] &\geq 0 \\ \text{Tr}[\hat{n}_i(1-\hat{n}_j)\Gamma_N] &\geq 0 \\ \text{Tr}[(1-\hat{n}_i)(1-\hat{n}_j)\Gamma_N] &\geq 0\end{aligned} \qquad (43)$$

Recall that $\text{Tr}[\Gamma_N] = 1$.

The (2, 2) constraints are associated with the first three Weinhold–Wilson constraints [23]. This is clear when we rewrite the inequalities in terms of the diagonal elements of the 2-matrix

$$\Gamma_{ij;ij} = \rho_{ij} \geq 0$$
$$\Gamma_{i;i} - \Gamma_{ij;ij} = \rho_i - \rho_{ij} \geq 0 \qquad (44)$$
$$1 - \Gamma_{i;i} - \Gamma_{j;j} + \Gamma_{ij;ij} = 1 - \rho_i - \rho_j + \rho_{ij} \geq 0$$

(The second and third inequalities in Eq. (43) are the same, except for a permutation of the indices.) Weinhold and Wilson use the fact that ρ_i represents the probability of observing an electron in orbital "i" and ρ_{ij} represents the probability that orbitals "i" and "j" are both occupied to show that each of these constraints has a straightforward probabilistic interpretation.

- $\rho_{ij} \geq 0$. The probability that orbitals "i" and "j" are both occupied is greater than or equal to zero.
- $\rho_i \geq \rho_{ij}$. The probability that orbital "i" is occupied is greater than or equal to the probability that orbital "i" *and* orbital "j" are both occupied. An alternative interpretation, based on the number operators representation in Eq. (43), is that the probability that orbital "i" is occupied *and* orbital "j" is empty is greater than or equal to zero.
- $1 \geq \rho_i + \rho_j - \rho_{ij}$. Based on Eq. (43), we infer that this inequality states that the probability that both orbital "i" and orbital "j" are empty is greater than or equal to zero. An alternative interpretation, due to Weinhold and Wilson, is that the probability that orbital "i" *or* orbital "j" is occupied is less than or equal to one.

Two other constraints are

- $\rho_{ij} = \rho_{ji}$ (symmetry).
- $\rho_{ii} = 0$ (Pauli exclusion relation).

These conditions follow directly from the orbital resolution of the reduced density matrix.

3. (3, 3) Conditions

The (3, 3) conditions are, in the orbital representation,

$$\begin{aligned}
\text{Tr}[\hat{n}_i \hat{n}_j \hat{n}_k \Gamma_N] \geq 0 &\rightarrow \rho_{ijk} \geq 0 \\
\text{Tr}[\hat{n}_i \hat{n}_j (1 - \hat{n}_k) \Gamma_N] \geq 0 &\rightarrow \rho_{ij} - \rho_{ijk} \geq 0 \\
\text{Tr}[\hat{n}_i (1 - \hat{n}_j)(1 - \hat{n}_k) \Gamma_N] \geq 0 &\rightarrow \rho_i - \rho_{ij} - \rho_{ik} + \rho_{ijk} \geq 0 \qquad (45) \\
\text{Tr}[(1 - \hat{n}_i)(1 - \hat{n}_j)(1 - \hat{n}_k) \Gamma_N] \geq 0 &\rightarrow \begin{pmatrix} 1 - \rho_i - \rho_j - \rho_k \\ + \rho_{ij} + \rho_{ik} + \rho_{jk} - \rho_{ijk} \end{pmatrix} \geq 0
\end{aligned}$$

The probabilistic interpretation of these conditions is very similar to the interpretation of the (2, 2) conditions. The analysis is readily extended to higher-order densities, where one obtains $R + 1$ unique constraints on the R-matrix.

D. Necessary Conditions from the Slater Hull; (Q, R) Conditions

As discussed in Section III. A, because the set of R-orbital Rth-degree polynomials contains the set of Q-orbital Rth-degree polynomials ($Q < R$), the (R, R) conditions contain the (Q, R) conditions. That is, if an R-matrix satisfies the (R, R) conditions, then the associated Q-matrix (obtained by partial tracing of Γ_R) satisfies the (Q, R) conditions. It should be possible, then, to use the (R, R) conditions in the previous section to derive the (Q, R) conditions [26].

To this end, note that adding the first two (2, 2) inequalities (cf. Eq. (44)) recovers the first (1, 1) inequality:

$$\rho_{ij} + (\rho_i - \rho_{ij}) = \rho_i \geq 0 \tag{46}$$

Similarly, adding the second two (2, 2) inequalities gives the second (1, 1) inequality:

$$(\rho_j - \rho_{ij}) + (1 - \rho_i - \rho_j + \rho_{ij}) = 1 - \rho_i \geq 0 \tag{47}$$

Since the (1, 1) conditions were sufficient to ensure the N-representability of the 1-matrix, it is reassuring that no new constraints were obtained.

It is not difficult to see that the (2, 2) inequalities can be derived from appropriate combinations of the (3, 3) inequalities in Eq. (45). However, there are (2, 3) inequalities that are not contained in the (2, 2) inequalities. For example, adding the first and fourth inequalities in Eq. (45) provides the new constraint

$$\rho_{ijk} + (1 - \rho_i - \rho_j - \rho_k + \rho_{ij} + \rho_{ik} + \rho_{jk} - \rho_{ijk}) = 1 - \rho_i - \rho_j - \rho_k + \rho_{ij} + \rho_{ik} + \rho_{jk} \geq 0 \tag{48}$$

Similarly, adding the second and third (3, 3) inequalities gives

$$(\rho_{jk} - \rho_{ijk}) + (\rho_i - \rho_{ij} - \rho_{ik} + \rho_{ijk}) = \rho_i - \rho_{ij} - \rho_{ik} + \rho_{jk} \geq 0 \tag{49}$$

These inequalities are the fourth and fifth Weinhold–Wilson constraints. These (2, 3) inequalities have a probabilistic interpretation very similar to the (2, 2) inequalities. Returning to the fundamental number operators form of the (3, 3) inequalities (cf. Eq. (45)), we see that Eq. (48) was generated from

$$\text{Tr}\left[\left(\hat{n}_i \hat{n}_j \hat{n}_k + (1 - \hat{n}_i)(1 - \hat{n}_j)(1 - \hat{n}_k)\right) \Gamma_N\right] \geq 0 \tag{50}$$

so it is the probability that all three orbitals are filled *or* all three orbitals are empty. Similarly, Eq. (49) comes from

$$\text{Tr}\big[\big((1-\hat{n}_i)\hat{n}_j\hat{n}_k + \hat{n}_i(1-\hat{n}_j)(1-\hat{n}_k)\big)\Gamma_N\big] \geq 0 \tag{51}$$

So it is the probability that orbital "i" is empty and orbitals "j" and "k" are filled *or* that orbital "i" is filled and orbitals "j" and "k" are empty. It is interesting that these particular three-electron correlations can be evaluated using the two-electron reduced density matrix.

Inequalities (49) and (48), along with the (2, 2) inequalities in Eq. (44), constitute the (2, 3) conditions. This is the entire family of restrictions on the diagonal elements of the 2-matrix that can be derived from polynomials containing no more than three distinct orbital indices.

To derive the (2, 4) conditions from the (4, 4) conditions, one first derives the (3, 4) conditions by adding pairs of (4, 4) inequalities. The (2, 4) conditions are then derived by taking appropriate linear combinations of the (3, 4) inequalities [26]. The (2, 5) conditions can be derived in a similar way [26]. Compared to the mathematically unwieldy and computationally costly alternative methods for obtaining (Q, R) conditions [24, 25, 43], deriving (Q, R) conditions directly from the (R, R) conditions seems much simpler. The (2, 5) conditions are quite complicated, however, and as R increases further the number of distinct (Q, R) conditions burgeons toward impracticality [25].

E. Sufficiency of the Slater Hull Constraints for Diagonal Elements of Density Matrices

The Slater hull constraints represent the entire family of N-representability constraints that can be expressed using only the diagonal elements of the reduced density matrix [25, 43]. That is, the complete set of (Q, K) conditions is necessary and sufficient to ensure the N-representability of the Q-density.

- $\rho_{i_1\ldots i_Q}$ is an N-representable Q-density if and only if it satisfies all of the Slater hull constraints. That is, $\rho_{i_1\ldots i_Q}$ is N-representable if and only if it satisfies the (Q, K) conditions, where K is the number of spin orbitals in the basis set.

People rarely discuss the N-representability of the diagonal of a density matrix. The diagonal elements of a density matrix are N-representable if and only if there exists some N-electron ensemble that reproduces those elements. The preceding statement asserts that if the diagonal elements of the Q-matrix, $\Gamma_{i_1\ldots i_Q;i_1\ldots i_Q}$, satisfy the (Q, K) conditions, then there always exists an N-electron ensemble with these diagonal elements. Thus, if $\rho_{i_1\ldots i_Q}$ satisfies the (Q, K) conditions, then there must be an ensemble-N-representable Q-matrix with those diagonal elements [25, 43]:

$$\rho_{i_1\ldots i_Q} = \int\int\int\int \left(\phi_{i_1}^*(\mathbf{r}_1)\cdots\phi_{i_Q}^*(\mathbf{r}_Q)\Gamma_Q(\mathbf{r}_1,\ldots,\mathbf{r}_Q;\mathbf{r}'_1,\ldots,\mathbf{r}'_Q)\phi_{i_1}(\mathbf{r}'_1)\right.$$
$$\left.\cdots\phi_{i_Q}(\mathbf{r}'_Q)\right)d\mathbf{r}_1\,d\mathbf{r}'_1\cdots d\mathbf{r}_Q\,d\mathbf{r}'_Q \tag{52}$$

In fact, we have a stronger result: if $\rho_{i_1\ldots i_Q}$ satisfies the (Q, K) conditions, then the Q-matrix constructed from $\rho_{i_1\ldots i_Q}$ by setting all the off-diagonal matrix elements equal to zero,

$$\Gamma_{i_1\ldots i_Q;j_1\ldots j_Q} = \begin{cases} \rho_{i_1\ldots i_Q}, & i_1=j_1,\ldots,i_Q=j_Q \\ 0, & \text{otherwise} \end{cases} \tag{53}$$

is N-representable. This is true because any $\rho_{i_1\ldots i_Q}$ that satisfies the (Q, K) conditions can always be derived from an ensemble of Slater determinants,

$$\rho_{i_1\ldots i_Q} = \frac{1}{Q!}\sum_i p_i \left\langle \Phi_i \left| a_{i_1}^+ a_{i_2}^+ \cdots a_{i_Q}^+ a_{i_Q} a_{i_{Q-1}} \cdots a_{i_1} \right| \Phi_i \right\rangle \tag{54}$$

and the Q-matrix from an ensemble of Slater determinants has no off-diagonal elements. Because the wavefunction $|\Psi\rangle = \sum_i \sqrt{p_i} e^{i\phi_i} |\Phi_i\rangle$ (ϕ_i is an arbitrary phase factor) also gives $\rho_{i_1\ldots i_Q}$, every Q-density that satisfies the (Q, K) conditions is actually pure-state N-representable. (Note that the pure-state and ensemble precursors of the Q-density give different off-diagonal elements for the Q-matrix.)

To derive this key result, we construct the set of Q-matrices associated with the Slater determinants in this orbital basis, $\mathbf{S}^{(Q,\mathbf{I}_K)}$. Every element in $\mathbf{S}^{(Q,\mathbf{I}_K)}$ has the form of Eq. (52), where the reduced density matrix is written in terms of Slater determinants (cf. Eq. (32)). $\mathbf{S}^{(Q,\mathbf{I}_K)}$ is convex because weighted averages of elements in $\mathbf{S}^{(Q,\mathbf{I}_K)}$ are also in $\mathbf{S}^{(Q,\mathbf{I}_K)}$: this follows directly from the fact that Eq. (31) is a convex linear combination of Q-densities from Slater determinants. Geometrically, $\mathbf{S}^{(Q,\mathbf{I}_K)}$ is a polyhedron and its vertices correspond to pure states.

The Qth-degree, K-orbital Hamiltonians (cf. Eq. (22)) generate hyperplanes,

$$\text{Tr}\left[\hat{L}_{Q,N}^{(Q,\mathbf{I}_K)}\Gamma_Q\right] \geq E_{\text{g.s.}}\left(\hat{L}_N^{(Q,\mathbf{I}_K)}\right) \tag{55}$$

that are tangent to the surface of $\mathbf{S}^{(Q,\mathbf{I}_K)}$. Any Q-density that is not in $\mathbf{S}^{(Q,\mathbf{I}_K)}$ must lie on the "wrong side" of at least one of these hyperplanes. If it does so, then one of the (Q, K) conditions will not be met. Thus, if a Q-density meets all the (Q, K) conditions, it must be expressible in the form of Eq. (54). This establishes that the (Q, K) conditions are necessary and sufficient for the N-representability of the Q-electron distribution function. Although the implementation of the (Q, R) conditions with large R is intractable, a sequence of calculations based on the (Q, Q), $(Q, Q+1)$, $(Q, Q+2)$, ... conditions will converge to the correct result from below (if the kinetic energy functional is exact). The convergence is monotonic because, as shown in Section III.A, the (Q, R) conditions are contained in the (Q, S) conditions whenever $S > R$. This sequence of calculations converges from below because the imposed N-representability conditions are necessary, but not sufficient. One would hope that rapid convergence to the correct result might be achieved.

This result suggests that building Q-densities from Slater determinants may be a useful computational technique.

Although every Q-density in the Slater hull is N-representable, not every Q-matrix that satisfies the Slater hull constraints is N-representable. That is, if $\rho_{i_1...i_Q} = \Gamma_{i_1...i_Q;i_1...i_Q}$ satisfies the (Q, K) conditions, some choices for the off-diagonal elements of the density matrix $\Gamma_{i_1...i_Q;j_1...j_Q}$ are N-representable, but most are not. (As shown in Eq. (53), one of the N-representable choices is to choose all the off-diagonal elements to be zero.) Simply stated, when only the diagonal elements of the reduced density matrix are considered, the Slater hull is equal to the set of N-representable Q-densities, which is equal to the set of Q-densities from ensembles of Slater determinants. However, when off-diagonal elements of the reduced density matrix are considered, one finds that the Slater hull contains the set of N-representable Q-matrices, which contains the set of Q-matrices from ensembles of Slater determinants. These results expose the need for constraints on the off-diagonal elements of reduced density matrices.

F. Constraints on Off-Diagonal Elements from Unitary Transformation of the Slater Hull Constraints

Note that the (Q, R) conditions apply for any choice of the spin-orbital basis. Because the diagonal elements of the Q-matrix in one orbital basis depend on the off-diagonal elements of the Q-matrix in another orbital basis,

$$\rho_{a_1,...a_Q} = \sum_{\substack{i_1...i_Q \\ j_1...j_Q}} \left(U_{a_1 i_1} U_{a_2 i_2} \cdots U_{a_Q i_Q} \right) \Gamma_{i_1...i_Q;j_1...j_Q} \left(U_{j_1 a_1} U_{j_2 a_2} \cdots U_{j_Q a_Q} \right) \qquad (56)$$

the (Q, R) conditions actually imply constraints on the off-diagonal elements of the reduced density matrix. In Eq. (56), U_{ai} denotes the unitary transformation of spin orbitals from one basis to another,

$$\psi_a(\mathbf{r}, \sigma) = \sum_{i=1}^{K} U_{ai} \phi_i(\mathbf{r}, \sigma) \qquad (57)$$

Equation reveals that it is impossible to determine the Q-density in one orbital basis directly from the Q-density in a different orbital basis. Given one N-representable Q-density, $\rho_{i_1\ldots i_Q}$, it is reasonable to choose the off-diagonal elements of the Q-matrix, $\Gamma_{i_1\ldots i_Q j_1\ldots j_Q}$, so that the Q-densities in other orbital basis sets are also N-representable. That is, it is reasonable to require that $\Gamma_{i_1\ldots i_Q j_1\ldots j_Q}$ satisfy the (Q, K) conditions in *every* orbital basis set. This is clearly a necessary condition for the N-representability of $\Gamma_{i_1\ldots i_Q j_1\ldots j_Q}$. Unfortunately, even the complete set of (Q, K) conditions in every orbital basis is still insufficient to ensure the N-representability of the Q-matrix. This contrasts with the pleasing result from the last section, where imposing the (Q, K) conditions in *any* orbital basis sufficed to ensure the N-representability of the Q-density.

G. Constraints on Off-Diagonal Elements from Other Positive-Definite Hamiltonians

Because the Slater hull constraints are insufficient to ensure N-representability, it is important to find additional methods for constraining the off-diagonal elements of the density matrix. Obtaining constraints that supersede the Slater hull requires considering Hamiltonians with a more general form than polynomials of number operators. As discussed in Section III.A, matters are especially simple if the Hamiltonian has nonnegative eigenvalues, because then the necessary conditions for N-representability take the form

$$\mathrm{Tr}\left[\hat{P}_N \Gamma_N\right] \geq 0 \qquad (58)$$

If the Hamiltonian $\hat{P}_N$ contains at most Q-body terms, then Eq. (58) will produce a necessary condition on the Q-matrix.

Positive semidefinite Hamiltonians can be constructed by taking an operator that depends on Q creation/annihilation operators and multiplying the operator by its Hermitian transpose, $\hat{P}_Q = \hat{A}_Q^\dagger \hat{A}_Q$. Q-body Hamiltonains can be constructed, for example, by choosing

$$\hat{A}_{2,p} = \sum_{i=1}^{K}\sum_{j=1}^{K} p_{ij} a_j a_i \qquad (59)$$

$$\hat{A}_{2,q} = \sum_{i=1}^{K}\sum_{j=1}^{K} q_{ij} a_j^+ a_i^+ \qquad (60)$$

$$\hat{A}_{2,g} = \sum_{i=1}^{K}\sum_{j=1}^{K} g_{ij} a_j^+ a_i \qquad (61)$$

hence

$$\sum_{i,j=1}^{K} \sum_{i',j'=1}^{K} \left(p_{i'j'}^* \mathrm{Tr}\left[a_{j'}^+ a_{i'}^+ a_i a_j \Gamma_N \right] p_{ij} \right) \geq 0 \tag{62}$$

$$\sum_{i,j=1}^{K} \sum_{i',j'=1}^{K} \left(q_{i'j'}^* \mathrm{Tr}\left[a_{j'} a_{i'} a_i^+ a_j^+ \Gamma_N \right] q_{ij} \right) \geq 0 \tag{63}$$

$$\sum_{i,j=1}^{K} \sum_{i',j'=1}^{K} \left(g_{i'j'}^* \mathrm{Tr}\left[a_{j'}^+ a_{i'} a_i^+ a_j \Gamma_N \right] g_{ij} \right) \geq 0 \tag{64}$$

These constraints are the P condition, Q condition, and G condition, respectively [22, 40]. Because $\mathrm{Tr}\left[a_{j'}^+ a_{i'}^+ a_i a_j \Gamma_N \right]$ is just the 2-matrix, the P condition reduces to the requirement that $\mathbf{p}^\dagger \Gamma_2 \mathbf{p} \geq 0$ for all choices of the vector $\mathbf{p}$. Thus the P condition reduces to the requirement that the 2-matrix is positive semidefinite. Similarly, the Q condition states that the "hole" 2-matrix (i.e., the 2-matrix for the unfilled orbitals) is positive semidefinite. The G condition can be interpreted if one recognizes that

$$\sum_{i=1}^{K} \sum_{j=1}^{K} a_j^+ g_{ij} a_i = \sum_{i=1}^{K} \sum_{j=1}^{K} a_j^+ \langle \phi_j | \hat{g} | \phi_i \rangle a_i \tag{65}$$

is the operator associated with a one-electron Hamiltonian, $\hat{g}(\mathbf{x})$. The G condition then states that the square of a one-electron Hamiltonian is positive semidefinite,

$$\hat{H} \equiv \left| \sum_{i=1}^{N} \hat{g}(\mathbf{x}_i) \right|^2 \succeq 0 \tag{66}$$

Recall that the (2, 2) conditions contain all the N-representability constraints on the diagonal element of the 2-matrix that can be stated without reference to more than two orbitals.

Accordingly, the diagonal elements of the P, Q, and G Hamiltonians should give results consistent with the (2, 2) conditions. This occurs. For example, using the commutation rules for creation and annihilation operators, the diagonal elements of the P Hamiltonian become

$$a_j^+ a_i^+ a_i a_j = -a_j^+ a_i^+ a_j a_i = -a_j^+ \left(\delta_{ij} - a_j a_i^+ \right) a_i = \hat{n}_j \hat{n}_i - \delta_{ij} \hat{n}_i = \hat{n}_j \hat{n}_i \tag{67}$$

Because the diagonal elements of a positive semidefinite matrix are never negative, this implies the first (2, 2) condition, $\mathrm{Tr}\left[\hat{n}_j \hat{n}_i \Gamma_N \right] \geq 0$.

The other (2, 2) conditions are similarly obtained. In particular, the Q condition implies that $\text{Tr}\left[(1-\hat{n}_j)(1-\hat{n}_i)\Gamma_N\right] \geq 0$; the G condition implies that $\text{Tr}\left[\hat{n}_j(1-\hat{n}_i)\Gamma_N\right] \geq 0$.

While the P, Q, and G conditions give no new constraints on the diagonal elements of the 2-matrix, they provide important constraints for the off-diagonal elements.

Based on the preceding analysis, one can develop an algorithm for generalizing any of the (Q, R) conditions to constrain off-diagonal elements of the Q-matrix. In general, an operator with the form of Eq. (29) becomes

$$\prod_{i=1}^{Q} \left[w_i \hat{n}_i + (1-w_i)(1-\hat{n}_i)\right] \to a_{Q'}^{\text{sgn}(w_i)} \cdots a_{2'}^{\text{sgn}(w_i)} a_{1'}^{\text{sgn}(w_i)} a_1^{\text{sgn}(1-w_i)} a_2^{\text{sgn}(1-w_i)} \cdots a_Q^{\text{sgn}(1-w_i)} \tag{68}$$

where the choice between creation and annihilation operators is made according to the rule

$$a_i^{\text{sgn}(x)} \to \begin{cases} a_i, & x = 0 \\ a_i^+, & x = 1 \end{cases} \tag{69}$$

Recall that in Eq. (68), the w_i are all either zero or one.

Using this method, the first (2, 3) condition (cf. Eq. (50)) implies that

$$\sum_{i,j,k=1}^{K} \sum_{i',j',k'=1}^{K} \left(\left(t_{i'j'k'}^{(1)}\right)^* \text{Tr}\left[\left(a_{i'}^+ a_{j'}^+ a_{k'}^+ a_k a_j a_i + a_{i'} a_{j'} a_{k'} a_k^+ a_j^+ a_i^+\right) \Gamma_2\right] t_{ijk}^{(1)}\right) \geq 0 \tag{70}$$

and the second (2, 3) condition (cf. Eq. (51)) implies that

$$\sum_{i,j,k=1}^{K} \sum_{i',j',k'=1}^{K} \left(\left(t_{i'j'k'}^{(2)}\right)^* \text{Tr}\left[\left(a_{i'}^+ a_{j'}^+ a_{k'} a_k^+ a_j a_i + a_{i'}^+ a_{j'} a_{k'} a_k^+ a_j^+ a_i\right) \Gamma_2\right] t_{ijk}^{(2)}\right) \geq 0 \tag{71}$$

These N-representability constraints are called the $T1$ (Eq. (70)) and $T2$ (Eq. (71)) conditions [52]. Calculations with these constraints give dramatically better results than calculations using only the P, Q, and G conditions [35, 53]. From the standpoint of conventional quantum chemistry, this is not that surprising: one would expect good results from constraints that include three-electron operators, since these constraints help ensure that the form of the 2-matrix is consistent with a proper representation of three-electron correlations.

Although the same algorithm suffices to derive off-diagonal analogues for any of the $(2, R)$ conditions, the complexity of the resulting expressions increases as R becomes larger. For example, one of the generalized (2, 4) constraints is

$$\sum_{i,j,k=1}^{K} \sum_{i',j',k'=1}^{K} \left(q_{i'j'k'}^{(1)}\right)^* \text{Tr} \left[\begin{pmatrix} 3a_{l'}^+ a_{k'}^+ a_{j'}^+ a_{i'}^+ a_i a_j a_k a_l + a_{l'} a_{k'}^+ a_{j'} a_{i'}^+ a_i^+ a_j^+ a_k^+ a_l^+ \\ + a_{l'} a_{k'}^+ a_{j'}^+ a_{i'}^+ a_i a_j a_k a_{l}^+ + a_{l'}^+ a_{k'} a_{j'}^+ a_{i'}^+ a_i a_j a_k^+ a_l \\ + a_{l'}^+ a_{k'}^+ a_{j'} a_{i'}^+ a_i a_j^+ a_k a_l + a_{l'}^+ a_{k'}^+ a_{j'}^+ a_{i'} a_i^+ a_j a_k a_l \end{pmatrix} \Gamma_2 \right] q_{ijk}^{(1)} \geq 0 \tag{72}$$

The other four generalized (2, 4) constraints have similarly daunting complexity.

IV. LINEAR INEQUALITIES FROM THE SPATIAL REPRESENTATION

The orbital representation is not used in most of the recent work on computational methods based on diagonal elements of density matrices. This is partly for historical reasons—most of the work has been done by people trained in density functional theory—and partly this is because most of the available kinetic energy functionals are known only in first-quantized form. For example, the popular generalized Weisacker functional [2, 7–11],

$$T_w^{(Q)} = \frac{(N-Q)!Q!}{2(N-1)!} \iint \cdots \int \left|\nabla_1 \sqrt{\rho_Q(\mathbf{r}_1, \mathbf{r}_2, \ldots, \mathbf{r}_Q)}\right|^2 d\mathbf{r}_1 d\mathbf{r}_2 \cdots d\mathbf{r}_Q \tag{73}$$

is not readily expressible in second-quantized form.

Unfortunately, the N-representability constraints from the orbital representation are not readily generalized to the spatial representation. A first clue that the N-representability problem is more complicated for the spatial basis is that while every N-representable Q-density can be written as a weighted average of Slater determinantal Q-densities in the orbital resolution (cf. Eq. (54)), this is clearly not true in the spatially resolved formulation. For example, the pair density ($Q = 2$) of any real electronic system will have a cusp where electrons of opposite spin coincide; but a weighted average of Slater determinantal pair densities,

$$\rho_Q(\mathbf{y}_1, \mathbf{y}_2) = \sum_i p_i \left(\frac{1}{2} \left\langle \Phi_i \left| \sum_{j_1=1}^{N} \sum_{\substack{j_2=1 \\ j_2 \neq j_1}}^{N} \delta(\mathbf{x}_{j_1} - \mathbf{y}_1) \delta(\mathbf{x}_{j_2} - \mathbf{y}_2) \right| \Phi_i \right\rangle \right) \tag{74}$$

cannot reproduce this behavior.

A second clue arises when one realizes that among the entire set of (Q, Q) conditions, only the simplest condition,

$$\text{Tr}\left[\hat{n}_{i_1} \hat{n}_{i_2} \cdots \hat{n}_{i_Q} \Gamma_Q\right] \geq 0 \tag{75}$$

works for spatial basis functions. (Replacing $1 - a_i^+ a_i$ with its analogue using the field operators, $1 - \psi^+(\mathbf{y}_i)\psi(\mathbf{y}_i)$, yields no new insight because the probability of observing an electron at any one point, $\mathbf{x}_i$, is always infinitesimal.) If one simply replaces the number operator for orbitals with the number operator for points in space, Eq. (75) produces the trivial nonnegativity constraint

$$\rho_Q(\mathbf{x}_1, \mathbf{x}_2, \ldots, \mathbf{x}_Q) \geq 0 \tag{76}$$

Recall that the spatial representation of the Q-density actually depends on the *off-diagonal* elements of the density matrix in the orbital representation. (See Eqs. (18)–(21) and the surrounding discussion.) This suggests that some progress can be made by using the N-representability constraints for *off-diagonal* elements in the density matrix. If one chooses the one-electron Hamiltonian associated with the G condition to be a simple function, then one finds that [22, 28, 54]

$$\sum_i p_i \left\langle \Psi_i \left\| \sum_{j=1}^N g(\mathbf{x}_j) \right\|^2 \Psi_i \right\rangle \geq 0$$

$$\iint g^*(\mathbf{x}_2)\rho_2(\mathbf{x}_1, \mathbf{x}_2) g(\mathbf{x}_1) d\mathbf{x}_1 d\mathbf{x}_2 - \int |g(\mathbf{x}_1)|^2 \rho(\mathbf{x}_1) d\mathbf{x}_1 \geq 0 \tag{77}$$

$$\iint g^*(\mathbf{x}_2)[\rho_2(\mathbf{x}_1, \mathbf{x}_2)(1 - \delta(\mathbf{x}_2 - \mathbf{x}_1))] g(\mathbf{x}_1) d\mathbf{x}_1 d\mathbf{x}_2 \geq 0$$

The argument in Eq. (77) can be generalized to higher-order electron distribution functions [28]. Unfortunately, the other N-representability conditions in Section III.G do not seem amenable to this approach.

V. LINKING THE ORBITAL AND SPATIAL REPRESENTATIONS

A. Review of Orbital-Based Density Functional Theory

It is clear from the preceding sections that the powerful N-representable constraints from the orbital representation do not extend to the spatial representation. This suggests reformulating the variational principle in Q-density functional theory in the orbital representation.

To motivate the form of orbital-based Q-density functional theory, it is useful to start with the familiar case of 1-density functional theory, where the orbital representation is well established [55].

One of the simplest variational methods in density functional theory was already mentioned in Section I: define the pseudo-wavefunction

$\chi(\mathbf{x}) = (1/\sqrt{N})\sqrt{\rho(\mathbf{x})}$ and then minimize the energy expression [12–16]

$$E_{\text{g.s.}}[v;N] = \min_{\langle\chi|\chi\rangle=1}\left(N\left\langle\chi\left|-\frac{\nabla^2}{2}\right|\chi\right\rangle + \int \rho[\chi,\mathbf{x}]v(\mathbf{x})d\mathbf{x} + J[\rho[\chi]] + E_{xc}[\rho[\chi]] + T_{\text{Pauli}}[\rho[\chi]]\right)$$

$$\rho[\chi,\mathbf{x}] = N|\chi(\mathbf{x})|^2 \tag{78}$$

The kinetic energy functional in this case is the Weisacker kinetic energy [56],

$$T_w^{(1)}[\rho] = N\int \frac{1}{2}\left|\nabla\sqrt{\rho(\mathbf{x})}\right|^2 d\mathbf{x} \tag{79}$$

which is inconsistent with the Pauli principle because it assumes all the electrons occupy the same orbital. The so-called Pauli kinetic energy term, $T_{\text{Pauli}}[\rho]$, corrects the Weisacker kinetic energy for the effects of the Pauli exclusion principle. Unfortunately, it is hard to find good approximations for $T_{\text{Pauli}}[\rho]$ and neglecting this term altogether gives terrible results.

The method based on the Weisacker functional can be improved if one chooses a computational ansatz that imposes the Pauli principle. For every N-representable electron density, $\rho(\mathbf{x})$, there exists some Slater determinant with that density. More generally, there exists an ensemble of N-electron Slater determinants,

$$\Gamma_N^{\text{Sl}} = \sum_i p_i|\Phi_i\rangle\langle\Phi_i| \tag{80}$$

with that density,

$$\rho[\Gamma_N^{\text{Sl}},\mathbf{x}] = \sum_i p_i\left\langle\Phi_i\left|\sum_{j=1}^N \delta(\mathbf{x}_j-\mathbf{x})\right|\Phi_i\right\rangle \tag{81}$$

This suggests replacing Eq. (78) with

$$E_{\text{g.s.}}[v;N] = \min_{\Gamma_N^{\text{Sl}}}\left(\text{Tr}[\hat{T}\Gamma_N^{\text{Sl}}] + \int \rho[\Gamma_N^{\text{Sl}},\mathbf{x}]v(\mathbf{x})d\mathbf{x} + J[\rho[\Gamma_N^{\text{Sl}}]] + E_x[\rho[\Gamma_N^{\text{Sl}}]] + E_c[\rho[\Gamma_N^{\text{Sl}}]]\right) \tag{82}$$

The ensemble search in Eq. (82) is the Kohn–Sham procedure, generalized to allow fractional orbital occupation numbers [55, 57–59]. Equation (82) can thus be rewritten in the form

$$E_{\text{g.s.}}[v; N] = \min_{\substack{0 \leq \rho_i \leq 1 \\ \langle \phi_i | \phi_j \rangle = \delta_{ij}}} \left(\sum_{i=1}^{\infty} \rho_i \left\langle \phi_i \left| -\frac{\nabla^2}{2} \right| \phi_i \right\rangle + \int \rho[\rho_i, \phi_i; \mathbf{x}] v(\mathbf{x}) d\mathbf{x} + J[\rho_i, \phi_i] \right.$$

$$\left. + E_x[\rho_i, \phi_i] + E_c[\rho_i, \phi_i] \right) \tag{83}$$

where

$$\rho[n_i, \phi_i; \mathbf{x}] = \sum_{i=1}^{\infty} \rho_i |\phi_i(\mathbf{x})|^2 \tag{84}$$

Note that the energy is minimized with respect to all choices of the orbital basis and subject to the $(1, K)$ conditions on $\rho_i = \Gamma_{i;i}$; this ensures that there exists an ensemble of Slater determinants with the desired electron density. Because an ensemble average of Slater determinants does not describe electron correlation, these variational energy expressions include a correlation functional, $E_c[\rho]$, which corrects the energy for the effects of electron correlation. Reasonable approximations for $E_c[\rho]$ exist, though they tend to work only in conjunction with approximate exchange-energy functionals, $E_x[\rho]$.

The computational procedure in Eq. (82) can also be written from a dual perspective, in which the Kohn–Sham potential is the fundamental descriptor [60]. In this perspective, one solves the Kohn–Sham equations,

$$\left(-\frac{\nabla^2}{2} + w(\mathbf{x}) \right) \phi_i(\mathbf{x}) = \varepsilon_i \phi_i(\mathbf{x}) \tag{85}$$

and writes the electron density in terms of the Kohn–Sham orbitals and orbital occupation numbers as

$$\rho[w; \mathbf{x}] = \sum_{i=1}^{\infty} n_i |\phi_i(\mathbf{x})|^2 \tag{86}$$

The variational principle is then restated as

$$E_{\text{g.s.}}[v; N] = \min_{\substack{w(\mathbf{x}) \\ 0 \leq n_i \leq 1}} \left(\sum_{i=1}^{\infty} n_i \left\langle \phi_i \left| -\frac{\nabla^2}{2} \right| \phi_i \right\rangle + \int \rho[w; \mathbf{x}] v(\mathbf{x}) d\mathbf{x} + J[\rho[w]] \right.$$

$$\left. + E_x[\rho[w]] + E_c[\rho[w]] \right) \tag{87}$$

In conventional 1-density functional theory, the approaches based on Eqs. (82), (83), and (87) all give identical results [61, 62].

Before generalizing these results to the k-density, the interpretation of $E_c[\rho]$ should be scrutinized carefully. Every N-representable electron density corresponds to many possible N-electron ensembles. One of these N-electron ensembles, $\Gamma_N^{\text{g.s.}}$, corresponds to the ground state of the system of interest. However, for computational expediency, we do not use the (unknown) ground-state ensemble to compute the kinetic energy but, instead, we select a Slater determinantal density matrix, Γ_N^{Sl}, with the same electron density, which corresponds to the ground state of a system of noninteracting electrons in some different potential, $w(\mathbf{r})$ (cf. Eq. (85)). The energy of Γ_N^{Sl} is greater than the true ground-state energy, so obtaining the correct energy requires adding a correction,

$$E_c[\rho] = E[\Gamma_N^{\text{g.s.}}] - E[\Gamma_N^{\text{Sl}}] \tag{88}$$

The Kohn–Sham scheme then provides a mapping from the true interacting system to a Slater determinantal approximation,

$$\Gamma_N^{\text{g.s.}} \to \rho(\mathbf{r}) \to \Gamma_N^{\text{Sl}} \tag{89}$$

Obtaining a useful functional for $E_c[\rho]$ requires (approximately) inverting this mapping using, for example, the adiabatic connection [63–65].

B. Orbital-Based Q-Density Functional Theory

Each of these approaches to the density functional theory can be generalized to Q-density functional theory. In Section I, we mentioned the commonly considered generalization of the Weisacker ansatz, namely,

$$E_{\text{g.s.}}[v; N] = \min_{\langle \chi | \chi \rangle = 1} \left(N \left\langle \chi \left| -\frac{\nabla_1^2}{2} \right| \chi \right\rangle + T_{\text{Pauli}}^{(Q)}[\rho_Q[\chi]] + \frac{(N-Q)!Q!}{2(N-1)!} \right.$$

$$\left. \int \cdots \int \rho_Q[\chi; \mathbf{x}_1 \ldots, \mathbf{x}_Q] \left(v(\mathbf{x}_1) + v(\mathbf{x}_2) + \frac{N-1}{|\mathbf{x}_1 - \mathbf{x}_2|} \right) d\mathbf{x}_1 \cdots d\mathbf{x}_Q \right)$$

$$\rho_Q[\chi, \mathbf{x}_1, \ldots, \mathbf{x}_Q] = \frac{N!}{(N-Q)!Q!} |\chi(\mathbf{x}_1, \ldots, \mathbf{x}_Q)|^2 \tag{90}$$

This formula for the kinetic energy corresponds to assuming that the Q-matrix has the form

$$\Gamma_Q(\mathbf{x}_1, \ldots, \mathbf{x}_Q; \mathbf{x}'_1, \ldots, \mathbf{x}'_Q) = \sqrt{\rho_Q(\mathbf{x}_1, \ldots, \mathbf{x}_k)\rho_Q(\mathbf{x}'_1, \ldots, \mathbf{x}'_k)} \tag{91}$$

Even if the Q-density is N-representable, this Q-matrix is not N-representable because its largest eigenvalue exceeds the upper bound $N!/Q!(N - Q + 1)!$. (That is, this Q-matrix violates the Pauli exclusion principle for Q-tuples of electrons.) Approximating the correction term, $T^{(Q)}_{\text{Pauli}}[\rho_Q[\chi]]$, seems difficult, and neglecting this term would give poor results, although the results improve with increasing Q [2, 10].

One can refine the approximation in Eq. (91) by writing the Q-density in terms of the eigenfunctions of the Q-matrix.

$$\rho_Q(\mathbf{x}_1,\ldots,\mathbf{x}_Q) = \sum_{i=1}^{\infty} n_i^{(Q)} |\phi_i(\mathbf{x}_1,\ldots,\mathbf{x}_Q)|^2 \tag{92}$$

$$0 \leq n_i^{(Q)} \leq \frac{N!}{Q!(N - Q + 1)!} \tag{93}$$

If one obtains the eigenfunctions by solving a Q-electron Schrödinger equation,

$$\left(\sum_{i=1}^{Q} -\frac{\nabla_i^2}{2} + w_Q(\mathbf{x}_1,\ldots,\mathbf{x}_Q)\right)\phi_i(\mathbf{x}_1,\ldots,\mathbf{x}_Q) = \varepsilon_i \phi_i(\mathbf{x}_1,\ldots,\mathbf{x}_Q) \tag{94}$$

then the potential functional method from Section V.A (cf. Eq. (87)) can be generalized to

$$E_{\text{g.s}}[v; N] = \min_{\substack{w_Q(\mathbf{x}_1,\ldots,\mathbf{x}_Q) \\ 0 \leq n_i^{(Q)} \leq \frac{N!}{Q!(N-Q+1)!}}} \frac{(N-Q)!Q!}{2(N-1)!}\left(\sum_{i=1}^{\infty} n_i^{(Q)} \left\langle \phi_i \left| -\frac{\nabla_1^2}{2} - \frac{\nabla_2^2}{2} \right| \phi_i \right\rangle + \frac{2(N-1)!}{(N-Q)!Q!}\right.$$

$$\left. T_N^{(Q)}[\rho_Q[w_Q]] + \int \cdots \int \rho_Q[w_Q; \mathbf{x}_1,\ldots,\mathbf{x}_Q]\left(v(\mathbf{x}_1) + v(\mathbf{x}_2) + \frac{N-1}{|\mathbf{x}_1 - \mathbf{x}_2|}\right)d\mathbf{x}_1 \cdots d\mathbf{x}_k\right) \tag{95}$$

While Eq. (95) is indubitably more accurate than the generalized Weisacker approach, merely imposing the bounds, Eq. (93), on the eigenvalues of the Q-matrix is insufficient to ensure the N-representability of the Q-matrix or the corresponding Q-density. The correction functional, $T_N^{(Q)}[\rho_Q]$, corrects for the approximations inherent in the model. Specifically, $T_N^{(Q)}[\rho_Q]$ corrects for (i) the constraint $0 \leq n_i^{(Q)} \leq N!/Q!(N - Q + 1)!$ which is insufficient to ensure the N-representability of the Q-matrix and (ii) the restriction to $\phi_i(\mathbf{x}_1,\ldots,\mathbf{x}_k)$ that are eigenfunctions of the same Q-electron Hamiltonians (Eq. (94)). Neither approximation (i) (which gives too much flexibility in the variational procedure) nor approximation (ii) (which gives too little flexibility) are expected to be useful by themselves. (The results of Bopp and Bender, Davidson, and Peat are germane to the question of how well the 2-matrix can be approximated using the eigenfunctions of the two-electron reduced Hamiltonian [66, 67].) Taken together, though, approximations (i) and (ii) *might* be useful. Unfortunately, the correction functional $T_N^{(Q)}[\rho_Q]$ seems difficult to approximate [3].

Nonetheless, Eq. (95) is perhaps the most natural generalization of the Kohn–Sham formulation to Q-density functional theory. Indeed, Ziesche's first papers on 2-density functional theory feature an algorithm based on Eq. (95), although he did not write his equations in the potential functional formulation [1, 4]. The early work of Gonis and co-workers [68, 69] is also of this form.

Equation (95) reveals an interesting link between Q-density functional theory and Q-matrix functional theory. Consider rewriting Eq. (95) in a form analogous to Eq. (83),

$$E_{\text{g.s.}}[v; N] = \min_{\substack{0 \le n_i^{(Q)} \le \frac{N!}{Q!(N-Q+1)!} \\ \langle \phi_i | \phi_j \rangle = \delta_{ij}}} \frac{(N-Q)!Q!}{2(N-1)!} \left(\sum_{i=1}^{\infty} n_i^{(Q)} \left\langle \phi_i \left| -\frac{\nabla_1^2}{2} - \frac{\nabla_2^2}{2} \right| \phi_i \right\rangle + \frac{2(N-1)!}{(N-Q)!Q!} \right.$$

$$\times T_N^{(Q)}\left[\rho_Q\left[n_i^{(Q)}, \phi_i\right]\right] + \int \cdots \int \rho_Q\left[n_i^{(Q)}, \phi_i; \mathbf{x}_1 \ldots, \mathbf{x}_Q\right]$$

$$\left. \times \left(v(\mathbf{x}_1) + v(\mathbf{x}_2) + \frac{N-1}{|\mathbf{x}_1 - \mathbf{x}_2|} \right) d\mathbf{x}_1 \cdots d\mathbf{x}_Q \right) \quad (96)$$

This algorithm is essentially a variational optimization of the Q-matrix,

$$\Gamma_Q(\mathbf{x}_1, \ldots, \mathbf{x}_Q; \mathbf{x}'_1, \ldots, \mathbf{x}'_Q) = \sum_{i=1}^{\infty} n_i^{(Q)} \phi_i(\mathbf{x}_1, \ldots, \mathbf{x}_Q) \phi_i^*(\mathbf{x}'_1, \ldots, \mathbf{x}'_Q) \quad (97)$$

subject to the eigenvalue constraint, Eq. (93). The eigenvalue constraint is, unfortunately, rather weak. (For the 2-matrix, the lower bound is equivalent to the P condition and the upper bound is implied by the G condition [22, 40].) This raises questions about the utility of Eq. (96) as a computational approach.

The computational procedure could be improved by imposing addition N-representability constraints. For example, the minimization in Eq. (96) could be performed subject to the $(Q, Q+1)$ conditions or their off-diagonal generalizations from Eq. (68). What results is a (rather complicated) restatement of the direct optimization procedure for the Q-matrix, except that in this formulation one can attempt to correct for the non-N-representability error with the functional $T_N^{(Q)}[\rho_Q]$. Enormous difficulties seem to be associated with approximating $T_N^{(Q)}[\rho_Q]$, however. Specifically, $T_N^{(Q)}[\rho_Q]$ is discontinuous and should be infinite for non-N-representable Q-densities and zero for N-representable ρ_Q [3].

A different—and arguably better—approach is based on the realization that every N-representable Q-density can be associated with an ensemble average of Slater determinants. (Recall Section III.E, and especially the discussion surrounding Eq. (54). For a Q-density built from Slater determinants, it follows from the Slater–Condon rules that the one-electron contributions to the energy can be written

$$F_1^{(Q)}[\rho_{i_1\ldots i_Q}] = \frac{(N-Q)!Q!}{(N-1)!} \sum_{i_1,\ldots,i_Q=1}^{K} \rho_{i_1\ldots i_Q} \left\langle \phi_{i_1} \left| -\frac{\nabla^2}{2} + v(\mathbf{x}) \right| \phi_{i_1} \right\rangle \tag{98}$$

and the electron–electron repulsion contribution can be written

$$F_2^{(Q)}[\rho_{i_1\ldots i_Q}] = \frac{(N-Q)!Q!}{(N-2)!} \sum_{i_1,\ldots,i_Q=1}^{K} \rho_{i_1\ldots i_Q} \left(\left\langle \phi_{i_1} \phi_{i_2} \left| \frac{1}{2|\mathbf{x}_1 - \mathbf{x}_2|} \right| \phi_{i_1} \phi_{i_2} \right\rangle \right.$$
$$\left. - \left\langle \phi_{i_1} \phi_{i_2} \left| \frac{1}{2|\mathbf{x}_1 - \mathbf{x}_2|} \right| \phi_{i_2} \phi_{i_1} \right\rangle \right) \tag{99}$$

The variational procedure becomes

$$E_{\text{g.s.}}[v;N] = \underset{\substack{(k,K) \text{ conditions} \\ \langle \phi_j | \phi_i \rangle = \delta_{ij}}}{\min} \left(F_1^{(Q)}[\rho_{i_1\ldots i_Q}] + F_2^{(Q)}[\rho_{i_1\ldots i_Q}] + E_c^{(Q)}[\rho_{i_1\ldots i_Q}] \right) \tag{100}$$

Here, the minimization is over all sets of orthonormal orbitals, subject to the requirement that $\rho_{i_1\ldots i_Q}$ satisfies the (Q, K) conditions. Because an ensemble of Slater determinants is incapable of describing electron correlation, one must include the additional correlation energy functional, $E_c^{(Q)}[\rho_{i_1\ldots i_Q}]$. No form for $E_c^{(Q)}[\rho_{i_1\ldots i_Q}]$ has ever been derived, and omitting this term is undesirable, since this procedure reduces to a (very complicated!) formulation of the Hartree–Fock method when $E_c^{(Q)}[\rho_{i_1\ldots i_Q}] = 0$. Computational methods based on Eq. (100) are members of the "Hartree–Fock plus corrections" family of methods. It should be noted that, in practical calculations, it will be impossible to use the (Q, K) conditions in Eq. (100). However, the ground-state energies from a sequence of calculations using the (Q, Q), $(Q, Q+1)$, ... conditions will converge monotonically to the (Q, K)-based result.

The variational procedure in Eq. (100) is in the spirit of the Kohn–Sham ansatz. Since $\rho_{i_1\ldots i_Q}$ satisfies the (Q, K) conditions, it is N-representable. In general, $\rho_{i_1\ldots i_Q}$ corresponds to many different N-electron ensembles and one of them, $\Gamma_N^{\text{g.s.}}$, corresponds to the ground state of interest. However, for computational expediency in computing the energy, a Slater determinantal density matrix,

Γ_N^{SI}, is selected instead. The energy of Γ_N^{SI} is greater than the true ground-state energy, and so obtaining the correct energy requires adding a correction,

$$E_c^{(Q)}\left[\rho_{i_1\ldots i_Q}\right] = E[\Gamma_N^{g.s.}] - E[\Gamma_N^{SI}] \tag{101}$$

Approximating $E_c^{(Q)}\left[\rho_{i_1\ldots i_Q}\right]$ requires (approximately) inverting the mapping

$$\Gamma_N^{g.s.} \to \rho_{i_1\ldots i_Q} \to \Gamma_N^{SI} \tag{102}$$

The preceding approach can be viewed as an orbital representation analogue for a recently proposed Kohn–Sham-based pair-density functional theory [17].

VI. CONCLUSION

The preceding is a rather comprehensive—but not exhaustive—review of N-representability constraints for diagonal elements of reduced density matrices. The most general and most powerful N-representability conditions seem to take the form of linear inequalities, wherein one states that the expectation value of some positive semidefinite linear Hermitian operator is greater than or equal to zero, $\text{Tr}[\hat{P}_N \Gamma_N] \geq 0$. If $\hat{P}_N$ depends only on Q-body operators, then it can be reduced into a Q-electron reduced operator, $\hat{P}_{Q,N}$, and $\text{Tr}[\hat{P}_{Q,N}\Gamma_Q] \geq 0$ provides a constraint for the N-representability of the Q-electron reduced density matrix, or Q-matrix. Requiring that $\text{Tr}[\hat{P}_{Q,N}\Gamma_Q] \geq 0$ for *every* Q-body positive semidefinite linear operator is necessary and sufficient for the N-representability of the Q-matrix [22].

Since it is obviously impossible to require that $\text{Tr}[\hat{P}_{Q,N}\Gamma_Q] \geq 0$ for *every* choice of $\hat{P}_{Q,N}$, one imposes this constraint only for a few operators. Moreover, because one needs to be able to prove that the operators are positive semidefinite, the operators that are selected for use as constraints are typically much simpler than a molecular Hamiltonian. This is unfortunate, because if one could ensure that $\text{Tr}[\hat{H}_{Q,N}\Gamma_Q] \geq E_{g.s.}(\hat{H}_N)$ for the Hamiltonian of interest, then the computational procedure would be exact. Future research in N-representability might focus on developing constraints based on "molecular" considerations.

Among the simple linear operators that are commonly used to construct constraints, polynomials of the number operators (cf. Eq. (22)) are particularly useful. Polynomials of number operators are convenient because (i) the ground-state wavefunction of number-operator polynomials is a Slater determinant and (ii) the number-operator constraints depend only on the diagonal elements of the Q-matrix, $\rho_{i_1\ldots i_Q} = \Gamma_{i_1\ldots i_Q;i_1\ldots i_Q}$. The (Q, R) conditions for N-representability are based on the requirement that $\text{Tr}\left[\hat{L}_N^{(Q,\mathbf{I}_R)}\Gamma_N\right] \geq E_{g.s.}\left(\hat{L}_N^{(Q,\mathbf{I}_R)}\right)$ for every possible Qth-degree polynomial of number operators for R orbitals, $\hat{L}_N^{(k,\mathbf{I}_R)}$ [24–26].

In general, it is difficult to derive the (Q, R) conditions directly. An exception occurs for the (R, R) constraints, which have an especially simple form based on the positive semidefinite Hamiltonian in Eq. (29). Fortunately, the (Q, R) conditions ($Q < R$) are easily derived from the (R, R) conditions [26]. In Section III.D we used this result to derive the Weinhold–Wilson constraints on the diagonal elements of the 2-matrix [23]. (The Weinhold–Wilson constraints are identical to the (2, 3) conditions.)

In a basis set that contains K spin orbitals, the (Q, K) conditions are necessary and sufficient for the N-representability of the diagonal elements of the Q-matrix, but only necessary for the off-diagonal elements, $\Gamma_{i_1 \ldots i_Q; j_1 \ldots j_Q}$. This is of academic interest, but for reasonable basis sets, implementing the (Q, K) conditions is intractable. The (Q, K) constraints define the Slater hull [29].

Because polynomials of orbital occupation number operators do not depend on the off-diagonal elements of the reduced density matrix, it is important to generalize the (Q, R) constraints to off-diagonal elements. This can be done in two ways. First, because the (Q, R) conditions must hold in any orbital basis, unitary transformations of the orbital basis set can be used to constrain the off-diagonal elements of the density matrix. (See Eq. (56) and the surrounding discussion.) Second, one can replace the number operators by creation and annihilation operators on different orbitals according to the rule

$$\hat{n}_i \to a_{i'}^+ a_i$$
$$1 - \hat{n}_i \to a_{j'} a_j^+ \quad (103)$$

(cf. Eqs. (68) and (69)). If one applies this procedure to the (2, 2) conditions, one derives the P (Eq. (62)), Q (Eq. (63)), and G (Eq. (64)) conditions for the N-representability of the 2-matrix. The P, Q, and G conditions were originally formulated in their off-diagonal form [22, 40]. Applying this procedure to the (2, 3) conditions produces the $T1$ and $T2$ conditions (in addition to the P, Q, and G conditions). The diagonal elements of the $T1$ and $T2$ conditions were derived in "diagonal form" by Weinhold and Wilson [23] before Erdahl [52] derived the off-diagonal form. This general procedure—use the (R, R) conditions to derive the (Q, R) conditions for the diagonal elements of the Q-matrix; then use Eq. (68) to generalize those constraints to include off-diagonal elements of the Q-matrix—seems to be an especially easy method for deriving N-representability constraints on the Q-matrix and/or its diagonal elements.

Almost all of the available N-representability constraints are based on the orbital representation of the reduced density matrix, $\Gamma_{i_1 \ldots i_Q; j_1 \ldots j_Q}$, instead of the spatial representation, $\Gamma_Q(\mathbf{x}_1, \ldots, \mathbf{x}_Q; \mathbf{x}'_1, \ldots, \mathbf{x}'_Q)$. This is not problematic when the reduced density matrix is available, because it is easy to convert Q-matrices to and from the spatial representation (cf. Eqs. (18) and (19)). There has been a lot of recent interest in developing computational algorithms based on the Q-electron distribution function, which comprises the diagonal elements of the Q-matrix. Unfortunately, there is no analogous method for converting the Q-electron distribution function, or Q-density, from the orbital to the spatial representation. (The orbital representation of the Q-density, $\rho_{i_1\ldots i_Q}$, depends on off-diagonal elements of the spatially resolved Q-matrix. Conversely, the spatial representation of the Q-density, $\rho_Q(\mathbf{x}_1,\ldots,\mathbf{x}_Q)$, depends on the off-diagonal elements of the orbital-resolved Q-matrix.) This is problematic because computational approaches based on the Q-density are usually based on the spatial representation, $\rho_Q(\mathbf{x}_1,\ldots,\mathbf{x}_Q)$, while the (Q, R) N-representability conditions are based on the orbital representation, $\rho_{i_1\ldots i_Q}$.

There are two ways to "fix" this problem. First, one can attempt to derive N-representability conditions for the Q-density in the spatial representation. This seems hard to do, although one constraint (basically a special case of the G condition for the density matrix) of this type is known, see Eq. (77). Deriving additional constraints is a priority for future work.

The second approach to this problem is to derive orbital-based reformulations of existing algorithms based on the spatial representation of the Q-density. The resulting formulations are in the spirit of the "orbital-resolved" Kohn–Sham approach to density functional theory.

It is fair to say that neither of these two approaches works especially well: N-representability conditions in the spatial representation are virtually unknown and the orbital-resolved computational methods are promising, but untested. It is interesting to note that one of the most common computational algorithms (cf. Eq. (96)) can be viewed as a density-matrix optimization, although most authors consider only a weak N-representability constraint on the occupation numbers of the Q-matrix [1, 4, 69]. Additional N-representability constraints could, of course, be added, but it seems unlikely that the resulting Q-density functional theory approach would be more efficient than direct methods based on semidefinite programming [33, 35–37].

The N-representability constraints presented in this chapter can also be applied to computational methods based on the variational optimization of the reduced density matrix subject to necessary conditions for N-representability. Because of their hierarchical structure, the (Q, R) conditions are also directly applicable to computational approaches based on the contracted Schrödinger equation. For example, consider the (2, 4) contracted Schrödinger equation. Requiring that the reconstructed 4-matrix in the (2, 4) contracted Schrödinger equation satisfies the (4, 4) conditions is sufficient to ensure that the 2-matrix satisfies the rather stringent (2, 4) conditions. Conversely, if the 2-matrix does not satisfy the (2, 4) conditions, then it is impossible to construct a 4-matrix that is consistent with this 2-matrix and also satisfies the (4, 4) conditions. It seems that the (Q, R) conditions provide important constraints for maintaining consistency at different levels of the contracted Schrödinger equation hierarchy.

Compared to Q-density functional theory, Q-matrix theory is a relatively mature field. The biggest impediment to widespread adoption of reduced-density-matrix optimization for quantum chemical calculations seems to be the uncompetitive computational cost of these methods compared to existing ab initio techniques. Recent algorithmic advances are closing the gap, and developing even better algorithms is an active area of research in the density-matrix research community. In addition to the development of improved constrained optimization algorithms, research into new N-representability conditions continues unabated. Because it is easy to derive N-representability constraints using the Slater hull, we believe that the material reviewed here may be helpful.

REFERENCES

1. P. Ziesche, Pair density-functional theory — a generalized density-functional theory. *Phys. Lett. A* **195**, 213–220 (1994).
2. P. W. Ayers, Generalized density functional theories using the k-electron densities: development of kinetic energy functionals. *J. Math. Phys.* **46**, 062107 (2005).
3. P. W. Ayers, S. Golden, and M. Levy, Generalizations of the Hohenberg–Kohn theorem: I. Legendre transform constructions of variational principles for density matrices and electron distribution functions. *J. Chem. Phys.* **124**, 054101 (2006).
4. P. Ziesche, Attempts toward a pair density functional theory. *Int. J. Quantum Chem.* **60**, 149–162 (1996).
5. P. W. Ayers and M. Levy, Generalized density-functional theory: conquering the N-representability problem with exact functionals for the electron pair density and the second-order reduced density matrix. *J. Chem. Sci.* **117**, 507–514 (2005).
6. M. E. Pistol, Characterization of N-representable n-particle densities when N is infinite. *Chem. Phys. Lett.* **417**, 521–523 (2006).
7. A. Nagy, Density-matrix functional theory. *Phys. Rev. A* **66**, 022505 (2002).
8. J. Y. Hsu, Derivation of the density functional theory from the cluster expansion. *Phys. Rev. Lett.* **91**, 133001 (2003).
9. A. Nagy and C. Amovilli, Effective potential in density matrix functional theory. *J. Chem. Phys.* **121**, 6640–6648 (2004).
10. F. Furche, Towards a practical pair density-functional theory for many-electron systems. *Phys. Rev. A* **70**, 022514 (2004).
11. J. Y. Hsu, C. H. Lin, and C. Y. Cheng, Pair-correlation effect and virial theorem in the self-consistent density-functional theory. *Phys. Rev. A* **71**, 052502 (2005).
12. P. K. Acharya, L. J. Bartolotti, S. B. Sears, and R. G. Parr, An atomic kinetic energy functional with full Weizsacker correction. *Proc. Natl. Acad. Sci. USA* **77**, 6978–6982 (1980).
13. M. R. Nyden, An orthogonality constrained generalization of the Weizacker density functional model. *J. Chem. Phys.* **78**, 4048–4051 (1983).
14. M. Levy, J. P. Perdew, and V. Sahni, Exact differential-equation for the density and ionization-energy of a many-particle system. *Phys. Rev. A* **30**, 2745–2748 (1984).
15. N. H. March, Differential-equation for the electron-density in large molecules. *Int. J. Quantum Chem.* **13**, 3–8 (1986).

16. N. H. March, The local potential determining the square root of the ground-state electron-density of atoms and molecules from the schrödinger equation. *Phys. Lett. A* **113**, 476–478 (1986).
17. P. W. Ayers and M. Levy, Using the Kohn–Sham formalism in pair density-functional theories. *Chem. Phys. Lett.* **416**, 211–216 (2005).
18. T. L. Gilbert, Hohenberg Kohn theorem for nonlocal external potentials. *Phys. Rev. B* **12**, 2111–2120 (1975).
19. J. E. Harriman, Orthonormal orbitals for the representation of an arbitrary density. *Phys. Rev. A* **24**, 680–682 (1981).
20. Y. A. Wang and E. A. Carter, in *Theoretical Methods in Condensed Phase Chemistry* (S. D. Schwartz, ed.), Kluwer, Dordrecht, 2000, p. 117.
21. G. K. L. Chan and N. C. Handy, An extensive study of gradient approximations to the exchange-correlation and kinetic energy functionals. *J. Chem. Phys.* **112**, 5639–5653 (2000).
22. C. Garrod and J. K. Percus, Reduction of the N-particle variational problem. *J. Math. Phys.* **5**, 1756–1776 (1964).
23. F. Weinhold and E. B. Wilson, Reduced density matrices of atoms and molecules. II. On the N-representability problem. *J. Chem. Phys.* **47**, 2298–2311 (1967).
24. E. R. Davidson, Linear inequalities for density matrices. *J. Math. Phys.* **10**, 725–734 (1969).
25. W. B. McRae and E. R. Davidson, Linear inequalities for density matrices II. *J. Math. Phys.* **13**, 1527–1538 (1972).
26. E. R. Davidson, Linear inequalities for density matrices: III. *Int. J. Quantum Chem.* **91**, 1–4 (2003).
27. E. R. Davidson, N-representability of the electron-pair density. *Chem. Phys. Lett.* **246**, 209–213 (1995).
28. P. W. Ayers and E. R. Davidson, Necessary conditions for the N-representability of pair distribution functions. *Int. J. Quantum Chem.* **106**, 1487–1498 (2006).
29. H. Kummer, N-representability problem for reduced density matrices. *J. Math. Phys.* **8**, 2063–2081 (1967).
30. P.-O, Lowdin, Quantum theory of many-particle systems. I. Physical interpretation by means of density matrices, natural spin-orbitals, and convergence problems in the method of configuration interaction. *Phys. Rev.* **97**, 1474–1489 (1955).
31. E. R. Davidson, *Reduced Density Matrices in Quantum Chemistry*, Academic Press, New York, 1976.
32. A. J. Coleman and V. I. Yukalov, *Reduced Density Matrices: Coulson's Challenge*, Springer, Berlin, 2000.
33. M. Nakata, M. Ehara, and H. Nakatsuji, Density matrix variational theory: application to the potential energy surfaces and strongly correlated systems. *J. Chem. Phys.* **116**, 5432–5439 (2002).
34. M. Nakata, H. Nakatsuji, M. Ehara, M. Fukuda, K. Nakata, and K. Fujisawa, Variational calculations of fermion second-order reduced density matrices by semidefinite programming algorithm. *J. Chem. Phys.* **114**, 8282–8292 (2001).
35. Z. J. Zhao, B. J. Braams, M. Fukuda, M. L. Overton, and J. K. Percus, The reduced density matrix method for electronic structure calculations and the role of three-index representability conditions. *J. Chem. Phys.* **120**, 2095–2104 (2004).
36. D. A. Mazziotti, First-order semidefinite programming for the direct determination of two-electron reduced density matrices with application to many-electron atoms and molecules. *J. Chem. Phys.* **121**, 10957–10966 (2004).
37. D. A. Mazziotti, Realization of quantum chemistry without wave functions through first-order semidefinite programming. *Phys. Rev. Lett.* **93**, 213001 (2004).

38. G. Gidofalvi and D. A. Mazziotti, Application of variational reduced-density-matrix theory to organic molecules. *J. Chem. Phys.* **122**, 094107 (2005).
39. D. A. Mazziotti, Quantum chemistry without wave functions: two electron reduced density matrices. *Acc. Chem. Res.* **39**, 207–215 (2006).
40. J. Coleman, Structure of fermion density matrices. *Rev. Mod. Phys.* **35**, 668–687 (1963).
41. R. Larsen, *Functional Analysis: An Introduction*, Marcel Dekker, New York, 1973.
42. H. W. Kuhn, *Proc. Symp. Appl. Math.* **10**, 141 (1960).
43. M. L. Yoseloff and H. W. Kuhn, Combinatorial approach to the N-representability of P-density matrices. *J. Math. Phys.* **10**, 703–706 (1969).
44. M. Deza and M. Laurent, Applications of cut polyhedra. 2. *J. Comput. Appl. Math.* **55**, 217–247 (1994).
45. L. Cohen and C. Frishberg, Hierarchy equations for reduced density matrices. *Phys. Rev. A* **13**, 927–930 (1976).
46. H. Nakatsuji, Equation for the direct determination of the density matrix. *Phys. Rev. A* **14**, 41–50 (1976).
47. C. Valdemoro, L. M. Tel, and E. Pérez-Romero, The contracted Schrödinger equation: some results. *Adv. Quantum Chem.* **28**, 33–46 (1997).
48. D. A. Mazziotti, Contracted Schrödinger equation: determining quantum energies and two-particle density matrices without wave functions. *Phys. Rev. A* **57**, 4219–4234 (1998).
49. C. Valdemoro, Electron correlation and reduced density matrices. *Correlation Localization* **203**, 187–200 (1999).
50. D. A. Mazziotti, Variational minimization of atomic and molecular ground-state energies via the two-particle reduced density matrix. *Phys. Rev. A* **65**, 062511 (2002).
51. D. K. Lee, H. W. Jackson, and E. Feenberg, *Ann. Phy.* **44**, 84–104 (1967).
52. R. M. Erdahl, Representability. *Int. J. Quantum Chem.* **13**, 697–718 (1978).
53. J. R. Hammond and D. A. Mazziotti, Variational two-electron reduced-density-matrix theory: partial 3-positivity conditions for N-representability. *Phys. Rev. A* **71**, x (2005).
54. J. K. Percus, At the boundary between reduced density-matrix and density-functional theories. *J. Chem. Phys.* **122**, 234103 (2005).
55. W. Kohn and L. J. Sham, Self-consistent equations including exchange and correlation effects. *Phys. Rev.* **140**, A1133–A1138 (1965).
56. C. F. Weizsacker, Zur theorie dier kernmassen. *Z. Phys.* **96**, 431–458 (1935).
57. S. M. Valone, A one-to-one mapping between one-particle densities and some normal-particle ensembles. *J. Chem. Phys.* **73**, 4653–4655 (1980).
58. H. Englisch and R. Englisch, Exact density functionals for ground-state energies. 2. Details and remarks. *Phys. Status Solidi B* **124**, 373–379 (1984).
59. H. Englisch and R. Englisch, Exact density functionals for ground-state energies. I. General results. *Phys. Status Solidi B* **123**, 711–721 (1984).
60. W. T. Yang, P. W. Ayers, and Q. Wu, Potential functionals: dual to density functionals and solution to the upsilon-representability problem. *Phys. Rev. Lett.* **92**, 146404 (2004).
61. E. H. Lieb, Density functionals for Coulomb systems. *Int. J. Quantum Chem.* **24**, 243–277 (1983).
62. P. W. Ayers, Axiomatic formulations of the Hohenberg–Kohn functional. *Phys. Rev. A* **73**, 012513 (2006).
63. J. Harris and R. O. Jones, The surface energy of a bounded electron gas. *J. Phys. F* **4**, 1170–1186 (1974).

64. O. Gunnarsson and B. I. Lundqvist, Exchange and correlation in atoms, molecules and solids by the spin-density-functional formalism. *Phys. Rev.* B **13**, 4274–4298 (1976).
65. D. C. Langreth and J. P. Perdew, Exchange-correlation energy of a metallic surface: wave-vector analysis. *Phys. Rev.* B **15**, 2884–2901 (1977).
66. F. Bopp, *Z. Phys.* **156**, 348 (1959).
67. C. F. Bender, E. R. Davidson, and F. D. Peat, Application of geminal methods to molecular calculations. *Phys. Rev.* **174**, 75–80 (1968).
68. A. Gonis, T. C. Schulthess, P. E. A. Turchi, and J. Vanek, Treatment of electron–electron correlations in electronic structure calculations. *Phys. Rev.* B **56**, 9335–9351 (1997).
69. A. Gonis, T. C. Schulthess, J. Vanek, and P. E. A. Turchi, A general minimum principle for correlated densities in quantum many-particle systems. *Phys. Rev. Lett.* **77**, 2981–2984 (1996).

PART V

CHAPTER 17

PARAMETERIZATION OF THE 2-RDM

A. JOHN COLEMAN

*Department of Mathematics and Statistics, Queen's University,
Kingston, Ontario K7L 3N6, Canada*

CONTENTS

I. Calculating Energy Levels and 2-Matrices
   A. Notation
   B. Algorithm
   C. Proof of Algorithm
II. Concluing Remarks
References

I. CALCULATING ENERGY LEVELS AND 2-MATRICES

A new algorithm is presented for the calculation of energy levels and their associated second-order density matrices, which aims to produce the exact energy as in full configuration interaction but without the N-particle wavefunction.

The efforts by several very able quantum scientists in four countries in the period preceding 1972 had failed to obtain a complete solution of the N-representability problem. It was assumed that we would never find one. My announcement of the solution in June of that year at a Conference in Boulder was therefore greeted with incredulity except by Ernie Davidson who understood my argument immediately.

The proof, as I understood it in 1972, can be found in Sections 2.3–2.6 of my book with V. I. Yukalov [1], which contains a misleading although minor error in Eq. (2.47). This mistake was corrected in 2002 in the paper [2], which also throws exciting light on the application of the RDM approach to the theory of condensed matter and to the complex geometry of the graph of the equation

$|B^N| = 0$. Essential for understanding the present approach is the basic paper [3] of Hans Kummer.

When I recently reread these papers, I myself found them difficult to understand! I believe this is partly because for these early attempts to write the proof I had not fully digested the argument nor recognized the generality of the result. My attempt to cover both fermions and bosons, not merely D^2 but also the general case of D^p, did not help.

So, here, I shall outline an algorithm and then comment. I claim proprietary rights to this algorithm in the sense that it may be used freely but never for financial gain. I believe that recent increases in computing power and improvements in programming make the use of my algorithm feasible, as it was not in 1972. Of course, the best test of my conviction will be for it to be used.

I suggest that the reader begin by getting clearly in mind the definitions in Section 2 of Ref. [2], making a rough sketch depicting the mutual relation of the key cones involved and proceed.

A. Notation

I have in mind a system of N electrons in a molecule or solid in Born–Oppenheimer approximation governed by a Hamiltonian H such that

$$H = \sum_{i=1}^{N} H(i) + \sum_{1 \leq i < j \leq N} H(ij)$$

Recall that in his Theorems 3 and 4 Hans Kummer [3] defined a *contraction* operator, L_N^p, which maps a linear operator on N-space onto an operator on p-space and an *expansion* operator, Γ_p^N, which maps an operator on p-space onto an operator on N-space. Note that the contraction and expansion operators are "super operators" in the sense that they act not on spaces of wavefunctions but on linear spaces consisting of linear operators on wavefunction spaces. If the two-particle reduced Hamiltonian is defined as

$$K^2 = H(1) + H(2) + (N-1)H(12) \tag{1}$$

then the N-particle Hamiltonian is

$$H = \frac{N}{2} \Gamma_2^N(K^2) \tag{2}$$

As is common in papers on quantum mechanics, we assume that a choice has been made of a fixed set of one-particle functions, $\{\varphi_i\}$, $0 \leq i \leq r$, in terms of which all functions occurring in our argument are expanded.

If B^2 is a Hermitian operator acting on the space of antisymmetric two-particle functions, then we define

$$B^N = \Gamma_2^N(B^2) = B^2 \wedge I^{N-2} \tag{3}$$

Thus the $s_N \times s_N$ matrix B^N is a function of B^2 and therefore of r^4 real numbers, which in our approach play the role of the parameters for N-representable 2-matrices within the limitations of the given one-particle basis set. Compare this with the $s_N = \binom{r}{N}$ parameters of the FCI approach. Recall Kummer's basic theorem [1, Theorem 2.8, p. 56] that B^2 could be a second-order RDM if and only if $\Gamma_2^N(B^2)$ is a positive operator on N-space. For λ, μ real and $\lambda > 0$, we set

$$B^2 = \lambda I^2 + \mu \frac{N}{2} K^2 \tag{4}$$

Therefore

$$B^N = \Gamma_2^N \left(\lambda I^2 + \mu \frac{N}{2} K^2 \right) \tag{5}$$

$$= \lambda^{s_N}(I^N + \sigma H) \tag{6}$$

The smallest value of σ for which the determinant $|B^N|$ vanishes is such that there exists an N-particle wavefunction, ψ, for which $B^N \psi = B^N P_\psi = 0$ and therefore $(I^N + \sigma H)\psi = 0$. It follows from this that

$$H\psi = -\frac{1}{\sigma} \psi \tag{7}$$

So the smallest positive value of σ corresponds to the ground state. The second smallest similarly provides the energy of the first excited state and so on. The possible occurrence and significance of multiple roots are discussed in Refs. [1, 2].

To discuss the 2-matrix it is convenient to use pairs of natural numbers as indices. We shall assume that lowercases Greek letters denote such pairs. For example, we define the *standard index set*: $\Omega = \{(ij)|\ i < j, 1 \leq i,j \leq r\}$. Then α, in Ω, could take any of $r(r-1)/2$ values. We shall define $[\alpha] = 1/\sqrt{2}[\varphi_i, \varphi_j]$, when $\alpha = (ij)$. Thus the set $\{[\alpha]|, \alpha \in \Omega\}$ is a complete orthonormal basis for 2-space.

B. Algorithm

1. Suppose that B^2 is a two-particle Hermitian operator.

2. Define

$$B^N = \Gamma_2^N(B^2) = B^2 \wedge I^{N-2} \tag{8}$$

3. Let

$$\Delta(B^2) = |B^N| \tag{9}$$

4. Set

$$B^2 = \lambda I^2 + \mu \frac{N}{2} K^2 \tag{10}$$

$$= \lambda \left(I^2 + \sigma \frac{N}{2} K^2 \right) \tag{11}$$

then a positive value, γ, of σ for which $\Delta(B^2) = 0$ implies the existence of an N-particle wavefunction ψ such that

$$H\psi = -\frac{1}{\gamma}\psi \tag{12}$$

Since this eigenvalue is negative it corresponds to a bound state and, if γ is the smallest positive zero of the determinant, to the ground state. The next smallest zero of the determinant will correspond to the first excited state and so on.

5. For any such ψ and γ, the corresponding 2-matrix is

$$D^2 = \frac{\partial \Delta}{\partial B^2} \tag{13}$$

$$= L_N^2(P_\psi) \tag{14}$$

Here, P_ψ denotes the projector onto the state ψ.

C. Proof of Algorithm

I am tempted to say "the proof is obvious" if it were not that the average chemist or physicist would then have all the evidence needed to consign me to his/her category of typical mathematicians who make lots of abstract assertions but seldom say anything of real use! Once, I did receive a letter addressed "Dear fellow quantum chemist." That gave me a really warm feeling—that I had finally arrived and may be of some slight use in the world!

In fact, the algorithm will be clear to anyone who understands Chapter 2 Ref. [1] and the pages following p. 26 about Grassmann algebra. In the following, numbers refer to steps in the algorithm.

1. We assume the familiar summation convention often ascribed to Einstein, so the repeated β is understood as running through the standard index set for pairs.
2. The condition that B^N is a positive semidefinite operator ensures that B^2 belongs to the *Kummer cone*. Unfortunately, the Grassmann wedge product of two operators is not explained explicitly in Ref. [1] but the section, p.79 may be sufficient. Here are two other references [4, 5] that may be helpful. Be sure that you understand Exercise 4 on p.73 of Ref. [1].
3. The determinant is of the same order as those occurring in an FCI calculation, but there is only one in contrast to billions which arise in contemporary FCI and it is of rather different structure. Although calculated on N-particle space, its components are functions of a two-particle operator.
4. Step 4 was suggested to me independently by Davidson and Erdahl. It is this step that simplifies the procedure and avoids a difficult variational approach.
5. The complete proof of these claims is on p. 65 of Ref. [1].

II. CONCLUDING REMARKS

1. Since this book is aimed at chemists, I think of electrons but, with only minor changes of language, most of it is immediately applicable to any type of fermion or boson. This is evident from ref. [1] and many of my other recent papers.

2. In contrast to methods based on variation accessible only to the ground state, my approach deals with excited states with exactly the same ease or difficulty as with the ground state.

3. The parameters for any calculation are the components of a Hermitian 2-matrix which will establish a basic pattern that will appear *ad nauseum* in B^N as N increases. I hesitate to be dogmatic since I have had so little hands-on experience in programming, but it would seem to me that one could use an off-the-shelf program for any particular configuration of nuclei, which could automatically be adjusted for different r and N. If so, numerical experiment might provide us with a clue as to what happens to the solution for a Hamiltonian such as that in ref. [1], Eq. (7.8), p. 256 when the number of electrons tends to infinity. This could be of great importance for condensed matter physics and/or BEC.

4. I draw the reader's attention to the summary of Yukalov's criteria for an adequate algorithm on p.7 and to my evaluation of present methods in the notes on p. 8 of Ref. [2].

5. With $\mu = 0$, the probe B^2 would be at the identity operator in the "center" of the Kummer cone and with increasing μ would be associated with increasing negative real eigenvalues, that is, smaller absolute value.

References

1. A. J. Coleman and V. I. Yukalov, *Reduced Density Matrices: Coulson's Challenge*, Springer-Verlag, New York, 2000.
2. A. J. Coleman, Kummer variety. *Phys. Rev. A* **66**, 022503 (2002).
3. H. Kummer, *J. Math. Phys.* **5**, 1756 (1964).
4. A. J. Coleman and I. Absar, *Int. J. Quantum Chem.* **18**, 1279 (1980).
5. D. A. Mazziotti, *Phys. Rev. A* **57**, 4219 (1998).

CHAPTER 18

ENTANGLEMENT, ELECTRON CORRELATION, AND DENSITY MATRICES

SABRE KAIS

Department of Chemistry, Purdue University, West Lafayette, IN 47907 USA

CONTENTS

I. Introduction
   A. Entanglement of Formation and Concurrence
   B. Entanglement Measure for Fermions
   C. Entanglement and Ranks of Density Matrices
II. Entanglement for Spin Systems
   A. Entanglement for Two-Spin Systems
   B. Entanglement for One-Dimensional N-Spin Systems
   C. Numerical Solution of the One-Dimensional Spin-$\frac{1}{2}$ Systems
   D. Entanglement and Spin Reduced Density Matrices
   E. Some Numerical Results
   F. Thermal Entanglement and the Effect of Temperature
   G. Entanglement for Two-Dimensional Spin Systems
III. Entanglement for Quantum Dot Systems
   A. Two-Electron Two-Site Hubbard Model
      1. Exact Solution
      2. Hartree–Fock Approximation
      3. Correlation Entropy
      4. Entanglement
   B. One-Dimensional Quantum Dots System
   C. Two-Dimensional Array of Quantum Dots
IV. Ab Initio Calculations and Entanglement
V. Dynamics of Entanglement and Decoherence
VI. Entanglement and Density Functional Theory
VII. Future Directions
Acknowledgments
References

Reduced-Density-Matrix Mechanics: With Application to Many-Electron Atoms and Molecules,
A Special Volume of Advances in Chemical Physics, Volume 134, edited by David A. Mazziotti.
Series editor Stuart A. Rice. Copyright © 2007 John Wiley & Sons, Inc.

I. INTRODUCTION

In quantum chemistry calculations, the correlation energy is defined as the energy error of the Hartree–Fock wavefunction, that is, the difference between the Hartree–Fock limit energy and the exact solution of the nonrelativistic Schrödinger equation [1]. Different types of electron correlation are often distinguished in quantum chemistry such as dynamical and nondynamical [2], radial versus angular correlation for atoms, left–right, in–out and, radial correlation for diatomic molecules, and weak and strong correlation for solids. There also exists other measures of electron correlation in the literature such as the statistical correlation coefficients [3] and more recently the Shannon entropy as a measure of the correlation strength [4–8]. Correlation of a quantum many-body state makes the one-particle density matrix nonidempotent. Therefore the Shannon entropy of the natural occupation numbers measures the correlation strength on the one-particle level [7]. Electron correlations have a strong influence on many atomic, molecular [9], and solid properties [10]. The concept of electron correlation as defined in quantum chemistry calculations is useful but not directly observable; that is, there is no operator in quantum mechanics that its measurement gives the correlation energy. Moreover, there are cases where the kinetic energy dominates the Coulomb repulsion between electrons, so the electron correlation alone fails as a correlation measure [6].

Entanglement is a quantum mechanical property that describes a correlation between quantum mechanical systems and has no classical analogue [11–15]. Schrödinger was the first to introduce these states and gave them the name "Verschränkung" to a correlation of quantum nature [16]: "For an entangled state the best possible knowledge of the whole does not include the best possible knowledge of its parts." Latter, Bell [17] defined entanglement as "a correlation that is stronger than any classical correlation." Thus it might be useful as an alternative measure of electron–electron correlation in quantum chemistry calculations.

Ever since the appearance of the famous EPR Gadanken experiment [18], the phenomenon of entanglement [19], which features the essential difference between classical and quantum physics, has received wide theoretical and experimental attention [17, 20–25]. Generally, if two particles are in an entangled state then, even if the particles are physically separated by a great distance, they behave in some respects as a single entity rather than as two separate entities. There is no doubt that the entanglement has been lying in the heart of the foundation of quantum mechanics.

A desire to understand quantum entanglement is fueled by the development of quantum computation, which started in the 1980s with the pioneering work of Benioff [26], Bennett [27], Deutsch [28], Feynman [29] and Landauer [30] but gathered momentum and research interest only after Peter Shor's revolutionary discovery [31] of a quantum computer algorithm in 1994 that would efficiently find the prime factors of composite integers. Since integer factorization is the basis for cryptosystems used for security nowadays, Shor's finding will have a profound effect on cryptography. The astronomical power of quantum computations has researchers all over the world racing to be the first to create a practical quantum computer.

Besides quantum computations, entanglement has also been at the core of other active research such as quantum teleportation [32, 33], dense coding [34, 35], quantum communication [36], and quantum cryptography [37]. It is believed that the conceptual puzzles posed by entanglement have now become a physical source of novel ideas that might result in applications.

A big challenge faced by all of the above-mentioned applications is to prepare the entangled states, which is much more subtle than classically correlated states. To prepare an entangled state of good quality is a preliminary condition for any successful experiment. In fact, this is not only an experimental problem but also poses an obstacle to theories, since how to quantify entanglement is still unsettled; this is now becoming one of the central topics in quantum information theory. Any function that quantifies entanglement is called an entanglement measure. It should tell us how much entanglement there is in a given mutipartite state. Unfortunately, there is currently no consensus as to the best method to define an entanglement for all possible multipartite states. And the theory of entanglement is only partially developed [13, 38–40] and for the moment can only be applied in a limited number of scenarios, where there is an unambiguous way to construct suitable measures. Two important scenarios are (i) the case of a pure state of a bipartite system, that is, a system consisting of only two components and (ii) a mixed state of two spin-$\frac{1}{2}$ particles.

When a bipartite quantum system AB describe by $H_A \otimes H_B$ is in a pure state, there is an essentially well-motivated and unique measure of the entanglement between the subsystems A and B given by the von Neumann entropy S. If we denote with ρ_A the partial trace of $\rho \in H_A \otimes H_B$ with respect to subsystem B, $\rho_A = \text{Tr}_B(\rho)$, the entropy of entanglement of the state ρ is defined as the von Neumann entropy of the reduced density operator ρ_A, $S(\rho) \equiv -\text{Tr}[\rho_A \log_2 \rho_A]$. It is possible to prove that, for the pure state, the quantity S does not change if we exchange A and B. So we have $S(\rho) \equiv -\text{Tr}[\rho_A \log_2 \rho_A] \equiv -\text{Tr}[\rho_B \log_2 \rho_B]$. For any bipartite pure state, if an entanglement $E(\rho)$ is said to be a good one, it is often required to have the following properties [14]:

- *Separability*: If ρ is separable, then $E(\rho) = 0$.
- *Normalization*: The entanglement of a maximally entangled state of two d-dimensional systems is given by $E = \log(d)$.

- *No Increase Under Local Operations*: Applying local operations and classically communicating cannot increase the entanglement of ρ.
- *Continuity*: In the limit of vanishing distance between two density matrices, the difference between their entanglement should tend to zero.
- *Additivity*: A certain number N of identical copies of the state ρ should contain N times the entanglement of one copy.
- *Subadditivity*: The entanglement of the tensor product of two states should not be larger that the sum of the entanglement of each of the states.
- *Convexity*: The entanglement measure should be a convex function, that is, $E(\lambda \rho + (1-\lambda)\sigma) \leq \lambda E(\rho) + (1-\lambda)E(\sigma)$ for $0 < \lambda < 1$.

For a pure bipartite state, it is possible to show that the von Neumann entropy of its reduced density matrix, $S(\rho_{\text{red}}) = -\text{Tr}(\rho_{\text{red}} \log_2 \rho_{\text{red}})$, has all the above properties. Clearly, S is not the only mathematical object that meets the requirement, but in fact, it is now basically accepted as the correct and unique measure of entanglement.

The strict definitions of the four most prominent entanglement measures can be summarized as follows [14]:

- *Entanglement of distillation E_D*.
- *Entanglement of cost E_C*.
- *Entanglement of formation E_F*.
- *Relative entropy of entanglement E_R*.

The first two measures are also called operational measures, while the second two don't admit a direct operational interpretation in terms of entanglement manipulations. Suppose E is a measure defined on mixed states that satisfy the conditions for a good measure mentioned above. Then we can prove that for all states $\rho \in (H^A \otimes H^B)$, $E_D(\rho) \leq E(\rho) \leq E_C(\rho)$, and both $E_D(\rho)$ and $E_C(\rho)$ coincide on pure states with the von Neumann reduced entropy as demonstrated earlier.

A. Entanglement of Formation and Concurrence

At the current time, there is no simple way to carry out the calculations with all these entanglement measures. Their properties, such as additivity, convexity, and continuity, and relationships are still under active investigation. Even for the best-understood entanglement of formation of the mixed states in bipartite systems AB, once the dimension or A or B is three or above, we don't know how to express it simply, although we have the general definitions given previously. However, for the case where both subsystems A and B are spin-$\frac{1}{2}$ particles, there exists a simple formula from which the entanglement of formation can be calculated [42].

Given a density matrix ρ of a pair of quantum systems A and B and all possible pure-state decompositions of ρ

$$\rho = \sum_i p_i |\psi_i\rangle\langle\psi_i| \qquad (1)$$

where p_i are the probabilities for ensembles of states $|\psi_i\rangle$, the entanglement E is defined as the entropy of either of the subsystems A or B:

$$E(\psi) = -\text{Tr}(\rho_A \log_2 \rho_A) = -\text{Tr}(\rho_B \log_2 \rho_B) \qquad (2)$$

The entanglement of formation of the mixed ρ is then defined as the average entanglement of the pure states of the decomposition [42], minimized over all decompositions of ρ:

$$E(\rho) = \min \sum_i p_i E(\psi_i) \qquad (3)$$

For a pair of qubits this equation can be written [42–44]

$$E(\rho) = \varepsilon(C(\rho)) \qquad (4)$$

where ε is a function of the "concurrence" C:

$$\varepsilon(C) = h\left(\frac{1 + \sqrt{1 - C^2}}{2}\right) \qquad (5)$$

where h in the binary entropy function [20]

$$h(x) = -x \log_2 x - (1 - x) \log_2(1 - x) \qquad (5)$$

In this case the entanglement of formation is given in terms of another entanglement measure, the concurrence C [42–44]. The entanglement of formation varies monotonically with the concurrence. From the density matrix of the two-spin mixed states, the concurrence can be calculated as follows:

$$C(\rho) = \max[0, \lambda_1 - \lambda_2 - \lambda_3 - \lambda_4] \qquad (6)$$

where λ_i are the eigenvalues in decreasing order of the Hermitian matrix $R \equiv \sqrt{\sqrt{\rho}\tilde{\rho}\sqrt{\rho}}$ with $\tilde{\rho} = (\sigma^y \otimes \sigma^y)\rho^*(\sigma^y \otimes \sigma^y)$. Here σ^y is the Pauli matrix of the spin in the y direction. The concurrence varies from $C = 0$ for a separable state to $C = 1$ for a maximally entangled state. The concurrence as a measure of entanglement will be used in Section II to discuss tuning and manipulating the entanglement for spin systems.

B. Entanglement Measure for Fermions

As we discussed in the previous section, for distinguishable particles, the most suitable and famous measure of entanglement is Wootters' measure [42], the entanglement of formation or concurrence. Recently, Schlieman and co-workers [45, 46] examined the influence of quantum statistics on the definition of entanglement. They discussed a two-fermion system with the Slater decomposition instead of Schmidt decomposition for the entanglement measure. If we take each of the indistinguishable fermions to be in the single-particle Hilbert space C^N, with $f_m, f_m^+ (m = 1, \ldots, N)$ denoting the fermionic annihilation and creation operators of single-particle states and $|\Omega\rangle$ representing the vacuum state, then a pure two-electron state can be written

$$\sum_{m,n} \omega_{mn} f_m^+ f_n^+ |\Omega\rangle,$$

where $\omega_{mn} = -\omega_{nm}$.

Analogous to the Schmidt decomposition, it can be proved that every $|\Psi\rangle$ can be represented in an appropriately chosen basis in C^N in a form of Slater decomposition [45],

$$|\Psi\rangle = \frac{1}{\sqrt{\sum_{i=1}^{K} |z_i|^2}} \sum_{i=1}^{K} z_i f_{a_1(i)}^+ f_{a_2(i)}^+ |\Omega\rangle \qquad (7)$$

where $f_{a_1(i)}^+ |\Omega\rangle, f_{a_2(i)}^+ |\Omega\rangle$, $i = 1, \ldots, K$, form an orthonormal basis in C^N. The number of nonvanishing coefficients z_i is called the Slater rank, which is then used for the entanglement measure. With similar technique, the case of a two-boson system is studied by Li et al. [47] and Paškauskas and you [48].

Gittings and Fisher [49] put forward three desirable properties of any entanglement measure: (i) invariance under local unitary transformations; (ii) noninvariance under nonlocal unitary transformations; and (iii) correct behavior as distinguishability of the subsystems is lost. These requirements make the relevant distinction between one-particle unitary transformation and one-site unitary transformations. A natural way to achieve this distinction [49] is to use a basis based on sites rather than on particles. Through the Gittings–Fisher investigation, it is shown that all of the above-discussed entanglement measures fail the tests of the three criteria. Only Zanardi's measure [50] survives, which is given in Fock space as the von Neumann entropy, namely,

$$E_j = -\text{Tr}\rho_j \log_2 \rho_j, \quad \rho_j = \text{Tr}_j |\psi\rangle\langle\psi| \tag{8}$$

where Tr_j denotes the trace over all but the jth site and ψ is the antisymmetric wavefunction of the studied system. Hence E_j actually describes the entanglement of the jth site with the remaining sites. A generalization of this one-site entanglement is to define an entanglement between one L-site block with the rest of the system [51],

$$E_L = -\text{Tr}(\rho_L \log_2 \rho_L) \tag{9}$$

C. Entanglement and Ranks of Density Matrices

In this section we review the known theorems that relate entanglement to the ranks of density matrices [52]. The rank of a matrix ρ, denoted as rank(ρ), is the maximal number of linearly independent row vectors (also column vectors) in the matrix ρ. Based on the ranks of reduced density matrices, one can derive necessary conditions for the separability of multiparticle arbitrary-dimensional mixed states, which are equivalent to sufficient conditions for entanglement [53]. For convenience, let us introduce the following definitions [54–56]. A pure state ρ of N particles $A_1, A_2, \ldots, A_N$ is called entangled when it cannot be written

$$\rho = \rho_{A_1} \otimes \rho_{A_2} \otimes \cdots \otimes \rho_{A_N} = \bigotimes_{i=1}^{N} \rho_{A_i} \tag{10}$$

where ρ_{A_i} is the single-particle reduced density matrix given by $\rho_{A_i} \equiv \text{Tr}_{\{A_j\}}(\rho)$ for $\{A_j | \text{all } A_j \neq A_i\}$. A mixed state ρ of N particles $A_1, A_2, \ldots, A_N$, described by M probabilities p_j and M pure states ρ^j as $\rho = \sum_{j=1}^{M} p_j \rho^j$, is called entangled when it cannot be written

$$\rho = \sum_{j=1}^{M} p_j \bigotimes_{i=1}^{N} \rho_{A_i}^j \tag{11}$$

where $p_j > 0$ for $j = 1, 2, \ldots, M$ with $\sum_{j=1}^{M} p_j = 1$.

Now we are in a position to list the separability conditions without proof. (The reader who is interested in the formal proofs can consult the paper by Chong, Keiter, and Stolze [53].)

Lemma 1 A state is pure if and only if the rank of its density matrix ρ is equal to 1, that is, $\text{rank}(\rho) = 1$.

Lemma 2 A pure state is entangled if and only if the rank of at least one of its reduced density matrices is greater than 1.

Lemma 3 Given a pure state ρ, if its particles are separated into two parts U and V, then $\text{rank}(\rho_U) = 1$ holds if and only if these two parts are separable, that is, $\rho = \rho_U \otimes \rho_V$.

Now we can discuss the necessary conditions for separable states. For convenience, we will use the following notation. For a state ρ of N particles $A_1, A_2, \ldots, A_N$, the reduced density matrix obtained by tracing ρ over particle A_i is written $\rho_{R(i)} = \text{Tr}_{A_i}(\rho)$, where $R(i)$ denotes the set of the remaining $(N-1)$ particles other than particle A_i. In the same way, $\rho_{R(i,j)} = \text{Tr}_{A_j}(\rho_{R(i)}) = \text{Tr}_{A_j}(\text{Tr}_{A_i}(\rho)) = \text{Tr}_{A_i}(\text{Tr}_{A_j}(\rho))$ denotes the reduced density matrix obtained by tracing ρ over particles A_i and A_j, $\rho_{R(i,j,k)} = \text{Tr}_{A_i}(\text{Tr}_{A_j}(\text{Tr}_{A_k}(\rho)))$, and so on. In view of these relations, ρ can be called the 1-level-higher density matrix of $\rho_{R(i)}$ and 2-level-higher density matrix of $\rho_{R(i,j)}$; $\rho_{R(i)}$ can be called the 1-level-higher density matrix of $\rho_{R(i,j)}$ and 2-level-higher density matrix of $\rho_{R(i,j,k)}$; and so on.

Now let us define the *separability condition theorem* [53]. If a state ρ of N particles $A_1, A_2, \ldots, A_N$ is separable, then the rank of any reduced density matrix of ρ must be less than or equal to the ranks of all of its 1-level-higher density matrices; that is,

$$\text{rank}(\rho_{R(i)}) \leq \text{rank}(\rho) \tag{12}$$

holds for any $A_i \in \{A_1, A_2, \ldots, A_N\}$; and

$$\text{rank}(\rho_{R(i,j)}) \leq \text{rank}(\rho_{R(i)}), \quad \text{rank}(\rho_{R(i,j)}) \leq \text{rank}(\rho_{R(j)}) \tag{13}$$

holds for any pair of all particles.

This will lead to the conditions for a mixed state to be entangled. *Given a mixed state ρ, if the rank of at least one of the reduced density matrices of ρ is greater than the rank of one of its 1-level-higher density matrices, then the state ρ is entangled.*

II. ENTANGLEMENT FOR SPIN SYSTEMS

A. Entanglement for Two-Spin Systems

We consider a set of N localized spin-$\frac{1}{2}$ particles coupled through exchange interaction J and subject to an external magnetic field of strength B. In this section we

ENTANGLEMENT, ELECTRON CORRELATION, AND DENSITY MATRICES 501

will demonstrate that (i) entanglement can be controlled and tuned by varying the anisotropy parameter in the Hamiltonian and by introducing impurities into the systems; (ii) for certain parameters, the entanglement is zero up to a critical point λ_c, where a quantum phase transition occurs, and is different from zero above λ_c; and (iii) entanglement shows scaling behavior in the vicinity of the transition point.

For simplicity, let us illustrate the calculations of entanglement for two spin-$\frac{1}{2}$ particles. The general Hamiltonian, in atomic units, for such a system is given by [57]

$$H = -\frac{J}{2}(1+\gamma)\sigma_1^x \otimes \sigma_2^x - \frac{J}{2}(1-\gamma)\sigma_1^y \otimes \sigma_2^y - B\sigma_1^z \otimes I_2 - BI_1 \otimes \sigma_2^z \quad (14)$$

where σ^a are the Pauli matrices ($a = x, y, z$) and γ is the degree of anisotropy. For $\gamma = 1$ Eq. (14) reduces to the Ising model, whereas for $\gamma = 0$ it is the XY model.

This model admits an exact solution; it is simply a (4×4) matrix of the form

$$H = \begin{pmatrix} -2B & 0 & 0 & -J\gamma \\ 0 & 0 & -J & 0 \\ 0 & -J & 0 & 0 \\ -J\gamma & 0 & 0 & 2B \end{pmatrix} \quad (15)$$

with the following four eigenvalues,

$$\lambda_1 = -J, \quad \lambda_2 = J, \quad \lambda_3 = -\sqrt{4B^2 + J^2\gamma^2}, \quad \lambda_4 = \sqrt{4B^2 + J^2\gamma^2} \quad (16)$$

and the corresponding eigenvectors,

$$|\phi_1\rangle = \begin{pmatrix} 0 \\ 1/\sqrt{2} \\ 1/\sqrt{2} \\ 0 \end{pmatrix}, \quad |\phi_2\rangle = \begin{pmatrix} 0 \\ -1/\sqrt{2} \\ 1/\sqrt{2} \\ 0 \end{pmatrix} \quad (17)$$

$$|\phi_3\rangle = \begin{pmatrix} \sqrt{\frac{\alpha+2B}{2\alpha}} \\ 0 \\ 0 \\ \sqrt{\frac{\alpha-2B}{2\alpha}} \end{pmatrix}, \quad |\phi_4\rangle = \begin{pmatrix} -\sqrt{\frac{\alpha-2B}{2\alpha}} \\ 0 \\ 0 \\ \sqrt{\frac{\alpha+2B}{2\alpha}} \end{pmatrix} \quad (18)$$

where $\alpha = \sqrt{4B^2 + J^2\gamma^2}$. In the basis set $\{|\uparrow\uparrow\rangle, |\uparrow\downarrow\rangle, |\downarrow\uparrow\rangle, |\downarrow\downarrow\rangle\}$, the eigenvectors can be written

$$|\phi_1\rangle = \frac{1}{\sqrt{2}}(|\downarrow\uparrow\rangle + |\uparrow\downarrow\rangle) \qquad (19)$$

$$|\phi_2\rangle = \frac{1}{\sqrt{2}}(|\downarrow\uparrow\rangle - |\uparrow\downarrow\rangle) \qquad (20)$$

$$|\phi_3\rangle = \sqrt{\frac{\alpha - 2B}{2\alpha}}|\downarrow\downarrow\rangle + \sqrt{\frac{\alpha + 2B}{2\alpha}}|\uparrow\uparrow\rangle \qquad (21)$$

$$|\phi_4\rangle = \sqrt{\frac{\alpha + 2B}{2\alpha}}|\downarrow\downarrow\rangle - \sqrt{\frac{\alpha - 2B}{2\alpha}}|\uparrow\uparrow\rangle \qquad (22)$$

Now we confine our interest to the calculation of entanglement between the two spins. For simplicity, we take $\gamma = 1$; Eq. (14) reduces to the Ising model with the ground-state energy λ_3 and the corresponding eigenvector $|\phi_3\rangle$. All the information needed for quantifying the entanglement in this case is contained in the reduced density matrix $\rho(i,j)$ [42–44].

For our model system in the ground state $|\phi_3\rangle$, the density matrix in the basis set $(\uparrow\uparrow, \uparrow\downarrow, \downarrow\uparrow, \downarrow\downarrow)$ is given by

$$\rho = \begin{pmatrix} \frac{\alpha + 2B}{2\alpha} & 0 & 0 & \sqrt{\frac{\alpha^2 - 4B^2}{4\alpha^2}} \\ 0 & 0 & 0 & 0 \\ 0 & 0 & 0 & 0 \\ \sqrt{\frac{\alpha^2 - 4B^2}{4\alpha^2}} & 0 & 0 & \frac{\alpha + 2B}{2\alpha} \end{pmatrix} \qquad (23)$$

The eigenvalues of the Hermitian matrix R needed to calculate the concurrence [42], C, Eq. (6), can be calculated analytically. We obtained $\lambda_2 = \lambda_3 = \lambda_4 = 0$ and therefore

$$C(\rho) = \lambda_1 = \sqrt{\frac{\lambda^2}{4 + \lambda^2}} \qquad (24)$$

where $\lambda = J/B$. Entanglement is a monotonically increasing function of the concurrence and is given by

$$E(C) = h(y) = -y\log_2 y - (1-y)\log_2(1-y); \quad y = \tfrac{1}{2} + \tfrac{1}{2}\sqrt{1-C^2} \qquad (25)$$

Substituting the value of the concurrence C, Eq. (24) gives

$$E = -\frac{1}{2}\log_2\left(\frac{1}{4} - \frac{1}{4+\lambda^2}\right) + \frac{1}{\sqrt{4+\lambda^2}}\log_2\frac{\sqrt{4+\lambda^2} - 2}{\sqrt{4+\lambda^2} + 2} \qquad (26)$$

This result for entanglement is equivalent to the von Neumann entropy of the reduced density matrix ρ_A. For our model system of the form AB in the ground state $|\phi_3\rangle$, the reduced density matrix $\rho_A = \text{Tr}_B(\rho_{AB})$ in the basis set $(\uparrow, \downarrow)$ is given by

$$\rho_A = \begin{pmatrix} \dfrac{\alpha + 2B}{2\alpha} & 0 \\ 0 & \dfrac{\alpha - 2B}{2\alpha} \end{pmatrix} \tag{27}$$

As we mentioned before, when a biparticle quantum system AB is in a pure state, there is essentially a unique measure of the entanglement between the subsystems A and B given by the von Neumann entropy $S \equiv -\text{Tr}[\rho_A \log_2 \rho_A]$. This approach gives exactly the same formula as the one given in Eq. (26). This is not surprising since all entanglement measures should coincide on pure bipartite states and be equal to the von Neumann entropy of the reduced density matrix (uniqueness theorem).

This simple model can be used to examine the entanglement for two-electron diatomic molecules. The value of J, the exchange coupling constant between the spins of the two electrons, can be calculated as half the energy difference between the lowest singlet and triplet states of the hydrogen molecule. Herring and Flicker [58] have shown that J for the H_2 molecule can be approximated as a function of the interatomic distance R. In atomic units, the expression for large R is given by

$$J(R) = -0.821\ R^{5/2} e^{-2R} + O(R^2 e^{-2R}) \tag{28}$$

Figure 1 shows the calculated concurrence $C(\rho)$ as a function of the distance between the two electronic spins R, using $J(R)$ of Eq. (28), for different values of the magnetic field strength B. At the limit $R \to \infty$ the exchange interaction J vanishes as a result of the two electronic spins being up and the wavefunction being factorizable; that is, the concurrence is zero. At the other limit, when $R = 0$ the concurrence or the entanglement is zero for this model because $J = 0$. As R increases, the exchange interaction increases, leading to increasing concurrence between the two electronic spins. However, this increase in the concurrence reaches a maximum limit as shown in the figure. For large distance, the exchange interaction decreases exponentially with R and thus the decrease of the concurrence. Figure 1 also shows that the concurrence increases with decreasing magnetic field strength. This can be attributed to effectively increasing the exchange interaction. This behavior of the concurrence as a function of the internuclear distance R is typical for two-electron diatomic molecules. We will show later in Section IV that by using accurate ab initio calculations we essentially obtain qualitatively the same curve for entanglement for the H_2 molecule as a function of the internuclear distance R.

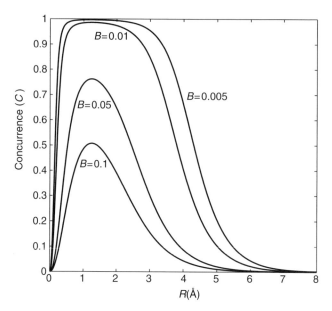

Figure 1. The concurrence (C) as a function of the distance R between the two spins for different values of the magnetic field strength B.

B. Entanglement for One-Dimensional N-Spin Systems

Now let us generalize it to a one-dimensional lattice with N sites in a transverse magnetic field and with impurities. The Hamiltonian for such a system is given by [59]

$$H = -\frac{1+\gamma}{2}\sum_{i=1}^{N} J_{i,i+1}\sigma_i^x \sigma_{i+1}^x - \frac{1-\gamma}{2}\sum_{i=1}^{N} J_{i,i+1}\sigma_i^y \sigma_{i+1}^y - \sum_{i=1}^{N} B_i \sigma_i^z \qquad (29)$$

where $J_{i,i+1}$ is the exchange interaction between sites i and $i+1$, B_i is the strength of the external magnetic field at site i, σ^a are the Pauli matrices ($a = x, y, z$), γ is the degree of anisotropy, and N is the number of sites. We assume cyclic boundary conditions, so that

$$\sigma_{N+1}^x = \sigma_1^x, \quad \sigma_{N+1}^y = \sigma_1^y, \quad \sigma_{N+1}^z = \sigma_1^z \qquad (30)$$

For $\gamma = 1$ the Hamiltonian reduces to the Ising model and for $\gamma = 0$ to the XY model. For the pure homogeneous case, $J_{i,i+1} = J$ and $B_i = B$, the system exhibits a quantum phase transition at a dimensionless coupling constant

$\lambda = J/2B = 1$. The magnetization $\langle \sigma^x \rangle$ is different from zero for $\lambda > 1$ and it vanishes at the transition point. The magnetization along the z direction $\langle \sigma^z \rangle$ is different from zero for any value of λ. At the phase transition point, the correlation length ξ diverges as $\xi \sim |\lambda - \lambda_c|^{-\nu}$ with $\nu = 1$ [60].

C. Numerical Solution of the One-Dimensional Spin-$\frac{1}{2}$ Systems

The standard procedure used to solve Eq. (29) is to transform the spin operators into fermionic operators [61]. Let us define the raising and lowering operators a_i^+, a_i^-,:

$$a_i^+ = \tfrac{1}{2}(\sigma_i^x + i\sigma_i^y); \quad a_i^- = \tfrac{1}{2}(\sigma_i^x - i\sigma_i^y)$$

Then we introduce the Fermi operators c_i, c_i^+ defined by

$$a_i^- = \exp\left(-\pi i \sum_{j=1}^{i-1} c_j^+ c_j\right) c_i; \quad a_i^+ = c_i^+ \exp\left(\pi i \sum_{j=1}^{i-1} c_j^+ c_j\right)$$

So the Hamiltonian assumes the following quadratic form:

$$H = -\sum_{i=1}^{N} J_{i,i+1}[(c_i^+ c_{i+1} + \gamma c_i^+ c_{i+1}^+) + \text{h.c.}] - 2\sum_{i=1}^{N} B_i\left(c_i^+ c_i - \frac{1}{2}\right) \quad (31)$$

$$\lambda = J/2B$$

We can write the parameters $J_{i,i+1} = J(1 + \alpha_{i,i+1})$, where α introduces the impurity in the exchange interactions, and the external magnetic field takes the form $B_i = B(1 + \beta_i)$, where β measures the impurity in the magnetic field. When $\alpha = \beta = 0$ we recover the pure XY case.

Introducing the matrices **A**, **B**, where **A** is symmetrical and **B** is antisymmetrical, we can rewrite the Hamiltonian:

$$H' = \sum_{i,j=1}^{N}\left[c_i^+ A_{i,j} c_j + \frac{1}{2}(c_i^+ B_{i,j} c_j^+ + \text{h.c})\right]$$

Introducing linear transformation, we have

$$\eta_k = \sum_{i=1}^{N} g_{ki} c_i + h_{ki} c_i^+; \quad \eta_k^+ = \sum_{i=1}^{N} g_{ki} c_i^+ + h_{ki} c_i$$

with the g_{ki} and h_{ki} real and which will give the Hamiltonian form

$$H = \sum_{k}^{N} \Lambda_k \eta_k^+ \eta_k + \text{constant}$$

From these conditions, we can get a set of equations for the g_{ki} and h_{ki}:

$$\Lambda_k g_{ki} = \sum_{j=1}^{N} (g_{kj} \mathbf{A}_{ji} - h_{kj} \mathbf{B}_{ji}) \qquad (32)$$

$$\Lambda_k h_{ki} = \sum_{j=1}^{N} (g_{kj} \mathbf{B}_{ji} - h_{kj} \mathbf{A}_{ji}) \qquad (33)$$

By introducing the linear combinations

$$\phi_{ki} = g_{ki} + h_{ki}; \quad \psi_{ki} = g_{ki} - h_{ki}$$

we can get the coupled equation

$$\phi_k(\mathbf{A} - \mathbf{B}) = \Lambda_k \psi_k \quad \text{and} \quad \psi_k(\mathbf{A} + \mathbf{B}) = \Lambda_k \phi_k$$

Then we can get both ϕ_k and ψ_k vectors from these two equations by the numerical method [62]. The ground state of the system corresponds to the state of "no-particles" and is denoted as $|\Psi_0\rangle$, and

$$\eta_k |\Psi_0\rangle = 0, \quad \text{for all } k$$

D. Entanglement and Spin Reduced Density Matrices

The matrix elements of the reduced density matrix needed to calculate the entanglement can be written in terms of the spin–spin correlation functions and the average magnetization per spin. The spin–spin correlation functions for the ground state are defined as [62]

$$S_{lm}^x = \tfrac{1}{4} \langle \Psi_0 | \sigma_l^x \sigma_m^x | \Psi_0 \rangle$$
$$S_{lm}^y = \tfrac{1}{4} \langle \Psi_0 | \sigma_l^y \sigma_m^y | \Psi_0 \rangle$$
$$S_{lm}^z = \tfrac{1}{4} \langle \Psi_0 | \sigma_l^z \sigma_m^z | \Psi_0 \rangle$$

and the average magnetization per spin is

$$M_i^z = \tfrac{1}{2} \langle \Psi_0 | \sigma_i^z | \Psi_0 \rangle \qquad (34)$$

These correlation functions can be obtained using the set ψ_k and ϕ_k from the previous section.

The structure of the reduced density matrix follows from the symmetry properties of the Hamiltonian. However, for this case the concurrence $C(i,j)$ depends on i,j and the location of the impurity and not only on the difference $|i-j|$ as for the pure case. Using the operator expansion for the density matrix and the symmetries of the Hamiltonian leads to the general form

$$\rho = \begin{pmatrix} \rho_{1,1} & 0 & 0 & \rho_{1,4} \\ 0 & \rho_{2,2} & \rho_{2,3} & 0 \\ 0 & \rho_{2,3} & \rho_{3,3} & 0 \\ \rho_{1,4} & 0 & 0 & \rho_{4,4} \end{pmatrix} \qquad (35)$$

with

$$\lambda_a = \sqrt{\rho_{1,1}\rho_{4,4}} + |\rho_{1,4}|, \quad \lambda_b = \sqrt{\rho_{2,2}\rho_{3,3}} + |\rho_{2,3}| \qquad (36)$$

$$\lambda_c = |\sqrt{\rho_{1,1}\rho_{4,4}} - |\rho_{1,4}||, \quad \lambda_d = |\sqrt{\rho_{2,2}\rho_{3,3}} - |\rho_{2,3}|| \qquad (37)$$

Using the definition $<A> = \mathrm{Tr}(\rho A)$, we can express all the matrix elements in the density matrix in terms of different spin–spin correlation functions [62]:

$$\rho_{1,1} = \tfrac{1}{2}M_l^z + \tfrac{1}{2}M_m^z + S_{lm}^z + \tfrac{1}{2} \qquad (38)$$

$$\rho_{2,2} = \tfrac{1}{2}M_l^z - \tfrac{1}{2}M_m^z - S_{lm}^z + \tfrac{1}{4} \qquad (39)$$

$$\rho_{3,3} = \tfrac{1}{2}M_m^z - \tfrac{1}{2}M_l^z - S_{lm}^z + \tfrac{1}{4} \qquad (40)$$

$$\rho_{4,4} = -\tfrac{1}{2}M_l^z - \tfrac{1}{2}M_m^z + S_{lm}^z + \tfrac{1}{4} \qquad (41)$$

$$\rho_{2,3} = S_{lm}^x + S_{lm}^y \qquad (42)$$

$$\rho_{1,4} = S_{lm}^x - S_{lm}^y \qquad (43)$$

E. Some Numerical Results

Let us show how the entanglement can be tuned by changing the anisotropy parameter γ by going from the Ising model ($\gamma = 1$) to the XY model ($\gamma = 0$). For the XY model the entanglement is zero up to the critical point λ_c and is different from zero above λ_c. Moreover, by introducing impurities, the entanglement can be tuned down as the strength of the impurity α increases [59]. First, we examine the change of the entanglement for the Ising model ($\gamma = 1$) for different values of the impurity strength α as the parameter λ, which induces the quantum phase transitions, varies. Figure 2 shows the change of the nearest-neighbor concurrence $C(1,2)$ with the impurity located at $i_m = 3$ as a

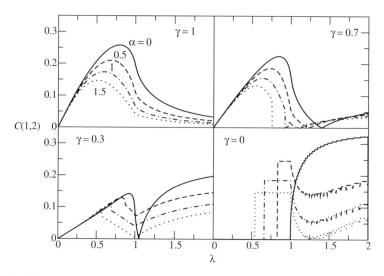

Figure 2. The nearest-neighbor concurrence $C(1,2)$ for different values of the anisotropy parameter $\gamma = 1$, 0.7, 0.3, 0 with an impurity located at $i_m = 3$ as a function of the reduced coupling constant $\lambda = J/2h$, where J is the exchange interaction constant and h is the strength of the external magnetic field. The curves correspond to different values of the impurity strength $\alpha = 0, 0.5, 1, 1.5$ with system size $N = 201$.

function of λ for different values of α. One can see clearly in Figure 2 that the entanglement can be tuned down by increasing the value of the parameter α. For $\alpha = 1.5$, the concurrence approaches zero above the critical $\lambda_c = 1$. The system size was taken as $N = 201$ based on finite size scaling analysis. Analysis of all the results for the pure case ($\alpha = 0$) for different system sizes ranging from $N = 41$ up to $N = 401$ collapse into a single curve. Thus all key ingredients of the finite size scaling are present in the concurrence. This holds true for the impurity problem as long as we consider the behavior of the value of λ for which the derivative of the concurrence attains its minimum value versus the system size. As expected, there is no divergence of the derivative $dC(1,2)/d\lambda$ for finite N, but there are clear anomalies. By examining $\ln(\lambda_c - \lambda_m)$ versus $\ln N$ for $\alpha = 0.1$, one obtains that the minimum λ_m scales as $\lambda_m \sim \lambda_c + N^{-0.93}$ and $dC(1,2)/d\lambda$ diverges logarithmically with increasing system size. For a system with the impurity located at larger distance $i_m = 10$ and the same $\alpha = 0.1$, $\lambda_m \sim \lambda_c + N^{-0.85}$, showing that the scaling behavior depends on the distance between the impurity and the pair of sites under consideration.

Figure 2 also shows the variation of nearest-neighbor concurrence as the anisotropy parameter γ decreases. For the XY model ($\gamma = 0$), the concurrence for $\alpha = 0$ is zero up to the critical point $\lambda_c = 1$ and different from zero above $\lambda_c = 1$. However, as α increases the concurrence develops steps and the results strongly depend on the system size. For small system size, such as $N = 101$, the steps and oscillations are large but become smaller as the system size increases as shown in Fig. 2 for $N = 201$. But they disappear in the limit $N \to \infty$. To examine the different behavior of the concurrence for the Ising model and the XY model, we took the system size to be infinite, $N \to \infty$, where the two models have exact solutions. However, the behavior is the same for a finite system with $N = 201$. For larger values of i_m the concurrence gets larger and approaches its maximum value, the pure case with $\alpha = 0$, at large values $i_m \gg 1$. It is worth mentioning that, for the Ising model, the range of entanglement [63], which is the maximum distance between spins at which the concurrence is different from zero, vanishes unless the two sites are at most next-nearest neighbors. For $\gamma \neq 1$, the range of entanglement is not universal and tends to infinity as γ tends to zero.

So far we have examined the change of entanglement as the degree of the anisotropy γ varies between zero and one and by introducing impurities at fixed sites. Rather than locating the impurity at one site in the chain, we can also introduce a Gaussian distribution of the disorder near a particular location [62]. This can be done by modifying α, the exchange interaction, where α introduces the impurity in a Gaussian form centered at $(N+1)/2$ with strength or height ζ:

$$\alpha_{i,i+1} = \zeta e^{-\epsilon(i-(N+1)/2)^2} \tag{44}$$

The external magnetic field can also be modified to take the form $h_i = h(1 + \beta_i)$, where β has the following Gaussian distribution [62]:

$$\beta_i = \xi e^{-\epsilon(i-(N+1)/2)^2} \tag{45}$$

where ϵ is a parameter to be fixed. Numerical calculations show that the entanglement can be tuned in this case by varying the strengths of the magnetic field and the impurity distribution in the system. The concurrence is maximum close to λ_c and can be tuned to zero above the critical point.

F. Thermal Entanglement and the Effect of Temperature

Recently, the concept of thermal entanglement was introduced and studied within one-dimensional spin systems [64–66]. The state of the system described by the Hamiltonian H at thermal equilibrium is $\rho(T) = \exp(-H/kT)/Z$, where $Z = \text{Tr}[\exp(-H/kT)]$ is the partition function and k is Boltzmann's constant. As $\rho(T)$ represents a thermal state, the entanglement in the state is called the thermal entanglement [64].

For a two-qubit isotropic Heisenberg model, there exists thermal entanglement for the antiferromagnetic case and no thermal entanglement for the ferromagnetic case [64]; while for the XY model the thermal entanglement appears for both the antiferromagnetic and ferromagnetic cases [67, 68]. It is known that the isotropic Heisenberg model and the XY model are special cases of the anisotropic Heisenberg model.

Now that the entanglement of the XY Hamiltonian with impurities has been calculated at $T = 0$, we can consider the case where the system is at thermal equilibrium at temperature T. The density matrix for the XY model at thermal equilibrium is given by the canonical ensemble $\rho = e^{-\beta H}/Z$, where $\beta = 1/k_B T$, and $Z = \text{Tr}\,(e^{-\beta H})$ is the partition function. The thermal density matrix is diagonal when expressed in terms of the Jordan–Wigner fermionic operators. Our interest lies in calculating the quantum correlations present in the system as a function of the parameters β, γ, λ, and α.

For the pure Ising model with $\alpha = 0$, the constructed two-site density matrices [66] are valid for all temperatures. By using these matrices, it is possible to study the purely two-party entanglement present at thermal equilibrium because the concurrence measure of entanglement can be applied to arbitrary mixed states. For this model the influence that the critical point has on the entanglement structure at nonzero temperatures is particularly clear. The entanglement between nearest-neighbor in the Ising model at nonzero temperature is shown in Fig. 3. The entanglement is nonzero only in a certain region in the $k_B T$–λ plane. It is in this region that quantum effects are likely to dominate the behavior of the system. The entanglement is largest in the vicinity of the critical point $\lambda = 1$, $k_B T = 0$.

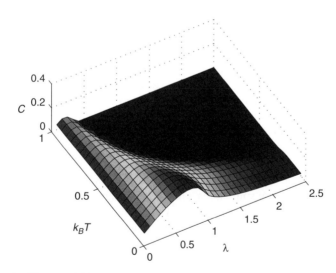

Figure 3. Nearest-neighbor concurrence C at nonzero temperature for the transverse Ising model.

Figure 3 shows that, for certain values of λ, the two-site entanglement can increase as the temperature is increased. Moreover, it shows the existence of appreciable entanglement in the system for temperatures $k_B T$ above the ground-state energy gap Δ. It has been argued that quantum systems behave classically when the temperature exceeds all relevant frequencies. For the transverse Ising model, the only relevant frequency is given by the ground-state energy gap $\Delta \equiv \hbar\omega$. The presence of entanglement in the system for temperatures above the energy gap indicates that quantum effects may persist past the point where they are usually expected to disappear.

The zero-temperature calculations of the previous section Section, the XY model with impurities, represent a highly idealized situation; however, it is unclear whether they have any relevance to the system at nonzero temperature. Since the properties of a quantum system for low temperatures are strongly influenced by nearby quantum critical points, it is tempting to attribute the effect of nearby critical points to persistent mixed-state entanglement in the thermal state.

G. Entanglement for Two-Dimensional Spin Systems

Quantum spin systems in two-dimensional lattices have been the subject of intense research, mainly motivated by their possible relevance in the study of high-temperature superconductors [69]. On the other hand, high magnetic field experiments on materials with a two-dimensional structure, which can be described by the Heisenberg antiferromagnetic model in frustrated lattices, have revealed novel phases as plateaus and jumps in the magnetization curves [70] and might be useful for quantum computations. Among the many different techniques that have been used to study such systems, the generalization of the celebrated Jordan–Wigner transformation [71] to two spatial dimensions [72] has some appealing features. It allows one to write the spin Hamiltonian completely in terms of spinless fermions in such a way that the $S = \frac{1}{2}$ single-particle constraint is automatically satisfied due to the Pauli principle, while the magnetic field enters as the chemical potential for the Jordan–Wigner fermions. This method has been applied to study the XXZ Heisenberg antiferromagnet [73–75].

For this case one can use the Jordan–Wigner transformation since it is a generalization of the well-known transformation in one dimension that we have used in previous sections. The Jordan–Wigner transformation is exact but the resulting Hamiltonian is highly nonlocal and some kind of approximation is necessary to proceed. One can use numerical methods such as Monte Carlo and variational approach to deal with the transformed Hamiltonian. This will allow us to explore the ground state of two-dimensional lattice spin $\frac{1}{2}$ systems, in a way that could be applied to arbitrary lattice topologies. The method can also be used in the presence of an external magnetic field, at finite temperature, and can even be applied to disordered systems. Once this is solved and we have the density matrix, we can follow the previous procedure to examine the entanglement as the parameters of external magnetic field, temperature, lattice topologies, and impurities vary.

III. ENTANGLEMENT FOR QUANTUM DOT SYSTEMS

A. Two-Electron Two-Site Hubbard Model

Many electron systems such as molecules and quantum dots show the complex phenomena of electron correlation caused by Coulomb interactions. These phenomena can be described to some extent by the Hubbard model [76]. This is a simple model that captures the main physics of the problem and admits an exact solution in some special cases [77]. To calculate the entanglement for electrons described by this model, we will use Zanardi's measure, which is given in Fock space as the von Neumann entropy [78].

1. Exact Solution

The Hamiltonian of the two-electron two-site Hubbard model can be written [77]

$$H = -\frac{t}{2} \sum_{i,\sigma} c_{i\sigma}^{\dagger} c_{i\sigma} + 2U \sum_{i} \hat{n}_{i\uparrow} \hat{n}_{i\downarrow} \tag{46}$$

where $c_{i\sigma}^{\dagger}$ and $c_{i\sigma}$ are the Fermi creation and annihilation operators at site i and with spin $\sigma = \uparrow, \downarrow$ and $\hat{n}_{i\sigma} = c_{i\sigma}^{\dagger} c_{i\sigma}$ is the spin-dependent occupancy operator at site i. For a two-site system $i = 1$ and 2, $\bar{i} = 3 - i$, $t/2$ is the hopping term of different sites, and $2U$ is the on-site interaction ($U > 0$ for repulsion in our case). The factors $t/2$ and $2U$ are chosen to make the following expressions for eigenvalues and eigenvectors as simple as possible. This Hamiltonian can be solved exactly in the basis set $|1\uparrow, 1\downarrow, 2\uparrow, 2\downarrow\rangle$; it is simply a (4×4) matrix of the form

$$H = \begin{pmatrix} 2U & -t/2 & -t/2 & 0 \\ -t/2 & 0 & 0 & -t/2 \\ -t/2 & 0 & 0 & -t/2 \\ 0 & -t/2 & -t/2 & 2U \end{pmatrix} \tag{47}$$

with the following four eigenvalues and eigenvectors,

$$\lambda_1 = U - \sqrt{t^2 + U^2}, \quad \lambda_2 = 0, \quad \lambda_3 = 2U, \quad \lambda_4 = U + \sqrt{t^2 + U^2} \tag{48}$$

and the corresponding eigenvectors,

$$|\phi_1\rangle = \begin{pmatrix} 1 \\ x+\sqrt{1+x^2} \\ x+\sqrt{1+x^2} \\ 1 \end{pmatrix}, \quad |\phi_2\rangle = \begin{pmatrix} 0 \\ -1 \\ 1 \\ 0 \end{pmatrix}, \quad (49)$$

$$|\phi_3\rangle = \begin{pmatrix} -1 \\ 0 \\ 0 \\ 1 \end{pmatrix}, \quad |\phi_4\rangle = \begin{pmatrix} 1 \\ x-\sqrt{1+x^2} \\ x-\sqrt{1+x^2} \\ 1 \end{pmatrix} \quad (50)$$

with $x = U/t$. The eigenvalue and eigenvector for the ground state are

$$E = U - \sqrt{t^2 + U^2} \quad (51)$$

and

$$|GS\rangle = |1, x+\sqrt{1+x^2}, x+\sqrt{1+x^2}, 1\rangle \quad (52)$$

2. Hartree–Fock Approximation

In quantum chemistry, the correlation energy E_{corr} is defined as $E_{\text{corr}} = E_{\text{exact}} - E_{\text{HF}}$. In order to calculate the correlation energy of our system, we show how to calculate the ground state using the Hartree–Fock approximation. The main idea is to expand the exact wavefunction in the form of a configuration interaction picture. The first term of this expansion corresponds to the Hartree–Fock wavefunction. As a first step we calculate the spin-traced one-particle density matrix [5] (1PDM) γ:

$$\gamma_{ij} = \langle GS| \sum_\sigma c_{i\sigma}^\dagger c_{j\sigma} |GS\rangle \quad (53)$$

We obtain

$$\gamma = \begin{pmatrix} 1 & 2\alpha\beta \\ 2\alpha & 1 \end{pmatrix} \quad (54)$$

where

$$\alpha = \frac{1}{\sqrt{2}}\sqrt{1 - \frac{x}{1+x^2}} \quad \text{and} \quad \beta = \frac{1}{\sqrt{2}}\sqrt{1 + \frac{x}{1+x^2}}$$

Diagonalizing this 1PDM, we can get the binding (+) and unbinding (−) molecular natural orbitals (NOs),

$$|\pm\rangle = \frac{1}{\sqrt{2}}(|1\rangle \pm |2\rangle) \qquad (55)$$

and the corresponding eigenvalues

$$n_\pm = 1 \pm \frac{1}{\sqrt{1+x^2}} \qquad (56)$$

where $|1\rangle$ and $|2\rangle$ are the spatial orbitals of sites 1 and 2, respectively. The NOs for different spins are defined as

$$|\pm\sigma\rangle = \frac{1}{\sqrt{2}}(c^\dagger_{1\sigma} \pm c^\dagger_{2\sigma})|0\rangle \equiv c^\dagger_{\pm\sigma}|0\rangle \qquad (57)$$

where $|0\rangle$ is the vacuum state. After we define the geminals $|\pm\pm\rangle = c^\dagger_{\pm\uparrow}c^\dagger_{\pm\downarrow}|0\rangle$, we can express $|GS\rangle$ in terms of NOs as

$$|GS\rangle = \sqrt{\frac{n_+}{2}}|++\rangle - \operatorname{sgn} U \sqrt{\frac{n_-}{2}}|--\rangle \qquad (58)$$

In the Hartree–Fock approximation, the GS is given by $|HF\rangle = |++\rangle$ and $E_{HF} = -t + U$. Let us examine the Hartree–Fock results by defining the ionic and nonionic geminals, respectively:

$$\begin{aligned}|A\rangle &= \frac{1}{\sqrt{2}}(c^\dagger_{1\uparrow}c^\dagger_{1\downarrow} + c^\dagger_{2\uparrow}c^\dagger_{2\downarrow})|0\rangle \\ |B\rangle &= \frac{1}{\sqrt{2}}(c^\dagger_{1\uparrow}c^\dagger_{2\downarrow} + c^\dagger_{2\uparrow}c^\dagger_{1\downarrow})|0\rangle\end{aligned} \qquad (59)$$

If $x \to 0$, the system is equally mixed between ionic and nonionic germinal, $|HF\rangle = |A\rangle + |B\rangle$. When $x \to +\infty$, $|GS\rangle \to |B\rangle$, which indicates that as x becomes large, our system goes to the nonionic state. Similarly, $|GS\rangle \to |C\rangle$, as $x \to -\infty$, where

$$|C\rangle = \frac{1}{\sqrt{2}}(c^\dagger_{1\uparrow}c^\dagger_{1\downarrow} - c^\dagger_{2\uparrow}c^\dagger_{2\downarrow})|0\rangle$$

Thus the HF results are a good approximation only when $x \to 0$. The unreasonable diverging behavior results from not suppressing the ionic state $|A\rangle$ in $|HF\rangle$

when $|x| \to \infty$. In order to correct this shortcoming of the Hartree–Fock method, we can combine different wavefunctions in different ranges to obtain a better wavefunction for our system. This can be done as follows:

Range	GS Energy	Correlation Energy	Wavefunction	n_+	n_-
$U > t$	0	$U - \sqrt{U^2 + t^2}$	$\|B\rangle$	1	1
$-t \geq U \leq t$	$-t + U$	$t - \sqrt{U^2 + t^2}$	$\frac{1}{\sqrt{2}}(\|A\rangle + \|B\rangle)$	2	0
$U < -t$	$2U$	$-U - \sqrt{U^2 + t^2}$	$\|C\rangle$	1	1

3. Correlation Entropy

The correlation entropy is a good measure of electron correlation in molecular systems [5, 7]. It is defined using the eigenvalues n_k of the one-particle density matrix 1PDM,

$$S = \sum_k n_k(-\ln n_k), \quad \sum_k n_k = N \tag{60}$$

This correlation entropy is based on the nonidempotency of the NONs n_k and proves to be an appropriate measure of the correlation strength if the reference state defining correlation is a single Slater determinant. In addition to the eigenvalues n_k of the "full" (spin-dependent) 1PDM, it seems reasonable to consider also the eigenvalues n_k of the spin-traced 1PDM. Among all the n_k there are a certain number N_0 of NONs n_{k_0} with values between 1 and 2 and all the other N_1 NONs n_{k_1} also have values between 0 and 1. So one possible measure of the correlation strength of spin-traced 1PDM is

$$S_1 = -\sum_{k_0}(n_{k_0} - 1)\ln(n_{k_0} - 1) - \sum_{k_1} n_{k_1} \ln n_{k_1} \tag{61}$$

Since all the $n_k/2$ have values between 0 and 1, there is another possible measurement of the correlation strength:

$$S_2 = -\sum_k \frac{n_k}{2} \ln \frac{n_k}{2} \tag{62}$$

4. Entanglement

The entanglement measure is given by the von Neumann entropy [78]

$$E_j = -\text{Tr}(\rho_j \log_2 \rho_j), \quad \rho_j = \text{Tr}_j(|\Psi\rangle\langle\Psi|) \tag{63}$$

where Tr_j denotes the trace over all but the jth site, $|\Psi\rangle$ is the antisymmetric wavefunction of the fermions system, and ρ_j is the reduced density matrix. Hence E_j actually describes the entanglement of the jth site with the remaining sites [79].

In the Hubbard model, the electron occupation of each site has four possibilities; there are four possible local states at each site, $|v\rangle_j = |0\rangle_j, |\uparrow\rangle_j, |\downarrow\rangle_j, |\uparrow\downarrow\rangle_j$. The reduced density matrix of the jth site with the other sites is given by [80, 81]

$$\rho_j = z|0\rangle\langle 0| + u^+|\uparrow\rangle\langle\uparrow| + u^-|\downarrow\rangle\langle\downarrow| + w|\uparrow\downarrow\rangle\langle\uparrow\downarrow| \quad (64)$$

with

$$w = \langle n_{j\uparrow} n_{j\downarrow}\rangle = \mathrm{Tr}(n_{j\uparrow} n_{j\downarrow} \rho_j) \quad (65)$$
$$u^+ = \langle n_{j\uparrow}\rangle - w, \quad u^- = \langle n_{j\downarrow}\rangle - w \quad (66)$$
$$z = 1 - u^+ - u^- - w = 1 - \langle n_{j\uparrow}\rangle - \langle n_{j\downarrow}\rangle + w \quad (67)$$

And the entanglement between the jth site and other sites is given by

$$E_j = -z \log_2 z - u^+ \log_2 u^+ - u^- \log_2 u^- - w \log_2 w \quad (68)$$

For the one-dimensional Hubbard model with half-filling electrons, we have $\langle n_\uparrow\rangle = \langle n_\downarrow\rangle = \frac{1}{2}$, $u^+ = u^- = \frac{1}{2} - w$, and the entanglement is given by

$$E_j = -2w \log_2 w - 2\left(\tfrac{1}{2} - w\right)\log_2\left(\tfrac{1}{2} - w\right) \quad (54)$$

For our case of a two-site two-electron system

$$w = \frac{1}{2 + 2[x + \sqrt{1 + x^2}]^2}$$

Thus the entanglement is readily calculated from Eq. (69). In Fig. 4, we show the entanglement between the two sites (top curve) and the correlation entropy S_1 and S_2 as a function of $x = U/t$. The entanglement measure is given by the von Neumann entropy in which the density matrix of the system is traced over the other site to get the reduced density matrix. The reduced density matrix describes the four possible occupations on the site: $|0\rangle, |\uparrow\rangle, |\downarrow\rangle, |\uparrow\downarrow\rangle$. The minimum of the entanglement is 1 as $x \to \pm\infty$. It can be understood that when $U \to +\infty$, all the sites are singly occupied; the only difference is the spin of the electrons at each site, which can be referred to as spin entanglement. As $U \to -\infty$, all the sites are either doubly occupied or empty, which is referred to as space entanglement. The maximum of the entanglement is 2 at $U = 0$; all four

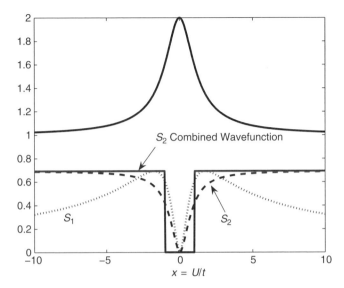

Figure 4. Two-site Hubbard model. Upper curve is the entanglement calculated by the von Newmann entropy. The curves S_1 and S_2 are the correlation entropies of the exact wavefunction as defined in the text. The dashed line is the S_2 for the combined wavefunction based on the range of x values. S_1 for the combined wavefunction is zero.

occupations are evenly weighted, which is the sum of the spin and space entanglements of the system. The correlation entropy S_1 vanishes for $x \to 0$ and $x \to \pm\infty$ and has a maximum near $|x| = 1$; the correlation entropy S_2 vanishes for $x \to 0$ and increases monotonically and approaches ln 2 for $x \to \pm\infty$. For $x \to +\infty$ it can be viewed as $t \to 0$ for fixed $U > 0$ or as $U \to +\infty$ for fixed t.

B. One-Dimensional Quantum Dots System

We consider an array of quantum dots modeled by the one-dimensional Hubbard Hamiltonian of the form [82]

$$H = -\sum_{\langle ij \rangle, \sigma} t_{ij}\, c_{i\sigma}^+ c_{j\sigma} + U \sum_i n_{i\uparrow} n_{i\downarrow} \qquad (70)$$

where t_{ij} stands for the hopping between the nearest-neighbor sites for the electrons with the same spin, i and j are the neighboring site numbers, σ is the electron spin, $c_{i\sigma}^+$ and $c_{j\sigma}$ are the creation and annihilation operators, and U is the Coulomb repulsion for the electrons on the same site. The periodic boundary condition is applied. The entanglement measure is given by the von Neumann entropy [78].

In the Hubbard model, the electron occupation of each site has four possibilities; there are four possible local states at each site, $|v\rangle_j = |0\rangle_j, |\uparrow\rangle_j, |\downarrow\rangle_j, |\uparrow\downarrow\rangle_j$. The dimensions of the Hilbert space of an L-site system is 4^L and $|v_1 v_2 \cdots v_L\rangle = \prod_{j=1}^{L} |v_j\rangle_j$ can be used as basis vectors for the system. The entanglement of the jth site with the other sites is given in the previous section by Eq. (65).

In the ideal case, we can expect an array of the quantum dots to have the same size and to be distributed evenly, so that the parameters t and U are the same everywhere. We call this the pure case. In fact, the size of the dots may not be the same and they may not be evenly distributed, which we call the impurity case. Here, we consider two types of impurities. The first one is to introduce a symmetric hopping impurity t' between two neighboring dots; the second one is to introduce an asymmetric electron hopping t' between two neighboring dots, the right hopping is different from the left hopping, while the rest of the sites have hopping parameter t.

Consider the particle–hole symmetry of the one-dimensional Hubbard model. One can obtain $w(-U) = \frac{1}{2} - w(U)$, so the entanglement is an even function of U, $E_j(-U) = E_j(U)$. The minimum of the entanglement is 1 as $U \to \pm\infty$. As $U \to +\infty$, all the sites are singly occupied; the only difference is the spin of the electrons on each site, which can be referred to as spin entanglement. As $U \to -\infty$, all the sites are either doubly occupied or empty, which is referred to as space entanglement. The maximum of the entanglement is 2 at $U = 0$, which is the sum of the spin and space entanglements of the system. The ground state of the one-dimensional Hubbard model at half-filling is metallic for $U < 0$, and insulating for $U > 0$; $U = 0$ is the critical point for the metal–insulator transition, where the local entanglement reaches its maximum. In Fig. 5 we show the

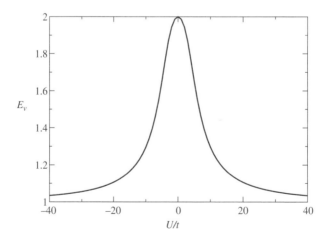

Figure 5. Local entanglement given by the von Neumann entropy, E_v, versus U/t in the pure case.

entanglement as a function U/t for six sites and six electrons. Our results are in complete agreement with the exact one obtained by Bethe ansatz [80].

C. Two-Dimensional Array of Quantum Dots

Using the Hubbard model, we can study the entanglement scaling behavior in a two-dimensional itinerant system. Our results indicate that, on the two sides of the critical point denoting an inherent quantum phase transition (QPT), the entanglement follows different scalings with the size just as an order parameter does. This fact reveals the subtle role played by the entanglement in QPT and points to its potential application in quantum information processing as a fungible physical resource.

Recently, it has been speculated that the most entangled systems could be found at the critical point [83] when the system undergoes a quantum phase transition; that is, a qualitative change of some physical properties takes place as an order parameter in the Hamiltonian is tuned [84]. QPT results from quantum fluctuations at the absolute zero of temperature and is a pure quantum effect featured by long-range correlations. So far, there have already been some efforts in exploring the above speculations, such as the analysis of the XY model about the single-spin entropies and two-spin quantum correlations [59, 85], the entanglement between a block of L contiguous sites and the rest of the chain [51], and also the scaling of entanglement near QPT [60]. But because there is still no analytical proof, the role played by the entanglement in quantum critical phenomena remains elusive. Generally, at least two difficulties exist in resolving this issue. First, until now, only two-particle entanglement is well explored. How to quantify the multiparticle entanglement is not clear. Second, QPT closely relates to the notorious many-body problems, which is almost intractable analytically. Until now, the only effective and accurate way to deal with QPT in critical region is the density-matrix renormalization group method [86]. Unfortunately, it is only efficient for one-dimensional cases because of the much more complicated boundary conditions for the two-dimensional situation [87].

In this chapter, we will focus on the entanglement behavior in QPT for the two-dimensional array of quantum dots, which provide a suitable arena for implementation of quantum computation [88, 89, 103]. For this purpose, the real-space renormalization group technique [91] will be utilized and developed for the finite-size analysis of entanglement. The model that we will be using is the Hubbard model [83],

$$H = -t \sum_{\langle i,j \rangle, \sigma} [c_{i\sigma}^+ c_{j\sigma} + \text{h.c.}] + U \sum_i \left(\tfrac{1}{2} - n_{i\uparrow}\right)\left(\tfrac{1}{2} - n_{i\downarrow}\right) \tag{71}$$

where t is the nearest-neighbor hopping term and U is the local repulsive interaction. $c_{i\sigma}^+(c_{i\sigma})$ creates(annihilates) an electron with spin σ in a Wannier orbital located at site i; the corresponding number operator is $n_{i\sigma} = c_{i\sigma}^{+} c_{i\sigma}$ and $\langle \rangle$ denotes the nearest-neighbor pairs; h.c. denotes the Hermitian conjugate.

For a half-filled triangular quantum lattice, there exists a metal–insulator phase transition with the tuning parameter U/t at the critical point 12.5 [92–94]. The corresponding order parameter for metal–insulator transition is the charge gap defined by $\triangle_g = E(N_e - 1) + E(N_e + 1) - 2E(N_e)$, where $E(N_e)$ denotes the lowest energy for an N_e-electron system. In our case, N_e is equal to the site number N_s of the lattice. Unlike the charge gap calculated from the energy levels, the Zanardi measure of the entanglement is defined based on the wavefunction corresponding to $E(N_e)$ instead. Using the conventional renormalization group method for the finite-size scaling analysis [92–94], we can discuss three schemes of entanglement scaling: single-site entanglement scaling with the total system size, E_{single}; single-block entanglement scaling with the block size, E_{block}; and block–block entanglement scaling with the block size, $E_{\text{block-block}}$. Our initial results of the single-site entanglement scaling indicate that E_{single} is not a universal quantity. This conclusion is consistent with the argument given by Osborne and Nielsen [85], who claim that the single-site entanglement is not scalable because it does not have the proper extensivity and does not distinguish between the local and the distributed entanglement. This implies that only a limited region of sites around the central site contributed significantly to the single-site entanglement. Using the one-parameter scaling theory, near the phase transition point, we can assume the existence of scaling function f for $E_{\text{block-block}}$ such that $E_{\text{block-block}} = q^{y_E} f(L/\xi)$, where $q = (U/t) - (U/t)_c$ measures the deviation distance of the system away from the critical state with $(U/t)_c = 12.5$, which is exactly equal to the critical value for metal–insulator transition when the same order parameter U/t is used [92–94]. $\xi = q^{-\nu}$ is the correlation length of the system with the critical exponent ν and $N = L^2$ for the two-dimensional systems.

In Fig. 6, we show the results of $E_{\text{block-block}}$ as a function of (U/t) for different system sizes. With proper scaling, all the curves collapse into one curve, which can be expressed as $E_{\text{block-block}} = f(qN^{1/2})$. Thus the critical exponents are $y_E = 0, \nu = 1$. It is interesting to note that we obtained the same ν as in the study of the metal–insulator transition. This shows the consistency of the initial results since the critical exponent ν is only dependent on the inherent symmetry and dimension of the investigated system. Another significant result lies in the finding that the metal state is highly entangled while the insulating state is only partly entangled.

It should be mentioned that the calculated entanglement here has a corresponding critical exponent $y_E = 0$. This means that the entanglement is constant at the critical point over all sizes of the system. But it is not a constant over all values of U/t. There is an abrupt jump across the critical point as $L \to \infty$. If we divide the regime of the order parameter into noncritical regime and critical

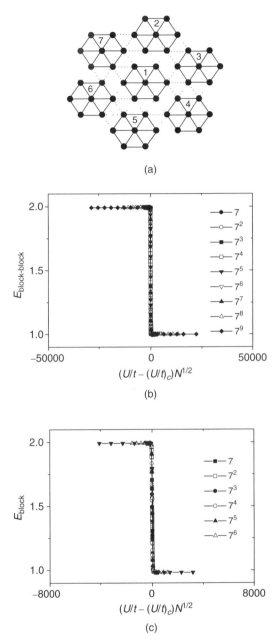

Figure 6. (a) Schematic diagram displays the lattice configuration with central block and the surrounding ones. (b) Scaling of block–block for various system size and (c) scaling of block entanglements with the block size.

regime, the results can be summarized as follows. In the noncritical regime, that is, U/t is away from $(U/t)_c$, as L increases, the entanglement will saturate onto two different values depending on the sign of $U/t - (U/t)_c$; at the critical point, the entanglement is actually a constant independent of the size L. These properties are qualitatively different from the single-site entanglement discussed by Osborne and Nielsen [85], where the entanglement with Zanardi's measure increases from zero to the maximum at the critical point and then decreases again to zero as the order parameter γ for the XY mode is tuned. These peculiar properties of the entanglement found here can be of potential interest to make an effective ideal "entanglement switch." For example, with seven blocks of quantum dots on a triangular lattice, the entanglement among the blocks can be regulated as "0" or "1" almost immediately once the tuning parameter U/t crosses the critical point. The switch errors will depend on the size of the blocks. Since it is already a well-developed technique to change U/t for the quantum dot lattice [95, 103], the above scheme should be workable. To remove the special confinement we have made upon the calculated entanglement, namely, only the entanglement of blocks 1 and 7 with the rest ones are considered, in the following, we will prove that the average pairwise entanglement also has the properties shown in Fig. 6. As we change the size of the central block, its entanglement with all the rest of the sites follows the same scaling properties as $E_{\text{block-block}}$. It is understandable if we consider the fact that only a limited region round the block contributes mostly to E_{block}. This result greatly facilitates the fabrication of realistic entanglement control devices, such as quantum gates for a quantum computer, since we don't need to consider the number of component blocks in fear that the next neighboring or the next-next neighboring quantum dots will influence the switching effect.

IV. AB INITIO CALCULATIONS AND ENTANGLEMENT

For a two-electron system in $2m$-dimensional spin-space orbital, with c_a and $c_a^\dagger$ denoting the fermionic annihilation and creation operators of single-particle states and $|0\rangle$ representing the vacuum state, a pure two-electron state $|\Phi\rangle$ can be written [57]

$$|\Phi\rangle = \sum_{a,b \in \{1,2,3,4,\ldots,2m\}} \omega_{a,b} c_a^\dagger c_b^\dagger |0\rangle \qquad (72)$$

where a, b run over the orthonormal single-particle states, and Pauli exclusion requires that the $2m \times 2m$ expansion coefficient matrix ω is antisymmetric: $\omega_{a,b} = -\omega_{b,a}$, and $\omega_{i,i} = 0$.

In the occupation number representation $(n_1 \uparrow, n_1 \downarrow, n_2 \uparrow, n_2 \downarrow, \ldots, n_m \uparrow, n_m \downarrow)$, where $\uparrow$ and $\downarrow$ mean α and β electrons, respectively, the subscripts denote the spatial orbital index and m is the total spatial orbital number. By tracing out all other spatial orbitals except n_1, we can obtain a (4×4) reduced density matrix for the spatial orbital n_1

$$\rho_{n_1} = \text{Tr}_{n_1} |\Phi\rangle\langle\Phi|$$

$$= \begin{pmatrix} 4\sum_{i,j=1}^{m-1} |\omega_{2i+1,2j+2}|^2 & 0 & 0 & 0 \\ 0 & 4\sum_{i=1}^{m-1} |\omega_{2,2i+1}|^2 & 0 & 0 \\ 0 & 0 & 4\sum_{i=2}^{m} |\omega_{1,2i}|^2 & 0 \\ 0 & 0 & 0 & 4|\omega_{1,2}|^2 \end{pmatrix} \tag{73}$$

The matrix elements of ω can be calculated from the expansion coefficient of the ab initio configuration interaction method. The CI wavefunction with single and double excitations can be written

$$|\Phi\rangle = c_0|\Psi_0\rangle + \sum_{ar} c_a^r |\Psi_a^r\rangle + \sum_{a<b,r<s} c_{a,b}^{r,s} |\Psi_{a,b}^{r,s}\rangle \tag{74}$$

where $|\Psi_0\rangle$ is the ground-state Hartree–Fock wavefunction, c_a^r is the coefficient for single excitation from orbital a to r, and $c_{a,b}^{r,s}$ is the double excitation from orbital a and b to r and s. Now the matrix elements of ω can be written in terms of the CI expansion coefficients. In this general approach, the ground-state entanglement is given by tbe von Neumann entropy of the reduced density matrix ρ_{n1} [57]:

$$S(\rho_{n_1}) = -\text{Tr}(\rho_{n_1} \log_2 \rho_{n_1}) \tag{75}$$

We are now ready to evaluate the entanglement for the H$_2$ molecule [57] as a function of R using a two-electron density matrix calculated from the configuration interaction wavefunction with single and double electronic excitations [96]. Figure 7 shows the calculated entanglement S for the H$_2$ molecule, as a function of the internuclear distance R using a minimal Gaussian basis set STO-3G (each Slater-type orbital fitted by 3 Gaussian functions) and a split valence Gaussian basis set 3-21G [96]. For comparison we included the usual electron correlation $(E_c = |E^{\text{exact}} - E^{\text{UHF}}|)$ and spin-unrestricted Hartree–Fock (UHF) calculations [96] using the same basis set in the figure. At the limit $R = 0$, the electron correlation for the He atom, $E_c = 0.0149$ (au) using the 3-21G basis set compared with the entanglement for the He atom $S = 0.0313$. With a larger basis set, $cc - pV5Z$ [97], we obtain numerically $E_c = 0.0415$ (au) and $S = 0.0675$. Thus qualitatively entanglement and absolute correlation have similar behavior.

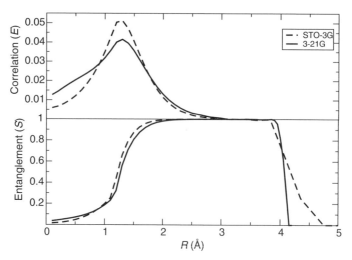

Figure 7. Comparison between the absolute value of the electron correlation $E_c = |E^{\text{Exact}} - E^{\text{UHF}}|$ and the von Neumann entropy (S) as a function of the internuclear distance R for the H_2 molecule using two Gaussian basis sets STO-3G and 3-21G.

At the united atom limit, $R \to 0$, both have small values, then rise to a maximum value, and finally vanish at the separated atom limit, $R \to \infty$. However, note that for $R > 3$ Å the correlation between the two electrons is almost zero but the entanglement is maximal until around $R \sim 4$ Å; the entanglement vanishes for $R > 4$ Å.

V. DYNAMICS OF ENTANGLEMENT AND DECOHERENCE

In this section, we investigate the dynamics of entanglement in one-dimensional spin systems with a time-dependent magnetic field. The Hamiltonian for such a system is given by [98]

$$H = -\frac{J}{2}(1+\gamma)\sum_{i=1}^{N}\sigma_i^x\sigma_{i+1}^x - \frac{J}{2}(1-\gamma)\sum_{i=1}^{N}\sigma_i^y\sigma_{i+1}^y - \sum_{i=1}^{N}h(t)\sigma_i^z \quad (76)$$

where J is the coupling constant, $h(t)$ is the time-dependent external magnetic field, σ^a are the Pauli matrices ($a = x, y, z$), γ is the degree of anisotropy, and N is the number of sites. We can set $J = 1$ for convenience and use periodic boundary conditions. Next, we transform the spin operators into fermionic operators. So the Hamiltonian assumes the following form:

$$H = \sum_{p=1}^{N/2} \alpha_p(t)[c_p^+ c_p + c_{-p}^+ c_{-p}] + i\delta_p[c_p^+ c_{-p}^+ + c_p c_{-p}] + 2h(t) = \sum_{p=1}^{N/2} \tilde{H}_p \quad (77)$$

where $\alpha_p(t) = -2\cos\phi_p - 2h(t)$, $\delta_p = 2\gamma\sin\phi_p$, and $\phi_p = 2\pi p/N$. It is easy to show $[\tilde{H}_p, \tilde{H}_q] = 0$, which means the space of $\tilde{H}$ decomposes into noninteracting subspace, each of four dimensions. No matter what $h(t)$ is, there will be no transitions among those subspaces. Using the following basis for the pth subspace, $(|0\rangle; c_p^+ c_{-p}^+|0\rangle; c_p^+|0\rangle; c_{-p}^+|0\rangle)$, we can explicitly get

$$\tilde{H}_p(t) = \begin{pmatrix} 2h(t) & -i\delta_p & 0 & 0 \\ i\delta_p & -4\cos\phi_p - 2h(t) & 0 & 0 \\ 0 & 0 & -2\cos\phi_p & 0 \\ 0 & 0 & 0 & -2\cos\phi_p \end{pmatrix} \qquad (78)$$

We only consider the systems that, at time $t = 0$, are in thermal equilibrium at temperature T. Let $\rho_p(t)$ be the density matrix of the pth subspace; we have $\rho_p(0) = e^{-\beta \tilde{H}_p(0)}$, where $\beta = 1/kT$ and k is Boltzmann's constant. Therefore, using Eq. (78), we have $\rho_p(0)$. Let $U_p(t)$ be the time-evolution matrix in the pth subspace, namely, ($\hbar = 1$): $i dU_p(t)/dt = U_p(t)\tilde{H}_p(t)$, with the boundary condition $U_p(0) = I$. Now the Liouville equation of this system is

$$i\frac{d\rho(t)}{dt} = [H(t), \rho(t)] \qquad (79)$$

which can be decomposed into uncorrelated subspaces and solved exactly. Thus, in the pth subspace, the solution of the Liouville equation is $\rho_p(t) = U_p(t)\rho_p(0)U_p(t)^\dagger$.

As a first step to investigate the dynamics of the entanglement, we can take the magnetic field to be a step function then generalize it to other relevant functional forms such as an oscillating one [98]. Figure 8 shows the results for nearest-neighbor concurrence $C(i, i+1)$ at temperature $T = 0$ and $\gamma = 1$ as a function of the initial magnetic field a for the step function case with final field b. For the $a < 1$ region, the concurrence increases very fast near $b = 1$ and reaches a limit $C(i, i+1) \sim 0.125$ when $b \to \infty$. It is surprising that the concurrence will not disappear when b increases with $a < 1$. This indicates that the concurrence will not disappear as the final external magnetic field increases at infinite time. It shows that this model is not in agreement with obvious physical intuition, since we expect that increasing the external magnetic field will destroy the spin–spin correlation functions and make the concurrence vanish. The concurrence approaches a maximum $C(i, i+1) \sim 0.258$ at $(a = 1.37, b = 1.37)$ and decreases rapidly as $a \neq b$. This indicates that the fluctuation of the external magnetic field near the equilibrium state will rapidly destroy the entanglement. However, in the region where $a > 2.0$, the concurrence is close to zero when $b < 1.0$ and maximum close to 1. Moreover, it disappear in the limit of $b \to \infty$.

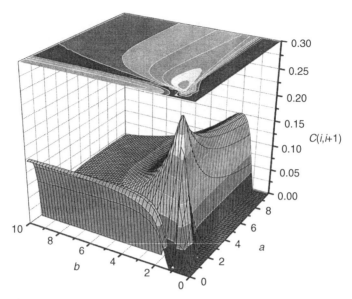

Figure 8. Nearest-neighbor concurrence C at zero temperature as a function of the initial magnetic field a for the step function case with final field b.

Now let us examine the system size effect on the entanglement with three different external magnetic fields changing with time t [99]:

$$h_{\mathrm{I}}(t) = \begin{cases} a, & t \leq 0 \\ b + (a-b)e^{-Kt}, & t > 0 \end{cases} \tag{80}$$

$$h_{\mathrm{II}}(t) = \begin{cases} a, & t \leq 0 \\ a - a\sin(Kt), & t > 0 \end{cases} \tag{81}$$

$$h_{\mathrm{III}}(t) = \begin{cases} 0, & t \leq 0 \\ a - a\cos(Kt), & t > 0 \end{cases} \tag{82}$$

where a, b, and K are varying parameters.

We have found that the entanglement fluctuates shortly after a disturbance by an external magnetic field when the system size is small. For larger system size, the entanglement reaches a stable state for a long time before it fluctuates. However, this fluctuation of entanglement disappears when the system size goes to infinity. We also show that in a periodic external magnetic field, the nearest-neighbor entanglement displays a periodic structure with a period related to that of the magnetic field. For the exponential external magnetic field, by varying the constant K, we have found that as time evolves, $C(i, i+1)$ oscillates but it does not reach its equilibrium value at $t \to \infty$.

This confirms the fact that the nonergodic behavior of the concurrence is a general behavior for slowly changing magnetic field. For the periodic magnetic field $h_{\mathrm{II}} = a(1 - \sin(-Kt))$, the nearest-neighbor concurrence is a maximum at $t = 0$ for values of a close to one, since the system exhibits a quantum phase transition at $\lambda_c = J/h = 1$, where in our calculations we fixed $J = 1$. Moreover, for the two periodic $\sin(-Kt)$ and $\cos(-Kt)$ fields the nearest-neighbor concurrence displays a periodic structure according to the periods of their respective magnetic fields [99].

For the periodic external magnetic field $h_{\mathrm{III}}(t)$, we show in Fig. 9 that the nearest-neighbor concurrence $C(i, i+1)$ is zero at $t = 0$ since the external magnetic field $h_{\mathrm{III}}(t = 0) = 0$ and the spins align along the x-direction: the total wavefunction is factorizable. By increasing the external magnetic field, we see the appearance of nearest-neighbor concurrence but very small. This indicates that the concurrence cannot be produced without a background external magnetic field in the Ising system. However, as time evolves one can see the periodic structure of the nearest-neighbor concurrence according to the periodic structure of the external magnetic field $h_{\mathrm{III}}(t)$ [99].

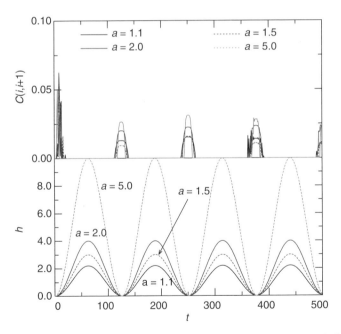

Figure 9. The nearest-neighbor concurrence $C(i, i+1)$ (upper panel) and the periodic external magnetic field $h_{\mathrm{III}}(t) = a(1 - \cos[Kt])$; see Eq. (14) in the text (lower panel) for $K = 0.05$ with different values of a as a function of time t.

Recently, interest in solid state systems has increased because they facilitate the fabrication of large integrated networks that would be able to implement realistic quantum computing algorithms on a large scale. On the other hand, the strong coupling between a solid state system and its complex environment makes it a significantly challenging mission to achieve the high coherence control required to manipulate the system. Decoherence is considered as one of the main obstacles toward realizing an effective quantum computing system [100–103]. The main effect of decoherence is to randomize the relative phases of the possible states of the isolated system as a result of coupling to the environment. By randomizing the relative phases, the system loses all quantum interference effects and may end up behaving classically.

In order to study the decoherence effect, we examined the time evolution of a single spin coupled by exchange interaction to an environment of interacting spin bath modeled by the XY-Hamiltonian. The Hamiltonian for such a system is given by [104]

$$H = -\frac{1+\gamma}{2}\sum_{i=1}^{N} J_{i,i+1}\sigma_i^x\sigma_{i+1}^x - \frac{1-\gamma}{2}\sum_{i=1}^{N} J_{i,i+1}\sigma_i^y\sigma_{i+1}^y - \sum_{i=1}^{N} h_i\sigma_i^z \qquad (83)$$

where $J_{i,i+1}$ is the exchange interaction between sites i and $i+1$, h_i is the strength of the external magnetic field at site i, σ^a are the Pauli matrices ($a = x, y, z$), γ is the degree of anisotropy, and N is the number of sites. We consider the centered spin on the lth site as the single-spin quantum system and the rest of the chain as its environment, where in this case $l = (N+1)/2$. The single spin directly interacts with its nearest-neighbor spins through exchange interaction $J_{l-1,l} = J_{l,l+1} = J'$. We assume exchange interactions between spins in the environment are uniform and simply set it as $J = 1$. The centered spin is considered as inhomogeneously coupled to all the spins in the environment by being directly coupled to its nearest neighbors and indirectly to all other spins in the chain through its nearest neighbors.

By evaluating the spin correlator $C(t)$ of the single spin at the jth site [104],

$$C_j(t) = \rho_j^z(t,\beta) - \rho_j^z(0,\beta) \qquad (84)$$

we observed that the decay rate of the spin oscillations strongly depends on the relative magnitude of the exchange coupling between the single spin and its nearest neighbor J' and coupling among the spins in the environment J. The decoherence time varies significantly based on the relative coupling magnitudes of J and J'. The decay rate law has a Gaussian profile when the two exchange couplings are of the same order, $J' \sim J$, but converts to exponential and then a power law as we move to the regimes of $J' > J$ and $J' < J$. We also show that the spin oscillations propagate from the single spin to the environment spins with a certain speed.

Moreover, the amount of saturated decoherence induced into the spin state depends on this relative magnitude and approaches a maximum value for a relative magnitude of unity. Our results suggest that setting the interaction within the environment in such a way that its magnitude is much higher or lower than the interaction with the single spin may reduce the decay rate of the spin state. The reason behind this phenomenon could be that the variation in the coupling strength along the chain at one point (where the single spin exits) blocks the propagation of decoherence along the chain by reducing the entanglement among the spins within the environment, which reduces its decoherence effect on the single spin in return [104]. This result might be applicable in general to similar cases of a centered quantum system coupled inhomogeneously to an interacting environment with large degrees of freedom.

VI. ENTANGLEMENT AND DENSITY FUNCTIONAL THEORY

Density functional theory is originally based on the Hohenberg–Kohn theorem [105, 106]. In the case of a many-electron system, the Hohenberg–Kohn theorem establishes that the ground-state electronic density $\rho(\mathbf{r})$, instead of the potential $v(\mathbf{r})$, can be used as the fundamental variable to describe the physical properties of the system. In the case of a Hamiltonian given by

$$H = H_0 + H_{\text{ext}} = H_0 + \sum_l \lambda_l \widehat{A}_l \tag{85}$$

where λ_l is the control parameter associated with a set of mutually commuting Hermitian operators $\{\widehat{A}_l\}$, the expectation values of $\widehat{A}_l$ for the ground state $|\psi\rangle$ are denoted by the set $\{a_l\} \equiv \{\langle\psi|\widehat{A}_l|\psi\rangle\}$. For such a Hamiltonian Wu et al. [107] linked entanglement in interacting many-body quantum systems to density functional theory. They used the Hohenberg–Kohn theorem on the ground state to show that the ground-state expectation value of any observable can be interchangeably viewed as a unique function of either the control parameter $\{\lambda_l\}$ or the associated operator representing the observable $\{a_l\}$.

The Hohenberg–Kohn theorem can be used to redefine entanglement measures in terms of new physical quantities: expectation values of observables, $\{a_l\}$, instead of external control parameters, $\{\lambda_l\}$. Consider an arbitrary entanglement measure M for the ground state of Hamiltonian (85). For a bipartite entanglement, one can prove a central lemma, which very generally connects M and energy derivatives.

Lemma *Any entanglement measure M can be expressed as a unique functional of the set of first derivatives of the ground-state energy [107]* :

$$M = M(\{a_l\}) = M\left(\left\{\frac{\partial E}{\partial \lambda_l}\right\}\right) \tag{86}$$

The proof follows from the fact that, according to the generalized Hohenberg–Kohn theorem, the ground-state wavefunction $|\Psi\rangle$ is a unique functional of $\{a_l\}$, and since $|\Psi\rangle$ provides a complete description of the state of the system, everything else is a unique functional of $\{a_l\}$ as well, including M. Wu et al. [107] use density functional theory concepts to express entanglement measures in terms of the first or second derivative of the ground-state energy. As a further application they discuss entanglement and quantum phase transitions in the case of mean field approximations for realistic models of many-body systems [107].

This interesting connection between density functional theory and entanglement was further generalized for arbitrary mixed states by Rajagopal and Rendell [108] using the maximum entropy principle. In this way they established the duality in the sense of Legendre transform between the set of mean values of the observables based on the density matrix and the corresponding set of conjugate control parameters associated with the observables.

VII. FUTURE DIRECTIONS

We have examined and reviewed the relation between electron–electron correlation, the correlation entropy, and the entanglement for two exactly solvable models: the Ising model and the Hubbard model for two sites. The ab initio calculation of the entanglement for the H_2 system is also discussed. Our results show that there is a qualitatively similar behavior between the entanglement and absolute standard correlation of electrons for the Ising model. Thus entanglement might be used as an alternative measure of electron correlation in quantum chemistry calculations. Entanglement is directly observable and it is one of the most striking properties of quantum mechanics.

Dimensional scaling theory [109] provides a natural means to examine electron–electron correlation, quantum phase transitions [110], and entanglement. The primary effect of electron correlation in the $D \to \infty$ limit is to open up the dihedral angles from their Hartree–Fock values [109] of exactly $90°$. Angles in the correlated solution are determined by the balance between centrifugal effects, which always favor $90°$, and interelectron repulsions, which always favor $180°$. Since the electrons are localized at the $D \to \infty$ limit, one might need to add the first harmonic correction in the $1/D$ expansion to obtain a useful density matrix for the whole system, thus the von Neumann entropy. The relation between entanglement and electron–electron correlation at the large-dimensional limit for the dimensional scaling model of the H_2 molecule [111] will be examined in future studies.

A new promising approach is emerging for the realization of quantum chemistry calculations without wavefunctions through first-order semidefinite programming [112]. Mazziotti has developed a first-order, nonlinear algorithm for the semidefinite programing of the two-electron reduced density matrix method that reduces memory and floating-point requirements by orders of magnitude [113, 114]. The electronic energies and properties of atoms and molecules are computable simply from an effective two-electron reduced density matrix $\rho_{(AB)}$ [115, 116]. Thus the electron–electron correlation can be calculated directly as effectively the entanglement between the two electrons, which is readily calculated as the von Neumann entropy $S = -\text{Tr}\rho_A \log_2 \rho_A$, where $\rho_A = \text{Tr}_B \rho_{(AB)}$. With this combined approach, one calculates the electronic energies and properties of atoms and molecules including correlation without wavefunctions or Hartree–Fock reference systems. This approach provides a natural way to extend the calculations of entanglement to larger molecules.

Quantum phase transitions are a qualitative change in the ground state of a quantum many-body system as some parameter is varied [84, 117]. Unlike classical phase transitions, which occur at a nonzero temperature, the fluctuations in quantum phase transitions are fully quantum and driven by the Heisenberg uncertainty relation. Both classical and quantum critical points are governed by a diverging correlation length, although quantum systems possess additional correlations that do not have a classical counterpart: this is the entanglement phenomenon. Recently, a new line of exciting research points to the connection between the entanglement of a many-particle system and the appearance of quantum phase transitions [60, 66, 118, 119]. For a class of one-dimensional magnetic systems, the entanglement shows scaling behavior in the vicinity of the transition point [60]. Deeper understanding of quantum phase transitions and entanglement might be of great relevance to quantum information and computation.

Acknowledgments

I would like to thank my collaborators, Dr. Jiaxiang Wang and Hefeng Wang, for their contributions in calculating the entanglement for quantum dot systems. Also, I thank Dr. Omar Osenda, Zhen Huang, and Gehad Sadiek for their contributions to the studies of entanglement of formation and dynamics of one-dimensional magnetic systems with defects.

I also acknowledge the financial support of The National Science Foundation, the US–Israel Binational Science Foundation, and the Purdue Research Foundation.

References

1. P. O. Löwdin, *Adv. Chem. Phys.* **II**, 207 (1959).
2. E. Clementi and G. Corongiu, in *Methods and Techniques in Computational Chemistry*, METECC-95, STEF, Cagliari, 1995.
3. W. Kutzelnigg, G. Del Re, and G. Berthier, *Phys. Rev.* **172**, 49 (1968).
4. N. L. Guevara, R. P. Sagar, and R. O. Esquivel, *Phys. Rev. A* **67**, 012507 (2003).
5. P. Ziesche, O. Gunnarsson, W. John, and H. Beck, *Phys. Rev. B* **55**, 10270 (1997).
6. P. Ziesche et al., *J. Chem. Phys.* **110**, 6135 (1999).
7. P. Gersdorf, W. John, J. P. Perdew, and P. Ziesche, *Int. J. Quantum Chem.* **61**, 935 (1997).
8. Q. Shi and S. Kais, *J. Chem. Phys.* **121**, 5611 (2004).
9. S. Wilson, *Electron Correlation in Molecules*, Clarendon Press, Oxford, 1984.
10. N. H. March, *Electron Correlation in Solid State*, Imperial College Press, London, 1999.
11. C. H. Bennett and D. P. DiVincenzo, *Nature* **404**, 247 (2000).
12. C. Macchiavello, G. M. Palma, and A. Zeilinger, *Quantum Computation and Quantum Information Theory*, World Scientific, 2000.
13. M. Nielsen and I. Chuang, *Quantum Computation and Quantum Communication*, Cambridge University Press, Cambridge, 2000.
14. D. BruB, *J. Math. Phys.* **43**, 4237 (2002).
15. J. Gruska, *Quantum Computing*, McGraw-Hill, New York, 1999.
16. E. Schrödinger, *Naturwissenschaften* **23**, 807 (1935).
17. J. S. Bell, *Physics* **1**, 195 (1964).
18. A. Einstein, B. Podolsky, and N. Rosen, *Phys. Rev.* **47**, 777 (1935).
19. E. Schrödinger, *Proc. Cambridge Philos. Soc.* **31**, 555 (1935).
20. C. H. Bennett, D. P. DiVincenzo, J. A. Smolin, and W. K. Wootters, *Phys. Rev. A* **54**, 3824 (1996).
21. E. Hagley, X. Maytre, G. Nogues, C. Wunderlich, M. Brune, J. M. Raimond, and S. Haroche, *Phys. Rev. Lett.* **79**, 1(1997).
22. Q. A. Turchette, C. S. Wood, B. E. King, C. J. Myatt, D. Leibfried, W. M. Itano, C. Monroe, and D. J. Wineland, *Phys. Rev. Lett.* **81**, 3631(1998).
23. D. Bouwmeester, J.-W. Pan, M. Daniell, H. Weinfurter, and A. Zeilinger, *Phys. Rev. Lett.* **82**, 1345–1349 (1999).
24. C. Monroe, D. M. Meekhof, B. E. King, and D. J. Wineland, *Science* **272**, 1131(1996).
25. C. A. Sackett, D. Kielpinksi, B. E. King, C. Langer, V. Meyer, C. J. Myatt, M. Rowe, Q. A. Turchette, W. M. Itano, D. J. Wineland, and C. Monroe, *Nature* **404**, 256–259 (2000).
26. P. Benioff, *J. Statist. Phys.* **22**, 563–591(1980); *J. Math. Phys.* **22**, 495–507(1981); *Int. J. Theor. Phys.* **21**, 177–201(1982).
27. C. H. Bennett and R. Landauer, *Sci. Am.* **253**, 48 (1985).
28. D. Deutsch, *Proc. R. Soc. A* **400**, 97 (1985); *Proc. R. Soc. A* **425**, 73 (1989).
29. R. P. Feynman, *Int. J. Theor. Phys.* **21**, 467–488(1982).
30. R. Landauer, *IBM J. Res. Dev.* **3**, 183 (1961).
31. *Proceedings of the 35th Annual Symposium on the Foundations of Computer Science*, IEEE Computer Society Press, Los Alamos, CA, 1994, p. 124; quant-ph/19508027.
32. D. Bouwmeester, K. Mattle, J.-W. Pan, H. Weinfurter, A. Zeilinger, and M. Zukowski, *Appl. Phys. B* **67**, 749 (1998).

33. D. Boyuwmeester, J. W. Pan, K. Mattle, M. Eibl, H. Weinfurter, and A. Zeilinger, *Nature* **390**, 575(1997).
34. C. H. Bennett and S. J. Wiesner, *J. Phys. Rev. Lett.* **69**, 2881 (1992).
35. K. Mattle, H. Weinfurter, P. G. Kwiat, and A. Zeilinger, *Phys. Rev. Lett.* **76**, 4546 (1996).
36. B. Schumacher, *Phys. Rev. A* **51**, 2738 (1995).
37. C. H. Bennett, G. Brassard, C. Creau, R. Jozsa, A. Peres, and W. K. Wootters, *Phys. Rev. Lett.* **70**, 1895(1993).
38. M. Horodecki, *Quant. Inf. Comp.* **1**, 3 (2001).
39. P. Horodecki and R. Horodecki, *Quant. Inf. Comp.* **1**, 45 (2001).
40. W. K. Wooters. *Quant. Inf. Comp.* **1**, 27 (2001).
41. C. H. Bennett, *Phys. Rev. A* **53**, 2046 (1996).
42. W. K. Wootters, *Phys. Rev. Lett.* **80**, 2245 (1998).
43. W. K. Wootters, *Quant. Inf. Comp.* **1**, 27 (2001).
44. S. Hill and W. K. Wooters, *Phys. Rev. Lett.* **78**, 5022 (1997).
45. J. Schliemann, J. Ignacio Cirac, M. Kus, M. Lewenstein, and D. Loss, *Phys. Rev. A* **64**, 022303 (2001).
46. J. Schliemann, D. Loss, and A. H. MacDonald, e-print, cond-mat/0009083.
47. Y. S. Li, B. Zeng, X. S. Liu, and G. L. Long, *Phys. Rev. A* **64**, 054302 (2001).
48. R. Paškauskas and L. You, *Phys. Rev. A* **64**, 042310 (2001).
49. J. R. Gittings and A. J. Fisher, *Phys. Rev. A* **66**, 032305 (2002).
50. P. Zanardi, *Phys. Rev. A* **65**, 042101 (2002).
51. G. Vidal, J. I. Latorre, E. Rico, and A. Kitaev, e-print, quant-ph/0211074.
52. A. C. Doherty, P. A. Parrilo, and F. M. Spedalieri, *Phys. Rev. A* **71**, 032333 (2005).
53. B. Chong, H. Keiter, and J. Stolze, arXiv:quant-ph/0512199 (2005).
54. J. Uffink, *Phys. Rev. Lett.* **88**, 230406 (2002).
55. M. Seevinck and G. Svetlichny, *Phys. Rev. Lett.* **89**, 060401 (2002).
56. R. F. Werner, *Phys. Rev. A* **40**, 4277 (1989).
57. Z. Huang and S. Kais, *Chem. Phys. Lett.* **413**, 1 (2005).
58. C. Herring and M. Flicker, *Phys. Rev.* **134**, A362 (1964).
59. O. Osenda, Z. Huang, and S. Kais, *Phys. Rev. A* **67**, 062321 (2003).
60. A. Osterloh, L. Amico, G. Falci, and R. Fazio, *Nature* **416**, 608 (2002).
61. E. Lieb, T. Schultz, and D. Mattis, *Ann. Phys.* **60**, 407 (1961).
62. Z. Huang, O. Osenda, and S. Kais, *Phys. Lett. A* **322**, 137 (2004).
63. D. Aharonov, *Phys. Rev. A* **62**, 062311 (2000).
64. M. C. Arnesen, S. Bose, and V. Vedral, *Phys. Rev. Lett.* **87**, 017901 (2001).
65. D. Gunlycke, V. M. Kendon, and V. Vedral, *Phys. Rev. A* **64**, 042302 (2001).
66. T. J. Osborne and M. A. Nielsen, *Phys. Rev. A* **66**, 032110 (2002).
67. X. Wang, H. Fu, and A. I. Solomon, *J. Phys. A* **34**, 11307 (2001).
68. X. Wang, *Phys. Rev. A* **66**, 044395 (2002).
69. P. W. Anderson, *Science* **235**, 1196 (1987).
70. K. Onizuka et al., *J. Phys. Soc. Japan* **69**, 1016 (2000).
71. P. Jordan and E. P. Wigner, *Z. Phys.* **47**, 631 (1928).

72. E. Fradkin, *Phys. Rev. Lett.* **63**, 322 (1989).
73. A. Lopez, A. G. Rojo, and E. Fradkin, *Phys. Rev. B* **49**, 15139 (1994).
74. Y. R. Wang, *Phys. Rev. B* **43**, 3786 (1991); **45**, 12604 (1992); **45**, 12608 (1992); K. Yang, L. K. Warman, and S. M. Girvin, *Phys. Rev. Lett.* **70**, 2641 (1993).
75. O. Derzhko, *J. Phys. Studies (L'viv)* **5**, 49 (2001).
76. J. Hubbard, *Proc. R. Soc. London A* **276**, 238 (1963).
77. F. H. L. Essler, H. Frahm, F. Gohmann, A. Klumper, and V. E. Korepin, *The One-Dimensional Hubbard Model*, Cambridge University Press, Cambridge, 2005.
78. P. Zanardi, *Phys. Rev. A* **65**, 042101 (2002).
79. J. Wang and S. Kais, *Phys. Rev. A* **70**, 022301 (2004).
80. S. Gu, S. Deng, Y. Li, and H. Lin, *Phys. Rev. Lett.* **93**, 086402 (2004).
81. H. Wang and S. Kais, *Chem. Phys. Lett.* x,xx (2006).
82. H. Wang and S. Kais, *Int. J. Quantum Info.* (submitted).
83. J. Wang and S. Kais, *Int. J. Quantum Info.* **1**, 375 (2003).
84. S. Sachdev, *Quantum Phase Transitions*, Cambridge University Press, Cambridge, 1999.
85. T. J. Osborne and M. A. Nielsen, *Phys. Rev. A* **66**, 032110 (2002); e-print, quant-ph/0109024.
86. S. R. White, *Phys. Rev. Lett.* **69**, 2863 (1992).
87. G. Vidal, quant-ph/0301063, 2003; quant-ph/0310089, 2003. J. I. Latorre, E. Rico, and G. Vidal, quant-ph/0304098, 2003.
88. D. Loss and D. P. DiVincenzo, *Phys. Rev. A* **57**, 120 (1998).
89. G. Burkard, D. Loss, and D. P. DiVincenzo, *Phys. Rev. B* **59**, 2070 (1999).
90. X. Hu and S. Das Sarma, *Phys. Rev. A* **61**, 062301-1 (2000).
91. *Real-Space Renormalization*, Topics in Current Physics (T. W. Burkhardt and J. M. J. van Leeuwen, eds., Springer-Verlag, New York 1982.
92. J. X. Wang, S. Kais, and R. D. Levine, *Int. J. Mol. Sci.* **3**, 4 (2002).
93. J. X. Wang and S. Kais, *Phys. Rev. B* **66**, 081101(R) (2002).
94. J. X. Wang, S. Kais, F. Remacle, and R. D. Levine, *J. Chem. Phys. B* **106**, 12847(2002).
95. F. Remacle and R. D. Levine, *CHEMPHYSCHEM* **2**, 20 (2001).
96. M. J. Frisch et al., *Gaussian 98, Revision A.11.3*, Gaussian, Inc., Pittsburgh, PA, 1998.
97. V. A. Rassolov, M. A. Ratner, and J. A. Pople, *J. Chem. Phys.* **112**, 4014 (2000).
98. Z. Huang and S. Kais, *Int. J. Quant. Inf.* **3**, 483 (2005).
99. Z. Huang and S. Kais, *Phys. Rev. A* **73**, 022339 (2006).
100. For a review, see W. H. Zurek, *Phys. Today* **44**, 36 (1991).
101. D. Bacon, J. Kempe, D. A. Lidar, and K. B. Whaley, *Phys. Rev. Lett.* **85**, 1758 (2000).
102. N. Shevni, R. de Sousa, and K. B. Whaley, *Phys. Rev. B* **71**, 224411 (2005).
103. R. de Sousa and S. Das Sarma, *Phys. Rev. B* **68**, 115322 (2003).
104. Z. Huang, G. Sadiek, and S. Kais, *J. Chem. Phys.* x,x (2006).
105. P. Hohenberg and W. Kohn, *Phys. Rev.* **136**, B864 (1964).
106. W. Kohn and L. J. Sham, *Phys. Rev.* **140**, A1133 (1965).
107. L. A. Wu, M. S. Sarandy, D. A. Lidar, and L. J. Sham, arXiv:quant-ph/0512031 (2005).
108. A. K. Rajagopal and R. W. Rendell, arXiv:quant-ph/0512102 (2005).
109. D. R. Herschbach, J. Avery, and O. Goscinski, *Dimensional Scaling in Chemical Physics*, Kluwer, Dordrecht, 1993.

110. S. Kais and P. Serra, *Adv. Chem. Phys.* **125**, 1 (2003).
111. A. A. Svidzinsky, M. O. Scully, and D. R. Herschbach, *Phys. Rev. Lett.* **95**, 080401 (2005).
112. D. A. Mazziotti, *Phys. Rev. Lett.* **93**, 213001 (2004).
113. D. A. Mazziotti, *J. Chem. Phys.* **121**, 10957 (2004).
114. D. A. Mazziotti and R. M. Erdahl, *Phys. Rev. A* **63**, 042113 (2001).
115. E. R. Davidson, *Reduced Density Matrices in Quantum Chemistry*, Academic Press, New York, 1976.
116. H. Nakatsuji, in *Many-Electron Densities and Reduced Density Matrices* (J. Cioslowski, ed.) Kluwer, Academic, New York, 2000.
117. S. L. Sondhi, S. M. Girvin, J. P. Carini, and D. Shahar, *Rev. Mod. Phys.* **69**, 315 (1997).
118. D. Larsson and H. Johanneson, *Phys. Rev. Lett.* **95**, 196406 (2005).
119. L. A. Wu, M. S. Sarandy, and D. A. Lidar, *Phys. Rev. Lett.* **93**, 250404 (2004).

AUTHOR INDEX

Numbers in parentheses are reference numbers and indicate that the author's work is referred to although his name is not mentioned in the text. Numbers in *italic* show the page on which the complete references are listed.

Absar, I., 123(23), 139(23), *162*, 185 (77–78), *203*, 208(89), 214(89), 215(89, 97–99), *258*, 271(56), *291*, 491(4), *492*
Acharya, P. K., 446(12), *480*
Adamo, C., 99(19), *101*
Aguado, A., 249(112, 113), *259*
Aharonov, D., 509(63), *533*
Al-Laham, M. A., 99(19), *101*, 417(89), *427*
Alcoba, D. R., 126(36, 38), 132-(48), 133(38, 50), 137(54), 138(54), 139(54), 140(48, 54), 146(54), 149(48), 151(38, 51), 153(79–80), 154(80), 160(48), *162–164*, 166(34–35), 167(34), 183(34–35), *201*, 206(18, 62, 64, 71–72, 77), 207(18, 71–72, 83, 87–88), 210(71–72, 83, 211(72), 214(77, 83), 217(71), 220(77), 221(77), 226(77), 227(71), 229(71–72), 230(18, 64, 71–72, 83, 87), 233(71, 83), 234(71, 83), 235(72), 239(72), 244(18, 6, 64, 83), 245(18, 62, 64, 71, 83, 87–88), 246(18), 247(18, 88), 249(88), 253(18, 62, 64, 87, 114), *254–257, 259*, 264(4), *290*, 331(24–25), 336(24), *341*, 393(64), *426*
Alizadeh, F., 110(18), *118*
Amico, L., 505(60), 519(60), 531(60), *533*
Amovilli, C., 446(9), 469(9), *480*
Anderson, Arlen, 348(37), *383*
Anderson, P. W., 511(69), *533*
Andersson, K., 345(18), *382*
Andres, J. L., 99(19), *101*, 417(89), *427*
Arai, T., 267(53), 271(53), *291*

Arias, T. A., 392(40, 46), *425*
Arnesen, M. C., 509(64), 510(64), *533*
Artemiev, A., 388(6), *424*
Asbrink, L., 419(92), *427*
Avery, J., 530(109), *534*
Ayala, P. Y., 99(19), *101*, 417(89), *427*
Ayers, P. W., 444(2–3, 5), 446(2–3, 7), 447(17, 28), 449(5), 456(2), 469(2), 470(28), 472(60), 473(62), 474(2), 475(3), 477(17), *480–482*
Ayres, R. U., 122(4), *161*, 206(11), *254*
Ažman, A., 392(50), *426*

Baboul, A. G., 99(19), *101*
Bacon, D., 528(101), *534*
Baerends, E. J., 391(36), 392(44), 393(71), 417(44), *425*
Baker, J., 417(89), *427*
Baldridge, K. K., 367(56), *384*
Bansil, A., 392(54), *426*
Barbiellini, B., 392(54), *426*
Barlett, R., 127(40), *163*, 209(90), *258*
Barone, V., 99(19), *101*
Bartlett, R. J., 262(40, 42), 268 (40–42), *291*, 332(35), 338(35), 339(35), *341*, 345(14), 346(27–28, 32–33), 352(33), 356(33, 50), 357(32–33, 50), 358(51), 372(60), 375(61), 377(33, 49, 61), *382–384*, 422(94), *427*
Bartolotti, L. J., 389(9), *424*, 446(12), *480*
Beck, H., 494(5), 513(5), 515(5), *532*

Reduced-Density-Matrix Mechanics: With Application to Many-Electron Atoms and Molecules, A Special Volume of *Advances in Chemical Physics*, Volume 134, edited by David A. Mazziotti. Series editor Stuart A. Rice. Copyright © 2007 John Wiley & Sons, Inc.

Beitia, F., 206(41), 226(41), *255*
Beiyan, Jin, 63(2), 67(2), 84(2), 90(2), *90*
Bell, J. S., 494(17), *532*
Ben-Tal, A., 104(5), 111(5), 113(5), 115(5), *117*
Benayoun, M. D., 166(33), *201*, 331(23), *341*
Bender, C. F., 475(67), *483*
Benioff, P., 494(26), *532*
Bennett, C. H., 494(11), 494(20, 27), 495(34, 36, 41), 496(41), *532–533*
Berrondo, M., 389(16), *425*
Berthier, G., 494(3), *532*
Beste, A., 127(40), *163*, 209(90), *258*
Bieri, G., 419(92), *427*
Bingel, W. A., 39(56), 44(56), *58*, 295(10), 296(10), *330*, 414(85), *427*
Binkley, J. S., 417(87, 89), *427*
Boatz, J. A., 367(56), *384*
Bochicchio, R., 206(44), 226(43–44), *255*, 262(37), 266(37), *291*, 393(64), *426*
Bopp, F., 12(5), 15, *16*, 123(22), *162*, 475(66), *483*
Borchers, B., 114(23), 115(23, 29), *118*
Borštnik, 392(50), *426*
Bose, S., 509(64), 510(64), *533*
Bouten, M., 15(26–27), *17*
Bouwmeester, D., 494(23), 495(32–33), *532–533*
Boyd, S., 45(63), 46(63), *58*, 103(3), 115(3), *117*
Braams, B. J., 23(27), 27(27), 29(27), 38(27), 44(27), 47(27–28), 55(27), *57*, 63(6), 66(6, 14), 90(6, 14), *91*, 94(5, 8), 96(5), 97(5), 98(5), 99(5, 8), 100(5), *100-101*, 106(14–16), 108(14–15), 109(14–16), 110(15), 114(14–16), 116(15), 116(15), *117–118*, 184(70), 198(70), *202*, 206(33), *255*, 448(35), 468(35), 479(35), *481*
Brassard, G., 495(37), *533*
Brink, D. M., 38(54), *58*
Bruant, J. C., 99(19), *101*
Bru, B. D., 494(14), *532*
Brune, M., 494(21), *532*
Buchowiecki, M., 393(65, 67), *426*
Buijse, M. A., PhD., 390(23), 391(23, 36), *425*
Burer, S., 47(68), *59*, 115(26–28), 116(26–28), *118*
Burkard, G., 519(89), *534*

Cammi, R., 99(19), *101*
Carini, J. P., 531(117), *535*

Carlson, D. C., 306(26), *330*
Carlsson, A. E., 391(29), *425*
Carter, E. A., 446(20), *481*
Casanyi, G., 392(46), *425*
Casida, M. E., 185(80), 186(80), *203*, 206(76), 214(76), *257*
Casquero, F. J., 44(62), *58*, 126(36), 151(51), 153(81), 160(81), *162–164*, 166(35), 184(35), *201*, 206(83–85, 88), 210(83), 230(83–85), 233(83), 244(83–85), 245(83–85, 88), 247(88), 249(88), *257*, 331(25), *341*
Castaño, F., 206(41), 226(41), *255*
Challacombe, M., 99(19), *101*, 417(89), *427*
Chan, G. K.-L., 332(30), 338(30), 339(30), *341*, 344(1, 7–11), 345(1), 346(28), *381–382*, 446(21), *481*
Chattopadhyay, S., 345(19), 356(19), *382*
Chaudhuri, R. K., 332(32), 338(32), *341*, 345(19), 356(19), *382*
Cheeseman, J. R., 99(19), *101*, 417(89), *427*
Chen, W., 99(19), *101*, 417(89), *427*
Cheng, C. Y., 446(11), 469(11), *480*
Cho, S., 122(13), 127(13), *161*, 244(103), *258*
Choi, C., 115(28), 116(27), *118*
Chong, B., 499(53), 500(53), *533*
Chuang, I., 494(13), 495(13), *532*
Cioslowski, J., 37(47), *58*, 99(19), *101*, 389(12–13), 391(31, 38), 392(39, 41, 56), 393(65, 67–69, 72), 400(12–13), 404(72), 405(13), 417(89), *424-427*
Cirac, J. Ignacio, 498(45), *533*
Cizek, J., 345(15–16), *382*
Clementi, E., 156(82), *164*, 494(2), *532*
Clifford, S., 99(19), *101*
Cohen, A. J., 392(44), 417(44), *425*
Cohen, L., 122(14–15), 127(14–15), *161*, 166(36), *201*, 244(104–105, *258*, 262(17), 265(17), *290*, 317(36), *330*, 455(45), 456(45), *482*
Coleman, A. J., 11(1), 15, *16*, 22(6–7), 23(6), 26(6), 31(42), 32(6–7), 33(6–7), 37(6, 48, 50), 38(6), 54(7), *56*, *58*, 62(1), 90(23), *90–91*, 94(1), 95(1, 12), *100–101*, 106(170, 107(17), *118*, 122(12), 123(23), 124(27), 125(27), 126(3), 127(12), 139(23), *161–162*, 166(3–4, 7–8, 10–11), 183(3–4), 184(3–4, 7), 185(3–4, 76–78), 188(7), 196(4), *200*, *203*, 206(1, 13, 73), 207(1), 208(1, 13, 89), 209(1, 91), 210(91),

AUTHOR INDEX

214(73, 89), 215(73, 89, 97–98), 226(73), 244(1), 245(1), *253–254*, *257–258*, 262(20, 24, 271(56), 273(24), 287(24), *290–291*, 304(21022), 306(21022), 318(22), *330*, 331(16), *340*, 388(1), 389(1), 396(1), 400(1), *424*, 433(15), 434(15–17), 435(17), *440*, 448(32), 450(40), 460(40), 467(40), 475(40), 478(40), *481–482*, 487(1–2), 488(2), 489(2), 491(4), 492(2), *492*

Colmenero, F., 23(8–9), *56*, 122(21), 133(53), 136(20), 137(20), *162–163*, 166(14–16), 167(16), 169(16) 170(14), 174(14), *200*, 206(46–48), 207(47–48), 244(46–48), 245(46–48), 246(46–48), *255–256*, 262(1–5, 25–26), 288(1–2, 25–26), *290*, 331(1–3), 333(1), *340*, 352(43–44), *383*

Corongiu, G, 494(2), *532*
Cossi, M., 99(19), *101*
Cotton, F. A., 40(58), *58*
Coulson, C. A., 122(5), *161*, 206(12), *254*
Creau, C., 495(37), *533*
Cremer, D., 432(10), *440*
Cruz, N., 392(52), *426*
Cruz, R., 392(52), *426*
Csanyi, G., 392(40), *425*
Cui, Q., 99(19), *101*
Curtiss, L. A., 432(6–7), *440*

Daniell, M., 494(23), *532*
Daniels, A. D., 99(19), *101*
Dapprich, S., 99(19), *101*
Das Sarma, S., 519(103), 522(103), 528(103), *534*
Davidson, E. R., 94(10), 95(10), *101*, 122(7), 146(75), *161*, *164*, 166(8), *200*, 206(2, 29), 208(2), 209(2), *253*, *255*, 262(23), *290*, 389(7), 403(84), *424*, *427*, 447(24–28), 448(31), 454(25–26), 455(24–26), 456(24–26), 459(25–26), 462(26), 463(24–26), 464(25), 470(28), 475(67), 477(24–26), 478(26), *481–483*, 531(115), *535*
Daw, M. S., 183(60), *202*
Day, O. W., 391(32), 403(32), *425*
DeFrees, D. J., 417(87, 89), *427*
Del Re, G., 494(3), *532*
Deng, S., 516(80), 519(80), *534*
Derzhko, O., 511(75), *534*

Deutsch, D., 494(28), *532*
Deza, M. M., 96(16), *101*, 455(44), *482*
Ditchfield, R., 366(54), *383*
DiVincenzo, D. P., 494(11, 20), 519 (88–89), *532*, *534*
Doherty, A. C., 499(52), *533*
Dominicis, C. de, 271(59), *291*
Donnelly, R. A., 389(17), 390(17–18), 403(18), 404(18), *425*
Dreizler, R. M., 391(30), *425*
Duch, W., 206(40), 226(40), *255*
Dunning, T. H., Jr., 194(87), 197(87), *203*, 335(41), *342*, 366(55), *383*, 437(24, 25), *441*
Dupuis, M., 367(56), *384*
Durand, P., 392(49), *426*

Ehara, M., 23(20, 22), 38(20, 22), 44(20), 47(20, 22), 55(20, 22), *56–57*, 65(12), 90(12), *91*, 96(14, 15), 98(18), *101*, 103(1), 105(1, 8), 108(1), *117*, 139(68–69), *164*, 184(66, 68), 198(66, 68), *202*, 206(29, 56), 244(56, 109), 245(56, 109), 252(56, 109), *255–256*, *258–259*, 262(9–10), 287(9), 288(9), *290*, 448(33–34), 479(33), *481*
Einstein, A., 494(17), *532*
Elbert, S. T., 344(3), 367(56), *381*, *384*
Ema, I., 249(112, 113), *259*
England, W. B., 392(51), *426*
Englisch, H., 471(58–59), *482*
Englisch, R., 471(58–59), *482*
Erdahl, R. M., 16(32), *18*, 23(16–17), 26(16–17), 27(16–17), 28(17), 29(38), 37(51), 38(17), 55(16–17), *56–58*, 63(3–5), 65(10), 66(3–4, 15), 67(3–4, 5), 77(10), 83(3–4) 84(5), 88(5), 90(3–5), *90–91*, 94(4), 95(4), 96(4), 97(4), 100(4), *100*, 122(6), 125(28), 126(28, 32–33), 146(75), *161–162*, *164*, 172(52), 184(63, 65), 188(63), 196(63), 198(63), *202*, 206(17, 20, 29, 36), 209(17), 210(17), 244(36), 245(36), *254–255*, 262(22), 287(66), *290*, *292*, 332(28), *341*, 468(52), 478(52), *482*, 531(114), *535*
Eschrig, H., 388(5), *424*
Esquivel, R. O., 494(4), *532*
Essler, F. H. L., 512(77), *534*

Falci, G., 505(60), 519(60), 531(60), *533*
Fan, P.-D., 352(41), *383*

Fano, G., 344(5–6), *381*
Farkas, O., 99(19), *101*
Fazio, R., 505(60), 519(60), 531(60), *533*
Feenberg, E., 456(51), *482*
Feynman, R. P., 494(29), *532*
Fisher, A. J., 498(49), *533*
Flicker, M., 503(58), *533*
Foresman, J. B., 99(19), *101*, 417(89), *427*
Fox, D. J., 99(19), *101*, 417(89), *427*
Fradkin, E., 511(72–73), *534*
Frahm, H., 512(77), *534*
Francl, M. M., 417(87), *427*
Freed, K. F., 345(19, 25), 346(25), 256(19, 25), 332(31–32), 338(31–32), *341*, *382*
Freeman, D. L., 275(64), *292*
Frisch, M. J., 99(19), *101*, 417(89), *427*, 523(96), *534*
Frishberg, C., 122(14–15), 127(14–15), *161*, 166(36), *201*, 244(104–105), *258*, 262(17), 265(17), *290*, 317(36), *330*, 455(45), 456(45), *482*
Fu, H., 510(67), *533*
Fujisawa, K., 23(20), 38(20), 44(20), 47(20), 55(20), *56*, 65(12), 90(12), *91*, 98(18), *101*, 103(1), 104(2), 105(1), 108(1), 113(21), 114(22), 115(2, 22, 34), 116(21), *117–118*, 184(66), 198(66), *202*, 206(29), *255*, 448(34), *481*
Fukuda, H., 23(27), 27(27), 29(27), 38(27), 44(27), 47(27–28), 55(27), *57*, 184(70), 198(70), *202*, 206(33), *255*
Fukuda, M., 23(20), 38(20), 44(20), 47(20), 55(20), *56*, 63(6), 65(12), 66(6, 14), 90(6, 12, 14), *91*, 94(5, 8), 96(5), 97(5), 98(5, 18), 99(5, 8), 100(5), *100–101*, 103(1), 105(1), 106(14–16), 108(1, 14–15), 109(14–16), 110(15), 114(14–16), 115(34), 116(15), *117–118*, 184(66), 198(66), *202*, 206(29), *255*, 448(34–35), 479(35), *481*
Fulde, P., 176(57), *202*, 214(93–94), *258*
Furche, F., 446(10), 469(10), 474(10), *480*
Fusco, M. A., 64(7, 9), *90*, 206(26), *254*
Futakata, Y., 115(34), *118*

Gardiner, C. W., 176(56), *202*
Garrod, C., 12(3), 15(3, 19–22), 16(32), *16–18*, 23(18–19), 31(18), 34(18), 55(19), *56*, 64(8–9), *91*, 94(2), 95(2, 13), *100–101*, 126(29), 126(39), 146(27), *162–163*, 166(5), 183(5), 184(64), 188(5), 196(5), *200*, *202*, 206(16, 22–24, 26), 209(16, 24), 210(16, 24), 236(24), *254*, 287(67), *292*, 391(32), 403(32), *425*, 447(22), 450(22), 460(22), 467(22), 470(22), 475(22), 477(22), 478(22), *481*
Gauss, J., 344(10), *381*
Gersdorf, P., 494(7), 515(7), *532*
Gidofalvi, G., 23(24–25, 30–32, 37), 36(24), 37(24), 38(24–25, 30–31), 44(32), 48(37), 49(31), 53(33), 55(24–25, 30–32, 37), *57*, 105(13), 109(12), *117*, 184(73), *203*, 206(31), *255*, 332(37, 39), 338(37, 39), *341–342*, 448(38), *481*
Gilbert, M. M., 344(3), *381*
Gilbert, T. L., 171(49), *202*, 389(15), 394(15), 402(15), 405(13), *425*, 446(18), *481*
Gill, P. M. W., 99(19), *101*, 417(89), *427*, 432(9), *440*
Girvin, S. M., 511(74), 531(117), *534–535*
Gittings, J. R., 498(49), *533*
Glazek, S. D., 345(24), *382*
Glenn, K. G., 419(93), *427*
Goedecker, S., 183(61), *202*, 392(46), 391(35), 394(76), 400(76), 410(76), 417(76), *425–426*
Gohmann, F., 512(77), *534*
Golden, S., 444(3), 446(3), 475(3), *480*
Goldstone, J., 315(33), *330*
Golli, B., 16(32), *18*, 126(33), *162*
Gómez, D., 388(6), *424*
Gomperts, R., 99(19), *101*, 417(89), *427*
Gonis, A., 479(69), *483*
Gonzalez, C., 99(19), *101*, 417(89), *427*
Gordon, M. S., 367(56), *384*, 417(87), *427*
Goscinski, O., 389(16), 392(49), *425–426*, 530(109), *534*
Grabowski, I., 372(60), *384*
Granovsky, Alex A., 194(85), *203*
Gritsenko, O., 393(71), *426*
Gross, E. K. U., 393(63), *426*
Gruska, J., 494(15), *532*
Gu, S., 516(80), 519(80), *534*
Guevara, N. L., 494(4), *532*
Gunlycke, D., 509(65), *533*
Gunnarsson, O., 473(64), *483*, 494(5), 513(5), 515(5), *532*
Guo H., 186(81), *203*
Gwaltney, S. R., 372(59), *384*

AUTHOR INDEX

Hagley, E., 494(21), *532*
Hall, Richard, 12(6), *16*
Hammond, J. R., 23(33, 35), 27(33), 28(33), 48(35), 55(35), *57* 94(6), 99(6), *100*, 184(74), *203*, 332(38), 338(38), *342*, 468(53), *482*
Handy, N. C., 446(21), *481*
Hariharan, P. C., 417(87), *427*
Haroche, S., 494(21), *532*
Harriman, J. E., 47(67), *59*, 122(18), 126(37), 129(47), 133(18), 139(67), *162–164*, 166(12–13, 30), 169(45), 171(50), 183(30), 85(79–80), 186(80), *200–203*, 206(65–66, 74–76), 207(86), 214(74–76, 86), 215(86), 230(86), 233(86), 244(65, 107), 245(107), *256-258*, 262(38), 263(43), 265(47–48), 266(38), 271(47–48), 275(43), 282(47), 284(43), *291*, 331(20), *341*, 392(47), 414(85), 417(47), *425*, *427*, 446(19), *481*
Harris, F. E., 139(66), *164*, 214(96), *258*, 272(62), 275(63), *292*
Harris, J., 473(63), *482*
Hartree, D. R., 432(8), *440*
Hay, P. J., 194(87), 197(87), *203*, 335(41), *342*
Hazra, A., 206(68), 244(68), 245(68), *256*, 262(44), 284(44), *291*
Head-Gordon, M., 99(19), *101*, 268(54), 271(54), *291*, 344(8–9), 352(42), 372(59), *381*, *383-384*, 417(89), *427*
Hehre, W. J., 366(54), *383*, 417(87), *427*
Helbig, N., 393(63), *426*
Helgaker, T., 38(55), *58*
Hennig, R. G., 391(29), *425*
Herbert, J. M., 139(67), *164*, 166(30), 183(30), *201*, 206(65–66), 244(65), *256*, 263(43), 265(47), 271(47), 275(43), 282(47), 284(43), 288(46, 69), 289(69), *291–292*, 331(20), *341*, 392(47), 417(47), *425*
Hernandez-Laguna, A., 141(72), 143(72), *164*
Herring, C., 503(58), *533*
Herschbach, D. R., 530(109), 531(111), *534–535*
Hess, B. A., 344(12–13), *381–382*, 393(62, 66), *426*
Hill, S., 497(44), *533*
Hino, O., 346(27–28), *382*
Hirao, K., 345(20), 367(20, 58), *382*, *384*
Hirata, S., 367(57), 372(60), *384*

Hoffmann, M R., 332(34), 338(34), *341*, 345(26), 346(26, 29), 375(29), *382*
Hohenberg, P., 170(46–47), *201*, 388(2), *424*, 435(19), *440*, 528(105), *534*
Holas, A., 391(37), *425*
Horodecki, M., 495(38), *533*
Horodecki, P., 495(39), *533*
Horodecki, R., 495(39), *533*
Hose, G., 345(19), 356(19), *382*
Hsu, J. Y., 446(8, 11), 469(8, 11), *480*
Huang, Z., 501(57), 504(59), *53*, 506(62), 507(59, 62), 509(62), 519(59), 522(57), 523(57), 524(98), 526(99), 527(99), 528(104), 529(104), *533–534*
Hubbard, J., 512(76), *534*
Hugenholtz, N. M., 315(32), *330*
Hummer, H., 488(3), *491*
Hurley, A. C., 429(1), *440*
Husimi, K., 122(1), *161*, 206(7), *254*

Itano, W. M., 494(22, 25), *532*

Jackson, H. W., 456(51), *482*
Jarzecki, A. A., 389(7), *424*
Jedziniak, K., 352(41), *383*
Jensen, J. J., 367(56), *384*
Jeziorski, B., 299(16), *330*
Jin, B., 23(16), 26(16), 27(16), 55(16), *56*, 63(3–4), 66(3–4), 67(3–4), 83(3–4), 84(3–4), 90(3–5), *90*, 206(20), *254*
Johanneson, H., 531(118), *535*
John, W., 494(5, 7), 513(5), 515(5, 7), *532*
Johnson, B., 99(19), *101*
Johnson, B. G., 417(89), *427*
Johnson, R. D., III, 417(90), 419(90), 421(90), 422(90), 423(90), *427*
Jones, R. O., 473(63), *482*
Jordan, P., 511(71), *533*
Jorgensen, P., 38(55), *58*
Jozsa, R., 495(37)*533*
Juhász, T., 23(26), 38(26), 47(26), 55(26–27), *57*, 206(32), *255*

Kais, S., 494(8), 501(57), 504(59), 506(62), 507(59, 62), 509(62), 156(79, 81), 517(82), 519(59, 83), 520(92–94), 522(57), 523(57), 524(98), 526(99), 527(99), 528(104), 529(104), 530(11), *532–535*
Ka'llay, M., 344(10), *381*
Kamiya, M., 367(58), *384*

Karasiev, V., 388(6), *424*
Karp, R. M., 96(17), *101*
Karwoski, J., 206(40), 226(40), *255*
Katriel, J., 403(84), *427*
Kawashima, Y., 367(58), *384*
Keiter, H., 499(53), 500(53), *533*
Keith, T., 99(19), *101*, 417(89), *427*
Keller, J. H., 306(26), *330*
Kempe, J., 528(101), *534*
Kendon, V. M., 509(65), *533*
Kielpinksi, D., 494(25), *532*
Kijewski, L. J., 15(19), *17*, 206(21), *254*
King, B. E., 494(22), 494(24–25), *532*
Kinoshita, T., 346(27–28), *382*
Kirtman, B., 346(36), *383*
Kitaev, A., 499(51), 519(51), *533*
Kladko, K., 176(57), *202*, 214(94), *258*
Klahn, B., 295(10), 296(10), *330*
Klein, A., 391(30), *425*
Klein, D. J., 44(61), *58*
Klumper, A., 512(77), *534*
Kobayashi, K., 115(34), *118*
Koch, S., 267(51), 271(51), *291*, 315(7), *329*
Kočvara, M., 115(24), 116(24), *118*
Kohn, W., 170(46–47), *201*, 388(2, 4), *424*, 435(20), *440*, 470(55), 471(55), *482*, 528(105–106), *534*
Kojima, M., 104(2), 113(21), 114(22), 115(2, 22, 34), 116(21), *117–118*
Koller, J., 392(50), *426*
Kollmar, C., 393(62, 66, 70), *426*
Komaromi, I., 99(19), *101*
Korepin, V. E., 512(77), *534*
Koseki, S., 367(56), *384*
Kou, H., 139(68), *164*, 206(56), 244(56), 245(56), 252(56), *255-256*, 262(10), *290*
Kowalski, K., 345(19), 352(41), 356(19), *382–383*
Kryachko, E. S., 388(3), *424*
Kubo, R., 138(60), *163*, 176(55), 177(55), *202*, 214(92), *258*, 262(39), 269(39), 272(39), *291*, 301(19), *330*
Kucharski, S. A., 332(35), 338(35), 339(35), *341*, 346(32), 356(50), 357(32, 50), 358(51), rski, S. A., 356(50), 357(32, 50), 358(51), *382–383*
Kudin, K. N., 99(19), *101*
Kuhn, H. W., 455(42–43), 463(43), 464(43), *482*

Kummer, H., 12(4), *16*, 31(41), *58*, 127(41), *163*, 166(6), *200*, 262(21), *290*, 304(23), *330*, 447(29), 455(29), 478(29), *481*
Kus, M., 498(45), *533*
Kutzelnigg, W., 28(39), 39(56), 44(56, 60), 57–58, 139(64), *164*, 166(27, 31, 39), 176(39), 183(27), 187(39), 198(27, 31), *201*, 206(61), 214(61, 95), *256*, *258*, 262(28–32), 263(29, 31), 265(29–31), 266(29–32), 267(32, 50–52), 271(50–52), 272(28, 32), 274(28), 277(30–31), 282(29–31), 283(30), *291*, 295(5–9), 296(6), 299(3, 5, 15), 300(17), 302(17), 303(17), 305(24–25), 307(17, 30), 310(8), 311(3, 8), 315(5–7, 30), 317(20), 318(38), 319(30), 320(20, 29), 322(25), 324(25), 326(25), 327(6), *329–330*, 331(13, 17, 21), 332(33), 333(13, 21), 338(33), 339(13, 33), *340–341*, 346(30–31), 351(38), *382–383*, 393(75, 77), 400(77), 410(77), 414(85), *426-427*, 494(3), *532*
Kwiat, P. G., 495(35), *533*

Laidig, W. D., 375(61), 377(61), *384*
Lain, L., 167(44), 169(44), *201*, 206(41–42), 226(41–42), *255*, 262(37), 266(37), *291*, 393(64), *426*
Landau, L. D., 21(1), *56*
Landauer, R., 494(27, 30), *532*
Langer, C., 494(25), *532*
Langreth, D. C., 473(65), *483*
Lara-Castells, M. P. de, 126(34–35), *162*, 206(44), 207(78–79), 210(79), 214(78–79), 226(43–44), 230(78–79), 233(78–79), 244(79), 245(79), *255*, *257*
Larson, R., 450(41), *482*
Larsson, D., 531(118), *535*
Lathiotakis, N. N., 393(63), *426*
Latorre, J. I., 499(51), 519(51, 87), *533–534*
Laurent, M., 96(16), *101*, 455(44), *482*
Lee, D. K., 456(51), *482*
Legeza, Ö., 344(12–13), *381–382*
Leibfried, D., 494(22), *532*
Leiva, P., 392(60), 394(78–80), *426*
Lennard-Jones, J., 429(1), *440*
Levine, I. N., 417(88), *427*
Levine, R. D., 520(92, 94), 522(95), *534*
Levy, M., 170(48), *202*, 271(57), *291*, 389(10–11, 14), 390(10), 391(33),

400(10–11), 403(33), *424–425*, 435(21), *441*, 444(3, 5), 446(3, 14, 17), 449(5), 475(3), 477(17), *480–481*
Lewenstein, M., 498(45), *533*
Li, X. Q., 186(82), *203*, 226(102), *258*, 345(17), 346(17), 351(17), *382*
Li, X.-P., 183(59), *202*
Li, Y. S., 498(47), 516(80), 519(80), *533–534*
Liashenko, A., 99(19), *101*
Lidar, D. A., 528(101), 529(107), 530(107), 531(119), *534–535*
Lieb, E., 271(57), *291*, 390(20–21), *425*, 473(61), *482*, 505(61), 519(60), 531(60), *533*
Lin, C. H., 446(11), 469(11), *480*
Lin, H., 516(80), 519(80), *534*
Lindgren, I., 296(11), *330*
Linguerri, R., 34(5–6), *381*
Liu, G., 99(19), *101*
Liu, X. S., 498(47), *533*
Long, G. L., 498(47), *533*
Lopez, A., 511(73), *534*
Lopez, R., 249(112, 113), *259*
Lopez-Boada, R., 391(31), *425*
López-Sandoval, R., 391(27), *425*
Loss, D., 498(45), 519(88–89), *533–534*
Löwdin, P. O., 22(3), *56*, 122(2), 124(2), *161*, 206(8), *254*, 309(31), *330*, 394(82), 396(82), 403(82), *426*, 448(30), *481*, 494(1), *532*
Lu, A. Y., 166(33), *201*, 331(23), *341*
Ludeña, E. V., 388(3, 6), 390(25), *424–425*
Lukman, B., 392(50), *426*
Lundqvist, B. I., 473(64), *483*

Macchiavello, C., 494(12), *532*
McRae, W. B., 94(10), 95(10), *101*, 447(25), 454(25), 455(25), 456(25), 459(25), 463(25), 464(25), 477(25), *481–482*
McWeeney, R., 44(59), *58*, 133(49), 141(72), 143(72), 156(49), *163–164*, 166(2), 183(2), 184(2), *200*, 206(10), 219(100), *254*, *258*, 399(83), *427*
Malick, D. K., 99(19), *101*
Malmqvist, P.-Å., 345(18), *382*
March, N. H., 44(61), *58*, 446(15–16), *480-481*, 494(10), *532*
Martin, P. C., 138(58), *163*, 271(59), *291*
Martin, R. L., 99(19), *101*, 344(4), *381*, 417(89), *427*
Martinez, A., 392(59), *426*

Maruani, J., 141(72), 143(72), *164*
Maschke, K., 390(24), *425*
Maslen, P. E., 268(54), 271(54), *291*
Matsunaga, N., 367(56), *384*
Mattis, D., 505(61), *533*
Mattle, K., 495(32, 35), *532–533*
Mattuck, R. D., 275(63), *292*
Mayer, J. E., 22(4), *56*, 95(11), *101*, 122(3), *161*, 206(9), *254*
Maytre, X., 494(21), *532*
Mazziotti, D. A., 23(13–15, 17, 21, 23–26, 28–37), 26(17), 27(17, 21, 33–34), 28(14–15, 17, 33, 40), 29(34), 30(21), 30(13, 29), 36(24), 37(24, 49), 38(17, 21, 23–26, 28–31), 43(13), 44(21, 32, 34), 47(13, 21, 23, 26, 28, 29), 48(28–29, 34–35, 37), 49(31), 50(28–29), 53(33), 54(28, 34), 55(17), 55(21, 23–26, 28–34, 37), *56–58*, 63(5), 65(13), 67(5, 17), 82(20–22), 84(5), 88(5), 90(3–5, 20–22), *90–91*, 94(3, 6–7), 99(6–7), *100–101*, 105(9–13), 108(9), 109(12–13), 115(10–11), 116(10–11), *117*, 128(45), 137(45), 138(59), 139(45, 54, 61–63), 141(45, 62), 145(45, 62–63), 146(75), 151(78), *163–164*, 166(20–22, 24–26, 28–29, 33, 40), 167(20, 28–29), 169(20), 170(20–21, 29), 174(20–22), 175(20), 176(21–22, 24, 26), 179(20, 24, 26), 180(20, 24, 26), 181(24), 183(62), 184(28–29, 62–63, 67, 69, 71–74), 187(21–22, 24, 26, 40), 188(63), 192(20), 196(63), 198(32, 63, 67, 69, 71–74), *200–203*, 206(30–32, 34, 36, 52–54, 57–59, 67, 69–70), 214(53), 215(70), 217(70), 219(67), 244(36, 52–54, 57–59, 70, 111), 245(36, 52–54, 57, 59, 70, 111), 246(52, 111), 249(70, 111), 252(70, 111), 253(34, 52), *255–258*, 262(13–16, 27, 33–34), 265(13), 266(27, 33–34), 270(13, 33), 271(33), 275(27), 287(34, 66), 288(4), *290–292*, 300(18), 318(18), *330*, 331(7, 9, 11–12, 14–15, 18–19, 22–23, 26–27), 332(27–28, 36–39), 333(7–9, 11, 14–15), 336(18), 337(36), 338(22, 36–39), 339(8–9, 11, 14–15), *340–342*, 352(47–48), *383*, 392(42–43), 393(74), 400(43), *425–426*, 435(18), 436(23), 437(23), *440–441*, 448(36–37, 39), 456(48, 50), 468(53), 479(36–39), *481–482*, 491(5), *492*, 531(112–114), *535*
Meekhof, D. M., 494(24), *532*

Mehler, E. L., 429(2), 431(2–4), *440*
Mennucci, B., 99(19), *101*
Meyer, V., 494(25), *532*
Meyer, W., 365(53), *383*
Mihailović, M. V., 14(7), 15(17, 18, 21, 25–27, 29), *16–17*, 23(19), 37(46), 47(69), 56, *58-59*, 64(8), 66(16), 67, 87, *91*, 184(64), 202, 206(22, 25, 27), 209(25), 210(25, 27), 236(25), *254*
Millam, J. M., 99(19), *101*
Millan, J., 206(42), 226(42), *255*
Mitrushenkov, A. O., 344(5–6), *381*
Mizuno, Y., 44(59), *58*, 133(49), 156(49), *163*, 219(100), *258*, 399(83), *427*
Møll, C., 432(11), *440*
Møller, L., 324(34), *330*
Monkhorst, H. J., 275(64), *292*
Monroe, C., 494(22, 24–25), *532*
Monteiro, R. D. C., 47(68), *59*,, 104(7), 111(7), 113(7), 115(7, 26–27), 116(26–27), *117–118*
Montero, L. A., 392(52), *426*
Montgomery, J. A., Jr., 99(19), *101*, 367(56), *384*, 417(89), *427*
Moré, J., 115(32), *118*
Morokuma, K., 99(19), *101*
Morrell, M. M., 391(33), 403(33), *425*
Morrison, J., 296(11), *330*
Morrison, R. C., 389(9), 419(91), *424*, *427*
Mukherjee, D., 28(39), *57*, 139(64), *164*, 166(27, 31, 39), 176(39), 183(27), 187(39), 198(27, 31), *201*, 206(61), 214(61, 95), *256*, *258*, 262(28–31), 263(29, 31), 265(29–31), 266(29–31), 272(28), 274(28), 277(30–31), 282(29–31), 283(30), *291*, 299(3), 300(17), 302(17), 303(17), 305(24–25), 307(17, 30), 311(2–3), 315(30), 317(20), 319(30), 320(20, 29), 322(25), 324(25), 326(25), *329–330*, 331(13, 17, 21), 333(13, 21), 339(13), *340–341*, 345(19), 346(35), 356(19), *382–383*, 393(75), *426*
Müller, A. M. K., 390(22), *425*
Munson, T., 115(32), *118*
Myatt, C. J., 494(20, 25), *532*

Nagy, A., 446(7, 9), 469(7, 9), *480*
Nakajima, T., 367(58), *384*
Nakano, H., 345(21), 367(21, 58), 375(21), *382*, *384*
Nakao, Y., 367(58), *384*

Nakata, J., 23(20, 22), 38(20, 22), 44(20), 47(20, 22), 55(20, 22, 28–31), *56–57*
Nakata, K., 23(20, 22), 38(20), 44(20), 47(20), 55(20), *56–57*, 65(12), 90(12), 98(18), *91*, 103(1), 105(1), 108(1), 113(21), 115(34), 116(21), *117–118*, 184(66), 198(66), *202*, 206(29), *255*, 448(34), *481*
Nakata, M., 65(12), 66(14), 90(12, 14), *91*, 94(8), 96(14, 15), 98(18), 99(8), *101*, 103(1), 105(1, 8), 106(14–15), 108(1, 15), 109(15–16), 110(15), 114(15–16), 115(34), 116(15), *117–118*, 139(68–69), *164*, 184(66, 68), 198(66, 68), *202*, 206(29, 56), 244(56, 109), 245(56, 109), 252(56, 109), *255–256*, *258*, 262(9–10), 287(9), *290*, 448(33), 479(33), *481*
Nakatsuji, H., 23(10–11, 20, 22), 38(20, 22), 44(20), 47(20, 22), 55(20, 22), *56-57*, 65(12), 90(12), *91*, 96(14, 15), 98(18), *101*, 103(1), 105(1, 8), 108(1), *117*, 122(16), 127(16), 129(16), 138(56–57), 139(68–69), 141(56–57), 142(56–57), *161*, *163–164*, 166(18–19, 37), 167(37), 169(19, 37), 170(18), 181(19), 184(66, 68), 198(66, 68), *200-202*, 206(29, 49, 51, 56), 207(49, 51), 244(49, 51, 56, 106, 109), 245(49, 51, 56, 106, 109), 246(49, 51, 252(49, 51, 56, 109), *255–256*, *258*, 262(7–10, 12, 18–19), 265(18), 277(7–8), 287(7, 9), 288(7–9), *290*, 317(35), 318(37), *330*, 331(5–6), 333(5), 339(5–6), *340*, 352(39, 45–46), *383*, 448(33–34), 456(46), 479(33), *481–482*, 531(116), *535*
Nanayakkara, A., 99(19), *101*, 417(89), *427*
Negele, J. W., 176(58), 177(58), *202*
Nemirovski, A., 104(5), 110(18), 111(5), 113(5), 115(5), *117–118*
Nesterov, Yu, 110(18), *118*
Nguyen, K. A., 367(56), *384*
Nielsen, M., 494(13), 495(13), *532*
Nielsen, M. A., 509(66), 510(66), 519(85), 520(85), 522(85), 531(66), *533–534*
Nocedal, J., 115(25), *118*
Noga, J., 332(35), 338(35), 339(35), *341*, 346(32), 356(50), 357(32, 50), 377(49), *382–383*
Nogues, G., 494(21), *532*
Nooijen, M., 139(70), *164*, 206(68), 244(68, 110), 245(68, 110), *256*, *258*, 263(44),

275(65), 284(44), *291–292*, 352(40), 372(60), *383–384*, 391(34), *425*
Noziérees, P., 296(11), *330*
Nunes, R. W., 183(59), *202*
Nyden, M. R., 446(13), *480*

Ochterski, J., 99(19), *101*
Oliphant, N., 422(94), *427*
Olsen, J., 38(55), *58*
Onizuka, K., 511(70), *533*
Orland, H., 176(58), 177(58), *202*
Ortiz, J. V., 99(19), *101*, 417(89), *427*
Ortiz de Zarate, A., 206(41), 226(41), *255*
Ortolani, F., 344(5), *381*
Osborne, T. J., 509(66), 510(66), 519(85), 520(85), 522(85), 531(66), *533–534*
Osenda, O., 504(59), 506(62), 507(59, 62), 509(62), 519(59), *533*
Oshsenfeld, C., 268(54), 271(54), *291*
Osterloh, A., 505(60), 519(60), 531(60), *533*
Otto, P., 392(52, 58–59), 393(61), *426*
Overton, M. L., 23(27), 27(27), 29(27), 38(27), 44(27), 47(27–28), 55(27), *57*, 63(6), 65(11), 66(6, 14), 77(11), 90(6, 14), *91*, 94(5, 8), 96(5), 97(5), 98(5), 99(5, 8), 100(5), *100–101*, 106(14–16), 108(14–15), 109(14–16), 110(15), 114(14–16), 116(15), *117–118*, 184(70), 198(70), *202*, 206(33), *255*, 448(35), 468(35), 479(35), *481*

Pal, S., 346(34–35), *383*
Paldus, J., 299(16), *330*, 345(16–17), 346(17), 351(17), *382*
Palma, G. M., 494(12), *532*
Palmieri, P., 344(5–6), *381*
Pan, J.-W., 494(23), 495(32–33), *532–533*
Papadimitriou, C. H., 96(17), *101*
Parr, R. G., 122(9), *161*, 388(3), 391(33), 403(33), *424–425*, 446(12), *480*
Parrilo, P. A., 49(52), *533*
Paškauskas, R., 498(48), *533*
Pastor, G. M., 391(27), *425*
Peat, F. D., 475(67), *483*
Peng, C. Y., 99(19), *101*, 417(89), *427*
Percus, J., 12(3), 15(3, 19), *16–17*, 23(18, 27), 27(17), 29(27), 31(18), 34(18), 38(27), 44(27), 47(27–28), 55(27), *56–57*, 63(6), 66(6, 14), 90(6, 14), *90–91*, 94(2, 5, 8), 95(2), 96(5), 97(5), 98(5), 99(5, 8), 100(5), *100–101*, 106(14–16), 108(14–15),
109(14–16), 110(15), 114(14–16), 116(15), *117–118*, 126(29, 31), 146(27), *162*, 166(5), 183(5), 184(70), 185(5), 196(5), 198(70), *200*, *202*, 206(16, 21, 33), 209(16), 210(16), *254-255*, 287(67), *292*, 447(22), 448(35), 448(35), 450(22), 460(22), 467(22), 468(35), 470(22, 54), 475(22), 477(22), 478(22), 479(35), *481–482*
Perdew, J. P., 389(8, 14), *424–425*, 446(14), 473(65), *480*, *483*, 494(7), 515(7), *532*
Peres, A., 495(37), *533*
Perez del Valle, C., 122(20), 136(20), 137(20), *162*, 166(14), 170(14), 174(14), *200*, 206(46), 245(46), 246(46), *255*, 262(25), 288(25), *290*, 331(1), 333(1), *340*
Pérez-Roman, E., 23(12), 39(57), 41(57), 42(57), 44(57, 62), *56*, *58*, 124(26), 125(26), 126(34–35), 128(46), 131(46), 132(46), 136(46), 137(46, 54), 138(54), 139(54), 140(26, 54), 141(73–74), 143(72–73), 146(26, 54), 147(26 73), 151(51), 153(81), 160(81), *162–164*, 166(35), 167(17), 170(17), 184(17, 35, 75), *200-201*, *203* 206(14–15, 18, 44), 206(44, 60), 207(18, 50, 60, 78–79, 81–85, 88), 209(15), 210(14, 79, 83), 214(78–79, 81–83), 219(14), 226(43–44), 230(18, 60, 78–79, 81–85), 233(15, 78–79, 81–83), 234(83), 244(15, 18, 50, 60, 79, 81–83), 245(15, 18, 50, 60, 79, 81–85, 88), 246(18, 50, 60, 81), 247(18, 88), 249(50, 88), 253(18), *254-257*, 262(3, 5–6), 288(6), *290*, 331(4, 25), *340–341*, 456(47), *482*
Pernal, K., 389(12), 391(38), 392(39, 41, 56–57), 393(65, 68–69, 71–72), 394(81), 400(12), 404(72), 405(81), 417(81), *424–426*
Peterson, G. A., 417(89), *427*
Peterson, K. A., 437(25), *441*
Petersson, G. A., 99(19), *101*
Petro, W. J., 417(87), *427*
Piecuch, P., 345(19), 352(41), 356(19), *382–383*
Piesche, P., 479(1, 4), *480*
Piris, M., 392(52–53, 58–60), 393(61, 73), 394(78–81), 408(53), 411(73), 413(53), *426*
Piskorz, P., 99(19), *101*
Pistol, M. E., 446(6), *480*
Pleaset, M. S., 324(34), *330*, 432(11), *440*
Podolsky, B., 494(18), *532*
Pomelli, C., 99(19), *101*

Pople, J. A., 99(19), *101*, 366(54), *383*, 417(87, 89), *427*, 429(1), 432(6–7), *440*
Post, H., 12(6), *16*
Potts, A. W., 419(93), *427*
Prasad, M. D., 346(35), *383*
Price, W. C., 419(93), *427*
Primas, H., 327(12), *330*
Purvis, G.D., 262(40), 268(40–41), *291*

Rabuck, A. D., 99(19), *101*
Raghavachari, K., 99(19), *101*, 417(89), *427*, 432(6–7), *440*
Raimond, J. M., 494(21), *532*
Rajagopal, A. K., 530(108), *534*
Ramírez, G., 249(112, 113), *259*
Rassolov, V. A., 431(5), 432(5), *440*
Redfern, P. C., 432(6–7), *440*
Redondo, A., 390(26), *425*
Remacle, F., 520(94), 522(95), *534*
Rendell, R. W., 530(108), *534*
Replogle, E. S., 99(19), *101*
Replogle, E. S., 417(89), *427*
Rico, E., 499(51), 519(51, 87), *533–534*
Rico, J. Fernandez, 249(112, 113), *259*
Rinderspacher, B. C., 438(26), 439(26), *441*
Robb, M. A., 99(19), *101*, 417(89), *427*
Röder, J., 344(12–13), *381–382*
Roetti, C., 156(82), *164*
Rojo, A. G., 511(73), *534*
Rolik, Z., 345(19), 356(19), *382*
Romanko, O., 115(33), *118*
Roos, B. O., 344(2), 345(18), *381–382*
Rosen, N., 494(18), *532*
Rosina, M., 14(7), 15(9, 10), 15(17–31), 16(32), *16–18*, 37(46, 51), 47(69), 55(19), *56*, *58–59*, 64(8), 66(16), 67, *91*, 95(13), *101*, 126(32, 39), *162–163*, 171(51), 179(51), 184(51, 64), *202*, 206(22–25, 27, 35), 209(24–25), 210(24–25, 27), 236(24–25), *254–255*
Rosta, E., 432(13–14), 433(14), *440*
Rowe, M., 494(25), *532*
Ruedenberg, K., 344(3), *381*, 429(2), 431(2–4), *440*
Runge, K., 127(40), *163*, 209(90), *258*
Ruskai, M. B., 172(53), 184(53), *202*

Saad, Y., 364(52), *383*
Sachdev, S., 519(84), 531(84), *534*

Sackett, C. A., 494(25), *532*
Sadiek, G., 528(104), 529(104), *534*
Sadlej, A. J., 345(18), *382*
Sagar, R. P., 494(4), *532*
Sahni, V., 446(14), *480*
Saigal, R., 81(9), *91*, 103(4), 111(4), 113(4), 115(4), *117*
Sarandy, M. S., 529(107), 530(107), 531(119), *534–535*
Sarich, J., 115(32), *118*
Sasaki, F., 12(2), 15, *16*, 38(53), *58*, 306(28), *330*
Satchler, G. R., 38(54), *58*
Schlegel, H., 99(19), *101*
Schlegel, H. B., 417(89), *427*
Schliemann, J., 498(45), *533*
Schlosser, H., 127(44), *162*
Schmidt, E., 306(27), *330*
Schmidt, M. W., 194(86), *203*, 335(42), *342*, 344(3), 367(56), *381*, *384*
Schork, T., 214(93), *258*
Schreiner, P. R., 438(26), 439(26), *441*
Schrödinger, E., 494(16, 19), *532*
Schulthess, T. C., 479(69), *483*
Schultz, M., 364(52), *383*
Schultz, T., 505(61), *533*
Schumacher, B., 495(36), *533*
Schwinger, J., 138(58), *163*
Scully, M. O., 531(111), *535*
Scuseria, G. E., 389(8), 392(45, 48), *424–426*
Sears, S. B., 446(12), *480*
Seevinck, M., 499(55), *533*
Sekino, H., 367(58), *384*
Serra, P., 530(110), *535*
Seuseria, G. E., 99(19), *101*
Shahar, D., 531(117), *535*
Sham, L. J., 170(47), *201*, 388(4), *424*, 435(20), *440*, 470(55), 471(55), *482*, 528(106), 529(107), 530(107), *534*
Shavitt, I., 345(16), *382*
Sheppard, M. G., 332(31), 338(31), *341*
Shevni, N., 528(102), *534*
Shi, Q., 494(8), *532*
Sierraalta, A., 390(25), *425*
Silver, D. M., 429(2), 431(2–4), *440*
Simons, J., 346(29), 375(29), *382*, 432(12), *440*
Simons, J. A., 332(34), 338(34), *341*
Slebodziński, W., 30(43), *58*, 173(54), *202*

Smith, D. W., 391(32), 403(32), *425*
Smolin, J. A., 494(20), *532*
Solomon, I., 510(67), *533*
Sondhi, S. L., 531(117), *535*
Sousa, R. de, 519(103), 522(103), 528(102–103), *534*
Spedalieri, F. M., 499(52), *533*
Staroverov, V. N., 389(8), 392(45, 48), *424–426*
Stefanov, B. B., 99(19), *101*, 417(89), *427*
Stevens, J. E., 332(32), 338(32), *341*
Stewart, J. P., 417(89), *427*
Stingl, M., 115(24), 116(24), *118*
Stolze, J., 499(53), 500(53), *533*
Strain, M. C., 99(19), *101*
Strang, G., 192(83), *203*
Stratmann, R. E., 99(19), *101*
Sturm, J. F., 115(30), *118*
Su, S., 367(56), *384*
Sun, C. C., 186(82), *203*, 226(102), *258*
Surján, P. R., 167(41), *201*, 219(101), *258*, 345(19), 356(19), *382*, 432(13–14), 433(14), *440*, 167(41), *201*
Svetlichny, G., 499(55), *533*
Svidzinsky, A. A., 531(111), *535*
Szabados, A., 345(19), 356(19), *382*

Tal, Y., 390(25), *425*
Tang, A. C., 186(81–82), *203*, 226(102), *258*
Tao, J., 389(8), *424*
Tasnádi, F., 262(36), 266(36), *291*
Tel, L. M., 23(12), 39(57), 41(57), 42(57), 44(57, 62), *56*, *58*, 124(26), 125(26), 126(35–36), 128(46), 131(46), 132(46), 133(50), 136(46), 137(46, 54), 138(54), 139(54), 140(26, 54), 141(73–74), 143(73), 146(26, 54), 147(26, 73), 151(51), 153(81), 160(81), *162-164*, 166(35), 167(17), 170(17), 184(17, 35, 75), *200-201*, *203*, 206(14–15, 18, 60, 71), 207(18, 50, 60, 71, 78–79, 81–85, 88), 209(15), 210(14, 71, 79, 83), 214(78–79, 81–83), 217(71), 219(14), 227(71), 229(71), 230(18, 60, 78–79, 81–85, 233(15, 71, 78–79, 81–83), 234(71, 83), 244(15, 18, 50, 60, 79, 81–85), 245(15, 18, 50, 60, 71, 79, 81–85, 88), 246(18, 50, 60, 81), 247(18, 88), 249(50, 88), 253(18), *254-257*, 262(3–6), 288(6), *290*, 331(4, 25), *340-341*, 456(47), *482*
Ter Haar, D., 166(1), *200*
Terlaky, T., 115(33), *118*

Theophilou, A. K., 44(61), *58*
Thouless, D. J., 296(11), 318(39), *330*
Todd, M. J., 81(81), *91*, 104(6), 111(6), 113(6), 115(6, 31), *117–118*
Toh, K.-C., 115(31), *118*
Tomasi, J., 99(19), *101*
Torre, A., 141(74), *164*, 167(44), 169(44), *201*, 206(42, 60), 207(60), 226(42), 230(60), 255-256, 262(6, 37), 266(37), 288(6), 290-291, 393(64), *426*
Torre, E., 244(60), 245(60), 246(60), *256*
Toth, G., 345(19), 356(19), *382*
Tredgold, 22(5), *56*
Trucks, G. W., 99(19), *101*, 346(33), 352(33), 356(33, 50), 357(33, 50), 377(33), *382-383*, 417(89), *427*
Tsundeda, T., 367(58), *384*
Tung Nguyen-Dang, T., 390(25), *425*
Turchette, Q. A., 494(22, 25), *532*
Turchi, P. E. A., 479(69), *483*
Tütüncii, R. H., 115(31), *118*

Uffink, J., 499(54), *533*
Umrigar, C. J., 391(35), 394(76), 400(76), 410(76), 417(76), *425-426*

Valdemoro, C., 23(8–9, 12), 29(57), 41(57), 42(57), 44(57, 62), *56*, *58*, 122(17, 19–21), 123(25), 124(26), 125(26), 126(34–36, 38), 127(17, 24, 26, 42–43), 128(46), 131(46), 132(46), 133(19, 38, 50, 53), 135(19), 136(19–20, 46), 137(20, 46, 54), 138(54), 139(54), 140(26, 54), 141(71, 73–74), 143(19, 73), 146(26, 54), 147(26, 73), 151(38, 51), 153(79, 81), 160(81), *162-164*, 166(14–17, 34–35, 38), 167(16–17, 34, 42–44), 169(16, 42–44), 170(14, 17), 174(14), 184(17, 34–35, 75), *200-201*, *203*, 206(14–15, 18, 38–41, 43–48, 50, 60, 62, 71–72), 207(18, 45, 47–48, 50, 60, 71–72, 78–85, 88), 208(37), 209(15), 210(14, 71–72, 79, 83), 211(72), 214(78–83), 217(71), 219(14), 226(37–39, 40–41, 43–45), 227(71), 229(71–72), 230(18, 60, 71–72, 78–85), 233(15, 71, 78–83), 234(71), 235(72), 239(72), 244(45, 108), 245(15, 18, 45–48, 50, 60, 62, 71, 79–85, 88, 108), 246(18, 46–48, 50, 60, 81, 108), 247(18, 88), 249(50, 88), 253(18, 62), *254–258*, 262(1–6, 25–26), 265(3),

288(1–2, 6, 25–26), *290*, 317(4), 318(4), *330*, 331(1–4, 24–25), 333(1), 336(24), *340–341*, 352(43–44), *383*, 391(28), *425*, 456(47, 49), *482*
Valone, S. M., 390(19), 400, 435(22), *425*, *440*, 471(57), *482*
Vandenberghe, L., 45(63), 46(63), *58*, 81(19), *91*, 104(3–4), 111(4), 113(4), 115(3–4), *117*
Vanderbilt, D., 183(59), *202*
Vanek, J., 479(69), *483*
Van Leuven, P., 15(26–27, 31), *17-18*
Van Voorhis, Troy, 344(11), *381*
Vedral, V., 509(64–65), 510(64), *533*
Vidal, G., 499(51), 519(51, 87), *533–534*
von Neumann, J., 21(2), *56*
von Niessen, W., 419(92), *427*
Voorhis, T. V., 352(42), *383*

Wang, J., 516(79, 81), 517(82), 519(83), 520(92–94), *534*
Wang, X., 510(67–68), *533*
Wang, Y. A., 446(20), *481*
Wang, Y. R., 511(74), *534*
Warman, L. K., 511(74), *534*
Watts, J., 352(33), 356(33), 357(33), *382*
Watts, J. D., 346(33), 356(50), 357(50), 377(3), *382–383*
Wegner, F., 345(23), *382*
Weinberg, S., 271(60–61), *292*
Weinfurter, H., 494(23), 495(32), 495(35), *532–533*
Weinhold, F., 94(9), 95(9), *101*, 109(18), *118*, 146(75), *164*, 447(23), 449(23), 460(23), 478(23), *481*
Weizsacer, C. F., 471(56), *482*
Werner, H.-J., 365(53), *383*
Werner, R. F., 499(56), *533*
Whaley, K. B., 528(101–102), *534*
White, C. A., 268(54), 271(54), *291*
White, S. K., 338(29), *341*
White, S. R., 332(29), 339(29), *341*, 344(4), 345(22), 349(22), 352(22), 356(22), 357(22), *381–382*, 519(86), *534*
Wick, W. C., 296(1), *329*
Wiesner, S. J., 495(34), *533*
Wigner, E. P., 511(71), *533*
Wilson, E. B., 146(75), *164*, 447(23), 449(23), 460(23), 478(23), *481*
Wilson, E. B., Jr., 94(9), 95(9), *101*, 109(18), *118*

Wilson, K. G., 345(24), *382*
Wilson, S., 143(72), *164*, 494(9), *532*
Windus, T. L., 367(56), *384*
Wineland, D. J., 494(22, 24, 25), *532*
Wladyslawski, M., 139(70), *164*, 206(68), 244(68), 245(68), *256*, 263(44), 284(44), *291*
Woch, M., 345(19), 356(19), *382*
Wolinski, K., 345(18), *382*
Wolkowicz, H., 81(19), *91*, 103(4), 111(4), 113(4), 115(4), *117*
Wong, M. W., 99(19), *101*, 417(89), *427*
Wood, C. S., 494(22), *532*
Woon, D., 437(25), *441*
Wootters, W. K., 494(20), 495(37, 40), 496(42), 497(42–44), 498(42), 502(42), *532–533*
Wright, S., 45(64), *59*
Wright, S. J., 115(25), *118*
Wu, L. A., 529(107), 530(107), 531(119), *534–535*
Wu, Q., 472(60), *482*
Wunderlich, C., 494(21), *532*

Yamashita, M., 66(14), 90(14), *91*, 99(8), *101*, 104(2), 106(15–16), 108(15), 109(15–16), 110(15), 114(15–16, 22), 115(2, 22 34), 116(15), *117–118*
Yanai, T., 332(30), 338(30), 339(30), *341*, 344(1), 345(1), 367(58), *381*, *384*
Yang, C. N., 38(52), *58*
Yang, K., 511(74), *534*
Yang, W., 122(9), *161*, 388(3), *424*, 472(60), *482*
Yasuda, K., 23(10–11), *56*, 133(52), 138(56–57), 139(68–69), 141(56–57), 142(56–57), *163–164*, 166(18–19, 23), 169(19), 170(18), 181(19) 183(23), *200*, 206(49, 51, 55–56, 63), 207(49, 51), 244(49, 51, 55–56, 63, 109), 245(49, 51, 55–56, 63 109), 246(49, 51), 249(49, 51, 56, 109), *256*, *258*, 262(7–11), 265(11, 49), 271(49), 277(7–8, 11), 282(11), 283(11), 287(7, 9), 288(7–9, 11), *290–291*, 318(37), *330*, 331(5–6, 10), 333(5), 339(5–6), *340*, 352(45–46), *383*, 392(55), *426*
Yoseloff, M. L., 455(43), 463(43), 464(43), *482*
You, L., 498(48), *533*
Young, J., 114(23), 115(23), *118*
Yukalov, V. I., 22(7), 32(7), 33(7), 54(7), *56*, 90(23), *91*, 122(12), 127(12), *161*, 166(4), 185(4), *200*, 206(1), 207(1), 208(1), *253*, 262(24), 273(24), 287(24), *290*, 304(22),

306(22), 318(22), *330*, 331(16), *340*, 434(17), 435(17), *440*, 448(32), *481*, 487(1), 489(1), 491(1), *492*

Zakrzewski, V. G., 99(19), *101*, 417(89), *427*
Zanardi, P., 499(50), 512(78), 515(78), 517(78), *533–534*
Zeilinger, A., 494(12, 23), 495(32, 35), *532*
Zeng, B., 48(47), *533*
Zhao, Z., 23(27), 27(27), 29(27), 38(27), 44(27), 47(27–28), 55(27), *57*, 63(6), 66(6, 14), 90(6, 14), *91*, 94(5), 96(5), 97(5) 98(5), 99(5, 8), 100(5), *100–101*, 106(14–16), 108(14–15), 109(14–16), 110(15), 114(14–16), 116(15), *117–118*, 184(70), 198(70), *202*, 206(33), *255*, 448(34–35), 468(35), 479(35), *481*
Zhu, L., 115(33), *118*, 139(65), *164*, 262(35–36), 266(35–36), 269(35, 55), 271(55), *291*
Ziesche, P., 389(12), 392(56–57), 393(67), 400(12), *424*, *426*, 44(1, 4), 475(1, 4), *480*, 494(5–7), 513(5), 515(5, 7), *532*
Zukowski, M., 495(32), *532*
Zumbach, G., 390(24), *425*
Zurek, W. H., 528(100), *534*

SUBJECT INDEX

Ab initio methods, 344, 390
Absolute correlation, 523–524
Absolute error, 338, 367, 375
Absolute value, 309
Accuracy, significance of, 84, 100
Active-active space correlations, 348–349, 367
Active-external space correlations, 348–351
Active NSOs, 302
Active orbitals, 347, 364
Active-space
  calculations, 364
  reference, 381
Adaptation
  spatial symmetry, 40–42
  spin-adapted operators, 38–42
Additivity
  characterized, 496
  additively separable RDMs, 266–269
  dditively separate cumulants, 302
Adjoint equations, 283
Alcoba-Valdemoro purification procedures, 237–239
Algebraic formulation, 284
Alkanes, 268
α/α blocks, 190–191, 195, 222, 225
α/β block
  characterized, 190–192, 222, 225
  CSE, 132–133, 195
  purification procedures, 217, 219–220, 224, 226–227
Amplitude
  equations, 363–365, 367, 370
  transition, 14–15
Angular correlation, 494

Anions, 421
Anisotropic Heisenberg model, 510
Anisotropy, 501, 504, 508, 524
Annihilation operators, 7, 23, 25, 28, 68, 85, 94, 97, 125, 137, 155, 169, 172–173, 176, 232, 269–270, 294, 332, 350, 432, 452, 466–467, 498, 512, 522
Ansatz
  Bethe, 519
  unrestricted antisymmetrized geminal products (UAGP), 437, 439
  Weisacker, 573
Antibonding orbitals, 372
Anticommutation
  relations, 37, 175, 184, 269
  rules, 125
Anti-Hermitian contracted Schrödinger equation (ACSE)
  applications, 335–338
  characterized, 198, 331–333, 338–339
  defined, 331
  future directions for, 339
  optimization of 2-RDM, 334–335, 339
  reconstruction of 3-RDM, 333–334, 336, 338–339
Antisymmetric/antisymmetries,
  see Antisymmetrization
  conditions, semidefinite programming problem, 109
  construction algorithms, 136–137
  functions, 8–9, 39
  matrices, 141, 157, 191
  nondegenerate ground state, 171
  operators, 266–267
  permutations, 177, 180, 199, 333–334
  second-order matrices, 215–216

---

*Reduced-Density-Matrix Mechanics: With Application to Many-Electron Atoms and Molecules,
A Special Volume of Advances in Chemical Physics, Volume 134*, edited by David A. Mazziotti.
Series editor Stuart A. Rice. Copyright © 2007 John Wiley & Sons, Inc.

Antisymmetric/antisymmetries, (*Continued*)
  spin-free two-electron functions, 290
  two-particle matrix, 186
  wavefunction, 516
Antisymmetrical geminal power of
  extreme type, 306
Antisymmetrization
  canonical transformation theory, 353
  implications of, 179
  operator, 198
  T1/T2 conditions, 96
Antisymmetrized geminal products (AGP)
  characterized, 37–38, 303, 306, 433–434
  strongly orthogonal (SOAGP), 429–433, 437
  unrestricted (UAGP), 437–439
Antisymmetrized logarithms, 300
Antisymmetrized product of strongly
  orthogonal geminals (APSG), 303, 393
Antisymmetrizer, 302, 306
Antisymmetry
  conditions, 28, 302
  functionals and, 396, 398
Applied mathematics, 63
Approximation, *see specific types
  of approximations*
  canonical transformation theory, 354
  cumulant 4-RDM, 179–182
  cumulant 3-RDMs, 179–182
  finite-order ladder-type, 288
  first-order, 331
  k-particle hierarchy, 294
  ladder-type, 288–289
  mean field, 177
  one-particle, 322–323
  two-particle, 323–324
Arbitrary reference function, 311–314
Arbitrary symmetry, 220–221
Asymmetric
  stretch, 337
  two-particle functions, 489
Asymmetry, dynamic *vs.* nondynamic
  correlations, 347–348
Augmented Hessian techniques, 365

B, 68
B^+, 226, 229–232, 241–243
Backer-Campbell-Hausdorff (BCH)
  expansion, 349, 351–352, 354, 363

Basis sets, 23–24, 43, 46–47, 49, 52, 55,
  180–181, 194, 263, 283, 334, 337,
  366–377, 417, 421, 455, 466, 523–524
Becke-3-Lee-Yang-Parr (B3LYP) density
  functional, 417
BeH_2, 143, 145, 147, 149, 194–197, 226,
  229–232, 241–244, 249–251, 337,
  376–377, 437
Beryllium (Be), 64, 66, 156–160, 228–229,
  231–232, 241–243, 332, 336, 375, 436
BeS, 418, 422–423
Beta (β), 491
β/β blocks, 132–133, 190
Bethe ansatz, 519
BH, 197, 332, 336, 375–378, 431, 436
Binary entropy function, 497
Binomial normalization, 273–274
Bipartite entanglement, 529
Bipartite quantum system, 495–496
Block-diagonal(s)
  antisymmetrized geminal products
    (AGP), 434
  matrices, semidefinite programming
    formulations, 104–105, 107,
    109–110, 114
Block entanglements, 520–521
Block equations, CSE, 131–133
Boltzmann's constant, 509, 525
Bond-breaking, simultaneous, of water
  molecule with 6–31G and cc-pVDZ
  basis sets,
Bonding, UAGP, 438
Boolean quadric polytope (BQP), 96
Bopp, Fritz, 12, 15
Born-Oppenheimer approximation, 488
Bose condensation, 37, 55
Boson(s)
  characterized, 36, 267, 329, 488
  creation operators for, 28
  distinguished from fermions, 3–4
  two-boson system, 498
Boundary conditions, N-representability, 262
Boundary methods, semidefinite
  programming, 82
Bounds
  correction, 146–147
  cumulants and, 303–305
  p-particle density matrix, 303–306
Brillouin conditions, 315–316, 320–321, 331
Brillouin theorem, 358

SUBJECT INDEX

Brueckner doubles(with triples) (BD(T)), 66
Buijse and Baerends (BB)
  functional (BBC), 393
  reconstruction 391

Canadian Mathematical Society, 7
CaNEOS Server for Optimization, 115
Canonical diagonalization (CD), 338–339, 355–356
Canonical transformation (CT)
  defined, 348
  multireference state, 356
  physical equivalence in, 349–350
  theory, *see* Canonical transformation (CT) theory
Canonical transformation (CT) theory, for dynamic correlations
  bond breaking of nitrogen molecules with 6–31G basis sets, 372–375
  coupled-cluster theory and, 345, 356–361
  defined, 345
  dynamic correlations, characterized, 343–351, 378–381
  linearized model, *see* Linearized canonical transformation theory, for dynamic correlations
  MR-CISD and MR-LCCM compared on two-configuration reference insertion of Be in H_2 molecule, 375
  perturbative analysis, 356–361, 377, 380
  reference function, 357, 361–363
  simultaneous bond breaking of water molecule with 6–31G and cc-pVDZ basis sets, 366–372
  single-function, 357
  theory overview, 351–355, 378–381
Carbon (C^{2+}), 226, 229–232, 241–243, 268
Carbon monoxide (CO)
  characterized, 194–197, 437
  closed-shell total energy, 418
  dipole moment, 421–422
  equilibrium geometries, 423
  SOAGP, 432
  variational 2-RDM method, 48, 50–52
  vertical ionization potential, 419
Cauchy-Schwartz inequalities, 305–306
cc-p-VDZ, 366–368
CCSD(T)
  characterized, 49–50, 52

natural orbital functionals, 417
  T1/T2 conditions, 99
CF, 99–100
CH_2, 418, 420–422
C_2H_2, 437
CH_4, 194–197, 337, 436
Ciampi, Antonio, 13
CI calculation, 139
Cl_2, 418
Classical RDM approach, 98
Closed-shell
  state, 294, 306, 309, 323–324, 327
  systems, 98, 400
Coefficients, geminal, 430
Coleman, A. J., 11, 13, 15, 26
Combinatorial optimization, 23, 47, 55
Commutation relations, 69–70, 172, 175, 199, 348, 402, 404
Complete-active-space (CAS)
  characterized, 372–373
  self-consistent field,
    see Complete-active-space self-consistent field (CASCF) theory
Complete-active-space self-consistent field (CASSCF) theory
  characterized, 344–345, 347, 356
  CTSD/L-CTSD applications, 367, 371–374, 376–380
  L-CTSD applications, 370
  wavefunction, 363
Complex numbers, interpretation of, 8
Composite system, RDM cumulants, 266–267
Computer software programs, 388
Concurrence, entanglement and, 497–498, 502–504, 507–510, 525–527
Condensed matter physics, 491
Condon and Shortley notation, 123, 208
Conferences, international, 12–15
Configuration interaction (CI)
  calculations, 15
  energy values, 247, 249
  full, *see* Full configuration interaction (FCI)
  functionals and, 393
  method, 388
Conjugation, *p*-particle density matrix, 306
Connected diagrams, 294
Connected equations, *see* Connectivity
  cancellation of unconnected terms, 282–286
  characterized, 262–263, 269, 286–288
  CSE, 182–183

Connected equations, *see* Connectivity (*Continued*)
  p-RDMs, 177–179
  2-RDM, 187
Connected-moments expansion, 37
Connectivity, extensivity and, 275
Construction algorithms
  comparative results, 147–150
  error matrix, 141–146
  higher-order RDMs, 136–140
  N-representability problem, 146–147
  overview of, 138–140
  2-RDM, 135–136
  spin-orbital basis, 136
  unifying, 140–141
Continuity, 496
Contracted power method, 193
Contracted Schrödinger equation (CSE)
  algorithm for solving, 193–194
  anti-Hermitian formulation,
    see Anti-Hermitian contracted
    Schrödinger equation (ACSE)
  applications, 194–197
  characterized, 89, 128–129, 165–166, 181, 262–263
  connected equations, 262–263, 282–290
  convergence, *see* CSE convergence
    with iterative solution, influential factors
  cumulants, 182–183, 266–277
  defined, 122
  density equation compared with, 127–128
  density functional theory, 479
  derivatization in second quantization, 167–169
  development of, 127–128
  eigenvalue equations, reduced, 263–266
  4-RDM reconstruction, 166, 170–183
  future directions for, 197–198
  Grassmann products, 198–199
  indeterminacy, exact formal solution to, 153–159
  irreducible, *see* Irreducible contracted
    Schrödinger equation (ICSE)
  iterative solution, 133–134, 157
  matrix representation, 128–130
  Nakatsuji's theorem, 167, 169–170
  N-representability conditions, 166–167
  p-order, 122
  purification of 2-RDM, 166–167, 183–192
  reduced density matrices construction
    algorithms, 134–150
  as reduced eigenvalue equations, 263–266
  second-order, purification within, 244–252
  second quantization, 167
  self-consistent iteration, 192–193
  spin, role of, 130–133
  stationarity conditions, 317–318
  3-RDM reconstruction, 170–183
  2-RDM, 166, 183–192
2-CSE convergence, influential factors
  on iterative solution
  algorithms, 150–151
  N-representability, 151–152, 248
  S-representability, 151–152, 248
Contraction
  contracted spin equation, 131–132
  contracting operations, 233–234
  defined, 314
  mappings, *see* Contract mapping (CM)
  operator, 184, 435, 488
  rules, 314
Contract mapping (CM), 28, 153
Control theory, 23, 55
Convergence
  canonical transformation theory, 346, 354, 380
  connected formulations with CSE, 141, 287
  dynamic correlations, 364–365
  implications of, 46, 48, 55, 63, 192–193
  iteration schemes, 321
  L-CTSD, 363, 365
  linear inequalities for diagonal elements, 465
  lower bound method, 89–90
  N-representability of 2-RDM, 190
  with order k, 86–88
  primal-dual interior-point method, 113
  purification procedures, 227, 239–240, 243, 253
  semidefinite programming, 83–84, 86–88
  second-order Schrödinger equation (2-CSE), 248
Convex analysis, lower bound method, 76
Convexity theory, 77, 496
Convex optimization, 104
Convex set
  characterized, 30–31
  k-matrices, 70
  lower bound method, 75
Coordinate-space representation, 399
Core Hamiltonian operator, 394
Core orbitals, 347–348, 364

SUBJECT INDEX

Corrected Hartree (CH) functionals, 392
Corrected Hartree-Fock (CHF) functionals, 392
Correlation(s)
  coefficients, 494
  Coulomb, 301
  energy, 65, 268–269, 280, 308–309, 476, 494
  entropy, 515–517
  functionals, 388–389, 472
  increment, 301
  intergeminal, 431
  matrix (CM), 125–126. See also Correlation matrices decomposition
  strength, 515
Correlation matrix decomposition, purification procedures
  Alcoba-Valdemoro purification, 237–239
  characterized, 229–231
  pure two-body correlation matrix, 231–237
  test calculation and results, 239–244
Coulomb
  correlation, 301
  energy, 280
  interactions, 394, 407, 416, 424, 512
  repulsion, 494
Coulson, Charles, 11
Coupled cluster
  method, 54, 324, 388
  theory (CCSD/CCSDT), 55, 268, 287, 344, 351–352, 367, 372–373, 380
  unitary, 338
Coupled cluster singles and doubles (CCSD)
  ACSE, 336
  approximation, 432
  CTSD/L-CTSD applications, 377–379
  functionals and, 421
  implications of, 89, 332, 337–338
  with perturbational treatment of triples (CCSD(T)), 66
Creation operators, 7, 23, 28, 39, 68, 85, 94, 97, 125, 137, 155, 169, 172–173, 176, 232, 269–270, 294, 332, 350, 432, 452, 466–467, 498, 512, 522
Cryptosystems, 495
Crystals, 54
Csanyi and Arias (CA) functional, 392
CSE-NS iterative process, 249–250
Cumulant(s)
  bounds, 306
  canonical transformation theory, 359–363, 381

correlation energy, 280, 308–309
  decomposition, 185–188, 352–355
  for degenerate states, 307–308
  density, properties of, 302–303
  diagrammatic representations, 277–281
  expansions, 139, 277–278, 333, 339, 353
  extensivity, 268, 272–275
  formalism, 262–263, 269–272
  4-RDM reconstruction, 176–183
  functionals and, 176, 407–408
  inequalities, 303–306
  independence of, 275–277
  of k-particle density matrices, 299–302
  reconstruction functionals, 288–290
  separability, additive vs. multiplicative, 266–269
  structure, 182–183
  3-RDMs, 176–183, 288
  2-RDMs, 405–409
  theory, 139, 166, 176–183, 193
Cut polytope, 96
Cutting plane method, 15

D^2, 4, 8, 488
Damping, 248, 253
Davidson, Ernest, 13
D-condition, 167, 188, 196–197, 209, 218, 237–238, 242, 304
Decay rate law, 528–529
Decoherence, 524, 528–529
Decomposition
  canonical transformation theory, 352–354
  characterized, 154–155, 185–188
  convergence of 2-CSE, 151
  cumulant, 271–273, 275, 279–280, 309, 359–363, 381
  dynamic correlations and, 349, 351
  entanglement and, 525
  linearized canonical transformation theory, 355–356
  pure states of, 497
  two-body correlation matrix, 232, 234
  2-RDM, 126, 214–229
Deexcitations, 233
Degeneration
  antisymmetrized geminal products (AGP), 434
  density cumulants, 307–308
  geminal functional theory (GFT), 435
  L-CTSD, 364

Degrees of freedom, 334, 344
Delta (Δ), 141–146, 175, 275–277, 490, 511.
　See also Dirac delta; Kronecker delta
Density cumulants, 294, 302–303
Density functional theory (DFT)
　characterized, 55, 183, 309, 388–389, 447, 469
　Kohn-Sham approach, 475, 479
　orbital-based, 470–473
　Q-, orbital-based, 473–477
　perturbation theory and, 391
　refinement of, 444
Density matrix, see specific density matrices
　four requirements for, 22
　historical perspectives, 21–22
　renormalization group, 344
　renormalization group (DMRG), 344–345, 347, 355
^1D-functional theory, 389–394, 423–424
Diagonal block(s)
　antisymmetrized geminal products, 430
　T1/T2 condition, 94–95, 97
Diagonal elements
　linear inequalities, see Linear inequalities for diagonal elements
　Q-matrix, 477
Diagonal representability problem, 96
Diagrammatic representation, 314–316, 277–281, 359
Diatomic molecules, 65, 83, 494, 503
Differential equations, 335
Differential geometry, 4
Dimensional scaling theory, 530
Dimensionless coupling, 504
Dipole moment, 98, 421
Dipole resonance, 14
Dirac, Paul, 5–6
Dirac delta function, 95
Dirac notation, 5, 266
Disjoint sets, 266
Dispersion condition, 169
Dissociation, 268–269, 344, 372, 391
^2D matrix
　ACSE, 337
　characterized, 388–389, 400
　contracted Schrödinger equation, 171–172, 248
　nonnegativity, 15
　positive semidefinite, 45–46
　purification procedures, 219–220, 222–224, 227, 242, 248

Doubled-value functions, 276
Double excitation, 365–366
Double-zeta basis set, 49–50, 52, 156, 194
D_p, 265, 267
D^p, 3
D-positivity, 190
DQGT2 conditions, 53–54
D-reduced Hamiltonians, 34–36
D-representability conditions, 146
Dual configuration interaction, 66
Dual SDP formulation, 45–46, 104–107, 109–110, 114, 116
Dual simplex, 15
Dual spectral optimization problem, 63–64, 74
Duality gap μ, 46
Dummy variables, 267, 269–270
Dunning DZP basis sets, 376–377
Dynamic correlation mechanism, 233
Dyson equation, 141

Eddington, 6
Effective Hamiltonian
　canonical transformation theory, 362
　theories, 344
Effective valence Hamiltonian method, 338–339
Effective valence-shell Hamiltonian (EVH), 356
Eigenenergy, 349
Eigenequation, 405
Eigenfunctions
　canonical transformation theory, 362
　characterized, 42, 305
　dynamic correlations, 348
　functionals and, 424
　one-particle approximations, 322
　Q-density, 453
　(R, R) conditions, 456
　SOAGP, 432
Eigenstates, 156, 171, 193, 264
Eigenvalue(s)
　ACSE, 337
　antisymmetrized geminal products (AGP), 434
　characterized, 33, 44, 72–74, 79–81, 96, 98, 104, 106, 108, 227–228, 237, 241
　dynamic correlations, 349
　entanglement measure, 501–502, 512–514
　equations, reduced, 263–266
　functionals and, 403
　generalized normal ordering, 299
　independent cumulants and, 276–277

natural orbital functionals, 411–412
off-diagonal matrix elements, 466
orbital-based density functional theory, 475
p-particle density matrix, 305
Q-density, 454
reduced Hamiltonians, 31
2-positivity conditions, 12, 25–26
Eigenvectors, 33, 185, 189, 237, 276, 501, 512–513
Einstein's theories, 4
Electron(s)
affinities, 420–421
correlations, 54–55, 67, 436–437, 472, 494, 530–531
excitation, construction algorithms, 142, 145, 150
integrals, 44
interaction energy, 308
interactions, multiple, 4
repulsion, 171, 400, 476, 494
self-repulsion, 233
Electron-electron
correlations, 494, 531
repulsion, 400, 476
Electron-hole density matrix, 411
Electronic Hamiltonian, 166, 287, 348
Electronic structure, 266
Electronic structure
problems, 79, 83, 90, 170
significance of, 266
structure theory, 72, 392, 449
Electrostatic operator, 85
Energy, *see specific types of energies*
gap, 73–75, 511
levels, calculation of, 487–490
lower bounds, 71–72
minimization problem, 64, 77–78, 81, 90
Entanglement
ab initio calculations, 503, 522–524
density functional theory, 529–530
dynamics of, 524–529
fermion measurement, 498–499
finite size, 519
of formation and concurrence, 496–498, 502–504, 507–510
future directions for, 530–531
implications of, 494–496
nearest-neighbor, 526
for quantum dot systems, 512–522
ranks of density matrices, 499–500

for spin systems, 500–512
switch, 522
Epsilon (ϵ), 334
Equilibrium
bond lengths, 423
geometries, 54, 67, 344, 365, 392, 421–423
significance of, 344
thermal, 525
Erdahl, Bob, 13, 16
Error matrices
characterized, 141–146
construction algorithms, 147–149
Euclidean scalar product, 76
Euler equations
characterized, 64, 76, 79, 401–402
derivation of, 390
functionals and, 416
Gilbert nonlocal potential, 402–403
natural orbital functionals and, 412
Pernal nonlocal potential, 404–405
relation with extended Koopmans' theorem, 403–404
Euler's method, 335
Exchange and time-inversion integral, 408, 413
Exchange-energy functionals, 280, 472
Exchange interaction, 503–504
Excitation, 14–15, 89, 139, 142, 145, 150, 168–169, 233, 295–298, 313, 339, 364–366, 370–371
Exclusion-principle violating (EPV) cumulants, 321
Existence theorems, 77–79
Expansion
ICSE, 289
ladder-type perturbation, 288
operators, 488
Expectation value(s), significance of, 69, 85, 168, 172, 182, 184, 229, 264, 266, 270, 299, 301, 311–312, 314, 353, 398
Extended Koopman's theorem (EKT), 391, 403–404, 419–420
External, generally
excitation, 371
external-external space correlations, 348, 350–351, 367
orbitals, 347, 364
potential, 399
Extreme B matrix, 32–33
Extreme conditions, 97
Extreme reduced Hamiltonians, 31, 35, 37–38

Factorization, 284
Fermi, generally
  correlation, 301
  operators, 505, 512
  vacuum, 348, 351, 357
Fermion(s)/fermionic
  anticommutation relation, 172–173
  anticommutation rules, 125
  bosons distinguished from, 3–4
  characterized, 36, 176, 184, 329, 488
  commutation relations, 69
  contracted Schrödinger equation, 167
  creation operators for, 28
  first-order relation, 20
  lower bound method, 88
  N identical, 8
  hoperators, 125, 154, 505, 524
  representability problem, 96
  system, 109, 516
Ferromagnetics, 510
Feynman, Richard, 5
Feynman perturbation theory, 181.
  See also Perturbation theory
Field operators, 269
Financial applications, variational 2-RDM, 55
Finite dimensional Fock space ($\mathfrak{F}$), 67–68
Finite Gaussian-orbital basis set, 43
First-order
  approximations, primal-dual interior-point method, 111–112
  corrections, 48, 325–326
  RDM method, *see* 1-RDM
  nonlinear algorithms, 48, 50
  reconstruction, 338
First-quantized formalism, 263
Fixed particle number, T1/T2 conditions, 97
5-CSE, 155
5-RDM functionals, 175
Floating-point operations, 46–48, 113, 116–117, 339
Flow-renormalization group, 345
Fluorine, 433
Fock-space, *see* Hartree-Fock
  Hamiltonian, 298
  many-body theory, *see* Fock-space formulation, many body theory
  operators, 97, 294, 314, 362
  positivity condition, 94
  unitary transformation, 329

Fock-space formulation, many-body theory
  excitation operators, 295–296
  k-particle density matrices, 296–298
  spin-free excitation operators, 297–298
Fock theory, 437
Formation, entanglement of, 496–498
Four-body problems, 12
4-CSE, 155
Four-electron
  CM, 154
  systems, geminal product theory (GPT), 436
4-positivity, 67, 88–89
4-RDM
  characterized, 129, 131, 133–135, 137–139
  construction algorithms, 144, 146–148, 150
  CSE, 160–161
  functionals, 175
  ICSE and, 288
  parameter value selection criterion, 140–141, 149
  purification strategies, 157, 245–247, 253
  reconstruction methods, *see* 4-RDM reconstruction methods
  reduced eigenvalue equations, 266
  2-CSE, 151, 248
4-RDM reconstruction methods
  approximation of cumulant 3-RDM, 179–182
  CSE, 193–194, 197
  cumulants, 176–179
  cumulant structure of CSE, 182–183
  particle-hole duality, 172–175
  Rosina's theorem, 170–171
Fourth-order, generally
  approximations, 88
  estimates, 88
  many-body perturbation methods (MP4), 89, 337–338
  modified contracted Schrödinger equation, 253
  perturbation theory, 65
  reduced-density matrices, *see* 4-RDM
  4×4 matrix, 501, 523
Free-electron gas, 392
Free labels, in diagrammatic representations, 315
Full configuration interaction (FCI)
  ACSE, 335–336, 338
  construction algorithms, 141, 143, 147–148, 150

SUBJECT INDEX 559

contracted Schrödinger equation, 156–157, 160, 194
geminal functional theory, 436–437
L-CTSD, 367, 376
lower bound method, 88
N-representabilities, 446, 489
parameterization of 2-RDM, 491
purification procedures, 229
T1/T2 conditions, 96, 99
wavefunction, 49–51, 53–54, 65
Full contractions, generalized normal ordering, 311, 314
Full-rank factorization, 115
Functionals, *see specific types of functionals*

GAMESS, 194, 335
Gamma (γ), 107–109, 324–325, 329, 490, 509, 513
Gap formula, 73–75
Garrod, Claude, 12–13, 15
Gaussian basis set, 375, 523–524
Gaussian distribution, 509
Gaussian functional, 194, 523
G conditions
  characterized, 27, 97, 99, 126, 196–197, 211, 287
  cumulants and, 304
  diagonal elements, 479
  off-diagonal matrix elements, 467–468
  purification procedures, 240, 242, 244
  semidefinite programming problem, 109–110, 116
Geminal functional theory (GFT)
  antisymmetrized geminal products (AGP), 433–436
  application of, 436–437
  characterized, 392, 439
  formal, 434–436
  strongly orthogonal antisymmetrized geminal products (SOAGP), 429–433
  unrestricted antisymmetrized geminal products (UAGP), 437–439
General Hermitian, 97
Generalized Brillouin conditions, 351
Generalized Fock operator, 301–302, 316
Generalized gradient-approximation (GGA), 391
Generalized Hellmann-Feynman theory, 355
Generalized matrix contraction mapping, 169

Generalized minimal residual (GM-RES) method, 364
Generalized normal ordering
  arbitrary reference function, 311–314
  diagrammatic representation, 314–316
  Hamiltonian in, 316
  particle-hole formalism, 309–312, 329
  Wick theorem, 311, 314
Generalized particles-holes (GP-H) separating approach, 138–147
General relativity theory (GRT), 4
Generalized Weisacker functional, 469
Generating functionals, 176
Gilbert nonlocal potential, 402–403
Global operators, 124
G matrix
  ACSE, 337
  characterized, 14, 126, 193, 196
  lower bound method, 66
  nonnegativity condition, 15
  positive semidefinite, 45–46, 64–65
  purification, 209–214
  radical calculation, 44
  spin-adapted, 40–41
  spin properties of, 211–212
  spin structure of, 211
GN-representability conditions, 146
Goedecker and Umrigar (UM) functional, 391–392
Goldstone-type diagrams, 315–316
Golli, Bojan, 16
Grassmann
  algebra, 173, 179, 183, 491
  field, 269
  product, 30, 43, 139, 198–199, 271–273, 279, 301. *See also* Wedge products
Gravitational theory, 5–6
G-reduced Hamiltonians, 34–38
Greeks, implications of, *see specific Greek symbols*
Green's function methods, 166, 176–177, 263, 271, 290
Ground-state, generally
  correlations, 14
  energies, *see* Ground-state energy
  entanglement, 523
  purification procedures, 240–242
  semidefinite programming, 83
  von Neumann density, 62
  wavefunction, 169, 192, 339

Ground-state energy, impact of
    ACSE, 336, 339
    construction algorithms and, 139
    CSE, 160
    density functional theory, 445
    entanglement and, 530
    functionals and, 416–417
    geminal functional theory (GFT), 434–435
    linear inequalities, 451, 458–459
    lower bound, 71–72
    orbital-based density functional theory, 476–477
    purification process, 250
    Q-densities, 446
    RDM method, 98
    reduced Hamiltonians, 32–33, 36–37
    semidefinite programming, 46–47, 109
    significance of, 4, 7–8, 14–15, 22–24, 39–40, 170
    variational 2-RDM method, 50, 52, 55
Group theory, 5
GU functional, 393

H_2CO, 418–419, 422
Hadamard, 7
Hall, Richard, 12–13
Hamiltonian (H), *see specific types of Hamiltonians*
    characterized, 4, 7–8, 15, 70, 84, 97, 316, 488
    eigenstate, 153
    entanglement measure and, 505–507, 529
    equation, 130
    Hermitian effective, 349–350
    many-body, 122–123
    many-electron, 23–24
    matrices, 23–24, 30–38, 80, 139, 172, 266
    methodology, spin-adapted reduced, 252
    operators, 23, 71–72, 122, 167–168, 170, 183, 332, 450, 501
    positive semidefinite, 459, 466–469
    Q-electron, 466, 474
    reduced, 454
    reduced, 456
    thermal entanglement, 510
    wavefunctions, 192
Harmonic(s)
    corrections, 530–531
    interactions, 36–37
Hartree functional, 416

Hartree-Fock (HF)
    approximations, 177–178, 406, 513–515
    basis, 156
    calculation, 251, 287, 335, 366, 372
    determinant, 49
    energy, 309
    equation, 160
    functions, 327, 418–419, 422–423, 437
    limit energy, 494
    method, 476
    operators, 362
    orbitals, 140, 239, 277
    purification process, 251
    reference, 50
    solution, 130
    state, 142
    theory, 301, 346, 352, 354, 438
    wavefunctions, 180–181, 193–197, 324, 332, 336, 339, 357, 375–378, 523, 531
Hartree-Fock-Bogoliubov (HFB) energy, 392
HCN, 436
Heisenberg, Werner, 7
Heisenberg
    antiferromagnetic model, 511
    representation, 339
Helium (He), 8, 65, 66
Hermitian-Lagrange multipliers, 401
Hermitian matrix
    characterized, 104, 126, 141, 157, 186, 215–216, 404, 502
    purification procedures, 209, 220–221
    reduced-density (HRDMs), 125–127, 175, 218–219, 228
    2-matrix, 491
Hermitian operators, 3–4, 67–68, 71, 85, 450, 489–490, 529
Hermiticity, 28, 302
Higher-occupied molecular orbital (HOMO), 421
Higher-order CSEs, 129–130
Higher-order equations, 134
Higher-order RDMs
    decomposition and cancellation relation, 153–154
    fourth-order MCSE, derivation of, 154–160
    indeterminancy of CSE solution, 153
Hilbert space
    characterized, 26, 264, 340, 518
    eigenvalue equation, 265, 283, 287
    operators, 263

SUBJECT INDEX

Historical perspectives, 389–394
HNO, 418, 422
Hohenberg-Kohn (HK) theorem, 170–171, 389–390, 529
Hole
   annihilation operators, 294
   creation operators, 294
   p-order, 124
   reduced density matrices (HRDMs), purification process, 207, 226, 230–231
   reduced Hamiltonians, 32–33
Homo orbitals, 143–144
Homogeneous electron gas (HEG), 391–392, 424
Hopf algebras, 329
Hopping, quantum dot systems, 517–519
HRVW/KSH/M direction, 112
Hubbard model, 55, 516–519, 530
Hugenholtz-type diagrams, 315
Husimi, 11
Hydrogen (H/H_2),
   atoms, 42–44, 503
   chain, variational 2-RDM method, 48, 53–54
   characterized, 375, 390, 523–524, 530–531
   Dirac's formula, 5
Hydrogen chloride (HCl), 418–419, 422–423
Hydrogen fluoride, 433
Hydroxide radical, variational 2-RDM method, 48, 53–54
Hyperplanes, Slater hulls, 457–458, 464

Idempotency, 273–274, 306, 309, 393
Identity
   matrix, 111, 173, 182, 184
   operator, 451
   representation, 3
I-MZ purification procedure, 230–231, 244
Independent-electron wavefunction, 273
Independent pair model, 226
Inequality, cumulants and, 303–305.
   See also Linear inequalities
Infeasible primal-dual path-following Mehrotra-type predictor-corrector interior-point method, 111
Infeld, Leopold, 5–6
Initial-value differential equations, 338
Inorganic molecules, variational 2-RDM method, 48, 52–53

Integrals, 44
Interelectronic interactions, 435
Interior points method, 80–81, 115
Internal, generally
   excitation, 365–366, 370–371
   operator, 278
Ionic geminals, 514
Ionization
   characterized, 43, 392, 403
   potentials, 419–420
Irreducible/irreducibility, generally
   Brillouin conditions, *see* Irreducible Brillouin conditions (IBC)
   contracted Schrödinger equation (ICSE), *See* Irreducible Schrödinger equation (ICSE)
   cumulants, 303
   representation (irrep), 307
   tensor components, 307
Irreducible Brillouin conditions (IBC)
   characterized, 294, 315–316, 320–321, 329
   solution strategies, 321–328
Irreducible contracted Schrödinger equation (ICSE)
   characterized, 263, 283–284, 286–287, 294, 319–320
   solution strategies, 321–328
Ising model, 507, 509–511, 527, 530
Isobaric spin, 16
Isotropic Heisenberg model, 509
Iteration/iterative
   algorithms, 167, 192–193
   diagonalization, natural orbital funtionals, 417
   geminal, 436
   ICSE, 288
   purification procedures, 239–242, 246–249
   self-consistent, 192–193, 246–249
   solutions, *see* Iterative solutions, 2-CSE purification
   2-CSE, 246–253, 288–289
   two-particle approximations, 323–324
   UAGP, 438–439
Iterative solutions, 2-CSE purification
   characterized, 157, 288–289
   N purification and S.purification, 249–253
   regulated self-consistent solutions, 246–249

Jeffery, R. L., 7
JK-only functional, 393
Jordan-Wigner transformation, 511

Karush-Kuhn-Tucker condition, 111
k-body operators, 85
k-densities, 71
"Killers," semidefinite programming, 83
Kinetic energy
   characterized, 171, 388, 399, 444, 494
   Q-densities, 446
   Weisacker, 471
Kinetic potential, geminal functional theory, 435
k-matrices
   lower bound method, 62–63, 68–74
   semidefinite programming, 83
Kohn-Sham (KS)
   determinant, 309
   equations, 472
   functionals, 388–389, 475, 479
   procedure, 471
Kollmar and Hess functional, 393
Koopman's theorem (KT), 419–420
k-particle density matrices
   cumulants, 299–302
   generalized normal ordering, 314
k-particle hierarchy, 321
Kronecker delta, 107, 135–137, 248
Kth-order approximations
   approximating states, 68
   characterized, 67–68
   energy lower bounds, 71–72
   expectation values, 69
   matrix representations, 68–71
   Pauli subspace, 69–70
Kubo cumulant expansion, 138
Kummer, Hans, 12–13, 488–489

Lagrange duality, lower bound method, 72–73
Lagrangian
   augmented function, 115
   multipliers, 47–48, 55, 319, 401–402
Lambda (λ) matrix, 314, 316, 324–326, 328–329, 334–335, 401–402, 404, 489–490, 497
Lamb shift, 5
Lattice(s)
   implications of, 391, 504, 521–522
   model, 68
   points/sites, 84–86
   topologies, 511–512
Least-squares norm, 195
Legendre transform, 530
Lemma 9, 76, 79, 306

Levy's functional, 390–391
Lie groups and algebras, 7
Li-Li bond, 250–251
Li_2, 226, 229–232, 241–244, 249–251
Lifted/lifting
   conditions, 27–28
   operator, 30
Light nuclei, giant dipole resonance, 14
LiH, 431, 437
Lindenberg, J., 13
Linear, generally
   algebra, 113
   conditions, lower bound method, 70
   contraction mapping, 180
   coupled cluster theory (L-CCSD), 377–378
   equation, primal-dual interior-point method, 112
   functionals, applications of, 22–23
   Hermitian operators, 448
   independence, 320
   inequalities, see Linear inequalites for diagonal elements
   mappings, 15, 25–26, 28, 172
   operators, 450, 488
   programming, 63–64, 80, 104
   relations, lower bound method, 74
   scale/scaling, 180–183, 333
   transformation, 505
Linear inequalities for diagonal elements
   characterized, 444–449, 477–480
   linking representations, orbital and spatial, 470–477
   n-representability, necessary and sufficient conditions for, 445–451
   from orbital representation, 451–469
   from spatial representation, 469–470
Linearization, primal-dual interior-point method, 112
Linearized canonical transformation theory (L-CTD/L-CTSD), for dynamic correlations
   characterized, 346, 355–356, 361, 373
   computational algorithm, 363–364
   computational scaling, 364
   excitation classes for exponential operator, 364–366
   single- and multireference for HF and BH molecules, 375–378
Linked-cluster theorem, 275
Liouville equation, 525

Lipkin model, 88–89
Lithium, 8
Local Hamiltonian, 84
Local operations, 496
Logarithm functions, antisymmetrized, 270–271
Logic, 4
Lorentz transformation, 5
Lower bound method
  algorithms, 79–83
  characterized, 23, 46, 62–64, 89–90
  energy, 71–72
  fundamental theorem, 75–79
  historical perspectives, 64–66
  Lipkin model, 88–89
  one-dimensional superconductor modeling, 83–890
  kth-order approximations, 67–72, 90
  semidefinite programming, 62–64, 72–75
  strong two-body forces, 66–67, 90
  variational 2-RDM methods and, 55
Lower-order RDMs, 156
Low-rank factorization method, 115
Lumo orbitals, 143–144

McWeeny, R., 13
McWeeny normalization, 353
Magnetic field
  entanglement and, 526–528
  strength, 500, 503, 508–509
  time-dependent, 524
Magnetic systems, one-dimensional, 531
Magnetization, 505–506
Many-body, generally electron dynamics, 381
  Hamiltonian matrices, 123, 208
  methods, 37
  perturbation methods, *see* Many-body perturbation theory (MBPT)
  problems, 11, 15, 519
  quantum system, 531
  theory, 263, 321
Many-body perturbation theory (MBPT), 53, 88–89, 178, 194–195, 197, 277, 288, 311, 315
Many-electron(s)
  atoms, 22–23
  densities, 456
  energy, 24
  quantum mechanics, 339
  wavefunctions, 55, 166
Mathematica, 137

Mathematical
  logic, 4
  programming, 104
Matrices, characterized, *see specific types of matrices*
Matrix contracting mapping (MCM), 127–128
Matrix theory, 77
Maximum absolute error, 47–48
Maximum entropy principle, 530
Mazziotti (M)
  formula, 194–197
  purification procedure, *see* Mazziotti (MZ) purification procedure
  reconstruction, 333–334, 336–338
Mazziotti (MZ) purification procedure
  characterized, 214–215
  improvement strategies, 222–226
  test calculations and results, 226–229
  unitary decomposition of arbitrary second-order matrices, 220–222
  unitary decomposition of antisymmetric second-order matrices, 215–220
Mazziotti, David, 9, 14, 16
MC-SCF theory, 320
Mean-field
  approximations, 177
  calculation, 335
Mehrotra-type predictor-corrector interior-point method, 111–113
Metal-insulator phase transition, 520
Metallic hydrogen, 54
M functional, 335
MgH, 418, 420, 422
Middle Ages, international conference agendas
  Density Matrices and Density Functionals, the A. John Coleman Symposium, 14
  Density Matrix Conference, 12–13
  Density Matrix Seminar, 13
  Density Matrix Seminar II, 13
  Reduced Density Matrices Conference, 13–15
  Reduced Density Matrix Workshop, 14
  Reduced Density Operators Conference, 13
Mihailović, Miodrag, 15
Millihartrees (mH), 52
Mixed state, entanglement measure, 499–500
Modified contracted Schrödinger equations (MCSEs)
  defined, 153
  fourth-order, derivation of, 154–160
  third-order, 160

SUBJECT INDEX

Molecular, generally
  conductivity, 55
  geometries, 65, 90
  ions, 98
  orbitals (MOs), 41, 143
  radical energies, 44
Moment-generating functionals, 176, 269
Momentum operator, 41
Monomials, lower bound method, 68, 71
Monte Carlo methods, 55, 511
MR-CISD, 375–376
MR-LCCM, 375–376
$m \times n$ matrix, 45
Mu (μ), 324–325, 489
Multiconfigurational reference, 347
Multiconfiguration self-constistent-field calculation, 332
Multiplicatively separable RDMs, 266–269
Multireference
  configuration interaction, 50
  perturbation theory, 344
  second-order perturbation theory (MRPT/MRPT2), 49–51, 55, 367, 372–374
  self-consistent-field calculation, 332, 339
  wavefunctions, 346, 356–357, 375–378

NaCl, 418, 422
Nakatsuji, Hiroshi, 9, 14, 16
Nakatsuji and Yasuda
  algorithm, 138–139
  correction, 181
  formula, 138–139
  reconstruction, 333, 336–338
Nakatsuji's theorem, 167, 169–170
Natural orbital (NO), 298, 389, 436, 514.
  See also Natural orbital function (NOF); Natural spin-orbital (NSO) functional
Natural orbital function (NOF)
  characterized, 393–394, 399–401, 423–424
  defined, 391
  dipole moment, 421
  energetics, 417–421
  -EKT, 420
  equilibrium geometries, 421–423
  numerical implementation, 416–417
  practical, 394, 415–416
  restricted closed-shell, 409–413
  restricted open-shell, 413–415
  vibrational frequencies, 423

Natural-order basis set, 180, 182
Natural spin-orbital (NSO) functional
  characterized, 302–303, 316, 319, 322–323, 402–403
  geminal functional theory, 433
  2-RDM cumulant and, 406
N-body eigenproblem,
  160 N-electron/N-electron systems
  density matrices, 22, 457
  Hilbert space eigenvalue equation, 265
  natural orbital functionals, 411–412
N-DM, 157
Nearest- neighbor
  characterized, 84
  concurrence, 507–508 510, 526–528
  entanglement, 526
  quantum dot systems, 517, 519
Necessary and sufficient conditions, *see specific types of functionals*
Negative semi-definite, purification procedures, 239
N electrons, 7
N-ensemble representable matrices, 171
NEOS Server, 115
Newton-Raphson procedure, 288
Next neighboring/next-next neighboring quantum dots, 522
NH, 418, 420, 422, 431
NH_3, 99–100, 194–197, 332, 336–337, 418–419, 422, 437
Nitrogen
  N_2, 48–51, 194–197, 332, 336, 374, 419, 423, 437
  N^{3+}, 158, 226, 230–232, 241–243
N_2O, 418, 422
Nonconvex optimization problem, 115
Noncovalent bonds, 432
Nondynamic correlation, 344–345, 347, 380–381
Noninteracting/noninteraction
  subsystems, 262, 266–268, 272, 300, 320
  supersystem, 302
Nonionic geminals, 514
Nonlinear/nonlinearity
  equations, 111, 262, 288
  optimization problem, 115
  semidefinite program, 47–48, 55
Nonlocal potential
  Gilbert, 402–403
  Pernal, 404–405

Nonparallelity error (NPE), 367, 374
Nonsmooth analysis, 77
Nonsymmetric matrix, 115
Normalization, 139, 168, 172, 184, 193,
   196, 199, 208, 264, 266, 270,
   237–275, 353, 396–397, 457, 460, 495
Normal ordering, 312, 326–327.
   See also Generalized normal ordering
Notation, 123–124. *See specific types of
   notations*
N-particle
   density matrix, 166, 183, 192, 325
   Hamiltonian, 32, 171
N-p electron variables, 124
N-purification process, 249
N-representability
   characterized, 3–4, 8, 294
   conditions, *see* N-representability
      conditions
   constraints, 23, 469–470
   convex set of two-particle reduced
      Hamiltonian matrices, 30–31
   corrections, 134
   decompositions, unitary and cumulant,
      185–18
   historical perspectives, 11–12
   ICSE, 287
   lower bound method, 62
   orbital-based density functional theory,
      473–474
   preservation of, 327
   problems, *see* N-representability problems
   pure-state, 304, 325
   purification procedures, 237
   semidefinite programming problem, 106, 110
   stationary conditions and, 323–324
   variational 2-RDM methods, 50–51
N-representability conditions
   ACSE, 336–339
   characterized, 22–23 124–127
   conditions, characterized, 206
   construction algorithms, 140, 147–148
   contracted Schrödinger equation, 132, 134,
      151–152, 159, 161, 166–167, 248
   diagonal elements, 478–479
   functional theory and, 388, 390
   geminal functional theory (GFT), 435–436
   natural orbital functionals, 412
   Q-densities, 445–446, 463, 477–478
   reduced eigenvalue equations, 265

Slater hull, 480
   two-body correlation matrix, 234, 236–237
N-representability problems
   construction algorithms, 146–148
   functionals, 400
   purification, 206, 208–210, 219–220,
      225–228
   solution methods, 487
$n \times n$ matrix, 45
Nuclear
   many-body problem, 14
   potential, geminal functional theory, 433
   shell model, 14
Nuclei, collective excitations of, 14
Nucleon-nucleon
   attraction, 67
   potential, 12
Number operator, 181–182
NY approximation, 143
NYM algorithm, 140, 145, 147–148

Occupancy, 8
Occupation number notation, 8
Off-diagonal elements
   density matrix, 447
   implications of, 453, 478–479
   linear inequalities, *see* Off-diagonal
      matrix elements, linear inequalities
Off-diagonal matrix elements, linear
   inequalities
   positive-definite Hamiltonians, 466–469
   Slater hull constraints, 464–466
OH, 371, 418, 420, 422
Ohrn, Yngve, 13
Old Ages, historical perspectives, 11
omega (Ω), 282, 284–386, 489
One-body
   conditions, 97
   operators, 67, 85, 94
1-density, 170–171
One-dimensional superconductor
   characterized, 83–84
   convergence with order k, 86–88
   details of, 84–85
   Lipkin model, 88–89
   results, 85–86
One-electron
   basis, 96
   cumulants, 279
   energy, 308

One-electron (*Continued*)
  Hamiltonian, 460
  operators, 132, 412–413
One-hole density matrices, 303–304, 314
1-HRDM, 135, 137, 142, 236
One-index, diagonal conditions, 95
One-parameter scaling theory, 520
One-particle
  density matrices, 186, 199, 294, 301–302, 494
  one-hole excitations, 15
  operators, 353, 362, 392
  unitary transformation, 498
1-RDM
  characterized, 106, 115, 130, 135, 139, 141, 156, 394–399
  closed-shell NOF, 409–410
  construction algorithms, 145, 147
  convergence of 2-CSE, 150–152
  correlation matrix, 233
  functional theory based on, 389–391
  geminal functional theory (GFT), 435–436
  natural orbital functionals, 402–403, 405
  N-representable, 37, 184–185
  purification procedures, 183–184, 245
  semidefinite programming problem, 108–109
  spin structure, 396–399
  T1/T2 conditions, 94, 96, 97
  two-body correlation matrix, 235–236
One-site unitary transformation, 498
1-TDRM, 127
Open-shell
  energies, 53
  molecules, 42–45, 344
  states, 322
  systems, 98, 277
Operational measures, 496
Operator diagrams, 278
Optimal values, 81. *See also* Optimization
Optimization
  antisymmetrized geminal products, 430–431
  density functional theory, 448, 479
  geminal product theory (GPT), 436
  natural orbital functionals, 417
  nonlinear, 47, 55
  self-consistent iteration, 192–198
  semidefinite programming, 45–48
Orbital(s)
  linear inequalities, 460–463
  representation, diagonal elements, 447, 449, 452–453

  rotation, 16
  spin-free, 297–299
  symmetry, 143–144
Order of magnitude, 52, 55, 89, 109, 337
Orthodox physicists, 6
Orthogonal/orthogonality
  characterized, 430–431
  complements, 70
  decomposition, 80
  spins, 190
Orthonormality conditions, 398
Orthonormalization, 404
Orthonormal one-particle functions (ϕ), 8
Oscillations, 248, 509, 525, 528
Oxygen
  O^{4+}, 158, 226, 229–232, 241–243
  O_3, 418, 422
  O_2, 99

p_2, 8, 418–419, 423
p, lower bound method, 67–68
Pair-density functional theory, 444, 447
Pairing relation, p-particle density matrix, 305
Pairwise
  entanglement, 522
  interactions, 22, 177, 288
Parameter value, 140–141, 149–150
Parametric algorithms, 140
Parametrization, 82, 487–491
Parent molecules, 44
Partially occupied NSOs, 302
Partial 3-positivity conditions
  lifting conditions, 27–28
  T_1/T_2 conditions, 27–29
Partial trace
  operators, 264–265
  relations, 303, 322
Particle distribution, cumulants and, 271–272
Particle-hole
  density matrix, 287, 303
  duality, 32, 39–41, 166, 172–175, 178
  formalism, *see* Particle-hole formalism
  symmetry, 303
Particle numbers, T1/T2 conditions, 94–95
Particle-hole formalism
  in arbitrary basis, 311–312
  characterized, 309–312, 329
Particle-reduced Hamiltonians, 33
Partition/partitioning
  antisymmetrized geminal products, 430

characterized, 222
cumulants and, 300–301
entanglement measure, 509
generalized normal ordering, 313
L-CTSD, 364
Pauli, generally
  exclusion relation, 461, 522
  kinetic energy system, 471
  matrices, 74, 497–498, 501, 524
  Principle, 62, 511
  space, 71, 73, 80
  subspace, 62, 79
p-body operators, 94
PC Gamess, 194
P conditions
  characterized, 99, 287
  off-diagonal matrix elements, 109–110, 116, 467–468
p-CSE, 129
Peat, David, 13
PEMCSCF, 393
Penalty barrier function, 64, 115
PENSDP, 116
Percus, Jerome, 12, 15
Permutation
  cumulants and, 274–275
  functionals, 391, 396, 408
  generalized normal ordering, 313
  implications, 3, 28, 125, 177, 209, 217, 220–221, 271, 284, 333–334
Pernal nonlocal potential, 404–405
Pertubation/perturbative
  analysis, 324–328
  canonical transformation theory and, 352
  expansion, 141
  natural orbital functionals, 417
  renormalization, 195
  theory, see Perturbation theory (PT)
Perturbation theory (PT)
  anti-Hermitian contracted Schrödinger equation (ACSE), 331, 334, 338–339
  characterized, 35, 37–38, 65, 178, 181, 275, 329
  diagrammatic, 279
  dynamic correlations and, 345, 356–361, 377, 380
  functionals, 391
  Møller-Plesset, 305, 324–328
  SOAGP, 432–433
Phase transition, 520

PH_3, 418, 422
Phi (Φ), 294, 313, 326–327
Physical vacuum, 294, 311
Physics Today, 5–6
pi (π), 199
Planck, 7
Polarization, 233, 392
Polymers, 54
Polynomials, 68, 85, 94, 98, 113, 453–456, 459, 463, 466, 477–478
Polytope, 64
Position-space kernels, 263, 265, 270
Positive-semidefinite
  conditions, 23, 39, 45, 68, 106, 116, 125–126, 185, 188–189, 287
  Hamiltonians, 32–33, 35–36
  Hermitian matrices, 104
  linear Hermitian operator, 477
  matrices, 22, 27–28, 70, 135, 157, 185, 188–189, 208–210, 435–436
  problem, CSE, 159
  programming, 77, 80
  purification procedures, 226, 249–250
  symmetric matrix, 104
Positivity conditions, 23, 185, 188–189. See also Positivity conditions for two-particle reduced Hamiltonian matrices
Positivity conditions for two-particle reduced Hamiltonian matrices
  1-RDM and, 32–33
  strength of, 35–38
  2-RDM and, 33–35
Positivity errors, 196
Post, H., 12
Potential energy
  characterized, 50
  curves (PECs) research, 250–253
  surface (PES), 23, 344, 432–433
  variational 2-RDM method, 53
Power
  contracted, 193
  geminal, 306
  method, see Wavefunction power method
  series, 269
Preconditioning, L-CTSD, 364–365
p-HRDM, 172
P-matrix
  lower bound method, 66
  positive semidefinite, 64–65

p-order
  moment, 138
  reduced density matrix (p-RDM), 124, 264
  transition density matrix (p-TRDM), 124
p-particle, 3, 24
p-positivity conditions, 23–24, 38
p-RDM, 30–31, 38, 127, 138, 140, 262
Primal-dual interior point
  algorithms, 46–48
  method, 82, 110–115
Primal SDP problem, 45–46, 48, 104–107, 110, 116
Principles of relativity (PR), 4
Probability applications, 462
Probability distribution, 25, 262
Probe variables, 269
p Schwinger variables, 138
Pseudo-two-electron wavefunction, 446
Pseudo-wavefunction, 470
Psi (ψ/Ψ), 4, 264, 267, 312–314, 335, 349–351, 355, 445, 489–490, 497, 516, 530
pth-order
  energy density matrix, 282
  RDM cumulant (p-RDMC), 262, 270, 272, 274
Pure-state density matrices, 171, 264
Pure two-body correlation matrix, 231–237
Purification procedures
  ACSE, 336–337, 339
  Alcoba-Valdemoro, 237–239
  based on correlation matrix decomposition, 229–244
  based on unitary decompositions, 214–229
  characterized, 135–136, 152, 160–161, 166–167, 183–192, 198–198, 206–207, 252–253, 336
  G-matrices problem, 210–214
  notations and definitions, 207
  N-representability problem, 208–210
  second-order CSE, 244–252
  S-representability problem, 210–214
  2-RDMs, 185

Q-body reduced operator, 448–449
Q condition
  characterized, 99, 126, 167, 188, 196–197, 209, 218, 287
  cumulants and, 304
  natural orbital functionals, 411

off-diagonal matrix elements, 467–468
purification procedures, 237–238, 242
semidefinite programming problems, 107–110, 116
Q-densities, linear inequalities, 469
Q-density functional theory, 444–445, 470–477, 480
Q-electron reduced density matrices, 447, 449
Q-matrix
  ACSE, 337
  characterized, 453–455
  diagonal elements, 477–479
  linear inequalities for diagonal elements, 462–466
  lower bound method, 66
  positive semidefinite, 45–46, 64–65
  radical calculation, 44
  spin-adapted, 40–41
  theory, 480
q-order matrix, 127
Q-positivity, 190
q-RDMs, 127, 264, 262
Q-reduced Hamiltonians, 34–36
Q-representability conditions, 146
Quadratic equations, 80
Quadratic penalty function, 47
Quadruple excitations, 169
Quantization formalism, SOAGP, 432–433
Quantum chemistry, 54–55, 104, 159, 252, 309, 344, 468, 480, 494, 513, 530
Quantum communication, 495
Quantum cryptography, 495
Quantum dot lattice, 522
Quantum dot system entanglement
  one-dimensional quantum dots system, 517–519
  two-dimensional array of quantum dots, 519–522
  two-electron two-site Hubbard model, 512–517
Quantum information theory, 55, 495, 519
Quantum interference effects, 528
Quantum many-body state, 494
Quantum mechanics, 11–12, 172, 262, 267–268, 272, 332, 339, 388, 489
Quantum phase transition (QPT), 82–83, 501, 504–505, 519, 531
Quantum statistics, 498
Quantum theory, 88
Quasi-Newton method, 115

SUBJECT INDEX

Radial correlation, 494
Radical, generally
  dissociated molecules, 42–44
  2-RDM, 43–44
Random phase approximation (RPA), 14
Rayleigh-Ritz variational principle, 22
RDM cumulant (RDMC), defined, 272.
  See also Cumulants
RDO News, 13
Reactivity research, 55
Reconstruction
  cumulants, 288–290, 329
  functionals, 178–179, 288, 294, 408
Reduced density matrices (RDM),
  see specific types of reduced density matrices
  characterized, 124
  fermions *vs.* bosons, 3–4
  mechanics, 55
Reduced density operator (RDO)
  defined, 3
  two-particle interactions, 8
Reduced equations, reconstruction
  and solution of, 288–290
Reduced Hamiltonian
  matrix, 247
  operator, 8
  two-particle, 30–38
Reduced operators, 264
Reduced two-body problems, 12
Renormalization, 135, 139, 148, 178,
  195, 238, 289, 345
Representability conditions, 98–100.
  See also N-representability conditions;
  N-representability problems
Repulsion, interelectron, 530
R matrix, 47–48, 455–460
Root mean square (RMS) deviation,
  141–142, 227–228, 240–241
Rosina, Mitja, 13
Rosina's theorem, 170–171
Roundoff coupling, 44
RRSDP, 115–117
RR', 115
Ruskai, Mary-Beth, 13

Sasaki, Fukashi, 12, 15
Scalar
  operators, 85
  products, 67, 69, 76, 85

Schmidt decomposition, 498
Schrödinger equation
  characterized, 62, 122, 169, 323
  clamped nuclei, 262
  contracted, *see* Contracted Schrödinger equation
  density equation and, 127
  lower bound method, 90
  nonrelativistic, 494
Schwinger probes, 176–177, 270
SDCI, T1/T2 conditions, 99
SDPA Online for Your Future, 115
SDPARA, 115
SDPLR, 116
Search direction, primal-dual interior-point
  method, 112
Second quantization, 7–8, 167–169, 176, 264
Second-order
  algebraic equation, 76
  approximations, 63, 65–67, 88, 90
  commutator, 125
  contracted Schrödinger equation,
    see Second-order contracted Schrödinger
    equation (2-CSE) purification
  corrections, 325–326, 334
  density equations, functionals and, 392
  estimates, 86–88
  ladder approximation, 289
  many-body perturbation theory
    (MP2/MBPT2), 53, 336
  matrices, 27
  perturbation theory, 65
Second-order contracted Schrödinger
  equation (2-CSE) purification
  characterized, 244–246
  indeterminancy problem, 245
  iterative, N and S-purification, 245, 249–250
  potential energy curve research, 250–252
  regulated iterative self-consistent solution,
    246–249
Second-quantized operators, 26, 170
Seifert, 7
Self-consistent field (SCF) theory, 365
Self-consistent iteration, 288
Self-contraction diagrams, 316
Semi-internal excitation, 365–366, 370–371
Semidefinite programming (SDP)
  algorithms, 79–83
  augmented Lagrangian method, 115
  characterized, 15, 23, 45–48, 104–105

Semidefinite programming (SDP) (*Continued*)
  density functional theory, 479
  dual, 104–107, 109–110, 114, 116
  existence theorems, 77–79
  first-order nonlinear algorithm, 50
  formulation as, 105–110
  fundamental theorems, 75–77
  general, 105
  historical perspective, 81
  large-scale, 47, 55, 115
  lower bound method and, 62–64, 72–75
  nonlinear, 47–48, 55
  one-dimensional superconductor modeling, 83–89
  primal, 104–107, 110, 116
  primal-dual interior-point method, 110–115
  search directions, 112
  software, 104, 115
  solving in practice, 115–116
  T1/T2 conditions, 95–96, 99
  two-electron reduced density matrix method, 531
Semidefinite representability condition, 97
Separability, 266–269, 495
Separation theorem, 76
Sequential quadratic programming (SQP), 417
SH, 418, 420, 422
Shannon entropy, 494
Shape vibrations, 14
Shor, Peter, 494–495
sigma (σ), 327–328, 517
Sign
  factors, 313
  rule, 316
Simulations, 388
Single, double, triple, and quadruple excitations (SDTQCI), 89
Single-double configuration interaction (MRCI) wavefunction, 49
Single-particle
  functions, 11
  Hilbert space, 498
  operators, 362
Single-reference
  canonical transformation theory, 357
  coupled cluster formalism, 329, 345, 375–378
  coupled-cluster theory, 364
Singlet pairs, 299
6-CSE-155
Six-body conditions, 97

Size-consistent/size-consistency
  canonical transformation theory, 354–355, 380
  equations, 262, 268
  formalism, 392
Slater condition, 105, 111
Slater-/condon rules, 476
Slater decomposition, 498
Slater determinantal density matrix, 476–477
Slater determinants, 53, 117, 141, 156, 183, 237, 248, 306, 311–314, 327, 329, 355, 388, 431, 434, 454, 456–457, 459, 464, 471–472, 515
Slater determination, 128
Slater hull
  characterized, 449–458, 480
  necessary conditions, *see* Slater hull necessary conditions
Slater hull necessary conditions
  off-diagonal matrix elements, 464–466
  Q, K conditions, 463–466
  Q, R conditions, 462–463, 465
  R, R conditions, 459–462
  sufficiency for diagonal elements, 463–465
Slater polynomials, 456
SMILES, 249
Smith, Darwin, 13
SO_2, 418, 422
s operator, 70
Space entanglement, 516
Spatial coordinates, 22
Spatial orbitals, 191, 339, 514, 523
Spatial representation
  characterized, 447
  diagonal elements, 449, 453
  from linear inequality, 469–470
Spatial symmetry, 40–42, 44, 109
Spectral optimization problem, 63–64, 74–75, 77–81
Spectral theorem, 83
Spin
  adaptation of, *see* Spin adaptation
  block, *see* Spin block(s)
  components, purification procedures, 225, 244
  coordinates, 22
  coupling, 323
  degeneracy, 307–308
  density matrices, 307
  entanglement, 516. *See also* Spin system entanglement

functions, 209
multiplicity eigenvalues, 98
operators, 524
orbitals, *see* Spin orbitals
significance of, 16
structure, two-body correlation matrix, 233–237
symmetries, semidefinite programming problem, 108–109
T1/T2 conditions, 98
unrestricted formalism, 398
Spin adaptation
 applications, 53–54
 characterized, 37–41
 open-shell molecules, 44
 operators, 38–42
 reduced Hamiltonian (SRH) theory, 226, 239
 S-representability, 41–42
Spin block(s)
 characterized, 190–192
 convergence of CSE, 151
 purification procedures, 216, 219, 237
 two-block correlation matrix, 235–236
Spin-compensated programs, 409
Spin-free
 cumulants, 307
 excitation operators, 297–298
 form, 319
 two particle excitation operators, 298
Spin-independent
 matrices, 409–410
 orbitals, 413–414
Spinless
 cumulant, 407–408
 density matrices, 399
Spin orbitals
 basis/basis sets, 23–24, 214, 337
 characterized, 108, 116, 140–141, 322, 347
 construction algorithms, 142
 CSE, 159–160
 diagonal elements and, 478
 generalized normal ordering, 311
 natural, 273, 276–277
 orthonormal, 266, 270
 representation, functionals, 394–397, 409
 2-CSE, 130–133, 246
Spin reduced density matrices, 506–507
Spin-spin correlation, 507
Spin system entanglement
 numerical results, 507–509

one-dimensional N spin system, 504–505
one-dimensional spin-½ systems, 500, 505–506
quantum phase transition (QPT), 501, 504–505, 519
spin reduced density matrices, 506–507
thermal entanglement, temperature effects, 509–511
two-dimensional spin systems, 511–512
two-spin systems, 500–504
Spin-traced one-particle density matrix (IPDM), 513–515
Spin-unrestricted Hartree-Fock (UHF), 523
Spin-up, spin-down electron pairs, 84–86, 431
Split-valence double-zeta basis set, 194
S-purification process, 249
Spurious solutions, 265
Standard index set, 489, 491
Stationarity conditions for energy
 contracted Schrödinger equation (CSE), 317–318
 irreducible contracted Schrödinger equation (ICSE), 319–320
 one-particle approximation, 322–323
 k-particle Brillouin conditions, 318–319
 perturbative analysis, 324–328
 solutions for IBC and CSE, 321–328
 two-particle approximation, 323–324, 328
Stationary state, functionals, 402
Stretched geometries, 53, 55
S-representability
 conditions, 206, 214, 230, 238–240
 convergence of 2-CSE, 151–152, 248
 purificaiton, 210–214, 219–220, 225–227, 237, 242–243
S-representable 2-RDM, 41–42
Strongly orthogonal antisymmetrized geminal products (SOAGP)
 components of, 429–431
 perturbation theory, 432–433
 singlet-type strongly orthogonal geminals, 431–432
 unrestricted antisymmetrized geminal products (UAGP), 437–438
Strong orthogonal functions, 267, 302
Strongly orthogonal geminal (SSG)
 approximation, 432
 geminal functional theory and, 435–436
 defined, 431
Subadditivity, 496

Substituted configuration interaction (SDCI), 66
Superconducting/superconductivity
  characterized, 38
  modeling, 83–89
  phase, 83
Super operators, 488
Symmetric matrices
  implications of, 80
  lower bound method, 72–73
  multiple, 104
  semidefinite programming problem, 107
Symmetric two-electron operators, 275
Symmetrization, 267
Symmetry
  adaptations, applications, 53–54, 308
  conditions, 28
  implications of, 16
Symmetry-adapted two-particle cumulants, 308

Tailored CC theory, 346
Tangency, 86
Taylor series expansion, 177
Tensor(s)
  characterized, 5, 274
  irreducible, 307
  two-body correlation matrix, 232–233
Thermal
  entanglement, 509–511
  equilibrium, 525
Third-order approximations, 63, 88
Third-order estimates, 84, 86, 88
Third-order reduced-density matrices,
  see 3-RDM
Three-body
  conditions, 97
  elemental excitation, construction
    algorithms, 142, 145
  problems, 12
Three-index conditions, 95–96
Three-nucleon nuclei, 12
Three-particle
  approximation, 324, 326
  correlations, 289
  operator, 353, 358–359
3-positivity (3POS) conditions
  characterized, 94
  density matrices, lower bound method, 65–67
  variational 2-RDM methods, see 3-positivity
    (3POS) conditions, variational
    2-RDM method

3-positivity conditions, variational 2-RDM
  characterized, 23, 26–27, 50–53
  partial, 23, 27–29
3-RDM
  ACSE, 332–333
  characterized, 129–131, 133, 135,
    138–139, 141, 143–145, 156
  construction algorithms, 146–147, 149–150
  CSE, 160–161
  purification procedure, 247
  reconstruction of, see 3-RDM
    reconstruction methods
  SOAGP, 436–437
  2-CSE, 153, 248
3-RDM reconstruction methods
  approximation of cumulant 3-RDM, 179–182
  CSE and, 193–194, 197
  cumulants, 176–179
  cumulant structure of CSE, 182–183
  overview of, 333–334, 336–338
  particle-hole duality, 172–175
  Rosina's theorem, 170–171
Threlfall, W., 7
Time-dependent
  magnetic field, 524, 528
  theories, 381
Time-evolution matrix, 525, 527
Time-inversion antiunitary operator, 408
T_1/T_2 conditions
  characterized, 94, 100
  classical RDM approach, 98–99
  Erdahl's, 96–98, 100
  historical perspectives, 94–96
  lower bound method, 66–67
  semidefinite programming problems, 109, 116
  variational 2-RDM methods, 27–29, 52
Trace
  conditions, 67–69, 157, 185–187, 191, 193,
    238, 434
  operators, 85
  relations, 302–303
  scalar products, 220
Traceless operators, 71, 80
Transition density matrix (TRDM), 124, 153
Transition metal chemistry, 344
Transition reduced density matrix
  (TRDM), 210–213
Transition-state structures, 23
Transpositions, 199
Transvection, 179–180, 182–183, 275

SUBJECT INDEX

Triatomic molecules, 65
Triple
  bonding, 372
  excitations, 169
  -zeta basis set, 54
Triplet pairs, 299
Truncation, canonical transformation theory, 351–352, 374–375
Two-body
  correlation matrix (CM), *see* Two-body correlation matrix (2-CM)
  forces, *see* Two-body forces
  interactions, 11, 36, 38, 63
  matrix, 206, 209
  operators, 67, 85, 94–95, 97, 334
  unitary transformations, 198
Two-body correlation matrix (2-CM)
  basic properties of, 233–234
  characterized, 231–233
  spin structure, 234–237
Two-body forces
  implications of, 84, 87
  lower bound methods, 62–63, 66–67, 88–90
Two-electron
  excitations, 168
  interactions, 273
  matrix, 123
  operators, 122, 208
  potential energy, 309
  systems, 321, 410
  two-site Hubbard model, 512–517
2-G matrix, 234–235, 249
Two-hole density matrices, 39, 303
2-HRDM, 136, 233–234, 239–240, 244, 249
Two-particle
  charge density (2-CD), 389
  density matrix, cumulants, 300–301
  electrons, 4, 7
  Hermitian operator, 490
  interactions, 4, 144, 171, 351, 363
  matrix, 186
  mean-field theory, 346
  operators, 353–355, 358, 362
  reduced Hamiltonian, 46, 186, 488.
    See also Two-particle reduced Hamiltonian matrices, convex set of
  system, 14
  two-hole interaction, 144
Two-particle reduced Hamiltonian matrices, convex set of

N-representable 2-RDMs, 30–31
  positivity and 1-RDM, 32–33
  positivity and 2-RDM, 33–35
  strength of positivity conditions, 2, 35–38
2-positivity (2POS) conditions
  density matrices, lower bound method, 65–67
  implications of, 94, 99
  variational 2-RDM methods, 23–26, 45, 50–52
2-RDM
  ACSE, 332–334
  characterized, 4, 6–7, 15, 156, 394–399
  chemistry applications, 9
  construction algorithms, 130–139, 147
  CSE and, 165–167
  cumulant, *see* 2-RDM cumulant
  defined, 7
  eigenvalues of, 12
  functionals and, 391
  geminal functional theory (GFT), 436
  generation of, 22
  ground-state energy, 22
  historical perspectives, 11–12
  mapping to N-particle wavelength, 15
  N-representability problem, 124–127.
    See also N-representability of 2-RDMS
  optimization of, 334–335, 338
  natural orbital functionals, 402–403, 414
  parameterization of, *see* Parametrization of 2
  purification process, 206
  spin structure, 396–399
  strongly orthogonal antisymmetrized geminal products (SOAGP), 429
  T1/T2 conditions, 95, 97
  2-CSE iterative solution, 157
  UAGP, 438
  unrestricted AGP, 437
  variational method, 12
2-RDM cumulant
  approximate cumulant, 407–409
  characterized, 405–407

Unconnected/unconnectedness
  equations, *see* Unconnected equations
  functionals, 177–179, 194
  p-RDMs, 179
Unconnected equations
  cancelled terms, 282–286
  characterized, 262–263, 268
Unifying algorithms, 144–145, 147–150

Unitary
  coupled-cluster theory (UCC), 351–352, 356, 379
  decomposition, 185–189, 220–222
  group, generators of the, 298
  operators, 126, 154
  transformation, 27, 323, 326–327, 329, 338–339, 351, 402, 465–466, 498
Unit trace, 325
Unity/unity operator, 232, 266
Un-occupied/occupied spin orbitals, 144
Unrestricted coupled-cluster singles-doubles (UCCSD), 53

Vacuum state, 498, 522
Valdemoro, Carmela, 9, 14, 16
Valdemoro (V) reconstruction, 333, 336, 339
Valence
  orbitals, 347, 364–365, 372
  shell Hamiltonian theory, 346, 351
Van Vleck theory, 351
Variational, *see* Variational two-electron reduced-density-matrix (2-RDM)
  approach, 511
  optimization, 198
  principle, 435, 472
Variational two-electron reduced-density-matrix (2-RDM)
  applications, 48–54
  energy as functional, 23–24
  future directions for, 54–56
  N-representability, 23–24, 50–51, 55
  open-shell molecules, 42–45
  overview of, 21–22
  partial 3-positivity conditions, 23, 27–29
  spatial symmetry adaptation, 40–42
  semidefinite programming, 45–48
  spin-adapted operators, 38–40, 41–42
  3-positivity conditions, 23, 26–27
  two-particle reduced Hamiltonian matrices, convex set of, 23, 30–38
  2-positivity conditions, 23–26, 45
VCP algorithms, 137–138, 140, 147
Vector(s)
  functionals and, 408
  kth-order approximations, 70
  semidefinite programming, 45–46, 83, 109
  spaces, 3-positivity conditions, 26
  T1/T2 conditions, 96–97

Vertex, diagrammatic representations, 278–279, 314–316
Vibrational frequencies, 392, 423
Virtual excitation, 233
Virtual orbitals, 180, 347–348, 364–365
von Neumann
  densities, 62, 67–69, 71–72, 74, 83, 85–86
  entropy, 495–496, 499, 503, 512, 515, 517–518, 523–524, 531
VTP algorithm, 143, 145

Wannier orbital, 519
Warsket, George, 13
Water (H_2O), 99–100, 194–197, 332, 336–337, 366–372, 418–419, 422, 437
Wavefunction, *see specific types of wavefunctions*
  antisymmetrical geminal power (AGP), 37–38
  characterized, 3, 11, 15
  correlated, 55
  ground-state, 24, 37–40, 170
  N-electron, 124
  N-particle, 15
  perturbation methods, 88–89
  power method, 192–193
  Slater, 171
  Ψ, 22, 24
  variational 2-RDM method, 49–50
Wavelength, many-electron, 22
Wavepacket dynamics, 15
Weak correlation, 344
Wedge products, 30, 37, 43, 173, 177–179, 182, 198–199, 301, 491
Weighting coefficients, 445, 456
Weinhold-Wilson constraints, 478
Weisacker
  ansatz, 473
  functional, 471
Wheeler, John A., 5
Whitehead, Alfred North, 4–6
Wick's theorem, 294, 311, 314
Wigner, Eugene, 5
Wootters' measure, 498
Workshops, international, 12–15

Zanardi entanglement measure, 512, 520, 522
Zero operator, 70
Zero-order wavefunction, 180
Zeroth-order energy, 432